Road Vehicle Dynamics

Other SAE titles of interest:

Fundamentals of Vehicle Dynamics
By Thomas D. Gillespie
(Product Code: R-114)

Race Car Vehicle Dynamics
By William F. Milliken and Douglas L. Milliken
(Product Code: R-146)

Tire and Vehicle Dynamics, Second Edition
By Hans B. Pacejka
(Product Code: R-372)

For more information or to order a book, contact SAE International at
400 Commonwealth Drive, Warrendale, PA 15096-0001;
phone (724) 776-4970; fax (724) 776-0790;
e-mail CustomerService@sae.org;
website http://store.sae.org.

Road Vehicle Dynamics

Rao Dukkipati

Jian Pang

Mohamad Qatu

Gang Sheng

Shuguang Zuo

SAE International™
Warrendale, Pa.

For permission and licensing requests, contact:

SAE Permissions
400 Commonwealth Drive
Warrendale, PA 15096-0001 USA
E-mail: permissions@sae.org
Tel: 724-772-4028
Fax: 724-772-4891

Library of Congress Cataloging-in-Publication Data

Road vehicle dynamics / Rao Dukkipati ... [et al.].
    p. cm.
ISBN 978-0-7680-1643-7
1. Automobiles—Dynamics. I. Dukkipati, Rao V.

TL243.R63 2008
629.2'31--dc22

2007018140

SAE International
400 Commonwealth Drive
Warrendale, PA 15096-0001 USA
E-mail: CustomerService@sae.org
Tel: 877-606-7323 (inside USA and Canada)
        724-776-4970 (outside USA)
Fax: 724-776-1615

SAE International is committed to preserving ancient forests and natural resources. We elected to print this title on 30% post consumer recycled paper, processed chlorine free. As a result, for this printing, we have saved: .

13 Trees (40' tall and 6-8" diameter)
4,830 Gallons of Wastewater
9 million BTU's of Total Energy
620 Pounds of Solid Waste
1,164 Pounds of Greenhouse Gases

SAE International made this paper choice because our printer, Thomson-Shore, Inc., is a member of Green Press Initiative, a nonprofit program dedicated to supporting authors, publishers, and suppliers in their efforts to reduce their use of fiber obtained from endangered forests.

For more information, visit www.greenpressinitiative.org

Environmental impact estimates were made using the Environmental Defense Paper Calculator. For more information visit: www.papercalculator.org

Copyright © 2008   SAE International

ISBN 978-0-7680-1643-7

SAE Order No. R-366

Printed in the United States of America.

Contents

Foreword

My career in vehicle dynamics began more than thirty years ago in what I now see with the benefit of hindsight to have been the Dark Ages. In those days, we were guided as much by intuition and experience as by science. Our ability to deductively model the physical laws of dynamics was rather limited, and our ability to measure the dynamics of a vehicle was somewhat primitive, as was our ability to predict the performance of a particular geometry and set of components. As a consequence, we had to rely heavily on tuning experts to iron out the problems that we had inadvertently designed into the vehicle—a luxury that as an industry we can no longer afford.

In the intervening years, the science of vehicle dynamics has undergone a revolution. A great deal of excellent work has been done in the field of geometry, and the availability of virtually high-powered numerically intensive computing capability has enabled us to vastly expand our understanding of vehicle dynamics through the construction of theoretical models. These analytical methods are a mix of deductive, empirical, and hybrid types that now can help us come much closer to the optimal design prior to the prototype stage. This includes not only the design concept but also the parameterization of the design to reduce error states. The result is that the design and development processes have become much more streamlined and efficient, and the end product has improved greatly. Today's cars undoubtedly are much safer and more enjoyable to drive than those of thirty years ago, and much of this improvement can be traced back to the more extensive understanding and use of analytical methods.

But we cannot be complacent. We still must do time-consuming tuning, and some failure modes are still detected after the point of initial release. Our quest to build safer cars in a world of increasing competition and cost constraints will be possible only with the application of analytical predictive methods. New product cycle times are shrinking, model ranges are expanding, and the cost of building prototypes is rising, making it imperative that we find faster and more efficient ways of working.

Despite all the technological advances we have made, vehicle dynamics remains an art as much as a science. It requires the exercise of good judgment in matters such as how to translate a customer's requirements into objective metrics that engineers can measure, and how to balance many competing requirements—such as ride, handling, and braking—within a package that is both affordable and consistent with the distinctive attributes of a particular brand.

This book will not teach you how to exercise good judgment—only first-hand experience can do that—but it will give you an excellent foundation in this fascinating and important field. A combination of theoretical fact and practical insight, this book will help you to play your part in meeting the challenges of vehicle dynamics in the twenty-first century.

Professor Richard Parry-Jones, CBE
Group Vice President, Global Product Development
Chief Technical Officer
Ford Motor Company

Preface

This book provides a fundamental understanding of how physical laws, human factor considerations, and design choices interact to determine the ride, handling, braking, and accelerating behavior of road vehicles used for personal transportation. It presents the dynamics of road vehicle systems from a perspective that unifies the treatment of the causes of physical events with a treatment of the reasons for physical functions. Emphasis is placed on the observation that driver vehicle system behavior is constrained by physical law, but the driver's objectives and the designer's goals ultimately will determine how the system is utilized and configured.

Currently, the driver's skill, situation awareness, and knowledge largely determine control qualities in service. However, systems aimed at aiding the driver in controlling the vehicle are being developed and sold, such as anti-lock braking systems, adaptive cruise control systems, and anti-spinout directional control systems. These new systems are based on control system concepts, and they contain sensors, units for processing sensor information, and actuators for implementing desired control actions. These control systems perform functions that are difficult or tedious for the driver to perform. In the context of this book, the point is that the reasons for these functions can be stated in analytical terms. Related to this point, a goal of this book is to provide a conceptual approach that can aid professionals and those in the academic world in using ideas similar to those used in control system design to plan and envision functional enhancements to the vehicle system. A key element of this approach lies in translating the reason for a new functional capability into analytical rules that can be employed in a control system that makes the vehicle simultaneously safer, more comfortable in which to ride, and/ or easier to drive.

Practical approaches involving pragmatic considerations of reason and cause are treated for each of the traditional vehicle dynamics areas—ride, handling (steering), braking, and accelerating behavior. In each of these areas, this book covers the analysis of vehicle dynamics issues using physical principles, and it considers vehicle design in the sense that a design is a plan based on an understanding of the physical properties of the vehicle, as well as an understanding of what the vehicle is expected to do.

In the chapters related to handling, braking, and accelerating, the individual control inputs are viewed as a means of communication between the driver, or an automatic control system, and the basic vehicle. From a systems perspective, the driver or control system activates the controls to obtain vehicle system dynamics that fit the driver's or the control system's plans and expectations. Accordingly, this book not only focuses inwardly on mechanisms within the vehicle but also looks outwardly at broader considerations involving what the driver desires and how to achieve it.

Even in areas such as ride and roll where the driver has no direct means of influencing the motions caused by hitting holes and bumps, emphasis is placed on driver concerns with control and safety. A common theme is to indicate how driver and passenger considerations influence the design characteristics of motor vehicle systems.

This book is written for vehicle designers, developers, evaluators, and both senior undergraduate students and graduate students. Previous knowledge of differential equations and dynamics systems will be helpful in reading this book. Other material is treated in the appendices.

There are probably a handful of books on vehicle dynamics. Some of these books are written at a fundamental level that may not meet ambitious engineering program requirements. Others are specialized in certain fields of vehicle dynamics, including modeling and simulation. In this book, we attempt to strike a balance between theory and practice, fundamentals and advanced subjects, and generality and specialization. However, the book focuses on road vehicles, and little treatment is made for off-road vehicles. Road vehicles that frequently are driven off-road (e.g., sport utility vehicles [SUVs]) are covered in this book. The book also emphasizes the use of cars and light trucks, with little treatment of vehicles with trailers. We had this emphasis in mind mainly for two reasons:

1. Engineers interested in off-road vehicle dynamics or the dynamics of multi-axle vehicles can build on the fundamentals given in this book.

2. Most vehicles produced for the public are classified as cars or light trucks (13 to 16 million such vehicles are sold annually). These also have the most stringent requirements for ride, safety, fuel economy, and other attributes.

We pay particular attention to the issue of safety in this book because it is among the leading (if not the leading) attributes in customers' minds when purchasing a vehicle. In addition, safety is the attribute loaded with governmental standards that must be met before the vehicle can be offered to the public. One complete chapter is written on accident reconstruction to address legal issues that result from an automotive accident.

This book contains ten chapters with an extended list of appendices. The appendices provide a review of the material needed before studying the book, or they provide other material that ensures consistency of the terms used in this book. Readers will find appendices covering matrix and vector algebra, Fourier series, Laplace transformation, vehicle dynamics terminology, direct numerical integration methods, and conversion of units.

Chapter 1 is an introduction and general review of dynamics. It covers vehicle system classification, dynamic systems and their models, generalized coordinates and degrees of freedom, discrete and continuous systems, vibration analysis, Newton's laws of motion, kinematics of rigid bodies, concepts of work and energy, impulse, D'Alembert's principle, Lagrange's equations of motion, holonomic and nonholonomic systems, and other subjects.

Chapter 2 focuses on the analysis of dynamic systems. Subjects treated include the classification of dynamic system models, constraints, generalized coordinates, degrees of freedom (DOF), classification of deterministic data, linear dynamic systems, free and forced vibrations, eigenvalues and eigenvectors, damping, nonlinear dynamic systems, random vibrations, and methods of analysis.

The forces acting on the vehicle are treated in Chapter 3, which covers tires and aerodynamic forces and moments. Particular emphasis is given to explaining how the tire produces the forces needed to move the vehicle laterally and longitudinally as well as to support the loads. The dynamic performance of the entire vehicle system is limited by what the tires can do. Every aspect of vehicle performance is constrained by what is happening at the tire/road interface. Hence, it is desirable that each subsystem of the vehicle aids in presenting the tire to the road as favorably as possible. No other component of the vehicle is singled out for separate treatment as that given to the tire because no other component is as important to so many different aspects of riding and driving performance. This book presents empirical and simplified theoretical approaches for representing the input-output characteristics of tires as needed for vehicle dynamic analyses.

Chapter 4 treats the subject of ride dynamics. Issues discussed include sources of vibrations, power spectral density, vehicle ride models, seat evaluation and modeling, discomfort evaluation, human body model, and material on active and semi-active control.

Chapter 5 covers roll dynamics and treats subjects on rollover, including rollover scenarios, rollover modeling, and rollover testing. Rollover modeling includes a suspended vehicle rollover model, steady-state rollover models, and a dynamic rollover model. Rollover testing includes the tilt table ratio (TTR) test and the side pull ratio (SPR) test. This chapter also includes studies on the contribution of tire deflection and suspension deflection to rollover. In addition, rollover control is discussed in this chapter.

Chapter 6 treats handling dynamics. Subjects include steady-state handling, understeer and oversteer, steering sensitivity, roll gain and roll damping, transient response, lateral acceleration and yaw rate response time, and others. Various handling models are presented, including a two-degrees-of-freedom yaw plane and nonlinear models. Safety considerations and overturning limits also are studied.

Chapter 7 discusses braking, spanning both drum and disk brakes. Torque distribution along the brake pads is discussed thoroughly, including the treatment of temperature rise.

Load transfer during braking under the effects of aerodynamic rolling and other forces, as well as the effect of grade, are studied. Optimal braking performance for single- and multi-axle braking is investigated, paying attention to safety. Front and rear lock-ups and anti-lock braking systems (ABS) also are discussed.

In Chapter 8, vehicle acceleration is studied, including load transfer during acceleration under the effect of various forces. Treatment of power and traction limited acceleration also is given. Studies are made on acceleration for various drivetrains, including front-wheel drive, rear-wheel drive, and four-wheel drive. Torque distribution for different drive axles and optimal tractive effort also are treated, as well as discussions on engine power and power transmission with manual, automatic, and other types of transmissions. Finally, safety features are presented, including limited slip axle and traction control.

Chapter 9 introduces total vehicle dynamics, covering both subjective and objective vehicle evaluations, dynamics of combined steering and braking, and combined steering and acceleration.

Chapter 10 presents the use of a wide variety of methodologies and software for accident reconstruction, based on the variation of physical evidence and accident investigative information.

Appendices A through G present a basic review of vector algebra, matrix analysis, Laplace transforms, a glossary of terms, direct numerical integration methods, units and conversions, and accident reconstruction formulae, respectively. Likewise, an extensive bibliography to guide the reader to further sources of information on road vehicle dynamics is provided at the end of the book. Both the S.I. and the U.S./English systems of units have been used throughout this book.

We wish to thank the many reviewers of this book for their reviews. We also thank SAE International and its staff, including Jeff Worsinger, for their help in editing and publishing this book. In addition, we would like to thank Richard Parry-Jones for his suggestions and for writing the Foreword for this book. Finally, we would like to thank our spouses and our children, who had to sacrifice family time while this book was being written.

Rao Dukkipati, Jian Pang, Mohamad Qatu,
Gang Sheng, and **Shuguang Zuo**

Chapter 1

Introduction

1.1 General

In this chapter, we introduce the classification of vehicles based on support and propulsion principles. The objective of an engineering analysis of a road vehicle system is prediction of the behavior or performance of that vehicle. Real vehicle dynamics systems are quite complex, and an exact analysis of the system often is impossible. However, simplifying assumptions can be made to reduce the system model to an idealized version of which the behavior or performance approximates that of the real system. The process by which a physical system is simplified to obtain a mathematically tractable situation is called modeling. The simplified version of the real system thus obtained is called the mathematical model or, quite simply, the model of the system.

The fundamental dynamics and modeling of mechanical systems are treated in this chapter. In general, mechanical systems are either in translational or rotational motion, or both. We present a brief review of the classification of dynamic system models; constraints; generalized coordinates; degrees of freedom; classification of vibrations; discrete and continuous systems; methods of vibration analysis; elements of vibrating systems, including spring elements, mass, or inertia elements; and damping elements.

The motion of vibrating systems is governed by the laws of mechanics and in particular by Newton's second law of motion. A brief review of rigid body dynamics—namely, Newton's laws of motion, the kinematics of rigid bodies, D'Alembert's principle, the principle of work and energy, the conservation of energy, and the principle of impulse and momentum—is presented to familiarize the reader with the basic notation and methods.

There are two approaches to the study of mechanics: (1) the formulation and solution of Newton's equations of motion, and (2) analytical dynamics. In analytical dynamics, we consider the system as a whole and formulate the expressions of energy in terms of the generalized coordinates.

We also discuss the concepts of generalized coordinates, the principle of virtual work, constraints, and D'Alembert's principle. In Section 1.10, the equations of motion for a dynamic system have been derived based on Lagrange's formulation. Lagrange's equations of motion consist of a system of simultaneous second-order differential equations, which generally are nonlinear in the generalized coordinates and the corresponding velocities.

1.2 Vehicle System Classification

Transportation can be considered as a means of conveyance from one point or place to another. There are many different kinds of transportation vehicles. These vehicles are based on different principles, they travel at different speeds, and they operate in diversified environments. In general, vehicles can be classified by their support and propulsion principles. The purpose of propulsion is to generate the forward speed, while the support mechanism balances the gravity acting on the vehicles, as shown schematically in Figure 1.1. Figure 1.2 shows the classification of vehicles based on support and propulsion principles. The reaction forces generated by the wheels, air cushions, or magnets support ground vehicles. Friction, flow, or magnetic forces drive all of these vehicles.

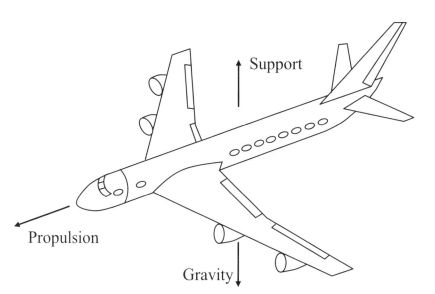

Figure 1.1 Propulsion and support of an airplane.

Static or dynamic lift forces generated by water or air support fluid vehicles, and flow forces generally propel them. Dynamic lift or inertia forces generated by air, jet propulsion, or orbital motion support inertia vehicles, and inertia forces accelerate them. Due to the nature of support and propulsion principles, quite a variation of maximum traveling speeds can be attained by these three classes of vehicles. Ground vehicles can attain a maximum speed of 600 km/h (375 mph). Fluid vehicles can attain up to 3000 km/h (865 mph), and inertia vehicles may have a traveling speed of 50,000 km/h (31,000 mph). The study of the dynamics of vehicles other than nonguided road vehicles is beyond the scope of this book.

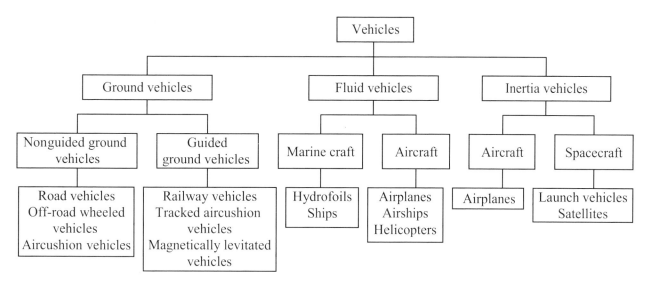

Figure 1.2 General classification of vehicles.

1.3 Dynamic System

Vibration is the motion of a particle, a body, or a system of connected bodies displaced from a position of equilibrium. Most vibrations are undesirable in machines, structures, and vehicles because they produce increased stresses and energy losses. They also can cause added wear, increase bearing loads, induce fatigue, create passenger discomfort in vehicles, and absorb energy from the system.

Vibration occurs when a system is displaced from the position of stable equilibrium. The system tends to return to this equilibrium position under the action of restoring forces (such as the elastic forces, as for a mass attached to a spring, or as the gravitational forces, as for a simple pendulum). The system keeps moving back and forth across its position of equilibrium. A system is considered as a combination of elements that are intended to act together to accomplish an objective. For example, an automobile is a system in which the elements are the wheels, suspension, car body, and so forth. A static element is one for which the output at any given time depends only on the input at that time, whereas a dynamic element is one for which the present output depends on past inputs. In a similar way, we also speak of static and dynamic systems. A static system contains all elements, whereas a dynamic system contains at least one dynamic element.

A physical system undergoing a time-varying interchange or dissipation of energy among or within its elementary storage or dissipative devices is said to be in a dynamic state. All of the elements in general are called passive (i.e., they are incapable of generating net energy). A dynamic system composed of a finite number of storage elements is said to be lumped or discrete, whereas a system containing elements that are dense in physical space is described as continuous. The analytical description of the dynamics of the discrete case is a set of ordinary differential equations. For the continuous case, it is a set of partial differential equations. The analytical formation of a dynamic system depends on the kinematic or geometric constraints and the physical laws governing the behavior of the system.

1.4 Classification of Dynamic System Models

To deal in an efficient and systematic way with problems involving time-dependent behavior, we must have a description of the objects or processes involved. Such a description is called a model. The model used most frequently is the mathematical model, which is a description in terms of mathematical relations and represents an idealization of the actual physical system. To describe a dynamic system, these relations will consist of differential or difference equations. Predicting performance from a model is called analysis. The purpose of the model partly determines its form; thus, the purpose influences the type of analytical techniques used to predict the behavior of the dynamic system.

Many types of analytical techniques are available, and their applicability depends on the purpose of the analysis. The physical properties, or characteristics, of a dynamic system are known as parameters. In general, real systems are continuous, and their parameters are distributed. However, in most cases, it is possible to replace the distributed characteristics of a system by discrete ones. In other words, many variables in a physical system are functions of location as well as time. If we ignore the spatial dependence by choosing a single representative value, then the process is called lumping, and the model of a lumped element or system is called a lumped parameter model. In a dynamic system, the independent variable in the model then would be only time. The model will be an ordinary differential equation that includes time derivatives but not spatial derivatives. If spatial dependence is included, then the resulting model is known as a distributed parameter model, in which the independent variables are the spatial coordinates as well as time. This model consists of one or a set of partial differential equations containing partial derivatives with respect to the independent variables. Discrete systems are simpler to analyze than distributed ones.

Vibrating systems are classified according to their behavior as either linear or nonlinear. If the dependent variables in the system differential equation(s) appear to only the first power and there are no cross products thereof, then the system is called linear. If there are fractional or higher powers, then the system is nonlinear. On the other hand, if the system contains terms in which the independent variables appear to powers higher than one or two fractional powers, then the system is known as a system with variable coefficients. Thus, the presence of a time-varying coefficient does not make a model nonlinear. Models with constant coefficients are known as time-invariant or stationary models, whereas those with variable coefficients are time-variant or nonstationary models. Figure 1.3 shows the relationships among the various model types. If there is uncertainty in the value of the model's coefficients or inputs, then stochastic models often are used. In a stochastic model, the inputs and coefficients are described in terms of probability distributions involving their means, variances, and so forth.

1.5 Constraints, Generalized Coordinates, and Degrees of Freedom

The position of a system of particles is called its configuration. Because of constraints on the system, actual coordinates usually need not be assigned to each particle. In a dynamic system, constraints may be at the boundary of the system or at points internal to the system. Constraints may be either static or kinematic in nature. The static constraints result from relationships among forces, whereas the kinematic constraints are due to relationships among displacements. In selecting the coordinates to describe a dynamic

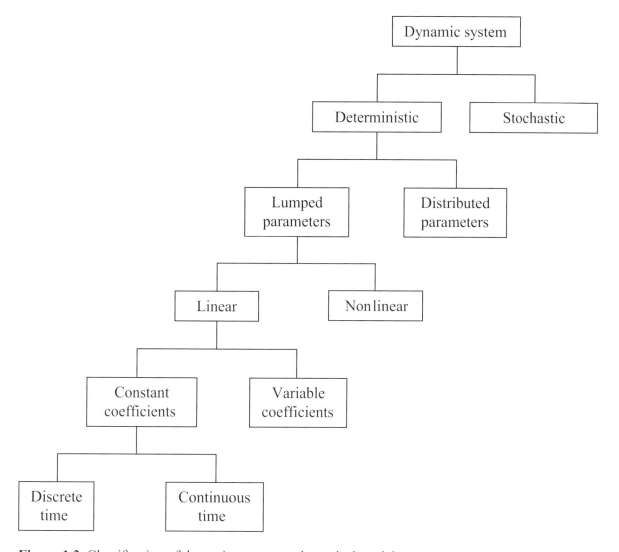

Figure 1.3 Classification of dynamic system mathematical models.

system, the static and kinematic constraints must be considered. Relationships among coordinates, which exist because of constraints on the system, are termed constraint equations. Based on this discussion, it can be said that systems of unconstrained or independent coordinates exist. In general, this is true in any dynamic system, and such a system can be described by a system of constrained coordinates.

For example, we can consider a dynamic system that is defined in terms of M coordinates. If there are R constrained displacements, then R coordinates can be expressed in terms of the remaining M-R coordinates, which are independent. Thus, if

$$N = M - R \tag{1.1}$$

N is the number of independent coordinates, and the forces and displacements are fully defined by these N coordinates. The independent coordinates required to specify completely the configuration of a dynamic system are called generalized coordinates. It is assumed that the generalized coordinates may be varied arbitrarily and independently without violating the constraints. Such a dynamic system is called a holonomic system. The number of generalized coordinates is called the number of degrees of freedom of a dynamic system.

To illustrate a dynamic system with constraint, we consider a rigid body attached to a point that is constrained to translate in the y direction, as shown in Figure 1.4. In three-dimensional space, five coordinates (i.e., two translations, one each along the x and z axes) and three rotations would describe the motion of the rigid body about the x, y, and z axes, respectively. In this case, the number of degrees of freedom for the system is five.

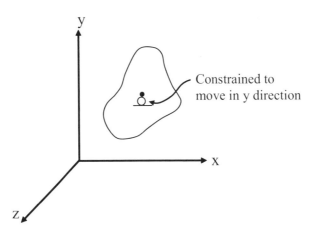

Figure 1.4 A rigid body in general motion (five degrees of freedom).

Let us suppose that the rigid body is further constrained and that it undergoes motion in only the x-y plane, as shown in Figure 1.5. The rigid body in a planar motion configuration would require two degrees of freedom to describe its motion. These degrees of freedom would correspond to the translation along the x axis and the rotation about the z axis.

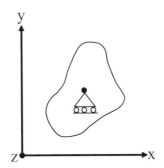

Figure 1.5 A rigid body in planar motion (two degrees of freedom).

The number of degrees of freedom of a vibratory system is the number of independent spatial coordinates necessary to define its configuration. A configuration is defined as the geometric location of all the masses of the system. A rigid body in space requires six coordinates for its complete identification: three coordinates to define the rectilinear positions, and three for the angular positions. Generally, the masses in a system are constrained to move only in a certain manner. Thus, the constraints limit the degrees of freedom of a system. The independent spatial coordinates describing the configuration of a dynamic system also are called generalized coordinates. Thus, the number of degrees of freedom of a dynamic system is equal to the number of generalized coordinates. Alternatively, the number of degrees of freedom of a system can be defined as the number of spatial coordinates required to completely specify its configuration minus the number of equations of constraint. Such a system is called a holonomic system.

Energy enters a dynamic system through the application of an excitation. The excitation varies according to a prescribed function of time. The vibratory behavior of a dynamic system is characterized by the motions caused by these excitations and is referred to as the system response. The response motion often is described by displacements, whereas the excitations could be in the form of initial displacements and velocities of externally applied forces.

We have already seen that the number of degrees of freedom of a dynamic system is the number of independent coordinates required to describe its motion completely. A discrete model of a dynamic system possesses a finite number of degrees of freedom, whereas a continuous model has an infinite number of degrees of freedom. Of the discrete mathematical models, the simplest one is the single-degree-of-freedom linear model. The following are the advantages of linear models:

1. Their response is proportional to input.

2. The principle of superposition is applicable.

3. They closely approximate the behavior of many dynamic systems.

4. Their response characteristics can be obtained from the form of system equations without a detailed solution.

5. A closed-form solution often is possible.

6. Numerical analysis techniques are well developed.

7. Linear models serve as a basis for understanding more complex nonlinear system behaviors.

However, note that in most nonlinear problems, it is impossible to obtain closed-form analytical solutions for the equations of motion. Therefore, a computer simulation often is used for the response analysis. Numerical analysis techniques used in computer simulations are discussed in Appendix E.

EXAMPLE E1.1

Determine the number of degrees of freedom for each of the four systems shown in Figure E.1.1, which are in equilibrium in the position shown.

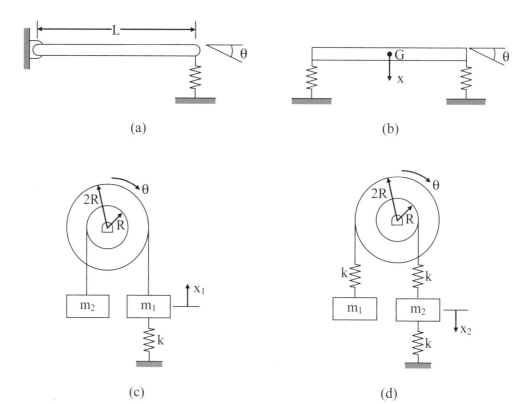

Figure E.1.1 Vibrating systems: (a) cantilever beam resting on a spring, (b) a beam resting on two springs, (c) a pulley system with one spring, and (d) a pulley system with three springs.

Solution:

a. Assuming θ is the clockwise angular displacement of the bar from the horizontal, a particle at a distance L from the fixed support of the system shown in Figure E1.1(a) has a horizontal position L cos θ and a vertical displacement L sin θ. Hence, the system has one degree of freedom, θ.

b. Denoting x as the displacement of the mass center of the bar as shown in Figure E1.1(b), measured from equilibrium, and θ the clockwise angular displacement of the bar, measured from the horizontal, then the displacement of a particle from equilibrium a distance ℓ to the right of the mass center is $x + \ell \sin \theta$.

c. Let θ be the counterclockwise angular displacement of the pulley measured from equilibrium, as shown in Figure E1.1(c). Assume there is no slip between the pulley and the cable. The displacement of the block of mass m_1 is $2R\theta$ vertically upward. Hence, the system has one degree of freedom, θ.

d. No kinematic relationship exists between the linear displacements of the blocks and the angular displacement of the pulley due to the fact that an elastic element connects each of the masses m_1 and m_2. A set of generalized coordinates (namely, the clockwise angular displacement of the pulley θ; the downward displacement of the block of mass m_2, x_2, and x_1; and the upward displacement of the block of mass m_1) can be chosen, as shown in Figure E1.1(d).

EXAMPLE E1.2

For the vibrating system shown in Figure E1.2(a),

a. Determine the number of degrees of freedom needed to specify the motion of the system.

b. Identify a set of generalized coordinates.

Solution:

a. The two springs k_1 and k_2 can be replaced by one equivalent spring. The system has two degrees of freedom, as shown in Figure E1.2(b).

b. One clear choice of a set of generalized coordinates is the downward displacement of the mass of the block m from the system equilibrium, x position, and the clockwise angular displacement of the rigid bar for the system equilibrium position, θ.

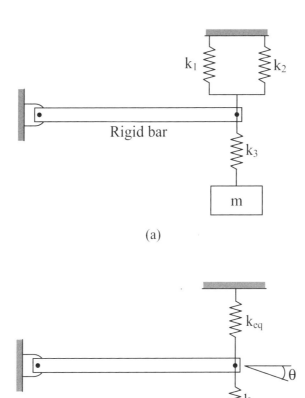

(a)

(b)

Figure E1.2 (a) A vibrating system, and (b) its equivalent system.

1.6 Discrete and Continuous Systems

Most of the mechanical and structural systems can be described using a finite number of degrees of freedom. However, some systems, especially those that include continuous elastic members, have an infinite number of degrees of freedom. Most mechanical and structural systems have elastic (deformable) elements or components as members and hence have an infinite number of degrees of freedom. Systems that have a finite number of degrees of freedom are known as discrete or lumped parameter systems. Systems that have an infinite number of degrees of freedom are called continuous or distributed systems. Continuous systems can be approximated as discrete systems by considering them as finite lumped masses, springs, and dampers. Better and accurate results can be obtained by increasing the number of masses, springs, and dampers, that is, by increasing the number of degrees of freedom as necessary.

1.7 Vibration Analysis

In general, the response (outputs) of a vibrating system depends on the initial conditions and the external excitations (inputs). A simplified mathematical model of the physical system can determine the overall complex behavior of the vibrating system. The vibration analysis of a physical system may be summarized by the following four steps:

1. **Mathematical modeling of a physical system:** The purpose of mathematical modeling is to determine the existence and nature of the system, its features and aspects, and the physical elements or components involved in the physical system. Necessary assumptions are made to simplify the modeling. The implicit assumptions that are used include the following:

   a. A physical system can be treated as a continuous piece of matter.

   b. Newton's laws of motion can be applied by assuming that the earth is an internal frame.

   c. Ignore or neglect the relativistic effects.

   d. All components or elements of the physical system are linear, elastic, homogenous, and isotropic.

   The resulting mathematical model may be linear or nonlinear, depending on the given physical system. Generally, all physical systems exhibit nonlinear behavior. Accurate mathematical modeling of any physical system will lead to nonlinear differential equations governing the behavior of the system. Often, these nonlinear differential equations have no solution, or it is difficult to find a solution. Assumptions are made to linearize a system, which permits quick solutions for practical purposes. When analyzing the results obtained from the mathematical model, realize that the mathematical model is only an approximation of the true or real physical system. Therefore, the actual behavior of the system may be different.

2. **Formulation of governing equations:** Once the mathematical model is developed, we can apply the basic laws of nature and the principles of dynamics to obtain the differential equations that govern the behavior of the vibrating system. Conservation of momentum, both linear and angular, is the only physical law that is of importance and of significance in application to vibrating systems. A basic law of nature is a

physical law that is applicable to all physical systems, regardless of the material from which the system is constructed. Several approaches to the principles of dynamics, such as Newton's second law of motion, D'Alembert's principle, and the principle of conservation of energy commonly are used to derive the governing equations of motion. Different materials behave differently under different operating conditions. Constitutive equations provide information about the materials from which a system is made. Application of geometric constraints—such as the kinematic relationship among displacement, velocity, and acceleration—often is necessary to complete the mathematical modeling of the physical system. The application of geometric constraints is necessary to formulate the required boundary and/or the initial conditions.

The resulting mathematical model may be linear or nonlinear, depending on the behavior of the elements of the vibrating system.

3. **Mathematical solution of the governing equations:** The mathematical modeling of a physical vibrating system results in the formulation of the governing equations of motion. Mathematical modeling of typical vibrating systems leads to differential equations of motion. Vibrations of a single-degree-of-freedom discrete system lead to one ordinary differential equation of motion. Vibrations of multiple-degrees-of-freedom systems are governed by a system of ordinary differential equations of motion. Similarly, the vibrations of continuous systems are governed by partial differential equations.

The governing equations of motion of a vibrating system are solved to find the response of the system. Many techniques are available for finding the solution, namely, the standard methods for the solution of ordinary differential equations, Laplace transformation methods, matrix methods, and numerical methods. The numerical methods of solving ordinary differential equations are treated in Appendix E. In addition, the solution of the partial differential equations that result for the vibration of continuous systems is much more involved than that of ordinary differential equations for the discrete systems.

In general, exact analytical solutions are available for many linear vibrating systems but for only a few nonlinear systems. Of course, exact analytical solutions always are preferable to numerical or approximate solutions.

4. **Physical interpretation of the results:** The solution of the governing equations of motion for the physical system generally gives the displacements, velocities, and accelerations of the various masses or inertias of the system. Physical interpretation of the results is an important and final step in the vibration analysis procedure. In some situations, this may involve the following:

a. Drawing general inferences from the mathematical solution

b. Development of design curves

c. Arriving at a simple arithmetic to reach a conclusion (for a typical or specific problem)

d. Recommendations regarding the significance of the results and changes, if any, required or desirable in the vibrating structure or system involved

EXAMPLE E1.3

Develop mathematical models of an automobile shown in Figure E1.3(a) traveling over a rough terrain, considering the critical elements of the vibrating system as follows:

a. The car body weight or chassis, passengers, seats, and front and rear wheels only

b. Suspension by considering the elasticity of the tires, main springs, and seats

c. Damping of the seats, shock absorbers, and tires

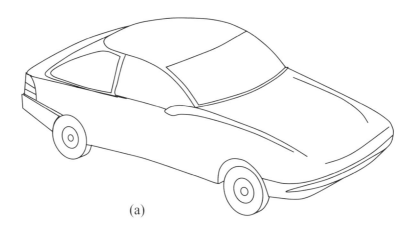

Figure E1.3 (a) Four-wheeled automobile.

(a)

Solution:

If we consider the equivalence values for the mass, stiffness, and damping of the system, a single-degree-of-freedom model of the automobile can be constructed as shown in Figure E1.3(b). In Figure E1.3(b), the equivalent stiffness (k_{eq}) includes the stiffness of the front and rear wheels, and the stiffness of the front and rear suspension. The equivalent damping constant (c_e) includes the damping of the front and rear wheels, and the damping of the front and rear suspension. The equivalent mass (m_{eq}) includes the mass of the automobile body and the mass of the passengers (m_p). This model is further refined by considering the mass of the wheels (m_w), the stiffness and damping of the tires (k_t and c_t), and the stiffness and damping of the front and rear suspension (k_s and c_s) as shown in Figures E1.3(c), (d), and (e). In Figure E1.3(e), the damping of the tires (k_t and c_t) and the stiffness and damping of the front and rear suspension of the automobile are assumed be the same for all the wheels and the front and rear suspensions.

In Figures E1.3(b) through (e), the subscripts are defined as

$$
\begin{aligned}
b &= \text{body} \\
eq &= \text{equivalent} \\
p &= \text{passengers} \\
s &= \text{suspension} \\
t &= \text{tire} \\
w &= \text{wheel}
\end{aligned}
$$

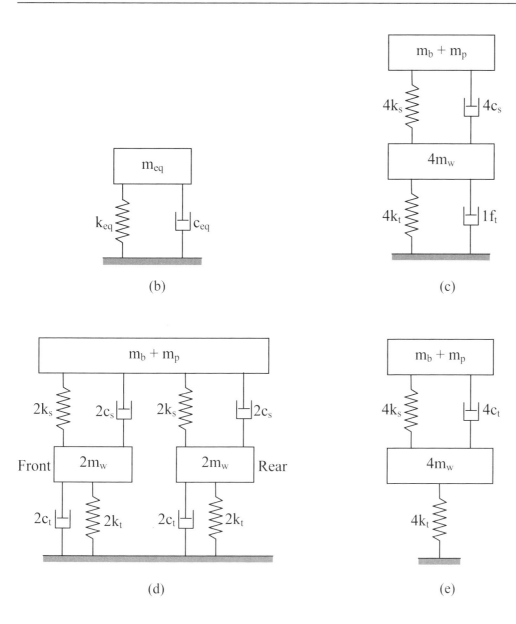

(b)

(c)

(d)

(e)

Figure E1.3 (continued) (b) A simplified automobile model; (c) another simplified automobile model; (d) a third simplified automobile model; and (e) a fourth simplified automobile model.

EXAMPLE E1.4

Figure E1.4(a) shows the overhead valve arrangement in an automobile engine. Develop a mathematical model for this arrangement as a lumped parameter or discrete parameter system. Assume that all components or elements of this system undergo no elastic deformation.

Solution:

The camshaft drives the cam with a constant angular velocity. When the operating speed is low, we can analyze the system consisting of rigid components. Thus, the simplified model considers the elasticity of the system components. Several models of varying complexity can be developed, which depend on the mass/inertia, stiffness, and damping characteristics of the system components. Figure E1.4(b) depicts a three-degrees-of-freedom model of the elastic body cam system, denoting the following:

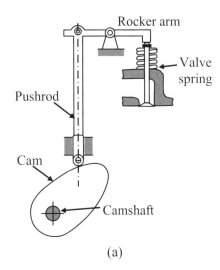

Figure E1.4 (a) Cam mechanism.

(a)

m_1 and m_3	$=$	lumped mass characteristics of the follower train
m_2	$=$	equivalent mass of the cam and a portion of the camshaft
k_1	$=$	stiffness of the follower retaining spring
k_2 and k_3	$=$	stiffness characteristics of the follower train
k_4	$=$	bending stiffness of the camshaft
$c_1, c_2, c_3,$ and c_4	$=$	dissipation of energy due to friction
m_{eq}	$=$	equivalent mass of the cam follower train
k_{eq}	$=$	equivalent stiffness of the pushrod
c_{eq}	$=$	equivalent damping of the system

Assuming the cam, camshaft, rocker arm, and valve springs are relatively stiff compared to the pushrod, a simple single-degree-of-freedom system as shown in Figure E1.4(c) is obtained.

Figure E1.4 (continued) (b) A three-degrees-of-freedom model, and (c) a single-degree-of-freedom model.

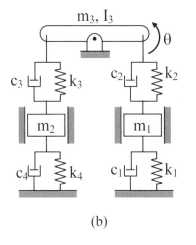

(b)

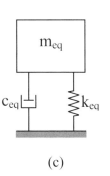

(c)

1.8 Elements of Vibrating Systems

Vibrations are the results of the combined effects of the inertia and elastic forces in a system. The inertia of moving or rotating components can be expressed in terms of the masses, moments of inertia, and time derivatives of the displacements. Elastic restoring forces are expressed in terms of the displacements and stiffness of the elastic members. Vibrations of systems occur with or without damping, and damping is another basic element in the analysis of vibrating systems.

1.8.1 Spring Elements

A linear spring is a flexible mechanical link between two particles in a mechanical system that is assumed to have negligible mass and damping.

Consider the spring connecting two masses m_1 and m_2, as shown in Figures 1.6(a), (b), and (c). The total deflection in the spring is given by

$$\Delta x = x_1 - x_2 \tag{1.2}$$

where

Δx = the total deflection of the spring due to the displacements of the masses m_1 and m_2

x_1 = the displacement of the mass m_1

x_2 = the displacement of the mass m_2

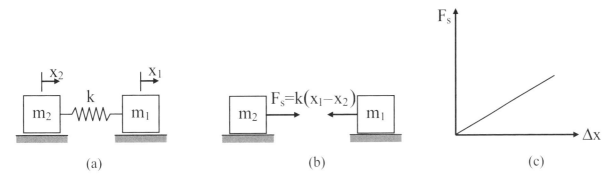

Figure 1.6 (a) A spring connecting two masses, (b) free body diagram, and (c) linear spring force.

Using Taylor's series, the new spring force can be expressed about the static equilibrium position x_0 as

$$F_s\left(x_0 + \Delta x\right) = F_s\left(x_0\right) + \frac{\partial F_s}{\partial x}\bigg|_{x=x_0} \Delta x + \frac{1}{2!}\frac{\partial^2 F_s}{\partial x^2}\bigg|_{x=x_0} \left(\Delta x\right)^2 + \dots \tag{1.3}$$

where F_s is the spring force, and x_0 may be defined as the pretension or precompression in the spring before the displacement Δx.

The spring force $F_s(x_0 + \Delta x)$ can be written as

$$F_s(x_0 + \Delta x) = F_s(x_0) + \Delta F_s \qquad (1.4)$$

where Δf_s is the change in the spring force as a result of the displacement. From Eqs. 1.3 and 1.4, we can write

$$\Delta F_s = \left. \frac{\partial F_s}{\partial x} \right|_{x=x_0} \Delta x + \frac{1}{2} \left. \frac{\partial^2 F_s}{\partial x^2} \right|_{x=x_0} (\Delta x)^2 + \dots \qquad (1.5)$$

If the displacement Δx is assumed to be small, higher-order terms in Δx can be neglected, and the spring force ΔF_s can be linearized. Hence,

$$\Delta F_s = \left. \frac{\partial F_s}{\partial x} \right|_{x=x_0} \Delta x \qquad (1.6)$$

or

$$\Delta F_s = k \, \Delta x = k(x_1 - x_2) \qquad (1.7)$$

where k is a proportionality constant called the spring constant, the spring coefficient, or the stiffness coefficient. The spring constant k is defined as

$$k = \left. \frac{\partial F_s}{\partial x} \right|_{x=x_0} \qquad (1.8)$$

The effect of the spring force F_s on the two masses is shown in Figure 1.6(b), and the linear relationship between the force and the displacement of the spring is shown in Figure 1.6(c). Springs commonly are used in many mechanical systems.

Continuous elastic elements such as rods, beams, and shafts produce restoring elastic forces. Figures 1.7(a) through (e) show some of these elastic elements that behave similarly to springs. In Figures 1.7(a) and (b), the rod produces a restoring elastic force that resists the longitudinal displacement in the system. If the mass of the rod is negligible compared to the mass m, we have

$$F = \frac{EA}{\ell} u \qquad (1.9)$$

where F is the force acting at the end of the rod, u is the displacement of the end point, and ℓ, A, and E are, respectively, the length, cross-sectional area, and modulus of elasticity of the rod. Equation 1.9 can be written as

$$F = ku \qquad (1.10)$$

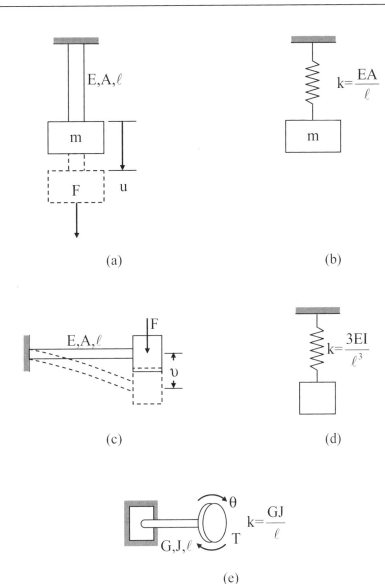

Figure 1.7 (a) Longitudinal vibration of rods, (b) an equivalent system for Figure 1.7(a), (c) transverse vibration of cantilever beams, (d) an equivalent system for Figure 1.7(c), and (e) a torsional system.

where k is the stiffness coefficient of the rod defined as

$$k = \frac{EA}{\ell} \tag{1.11}$$

Similarly, for the bending of the cantilever beam shown in Figures 1.7(c) and (d), we can write

$$F = \frac{3EI}{\ell^3} \upsilon$$

where F is the applied force, υ is the transverse deflection of the end point, and ℓ, I, and E are the length, second area moment of inertia, and modulus of elasticity of the beam. The beam stiffness is

$$k = \frac{3EI}{\ell^3} \tag{1.12}$$

The relationship between the torque T and the angular torsional displacement θ of the shaft shown in Figure 1.7(e) is given by

$$T = \frac{GJ}{\ell}\theta$$

where T is the torque, θ is the angular displacement of the shaft, and ℓ, J, and G are the length, polar moment of inertia, and modulus of rigidity, respectively. The torsional stiffness of the shaft is

$$k = \frac{GJ}{\ell} \tag{1.13}$$

1.8.2 Potential Energy of Linear Springs

The work done by the spring force (i.e., the spring connecting two masses m_1 and m_2) is

$$U_{1\to2} = \int_{x_1}^{x_2} kx\,dx = \frac{1}{2}kx_2^2 - \frac{1}{2}kx_1^2 \tag{1.14}$$

Now consider a path where the masses move between x_1 and $x_3 > x_2$ and then back to x_2. The total work done by the spring force is

$$\begin{aligned}
U_{1\to2} &= U_{1\to3} + U_{3\to2} \\
&= \int_{x_1}^{x_3} kx\,dx + \int_{x_3}^{x_2} kx\,dx \\
&= \frac{1}{2}kx_3^2 - \frac{1}{2}kx_1^2 + \frac{1}{2}kx_2^2 - \frac{1}{2}kx_3^2 \\
&= \frac{1}{2}kx_2^2 - \frac{1}{2}kx_1^2
\end{aligned} \tag{1.15}$$

Because the results of Eqs. 1.14 and 1.15 are the same, the work done by the force between the mass and the spring is independent of path. Thus, the force is conservative.

Therefore, a potential energy function can be defined for a linear spring as

$$V = \frac{1}{2}kx^2 \tag{1.16}$$

where x is the change in length of the spring from its unstretched length.

1.8.3 Equivalent Springs

In many practical applications, several linear springs are used in combination. It is quite possible to replace this combination of springs by a single linear spring of equivalent spring stiffness so that the behavior of the system with the equivalent spring is considered identical to the behavior of the actual system.

1.8.3.1 Springs in Parallel

Consider the springs in parallel connected to the mass m. We wish to find the elastic constant for a single spring, such that the behavior of the mechanical system shown in Figure1.8(b) is identical to that of the system shown in Figure 1.8(a).

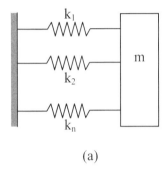

(a)

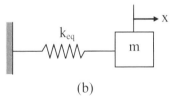

(b)

Figure 1.8 (a) Springs in parallel, and (b) a combination of springs in parallel replaced by a single equivalent spring.

This is accomplished by considering the force equilibrium condition of each system. In this case, the displacement of the mass of each system is the same, say x, when subject to the same resulting force F. The resulting force acting on the mass attached to the parallel combination of springs is the sum of the individual spring forces

$$F = k_1x + k_2x + k_3x + ... + k_nx = \left(\sum_{i=1}^{n} k_i \right) x \qquad (1.17)$$

The force acting on the mass attached to the spring of an equivalent stiffness is

$$F = k_{eq}x \qquad (1.18)$$

Equating the force F from Eqs. 1.17 and 1.18 gives

$$k_{eq} = \sum_{i=1}^{n} k_i \qquad (1.19)$$

The equivalent spring is one having an elastic spring constant equal to the sum of the constants of all the springs in the original system. Figures 1.9(a) and (b) show other examples of parallel systems.

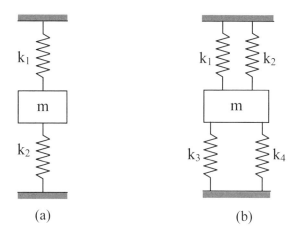

Figure 1.9 (a) Springs in a parallel system, and (b) other springs in a parallel system.

1.8.3.2 Springs in Series

The springs in the system shown in Figure 1.10(a) are said to be in series. The same force is developed in each spring and is equal to the force acting on the mass. In a series combination of elastic elements, the sum of the deformations of each element is equal to the total deformation. However, the change in length of each spring is different and is dependent on the spring stiffness. The displacement of the mass from equilibrium is the sum of the changes in the lengths of the springs

$$x = x_1 + x_2 + x_3 + ... + x_n = \sum_{i=1}^{n} x_i \qquad (1.20)$$

Because the force is the same in each spring, $x_i = \dfrac{F}{k_i}$, and

$$x = \sum_{i=1}^{n} \frac{F}{k_i} \qquad (1.21)$$

The displacement of the block attached to a single spring of an equivalent stiffness must equal the displacement of a block attached to the series combination of springs when each block is subject to the same resultant force. Equating Eqs. 1.20 and 1.21 gives

$$k_{eq} = \frac{1}{\displaystyle\sum_{i=1}^{n} \frac{1}{k_i}} \qquad (1.22)$$

Hence, when elastic elements are in series, the reciprocal of the equivalent elastic constant is equal to the reciprocals of the elastic constants of the elements in the original system. The combination of springs in series is replaced by a single equivalent spring, as shown in Figure 1.10(b).

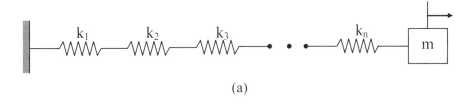

(a)

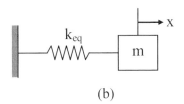

(b)

Figure 1.10 (a) Springs in series, and (b) a combination of springs in series replaced by a single equivalent spring.

EXAMPLE E1.5

Determine the equivalent spring stiffness for the system shown in Figure E1.5(a).

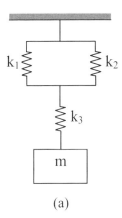

(a)

Figure E1.5 (a) Vibrating system.

Solution:

Because the springs k_1 and k_2 are parallel, they are equivalent to a single spring that has a stiffness coefficient k_{e1} given by

$$k_{e1} = k_1 + k_2$$

The system shown in Figure E1.5(a) then is equivalent to the system shown in Figure E1.5(b) or (c). We observe that the two springs k_{e1} and k_3 are connected in series.

Hence, they are equivalent to one spring that has a stiffness coefficient k_{eq} given by

$$\frac{1}{k_{eq}} = \frac{1}{k_{e1}} + \frac{1}{k_3}$$

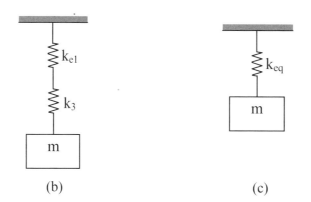

Figure E1.5 (continued) (b) An equivalent system, and (c) a final equivalent system.

(b)

(c)

or

$$k_{eq} = \frac{k_{e1} \, k_3}{k_{e1} + k_3} = \frac{(k_1 + k_2)k_3}{(k_1 + k_2 + k_3)}$$

EXAMPLE E1.6

A mass is suspended as shown in Figure E.1.6(a). Neglecting the mass of the beam, determine the equivalent stiffness of a linear spring when a linear spring is used to replace the system. Choose x as the generalized coordinate.

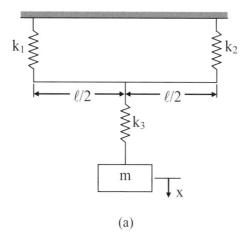

Figure E1.6 (a) Vibrating system.

(a)

Solution:

The beam behaves similarly to a linear spring. The equivalent spring stiffness of the two springs k_1 and k_1, which are parallel (Figure E1.6(b)), is

$$k_{12} = k_1 + k_2$$

(E1.6.1)

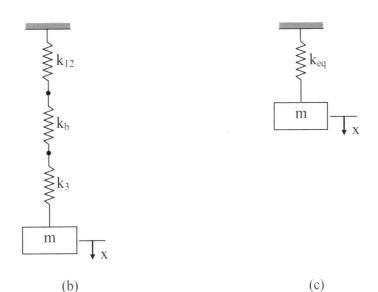

Figure E1.6 (continued)
(b) An equivalent system, and (c) a final equivalent system.

(b)

(c)

The stiffness of the beam is

$$k_b = \frac{48EI}{\ell^3} \tag{E1.6.2}$$

From Figures E1.6(b) and (c), in which the springs k_{12}, k_b, and k_3 are connected in series,

$$\frac{1}{k_{eq}} = \frac{1}{k_{12}} + \frac{1}{k_b} + \frac{1}{k_3} \tag{E1.6.3}$$

or

$$k_{eq} = \frac{k_{12}\,k_b\,k_3}{\left(k_{12}\,k_b + k_b\,k_3 + k_3 k_{12}\right)} \tag{E1.6.4}$$

where k_b and k_{12} are given in Eqs. E1.6.2 and E1.6.3, respectively.

EXAMPLE E1.7

Determine the equivalent spring constant for the system shown in Figure E1.7(a).

Solution:

Replacing the parallel combinations by springs of equivalent stiffness, we obtain the results shown in Figure E1.7(b). Noting that the springs on the left of the mass m are in series to one another, we can replace them by a spring of stiffness k_{e1}

$$\frac{1}{k_{e1}} = \frac{1}{5k} + \frac{1}{3k} + \frac{1}{k} + \frac{1}{7k} = \left(\frac{176}{105\,k}\right)$$

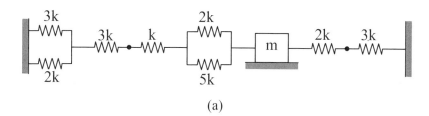

Figure E1.7 (a) Vibrating system.

(a)

or

$$k_{e1} = \left(\frac{105}{176}\right)k$$

Similarly, the springs attached to the right of the mass m are in series and can be replaced by a spring stiffness k_{e2}, as shown in Figure E1.7(c), as

$$\frac{1}{k_{e2}} = \frac{1}{2k} + \frac{1}{3k}$$

or

$$k_{e2} = \frac{6}{5}k$$

When the mass has an arbitrary displacement x, the displacements on each of the springs shown in Figure E1.7(c) are the same, and the total force acting on the mass is the sum of the forces developed in the springs. Hence, the springs will behave as if they are all parallel. We can replace them by a spring of stiffness k_E, as shown in Figure E1.7(d).

(b)

(c)

Figure E1.7 (continued) (b) Equivalent system, (c) further simplified equivalent system, and (d) a final equivalent system.

(d)

Hence,

$$\frac{105}{176}k + \frac{6}{5}k = k_E$$

or

$$k_E = \frac{1581}{880}k$$

1.8.4 Mass or Inertia Elements

The mass or inertia element is assumed to be a rigid body. Once the mathematical model of the physical vibrating system is developed, the mass or inertia elements of the system can easily be identified.

1.8.5 Damping Elements

In real mechanical systems, there is always energy dissipation in one form or another. The process of energy dissipation is referred to in the study of vibration as damping. Although the amount of energy dissipated into heat or sound is relatively small, the consideration of damping is important in the prediction of the vibration response of a system. A damper is considered to have neither mass nor elasticity. The three main forms of damping are viscous damping, Coulomb or dry friction damping, and hysteretic damping.

1.8.5.1 Viscous Damping

The most common type of energy-dissipating element used in vibration studies is the viscous damper, which also is referred to as a dashpot. The viscous damper or dashpot is characterized by the resistance force exerted on a body moving in a viscous fluid. In viscous damping, the damping force is proportional to the velocity of the body. The amount of dissipated energy depends on the size and shape of the vibrating body, the viscosity of the fluid, the frequency of vibration, and the velocity of the vibrating body. The desired damping characteristics can be obtained by changing these parameters. Examples of viscous damping include fluid flow around a piston in a cylinder and fluid film between sliding surfaces. Common motorcycle and automobile shock absorbers are viscous dampers. Consider the two masses m_1 and m_2 connected by a viscous damper c. If the velocity of mass m_1 is $\dot{x}_1$, then the velocity of m_2 is $\dot{x}_2$. Assuming $\dot{x}_1$ is greater than $\dot{x}_2$, the damping force F_d, which is proportional to the relative velocity, is given by

$$F_d \alpha \left(\dot{x}_1 - \dot{x}_2 \right)$$

or

$$F_d = c\left(\dot{x}_1 - \dot{x}_2 \right) \tag{1.23}$$

where c is the proportionality constant referred to as the coefficient of viscous damping.

Figures 1.11(a) and (b) show a block of a mass m sliding on a horizontal surface, where the two surfaces are separated by a liquid film. When a force F is applied to the block in the horizontal direction, the damping force f_c opposes the motion of the block, which depends on the nature of the fluid flow between the two surfaces. The magnitude of the damping force using a linear relationship for the viscous damper is given by

$$f_c = cV_{rel} \qquad \text{if } V_{rel} > 0 \tag{1.24}$$

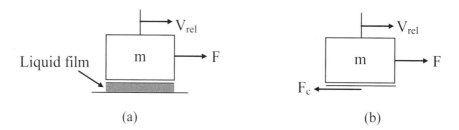

Figure 1.11 (a) Viscous damping, and (b) free body diagram.

This damping is called positive damping because it dissipates the energy of the system. The damping associated with self-excited vibration is called negative damping because it adds energy to the system and tends to make the system unstable.

Modeling of a system with viscous damping: Consider a single-degree-of-freedom spring-mass-damper system as shown in Figure 1.12(a). Figure 1.12(b) shows the free body diagram.

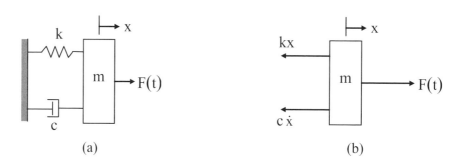

Figure 1.12 (a) A spring-mass-damper system, and (b) a free body diagram.

Applying Newton's second law in the x direction, we have

$$\Sigma F_x = ma_x \tag{1.25}$$

or

$$F(t) - c\dot{x} - kd = m\ddot{x} \tag{1.26}$$

Hence, the differential equation of motion is

$$m\ddot{x} = c\dot{x} + kx = F(t) \tag{1.27}$$

1.8.5.2 Coulomb Damping

Coulomb or dry friction damping occurs when sliding contact exists between surfaces in contact that are dry or that have insufficient lubrication. In this case, the damping force is constant in magnitude but is opposite in direction to that of the motion. In dry friction damping, energy is dissipated as heat.

Consider a block of mass m on a horizontal surface that is under the applied force F in the direction as shown in Figure 1.13(a). Figure 1.13(b) shows the friction force. The general friction force f_f is the static friction force f_s. That is, $f_f = f_s$. The range of f_s is from zero value to a maximum value given by

$$0 \le f_s \le \left[\left(f_s \right)_{max} = \mu_s N \right] \qquad \text{if } V_{rel} = 0 \tag{1.28}$$

where

F	=	applied force
f_f	=	general friction force
f_k	=	kinetic friction force
f_s	=	static friction force
$\left(f_s \right)_{max}$	=	maximum static friction force
N	=	normal force

Figure 1.13 (a) Coulomb damping, and (b) friction force.

The static force attains the maximum value at the transmission when the sliding is impending.

If the sliding occurs, then $f_f \to f_k$. For practical applications, the kinetic friction force is assumed to be a constant

$$f_k = \mu_k N \qquad \text{if } V_{rel} > 0 \tag{1.29}$$

where

$$\mu_k < \mu_s$$

where

μ_k = kinetic friction coefficient

μ_s = the static friction coefficient

Modeling of a system with Coulomb damping: Figure 1.14(a) shows a simple single-degree-of-freedom spring mass Coulomb damper system. The block of mass m slides on a dry surface. Figures 1.14(b) and (c) show the free body diagram.

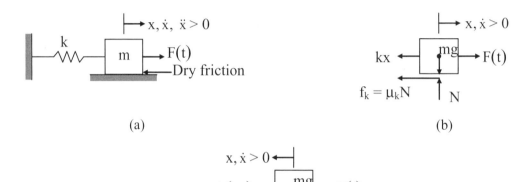

Figure 1.14 (a) A spring mass Coulomb damper system, (b) positive displacement and velocity, and (c) negative displacement and velocity.

Applying Newton's second law of motion in the x direction, we have

$$\Sigma F_x = m \, a_{cx} \tag{1.30}$$

For position displacement and velocity, we can write

$$F(t) - kx - \mu_k N = m\ddot{x} \qquad x > 0 \quad \text{and} \quad \dot{x} > 0 \tag{1.31}$$

or

$$m\ddot{x} + kx + \mu_k N = F(t) \qquad x > 0 \quad \text{and} \quad \dot{x} > 0 \tag{1.32}$$

For negative displacement and velocity, the equation becomes

$$F(t) + k(-x) + \mu_k N = m\ddot{x} \tag{1.33}$$

or

$$m\ddot{x} + kx + \mu_k N = F(t) \qquad x < 0 \quad \text{and} \quad \dot{x} < 0 \qquad (1.34)$$

Introducing the signum function,

$$\text{sgn}(\dot{x}) = \frac{\dot{x}}{|\dot{x}|} = +1 \qquad \text{if } \dot{x} > 0$$

$$\qquad\qquad\qquad\qquad (1.35)$$

$$\text{sgn}(\dot{x}) = \frac{\dot{x}}{|\dot{x}|} = -1 \qquad \text{if } \dot{x} < 0$$

Then, the equation for the system can be written as

$$m\ddot{x} + kx + \mu_k N \frac{\dot{x}}{|\dot{x}|} = F(t) \qquad \text{for all } \dot{x} \qquad (1.36)$$

Because $N = mg$, we can write Eq. 1.36 as

$$m\ddot{x} + kx + \mu_k mg \frac{\dot{x}}{|\dot{x}|} = F(t) \qquad \text{for all } \dot{x} \qquad (1.37)$$

1.8.5.3 Structural or Hysteretic Damping

Solid materials are not perfectly elastic. When they are deformed, energy is absorbed and is dissipated by the material. This effect is due to the internal friction that is caused by the relative motion between the internal planes of the material during the deformation process. Such materials are known as viscoelastic solids. The type of damping that they exhibit is called structural or hysteretic damping, or material or solid damping. When viscoelastic solid material is subjected to vibration, the stress-strain diagram shows a hysteresis loop as depicted in Figure 1.15(a). The area of the loop shown in Figure 1.15(b) shows the energy lost per unit volume of the body per cycle due to damping.

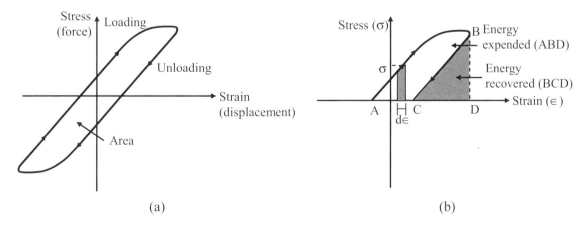

(a)   (b)

Figure 1.15 (a) Hysteresis loop for elastic materials, and (b) hysteresis in elastic materials.

1.8.5.4 Combination of Damping Elements

In many practical applications, several dashpots are used in combination. It is quite possible to replace this combination of dashpots by a single dashpot of an equivalent damping coefficient so that the behavior of the system with the equivalent dashpot is considered identical to the behavior of the actual system.

a. **Dashpots in parallel:** Consider the dashpots in parallel connected to the mass m. We wish to find the damping coefficient for a single dashpot, such that the behavior of the mechanical system shown in Figure 1.16(a) is identical to that of Figure 1.16(b). This is accomplished by considering the force equilibrium condition of each system. In this case, the displacement of the mass of each system is the same (say, x) when subjected to the same resulting force F. The resulting force acting on the mass attached to the parallel combination of dashpots is the sum of the individual damping forces

$$F = c_1\dot{x} + c_2\dot{x} + c_3\dot{x} + ... + c_n\dot{x} = \left(\sum_{i=1}^{n} c_i\right)x \tag{1.38}$$

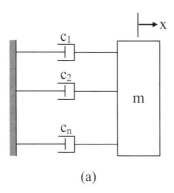

(a)

Figure 1.16 (a) Dashpots in parallel, and (b) a combination of dashpots in parallel replaced by a single equivalent dashpot.

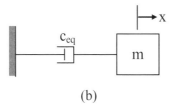

(b)

The force acting on the mass attached to the spring of an equivalent stiffness is

$$F = c_{eq}\dot{x} \tag{1.39}$$

Equating the force F from Eqs. 1.38 and 1.39 gives

$$c_{eq} = \sum_{i=1}^{n} c_i \tag{1.40}$$

The equivalent dashpot is one having a damping coefficient equal to the sum of the damping coefficients of all the dashpots in the original system. Other examples of parallel dashpot systems are shown in Figures 1.17(a), (b), and (c).

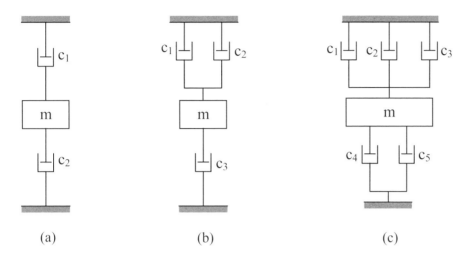

Figure 1.17 (a) A parallel system, (b) another parallel damper system, and (c) a complex parallel damper system.

b. **Dashpots in series:** The dashpots in the system shown in Figures 1.18(a) and (b) are said to be in series. The same force is developed in each dashpot and is equal to the force acting on the mass. In a series combination of dashpots, the sum of the deformations of each element is equal to the total deformation. However, the change in each deformation of each dashpot is different and is dependent on the dashpot coefficient. The displacement of the mass from equilibrium is the sum of the changes in the velocities of the dashpots

$$\dot{x} = \dot{x}_1 + \dot{x}_2 + \dot{x}_3 + \dots \dot{x}_n = \sum_{i=1}^{n} \dot{x}_n \qquad (1.41)$$

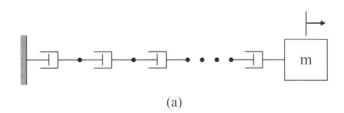

(a)

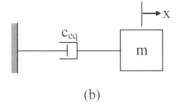

(b)

Figure 1.18 (a) Dashpots in series, and (b) a combination of dashpots in series is replaced by a single equivalent dashpot.

Because the force is the same in each dashpot, $\dot{x}_i = \dfrac{F}{c_i}$ and

$$\dot{x} = \sum_{i=1}^{n} \frac{F}{c_i} \tag{1.42}$$

The displacement of the block attached to the single dashpot of an equivalent damping coefficient must be equal to the displacement of a block attached to the series combination of dashpots when each block is subjected to the same resultant force.

Equating Eqs. 1.41 and 1.42, we obtain

$$c_{eq} = \frac{1}{\displaystyle\sum_{i=1}^{n} \frac{1}{c_i}} \tag{1.43}$$

Hence, when dashpots are in series, the reciprocal of the equivalent damping coefficient is equal to the reciprocals of the damping coefficients of the damping elements in the original system.

1.9 Review of Dynamics

1.9.1 Newton's Laws of Motion

Sir Isaac Newton (1642–1727) formulated three fundamental laws upon which the science of mechanics is based. These three laws are stated as follows:

First law: If the resultant force acting on a particle is zero, the particle will remain at rest (if originally at rest) or will move with constant speed in a straight line (if originally in motion).

Second law: If the resultant force acting on a particle is not zero, the particle will have acceleration proportional to the magnitude of the resultant force and in the direction of the resultant force.

When a particle of mass m is acted on by a force F, the force F and the acceleration a of the particle satisfies the relation

$$F = ma$$

When a particle is subjected simultaneously to several forces, then

$$\sum F = ma \tag{1.44}$$

where $\sum F$ represents the sum, or resultant, of all the forces acting on the particle.

Third law: The forces of action and reaction between bodies in contact have the same magnitude, the same line of action, and the opposite sense.

1.9.2 Kinematics of Rigid Bodies

The position of a particle P on a rigid body at any instant in time can be referred to a fixed rectangular or Cartesian coordinate system, as shown in Figure 1.19.

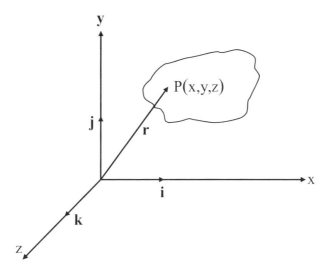

Figure 1.19 A Cartesian coordinate system.

The position vector of particle P is given by

$$\mathbf{r} = x\mathbf{i} + y\mathbf{j} + z\mathbf{k} \tag{1.45}$$

where the coordinates x, y, and z are functions of time t. Differentiating twice, we obtain the velocity and acceleration of the particle

$$\mathbf{V} = \frac{d\mathbf{r}}{dt} = \dot{x}\mathbf{i} + \dot{y}\mathbf{j} = \dot{z}\mathbf{k} \tag{1.46}$$

$$\mathbf{a} = \frac{d\mathbf{v}}{dt} \ddot{x}\mathbf{i} + \ddot{y}\mathbf{j} + \ddot{z}\mathbf{k} \tag{1.47}$$

where a dot above a quantity represents differentiation of that quantity with respect to time.

Note that the position $\mathbf{r}_B$ of particle B is the sum of the position vector $\mathbf{r}_A$ of particle A and of the position vector $\mathbf{r}_{B/A}$ of particle B relative to particle A, as shown in Figure 1.20.

Hence, we can write

$$\mathbf{r}_B = \mathbf{r}_A + \mathbf{r}_{B/A} \tag{1.48}$$

Differentiating Eq. 1.36 with respect to time within the fixed frame of references and using dots to indicate time derivatives, we have

$$\dot{\mathbf{r}}_B = \dot{\mathbf{r}}_A + \dot{\mathbf{r}}_{B/A} \tag{1.49}$$

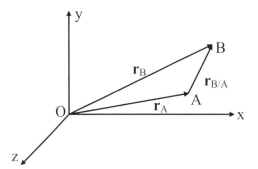

Figure 1.20 A vector representation.

or

$$\mathbf{V}_B = \mathbf{V}_A + \mathbf{V}_{B/A} \tag{1.50}$$

Differentiating Eq. 1.50 with respect to time, we obtain

$$\mathbf{a}_B = \mathbf{a}_A + \mathbf{a}_{B/A} \tag{1.51}$$

If the rigid body is rotating about an axis defined by a unit vector k with an angular velocity or speed ω, then the angular velocity vector is

$$\boldsymbol{\omega} = \omega \mathbf{k} \tag{1.52}$$

and the angular acceleration vector is given by

$$\boldsymbol{\alpha} = \frac{d\boldsymbol{\omega}}{dt} \tag{1.53}$$

Assuming two particles A and B are fixed to the rigid body, then

$$\mathbf{V}_{B/A} = \boldsymbol{\omega} \times \mathbf{r}_{B/A} \tag{1.54}$$

The acceleration of particle B relative to particle A is

$$\mathbf{a}_{B/A} = \boldsymbol{\alpha} \times \mathbf{r}_{B/A} + \boldsymbol{\omega} \times \left(\boldsymbol{\omega} \times \mathbf{r}_{B/A} \right) \tag{1.55}$$

EXAMPLE E1.8

Determine the angular acceleration of the crank at the instant shown in Figure E1.8(a).

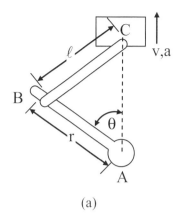

Figure E1.8 (a) A crank mechanism.

(a)

Solution:

Applying the law of sines (Figure E1.8(b)), we have

$$\frac{r}{\sin\phi} = \frac{\ell}{\sin\theta}$$

or

$$\sin\phi = \frac{r}{\ell}\sin\theta \qquad\qquad\text{(E1.8.1)}$$

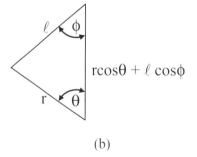

Figure E1.8 (continued) (b) The law of sines.

(b)

From trigonometry,

$$\cos\phi = \sqrt{1-\sin^2\phi}$$

$$= \sqrt{1-\left(\frac{r}{\ell}\sin\theta\right)^2} \qquad\qquad\text{(E1.8.2)}$$

From the relative velocity equation,

$$v_B = v_A + \omega_{AB} \times r_{B/A}$$

$$= \omega_{AB} k \times (-r \sin\theta i + r \cos\theta j)$$

$$= -r\omega_{AB}\left(\cos\dot{\theta}i - r\omega_{AB}\sin\theta_j\right)$$

and

$$v_c = vj = v_B = \omega_{BC} \times r_{C/B}$$

$$= v_B + \omega_{BC} k \times (\ell\sin\phi i + \ell\cos\phi j)$$

$$= \left(-r\omega_{AB}\sin\theta + \ell\omega_{BC}\sin\phi\right)j - \left(r\omega_{AB}\cos\theta + \ell\omega_{BC}\cos\phi\right)i$$

From the x component, we obtain

$$\omega_{BC} = -\frac{r}{\ell}\omega_{AB}\frac{\cos\theta}{\cos\phi} \tag{E1.8.3}$$

From the y component, we have

$$\omega_{AB} = -\frac{v}{r\left(\sin\theta + \cos\theta\tan\phi\right)} \tag{E1.8.4}$$

The relative acceleration equations give

$$a_B = a_A + \alpha_{AB} \times r_{B/A} + \omega_{AB} \times \left(\omega_{AB} \times r_{B/A}\right)$$

$$= \left(-r\alpha_{AB}\cos\theta + r\omega_{AB}^2\sin\theta\right)i + \left(-r\alpha_{AB}\sin\theta - r\omega_{AB}^2\cos\theta\right)j$$

and

$$a_C = aj = a_B + \alpha_{BC} \times r_{C/B} + \omega_{BC} \times \omega_{BC} \times r_{CB}$$

$$= \left(-r\alpha_{AB}\cos\theta + r\omega_{AB}^2\sin\theta - \ell\alpha_{BC}\cos\phi - \ell\omega_{BC}^2\sin\phi\right)i$$

$$+ \left(-r\alpha_{AB}\sin\theta + r\omega_{AB}^2\cos\theta + \ell\alpha_{BC}\sin\phi - \ell\omega_{BC}^2\cos\phi\right)j$$

The x component is used to obtain

$$\alpha_{BC} = -\frac{1}{\ell\cos\phi}\left(r\omega_{AB}^2\sin\theta + r\alpha_{AB}\cos\theta + \ell\omega_{BC}^2\sin\phi\right)$$

From the y component, we obtain

$$\alpha_{AB} = -\frac{a - r\omega_{AB}^2\cos\theta + \ell\omega_{BC}^2\cos\phi - r\omega_{AB}^2\sin\theta\tan\phi + \ell\omega_{BC}^2\sin\phi\tan\phi}{r(\sin\theta + \cos\theta\tan\phi)} \quad \text{(E1.8.5)}$$

Equation E1.8.5 can be used to find the angular acceleration of the crank using Eqs. E1.8.1 through E1.8.4.

1.9.3 Linear Momentum

If we replace the acceleration a by the derivative $\dfrac{d\mathbf{v}}{dt}$ in Newton's second law of motion, $\sum F = ma$, we can write

$$\sum F = m\frac{d\mathbf{v}}{dt} \quad (1.56)$$

because the mass of a particle is constant

$$\sum F = \frac{d}{dt}(mv) \quad (1.57)$$

The vector mv is called the linear momentum or simply the momentum of the particle, as shown in Figure 1.21. Equation 1.56 states that the resultant of the forces acting on the particle is equal to the rate of change of the linear momentum of the particle. Denoting by L the linear momentum of the particle,

$$L - mv$$

or

$$\dot{L} = \frac{d}{dt}(mv) = m\left(\frac{d\mathbf{v}}{dt}\right) = ma = \sum F \quad (1.58)$$

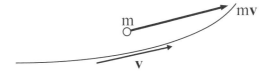

Figure 1.21 Linear momentum.

1.9.4 Principle of Conservation of Linear Momentum

It follows from Eq. 1.58 that the rate of change of the linear momentum mv is zero when $\sum F = 0$. Hence, if the resultant force acting on a particle is zero, the linear momentum of the particle remains constant in both magnitude and direction. This is known as the principle of conservation of linear momentum.

1.9.5 Angular Momentum

Consider a particle P of mass m moving with respect to a Newtonian frame of reference Oxyz as shown in Figure 1.22. The moment about O of the vector mv is called the moment of momentum or the angular momentum of the particle about O at that instant and is denoted by H_O. Denoting yr as the position vector of P, we have

$$\mathbf{H}_O = \mathbf{r} \times m\mathbf{v} \tag{1.59}$$

or

$$\mathbf{H}_O = \mathbf{r}\, mv \sin \phi \tag{1.60}$$

where ϕ is the angle between $\mathbf{r}$ and $m\mathbf{v}$ (Figure 1.22).

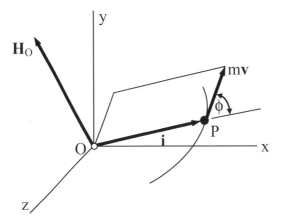

Figure 1.22 Angular momentum.

Differentiating Equation 1.59 with respect to time, we obtain

$$\dot{\mathbf{H}}_O = \dot{\mathbf{r}} \times m\mathbf{v} + \mathbf{r} \times m\dot{\mathbf{v}} = \mathbf{v} \times m\mathbf{v} + \mathbf{r} \times m\mathbf{a} \tag{1.61}$$

Because $\sum M_0$ equals the moment about O of the forces $\mathbf{r} \times \sum \mathbf{F}$, we can write

$$\sum M_0 = \dot{\mathbf{H}}_O \tag{1.62}$$

Equation 1.62 states that the sum of the moments about O of the forces acting on the particle is equal to the rate of change of the moment of momentum, or angular momentum, of the particle about O.

1.9.6 Equations of Motion for a Rigid Body

Figure 1.23 shows a rigid body acted on by several external forces F_1, F_2, F_3, ..., F_n.

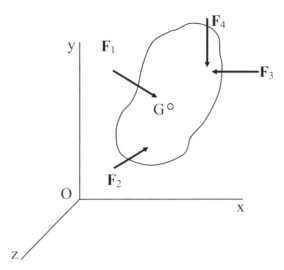

Figure 1.23 A rigid body acted on by external forces.

The motion of the mass center G of the body with respect to the Newtonian frame of reference Oxyz is given by

$$\sum F = ma \tag{1.63}$$

where m is the mass of the body, and a is the acceleration of the mass center G. The motion of the body relative to the centroidal frame of reference Gx'y'z' is given by

$$\sum M_G = \dot{H}_G \tag{1.64}$$

where $\dot{H}_G$ is the rate of change of H_G, the angular momentum about G of the system of particles forming the rigid body.

1.9.7 Angular Momentum of a Rigid Body

Assuming the rigid body in Figure 1.24 is made up of a large number of particles P_i of mass Δm_i, the angular momentum H_G of the rigid body about its mass center G can be obtained by taking the moment about G of the momenta of the particles of the body in their motion with respect to either of the frames Oxy or Gx'y'.

Hence,

$$H_G = \sum_{i=1}^{n} \left(r_i' \times v_i' \, \Delta m_i \right) \tag{1.65}$$

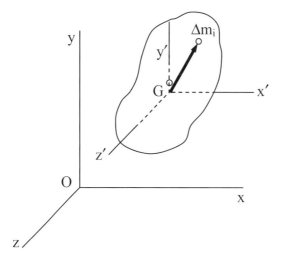

Figure 1.24 A rigid body with a particle of mass Δm_i.

where r_i' and $v_i' \Delta m_i$ are the position and the linear momentum of the particle P_i relative to the centroidal frame of reference $Gx'y'$, respectively. Because $v_i' = \omega \times r_i'$, where ω is the angular velocity of the body, we can write

$$H_G = \sum_{i=1}^{n} \left[\mathbf{r}_i' \times \left(\omega \times r_i' \right) \Delta m_i \right] \tag{1.66}$$

Noting that $\sum r_i'^2 \Delta m_i$ represents the moment of inertia $\overline{I}$ of the body about a centroidal axis normal to the body, the angular momentum H_G of the body about its mass center is

$$H_G = \overline{I}\omega \tag{1.67}$$

1.9.8 Principle of Work and Energy

Application of the principle of work and energy to the analysis of the motion of a rigid body gives

$$T_1 + U_{1 \to 2} = T_2 \tag{1.68}$$

where

T_1 and T_2 = initial and final values of the total kinetic energy of the particles forming the rigid body

$U_{1 \to 2}$ = work of all forces acting on various particles of the body

The total kinetic energy is

$$T = \frac{1}{2} \sum_{i=1}^{n} \Delta m_i v_i^2 \tag{1.69}$$

where the rigid body is assumed to be made of a large number n of particles of mass Δm_i.

The work done by a force F acting on a rigid body as the point of application of the force travels between two points described by position vectors r_A and r_B is given by

$$U_{1\rightarrow2} = \int_{r_A}^{r_B} \mathbf{F} \cdot d\mathbf{r} \tag{1.70}$$

where dr is a differential position vector in the direction of motion.

The work of a couple of moment M acting on a rigid body is

$$dU = M\,d\theta \tag{1.71}$$

where $d\theta$ is the small angle through which the body rotates. The work of the couple during a finite rotation of the rigid body from the initial value θ_A of the angle θ to the final value θ_B is given by

$$U_{1\rightarrow2} = \int_{\theta_A}^{\theta_B} M\,d\theta \tag{1.72}$$

A force is called conservative if the work of the force is independent of the path taken from 1 to 2. Spring force, gravity force, and normal force are some examples of conservative forces. The work done by a conservative force is equal to the difference in the potential energies

$$U_{A\rightarrow B} = V_A - V_B \tag{1.73}$$

where V_A and V_B are the potential energy functions for A and B, respectively. Because the system of external forces is equivalent to the system of effective forces, the total work done on a rigid body is given by

$$U_{A\rightarrow B} = \int_{r_A}^{r_B} ma \cdot dr + \int_{\theta_A}^{\theta_B} \bar{I}\alpha\,d\theta \tag{1.74}$$

1.9.9 Conservation of Energy

When a rigid body or a system of rigid bodies moves under the action of conservation forces, the principle of work and energy can be expressed as

$$T_1 + V_1 = T_2 + V_2 \tag{1.75}$$

Equation 1.75 states that when a rigid body or a system of rigid bodies moves under the action of conservation forces, the sum of the kinetic energy and of the potential energy of the system remains constant. This is known as the law of conservation of energy.

1.9.10 Principle of Impulse and Momentum

Considering that a rigid body is composed of a large number of particles P_i (Figure 1.25), we can write

$$\text{System Momenta}_1 + \text{System External Imp}_{1\rightarrow 2} = \text{System Momenta}_2 \qquad (1.76)$$

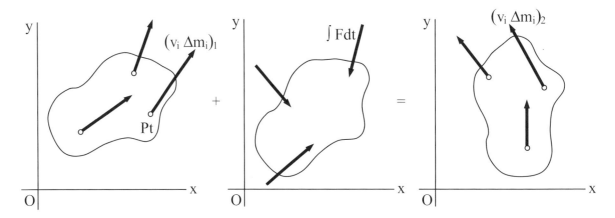

Figure 1.25 Principle of impulse and momentum.

But

$$L = \sum_{i=1}^{n} \mathbf{v}_i \Delta m_i \qquad (1.77)$$

and a couple of moments equal to the sum of their moments about G

$$H_G = \sum_{i=1}^{n} \mathbf{r}_i' \times v_i \, \Delta m_i \qquad (1.78)$$

where

$\quad$ L $\quad$ = $\quad$ the linear momentum about G of the system of particles forming the rigid body

$\quad$ H_G $\quad$ = $\quad$ the angular momentum about G of the system of particles forming the rigid body

Because $H_G = \bar{I}\omega$ from Eq. 1.67, we can conclude that the system of the moment $v_i\Delta m_i$ is equivalent to the linear momentum vector mv attached at G and to the angular momentum couple $\bar{I}\omega$ (Figure 1.26).

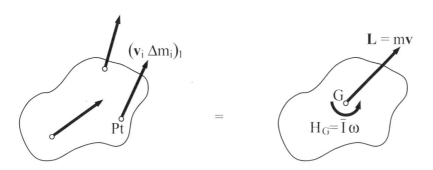

Figure 1.26 Angular momentum couple.

Figure 1.27 shows that the plane motion of a rigid body is symmetrical with respect to the reference plane. In the case of noncentroidal rotation, $\bar{v} = \bar{r}\omega$, where $\bar{r}$ represents the distance from the mass center to the fixed axis of rotation, and ω represents the angular velocity of the body and $m\,\bar{v} = m\,\bar{r}\,\omega$.

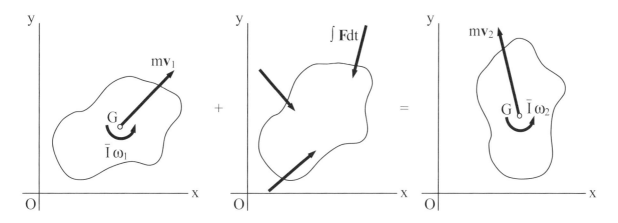

Figure 1.27 Plane motion of a rigid body.

Summing the moments about O of the momentum vector and momentum couple (Figure 1.28), we can write

$$\bar{I}\omega + \left(m\,\bar{r}\,\omega\right)\bar{r} = \left(\bar{I} + m\,\bar{r}^2\right)\omega = I_0\omega \tag{1.79}$$

Now, equating the moments about O of the momenta and impulses, we have

$$I_0\omega_1 + \sum \int_{t_1}^{t_2} M_0 dt = I_0\omega_2 \tag{1.80}$$

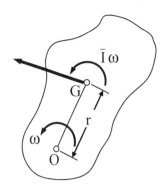

Figure 1.28 Momentum couple.

EXAMPLE **E1.9**

A sphere of radius r and mass m is moving along a rough horizontal surface with the initial linear velocity $\bar{v}_0$ and angular velocity $\bar{\omega}_0$, as shown in Figure E1.9(a). If the final velocity of the sphere is to be zero, determine the following:

a.   The required magnitude of ω_0 in terms of υ_0 and r

b.   The time required for the sphere to come to rest in terms of υ_0 and the coefficient of kinetic friction μ_k.

Solution:

Figure E1.9(b) shows the solution.

$$\Sigma y : Nt = Wt = 0$$

$$\text{or} \tag{E1.9.1}$$

$$N = W = mg$$

$$\Sigma x : m\bar{\upsilon}_0 - Ft = 0$$

$$\text{or} \tag{E1.9.2}$$

$$Ft = m\bar{\upsilon}_0$$

$$\Sigma M : \text{Moment about G}$$

$$\bar{I}\omega_0 - (Ft)r = 0 \tag{E1.9.3}$$

Note that

$$\bar{I} = \frac{2}{5}mr^2 \tag{E1.9.4}$$

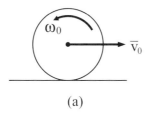

(a)

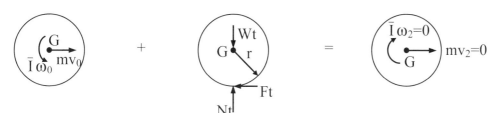

System Momenta₁ + System External Imp₁→₂ = System Momenta₂

(b)

Figure E1.9 (a) A sphere rolling on a horizontal surface, and (b) system momenta.

Using Eqs. E1.9.2 and E1.9.4, Eq. E1.9.3 becomes

$$\frac{2}{5}m^2\omega_0 - (m\bar{\upsilon}_0)r = 0$$

or   (E1.9.5)

$$\omega_0 = \frac{5}{2}\frac{\bar{\upsilon}_0}{r}$$

From Eq. E1.9.2,

$$t = \frac{m\bar{v}_0}{F} = \frac{m\bar{v}_0}{\mu_k mg} = \frac{\bar{\upsilon}_0}{\mu_k g}$$ (E1.9.6)

1.9.11 Mechanical Systems

The motion of a mechanical system can be described as

a. Pure translational motion

b. Pure rotational motion

c. Mixed: translational and rotational motion

There are three types of basic elements in a mechanical system. They are inertia elements, spring elements, and damper elements. For the purpose of modeling, the components of a mechanical system are assumed to contain only pure elements, that is, a spring is considered to be a pure spring with no mass and is assumed to dissipate no energy during extension or compression. Inertia elements include masses and moments of inertia. A damper is a mechanical element that dissipates energy in the form of heat instead of storing it. Inertia may be defined as the change in force (torque) needed to make a unit change in acceleration (angular acceleration).

1.9.12 Translational Systems

Newton's laws can be applied to develop a mathematical model of a mechanical system. For a translational motion of a mechanical system, Newton's second law states that the acceleration of any rigid body is directly proportional to the force acting on it and is inversely proportional to the mass of the body. For a system of n rigid bodies, Newton's second law can be written as

$$\sum_{i=1}^{N} \mathbf{F}_i = m\, a_c = \sum_{j=1}^{n} m_j a_{cj} \tag{1.81}$$

where

$$m = \sum_{j=1}^{n} m_j \tag{1.82}$$

in which

F_i = externally applied ith physical force vector

a_c = absolute acceleration vector of the mass center of the rigid body system

a_{cj} = absolute acceleration vector of the mass center of the jth rigid body

N = total number of externally applied physical forces

n = total number of rigid bodies

For a Cartesian coordinate system used for a translational system, the component form of Eq. 1.81 in the x and y directions can be written as

$$\sum_{i=1}^{N} F_{ix} = m a_{cx} = \sum_{j=1}^{n} m_j a_{cjx} \tag{1.83}$$

and

$$\sum_{i=1}^{N} F_{iy} = m a_{cy} = \sum_{j=1}^{n} m_j a_{cjy} \tag{1.84}$$

The line of action of the force acting on a mass must pass through the center of the mass.

1.9.13 Rotational Systems

Newton's second law for pure rotational motion about a fixed axis of a rigid body can be stated as

$$\Sigma T = J\alpha \tag{1.85}$$

where

ΣT = sum of all torques acting about a given axis

J = moment of inertia of a body about that given axis

α = angular acceleration

The general moment equation for a rigid body in three-dimensional arbitrary motion can be written as

$$\Sigma M_p = \dot{H}_p + m\, r_{c/p} \times a_p \tag{1.86}$$

where

ΣM_p = sum of all moments about point p

$\dot{H}_p$ = time rate of change of the angular momentum vector about point p

$r_{c/p}$ = position vector of the mass center C with respect to point p

a_p = absolute acceleration vector of point p

m = rigid body total mass

p = any arbitrary point fixed in the rigid body

In matrix form, we can write

$$H_P = I_P \omega \tag{1.87}$$

or

$$\{H\}_P = [I]_P \{\omega\} \tag{1.88}$$

Expanding Eq. 1.88, we have

$$\begin{Bmatrix} H_x \\ H_y \\ H_z \end{Bmatrix} = \begin{bmatrix} I_{xx} & I_{xy} & I_{xz} \\ I_{yx} & I_{yy} & I_{yz} \\ I_{zx} & I_{zy} & I_{zz} \end{bmatrix} \begin{Bmatrix} \omega_x \\ \omega_y \\ \omega_z \end{Bmatrix} \tag{1.89}$$

where

$(H_x)_p, (H_y)_p, (H_z)_p$ = three Cartesian components of the angular momentum vector H_p

ω = angular velocity of the rigid body (3×1 column matrix)

$$H_p \qquad = \text{angular momentum of the rigid body about point p } (3 \times 1 \text{ column matrix})$$

$$I_p \qquad = \text{mass moment of inertia of the rigid body about point p (square matrix).}$$

The time rate of change of the angular momentum vector for a rigid body in planar motion can be written as

$$\dot{\mathbf{H}}_p = I_p\,\alpha \tag{1.90}$$

where

$$I_P = I_c + md^2 \tag{1.91}$$

in which

I_p = mass moment of inertia about point p and the fixed axis of rotation

I_c = mass moment of inertia about the mass center C

d = distance from the mass center C to point p

Hence, the moment equation can be rewritten as

$$\Sigma\dot{\mathbf{H}}_p = I_p\,\alpha + m\,r_{c/p} \times a_p \tag{1.92}$$

where

$$I_P = I_c + md^2$$

Equation 1.92 also can be written as

$$\Sigma M_p = \dot{\mathbf{H}}_c + m\,r_{c/p} \times a_c \tag{1.93}$$

where c is the mass center.

1.9.14 Translational and Rotational Systems

For translational and rotational mechanical systems, first we use Newton's second law for the translational subsystem, and then we use the moment equation for the rotational subsystem. The differential equations governing the motion of the subsystems then are combined to give the complete system equations of motion of the mixed system.

1.9.15 Angular Momentum and Moments of Inertia

The angular momentum vector of a system of n particles rotating about a reference point P is defined as

$$H_p = \sum_{i=1}^{n} \mathbf{r}_{i/p} \times m_i\,v_{i/p} \tag{1.94}$$

where

$r_{i/p}$ = position vector of the ith particle with respect to point P

m_i = mass of the ith particle

$v_{i/p}$ = velocity vector of the ith particle with respect to point P

From the fundamentals of kinematics of rotating frames, if A is an arbitrary vector in a frame rotating with angular velocity ω, then

$$\frac{d\mathbf{A}}{dt} = \left(\frac{\partial \mathbf{A}}{\partial t}\right)_{rel} + \omega \times A \tag{1.95}$$

or

$$\dot{\mathbf{A}} = \left(\dot{\mathbf{A}}\right)_{rel} + \omega \times A \tag{1.96}$$

Thus,

$$v_{i/p} = \dot{r}_{i/p} = \left(\dot{r}_{i/p}\right)_{rel} + \omega \times r_{i/p} \tag{1.97}$$

For a rigid body, $\left(\mathbf{r}_{i/p}\right)_{rel} = 0$, the ith particle becomes the ith point, so the subscripts may be dropped for brevity. The velocity of the ith point with respect to point P becomes

$$v = \dot{\mathbf{r}} = \left(\dot{\mathbf{r}}\right)_{rel} + \omega \times r = \omega \times r \tag{1.98}$$

because $\left(\dot{\mathbf{r}}\right)_{rel} = 0$. In addition, the summation sign in the expression of the angular momentum becomes the integral; thus,

$$H_p = \int_V \mathbf{r} \times v_p \, dV = \int_V \mathbf{r} \times \omega \times r_p \, dV \tag{1.99}$$

The velocity vector of the ith point with respect to point P may be given as

$$r = xi + yj + jk \tag{1.100}$$

and the angular velocity vector is

$$\omega = \omega_x i + \omega_y j + \omega_z k \tag{1.101}$$

Performing the cross product,

$$\omega \times r = \left(\omega_x i + \omega_y j + \omega_z k\right) \times \left(xi + yj + zk\right) \tag{1.102}$$

or

$$\omega \times r = \begin{vmatrix} \mathbf{i} & \mathbf{j} & \mathbf{k} \\ \omega_x & \omega_y & \omega_z \\ x & y & z \end{vmatrix} = \left(z\omega_y - y\omega_z\right)\mathbf{i} + \left(z\omega_z - y\omega_x\right)\mathbf{j} + \left(z\omega_x - y\omega_y\right)\mathbf{k} \quad (1.103)$$

and

$$r \times \left(\omega \times r\right) = \begin{vmatrix} \mathbf{i} & \mathbf{j} & \mathbf{k} \\ x & y & z \\ z\omega_y - y\omega_z & x\omega_z - z\omega_x & y\omega_x - x\omega_y \end{vmatrix} \quad (1.104)$$

$$
\begin{aligned}
r \times \left(\omega \times r\right) &= \left[y\left(y\omega_x - x\omega_y\right) - z\left(x\omega_z - z\omega_x\right)\right]\mathbf{i} + \left[z\left(z\omega_y - y\omega_z\right) - x\left(y\omega_x - x\omega_y\right)\right]\mathbf{j} \\
&\quad + \left[x\left(x\omega_z - z\omega_x\right) - y\left(z\omega_y - x\omega_z\right)\right]\mathbf{k} \\
&= \left[\left(y^2 + z^2\right)\omega_x - xy\omega_y - xz\omega_z\right]\mathbf{i} + \left[-yx\omega_x + \left(z^2 + x^2\right)\omega_y - yz\omega_z\right]\mathbf{j} \\
&\quad + \left[-zx\omega_x - zy\omega_y + \left(x^2 + y^2\right)\omega_z\right]\mathbf{k}
\end{aligned}
\quad (1.105)
$$

Thus,

$$H_p = \int_V \mathbf{r} \times \left(\omega \times r\right)_\rho dV \quad (1.106)$$

$$
\begin{aligned}
H_p &= \left[\int_V \left(y^2 + z^2\right)\rho\, dV\omega_x - \int_V xy\rho\, dV\omega_y - \int_V xz\rho\, dV\omega_z\right]\mathbf{i} \\
&\quad + \left[-\int_V xy\rho\, dV\omega_x + \int_V \left(z^2 + x^2\right)\rho\, dV\omega_y - \int_V yz\rho\, dV\omega_z\right]\mathbf{j} \\
&\quad + \left[-\int_V zx\rho\, dV\omega_x - \int_V zy\rho\, dV\omega_y + \int_V \left(x^2 + y^2\right)\rho\, dV\omega_z\right]\mathbf{k}
\end{aligned}
\quad (1.107)
$$

The moments of inertia about point P may be defined as

$$\left(I_{xx}\right)_p = \int_V \left(y^2 + z^2\right)\rho\, dV \quad (1.108)$$

$$\left(I_{yy}\right)_p = \int_V \left(z^2 + x^2\right)\rho\, dV \quad (1.109)$$

$$\left(I_{zz}\right)_p = \int_V \left(x^2 + y^2\right)\rho\, dV \quad (1.110)$$

$$\left(I_{xy}\right)_p = \left(I_{yx}\right)_p = -\int_V xy\rho\, dV \quad (1.111)$$

$$\left(I_{yz}\right)_p = \left(I_{zy}\right)_p = -\left(I_{yz}\right)_p = \left(I_{zy}\right)_P$$

$$= -\left(I_{yz}\right)_p = \left(I_{zy}\right)_p = -\int_V yz\rho \, dV \tag{1.112}$$

$$\left(I_{zx}\right)_p = \left(I_{xz}\right)_p = -\int_V zx\rho \, dV \tag{1.113}$$

To avoid any confusion between the two different types of moments of inertia in mechanics, we use mass moments of inertia for dynamics and area moments of inertia for mechanics of materials.

The angular moment vector about point P thus is given as

$$H_P = \left(H_x\right)_P i = \left(H_y\right)_P j + \left(H_z\right)_P k \tag{1.114}$$

where

$$\left(H_x\right)_P = \left(I_{xx}\right)_P \omega_x + \left(I_{xy}\right)_P \omega_y + \left(I_{xz}\right)_P \omega_z \tag{1.115}$$

$$\left(H_y\right)_P = \left(I_{yx}\right)_P \omega_x + \left(I_{yx}\right)_P \omega_x + \left(I_{yy}\right)_P \omega_y + \left(I_{yz}\right)_P \omega_z \tag{1.116}$$

$$\left(H_z\right)_P = \left(I_{zx}\right)_P \omega_x + \left(I_{zy}\right)_P \omega_y + \left(I_{zz}\right)_P \omega_z \tag{1.117}$$

In terms of matrices,

$$H_P = I_P\omega \tag{1.118}$$

or

$$\begin{Bmatrix} H_x \\ H_y \\ H_z \end{Bmatrix}_p = \begin{bmatrix} I_{xx} & I_{xy} & I_{xz} \\ I_{yx} & I_{yy} & I_{yz} \\ I_{zx} & I_{zy} & I_{zz} \end{bmatrix}_p \begin{Bmatrix} \omega_x \\ \omega_y \\ \omega_z \end{Bmatrix} \tag{1.119}$$

Note that the word "thin" implies that the radius is negligible, $r \cong 0$. In fact, a uniform thin rod is simply a special case of a uniform circular cylinder.

Note that for a uniform circular disk of mass m and radius r, the moment of inertia about an axis through its mass center C and perpendicular to the plane of the disk is

$$I_c = \frac{1}{2}mr^2 \tag{1.120}$$

This is a direct result from the properties of a uniform circular cylinder. For brevity, this moment of inertia often is referred to as the centroidal mass moment of inertia in planar motion (2D).

1.9.16 Geared Systems

Examples of equivalent mechanical systems are given by devices called mechanical transformers. Some of these devices include the levers, cams, chains, and gears that transmit the motion at the input into a related motion at the output. Gear trains, levers, or timing belts over pulleys are mechanical devices that transmit energy from one part of the system to another in such a way that force, torque, speed, and displacement may be altered. Gear trains are used to reduce speed or to magnify torque, and so forth. Figure 1.29 shows a simple gear train system in which the gear train transmits motion and torque from the input shaft to the output shaft.

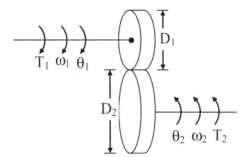

Figure 1.29 Gear train.

When the radii of gears 1 and 2 are r_1 and r_2, respectively, we have

$$\text{Gear ratio} = \frac{r_1}{r_2} = \frac{D_1}{D_2} = \frac{n_1}{n_2} \tag{1.121}$$

Because the surface speeds at the point of contact of the two gears must be equal, we have

$$r_1\omega_1 = r_2\omega_2 \tag{1.122}$$

where ω_1 and ω_2 are the angular velocities of gears 1 and 2, respectively.

$$\frac{\omega_2}{\omega_1} = \frac{r_1}{r_2} = \frac{n_1}{n_2} \tag{1.123}$$

where n_1 and n_2 are the number of gear teeth of gears 1 and 2, respectively. Neglecting the friction loss, the gear train transmits the same power. That is,

$$T_1\omega_1 = T_2\omega_2 \tag{1.124}$$

where T_1 and T_2 are the torques on gears 1 and 2, respectively,

or

$$\text{Torque ratio} = \frac{T_1}{T_2} = \frac{n_1}{n_2} = N \quad \text{(gear ratio)}$$

Similarly, we can write

$$\text{Displacement ratio} = \frac{D_1}{D_2} = \frac{n_1}{n_2} = N \quad \text{(1.125)}$$

$$\text{Velocity ratio} = \frac{\omega_1}{\omega_2} = \frac{n_1}{n_2} = \frac{1}{N} \quad \text{(1.126)}$$

EXAMPLE E1.10

For the single-degree-of-freedom spring mass vibrating system shown in Figure E1.10(a), derive the differential equation of motion.

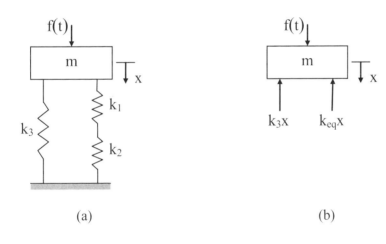

(a)   (b)

Figure E1.10 (a) Single-degree-of-freedom system, and (b) free body diagram.

Solution:

Referring to the free body diagram shown in Figure E1.10(b), the equivalent spring stiffness for stiffness k_1 and k_2 can be written as

$$k_{eq} = \frac{k_1 - k_2}{k_1 + k_2} \quad \text{(E.10.1)}$$

Applying Newton's second law of motion, we have

$$F_x = m a_{cx} \quad \text{(E.10.2)}$$

or

$$f(t) - k_3 x - k_{eq} x = m\ddot{x} \quad \text{(E.10.3)}$$

Hence, the differential equation of motion is

$$m\ddot{x} + \left(k_3 + \frac{k_1 k_2}{k_1 + k_2}\right)x = f(t) \qquad (E.10.4)$$

EXAMPLE E1.11

Figure E1.11 shows a homogeneous cylinder of radius r and mass M rolling without slipping on a rough surface. Obtain the natural frequency of oscillation of the system. The linear stiffness of the spring is k.

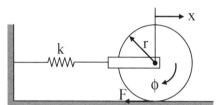

Figure E1.11 Cylinder rolling on a rough surface.

Solution:

Applying Newton's second law of motion, the equations of motion for the system are

$$M\ddot{x} = -kx - F$$

and

$$J\ddot{\phi} = Fr$$

where

$$x = r \quad \text{and} \quad J = \frac{1}{12}Mr^2$$

Hence,

$$M\ddot{x} = -kx - \frac{J\ddot{\phi}}{r} = -kx - \frac{1}{2}M\ddot{x}$$

or

$$\frac{3}{2}M\ddot{x} + kx = 0$$

The natural frequency of the system is

$$\omega_n = \sqrt{\frac{2k}{3m}}$$

EXAMPLE **E1.12**

Figure E1.12(a) shows an inverted pendulum with a cart of mass M and a uniform rod of mass m and length ℓ. The coefficient of viscous damping on the cart is c, and the applied force is F(t) as shown. The cart rolls without slipping. Determine the following:

a. The linearized equations of motion for small motions in matrix form

b. The transfer functions

c. The transfer matrix

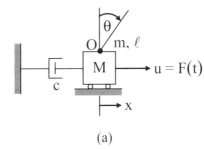

(a)

Figure E1.12 (a) An inverted pendulum.

Solution:

Refer to Figures E1.12(b), (c), and (d) for the solution.

a. Applying the moment equation for the rod, with the moment center at point O, we get

$$\Sigma \vec{M}_0 = I_0 \vec{\alpha} + m \, \vec{r}_{c/0} \times \vec{a}_0 \tag{E1.12.1}$$

$$L \sin \theta \, mg = \left(I_0 + mL^2\right)\ddot{\theta} + L\ddot{x}\cos\theta \tag{E1.12.2}$$

where

$$I_0 = \frac{1}{12}m\ell^2 \quad \text{and} \quad \sin(90 - \theta) = \cos\theta \tag{E1.12.3}$$

For the translational equation, see Figure E1.12(e).

Applying Newton's second law of motion for the system in the x direction, we can write

$$\sum_{i=1}^{n} F_{ix} = m\,a_{cx} = \sum_{j=1}^{2} m_j a_{cjx} = m_1 a_{c1x} + m_2 a_{c2x} \tag{E1.12.4}$$

$$u - c\dot{x} = M\ddot{x} + m\left(\ddot{x} + L\ddot{\theta}\cos\theta - L\dot{\theta}^2 \sin\theta\right) \tag{E1.12.5}$$

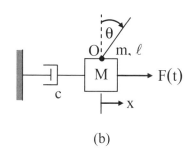

(b)

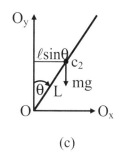

(c)

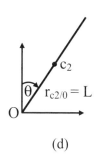

(d)

Figure E1.12 (continued)
(b) Free body diagram,
(c) another free body
diagram, (d) another free
body diagram, and
(e) another free body
diagram.

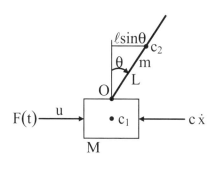

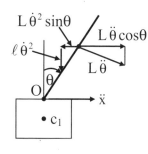

(e)

or

$$\left(I_c + mL^2\right)\ddot{\theta} + mL\ddot{x}\cos\theta - mgL\sin\theta = 0 \qquad (E1.12.6)$$

$$\left(M + m\right)\ddot{x} + mL\ddot{\theta}\cos\theta - mL\dot{\theta}^2\sin\theta + c\dot{x} = i = F(t) \qquad (E1.12.7)$$

For small motions,

$$\theta \ll 1\,\text{rad}, \quad \cos\theta = 1, \quad \text{and} \quad \sin\theta = \theta, \quad \dot{\theta}^2 = 0 \qquad (E1.12.8)$$

Hence, the equations of motion are

$$\left(I_c + mL^2\right)\ddot{\theta} + mL\ddot{x} - mgL\theta = 0 \qquad (E1.12.9)$$

$$mL\ddot{\theta} + (M + m)\ddot{x} + c\dot{x} = u = F(t) \qquad (E1.12.10)$$

or in matrix form,

$$\begin{bmatrix} Ic + mL^2 & mL \\ mL & M + m \end{bmatrix} \begin{Bmatrix} \ddot{\theta} \\ \ddot{x} \end{Bmatrix} + \begin{bmatrix} 0 & 0 \\ 0 & c \end{bmatrix} \begin{Bmatrix} \dot{\theta} \\ \dot{x} \end{Bmatrix} + \begin{bmatrix} -mgL & 0 \\ 0 & 0 \end{bmatrix} \begin{Bmatrix} \theta \\ x \end{Bmatrix} = \begin{Bmatrix} 0 \\ u \end{Bmatrix} \qquad (E1.12.11)$$

b. Taking the Laplace transform of Eq. E1.12.11 with zero initial conditions, we get

$$\begin{bmatrix} I_0 s^2 - m_0 g\ell & m_0 \ell s^2 \\ m_0 \ell s^2 & s(m_t s + c) \end{bmatrix} \begin{bmatrix} \theta(s) \\ X(s) \end{bmatrix} = \begin{Bmatrix} 0 \\ U(s) \end{Bmatrix} \qquad (E1.12.12)$$

Solving Eq. E1.12.12 for $\theta(s)$ and $X(s)$, we get

$$\theta(s) = \frac{\begin{vmatrix} 0 & m_0 \ell s^2 \\ U(s) & s(m_t s + c) \end{vmatrix}}{\Delta(s)} = -\frac{m_0 \ell s^2}{\Delta(s)} U(s)$$

and

$$X(s) = \frac{\begin{vmatrix} I_0 s^2 - m_0 g\ell & 0 \\ m_0 \ell s^2 & U(s) \end{vmatrix}}{\Delta(s)} = \frac{I_0 s^2 - m_0 g\ell}{\Delta(s)} U(s)$$

where

$$\Delta(s) = \left(I_0 s^2 - m_0 g\ell\right)(m_t s + c)s - \left(m_0 \ell s^2\right)^2$$

Hence, the two transfer functions are given as

$$G_1(s) = \frac{\theta(s)}{U(s)} = -\frac{m_0 \ell s^2}{\Delta(s)}$$

$$G_2(s) = \frac{X(s)}{U(s)} = \frac{I_0 s^2 - m_0 g\ell}{\Delta(s)}$$

c. The transfer matrix is

$$G(s) = \begin{Bmatrix} G_1(s) \\ G_2(s) \end{Bmatrix}$$

EXAMPLE **E1.13**

Develop a lumped inertia model of the impeller velocity as a function of the motor torque for the geared system shown in Figure 1.13. The elasticity of the gear teeth and shafts can be neglected. The friction and backlash in the gears are assumed to be small.

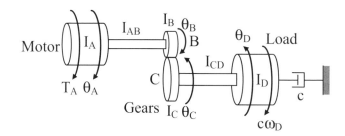

Figure E1.13 Geared system.

Solution:

Let N be the gear ratio. The angular displacements are related to the gear ratio as

$$\theta_B = N\theta_C$$

and

$$N = \frac{D_C}{D_B}$$

Because the shaft elasticity is negligible, the twist can be assumed to be zero. Hence,

$$\theta_A = \theta_B$$

$$\omega_A = \omega_B$$

$$\theta_C = \theta_D$$

$$\omega_C = \omega_D$$

$$\omega_A = N\omega_D$$

The lumped inertia on shaft AB is given by

$$I_1 = I_A + I_{AB} + I_B$$

The lumped inertia on shaft CD is given by

$$I_2 = I_C + I_{CD} + I_D$$

Choosing the motor shaft AB as the reference, the inertia equivalent to I_2 referenced to shaft AB is given by

$$I_{e2} = \frac{1}{N^2} I_2$$

The total equivalent inertia I is the sum of all the inertias with respect to shaft AB, or

$$I = I_1 + I_{e2}$$

The viscous damping acting on shaft CD has an equivalent value on shaft AB of

$$c_e = \frac{1}{N^2} c$$

Hence, the lumped inertia model is given by

$$I \frac{d\omega_A}{dt} = T_A - c_e \omega_A$$

1.10 Lagrange's Equation

There are two general approaches to classical dynamics: vectorial dynamics, and analytical dynamics. Vectorial dynamics is based directly on the application of Newton's second law of motion, concentrating on forces and motions. Analytical dynamics treats the system as a whole, dealing with scalar quantities such as the kinetic and potential energies of the system. Lagrange (1736–1813) proposed an approach that provides a powerful method for the formulation of the equations of motion for any dynamical system. Lagrange's equations are differential equations in which one considers the energies of the system and the work done instantaneously in time.

1.10.1 Degrees of Freedom

Application of Newton's laws considers the forces acting among particles or bodies within the system, their motion, and any constraints among the motions of individual particles. This is referred to as vectorial mechanics. The second approach considers the system as a whole and formulates expressions for the work and energy in terms of independently variable quantities, which are termed as generalized coordinates. This approach is referred to as analytical dynamics.

A particle in a three-dimensional space can move along any one of three directions. The complete motion can be expressed as a vector sum of the three independent motions; hence, the particle has three degrees of freedom. The three degrees of freedom need three independent quantities to describe the motion along these three degrees of freedom. A rigid body can translate along the three coordinate directions, which gives it three degrees of freedom. In addition, a rigid body also can rotate about any one of the axes. This ability gives it three more degrees of freedom. Totally, a rigid body needs six independent quantities to describe its motion in the six degrees of freedom.

1.10.2 Generalized Coordinates

The coordinates used to describe the motion in each degree of freedom of a system are termed as generalized coordinates. They may be Cartesian, polar, cylindrical, or spherical coordinates, provided any one of them can be used to describe the configuration of the system where the motion along any one coordinate direction is independent of the others. However, they sometimes may not have such simple physical or geometrical meaning. For example, the deflections of a string, stretched between two points, can be expressed in the form of trigonometric Fourier series, and the coefficients of all the terms in the series can be considered as a generalized coordinate set. This is true because each trigonometric function in the series may be considered as a unique degree of freedom, and the coefficients describe the extent of deflection in each degree of freedom.

It is possible to transform the coordinates from any one system to the generalized coordinate system, or vice versa, through coordinate transformation. Consider a mechanical system consisting of N particles whose positions are $(x_i, y_i, z_i,)$, where $i = 1, 2, N$, in a Cartesian coordinate system. The motion of the mechanical system is defined completely if the variations with time of these positions, that is, $x_i = x_i(t)$, $y_i = y_i(t)$, and $z_i = z_i(t)$ are known. These 3N coordinates completely define a representative space. If it is possible to find another set of generalized coordinates, q_i, $i = 1, 2, n$, where $n = 3N$, then these two coordinate systems are related by the following:

$$x_i(t) = x_i(q_1, q_2, ..., q_n, t)$$

$$y_i(t) = y_i(q_1, q_2, ..., q_n, t) \qquad (1.127)$$

$$z_i(t) = z_i(q_1, q_2, ..., q_n, t)$$

As an example, consider a Cartesian coordinate system and a polar coordinate system as shown in Figure 1.30. The coordinates of a point P are (x, y), which also are defined by (q_1, q_2). The transformation from Cartesian to polar coordinates has the form

$$x = q_1 \cos q_2$$
$$\qquad (1.128)$$
$$y = q_1 \sin q_2$$

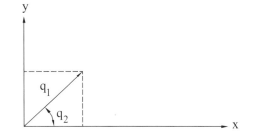

Figure 1.30 Cartesian and polar coordinate system.

1.10.3 Constraints

When two particles in space at (x_1, y_1, z_1) and (x_2, y_2, z_2) are connected by a rigid rod as in Figure 1.31, the length between them is a constant given by

$$(x_1 - x_2)^2 + (y_1 - y_2)^2 + (z_1 - z_2)^2 = L^2 = \text{ constant} \qquad (1.129)$$

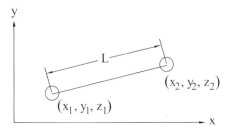

Figure 1.31 A rigid rod connected by two points.

By virtue of this constraint, the motion of particle A along the rod is the same as that of particle B. Hence, although the total number of coordinates needed to specify the position of the two particles is six, if five are known, the sixth can be obtained from the use of Eq. 1.129. Hence, the system has only five degrees of freedom, and only five generalized coordinates are needed to describe the configuration of the system.

In general, a system of N particles that are subject to c number of kinematic constraint relations such as Eq. 1.129 can be described uniquely by

$$n = 3N - c \qquad (1.130)$$

This can be imagined as a subspace of the configuration space of 3N dimensions. The use of generalized coordinates eliminates the need to consider the constraints in the system.

Constraints can be classified as follows:

1. Holonomic and nonholonomic

2. Scleronomic or rheonomic

3. Bilateral or unilateral

Holonomic constraints are those in which the generalized coordinates are related by a constraint equation of the form

$$\phi_j(q_1, q_2, ..., q_n, t) = 0 \qquad j = 1, 2, ..., k \qquad (1.131)$$

Using this constraint equation, it may be possible either to solve for one coordinate explicitly in terms of the other $(n - 1)$ coordinates, or the constraint may be an implicit relation among all the coordinates. If any differential expressions are involved in the equations, they must be integrable. Using k number of holonomic constraints, it is possible to eliminate k variables. Systems involving holonomic constraints are referred to as holonomic systems.

When the constraints are written as non-integrable differential expressions of the form

$$\sum_{i=1}^{n} a_{ji} dq_i + a_{jt} dt = 0 \qquad j = 1, 2, ..., k \qquad (1.132)$$

where a_{ji} and a_{jt} are, in general, functions of the q's and t. They are called nonholonomic constraints. Because these constraints are non-integrable in nature, it is not possible to express them in the form of the holonomic constraints given in Eq. 1.131 and to use them to eliminate variables. It also is not possible to find a set of independent generalized coordinates; hence, nonholonomic systems require a larger number of coordinates than their degrees of freedom to describe them.

As an illustration of a nonholonomic system, consider the motion of a vertical disk of radius a rolling on a rough horizontal surface, as shown in Figure 1.32. The contact point P of the disk moves along a curvilinear path as shown. The rough surface prevents slipping of the disk while rolling. When P is at (x, y), the tangent at P makes an angle φ with the OX axis. Let θ be the angular rotation of the disk when P has moved a distance s along the path. The four coordinates (x, y, θ, φ) specify the position of the disk at any instant. Suppose the position of the disk is varied from (x, y, θ, ϕ) to $(x + \delta x, y + \delta y, \theta + \delta\theta, \varphi + \delta\varphi)$. Because $s = a\theta$, we have $\delta s = a\delta\theta$ and

$$\delta x = \delta s \cos\varphi = a\cos\varphi\delta\theta$$
$$\delta y = \delta s \sin\varphi = a\sin\varphi\delta\theta \qquad (1.133)$$

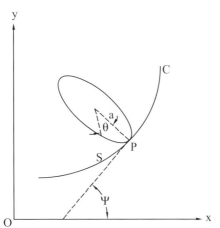

Figure 1.32 Vertical disk rolling on a rough horizontal surface.

The differential relations δx and δy in Eq. 1.133 are non-integrable because φ is arbitrary. Also, these relations show that the virtual displacements $(\delta x, \delta y)$ in the x and y depend on the virtual displacement $\delta\theta$ in θ. This means that (x, y, θ, φ) are not independent generalized coordinates; we need all four of them to specify the position of the disk. Hence, the system is nonholonomic.

When the constraint equations do not contain time, they are called scleronomic. For example, when two masses at (x_1, y_1, z_1) and (x_2, y_2, z_2) are connected by a rigid rod of length L, they are constrained by the relation

$$(x_2 - x_1)^2 + (y_2 - y_1)^2 + (z_2 - z_1)^2 = L^2 \qquad (1.134)$$

On the other hand, if the length L varies as a function of time, (say, $L(t)$), in the preceeding example, then the constraint is rheonomic. In the constraint of the type given in Eq. 1.134, both positive and negative displacements are allowed from any configuration of the system. Then the constraints are called bilateral constraints. If the constraint equation is of the type

$$\phi(q_1, q_2, ..., q_n, t) \leq 0 \qquad (1.135)$$

it is called a unilateral constraint. This implies that the configuration point is restricted to a certain region of the n-dimensional configuration space. If the configuration point lies on the boundary of the allowable region, the negative of a permitted small displacement will lie outside the allowable region. Hence, the unilateral nature of the constraint is obvious. Suppose that a free particle is constrained within a hollow sphere of radius r. With the origin of a Cartesian coordinate system fixed at the center of the sphere, the unilateral constraint is given by

$$x^2 + y^2 + z^2 - r^2 \leq 0 \qquad (1.136)$$

If the particle is inside the sphere and is not touching the inner surface, it can move freely with three degrees of freedom. However, if the particle moves on the inner surface, the equality condition applies. A unilateral constraint also can be holonomic in nature.

1.10.4 *Principle of Virtual Work*

The principle of virtual work, due to Johann Bernoulli, is essentially a statement of the static or dynamic equilibrium of a mechanical system. A virtual displacement, denoted by δr, is an imaginary displacement, and it occurs without the passage of time. The virtual displacement, being infinitesimal, obeys the rules of differential calculus. The virtual displacement takes place instantaneously, that is, it does not require any time to materialize, $\delta t = 0$.

Consider a mechanical system with N particles in a three-dimensional space, whose Cartesian coordinates are (x_1, y_1, z_1, z_n). Suppose the system is subject to k holonomic constraints

$$\phi_j(x_1, y_1, z_1, ..., z_n, t) = 0 \qquad j = 1, 2, ..., k$$

The virtual displacements $\delta x_1, \delta y_1, \delta z_1$, and so forth are said to be consistent with the system constraints if the constraint equations are still satisfied, that is,

$$\phi_j(x_1 + \delta x_1, y_1 + \delta y_1, z_1 + \delta z_1, ..., z_N + \delta z_N, t) = 0 \qquad (1.137)$$

Note that t is unchanged during the virtual displacement. Expanding Eq. 1.137 in Taylor's series about the original position, and neglecting the higher-order terms in δx_1, δy_1, δz_1, and so forth, we obtain

$$\phi_j\left(x_1, y_1, z_1, ..., z_N, t\right) + \sum_{i=1}^{N}\left(\frac{\partial \phi_j}{\partial x_i}\delta x_i + \frac{\partial \phi_j}{\partial y_i}\delta y_i + \frac{\partial \phi_j}{\partial z_i}\delta z_i\right) = 0 \qquad (1.138)$$

Hence,

$$\sum_{i=1}^{N}\left(\frac{\partial \phi_j}{\partial x_i}\partial x_i + \frac{\partial \phi_j}{\phi y_i}\partial y_i + \frac{\partial \phi_j}{\phi z_i}\partial z_i\right) = 0 \qquad (1.139)$$

as the condition for the virtual displacements to be consistent with constraints.

Let $\overline{F}_i$ be the force acting on particle i, which is subject to a virtual displacement δr_1. Then the virtual work of the system is

$$\delta W = \sum_{i=1}^{N}\overline{F}_i \cdot \delta \overline{r}_i \qquad (1.140)$$

Equation 1.140 represents the virtual work performed by the resultant force vector $\overline{F}_i$ over the virtual displacement vector δr_1 of particle i.

When the system is in equilibrium, the resultant force acting on each particle is zero. The resultant force is the sum of the applied force and the reaction force or the constraint force. Hence, under equilibrium conditions,

$$\overline{F}_i + \overline{R}_i = 0 \qquad (1.141)$$

Therefore, the virtual work done by all the forces in moving through an arbitrary virtual displacement consistent with the constraints is zero, that is,

$$\sum_{i=1}^{N}\left(\overline{F}_i + \overline{R}_i\right) \cdot \delta \overline{r}_i = \sum_{i=1}^{N}\overline{F}_i \cdot \delta \overline{r}_i + \sum_{i=1}^{N}\overline{R}_i \cdot \delta \overline{r}_i \qquad (1.142)$$

In Eq. 1.142, $\sum_{i=1}^{N}\overline{R}_i \cdot \delta \overline{r}_i$ is the work done by constraint forces. Many constraint forces that commonly occur do not do any work during a virtual displacement either because they are perpendicular to the displacement or because two equal and opposite reaction forces cancel the work done by each other.

The following are some examples of workless constraint forces:

1. A rigid rod connecting two particles (Figure 1.33(a)). Internal forces are equal in magnitude and opposite in direction. Hence, the network done by the internal forces is zero.

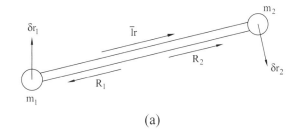

(a)

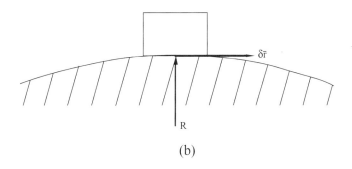

(b)

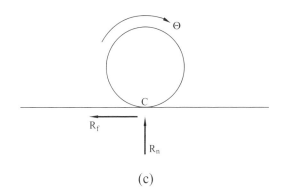

(c)

Figure 1.33 (a) A rigid rod connected by two points, (b) a rigid body sliding without friction, and (c) a disk rolling without slippage.

$$\bar{R}_2 = -R_1$$

$$\bar{e}_r \cdot \delta\bar{r}_1 = \bar{e}_r \cdot \delta\bar{r}_2$$

$$\delta W_c = \bar{R}_1 \cdot \delta\bar{r}_1 + \bar{R}_2 \cdot \delta\bar{r}_2 = 0$$

2. A body sliding without friction on a fixed surface (Figure 1.33(b)). The normal reaction is perpendicular to the direction of motion. Hence, the work done by the normal reaction is zero.

$$\bar{R} \text{ is normal to } \delta\bar{r}$$

hence,

$$\bar{R} \cdot \delta\bar{r} = 0$$

3. A circular disk rolling without slipping along a straight horizontal path (Figure 1.33(c)). Note that the instantaneous center does not move during a virtual

displacement (again, Figure 1.33(c)). Hence, the work of the friction force R_f is zero.

$$\overline{R}_f \cdot 0 = 0$$

Such constraints are called workless constraints. If the system considered has workless constraints, we have in Eq. 1.142

$$\sum_{i=1}^{N} \overline{R}_i \cdot \delta\overline{r}_i = 0 \qquad (1.143)$$

Hence, from Eq. 1.140, we have

$$\delta W = \sum_{i=1}^{N} \overline{F}_i \cdot \delta\overline{r}_i = 0 \qquad (1.144)$$

Equation 1.144 states that the work performed by the applied forces through infinitesimal virtual displacements compatible with the system constraints is zero. This is known as the principle of virtual work.

Now if we assume that the system is not in equilibrium, the system will start to move in the direction of the resultant force. Because any motion must be compatible with the constraints, we can always choose a virtual displacement in the direction of the actual motion at each point. In such a case, the virtual work is positive, that is,

$$\sum_{i=1}^{N} \overline{F}_i \cdot \delta\overline{r}_i + \sum_{i=1}^{N} \overline{R}_i \cdot \delta\overline{r}_i > 0 \qquad (1.145)$$

But the constraints are workless; hence,

$$\sum_{i=1}^{N} \overline{F}_i \cdot \delta\overline{r}_i > 0 \qquad (1.146)$$

These results are summarized in the principle of virtual work: The necessary and sufficient condition for the static equilibrium of the initially motionless scleronomic system that is subjected to workless constraints is that zero virtual work be done by the applied forces in moving through an arbitrary virtual displacement satisfying the constraints.

It sometimes is convenient to assume that a set of δx's conforming to the instantaneous constraints occurs during an interval of time δt. The corresponding ratios of the form $\frac{\delta x}{\delta y}$ have the dimensions of velocity and are known as virtual velocities. Accordingly, a principle of virtual power can be formulated: For a system at rest for a finite time, the total power of the system given by the product of the applied forces and the virtual velocity must vanish for all virtual velocities conforming to the constraints.

For the following holonomic and nonholonomic constraints,

$$\phi_j\left(x_1,\, y_1,\, z_1,\, ...,\, z_N\right) = 0 \tag{1.147}$$

$$\sum_{i=1}^{3N} a_{ji}dq_i + a_{ji}dt = 0 \tag{1.148}$$

the condition for the virtual displacements and virtual velocities to be consistent with the constraints are, respectively,

$$\frac{\partial\phi_j}{\partial t} = 0 \tag{1.149}$$

$$a_{jt} = 0 \tag{1.150}$$

A virtual displacement and virtual velocity also can be a possible real displacement and real velocity described by a set of dx's and assumed to occur during the time increment dt only if the conditions in Eqs. 1.149 and 1.150 are satisfied. Because these conditions are not met in the general case, a virtual displacement or virtual velocity is not, in general, a possible real displacement.

EXAMPLE E1.14

Figure E1.14(a) shows a system of a uniform rigid link of mass m and length ℓ and two linear springs of stiffness k_1 and k_2. The link is in the horizontal position when the springs are unstretched. Derive the equilibrium equation for the system using the principle of virtual work.

Solution:

Let C be the center of gravity of the link AB.

Referring to Figure E1.14(b), the virtual work principle can be written as

$$\delta W = -k_1 x\,\delta x - k_2 y\,\delta y + mg\frac{1}{2}\delta y = 0$$

The rectangular coordinates x and y, and the rectangular virtual displacements δx and δy in terms of the generalized coordinate θ and the generalized virtual displacement $\delta\theta$ are

$$x = \ell\left(1 - \cos\theta\right)$$

$$y = \ell\sin\theta$$

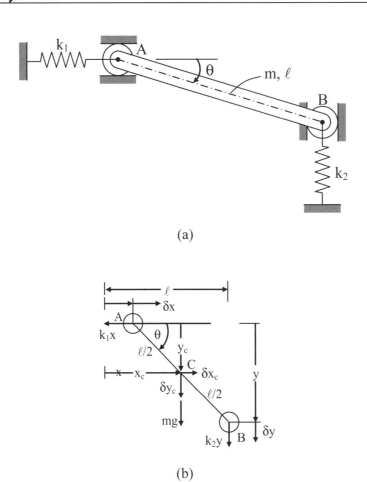

Figure E1.14 (a) Vibrating system, and (b) free body diagram.

(a)

(b)

$$\delta x = \ell \sin \theta \, \delta \theta$$

$$\delta y = \ell \cos \theta \, \delta \theta$$

Therefore, the virtual work principle in terms of the generalized displacement θ is

$$\delta W = \left[-k_1 \ell (1 - \cos \theta) \ell \sin \theta - k_2 \ell \sin \theta \ell \cos \theta + \frac{1}{2} mg \ell \cos \theta \right] \delta \theta = 0$$

Because $\delta \theta$ is arbitrary, the coefficient of $\delta \theta$ must be zero.

Hence, the equilibrium equation is

$$\frac{mg}{2\ell} = k_1 \tan \theta + (k_2 - k_1) \sin \theta$$

1.10.5 D'Alembert's Principle

Here, the principle of virtual work is extended to dynamics, in which form is known as D'Alembert's principle. The principle of virtual work is extended to the dynamic case by considering the inertia forces and by considering the systems to be in dynamic equilibrium. Consider a system consisting of N particles. We assume that a typical mass particle m_i in a system of particles (i = 1, 2, N) is acted on by the applied force $\overline{F}_i$ and the constraint force $\overline{R}_i$. If any inertial forces are negligibly small, we can apply Newton's second law for particle m_i.

The equation of motion can be written as

$$\overline{F}_i + \overline{R}_i - m_i\ddot{\overline{r}}_i = 0 \qquad i = 1, 2, N \qquad (1.151)$$

where $\overline{F}_i$ is the applied force, $\overline{R}_i$ is the constraint force, and $-m_i\ddot{r}_i$ is the reversed effective force or the inertia force. Equation 1.151 states that the sum of all the forces, external and inertial, acting on each particle of the system is zero. This is known as D'Alembert's principle. Extending the principle of virtual work to dynamic equilibrium state results in

$$\delta W = \sum_{i=1}^{N}\left(\overline{F}_i + \overline{R}_i - m_i\ddot{r}_i\right) \cdot \delta\overline{r}_i = 0 \qquad (1.152)$$

Equation 1.152 embodies both the virtual work principle of statics and D'Alembert's principle, and it often is referred to as the generalized principle of D'Alembert or the Lagrange version of D'Alembert's principle.

If the system has workless constraints, and if we choose $\delta\overline{r}_i$ to be reversible virtual displacements consistent with the constraints, we have

$$\sum_{i=1}^{N}\left(\overline{F}_i - m_i\ddot{r}_i\right) \cdot \delta\overline{r}_i = 0 \qquad (1.153)$$

In Eq. 1.153, the sum of the applied force $\overline{F}_i$ and the inertia force $-m_i\ddot{r}_i$ or $\left(\overline{F}_i - m_i\ddot{r}_i\right)$ is called the effective force acting on particle m_i.

The generalized principle of D'Alembert states that the virtual work performed by the effective forces through infinitesimal virtual displacements compatible with the system constraints is zero.

EXAMPLE E1.15

Derive the equation of motion for the system of Example E1.14 by means of D'Alembert's principle.

Solution:

Referring to Figure E1.14(b), we have

$$x_C = \frac{1}{2}x, \quad \ddot{x}_C = \frac{1}{2}\ddot{x}, \quad \text{and} \quad \delta x_C = \frac{1}{2}\delta x$$

$$y_C = \frac{1}{2}y, \quad \ddot{y}_C = \frac{1}{2}\ddot{y}, \quad \text{and} \quad \delta y_C = \frac{1}{2}\delta y$$

The generalized principle of D'Alembert for the system can be written as

$$-k_1 x \delta x + mg\delta y_C - k_2 y \delta y - m\ddot{x}_C \delta x_C - m\ddot{y}_C \delta x_C - m\ddot{y}_C \delta y_C - I_C \ddot{\theta}\delta\theta$$

$$= \left(-k_1 x - \frac{1}{4}m\ddot{x}\right)\delta x + \left(-k_2 y + \frac{1}{2}mg - \frac{1}{4}m\ddot{y}\right)\delta y - I_C \ddot{\theta}\delta\theta = 0$$

The relations between rectangular coordinates x and y and the generalized coordinate θ are given by

$$x = \ell(1 - \cos\theta) \Rightarrow \dot{x} = \ell\sin\theta\,\dot{\theta}, \ \ddot{x} = \ell\left(\cos\theta\dot{\theta}^2 + \sin\theta\ddot{\theta}\right), \ \delta x = \ell\sin\theta\delta\theta$$

$$y = \ell\sin\theta \Rightarrow \dot{y} = \ell\cos\theta\dot{\theta}, \ \ddot{y} = \ell\left(-\sin\theta\dot{\theta}^2 + \cos\theta\ddot{\theta}\right), \ \delta y = \ell\cos\theta\delta\theta$$

Also,

$$I_c = \frac{1}{12}m\ell^2$$

Therefore,

$$\left\{\left[-k_1\ell(1 - \cos\theta) - \frac{1}{4}\ell m\left(\cos\theta\dot{\theta}^2 + \sin\theta\ddot{\theta}\right)\right]\ell\sin\theta\right.$$

$$+ \left[-k_2\ell\sin\theta - \frac{1}{4}\ell m\left(-\sin\theta\dot{\theta}^2 + \cos\theta\ddot{\theta}\right) + \frac{1}{2}mg\right]\ell\cos\theta - \frac{1}{12}m\ell^2\ddot{\theta}\right\}\delta\theta$$

$$= \left[-k_1\ell^2(1 - \cos\theta)\sin\theta - k_2\ell^2\sin\theta\cos\theta + \frac{1}{2}mg\ell\cos\theta - \frac{1}{3}m\ell^2\ddot{\theta}\right]\delta\theta = 0$$

Because $\delta\theta$ is arbitrary, its coefficient must be zero.

Hence, the equation of motion is

$$\frac{1}{3}m\ddot{\theta} + k_1(1 - \cos\theta)\sin\theta + k_2\sin\theta\cos\theta - \frac{1}{2}m\frac{g}{L}\cos\theta = 0$$

EXAMPLE **E1.16**

Derive the equations of motion of a simple pendulum shown in Figure E1.16(a) using D'Alembert's principle.

Solution:

The generalized principle of D'Alembert (Eq. 1.153) for the system shown in Figure E1.16(b) is

$$\left(-mg\sin\theta - m\ell\ddot{\theta}\right)\delta\theta = 0 \qquad\qquad (E1.16.1)$$

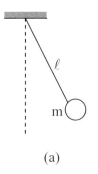

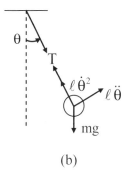

(a)　　　　　　　　　　　　(b)

Figure E1.16 (a) Vibrating system, and (b) free body diagram.

Because $\delta\theta$ is arbitrary, its coefficient must be zero, or

$$-mg\sin\theta - m\ell\ddot{\theta} = 0 \Rightarrow \ddot{\theta} + \frac{g}{\ell}\sin\theta = 0 \qquad\qquad (E1.16.2)$$

Equation E1.16.2 represents the equation of motion for the simple pendulum.

1.10.6 Generalized Force

If the given set of forces $\bar{F}_1$, $\bar{F}_2$, ..., $\bar{F}_N$ is applied to a system of N particles, the virtual work of these forces is

$$\delta W = \sum_{j=1}^{N} \bar{F}_j \cdot \delta\bar{r}_j = \sum_{j=1}^{N} F_{xj}\delta x_j + F_{yj}\delta y_j + F_{zj}\delta_{zj} \qquad\qquad (1.154)$$

Let us suppose that the 3N Cartesian coordinates $x_1,\ y_1,\ z_1,\ \ldots,\ z_N$ are related to the n generalized coordinates $q_1,\ q_2,\ \ldots,\ q_n$ (n = 3N) by the relations

$$x_1 = x_1\left(q_1, q_2, \ldots, q_n, t\right)$$

$$y_1 = y_1\left(q_1, q_2, \ldots, q_n, t\right) \qquad\qquad (1.155)$$

$$z_N = z_N\left(q_1, q_2, \ldots, q_n, t\right)$$

If we differentiate this expression and put $\delta t = 0$ (consistent with virtual displacements), we have

$$\delta x_j = \sum_{i=1}^{n} \frac{\partial x_j}{\partial q_i} \delta q_i$$

$$\delta y_j = \sum_{i=1}^{n} \frac{\partial y_j}{\partial q_i} \delta q_i \qquad j = 1, 2, N \qquad (1.156)$$

$$\delta z_j = \sum_{i=1}^{n} \frac{\partial z_j}{\partial q_i} \delta q_i$$

where δq_i are virtual generalized displacements, and they are all independent.

Note that the quantities $\frac{\partial x_j}{\partial q_i}$, $\frac{\partial y_j}{\partial q_i}$, and $\frac{\partial z_j}{\partial q_i}$ are functions of q and t. Using Eq. 1.156 into Eq. 1.154, we get the virtual work as

$$\delta W = \sum_{j=1}^{N} \sum_{i=1}^{n} \left(F_{xj} \frac{\partial x_j}{\partial q_i} + F_{yj} \frac{\partial y_j}{\partial q_i} + F_{zj} \frac{\partial z_j}{\partial q_i} \right) \delta q_i \qquad (1.157)$$

Interchanging the order of summation in Eq. 1.154, we get

$$\delta W = \sum_{i=1}^{n} \left(\sum_{j=1}^{N} F_{xj} \frac{\partial x_j}{\partial q_i} + F_{yj} \frac{\partial y_j}{\partial q_i} + F_{zj} \frac{\partial z_j}{\partial q_i} \right) \delta q_i \qquad (1.158)$$

We can write Eq. 1.158 in the form

$$\partial W = \sum_{i=1}^{n} Q_i \delta q_i \qquad (1.159)$$

where $Q_i = \sum_{j=1}^{N} F_{xj} \frac{\partial x_j}{\partial q_i} + F_{yj} \frac{\partial y_j}{\partial q_i} + F_{zj} \frac{\partial z_j}{\partial q_i}$ is defined as the generalized force Q_i along the ith generalized coordinate direction. It is assumed that the δq_i's are independent of each other and hence are entirely arbitrary. Also, they are small enough such that the forces remain constant during the virtual displacement. The equilibrium conditions are given by $Q_i = 0$ for i = 1, 2, N. The dimensions of a generalized force are not always that of a force. They depend on the dimensions of the corresponding generalized coordinate. However, dimensions of the product $(Q_i \delta q_i)$ have the dimensions of work or energy. Hence, if q_i represents linear displacement, Q_i has the dimensions of a force. If q_i is the angle of rotation, then Q_i has the dimensions of a moment.

Generalized coordinates are chosen such that they are independent. If the configuration of a holonomic system is expressed in terms of a set of independent generalized coordinates, then the necessary and sufficient condition for static equilibrium is that all the Q_i's must be independently equal to zero.

EXAMPLE E1.17

Figure E1.17(a) shows a damped single-degree-of-freedom system. Derive the generalized forces due to the spring, damper, and the external force $F(t)$.

Solution:

The generalized coordinate is $q_1 = x$, that is, $\delta x = \delta q_1$.

The forces that act on this system are shown in the Figure E1.17(b) and are given by

$$f = F(t) - kx - c\dot{x}$$

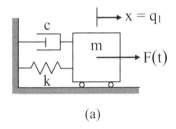
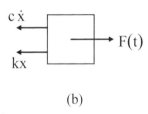

(a)

(b)

Figure E1.17 (a) Single-degree-of-freedom system, and (b) free body diagram.

The virtual work is then defined as

$$\delta W = f\delta x = \left(F(t) - kx - c\dot{x}\right)\delta x$$

Because $x = q_1$ and $\dot{x} = \dot{q}_1$, in terms of the generalized coordinate q_1, the virtual work δW can be written as

$$\delta W = \left(F(t) - kq_1 - c\dot{q}_1\right)\delta q_1 = Q_1\delta q_1$$

where Q_1 is the generalized force associated with the generalized coordinate q_1 and is defined as

$$Q_1 = F(t) - kq_1 = c\dot{q}_1$$

Generalized forces Q_j are

$$\delta W = Q_1\delta q_1 + Q_2\delta q_2 + \dots + Q_n\delta q_n = \sum_{j=1}^{n} Q_j\delta q_j$$

which can be written in a vector form as

$$\delta W = Q^T \delta q$$

where Q is the vector of generalized forces defined as

$$Q = \left[Q_1, Q_2, ..., Q_n\right]^T$$

1.10.7 Lagrange's Equations of Motion

We derive Lagrange's equations of motion for a dynamic system using the application of D'Alembert's principle and the principle of virtual work. First, we consider the holonomic systems, and later we generalize the results to nonholonomic systems.

1.10.8 Holonomic Systems

Consider a dynamic system of N particles. Using D'Alembert's principle and the principle of virtual work, we have

$$\sum_{k=1}^{N} \left(\vec{F}_k - m_k \ddot{\vec{r}}_k\right) \delta \vec{r}_k = 0 \tag{1.160}$$

where $\vec{F}_K$ is the impressed force on the ith particle of mass m_k. Let the system have n degrees of freedom. For this holonomic system, if we choose $q_1, q_2, ..., q_n$ as n generalized coordinates, then we have the transformation equation

$$\vec{r}_k = \vec{r}_k\left(q_1, q_2, ..., q_n, t\right) \tag{1.161}$$

between the vector coordinates of the particles and the n generalized coordinates. The velocities of the particles are then

$$\dot{\vec{r}}_k = \sum_{i=1}^{n} \frac{\partial r_k}{\partial q_i} \dot{q}_i + \frac{\partial \vec{r}_k}{dt} \tag{1.162}$$

Expressing the virtual displacements $\delta \vec{r}_k$ in terms of the n generalized coordinates,

$$\delta \vec{r}_k = \sum_{i=1}^{n} \frac{\partial \vec{r}_k}{\partial q_i} \delta q_i \tag{1.163}$$

Hence, Eq. 1.160 becomes

$$\sum_{k=1}^{N} \sum_{i=1}^{n} \left(\vec{F}_k \frac{\partial \vec{r}_k}{\partial q_i} - m_k \ddot{\vec{r}}_k \frac{\partial \vec{r}_k}{\partial q_i}\right) \delta q_i = 0 \tag{1.164}$$

In Eq. 1.164,

$$\sum_{k=1}^{N} F_k \frac{\partial r_k}{\partial q_i} = Q_i \qquad (1.165)$$

Q_i is called the generalized force in the direction of the ith generalized coordinate.

The other term involving the accelerations, that is, $m_k \ddot{\bar{r}}_k \frac{\partial r_k}{\partial q_i}$, may be related to the kinetic energy of the system as follows.

From Eqs. 1.160 and 1.162, we obtain

$$\frac{\partial \dot{\bar{r}}_k}{\partial \dot{q}_i} = \frac{\partial \bar{r}_k}{\partial q_i} \qquad (1.166)$$

The expression for kinetic energy is

$$T = \frac{1}{2} \sum_{k=1}^{N} m_k \dot{\bar{r}}_k^{\,2} \qquad (1.167)$$

and

$$\frac{\partial T}{\partial \dot{q}_i} = \sum_{k=1}^{N} m_k \dot{\bar{r}}_k \frac{\partial \dot{\bar{r}}_k}{\partial \dot{q}_i} \qquad (1.168)$$

$$\frac{d}{dt}\left(\frac{\partial T}{\partial \dot{q}_i}\right) = \sum_{k=1}^{N} m_k \ddot{\bar{r}}_k \frac{\partial \dot{\bar{r}}_k}{\partial \dot{q}_i} + \sum_{k=1}^{N} m_k \dot{\bar{r}}_k \frac{d}{dt}\left(\frac{\partial \dot{\bar{r}}_k}{\partial \dot{q}_i}\right) \qquad (1.169)$$

Substituting from Eq. 1.166 and noting that the order of differentiation can be changed,

$$\frac{d}{dt}\left(\frac{\partial T}{\partial \dot{q}_i}\right) = \sum_{k=1}^{N} m_k \ddot{\bar{r}}_k \frac{\partial \bar{r}_k}{\partial q_i} + \sum_{k=1}^{N} m_k \dot{\bar{r}}_k \frac{\partial \dot{\bar{r}}_k}{\partial q_i} \qquad (1.170)$$

From Eq. 1.167, we have

$$\frac{\partial T}{\partial \dot{q}_i} = \sum_{k=1}^{N} m_k \dot{\bar{r}}_k \frac{\partial \dot{\bar{r}}_k}{\partial \dot{q}_i} \qquad (1.171)$$

Substituting from Eq. 1.171 in Eq. 1.170,

$$\frac{d}{dt}\left(\frac{\partial T}{\partial \dot{q}_i}\right) = \sum_{k=1}^{N} m_k \ddot{\bar{r}}_k \frac{\partial \bar{r}_k}{\partial q_i} + \frac{\partial T}{\partial q_i} \qquad (1.172)$$

Substituting from Eqs. 1.136 and 1.163 into Eq. 1.162, we obtain

$$\sum_{i=1}^{n} \left[Q_i - \frac{d}{dt}\left(\frac{\partial T}{\partial \dot{q}_i}\right) + \frac{\partial T}{\partial q_i} \right] \delta q_i = 0 \tag{1.173}$$

Because q_i's are generalized coordinates for a holonomic system, they are independent. Hence, to satisfy Eq. 1.173, the coefficients of δq_i must be zero, that is,

$$\frac{d}{dt}\left(\frac{\partial T}{\partial \dot{q}_i}\right) - \frac{\partial T}{\partial q_i} - Q_i = 0 \tag{1.174}$$

For conservative systems,

$$Q_i = -\frac{\partial V}{\partial q_i} \tag{1.175}$$

where V is the potential energy of the system; hence,

$$\frac{d}{dt}\left(\frac{\partial T}{\partial \dot{q}_i}\right) - \frac{\partial T}{\partial q_i} + \frac{\partial V}{\partial q_i} = 0 \tag{1.176}$$

This is Lagrange's equation for a conservative system. Expressing $T - V = L$, known as the Lagrangian, we have

$$\frac{d}{dt}\left(\frac{\partial L}{\partial \dot{q}_i}\right) - \frac{\partial L}{\partial q_i} = 0 \tag{1.177}$$

If there are forces not derivable from a potential function V, that is, for nonconservative systems,

$$Q_i = -\frac{\partial V}{\partial q_i} + Q_i' \tag{1.178}$$

then we have

$$\frac{d}{dt}\left(\frac{\partial L}{\partial \dot{q}_i}\right) - \frac{\partial L}{\partial q_i} = Q_i' \tag{1.179}$$

1.10.9 Nonholonomic Systems

The derivation of Lagrange's equations for holonomic systems requires that the generalized coordinates be independent. For a nonholonomic system, however, there must be a greater number of generalized coordinates than the number of degrees of freedom. Therefore, the δq's are not independent if we assume a virtual displacement consistent with the constraints. Let n coordinates q_1, q_2, ..., q_n be chosen to describe the motion.

If there are m nonholonomic constraint equations of the form

$$\sum_{i=1}^{n} a_{ji}dq_i + a_{jt}dt = 0 \qquad j = 1, 2, ..., m \qquad (1.180)$$

where a_{ji}, $I = 1, 2, ..., n$ are functions of q_i. The degrees of freedom are given by $(n - m)$, and the coordinate q_i's are not all independent. The δq's must meet the following conditions:

$$\sum_{i=1}^{n} a_{ji}\delta q_i = 0 \qquad j = 1, 2, ..., m \qquad (1.181)$$

Let us assume that each generalized applied force Q_i is obtained from a potential function, and assume workless constraints. The generalized constraint force C_i must meet the condition

$$\sum_{i=1}^{N} C_i \delta q_i = 0 \qquad (1.182)$$

for any virtual displacement consistent with constraints.

Multiply Eq. 1.181 by factor λ_j, known as the Lagrange multiplier, and we obtain the m equations

$$\lambda_j \sum_{i=1}^{n} a_{ji}\delta q_i = 0 \qquad j = 1, 2, ..., m \qquad (1.183)$$

Subtract the sum of these m equations from Eq. 1.182. Then by interchanging the order of summation, we obtain

$$\sum_{i=1}^{n} \left(C_i - \sum_{j=1}^{m} \lambda_j a_{ji} \right) \delta q_i = 0 \qquad (1.184)$$

Now if we choose λ's in such a way that

$$C_i = \sum_{j=1}^{m} \lambda_j a_{ji} \qquad i = 1, 2, ..., n \qquad (1.185)$$

then the coefficient of δq's is zero, and Eq. 1.184 will apply for any δq's, that is, the δq's are independent.

Now the generalized force C_i can be equated to Q_i' or the force not derivable by a potential function

$$\frac{d}{dt}\left(\frac{\partial L}{\partial \dot{q}_i}\right) - \frac{\partial L}{\partial q_i} = \sum_{j=1}^{m} \lambda_j a_{ji} \qquad i = 1, 2, ..., n \qquad (1.186)$$

In addition to these n equations, we have m equations of constraints to solve for $(n + m)$ unknowns, that is, n q's and m λ's. The Lagrange multiplier relates the constraints to constraint forces (Eq. 1.185).

1.10.10 Rayleigh's Dissipation Function

The generalized force Q_i in Eq. 1.174 includes nonconservative forces that are not derivable from a potential function. The frictional forces acting on a particle can be separated into viscous and nonviscous friction forces. Viscous friction forces are proportional to the velocity of a given particle and resist the motion because they act in a direction opposite to that of the velocity. Nonviscous forces are nonlinear functions of the velocity and resist the motion. Viscous friction forces are known as dissipative forces because the system loses energy due to their action. It is possible to define a single scalar function F, known as Rayleigh's dissipation function, in terms of the generalized coordinates and velocities to represent these forces in Lagrange's equations of motion.

Denoting the dissipative force components acting on particle i as

$$F_{xk} = -c_{xk}\dot{x}_i$$

$$F_{yk} = -c_{yk}\dot{y}_i \qquad i = 1, 2, ..., N \qquad (1.187)$$

$$F_{zk} = -c_{zk}\dot{z}_i$$

where c_{xk}, c_{yk}, and c_{zk} are the functions independent of the velocities and are dependent only on the coordinates. The virtual work done by these forces can be written as

$$\sum_{k=1}^{N} \overline{F}_k \cdot \delta \overline{r}_k = \sum_{k=1}^{N}\left(F_{xk}\delta x_k + F_{yk}\delta y_k + F_{zk}\delta z_k\right)$$
$$= -\sum_{k=1}^{N}\left(c_{xk}\dot{x}_k\delta x_k + c_{yk}\dot{y}_k\delta y_k + c_{zk}\dot{z}_k\delta z_k\right) \qquad (1.188)$$

Referring to Eqs. 1.162 and 1.163, we have

$$\delta \overline{r}_k = \sum_{i=1}^{n}\frac{\delta \overline{r}_k}{\partial q_i}\delta q_i = \sum_{i=1}^{n}\frac{\partial \dot{\overline{r}}_k}{\partial \dot{q}_i}\delta q_i \qquad (1.189)$$

and introducing the components δx_k, δy_k, and δz_k into Eq. 1.188, we obtain

$$
\begin{aligned}
\sum_{k=1}^{N} \overline{F}_k \cdot \delta \overline{r}_k &= - \sum_{k=1}^{N} \left(c_{xk} \dot{x}_k \delta x_k + c_{yk} \dot{y}_k \delta y_k + c_{zk} \dot{z}_k \delta z_k \right) \\
&= - \sum_{k=1}^{n} \left(\sum_{i=1}^{N} \left(c_{xk} \dot{x}_k \frac{\partial \dot{x}_k}{\partial \dot{q}_i} + c_{yk} \dot{y}_k \frac{\partial \dot{y}_k}{\partial \dot{q}_i} + c_{zk} \dot{z}_k \frac{\partial \dot{z}_k}{\partial \dot{q}_i} \right) \right) \delta q_i \\
&= - \sum_{k=1}^{n} \left(\sum_{i=1}^{N} \frac{1}{2} \frac{\partial}{\partial \dot{q}_i} \left(c_{xk} \dot{x}_k^2 + c_{yk} \dot{y}_k^2 + c_{zk} \dot{z}_k^2 \right) \right) \delta q_i \\
&= - \sum_{i=1}^{n} \frac{\partial F}{\partial \dot{q}_i} \delta \dot{q}_i
\end{aligned}
\tag{1.190}
$$

where

$$
F = \sum_{k=1}^{N} \frac{1}{2} \left(c_{xk} \dot{x}_k^2 + c_{yk} \dot{y}_k^2 + c_{zk} \dot{z}_k^2 \right)
\tag{1.191}
$$

is known as Rayleigh's dissipation function. The virtual work for the case when only nonconservative forces present are of the dissipative type is given by

$$
\delta W_{nc} = \sum_{i=1}^{n} Q_i \delta q_i = - \sum_{i=1}^{n} \frac{\partial F}{\partial \dot{q}_i} \delta q_i
\tag{1.192}
$$

From Eq. 1.192, we note that the dissipative generalized forces can be obtained from Rayleigh's dissipation function as

$$
Q_i = - \frac{\partial F}{\partial \dot{q}_i} \qquad i = 1, \, 2, \, ..., \, n
\tag{1.193}
$$

Therefore, from Eqs. 1.193 and 1.174, Lagrange's equations for a system with dissipative forces are given by

$$
\frac{d}{dt} \left(\frac{\partial L}{\partial \dot{q}_i} \right) - \frac{\partial L}{\partial q_i} + \frac{\partial F}{\partial \dot{q}_i} = 0 \qquad i = 1, \, 2, \, ..., \, n
\tag{1.194}
$$

Equation 1.194 indicates clearly that all the equations can be obtained from two scalar functions—the Lagrangian (L) function, and Rayleigh's dissipation function (F). A more general form of the dissipation function is given by

$$
F = \frac{1}{2} \sum_{r=1}^{n} \sum_{s=1}^{n} c_{rs} \dot{q}_r \dot{q}_s
\tag{1.195}
$$

EXAMPLE **E1.18**

For the simple pendulum shown in Figure E1.18(a):

a. Set up the Lagrangian

b. Obtain the equation describing its motion

Solution:

a. The angle θ made by the string OB of the pendulum and the vertical OA is selected as the generalized coordinate. The kinetic energy of the system from Figure E1.18(b) is given by

$$T = \frac{1}{2}mV^2 = \frac{1}{2}m(\ell\dot{\theta})^2 = \frac{1}{2}m\ell^2\dot{\theta}^2 \qquad (E1.18.1)$$

where m is the mass of the body, and ℓ is the length of OB.

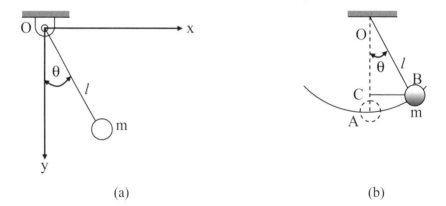

Figure E1.18 (a) Simple pendulum, and (b) free body diagram.

(a)   (b)

The potential energy of mass m is given by

$$V = mg(OA - OC) - mg(\ell - \ell\cos\theta) = mg\ell(1 - \cos\theta) \qquad (E1.18.2)$$

The Lagrangian is

$$L = T - V = \frac{1}{2}m\ell^2\dot{\theta}^2 - mg\ell(1 - \cos\theta) \qquad (E1.18.3)$$

b. Lagrange's equation is

$$\frac{d}{dt}\left(\frac{\partial L}{\partial \dot{\theta}}\right) - \frac{\partial L}{\partial \theta} = 0 \qquad (E1.18.4)$$

From Eq. E1.18.3, we have

$$\frac{\partial L}{\partial \theta} = -mg\ell \sin \theta$$

and

$$\frac{\partial L}{\partial \dot{\theta}} = m\ell^2 \dot{\theta} \qquad\qquad (E1.18.5)$$

Substituting Eq. E1.18.5 into Eq. E1.18.4, we obtain the equation of motion for the system as

$$m\ell^2 \ddot{\theta} + mg\ell \sin \theta = 0$$

$$\text{or} \qquad\qquad (E1.18.6)$$

$$\ddot{\theta} + \frac{g}{\ell} \sin \theta = 0$$

EXAMPLE E1.19

Derive the equations of motion for the system shown in Figure E1.19(a) using Lagrange's equation and the generalized coordinates $q_1 = x_1$ and $q_2 = x_2$.

Solution:

The kinetic energy of the system from Figure E1.19(b) is given by

$$T = \frac{1}{2} m\dot{x}_G^2 + \frac{1}{2} I_G \dot{\theta}^2$$

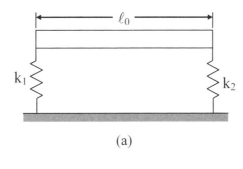

(a)

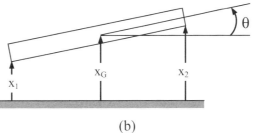

(b)

Figure E1.19 (a) Two-degrees-of-freedom vibrating system, and (b) free body diagram.

The potential energy is given by

$$V = \frac{1}{2}k_1\left(x_1 - L_1\right)^2 + \frac{1}{2}k_2\left(x_2 - L_2\right)^2 + mg\,x_G$$

where L_1 and L_2 are the undeformed lengths of springs 1 and 2, respectively. For small oscillations, we can write

$$x_G = \frac{x_1 + x_2}{2} \quad \text{and} \quad \theta = \frac{x_2 - x_1}{L_0}$$

Then, the Lagrangian is

$$L = T - V = \frac{1}{2}m\left(\frac{\dot{x}_1 + \dot{x}_2}{2}\right)^2 + \frac{1}{2}I_G\left(\frac{\dot{x}_2 - \dot{x}_1}{L_0}\right)^2$$

$$-\frac{1}{2}k_1\left(x_1 - L_1\right)^2 - \frac{1}{2}k_2\left(x_2 - L_2\right)^2 - mg\frac{x_1 + x_2}{2}$$

Because there are no nonconservative forces, Lagrange's equations are

$$\frac{d}{dt}\left(\frac{\partial L}{\partial \dot{x}_i}\right) - \frac{\partial L}{\partial x_i} = 0 \qquad i = 1, 2$$

For i = 1, performing the indicated differentiation gives

$$\frac{1}{2}m\left(\ddot{x}_1 + \ddot{x}_2\right) - \frac{I_G}{L_0}\left(\ddot{x}_2 - \ddot{x}_1\right) + k_1\left(x_1 - L_1\right) + \frac{mg}{2} = 0$$

while for i = 2,

$$\frac{1}{2}m\left(\ddot{x}_1 + \ddot{x}_2\right) + \frac{I_G}{L_0}\left(\ddot{x}_2 - \ddot{x}_1\right) + k_2\left(x_1 - L_2\right) + \frac{mg}{2} = 0$$

EXAMPLE E1.20

For the system shown in Figure E1.20:

a. Write the expressions for the kinetic and potential energy

b. Derive the equations of motion using Lagrange's equation.

Solution:

The kinetic energy is

$$T = \frac{1}{2}m_1\dot{x}_1^2 + \frac{1}{2}m_2\dot{x}_2^2 + \frac{1}{2}m_3\dot{x}_3^2$$

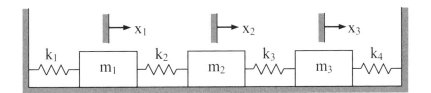

Figure E1.20 Three-degrees-of-freedom vibrating system.

The potential energy is

$$V = \frac{1}{2}k_1x_1^2 + \frac{1}{2}k_2(x_2 - x_1)^2 + \frac{1}{2}k_3(x_3 - x_2)^2 + \frac{1}{2}k_4x_3^2$$

Then

$$m_1\ddot{x}_1 + k_1x_1 + k_2(x_2 - x_1)(-1) = 0$$

$$m_2\ddot{x}_2 + k_2(x_2 - x_1)(1) + k_3(x_3 - x_2)(-1) = 0$$

$$m_3\ddot{x}_3 + k_3(x_2 - x_1)(1) + k_4x_3 = 0$$

$$\begin{bmatrix} m_1 & 0 & 0 \\ 0 & m_2 & 0 \\ 0 & 0 & m_3 \end{bmatrix} \begin{Bmatrix} \ddot{x}_1 \\ \ddot{x}_2 \\ \ddot{x}_3 \end{Bmatrix} + \begin{bmatrix} (k_1 + k_2) & -k_2 & 0 \\ -k_2 & (k_2 + k_3) & -k_3 \\ 0 & -k_3 & (k_3 + k_4) \end{bmatrix} \begin{Bmatrix} x_1 \\ x_2 \\ x_3 \end{Bmatrix} = \begin{Bmatrix} 0 \\ 0 \\ 0 \end{Bmatrix}$$

EXAMPLE E1.21

Derive the differential equation governing the motion of the four-degrees-of-freedom torsional system shown in Figure E1.21 using Lagrange's equation and the generalized coordinates θ_1, θ_2, θ_3, and θ_4. The k's represent the torsional spring constants for the indicated shaft portions, and the I's are the mass moment of inertia of the disks.

Solution:

The kinetic energy of the system is

$$T = \frac{1}{2}I_1\dot{\theta}_1^2 + \frac{1}{2}I_2\dot{\theta}_2^2 + \frac{1}{2}I_3\dot{\theta}_3^2 + \frac{1}{2}I_4\dot{\theta}_4^2$$

The potential energy of the system is equal to the work done by the shaft as it returns from the dynamic configuration to the reference equilibrium position. Hence,

$$V = \int_\theta^0 (-k\theta)d\theta = \frac{k\theta^2}{2}$$

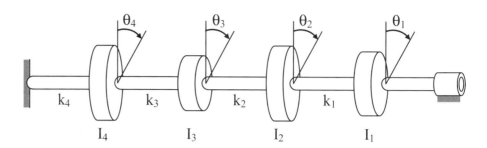

Figure E1.21 Four-degrees-of-freedom torsional system.

Applying this, the potential energy function is given by

$$V = \frac{1}{2}k_1(\theta_1 - \theta_2)^2 + \frac{1}{2}k_2(\theta_2 - \theta_3)^2 + \frac{1}{2}k_3(\theta_3 - \theta_4)^2 + \frac{1}{2}k_4\theta_4^2$$

For the subscript k = 1, the following are obtained using Lagrange's equation:

$$\frac{\partial T}{\partial \dot{\theta}_1} = I_1\dot{\theta}_1 \quad \text{and} \quad \frac{d}{dt}\left(\frac{\partial T}{\partial \dot{\theta}_1}\right) = I_1\ddot{\theta}_1$$

$$\frac{\partial T}{\partial \theta} = 0$$

Because T is independent of θ_1, and

$$\frac{\partial V}{\partial \theta_1} = k_1(\theta_1 - \theta_2)$$

Hence,

$$\frac{d}{dt}\left(\frac{\partial T}{\partial \dot{\theta}_1}\right) - \frac{\partial T}{\partial \theta} + \frac{\partial V}{\partial \theta_1} = 0$$

which gives

$$I_1\ddot{\theta}_1 + k_1(\theta_1 - \theta_2) = 0$$

Similarly, letting subscript k = 2, 3, 4 in Lagrange's equation gives

$$I_2\ddot{\theta}_2 - k_1(\theta_1 - \theta_2) + k_2(\theta_2 - \theta_3) = 0$$

$$I_3\ddot{\theta}_3 - k_2(\theta_2 - \theta_3) + k_2(\theta_3 - \theta_4) = 0$$

$$I_4 \ddot{\theta}_4 - k_3 \left(\theta_3 - \theta_4 \right) + k_4 \theta_4 = 0$$

Rearranging and writing these in matrix form, we obtain

$$\begin{bmatrix} I_1 & 0 & 0 & 0 \\ 0 & I_2 & 0 & 0 \\ 0 & 0 & I_3 & 0 \\ 0 & 0 & 0 & I_4 \end{bmatrix} \begin{Bmatrix} \ddot{\theta}_1 \\ \ddot{\theta}_2 \\ \ddot{\theta}_3 \\ \ddot{\theta}_4 \end{Bmatrix} + \begin{bmatrix} k_1 & -k_1 & 0 & 0 \\ -k_1 & \left(k_1 + k_2 \right) & -k_2 & 0 \\ 0 & -k_2 & \left(k_2 + k_3 \right) & -k_3 \\ 0 & 0 & -k_3 & \left(k_3 + k_4 \right) \end{bmatrix} \begin{Bmatrix} \theta_1 \\ \theta_2 \\ \theta_3 \\ \theta_4 \end{Bmatrix} = \begin{Bmatrix} 0 \\ 0 \\ 0 \\ 0 \end{Bmatrix}$$

1.11 Summary

The study of vibrations is concerned with the motion of a particle or a body or system of connected bodies displaced from a position of equilibrium. Vibration occurs when a system is displaced from the position of stable equilibrium. The system tends to return to this equilibrium position under the action of restoring forces. A brief review of the classification of dynamic system models, constraints, generalized coordinates, degrees of freedom, classification of vibrations, discrete and continuous systems, methods of vibration analysis, elements of vibrating systems including spring elements, mass or inertia elements, and damping elements is presented.

The motion of vibrating systems is governed by the laws of mechanics and in particular by Newton's second law of motion. A brief review of rigid body dynamics—namely, Newton's laws of motion, the kinematics of rigid bodies, the rigid body D'Alembert's principle, the principle of work and energy, the conservation of energy, and the principle of impulse and momentum—is presented to familiarize the reader with the basic notation and methods.

There are two approaches to the study of mechanics—the formulation and solution of Newton's equations of motion, and analytical dynamics. In analytical dynamics, we consider the system as a whole and formulate the expressions of energy in terms of the generalized coordinates.

We discussed the concepts of generalized coordinates, the principle of virtual work, the constraints, and D'Alembert's principle. Although the analytical dynamics approach is conceptually abstract, the analysis of complex systems becomes quite simple because we deal with the scalar energy expressions.

The equations of motion for a dynamic system have been derived based on Lagrange's formulation. Because the selection of generalized coordinates instead of the physical coordinates was made, Lagrange's formulation is quite versatile. Lagrange's equations of motion consist of a system of simultaneous second-order differential equations, which generally are nonlinear in the generalized coordinates and the corresponding velocities. The Lagrange formulation is a powerful approach compared to the method based on the direct application of Newton's second law in obtaining the equations of motion for a dynamic system. In the method based on the direct application of Newton's second law, the constraint forces appear in the equations of motion and must be eliminated. Because constraint forces do not perform work in virtual displacement, they are ignored in the Lagrangian formulation. With a suitable choice of coordinate systems, the equations of motion can be obtained in a simple and straightforward manner.

1.12 References

1. Bastow, D., *Car Suspension and Handling*, 2nd Ed., Pentech Press, London, 1990.

2. Bhat, R. and Dukkipati, R.V., *Advanced Dynamics*, Narosa Publishing House, New Delhi, India, 2001.

3. Campbell, C., *New Directions in Suspension Design*, Robert Bentley, Inc., Cambridge, MA, 1981.

4. Chaplin, H.R. and Ford, A., *Design Principles of Ground Effect Machines*, Section B, Maritime Press, Centerville, MD, 1965.

5. Costin, M. and Phillips, D., *Racing and Sports Car Chassis Design*, 2nd Ed., Robert Bentley, Inc., Cambridge, MA, 1965.

6. Dukkipati, R.V., *Advanced Engineering Analysis*, Narosa Publishing House, New Delhi, India, 2006.

7. Dukkipati, R.V., *Advanced Mechanical Vibrations*, Narosa Publishing House, New Delhi, India, 2006.

8. Dukkipati, R.V., *Analysis and Design of Control Systems Using MATLAB*, Narosa Publishing House, New Delhi, India, 2005.

9. Dukkipati, R.V., *Control Systems*, Narosa Publishing House, New Delhi, India, 2005.

10. Dukkipati, R.V., *Engineering System Dynamics*, Narosa Publishing House, New Delhi, India, 2004.

11. Dukkipati, R.V., *MATLAB for Mechanical Engineers*, New Age International Publishers, New Delhi, India, 2007.

12. Dukkipati, R.V., *Solving Engineering Mechanics Problems Using MATLAB*, New Age International Publishers, New Delhi, India, 2007.

13. Dukkipati, R.V., *Solving Engineering System Dynamics Problems with MATLAB*, New Age International Publishers, New Delhi, India, 2006.

14. Dukkipati, R.V., *Solving Vibration Analysis Problems Using MATLAB*, New Age International Publishers, New Delhi, India, 2006.

15. Dukkipati, R.V., *Spatial Mechanisms: Analysis and Synthesis*, Narosa Publishing House, New Delhi, India, 2001.

16. Dukkipati, R.V., *Vehicle Dynamics*, Narosa Publishing House, New Delhi, India, 2000.

17. Dukkipati, R.V., *Vibration Analysis*, Narosa Publishing House, New Delhi, India, 2004.

18. Dukkipati, R.V. and Amyot, J.R., *Computer Aided Simulation in Railway Dynamics*, Marcel Dekker Inc, New York, 1988.

19. Dukkipati, R.V., Ananda Rao, M., and Bhat, R., *Computer Aided Analysis and Design of Machine Elements*, New Age International Publishers, New Delhi, India, 2000.

20. Dukkipati, R.V. and Ray, P.K., *Product and Process Design for Quality, Economy and Reliability*, Narosa Publishing House, New Delhi, India, 2007.

21. Dukkipati, R.V. and Ray, P.K., *Quality Control, Improvement and Management*, Narosa Publishing House, New Delhi, India, 2007.

22. Dukkipati, R.V. and Srinivas, J.S., *Mechanical Vibrations*, Tata McGraw-Hill, New Delhi, India, 2005.

23. Dukkipati, R.V. and Srinivas, J.S., *Vibrations: Problem Solving Companion*, Narosa Publishing House, New Delhi, India, 2007.

24. Dukkipati, R.V., Srinivas, J.S., RaviKanth, A., and Ramji, K., *Applied Numerical Methods*, Narosa Publishing House, New Delhi, India, 2007.

25. Ellis, J.R., *Road Vehicle Dynamics*, John R. Ellis, Inc., Akron, OH, 1988.

26. Ellis, J.R., *Vehicle Dynamics*, Business Books Limited, London, 1969.

27. Elsley, G.H. and Devereux, A.J., *Hovercraft Design and Construction*, Cornell Maritime Press, Centerville, MD, 1968.

28. Firch, J.W., *Motor Truck Engineering Handbook*, 3rd Ed., James W. Fitch, Anacortes, WA, 1984.

29. Garg, V.K. and Dukkipati, R.V., *Dynamics of Railway Vehicle Systems*, Academic Press, New York, 1984.

30. Geary, P.J., *Magnetic and Elastic Suspensions—A Survey of Their Design, Construction and Use*, Taylor and Francis, London, England, 1964.

31. Giles, J.G., *Steering, Suspension and Tyres*, Iliffe Books Ltd., London, 1968.

32. Gillespie, T.D., *Fundamentals of Vehicle Dynamics*, Society of Automotive Engineers, Warrendale, PA, 1992.

33. Goodsell, D., *Dictionary of Automotive Engineering*, Butterworths, London, 1989.

34. Hay, W.W., *An Introduction to Transportation Engineering*, Wiley, New York, 1961.

35. Hennes, R.G. and Ekse, M., *Fundamentals of Transportation Engineering*, McGraw-Hill, New York, 1969.

36. Huang, M., *Vehicle Crash Mechanics*, CRC Press, Boca Raton, FL, 2002.

37. ISO, "Road Vehicles—Lateral Transient Response Test Methods," ISO 7401-1988, International Organisation for Standardization, 1988.

38. ISO, "Road Vehicles—Steady State Circular Test Procedure," ISO 4138-1982 (E), International Organisation for Standardization, 1982.

39. Jones, B., *Elements of Practical Aerodynamics*, Wiley, New York, 1942.

40. Laithwaite, E.R., *Transport Without Wheels*, Westview Press, Boulder, CO, 1977.

41. Magnus, K., ed., *Dynamics of Multibody Systems*, Springer-Verlag, Berlin-Heidelberg, 1978.

42. Milliken, W.F. and Milliken, D.L., *Race Car Vehicle Dynamics*, Society of Automotive Engineers, Warrendale, PA, 1995.

43. Newton, K., Steeds, W., and Garrett, T.K., *The Motor Vehicle*, 10th Ed., Butterworths, London, 1983.

44. Pacejka, H.B., ed., *The Dynamics of Vehicles on Roads and Railway Tracks*, Swets & Zeitlinger, Lisse, The Netherlands, 1976.

45. Puhn, F., *How to Make Your Car Handle*, H.P. Books, Los Angeles, CA, 1981.

46. Rao, J.S. and Dukkipati, R.V., *Mechanism and Machine Theory*, 2nd Ed., Wiley Eastern Limited, New Delhi, India, 1992.

47. Richardson, H.H., *et al.,* "Special Issue on Ground Transportation," *ASME Journal of Dynamic Systems, Measurement and Control*, Vol. 96, No. 2, 1974.

48. Roberts, P., *Collector's History of the Automobile*, Bonanza Books, New York, 1978.

49. SAE International, *Automobile Handbook*, 2nd Ed., Robert Bosch GmbH, Stuttgart, 1994.

50. SAE International, "Laboratory Testing Machines for Measuring the Steady State Force and Moment Properties of Passenger Car Tyres," SAE J1107, Society of Automotive Engineers, Warrendale, PA, September 1975.

51. SAE International, "Proposed Passenger Car and Light Truck Directional Control Response Test Procedures," SAE XJ266, Society of Automotive Engineers, Warrendale, PA, 1985.

52. SAE International, "Vehicle Aerodynamics Terminology," SAE J1594, Society of Automotive Engineers, Warrendale, PA, June 1987.

53. SAE International, "Vehicle Dynamics Terminology," SAE J670e, Appendix E, Society of Automotive Engineers, Warrendale, PA, July 1976.

54. Scibor-Rylski, A.J., *Road Vehicle Aerodynamics*, Halstead Press, Wiley, New York, 1975.

55. Segel, L., "Research in the Fundamentals of Automobile Control and Stability," *Transactions of the Society of Automotive Engineers*, Society of Automotive Engineers, Warrendale, PA, 1985.

56. Slibar, A. and Springer, H., "The Dynamics of Vehicles on Roads and Tracks," Swets & Zeitlinger, Lisse, The Netherlands, 1978.

57. Steeds, W., *Mechanics of Road Vehicles*, Iliffe and Sons, Ltd., London, 1960.

58 . Taborek, J.J., *Mechanics of Vehicles*, Towmotor Corporation, Cleveland, OH, 1957.

59. Terry, L. and Baker, A., *Racing Car Design and Development*, Robert Bentley, Inc., Cambridge, MA, 1973.

60. Van Eldik, H.C.A. and Pacejka, H.B., "The Tire as a Vehicle Component," Delft University of Technology, Delft, The Netherlands, 1971.

61. Willumeit, H.P., ed., "The Dynamics of Vehicles on Roads and Tracks," Swets & Zeitlinger, Lisse, The Netherlands, 1980.

62. Wong, J.Y., *Theory of Ground Vehicles*, 3rd Ed., Wiley, New York, 2001.

Chapter 2

Analysis of Dynamic Systems

2.1 Introduction

In this chapter, we briefly discuss the analytical techniques that are used with dynamic systems. The initial portion of the chapter is devoted to the analysis of a linear single-degree-of-freedom system. Both free and forced responses of the system are discussed. In subsequent sections of the chapter, the analytical methods for a multiple-degrees-of-freedom system are presented. Dynamic systems with and without damping are considered. Eigenvalue problems for these systems are formulated. The modal superposition technique for calculating the response of a multiple-degrees-of-freedom system also is presented. A single general analysis of a one-degree-of-freedom nonlinear system is given to provide a perspective of the problem. The characteristics of trajectories in a phase plane are discussed and illustrated. The isocline and Pell's method for the graphical construction of trajectories in the phase plane are designed. Some of the exact methods, a few approximate methods, and graphical procedures for nonlinear system are presented. Direct numerical integration methods for nonlinear multiple-degrees-of-freedom systems are described and presented in Appendix E. Finally, a brief discussion of the theory of random vibration is given, and a method for calculating the response of a linear system subjected to stationary random excitations is presented.

2.2 Classification of Vibrations

Vibrations can be classified into three categories: free, forced, and self-excited. Free vibration of a system is vibration that occurs in the absence of forced vibration, where damping may or may not be present. In the absence of damping, the total mechanical energy due to the initial conditions is conserved, and the system can vibrate forever because of the continuous exchange between the kinetic and potential energies. Because almost all mechanical systems exhibit some form of damping, the application of such free vibration theories lies in the areas of celestial mechanics, space dynamics, and structural dynamics problems, in which the amount of damping is so small that the system can be treated as an undamped system.

An external force that acts on the system causes forced vibrations. In this case, the exciting force continuously supplies energy to the system to compensate for that dissipated by damping. Forced vibrations may be either deterministic or random. The differential equations of motion of the dynamic systems considered in this book are all deterministic; that is, the parameters are not randomly varying with time. However, the exciting force may be either a deterministic or a random function of time. In deterministic vibrations, the amplitude and frequency at any designated future time can be completely predicted from past history, whereas random forced vibrations are defined in statistical terms, and only the probability of occurrence of designated magnitudes and frequencies can be predicted.

Self-excited vibrations are periodic and deterministic oscillations. Under certain conditions, the equilibrium state in such a vibration system becomes unstable, and any disturbance causes the perturbations to grow until some effect limits any further growth. The energy required to sustain these vibrations is obtained from a non-alternating power source. In self-excited vibrations, the vibrations create the periodic force that excites the vibrations themselves. If the system is prevented from vibrating, then the exciting force disappears. In contrast, in the case of forced vibrations, the exciting force is independent of the vibrations and can persist even when the system is prevented from vibrating.

2.3 Classification of Deterministic Data

Deterministic data can be classified as being either periodic or nonperiodic. Periodic data can be further classified as being either sinusoidal or complex periodic. Nonperiodic data can be further classified as being either almost periodic or transient. These various classifications of deterministic data are illustrated schematically in Figure 2.1. Any combination of these forms also may occur in practice. Each of these types of deterministic data will be reviewed briefly.

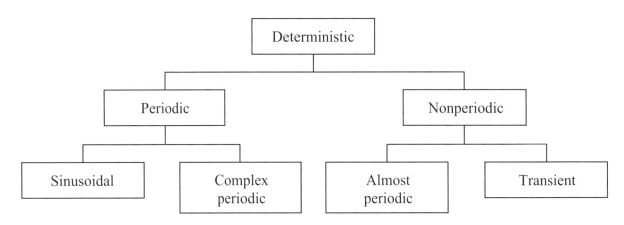

Figure 2.1 Classification of deterministic data.

2.3.1 Sinusoidal Periodic Data

Sinusoidal periodic data can be expressed by a time-varying function of the form

$$x(t) = X \sin(2\pi f_0 t + \theta) \tag{2.1}$$

where

$\quad X \quad$ = amplitude

$\quad f_0 \quad$ = cyclical frequency in cycles per unit time

$\quad \theta \quad$ = initial phase angle with respect to the time origin in radians

$\quad x(t) \quad$ = instantaneous value at time t

The sinusoidal time history described by Eq. 2.1 is commonly referred to as a sine wave. In practice, when analyzing sinusoidal data, the phase angle θ often is ignored. Therefore,

$$x(t) = X \sin 2\pi f_0 t \tag{2.2}$$

Equation 2.2 can be pictured by a time history plot or by an amplitude frequency plot (frequency spectrum), as illustrated in Figure 2.2. The time interval required for one full fluctuation or cycle of sinusoidal data is called the period T_P. The number of cycles per unit time is called the frequency f_0. The frequency and period are related by

$$T_p = \frac{1}{f_0} \tag{2.3}$$

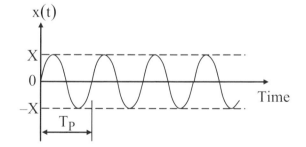
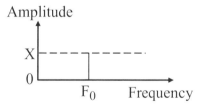

Figure 2.2 Time history and spectrum of sinusoidal data.

The frequency spectrum in Figure 2.2 is composed of an amplitude component at a specific frequency, as opposed to a continuous plot of amplitude versus frequency. Such spectra are called discrete spectra or line spectra. Many physical phenomena produce approximately sinusoidal data in practice. Sinusoidal data represent one of the simplest forms of time-varying data from the analysis point of view.

2.3.2 Complex Periodic Data

Complex periodic data can be defined mathematically by a time-varying function whose waveform exactly repeats itself at regular intervals, such that

$$x(t) = x(t \pm nT_P) \qquad n = 1, 2, 3 \tag{2.4}$$

As for sinusoidal data, the time interval required for one full fluctuation is called the period T_P. The number of cycles per unit time is called the fundamental frequency f. A special case for complex periodic data is clearly sinusoidal data, where $f_1 = f_0$.

Complex periodic data may be expanded into a Fourier series as

$$x(t) = \frac{a_0}{2} + \sum_{n=1}^{\infty} \left(a_n \cos 2\pi n f_1 t + b_n \sin 2\pi n f_1 t \right) \tag{2.5}$$

where

$$f_1 = \frac{1}{T_P}$$

$$a_n = \frac{2}{T_P} \int_0^{T_P} x(t) \cos 2\pi n f_1 t \, dt \qquad n = 0, 1, 2, \dots$$

$$b_n = \frac{2}{T_P} \int_0^{T_P} x(t) \sin 2\pi n f_1 t \, dt \qquad n = 0, 1, 2, \dots$$

Another way to express the Fourier series for complex periodic data is

$$x(t) = X_0 + \sum_{n=1}^{\infty} X_n \cos(2\pi n f_1 t - \theta_n) \tag{2.6}$$

where

$$X = \frac{a_0}{2}$$

$$X_n = \sqrt{a_n^2 + b_n^2} \qquad n = 1, 2, 3, \dots$$

$$\theta_n = \tan^{-1}\left(\frac{b_n}{a_n} \right) \qquad n = 1, 2, 3, \dots$$

Equation 2.6 implies that complex periodic data consist of a static component, X_0, and an infinite number of sinusoidal components called harmonics, which have amplitudes of X_n and phase of θ_n. The frequencies of the harmonic components are all integral multiples of f_1. The phase angles often are ignored when periodic data are analyzed in practice. For this case, Eq. 2.6 can be characterized by a discrete spectrum, as shown in Figure 2.3. Often, complex periodic data will include only a few components. In other cases, the fundamental component may be absent. The classification of data as being sinusoidal often is only an approximation of data, which are actually complex. Intense harmonic components can be present in periodic physical data.

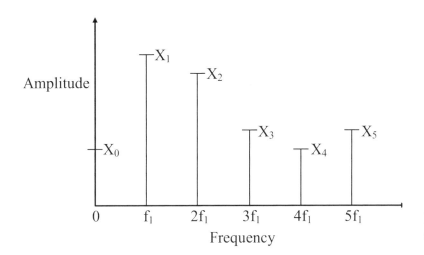

Figure 2.3 Spectrum of complex periodic data.

2.3.3 Almost Periodic Data

The data formed by summing two or more sine waves with arbitrary frequencies generally will not be periodic. More specifically, the sum of two or more sine waves will be periodic only when the ratios of all possible pairs of frequencies form rational numbers. Almost periodic data can be defined by a time-varying function of the form

$$x(t) = \sum_{n=1}^{\infty} X_n \sin(2\pi n f_n t + \theta_n) \tag{2.7}$$

Physical phenomena producing almost periodic data frequently occur in practice when the effects of two or more unrelated periodic phenomena are mixed. The vibration response in a multiple engine propeller airplane when the engines are out of synchronization is one example.

If the phase angles θ_n are ignored, Eq. 2.7 can be characterized by a discrete frequency spectrum similar to that for complex periodic data. Their only difference is that the frequencies of the components are not related by rational numbers, as illustrated in Figure 2.4.

2.3.4 Transient Nonperiodic Data

Transient data are defined as all nonperiodic data other than the almost periodic data previously discussed. Transient data include all data that can be described by some suitable time-varying function. Figure 2.5 gives three simple examples of transient data. Physical phenomena that produce transient data are numerous and diverse. An important characteristic of transient data is that a discrete spectral representation is

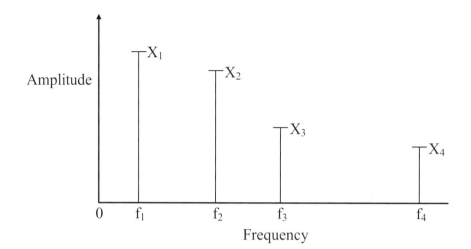

Figure 2.4 Spectrum of almost periodic data.

not possible. However, a continuous spectral representation for transient data can be obtained in most cases from a Fourier integral given by

$$X(f) = \int_{-\infty}^{\infty} x(t)e^{-j2\pi ft}dt \qquad (2.8)$$

$$X(t) = \begin{cases} Ae^{-at} & t \geq 0 \\ 0 & t < 0 \end{cases}$$

(a)

$$X(t) = \begin{cases} Ae^{-at}\cos bt & t \geq 0 \\ 0 & t < 0 \end{cases}$$

(b)

Figure 2.5 (a) Illustration of transient data, (b) another illustration of transient data, and (c) a third illustration of transient data.

$$X(t) = \begin{cases} A & c \geq t \geq 0 \\ 0 & c < t < 0 \end{cases}$$

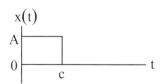

(c)

The Fourier spectrum $X(f)$ generally is a complex number that can be expressed in complex polar notation as

$$(f) = |X(f)|e^{-j\theta(f)}$$

where $|X(f)|$ is the magnitude of $X(f)$, and $\theta(f)$ is the argument. The Fourier spectra of the three transient time histories of Figure 2.5 are presented in Figure 2.6.

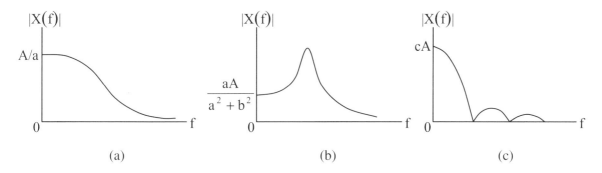

Figure 2.6 (a) Spectra of transient data of Figure 2.5(a), (b) spectra of transient data of Figure 2.5(b), and (c) spectra of transient data of Figure 2.5(c).

2.4 Linear Dynamic Systems

For small amplitude linear dynamic systems, the equation of motion for a single-degree-of-freedom system is of the form

$$m\ddot{x} + c\dot{x} + kx = F(t)$$

In this expression, m, c, and k represent the mass, damping coefficient, and stiffness of the system, respectively; $F(t)$ represents the excitation function; and x represents the response of the system due to the excitation force. The parameters m, c, and k are taken as constants, representing the proportionality coefficients of the inertia force, the damping force, and the stiffness force, respectively, which are taken directly proportional to the acceleration, velocity, and displacement, respectively. This assumption is valid for many vehicle dynamic systems for small amplitude vibration. Because all the forces are linear functions of the independent parameters, the differential equation of motion for the system is linear. Such a theory of vibration is useful in predicting the natural frequencies according to linear analysis, which is relatively simple, where large amplitudes of vibration occur. The linear theory also is useful in predicting the dynamic behavior of a system at frequencies away from resonance. Thus, the linear theory serves most of the requirements for design and analysis purposes as long as the system is detuned from resonance.

2.4.1 Linear Single-Degree-of-Freedom System

We now consider a single-degree-of-freedom model of a linear dynamic system, as shown in Figure 2.7. From Newton's third law, we write

$$F(t) - F_s(t) - F_d(t) = m\ddot{x}(t) \tag{2.9}$$

where $F(t)$, $F_s(t)$, and $F_d(t)$ are the exciting, spring, and damping forces, respectively; m denotes the mass of the body; and $\ddot{x}(t)$ denotes its acceleration. Because $F_s(t) = kx(t)$ and $F_d(t) = c\dot{x}(t)$, Eq. 2.9 becomes

$$m\ddot{x}(t) + c\dot{x}(t) + kx(t) = F(t) \tag{2.10}$$

where c and k are the damping and stiffness coefficients, respectively.

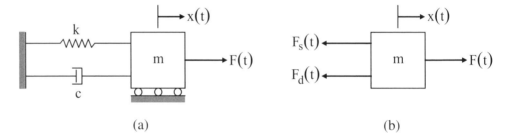

Figure 2.7 (a) Linear single-degree-of-freedom system, and (b) free body diagram.

Equation 2.10 is the equation of motion of the linear single-degree-of-freedom system and is a second-order linear differential equation with constant coefficients.

2.4.2 Free Vibration of a Single-Degree-of-Freedom System

In the case of the free vibration of a single-degree-of-freedom system, the exciting force $F(t) = 0$ and the equation of motion is

$$m\ddot{x}(t) + c\dot{x}(t) + kx(t) = 0m\ddot{x}(t) + c\dot{x}(t) + kx(t) = 0 \tag{2.11}$$

If we define $\omega_n^2 = \dfrac{k}{m}$ and $\xi = \dfrac{c}{2}m\omega_n$, Eq. 2.11 can be written as

$$\ddot{x}(t) + 2\xi\omega_n\dot{x}(t) + \omega_n^2 x(t) = 0 \tag{2.12}$$

To solve Eq. 2.12, we assume that

$$x(t) = Ae^{st} \tag{2.13}$$

where A is a constant, and s is a parameter that remains to be determined. By substituting Eq. 2.13 into Eq. 2.12, we obtain

$$\left(s^2 + 2\xi\omega_n s + \omega_n^2\right)Ae^{st} = 0 \tag{2.14}$$

Because $Ae^{st} \neq 0$, then

$$s^2 + 2\xi\omega_n s + \omega_n^2 = 0 \tag{2.15}$$

Equation 2.15 is known as the characteristic equation of the system. This equation has the two roots of

$$s_1, s_2 = \left(-\xi \pm \sqrt{\xi^2 - 1}\right)\omega_n \tag{2.16}$$

For $\xi < 1$ (underdamped condition),

$$s_1, s_2 = \left(-\xi \pm i\sqrt{\xi^2 - 1}\right)\omega_n \tag{2.17}$$

$$x(t) = A\exp(-i\omega_n t)\cos\left(\omega_n\sqrt{1 - \xi^2}\,t - \phi\right) \tag{2.18}$$

$$x(t) = A\exp(-i\omega_n t)\,\cos(\omega_d t - \phi) \tag{2.19}$$

where ω_n is the natural circular frequency, ξ is the damping factor, and $\omega_d = \omega_n\sqrt{1 - \xi^2}$ is the damped frequency of the system. Constants A and ϕ are determined from the initial conditions.

For $\xi > 1$ (overdamped condition),

$$s_1, s_2 = \left(-\xi \pm \sqrt{\xi^2 - 1}\right)\omega_n$$

$$x(t) = A_1\exp\left(-\xi + \sqrt{\xi^2 - 1}\right)\omega_n t + A_2\exp\left(-\xi - \sqrt{\xi^2 - 1}\right)\omega_n t \tag{2.20}$$

The motion is aperiodic and decays exponentially with time. Constants A_1 and A_2 are determined from the initial conditions.

For $\xi = 1$ (critically damped condition),

$$s_1 = s_2 = -\omega_n$$

$$x(t) = \left(A_1 + A_2\right)\exp(-\omega_n t) \tag{2.21}$$

Equation 2.21 represents an exponentially decaying response. The constants A_1 and A_2 depend on the initial conditions.

For this case, the coefficient of viscous damping has the value

$$c_c = 2m\omega_n = 2\sqrt{km}$$

Hence,

$$\xi = \frac{c}{c_c} \tag{2.22}$$

The locus of the roots s_1 and s_2 can be represented on a complex plane, as shown in Figure 2.8. This permits an instantaneous view of the effect of the parameter ξ on the system response. For an undamped system with $\xi = 0$, the imaginary roots are $\pm i\omega_n$.

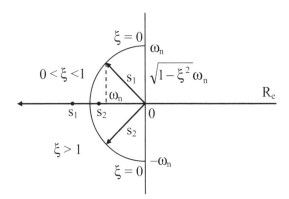

Figure 2.8 Complex planar representation of roots s_1, s_2.

For a system $0 < \xi < 1$, the roots s_1 and s_2 are complex conjugates that are located symmetrically with respect to the real axis on a circle of radius ω_n. For $\xi = 1$, $s_1 = s_2 = -\omega_n$, and as $\xi \to \infty$, $s_1 \to 0$ and $s_2 \to -\infty$.

We further consider the undamped condition in which t_1 and t_2 denote the times corresponding to the consecutive displacements x_1 and x_2, measured one cycle apart, as shown in Figure 2.9. By using Eq. 2.19, we can write

$$\frac{x_1}{x_2} = \frac{A\exp(-i\omega_n t_1)\cos(\omega_d t_1 - \phi)}{A\exp(-i\omega_n t_2)\cos(\omega_d t_2 - \phi)} \tag{2.23}$$

Because

$$t_2 = t_1 + T = t_1 + \frac{2\pi}{\omega_d}$$

then

$$\cos(\omega_d t_1 - \phi) = \cos(\omega_d t_2 - \phi)$$

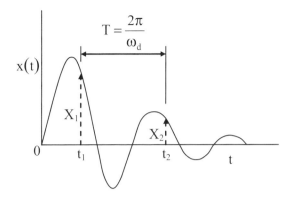

Figure 2.9 Response of an underdamped system.

Equation 2.23 then reduces to

$$\frac{x_1}{x_2} = \exp(\xi\omega_n T) \qquad (2.24)$$

If we define

$$\delta = \ln(x_1 - x_2) = \xi\omega_n T = \frac{2\pi\xi}{\sqrt{1 - \xi^2}} \qquad (2.25)$$

then δ is known as the logarithmic decrement. Thus, to determine the amount of damping in the system, it is sufficient to measure any two consecutive displacement x_1 and x_2 and obtain ξ from

$$\xi = \frac{\delta}{\sqrt{(2\pi)^2 + \delta^2}} \qquad (2.26)$$

For small damping, δ is a small quantity, and Eq. 2.26 becomes

$$\xi \cong \frac{\delta}{2\pi} \qquad (2.27)$$

EXAMPLE E2.1

For a spring mass damper system, m = 220.46 lb (100 kg) and k = 28.55 lb/in. (5000 N/m). Find the following:

a. Critical damping constant

b. Damped natural frequency when $c = \dfrac{c_c}{2}$

c. Logarithmic decrement

Solution:

The coefficient of viscous damping has the value

a. $\quad c_c = 2\sqrt{km} = 2\sqrt{\dfrac{28.55(220.46)}{12(32.2)}} = 8.07 \text{ lb sec/in.} \qquad (1414.21)\text{ N-s/m}$

b. $\quad c = \dfrac{c_c}{2} = \dfrac{8.07}{2} = 4.035 \text{ lb sec/in.} \qquad (707.1068)\text{ N-s/m}$

$$\omega_d = \omega_n\sqrt{1-\xi^2} = \sqrt{\dfrac{k}{m}}\sqrt{1-\left[\dfrac{c}{c_c}\right]^2} = 6.1235 \text{ rad/sec}$$

c. $\quad \delta = \dfrac{2\pi}{\omega_d}\left[\dfrac{c}{2m}\right] = 5.4416$

2.4.3 Forced Vibration of a Single-Degree-of-Freedom System

We now consider the response of a single-degree-of-freedom system to a harmonic excitation, for which the equation of motion is

$$m\ddot{x}(t) + c\dot{x}(t) + kx(t) = F_0\cos\omega t \tag{2.28}$$

where F_0 is the amplitude and ω is the frequency of the excitation.

Equation 2.28 can be simplified as

$$\ddot{x}(t) + 2\xi\omega_n\dot{x}(t) + \omega_n^2 x(t) = \left(\dfrac{F_0}{k}\right)\omega_n^2\cos\omega t \tag{2.29}$$

The solution of Eq. 2.29 consists of two parts: the complementary function, which is the solution of the homogeneous equation, and the particular integral. The complementary function dies out with time for $\xi > 0$ and often is called the transient solution, whereas the particular solution does not vanish for a large t and is referred to as the steady-state solution to the harmonic excitation. We assume a solution of the form

$$x(t) = X\cos(\omega t - \phi) \tag{2.30}$$

where X and ϕ are the amplitude and phase angle of response, respectively.

By substituting Eq. 2.30 into Eq. 2.29, we obtain

$$X\left[\left(\omega_n^2 - \omega^2\right)\cos(\omega t - \phi) - 2\xi\omega_n\omega\sin(\omega t - \phi)\right] = \left(\dfrac{F_0}{k}\right)\omega_n^2\cos\omega t \tag{2.31}$$

However,

$$\cos(\omega t - \phi) = \cos\omega t \cos\phi + \sin\omega t \sin\phi$$

and

$$\sin(\omega t - \phi) = \sin\omega t \cos\phi + \cos\omega t \sin\phi$$

By inserting these relationships into Eq. 2.31 and equating the coefficients of $\cos\omega t$ and $\sin\omega t$ on both sides of Eq. 2.31, we obtain

$$X\left[\left(\omega_n^2 - \omega^2\right)\cos\phi + 2\xi\omega_n\omega\sin\phi\right] = \left(\frac{F_0}{k}\right)\omega_n^2$$

$$X\left[\left(\omega_n^2 - \omega^2\right)\sin\phi - 2\xi\omega_n\omega\cos\phi\right] = 0$$

(2.32)

By solving Eq. 2.32, we obtain

$$\frac{X}{\left(\dfrac{F_0}{k}\right)} = \left\{\left[1 - \left(\frac{\omega}{\omega_n}\right)^2\right]^2 + \left[2\xi\left(\frac{\omega}{\omega_n}\right)^2\right]\right\}^{-1/2}$$

(2.33)

and

$$\phi = \tan^{-1}\left[\frac{2\xi\left(\dfrac{\omega}{\omega_n}\right)}{1 - \left(\dfrac{\omega}{\omega_n}\right)^2}\right]$$

(2.34)

Equations 2.33 and 2.34 indicate that the nondimensional amplitude $\dfrac{X}{\left(\dfrac{F_0}{k}\right)}$ and the phase angle ϕ are functions of the frequency ratio $\dfrac{\omega}{\omega_n}$ and the damping ratio ξ. For $\dfrac{\omega}{\omega_n} \ll 1$, both the inertia and damping forces are small, and this results in a small phase angle ϕ, with $\dfrac{X}{\left(\dfrac{F_0}{k}\right)} \cong 1$. However, for $\dfrac{\omega}{\omega_n} \gg 1$, the phase angle $\phi \to 180°$ and $\dfrac{X}{\left(\dfrac{F_0}{k}\right)} \to 0$. For $\dfrac{\omega}{\omega_n} = 1$, the phase angle $\phi = 90°$ and $\dfrac{X}{\left(\dfrac{F_0}{k}\right)} = \dfrac{1}{2}\xi$.

In summary, the complete solution of Eq. 2.29 is given as

$$x(t) = A_1 \exp(-i\omega_n t)\cos(\omega_d t + \phi_1) \frac{F_0}{k} \frac{\cos(\omega t - \phi)}{\sqrt{\left[1 - \left(\frac{\omega}{\omega_n}\right)^2\right]^2 + \left[2\xi\left(\frac{\omega}{\omega_n}\right)\right]^2}} \qquad (2.35)$$

where the constants A_1 and ϕ_1 are determined by the initial conditions.

Let us reconsider Eq. 2.29 and represent the excitation by the complex form

$$\left(\frac{F_0}{k}\right)\omega_n^2 e^{-i\omega t} = X_s \omega_n^2 e^{-i\omega t} \qquad (2.36)$$

where $X_s = \dfrac{F_0}{k}$ and is referred to as a static response.

We assume a solution in the form

$$x(t) = X e^{-i\omega t} \qquad (2.37)$$

By substituting Eq. 2.37 into Eq. 2.29, we obtain

$$\left[\omega_n^2 - \omega^2 - 2i\xi\omega_n\omega\right]X e^{-i\omega t} = X_s \omega_n^2 e^{-i\omega t} \qquad (2.38)$$

$$\frac{X}{X_s} = \left[1 - \left(\frac{\omega}{\omega_n}\right)^2 - 2i\xi\left(\frac{\omega}{\omega_n}\right)\right]^{-1} = H(\omega) \qquad (2.39)$$

where $H(\omega)$ is known as the complex frequency response function. Its magnitude $|H(\omega)|$ refers to a magnification factor and is given as

$$|H(\omega)| = \left\{\left[1 - \left(\frac{\omega}{\omega_n}\right)^2\right]^2 + \left[2\xi\left(\frac{\omega}{\omega_n}\right)^2\right]\right\}^{-1/2} \qquad (2.40)$$

The phase angle ϕ will be

$$\phi = \tan^{-1}\left[\frac{2\xi\left(\frac{\omega}{\omega_n}\right)}{1 - \left(\frac{\omega}{\omega_n}\right)^2}\right] \qquad (2.41)$$

The excitation considered thus far has been a simple harmonic force. We can generalize the results when the exciting force is periodic, because periodic force can be expanded in the terms of the Fourier series, as

$$F(t) = a_1 \sin \omega t + b_1 \cos \omega t + a_2 \sin 2\omega t$$
$$+ b_2 \cos 2\omega t + ... + a_n \sin n\omega t + b_n \cos n\omega t$$

(2.42)

where a_n and b_n are the coefficients of the Fourier series expansion, and it has been assumed that the constant $b_0 = 0$. Because

$$a_2 \sin n\omega t + b_n \cos n\omega t - f_n \sin(n\omega t + \alpha_n)$$

(2.43)

where

$$f_n = \left[a_n^2 + b_n^2 \right]^{1/2}$$

$$\alpha_n = \tan^{-1}\left(\frac{b_n}{a_n} \right)$$

It follows that

$$F(t) = f_1 \sin(\omega t + \alpha_1) + f_2 \sin(2\omega t + \alpha_2) + ... + f_n \sin(n\omega t + \alpha_n)$$

(2.44)

Because superposition is valid, we can consider each term on the right side of Eq. 2.44 as a separate forcing function and obtain the steady-state response by adding individual responses, due to each forcing function acting separately. Hence, it follows that

$$x(t) = X_1 \cos(\omega t + \alpha_1 - \phi_1) + X_2 \cos(2\omega t + \alpha_2 - \phi_2)$$
$$+ ... + X_n \cos(n\omega t + \alpha_n - \phi_n)$$

(2.45)

where

$$X_n = \frac{\dfrac{f_n}{k}}{\left\{ \left[1 - \left(\dfrac{n\omega}{w_n} \right)^2 \right]^2 + \left[2\xi\left(\dfrac{n\omega}{w_n} \right) \right]^2 \right\}^{1/2}}$$

and

$$\phi_n = \tan^{-1}\frac{2\xi\left(\dfrac{n\omega}{\omega_n} \right)}{\left[1 - \left(\dfrac{n\omega}{\omega_n} \right)^2 \right]} \qquad n = 1, 2, ...$$

Hence, the steady-state response is also periodic, with the same period as the forcing function but with a different amplitude and an associated phase lag.

We now discuss an alternative method for the analysis of the forced vibrations of a single-degree-of-freedom damped system. The approach uses a frequency domain analysis and is based on the concept of transfer and harmonic response functions. By taking the Laplace transformation of Eq. 2.11, we obtain

$$m\left[s^2\,\overline{x}(s) - x(0)s - \dot{x}(s)\right] + c\left[s\overline{x}(s) - x(0)\right] + k\overline{x}(s) = \overline{F}(s) \tag{2.46}$$

and it follows that

$$\overline{x}(s) = \frac{\overline{F}(s)}{ms^2 + cs + k} + \frac{(ms + c) \times (0) + m\dot{x}(0)}{ms^2 + cs + k} \tag{2.47}$$

Equation 2.47 can be written as

$$\overline{x}(s) = \frac{A(s)}{B(s)} \tag{2.48}$$

where $A(s)$ and $B(s)$ are polynomials, with $B(s)$ of a higher order.

The response $x(t)$ is found by taking the inverse Laplace transformation of Eq. 2.48. If only the forced solution is considered, we can define the impedance transform as

$$Z(s) = \left[\frac{\overline{F}(s)}{\overline{x}(s)}\right] = ms^2 + cs + k \tag{2.49}$$

A transfer function $H(s)$ is defined by

$$H(s) = \frac{1}{Z(s)} = \left(ms^2 + cs + k\right)^{-1} \tag{2.50}$$

and relates the input to the output of the system as (Figure 2.10)

$$\overline{x}(s) = H(s)\overline{F}(s) \tag{2.51}$$

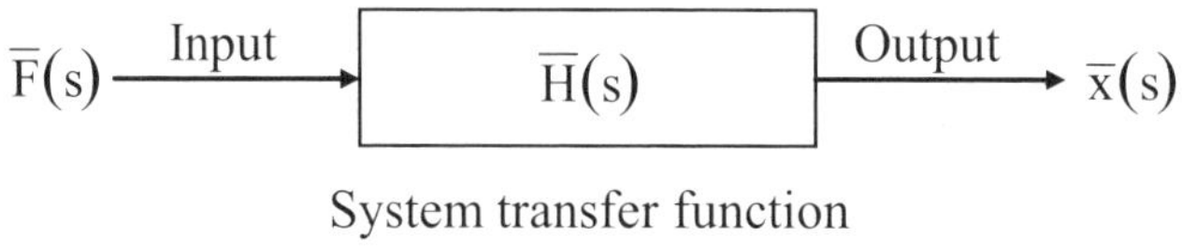

Figure 2.10 Input-output relationship through the transfer function.

System transfer function

The inverse Laplace transformation of $\bar{x}(s)$ in Eq. 2.51 yields

$$x(t) = \frac{1}{m\omega_d} \int_0^t F(\tau) \exp\left[-\xi\omega_n(t-\tau)\right] \sin\omega_d(t-\tau)d\tau \qquad (2.52)$$

EXAMPLE E2.2

Show that for a spring mass damper system, the peak amplitude occurs at the frequency ratio $r = \sqrt{1 - 2\xi^2}$. Also show that the peak amplitude is equal to

$$\left(\frac{X}{\delta_{st}}\right)_{max} = \frac{1}{2\xi\sqrt{1-\xi^2}}$$

Solution:

We know that

$$X = \frac{\delta_{st}}{\left[\left(1-r^2\right)^2 + \left(2\xi r\right)^2\right]^{1/2}}$$

For maximum X,

$$\frac{dX}{dr} = -\delta_{st}\frac{1}{2}\left[\frac{1}{\left[\left(1-r^2\right)^2 + \left(2\xi r\right)^2\right]^{3/2}}\right]\left[2\left(1-r^2\right)(-2r) + 2(2\xi r)(2\xi)\right] = 0$$

That is,

$$-4r\left(1-r^2\right) + 8r\xi^2 = 0$$

or

$$r = \sqrt{1-2\xi^2}$$

$$X\big|_{at\ r=\sqrt{1-2\xi^2}} = \frac{\delta_{st}}{\left[\left(1-\left(1-2\xi^2\right)\right)^2 + \left(2\xi\sqrt{1-2\xi^2}\right)^2\right]^{1/2}} = \frac{\delta_{st}}{2\xi\sqrt{1-\xi^2}}$$

Therefore,

$$\left(\frac{X}{\delta_{st}}\right)_{max} = \frac{1}{2\xi\sqrt{1-\xi^2}}$$

EXAMPLE E2.3

For a vibration system,

$$m = 10 \text{ kg } (0.69 \text{ lb s}^2/\text{ft})$$

$$k = 2500 \text{ N/m } (14.28 \text{ lb/in.})$$

$$c = 45 \text{ N-s/m } (0.26 \text{ lb-s/in.})$$

A harmonic force of amplitude 180 N (40.47 lb$_f$) and frequency 3.5 Hz acts on the mass. If the initial displacement and velocity of the mass are 15 mm (0.60 in.) and 5 m/s (16.40 ft/s), determine the complete solution representing the motion of the mass.

Solution:

The complete solution is

$$x(t) = X_0 e^{-\xi\omega_n t} \cos(\omega_d t + \phi_0) + X\cos(\omega t - \phi)$$

where

$$\omega = 2\pi(3.5) = 21.99 \text{ rad/sec}$$

$$\omega_n = \sqrt{\frac{k}{m}} = \sqrt{\frac{2500}{10}} = 15.81 \text{ rad/sec}$$

$$\delta_{st} = \frac{F_0}{k} = \frac{180}{2500} = 0.072 \text{ m } (2.83 \text{ in.})$$

$$r = \frac{\omega}{\omega_n} = \frac{21.99}{15.81} = 1.39$$

Also,

$$\xi = \frac{c}{2m\omega_n} = \frac{45}{2(10)(15.81)} = 0.1423$$

$$\xi\omega_n = 2.25$$

$$\omega_d = \sqrt{1-\xi^2}$$

$\omega_n = 15.65$

$$X = \frac{\delta_{st}}{\sqrt{\left(1 - r^2\right)^2 + \left(2\xi r\right)^2}} = \frac{0.072}{\left[\left(1 - 1.39^2\right)^2 + \left(2 \times 1.43 \times 1.39\right)^2\right]^{1/2}} = 0.07095 \text{ m } (2.73 \text{ in.})$$

$$\phi = \tan^{-1}\left[\frac{2\xi r}{1 - r^2}\right] = \tan^{-1}\left[\frac{0.3958}{-0.9343}\right] = -22.9591°$$

$$x(t) = X_0\, e^{-2.25t} \cos\left(15.6505\, t + \phi_0\right) + 0.07095 \cos\left(21.9912\, t + 22.9591°\right)$$

$$\dot{x}(t) = -2.25\, X_0\, e^{-2.25t} \cos\left(15.6505\, t + \phi_0\right) - 15.6505\, X_0\, e^{-2.25t} \sin\left(15.6505\, t + \phi_0\right)$$

$$- 21.9912(0.07095)\sin\left(21.9912\, t + 22.9591°\right)$$

$$x(0) = 0.015 = X\cos\phi_0 + 0.07095\cos 22.9591°$$

$$X\cos\phi_0 = -0.0503 \tag{E2.3.1}$$

$$\dot{x}(0) = -2.25 X_0 \cos\phi_0 - 15.6505 X_0 \sin\phi_0 - 1.5603\sin 22.9591°$$

$$X_0 \sin\phi_0 \frac{-0.6086 - 2.25 X_0 \cos\phi_0 - 5}{15.6505} = -0.3511 \tag{E2.3.2}$$

Therefore, from Eqs. E2.3.1 and E2.3.2,

$$X_0 = \sqrt{\left(X\cos\phi_0\right)^2 + \left(X\sin\phi_0\right)^2}$$

$$= \sqrt{\left(-0.0503\right)^2 + \left(-0.3511\right)^2}$$

$$= 0.3547 (13.96 \text{ in.})$$

and

$$\phi_0 = \tan^{-1}\left[\frac{-0.3511}{-0.0503}\right] = \tan^{-1}[6.9760] = 81.8423°$$

2.4.4 Linear Multiple-Degrees-of-Freedom System

The general equations of motion for a multiple-degrees-of-freedom discrete system with n degrees of freedom are written as

$$[m]\{\ddot{x}\} + [c]\{\dot{x}\} + [k]\{x\} = \{F(t)\} \qquad (2.53)$$

where $\{F(t)\}$ denotes the externally applied force(s). In Eq. 2.53, $[m]$, $[c]$, and $[k]$ are the n × n mass, damping, and stiffness matrices, respectively. For linear systems, these matrices are constants for nonlinear systems, and the elements of these matrices are functions of generalized displacements and velocities that are time dependent.

The response $\{x(t)\}$ of Eq. 2.53 consists of two parts: first, $\{x_h(t)\}$ is the transient response, that is, the homogeneous or complementary solution of the equations of motion; and, second, $\{x_p(t)\}$ is the steady-state or forced response, that is, the particular solution of the equations of motion.

2.4.5 Eigenvalues and Eigenvectors: Undamped System

The equations of motion for the free vibration of an undamped discrete multiple-degrees-of-freedom system are obtained by setting $[c]$ and $\{F(t)\}$ in Eq. 2.53 to zero. Then it follows that

$$[m]\{\ddot{x}\} + [k]\{x\} = \{0\} \qquad (2.54)$$

We use a linear transformation to replace $\{x\}$ by

$$\{x\} = [\phi]\{y\} \qquad (2.55)$$

where $[\phi]$ is a constant nonsingular square matrix and is referred to as a transformation matrix, and

$$\{\ddot{x}\} = [\phi]\{\ddot{y}\} \qquad (2.56)$$

By substituting Eqs. 2.55 and 2.56 into Eq. 2.54, we obtain

$$[m][\phi]\{\ddot{y}\} + [k][\phi]\{y\} = \{0\} \qquad (2.57)$$

Premultiply both sides of Eq. 2.57 by $[\phi]^T$ to yield

$$[\phi]^T[m][\phi]\{\ddot{y}\} + [\phi]^T[k][\phi]\{y\} = \{0\} \qquad (2.58)$$

From Eq. 2.58, it follows that

$$\left[M^*\right]\{\ddot{y}\} + \left[K^*\right]\{y\} = \{0\} \qquad (2.59)$$

where $\left[M^*\right]$ and $\left[K^*\right]$ are diagonal matrices known as the generalized mass and the stiffness matrix, respectively.

Equation 2.59 refers to the uncoupled homogeneous equations of motion of the system. It follows that the uncoupled equation of motion for the ith degree of freedom is

$$\ddot{y}_i + \omega_i^2 y_i = 0 \tag{2.60}$$

where ω_i is the frequency corresponding to the ith mode of vibration. The solution of Eq. 2.60 is given as

$$y_i(t) = A_i \sin \omega_i t + A_i^* \cos \omega_i t \tag{2.61}$$

where arbitrary constants A_i and A_i^* are determined by the initial conditions $x_i(0)$ and $\dot{x}_i(0)$.

We now consider Eq. 2.54 and premultiply both sides by $[m]^{-1}$ to yield

$$[m]^{-1}[m]\{\ddot{x}\} + [m]^{-1}[k]\{x\} = \{0\} \tag{2.62}$$

Equation 2.62 can be written as

$$[I]\{\ddot{x}\} + [D]\{x\} = \{0\} \tag{2.63}$$

where $[I]$ is the unit matrix and $[D] = [m]^{-1}[k]$, which is known as the dynamic matrix.

Let us assume a harmonic motion so that

$$\{x\} = \{A\}e^{i\omega t} \tag{2.64}$$

Equation 2.64 yields

$$\{\ddot{x}\} = -\omega^2 \{A\}e^{i\omega t} = -\lambda\{x\} \tag{2.65}$$

where $\lambda = \omega^2$. Substituting Eq. 2.65 into Eq. 2.63 results in

$$[[D] - \lambda[I]]\{x\} = \{0\} \tag{2.66}$$

The characteristic equation of the system is then the determinant equated to zero, that is,

$$[[D] - \lambda[I]] = \{0\} \tag{2.67}$$

The roots λ_i of the characteristic equation (Eq. 2.67) are called eigenvalues. The natural frequencies of the system are determined from

$$\lambda_i = \omega_I^2 \tag{2.68}$$

By substituting λ_i into the matrix Eq. 2.66, we obtain the corresponding mode shapes, which are called the eigenvectors. Thus, for an n-degrees-of-freedom system, there will be n eigenvalues and n eigenvectors.

Let us consider two distinct solutions corresponding to the rth and the sth modes, respectively, ω_r^2, $\left\{\phi^{(r)}\right\}$, and ω_s^2, $\left\{\phi^{(s)}\right\}$ of the eigenvalue problem. Because these solutions satisfy Eq. 2.54, it follows that

$$[k]\left\{\phi^{(r)}\right\} = \omega_r^2[m]\left\{\phi^{(r)}\right\} \tag{2.69}$$

and

$$[k]\left\{\phi^{(s)}\right\} = \omega_s^2[m]\left\{\phi^{(s)}\right\} \tag{2.70}$$

We premultiply both sides of Eq. 2.69 by $\left\{\phi^{(s)}\right\}^T$ and both sides of Eq. 2.70 by $\left\{\phi^{(r)}\right\}^T$ to obtain

$$\left\{\phi^{(s)}\right\}^T[k]\left\{\phi^{(r)}\right\} = \omega_r^2\left\{\phi^{(s)}\right\}^T[m]\left\{\phi^{(r)}\right\} \tag{2.71}$$

$$\left\{\phi^{(r)}\right\}^T[k]\left\{\phi^{(s)}\right\} = \omega_s^2\left\{\phi^{(r)}\right\}^T[m]\left\{\phi^{(s)}\right\} \tag{2.72}$$

Now we take the transpose of Eq. 2.72 to obtain

$$\left\{\phi^{(s)}\right\}^T[k]\left\{\phi^{(r)}\right\} = \omega_s^2\left\{\phi^{(s)}\right\}^T[m]\left\{\phi^{(r)}\right\} \tag{2.73}$$

By subtracting Eq. 2.73 from Eq. 2.71, we obtain

$$\left\{\omega_r^2 - \omega_s^2\right\}\left\{\phi^{(s)}\right\}^T[m]\left\{\phi^{(r)}\right\} = 0 \tag{2.74}$$

Because $\omega_r \neq \omega_s$, we conclude that

$$\left\{\phi^{(s)}\right\}^T[m]\left\{\phi^{(r)}\right\} = 0 \qquad r \neq s \tag{2.75}$$

Equation 2.75 represents the orthogonality condition of modal vectors. It also can be shown that

$$\left\{ \phi^{(s)} \right\}^T [k] \left\{ \phi^{(r)} \right\} = 0 \qquad r \neq s \tag{2.76}$$

If each column of the modal matrix $[\phi]$ is divided by the square root of the generalized mass M_i^*, the new matrix $[\bar\phi]$ is called the weighted modal matrix. It can easily be seen that

$$\left[\bar\phi\right]^T [m] \left\{\bar\phi\right\} = [I] \tag{2.77}$$

and

$$[k]\left[\bar\phi\right] = [m]\left[\bar\phi\right]\left[\omega^2\right] \tag{2.78}$$

Premultiplying Eq. 2.78 by $\left[\bar\phi\right]^T$ results in

$$\left[\bar\phi\right]^T [k]\left[\bar\phi\right] = \left[\bar\phi\right]^T [m]\left[\bar\phi\right]\left[\omega^2\right] = \left[\omega^2\right] \tag{2.79}$$

EXAMPLE E2.4

The mass matrix $[m]$ and the stiffness matrix $[k]$ of a vibration system are given by

$$[m] = m \begin{bmatrix} 1 & 0 & 0 \\ 0 & 1 & 0 \\ 0 & 0 & 1 \end{bmatrix}$$

and

$$[k] = k \begin{bmatrix} 2 & -1 & 0 \\ -1 & 2 & -1 \\ 0 & -1 & 2 \end{bmatrix}$$

Determine the eigenvalues and eigenvectors of the system.

Solution:

The equations of motion for the vibration system are

$$m \begin{bmatrix} 1 & 0 & 0 \\ 0 & 1 & 0 \\ 0 & 0 & 1 \end{bmatrix} \begin{Bmatrix} \ddot{x}_1 \\ \ddot{x}_2 \\ \ddot{x}_3 \end{Bmatrix} + k \begin{bmatrix} 2 & -1 & 0 \\ -1 & 2 & -1 \\ 0 & -1 & 2 \end{bmatrix} \begin{Bmatrix} x_1 \\ x_2 \\ x_3 \end{Bmatrix} = \begin{Bmatrix} 0 \\ 0 \\ 0 \end{Bmatrix}$$

For harmonic motion,

$$\begin{bmatrix} -m\omega^2 + 2k & -k & 0 \\ -k & -m\omega^2 + 2k & -k \\ 0 & -k & -m\omega^2 + 2k \end{bmatrix} \begin{Bmatrix} x_1 \\ x_2 \\ x_3 \end{Bmatrix} = \begin{Bmatrix} 0 \\ 0 \\ 0 \end{Bmatrix} \qquad \text{(E2.4.1)}$$

The frequency equation is

$$\left(-m\omega^2 + 2k\right)\left[\left(-m\omega^2 + 2k\right)^2 - k^2\right] + k\left[-k\left(-m\omega^2 + 2k\right)\right] = 0$$

or

$$\left(-\alpha + 2\right)\left(\alpha^2 - 4\alpha + 2\right) = 0$$

where

$$\alpha = \frac{m\omega^2}{k}$$

This gives

$$-\alpha_1 = 2 - \sqrt{2} = 0.5858$$

$$\alpha_2 = 2$$

$$\alpha_3 = 3.4142$$

Therefore,

$$\omega_1 = 0.7653\sqrt{\frac{k}{m}}$$

$$\omega_2 = 1.414\sqrt{\frac{k}{m}}$$

$$\omega_3 = 1.847\sqrt{\frac{k}{m}}$$

Equation E2.4.1 gives

$$x_2^{(j)} = \left(\frac{-m\omega_j^2 + 2k}{k}\right)x_1^{(j)} - kx_1^{(j)} + \left(-m\omega_j^2 + 2k\right)x_2^{(j)} - kx_3^{(j)} = 0$$

or

$$\left[-k + \left(-m\omega_j^2 + 2k\right)^2 \frac{1}{k}\right] x_1^{(j)} - kx_3^{(j)} = 0$$

or

$$x_3^{(j)} = \left[\frac{\left(-m\omega_j^2 + 2k\right)^2 - k^2}{k^2}\right] x_1^{(j)}$$

$$\text{jth mode} = \vec{x}^{(j)} = \begin{Bmatrix} x_1^{(j)} \\ x_2^{(j)} \\ x_3^{(j)} \end{Bmatrix} = x_1^{(j)} \begin{Bmatrix} 1 \\ \left(-m\omega_j^2 + 2k\right)/k \\ \left[\left(-m\omega_j^2 + 2k\right)^2 - k^2\right]/k^2 \end{Bmatrix}$$

The mode shapes are

$$\vec{x}^{(1)} = \begin{Bmatrix} 1 \\ 1.4142 \\ 1 \end{Bmatrix} x_1^{(1)}$$

$$\vec{x}^{(2)} = \begin{Bmatrix} 1 \\ 0 \\ -1 \end{Bmatrix} x_1^{(2)}$$

$$\vec{x}^{(3)} = \begin{Bmatrix} 1 \\ -1.4142 \\ 1 \end{Bmatrix} x_1^{(3)}$$

2.4.6 Eigenvalues and Eigenvectors: Damped System

The equation of motion for the free vibration of a damped discrete multiple-degrees-of-freedom system is given as

$$[m]\{\ddot{x}\} + [c]\{\dot{x}\} + [k]\{x\} = 0 \qquad (2.80)$$

We define a vector

$$\{y\} = \begin{Bmatrix} \{\dot{x}\} \\ \{x\} \end{Bmatrix}$$

and use an identity $\left([m]\{\dot{x}\} - [m]\{\dot{x}\} = 0\right)$ to obtain

$$\begin{bmatrix} [0] & | & [m] \\ \hline [m] & | & [c] \end{bmatrix} \begin{Bmatrix} \{\ddot{x}\} \\ \{\dot{x}\} \end{Bmatrix} + \begin{bmatrix} -[m] & | & [0] \\ \hline [0] & | & [k] \end{bmatrix} \begin{Bmatrix} \{\dot{x}\} \\ \{x\} \end{Bmatrix} = \begin{Bmatrix} \{0\} \\ \{0\} \end{Bmatrix} \qquad (2.81)$$

$$2n \times 2n \qquad 2n \times 1 \qquad 2n \times 2n \qquad 2n \times 1$$

Equation 2.81 can be rewritten as

$$[A]\{\dot{y}\} + [B]\{y\} = \{0\} \qquad (2.82)$$

where matrices $[A]$ and $[B]$ are defined as

$$[A] = \begin{bmatrix} [0] & | & [m] \\ \hline [m] & | & [c] \end{bmatrix} [B] \begin{bmatrix} -[m] & | & [0] \\ \hline [0] & | & [k] \end{bmatrix}$$

By premultiplying both sides of Eq. 2.82 with $[A]^{-1}$, we obtain

$$\{\dot{y}\} - [H]\{y\} = \{0\} \qquad (2.83)$$

where $[H] = -[A]^{-1}[B]$.

Let us assume a solution of Eq. 2.83 as

$$\{y\} = \{\Psi\}e^{\gamma t} \qquad (2.84)$$

in which γ is a complex number, and $\{\Psi\}$ is a modal vector with complex elements. Substituting Eq. 2.84 into Eq. 2.83 yields

$$\left[\gamma[I] - [H]\right]\{\Psi\} = \{0\} \qquad (2.85)$$

where $[I]$ is the unit matrix.

Hence, the characteristic equation of the system is

$$\left|\gamma[I] - [H]\right| = 0 \qquad (2.86)$$

The roots γ_i of the characteristic equation represent 2n eigenvalues, which are complex conjugates. By substituting γ_i into Eq. 2.85, we can obtain corresponding eigenvectors, which also are complex conjugates. The modal matrix $[\Psi]$ is given as

$$[\Psi] = \left[\{\Psi\}_1 \{\Psi\}_2 \cdots \{\Psi\}_{2n}\right] \tag{2.87}$$

By employing simple matrix manipulations, the orthogonality condition of eigenvectors can be established, and it can be shown that the following relationships are valid:

$$\{\Psi\}_r^T [A]\{\Psi\}_s = 0 \qquad \text{for } r \neq s \tag{2.88a}$$

$$\{\Psi\}_r^T [B]\{\Psi\}_s = 0 \qquad \text{for } r \neq s \tag{2.88b}$$

From Eqs. 2.88a and 2.88b, it follows that

$$[\Psi]^T [A][\Psi] = [A] \tag{2.89a}$$

$$[\Psi]^T [B][\Psi] = [B] \tag{2.89b}$$

where $[A]$ and $[B]$ are the diagonal matrices.

Example E2.5

The mass matrix of a vibration system is given by

$$[m] = \begin{bmatrix} 1 & 0 & 0 \\ 0 & 2 & 0 \\ 0 & 0 & 1 \end{bmatrix}$$

and the eigenvectors are given by

$$\begin{Bmatrix} 1 \\ -1 \\ 1 \end{Bmatrix}, \begin{Bmatrix} 1 \\ 1 \\ 1 \end{Bmatrix}, \text{ and } \begin{Bmatrix} 0 \\ 1 \\ 2 \end{Bmatrix}$$

Find the $[m]$-orthonormal modal matrix of the system.

Solution:

For orthonormalization,

$$\vec{x}^{(i)T}[m]\vec{x}^{(i)} = 1 \qquad i = 1, 2, 3$$

Let new

$$\vec{x}^{(1)} = a_1 \begin{Bmatrix} 1 \\ -1 \\ 1 \end{Bmatrix}$$

$$\vec{x}^{(2)} = a_2 \begin{Bmatrix} 1 \\ 1 \\ 1 \end{Bmatrix}$$

$$\vec{x}^{(3)} = a_3 \begin{Bmatrix} 0 \\ 1 \\ 2 \end{Bmatrix}$$

Then

$$\vec{x}^{(1)T}[m]\vec{x}^{(1)} = a_1^2 \begin{pmatrix} 1 & -1 & 1 \end{pmatrix} \begin{bmatrix} 1 & 0 & 0 \\ 0 & 2 & 0 \\ 0 & 0 & 1 \end{bmatrix} \begin{Bmatrix} 1 \\ -1 \\ 1 \end{Bmatrix} = 4a_1^2 = 1$$

Therefore,

$$a_1 = \frac{1}{2}$$

$$\vec{x}^{(2)T}[m]\vec{x}^{(2)} = a_2^2 \begin{pmatrix} 1 & 1 & 1 \end{pmatrix} \begin{bmatrix} 1 & 0 & 0 \\ 0 & 2 & 0 \\ 0 & 0 & 1 \end{bmatrix} \begin{Bmatrix} 1 \\ 1 \\ 1 \end{Bmatrix} = 4a_2^2 = 1$$

Therefore,

$$a_2 = \frac{1}{2}$$

$$\vec{x}^{(3)T}[m]\vec{x}^{(3)} = a_3^2 \begin{pmatrix} 0 & 1 & 2 \end{pmatrix} \begin{bmatrix} 1 & 0 & 0 \\ 0 & 2 & 0 \\ 0 & 0 & 1 \end{bmatrix} \begin{Bmatrix} 0 \\ 1 \\ 2 \end{Bmatrix} = 6a_3^2 = 1$$

which gives

$$a_3 = \frac{1}{\sqrt{6}}$$

Hence,

$$[m]\text{-orthonormal modal matrix} = [x] = \frac{1}{2}\begin{bmatrix} 1 & 1 & 0 \\ -1 & 1 & \sqrt{2/3} \\ 1 & 1 & \sqrt{8/3} \end{bmatrix}$$

2.4.7 Forced Vibration Solution of a Multiple-Degrees-of-Freedom System

We consider Eq. 2.53 and first solve the undamped free vibration problem to obtain the eigenvalues and eigenvectors, which describe the normal modes of the system and the weighted modal matrix $[\bar{\phi}]$. Let

$$\{x\} = [\bar{\phi}]\{y\} \qquad (2.90)$$

Substituting Eq. 2.90 into Eq. 2.53 yields

$$[m][\bar{\phi}]\{\ddot{y}\} + [c][\bar{\phi}]\{\dot{y}\} + [k][\bar{\phi}]\{y\} = \{F(t)\} \qquad (2.91)$$

Premultiply both sides of Eq. 2.91 by $[\bar{\phi}]^T$ to obtain

$$[\bar{\phi}]^T[m][\bar{\phi}]\{\ddot{y}\} + [\bar{\phi}]^T[c][\bar{\phi}]\{\dot{y}\} + [\bar{\phi}]^T[k][\bar{\phi}]\{y\} = [\bar{\phi}]^T\{F(t)\} \qquad (2.92)$$

Note that the matrices $[\bar{\phi}]^T[m][\bar{\phi}]$ and $[\bar{\phi}]^T[k][\bar{\phi}]$ on the left side of Eq. 2.92 are diagonal matrices that correspond to the matrices $[I]$ and $[\omega^2]$, respectively. However, the matrix $[\bar{\phi}]^T[c][\bar{\phi}]$ is not a diagonal matrix. If $[c]$ is proportional to $[m]$ or $[k]$ or both, then $[\bar{\phi}]^T[c][\bar{\phi}]$ becomes diagonal, in which case we can say that the system has proportional damping. The equations of motion then are completely uncoupled, and the ith equation will be

$$\ddot{y}_i + 2\xi_i\omega_i\dot{y}_i + \omega_i^2 y_i = \bar{f}_i(t) \qquad i = 1, 2, ..., n \qquad (2.93)$$

where

$$\bar{f}_i(t) = \{\bar{\phi}^{(i)}\}\{F(t)\}$$

Thus, instead of n-coupled equations, we will have n-uncoupled equations.

Let $[c] = \alpha[m] + \beta[k]$, in which α and β are proportionality constants. Therefore,

$$[\bar{\phi}]^T[c][\bar{\phi}] = [\bar{\phi}]^T(\alpha[m] + \beta[k])[\bar{\phi}] = \alpha[I] + \beta[\omega^2] \qquad (2.94)$$

This will yield the uncoupled ith equation of motion as

$$\ddot{y}_i + \left(\alpha + \beta\omega_i^2\right)\dot{y}_i + \omega_i^2 y_i = \bar{f}_i(t) \tag{2.95}$$

and the modal damping can be defined as

$$2\xi_i\omega_i = \alpha + \beta\omega_i^2 \tag{2.96}$$

Equation 2.96 represents Rayleigh's damping.

The solution of Eq. 2.93 is obtained by using Eq. 2.52 with initial conditions $y_i(0)$ and $\dot{y}_i(0)$ given as

$$
y_i(t) = \frac{1}{\omega_d}\int_0^t \bar{f}_i(\tau)\exp\left[-\xi_i\omega_i(t-\tau)\right]\sin\omega_{di}(t-\tau)d\tau
$$
$$
+ \frac{y_i(0)\exp\left(-\xi_i\omega_i t\right)}{\left(1-\xi_i^2\right)^{1/2}}\cos\left(\omega_{di}t - \psi_i\right) + \frac{\dot{y}_i(0)}{\omega_{di}}\exp\left(-\xi_i\omega_i t\right)\sin\omega_{di}t \tag{2.97}
$$

where

$$\omega_{di} = \left(1-\xi_i^2\right)^{1/2}\omega_i$$

and

$$\psi_i = \tan^{-1}\left[\frac{\xi_i}{\left(1-\xi_i^2\right)^{1/2}}\right]$$

Similarly, the contribution from each normal mode is calculated and substituted in Eq. 2.90 to obtain the complete response of the system. This is known as the normal mode summation method. The contributions of the higher vibration modes to the system response often are quite small and for all practical purposes may be ignored in the summation procedure by considering fewer modes of vibration.

EXAMPLE E2.6

Two of the eigenvalues of a vibrating system are given as $\begin{Bmatrix} 0.2755 \\ 0.3994 \\ 0.4491 \end{Bmatrix}$ and $\begin{Bmatrix} 0.6917 \\ 0.2974 \\ -0.3389 \end{Bmatrix}$.

a. Show that these are orthogonal with respect to the mass matrix $[m]$, and find the remaining $[m]$-orthogonal eigenvector, where $[m] = \begin{bmatrix} 1 & 0 & 0 \\ 0 & 2 & 0 \\ 0 & 0 & 3 \end{bmatrix}$.

b. Using the eigenvectors of part (a) of this example, find all the natural frequencies of
the system if the stiffness matrix $[k]$ of the system is given by $[k] = \begin{bmatrix} 6 & -4 & 0 \\ -4 & 10 & 0 \\ 0 & 0 & 6 \end{bmatrix}$.

Solution:

a. We are given that

$$\vec{x}^{(1)} = \begin{Bmatrix} 0.2755 \\ 0.3994 \\ 0.4491 \end{Bmatrix}$$

$$\vec{x}^{(2)} = \begin{Bmatrix} 0.6917 \\ 0.2974 \\ -0.3389 \end{Bmatrix}$$

$$[m] = \begin{bmatrix} 1 & 0 & 0 \\ 0 & 2 & 0 \\ 0 & 0 & 3 \end{bmatrix}$$

Therefore,

$$\vec{x}^{(1)T}[m]\vec{x}^{(2)} = \begin{pmatrix} 0.2755 & 0.3994 & 0.4491 \end{pmatrix} \begin{Bmatrix} 0.6917 \\ 0.5948 \\ -1.0168 \end{Bmatrix} = 0$$

Let

$$\vec{x}^{(3)} = \begin{Bmatrix} a_1 \\ a_2 \\ a_3 \end{Bmatrix}$$

Then

$$\vec{x}^{(3)T}[m]\vec{x}^{(3)} = 1$$

$$\vec{x}^{(1)T}[m]\vec{x}^{(3)} = 0$$

$$\vec{x}^{(2)T}[m]\vec{x}^{(3)} = 0$$

These relations give

$$a_1^2 + 2a_2^2 + 3a_3^2 = 1 \qquad\qquad \text{(E2.6.1)}$$

$$0.2755a_1 + 0.7989a_2 + 1.3472a_3 = 0 \qquad\qquad \text{(E2.6.2)}$$

$$0.6917a_1 + 0.5949a_2 + 1.0168a_3 = 0 \qquad \text{(E2.6.3)}$$

From Eqs. E2.6.2 and E2.6.3,

$$a_1 = -2.9a_2 - 4.89a_3 = -0.86a_2 + 1.47a_3$$

or

$$a_2 = 3.1176a_3 \qquad \text{(E2.6.4)}$$

and

$$a_1 = -2.9(-3.1176a_3) - 4.89a_3 = 4.1512a_3 \qquad \text{(E2.6.5)}$$

Equations E2.6.1, E2.6.4, and E2.6.5 give

$$a_3^2(17.2323 + 9.7197 + 1) = 1$$

or

$$a_3 = \pm 0.1891$$

Hence,

$$a_2 = \mp 0.5897$$

$$a_1 = \mp 0.7852$$

Therefore,

$$\vec{x}^{(3)} = \begin{Bmatrix} 0.7852 \\ -0.5897 \\ 0.1891 \end{Bmatrix}$$

b.

$$\omega_i^2 = \vec{x}^{(i)\text{T}}[k]\vec{x}^{(i)}$$

and

$$[k] = \begin{bmatrix} 6 & -4 & 0 \\ -4 & 10 & 0 \\ 0 & 0 & 6 \end{bmatrix}$$

$$\omega_1^2 = (0.2755 \quad 0.3995 \quad 0.4491)\begin{Bmatrix} 0.0551 \\ 2.8927 \\ 2.6943 \end{Bmatrix} = 2.3806$$

$$\omega_2^2 = \begin{pmatrix} 0.6917 & 0.2974 & -0.3389 \end{pmatrix} \begin{Bmatrix} 2.9605 \\ 0.2075 \\ -2.0336 \end{Bmatrix} = 2.7987$$

$$\omega_3^2 = \begin{pmatrix} 0.7852 & -0.5897 & 0.1891 \end{pmatrix} \begin{Bmatrix} 7.0698 \\ -9.0375 \\ 1.1348 \end{Bmatrix} = 11.0949$$

Therefore,

$$\omega_1 = 1.5429$$

$$\omega_2 = 1.6729$$

$$\omega_3 = 3.3309$$

2.5 Nonlinear Dynamic Systems

In this section, we briefly discuss the analytical techniques that are used with nonlinear dynamic systems. A general analysis of a single-degree-of-freedom nonlinear system is given to provide a perspective of the problem. The characteristics of trajectories in a phase plane are discussed and illustrated. The isocline and Pell's method for the graphical construction of trajectories in the phase plane are described. Some of the exact methods, four approximate methods, graphical procedures, and the stability of equilibrium for nonlinear systems are presented.

In most dynamic systems, nonlinear relationships exist, particularly when the deformations are large. In addition to the nonlinear behavior of coefficients of inertia force, damping force, and stiffness force, we come across nonlinearities due to geometry. Such nonlinearities in a dynamic system give rise to entirely different dynamic behavior when compared to linear theories. One of the main reasons for modeling a dynamic system as a nonlinear one is that totally unexpected phenomena sometimes occur in nonlinear systems—phenomena that are not predicted or even hinted at by linear theory. The characteristics of dynamic behavior of a nonlinear system may be summarized as follows:

a. A nonlinear system is governed by nonlinear differential equations. The system may have more than one equilibrium point, and the equilibrium point may be stable or unstable.

b. The steady-state behavior, when it exists, depends on the initial conditions.

c. The frequency of vibration may be dependent on the amplitude of vibration.

d. While the frequency of excitation is increased, the amplitude of vibration may have a significant jump; the amplitude suddenly may fall after reaching a peak value and suddenly may jump upward when the frequency of excitation is decreased.

e. When a harmonic force excites a system, the response will not only have the basic harmonic component but may consist of higher harmonics or subharmonics.

f. The system can become self-excited, and amplitudes of vibration may grow, even without any external disturbance.

g. The system can become unstable under certain conditions.

h. The principle of linear-super position is not applicable to analyze a nonlinear system subjected to a multiple frequency excitation.

i. Depending on the combinations of natural frequencies, internal resonances can exist in a multiple-degrees-of-freedom system.

j. A periodic excitation may result in a nonperiodic response in a nonlinear system.

2.5.1 Exact Methods for Nonlinear Systems

An exact solution is possible for only relatively few nonlinear systems. The solutions are exact in the sense that they are obtained in closed form or as expressions that can be evaluated numerically. Consider a simple nonlinear system for which the exact solution is possible. For a single-degree-of-freedom system with a general restoring (spring) force $F(x)$, the free vibration equation can be expressed as

$$\ddot{x} + a^2 F(x) = 0 \tag{2.98}$$

where a^2 is a constant. Equation 2.98 can be written as

$$\frac{d}{dx}\left(\dot{x}^2\right) + 2a^2 F(x) = 0 \tag{2.99}$$

Integration of Eq. 2.99 yields, assuming $x_0 = \dot{x}_0 = 0$,

$$\dot{x}^2 = 2a^2 \int_x^{x_0} F(\eta)\,d\eta$$

or $\hspace{4cm}$ (2.100)

$$|\dot{x}| = \sqrt{2}a\left\{\int_x^{x_0} F(\eta)\,d\eta\right\}^{1/2}$$

where η is the integration variable, and x_0 is the value of x when $\dot{x} = \frac{(dx)}{(dt)} = 0$. Equation 2.100, upon integration again, gives

$$t - t_0 = \frac{1}{\sqrt{2}a}\int_0^x \frac{d\xi}{\left\{\int_\xi^{x_0} F(\eta)\,d\eta\right\}^{1/2}} \tag{2.101}$$

where ξ is the new integration variable, and t_0 corresponds to the time when $x = 0$. Equation 2.101 gives the exact solution of Eq. 2.98 where the integrals of Eq. 2.101 can be evaluated in closed form. After the integrals of Eq. 2.101, we can obtain the displacement-time relation. If $F(x)$ is an odd function, then

$$F(-x) = -F(x) \tag{2.102}$$

Assuming Eq. 2.101 from zero displacement to maximum displacement, the period of vibration τ can be obtained as

$$\tau = \frac{4}{\sqrt{2}a} \int_0^x \frac{d\xi}{\left\{ \displaystyle\int_\xi^{x_0} F(\eta)d\eta \right\}^{1/2}} \tag{2.103}$$

As an example, let $F(x) = x^n$. Equations 2.101 and 2.103 give

$$t - t_0 = \frac{1}{a}\sqrt{\frac{n+1}{2}} \int_0^{x_0} \frac{d\xi}{\left(x_0^{n+1} - \xi^{n+1} \right)^{1/2}} \tag{2.104}$$

and

$$\tau = \frac{4}{a}\sqrt{\frac{n+1}{2}} \int_0^{x_0} \frac{d\xi}{\left(x_0^{n+1} - \xi^{n+1} \right)^{1/2}} \tag{2.105}$$

Now denoting $y = \dfrac{\xi}{x_0}$, Eq. 2.105 becomes

$$\tau = \frac{4}{a} \frac{1}{\left(x_0^{n-1} \right)^{1/2}} \sqrt{\frac{n+1}{2}} \int_0^1 \frac{dy}{\left(1 - y^{n+1} \right)^{1/2}} \tag{2.106}$$

Equation 2.106 can be used to compute values to any desired level of accuracy.

EXAMPLE E2.7

Consider the single-degree-of-freedom system shown in Figure E2.7(a). Derive the equation of motion, assuming the spring k is linear and the system is shown in its static equilibrium position.

Solution:

The kinetic energy of the system is

$$T = \frac{1}{2}J_0\dot{\phi}^2 + \frac{1}{2}m\dot{x}^2$$

The potential energy of the system is (Figure E2.7(b))

$$U = \frac{1}{2}k\left(\ell_1 - \ell\right)^2$$

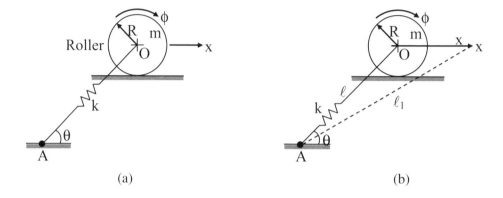

Figure E2.7 (a) Single-degree-of-freedom system, and (b) free body diagram.

(a)

(b)

Now

$$\frac{\partial U}{\partial x} = k\left(\ell_1 - \ell\right)\frac{\partial \ell_1}{\partial x}$$

and

$$\ell_1^2 = \left(\ell\cos\theta + x\right)^2 + \left(\ell\sin\theta\right)^2 = \ell^2 + x^2 + 2\ell x\cos\theta$$

$$\ell_1 = \frac{\partial \ell_1}{\partial x} = x + \ell\cos\theta$$

Applying Lagrange's equation, the equation of motion can be written as

$$\left(\frac{J_0}{R^2} + m\right)\ddot{x} + k\left[1 + \frac{\ell}{\sqrt{\ell^2 + x^2 + 2\ell x\cos\theta}}\right]\left(x + \ell\cos\theta\right) = 0$$

which is a nonlinear equation.

If x is small, then the linearized equation can be written as

$$\left(\frac{J_0}{R^2} + m\right)\ddot{x} + k\left(\cos^2\theta\right)x = 0$$

EXAMPLE E2.8

Figure E2.8(a) shows a mass m attached to the midpoint of a string of length 2ℓ. Assuming the tension in the string as T, obtain the governing differential equation of motion for the system.

Solution:

Newton's second law of motion to the system in Figure E2.8 gives

$$\Sigma F_x = -2T\sin\theta = m\ddot{x}$$

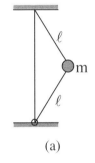
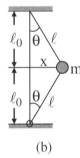

(a)   (b)

Figure E2.8 (a) Vibrating system, and (b) free body diagram.

We have

$$\sin\theta = \frac{x}{\ell} = \frac{x}{\sqrt{\ell_0^2 + x^2}} \approx \frac{x}{\ell_0}\left[1 - \frac{1}{2}\left(\frac{x}{\ell_0}\right)^2\right]$$

Also,

$$T = T_0 + k\left(\ell - \ell_0\right) = T_0 + k\left[\ell_0\left(1 + \frac{x^2}{\ell_0^2}\right)^{1/2} - \ell_0\right] \approx T_0 + k\left(\frac{1}{2}\right)\left(\frac{x}{\ell_0}\right)^2$$

Hence,

$$m\ddot{x} + 2\left[T_0 + \frac{k}{2}\left(\frac{x}{\ell_0}\right)^2\right]\frac{x}{\sqrt{\ell_0^2 + x^2}} = 0$$

or

$$m\ddot{x} + \frac{2}{\ell_0}\left[T_0 + \frac{k}{2}\left(\frac{x}{\ell_0}\right)^2\right]\left[1 + \frac{1}{2}\left(\frac{x}{\ell_0}\right)^2\right]x = 0$$

which is a nonlinear differential equation.

2.5.2 Approximate Methods for Nonlinear Systems

Both analytical and numerical methods are available for approximate solution of nonlinear dynamics problems. In this section, we consider four analytical techniques: the iterative method, the Ritz averaging method, the perturbation method, and the variation of parameter method.

2.5.2.1 Iterative Method

We present the iterative method for determining the periodic solutions of Duffing's equation. Duffing's equation refers to the equation of motion of a damped, harmonically excited, single-degree-of-freedom system with a nonlinear spring. We consider the solution of the undamped and damped equations.

Undamped equation: Neglecting damping, Duffing's equation can be written as

$$\ddot{x} + \alpha x \pm \beta x^3 = F\cos\omega t$$

or (2.107)

$$\ddot{x} = -\alpha x \pm \beta x^3 = F\cos\omega t$$

Assume the solution

$$x_1(t) = A\cos\omega t \tag{2.108}$$

where A is an unknown constant. From Eqs. 13.107 and 13.108, the differential equation for the second approximation is obtained as

$$\ddot{x}_2 = -A\alpha\cos\omega t \mp A^3\beta\cos^3\omega t + F\cos\omega t \tag{2.109}$$

Assuming the identity

$$\cos^3\omega t = \frac{3}{4}\cos\omega t + \frac{1}{4}\cos 3\omega t \tag{2.110}$$

Equation 2.109 becomes

$$\ddot{x}_2 = -\left(A\alpha \pm \frac{3}{4}A^3\beta - F\right)\cos\omega t \mp \frac{1}{4}A^3\beta\cos 3\omega t \tag{2.111}$$

Integrating Eq. 2.111 and taking the integration constants as zero, we obtain

$$x_2(t) = \frac{1}{\omega^2}\left(A\alpha \pm \frac{3}{4}A^3\beta - F\right)\cos\omega t \pm \frac{A^3\beta}{36\omega^2}\cos 3\omega t \qquad (2.112)$$

Assuming $x_1(t)$ and $x_2(t)$ as approximations to the solution $x(t)$, equating these coefficients from Eqs. 2.108 and 2.112, we obtain

$$A = \frac{1}{\omega^2}\left(A\alpha \pm \frac{3}{4}A^3\beta - F\right)$$

or $\qquad\qquad\qquad\qquad\qquad\qquad\qquad\qquad (2.113)$

$$\omega^2 = \alpha \pm \frac{3}{4}A^2\beta - \frac{F}{A}$$

It can be shown that this process gives the exact solution for the special case of a linear spring (with $\beta = 0$) as

$$A = \frac{F}{\alpha - \omega^2} \qquad (2.114)$$

where A is the amplitude of the harmonic response of the linear system.

Equation 2.113 indicates that the frequency ω is a function of β, A, and F. Here, A represents the coefficient of the first term of its solution or the amplitude of the harmonic response of the system. For the free vibration problem, we have $F = 0$; hence, Eq. 2.113 gives

$$\omega^2 = \alpha \pm \frac{3}{4}A^2\beta \qquad (2.115)$$

Equation 2.115 shows that the frequency of the response increases with the amplitude A for the hardening spring and decreases for the softening spring.

When $F \neq 0$, there exist two values for ω for any given amplitude $|A|$. One value of ω is smaller and the other larger than the corresponding frequency of free vibration at that amplitude. For the smaller value of ω, $A > 0$, and the harmonic response of the system will be in phase with the external force. For the larger value of ω, $A < 0$, and the response will be $180°$ out of phase with the external force.

Damped equation: Consider Duffing's equation (with viscous damping)

$$\ddot{x} + c\dot{x} + \alpha x \pm \beta x^3 = F\cos\omega t \qquad (2.116)$$

There will be a phase difference between the applied force and the response. Here, we fix the phase of the solution and keep the phase of the applied force as a quantity to be obtained. Equation 2.116 can be rewritten as

$$\ddot{x} + c\dot{x} + \alpha x \pm \beta x^3 = F\cos(\omega t + \phi)$$

$$= A_1 \cos \omega t - A_2 \sin \omega t \qquad (2.117)$$

where the amplitude $F = \left(A_1^2 + A_2^2\right)^{1/2}$ of the applied force is considered fixed, and the ratio $\dfrac{A_1}{A_2} = \tan^{-1}\phi$ is to be obtained. We assume that c, A_1, and A_2 are small, of order β. As with Eq. 2.107, we assume the first approximation to the solution to be

$$x_1 = A\cos\omega t \qquad (2.118)$$

where A is assumed fixed and ω is to be found.

From Eqs. 2.117, 2.118, and 2.110, we obtain

$$\left[\left(\alpha - \omega^2\right)A \pm \frac{3}{4}\beta A^3\right]\cos\omega t - c\omega A\sin\omega t \pm \frac{\beta A^3}{4}\cos 3\omega t$$

$$= A_1\cos\omega t - A_2\sin\omega t \qquad (2.119)$$

Ignoring the term involving $\cos 3\omega t$ and equating the coefficients of $\cos\omega t$ and $\sin\omega t$ on both sides of Eq. 2.119, we obtain

$$\left[\left(\alpha - \omega^2\right)A \pm \frac{3}{4}\beta A^3\right] = A_1$$

$$c\omega A = A_2 \qquad (2.120)$$

Squaring and adding Eq. 2.120, we obtain

$$\left[\left(\alpha - \omega^2\right)A \pm \frac{3}{4}\beta A^3\right]^2 + \left(c\omega A\right)^2 = A_1^2 + A_2^2 = F^2 \qquad (2.121)$$

Equation 2.121 can be expressed as

$$S^2(\omega, A) + c^2\omega^2 A^2 = F^2 \qquad (2.122)$$

where

$$S(\omega, A) = \left(\alpha - \omega^2\right)A \pm \frac{3}{4}\beta A^3 \qquad (2.123)$$

It can be observed that for $c = 0$, Eq. 2.122 reduces to $S(\omega, A) = F$, which is identical to Eq. 2.113. The response curves from Eq. 2.122 are depicted in Figure 2.11.

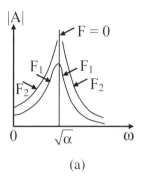 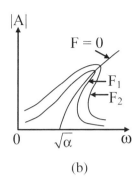 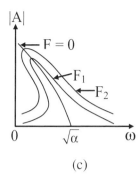

(a)   (b)   (c)

Figure 2.11 Response curves for (a) $\beta = 0$, (b) $\beta > 0$, and (c) $\beta < 0$.

2.5.2.2 Ritz Averaging Method

Here, an approximation solution is obtained by satisfying the governing nonlinear equation in the average. Consider the nonlinear differential equation given by

$$E(x) = 0 \tag{2.124}$$

An approximate solution of Eq. 2.124 can be assumed as

$$x(t) = a_1\phi_1(t) + a_2\phi_2(t) + \ldots + a_n\phi_n(t) \tag{2.125}$$

where $\phi_1(t)$, $\phi_2(t)$, ..., $\phi_n(t)$ are prescribed functions of time, and $a_1, a_2, ..., a_n$ are the weighting factors to be determined. The weighting factors are obtained by substituting Eq. 2.125 into Eq. 2.124, multiplying the differential equation by each of the functions $\phi_1(t)$, and then integrating the product over a period of the motion. Thus,

$$\int_0^\tau E(x)\phi_i(t)dt = 0 \qquad i = 1, 2, ..., n \tag{2.126}$$

Equation 2.126 represents a system of n algebraic equations in n unknowns, namely, the weighting factors $a_1, a_2, ..., a_n$, where $\phi_1(t)$, $\phi_2(t)$, ..., $\phi_n(t)$ are prescribed functions of time. From Eqs. 2.124 and 2.125, we obtain a function $E[x(t)]$. Because $x(t)$ is not the exact solution of Eq. 2.124, $E(t) = E[x(t)]$ will not be equal to zero. The value of $E[t]$ is a measure of the accuracy of the approximation.

The weighting factors a_i are obtained by minimizing the integral

$$\int_0^\tau E^2[t]dt \tag{2.127}$$

where τ is the period of the motion. The minimization process in Eq. 2.127 becomes

$$\frac{\partial}{\partial a_i}\left(\int_0^\tau \underset{\sim}{E}^2[t]dt\right) = 2\int_0^\tau \underset{\sim}{E}[t]\frac{\partial \underset{\sim}{E}[t]}{\partial a_i}dt = 0 \qquad i = 1, 2, ..., n \qquad (2.128)$$

Equation 2.128 represents a system of n algebraic equations that are solved simultaneously to obtain the values of a_1, a_2, ..., a_n.

EXAMPLE E2.9

Determine the natural time period of oscillation of the simple pendulum system shown in Figure E2.9 for $\frac{\pi}{2} < 0 < \frac{\pi}{2}$.

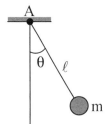

Figure E2.9 Simple pendulum.

Solution:

The differential equation governing the free vibration of the pendulum can be obtained as

$$m\ell^2\ddot{\theta} + \frac{g}{\ell}\sin\theta = 0 \qquad (E2.9.1)$$

or

$$\ddot{\theta} + \frac{g}{\ell}\sin\theta = 0 \qquad (E2.9.2)$$

Equation E2.9.2 can be rewritten as

$$\dot{\theta}\frac{d\dot{\theta}}{d\theta} + \frac{g}{\ell}\sin\theta = 0 \qquad (E2.9.3)$$

where we used

$$\ddot{\theta} = \frac{d\dot{\theta}}{dt} = \frac{d\dot{\theta}}{d\theta}\frac{d\theta}{dt} = \dot{\theta}\frac{d\dot{\theta}}{d\theta}$$

Integrating Equation E2.9.3, we obtain

$$\frac{\dot{\theta}^2}{2} - \frac{g}{\ell}\cos\theta = c \qquad\text{(E2.9.4)}$$

The initial conditions are $t = 0$, $\dot{\theta} = 0$, and $\theta = \theta_0$.

Hence,

$$c = -\frac{g}{\ell}\cos\theta_0$$

Equation E2.9.4 becomes

$$\left(\frac{d\theta}{dt}\right)^2 = \dot{\theta}^2 = 2\left(c + \frac{g}{\ell}\cos\theta\right)$$

$$dt = \frac{d\theta}{\sqrt{2\left(c + \dfrac{g}{\ell}\cos\theta\right)}} = \frac{d\theta}{\sqrt{2\dfrac{g}{\ell}(\cos\theta - \cos\theta_0)}}$$

or

$$t = \int \frac{d\theta}{\sqrt{2\dfrac{g}{\ell}(\cos\theta - \cos\theta_0)}} \qquad\text{(E2.9.5)}$$

Noting

$$\cos\theta - \cos\theta_0 = \left(1 - 2\sin^2\frac{\theta}{2}\right) - \left(1 - 2\sin^2\frac{\theta_0}{2}\right)$$

$$= 2\sin^2\frac{\theta_0}{2}\left[1 - \frac{\sin^2\dfrac{\theta}{2}}{\sin^2\dfrac{\theta_0}{2}}\right] \qquad\text{E2.9.6)}$$

Denoting $k = \sin\dfrac{\theta_0}{2}$, we have

$$\frac{\sin\dfrac{\theta}{2}}{k} = \sin\phi \qquad\text{(E2.9.7)}$$

Differentiating Eq. E2.9.7, we obtain

$$\frac{1}{k} \cos \frac{\theta}{2} \frac{d\theta}{2} = \cos \phi \, d\phi$$

or

$$d\theta = \frac{2k \cos \phi \, d\phi}{\cos \frac{\theta}{2}}$$

(E2.9.8)

From Eqs. E2.9.5 to E2.9.8, we can write

$$t = \int \sqrt{\frac{\left(\dfrac{2k \cos \phi \, d\phi}{\cos \dfrac{\theta}{2}} \right)}{\left(2\sqrt{\dfrac{g}{\ell}} \sin \dfrac{\theta_0}{2} \cos \phi \right)}}$$

$$= \sqrt{\frac{\ell}{g}} \int \frac{d\phi}{\cos \dfrac{\theta}{2}} = \sqrt{\frac{\ell}{g}} \int \frac{d\phi}{\sqrt{1 - \sin^2 \dfrac{\theta}{2}}} = \sqrt{\frac{\ell}{g}} \int \frac{d\phi}{\sqrt{1 - k^2 \sin^2 \phi}}$$

Hence, the time period τ is given by

$$\tau = 4\sqrt{\frac{\ell}{g}} \int_0^{\pi/2} \frac{d\phi}{\sqrt{1 - k^2 \sin^2 \phi}}$$

(E2.9.9)

where

$$k = \sin \frac{\theta_0}{2}$$

Eq. E2.9.9 can be solved for any value of θ_0 that can be determined from the elliptic integral mathematical tables.

2.5.2.3 Perturbation Method

The perturbation method assumes that the solution of a nonlinear equation can be written as a power series that consists of a generating solution and the added corrective terms.

The equation of motion of an undamped single-degree-of-freedom system vibrating under free conditions, as shown in Figure 2.12, can be written as

$$m \frac{d^2 x}{dt^2} + \alpha x + \beta x^3 = 0$$

(2.129)

Let

$$p = \sqrt{\frac{\alpha}{m}}$$

$$T = pt$$

$$dt = \frac{1}{p}dT \qquad\qquad (2.130)$$

$$\mu = \frac{\beta}{mp^2}$$

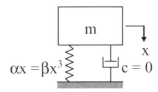

Figure 2.12 Undamped nonlinear system in free vibration.

Then

$$\ddot{x} + x + \mu x^3 = 0 \qquad\qquad (2.131)$$

where

$$\cdot = \frac{d}{dt}$$

Equation 2.129 is known as Duffing's equation. Equation 2.131 is nonlinear with respect to the restoring force defined by the cubic function. An exact solution to Eq. 2.131 is not possible. Hence, we try to find an approximation solution. Assuming the initial conditions for the system are given by

$$x(0) = A$$
$$\dot{x}(0) = 0 \qquad\qquad (2.132)$$

Let $x_0(t)$ be the linear solution ($\beta = 0$). We then can perturb the solution about the linear solution in a series form given by

$$x(T) = x_0(T) + \mu x_1(T) + \mu^2 x_2(T) + \dots \qquad\qquad (2.133)$$

The coefficient μ in Eq. 2.129 is a nonlinear parameter, and the solution assumed in Eq. 2.133 converges quickly if $\mu \ll 1$. From Eqs. 2.133 and 2.129, we obtain

$$\left(\ddot{x}_0 + \mu\ddot{x}_1 + \mu^2\ddot{x}_2 + ...\right) + \left(x_0 + \mu x_1 + \mu^2 x_2 + ...\right)$$
$$+ \mu\left(x_0 + \mu x_1 + \mu^2 x_2 + ...\right)^3 = 0 \tag{2.134}$$

Because μ is a parameter, separating the terms of μ^0, μ^1, μ^2, and so forth, we get

$$\ddot{x}_0 + x_0 = 0$$
$$\ddot{x}_1 + x_1 = -x_0^3 \text{ ... and so forth} \tag{2.135}$$

Noting that the first part of Eq. 2.135 is the linear part, and considering Eq. 2.132, we have the solution of the form

$$x_0 = a_1 \cos T + b_1 \sin T = A \cos T \tag{2.136}$$

From Eqs. 2.135 and 2.132, we obtain

$$\ddot{x}_1 + x_1 = -A^3 \cos^3 T = -\frac{3}{4}A^3 \cos^3 T - \frac{1}{4}A^3 \cos 3T \tag{2.137}$$

Equation 2.137 shows that there is an exciting term with the third harmonic of the natural frequency. The solution of Eq. 2.133 contains higher harmonic terms. We can separate Eq. 2.137 into two parts: one belonging to the harmonic solution, and the other belonging to the higher harmonic. Hence,

$$\ddot{x}_{11} + x_{11} = -\frac{3}{4}A^3 \cos T$$
$$\ddot{x}_{12} + x_{11} = -\frac{1}{4}A^3 \cos 3T \tag{2.138a}$$
$$x_1 = x_{11} + x_{12}$$

The terms on the right side of Eq. 2.138a are due to the perturbation of the nonlinear free vibration problem. Here, we consider both the transient and particular integral parts of Eq. 2.138a for the final solution. Because the first equation in Eq. 2.138a is excited by the forcing term at resonant frequency, we have

$$x_{11} = T\left(a_2 \cos T + b_2 \sin T\right)$$

or $\hspace{8cm}$ (2.138b)

$$\ddot{x}_{11} = -2a_2 \sin T + 2b_2 \cos T$$

From Eqs. 2.138a and b, we obtain

$$x_{11} = -\frac{3}{8}A^3 T \sin T \tag{2.139}$$

Similarly for the higher harmonic, we obtain

$$x_{12} = a_3 \cos T + b_3 \sin T + \frac{A^3}{32} \cos 3T$$

Noting

$$x_{12} = 0$$

and

$$\dot{x}_{12} = 0 \text{ at } T = 0$$

we have

$$x_{12} = -\frac{1}{32} A^3 (\cos T - \cos 3T) \qquad (2.140)$$

Thus, the complete solution for x_1 is

$$x_1 = -x_1 = -\frac{3}{8} A^3 T \sin T - \frac{1}{32} A^3 (\cos T - \cos 3T) \qquad (2.141)$$

We observe that the first term on the right side of Eq. 2.141 grows with time. This term is called the secular term. Here, we introduce the frequency amplitude interaction by the Lindstedt-Poincare method as follows:

For the linear system with $\mu = 0$, in Eq. 2.131, the solution becomes periodic with period 2π.

For $\mu \neq 0$, we introduce an unspecified function ω as

$$\tau = \omega T \qquad (2.142)$$

We select ω such that the secular term in Eq. 2.141 can be avoided. Equation 2.141 becomes

$$\omega^2 \ddot{x} + x + \mu x^3 = 0 \qquad (2.143)$$

where

$$\cdot = \frac{d}{d\tau}$$

The solution to Eq. 2.143 can be written as

$$x(t) = x_0(t) + \mu x_1(t) + \mu^2 x_2(t) + \dots \qquad (2.144)$$

We also select the function for ω in series as

$$\omega = \omega_0 + \mu\omega_1 + \mu^2\omega_2 + \dots \qquad (2.145)$$

Hence,

$$\left(\omega_0 + \mu\omega_1 + \mu^2\omega_2 + \dots\right)^2 \left(\ddot{x}_0 + \mu\ddot{x}_1 + \mu^2\ddot{x}_2 + \dots\right)$$
$$+ \left(x_0 + \mu x_1 + \mu^2 x_2 + \dots\right) + \mu\left(x_0 + \mu x_1 + \mu^2 x_2 + \dots\right) = 0 \qquad (2.146)$$

Separate the terms of μ^0, μ^1, μ^2, and so forth, and we obtain

$$\omega_0^2 \ddot{x}_0 + x_0 = 0$$
$$\omega_0^2 \ddot{x}_1 + x_1 = -2\omega_0\omega_2\ddot{x}_0 - x_0^3 \qquad (2.147)$$
$$\omega_0^2 \ddot{x}_2 + x_2 = -\left(2\omega_0\omega_2 + \omega_1^2\right)\ddot{x}_0 - 2\omega_0\omega_1\ddot{x}_1 - 3x_0^2 x_1 \dots \text{ and so forth.}$$

In addition to the initial condition of Eq. 2.132, we now introduce

$$x_i\left(\tau + 2\pi\right) = x_i\left(\tau\right) \qquad (2.148)$$

The solution of the first equation in Eq. 2.147 yields

$$x_0 = a_1 \cos\frac{\tau}{\omega_0} + b_1 \sin\frac{\tau}{\omega_0}$$

From Eqs. 2.132 and 2.148, we have

$$x_0 = A\cos\tau \qquad (2.149)$$

$$\omega_0 = 1 \qquad (2.150)$$

Now the second equation in Eq. 2.147 gives

$$\ddot{x}_1 + x_1 = -2\omega_1 A\cos\tau - A^3\cos^3\tau_1 + x_1$$
$$= -2\omega_1 A\cos\tau - A^3\cos^3\tau_1 \qquad (2.151)$$
$$\left(2\omega_1 A - \frac{3}{4}A^3\right)\cos\tau - \frac{1}{4}A^3\cos 3\tau$$

The first term on the right side of Eq. 2.151 gives rise to the secular term in the solution. To avoid the secular term in the solution, we select

$$\omega_1 = \frac{3}{8}A^2 \tag{2.152}$$

This would ensure that there is no term in the solution that grows with time.

Here, the solution of Eq. 2.151 is governed by the third harmonic excitation term, which can be expressed as

$$x_1 = a_2 \cos\tau + b_2 \sin\tau + \frac{A^3}{32}\cos 3\tau = \frac{A^3}{32}\left(-\cos\tau + \cos 3\tau\right) \tag{2.153}$$

From Eqs. 2.153 and 2.147, we can obtain ω_2 and x_{13}. The solution of Eq. 2.143 is given by

$$x(\tau) = \left(A - \frac{1}{32}\mu A^3 + \frac{23}{1024}\mu^2 A^5\right)\cos\tau$$

$$+ \left(\frac{1}{32}\mu A^3 - \frac{3}{128}\mu^2 A^5\right)\cos 3\tau + \frac{1}{1024}\mu^2 A^5 \cos 5\tau \tag{2.154}$$

$$\omega = 1 + \frac{3}{8}\mu A^2 - \frac{21}{256}\mu^2 A^4$$

Equation 2.154 shows the dependency of frequency ω on amplitude A given initially. If $\mu > 0$, we have a hardening spring. If $\mu < 0$, then the spring is the softening type, as shown in Figure 2.13. Equation 2.154 is plotted in Figure 2.14 for both cases.

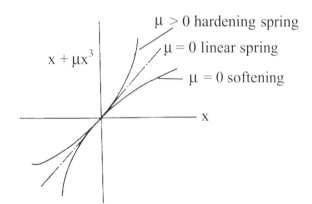

Figure 2.13 Hardening and softening spring characteristics.

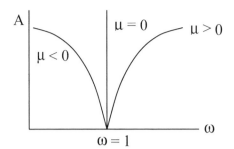

Figure 2.14 Amplitude-frequency relations.

Therefore, the free vibration response of the system shown in Figure 2.2 is given by

$$x(t) = \left(A - \frac{1}{32}\mu A^3 + \frac{23}{1024}\mu^2 A^5 \right) \cos \omega \sqrt{\frac{\alpha}{m}}\, t$$

$$+ \left(\frac{1}{32}\mu A^3 - \frac{3}{128}\mu^2 A^5 \right) \cos 3\omega \sqrt{\frac{\alpha}{m}}\, t + \frac{1}{1024}\mu^2 A^5 \cos 5\omega \sqrt{\frac{\alpha}{m}}\, t \tag{2.155}$$

For the linear system, Eq. 2.155 reduces to the solution

$$x(t) = A \cos \sqrt{\frac{\alpha}{m}}\, t \tag{2.156}$$

2.5.2.4 Variation of Parameter Method

The perturbation method gives a steady-state solution, whereas the variation of parameter method allows small changes in amplitude and phase angle in the time solution.

Consider the equation of a dynamic system whose equation of motion is given by

$$\ddot{x} + \omega_0^2 x + \mu\phi(x,\, \dot{x},\, t) = 0 \tag{2.157}$$

where $\phi(x,\, \dot{x},\, t)$ is the nonlinearity, and μ is a constant. If we neglect $\phi(x,\, \dot{x},\, t)$, the solution corresponding to the linear equation $\ddot{x} + \omega_0^2 x = 0$ is given by

$$x = A \cos(\omega_0 t + \theta) \tag{2.158}$$

$$\dot{x} = -\omega_0 A \sin(\omega_0 t + \theta) \tag{2.159}$$

Differentiating Eq. 2.158 with respect to time t, we obtain

$$\dot{x} = \dot{A}\cos\psi - (A\sin\psi)(\omega_0 + \dot{\theta}) \tag{2.160}$$

where

$$\psi = \omega_0 t + \theta$$

and

$$\dot{\psi} = \omega_0 + \dot{\theta}$$

From Eqs. 2.159 and 2.160, we obtain

$$\dot{A}\cos\psi - \dot{\theta}A\sin\psi = 0 \qquad (2.161)$$

Differentiating Eq. 2.161, we get with respect to t

$$\ddot{x} = -\omega_0\dot{A}\sin\psi - (\omega_0 A\cos\psi)(\omega_0 + \dot{\theta}) \qquad (2.162)$$

From Eqs. 2.157, 2.158, 2.159, and 2.162, we obtain

$$-\dot{A}\omega_0\sin\psi - \dot{\theta}\omega_0 A\cos\psi = -\mu\phi(A\cos\psi, -\omega_0 A\sin\psi, t) \qquad (2.163)$$

From Eqs. 2.161 and 2.163, we obtain $\dot{A}$ and $\dot{\theta}$ as

$$\dot{A} = \frac{\mu}{\omega_0}\sin\psi\,\phi(A\cos\psi, -\omega_0 A\sin\psi, t)$$

$$\dot{\theta} = \frac{\mu}{\omega_0 A}\cos\psi\,\phi(A\cos\psi, -\omega_0 A\sin\psi, t) \qquad (2.164)$$

An approximate solution of Eq. 2.164 is obtained by assuming μ is small and that A and θ do not change rapidly. In addition, if the changes in A and θ are considered small over one cycle, then we can consider the average values of $\dot{A}$ and $\dot{\theta}$ instead of the instantaneous values.

Therefore,

$$\dot{A}_{av} = \frac{\mu}{2\pi\omega_0}\int_0^{2\pi}\sin\psi\,\phi(A\cos\psi, -\omega_0 A\sin\psi, t)\,d\psi$$

$$\dot{\theta}_{av} = \frac{\mu}{2\pi\omega_0 A}\int_0^{2\pi}\cos\psi\,\phi(A\cos\psi, -\omega_0 A\sin\psi, t)\,d\psi \qquad (2.165)$$

where A is a constant in the integrand.

2.5.3 Graphical Method

2.5.3.1 Phase Plane Representation

Graphical methods are useful in determining the qualitative information about the behavior of a nonlinear dynamic system and integration of the equations of motion. A single-degree-of-freedom system requires two parameters to describe completely the state of motion. The two parameters are the displacement and the velocity of the system. When these two parameters are used as coordinate axes, the resulting graphical representation of the motion is called the phase plane representation. Here, each point in the phase plane represents a possible state of the system. As time changes, the state of the system changes. A typical representative point in the phase plane moves and traces a curve known as the trajectory. The trajectory shows how the solution of the system varies with time.

Consider a single-degree-of-freedom nonlinear oscillatory system that has a governing equation given by

$$\ddot{x} + f\left(x, \dot{x}\right) = 0 \qquad (2.166)$$

By denoting

$$\frac{dx}{dt} = \dot{x} = y \qquad (2.167)$$

and

$$\frac{dy}{dt} = \dot{y} = -f\left(x, y\right) \qquad (2.168)$$

we have

$$\frac{dy}{dx} = \frac{\left(\dfrac{dy}{dt}\right)}{\left(\dfrac{dx}{dt}\right)} = -\frac{f\left(x, y\right)}{y} = \phi\left(x, y\right) \qquad (2.169)$$

We observe that there exists a unique slope of the trajectory at every point $\left(x, y\right)$ in the phase plane, provided that $\phi\left(x, y\right)$ is not indeterminate. If $y = 0$ and $f \neq 0$, the slope of the trajectory is infinite. This indicates that all trajectories must cross the x-axis at right angles. If $y = 0$ and $f = 0$, the point is called a singular point, and the slope is indeterminate at such points. A singular point corresponds to a state of equilibrium of the system—the velocity $y = \dot{x}$ and the force $\ddot{x} = -f$ are zero at a singular point.

2.5.3.2 Phase Velocity

The velocity $\bar{v}$ with which a representative point moves along a trajectory is known as the phase velocity. The components of the phase velocity parallel to the x and y axes are given by

$$v_x = \dot{x}$$

$$v_y = \dot{y}$$

The magnitude of $\bar{v}$ becomes

$$|\bar{v}| = \sqrt{v_x^2 + v_y^2} = \sqrt{\left(\frac{dx}{dt}\right)^2 + \left(\frac{dy}{dt}\right)^2} \qquad (2.170)$$

Note here that if the system has a periodic motion, its trajectory in the phase plane is a closed curve. This is due to the fact that the representative point, having started its motion along a closed trajectory at an arbitrary point $(x,\,y)$ will return to the same point after one period. The time required to go around the closed trajectory is finite because the phase velocity is bounded away from zero at all points of the trajectory.

EXAMPLE E2.10

The equation of motion of a single-degree-of-freedom system is given by

$$2\ddot{x} + 0.8\dot{x} + 1.6x = 0$$

with initial conditions $x(0) = -1$ and $\dot{x}(0) = 2$.

a.   Plot $x(t)$ versus t for $0 \le t \le 8$.

b.   Plot a trajectory in the phase plane.

Solution:

The equation of motion is

$$2\ddot{x} + 0.8\dot{x} + 1.6x = 0$$

This is of the standard form

$$\ddot{x} + 2\zeta\omega_n\dot{x} + \omega_n^2 = 0$$

with

$$\omega_n = \sqrt{0.8} = 0.8945$$

and

$$\xi = 0.2237$$

$$\omega_d = \omega_n\sqrt{1 - \zeta^2} = 0.8718$$

The solution is given by

$$x(t) = e^{-\zeta\omega_n t}\left(A_1 \cos\omega_d t + A_2 \sin\omega_d t\right)$$

$$\dot{x} = e^{-\zeta\omega_n t}\left[\left(-\zeta\omega_n A_1 + \omega_d A_2\right)\cos\omega_d t - \left(\omega_d A_1 + \zeta\omega_n A_2\right)\sin\omega_d t\right]$$

Using the initial conditions

$$x(0) = A_1 = -1$$

and

$$\dot{x}(0) = -\zeta\omega_n A_1 + \omega_d A_2 = 2$$

we get

$$A_1 = -1$$

and

$$A_2 = \frac{2 + \xi\omega_n A}{\omega_d} = 1.7706$$

Hence,

$$x(t) = e^{-0.2t}\left(-\cos 0.8718t + 1.7706\ \sin 0.8718t\right)$$

and

$$\dot{x}(t) = e^{-0.2t}\left(1.7436\ \cos 0.8717t + 0.5177\ \sin 0.8718t\right)$$

The plots are shown in Figures E2.10(a) and (b).

2.5.3.3 Pell's Method

The isocline method already described is a general method but a rather tedious one. In Pell's method, we consider a second-order autonomous equation of the form

$$+\phi(\dot{x}) + \psi(x) = 0 \tag{2.171}$$

or

$$\frac{d\dot{x}}{dx} = \frac{-\left[\phi(\dot{x}) + \psi(x)\right]}{\dot{x}} \tag{2.172}$$

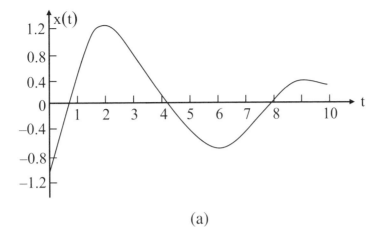

(a)

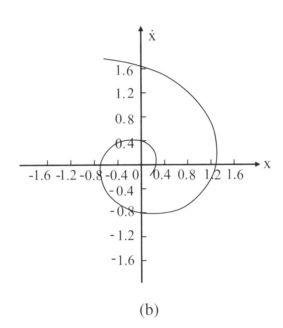

(b)

Figure E2.10 (a) Plot of $x(t)$ versus t, and (b) trajectory in the phase plane.

where $\phi(\dot{x})$ and $\psi(x)$ are known nonlinear functions. The slope of the trajectory is $\dfrac{d\dot{x}}{dx}$. The Pell's method is a procedure to (1) construct the slope of a trajectory and (2) extrapolate from a point at $(\dot{x}, x)$. Because the slope of the trajectory from Eq. 2.172 is $\dfrac{-\left[\phi(\dot{x}) + \psi(x)\right]}{\dot{x}}$, from simple geometry, a line segment perpendicular to

$$\frac{\dot{x}}{\left[\phi(\dot{x}) + \psi(x)\right]}$$ is the slope of the trajectory.

Referring to Figure 2.15, the construction procedure includes the following steps:

1. Plot the given $\psi(x)$ versus x in the fourth quadrant.

2. Plot $\phi(\dot{x})$ versus $\dot{x}$ in the second quadrant.

3. For the given point P at $(\dot{x}, x)$, follow the dashed line to obtain $\psi(x)$ and the dotted line for $\phi(\dot{x})$, as shown in Figure 2.15.

4. $\phi(\dot{x})$ is added to $\psi(x)$ on the x-axis by the parallel dashed-dotted lines. This gives $\left[\phi(\dot{x}) + \psi(x)\right]$. The slope of the line PA is $\dfrac{\dot{x}}{\left[\phi(\dot{x}) + \psi(x)\right]}$.

5. From P, draw a line segment PP_1 perpendicular to the line PA. This represents a small line segment of the trajectory.

6. Using P_1 as a new initial point, the procedure is repeated to obtain the full trajectory.

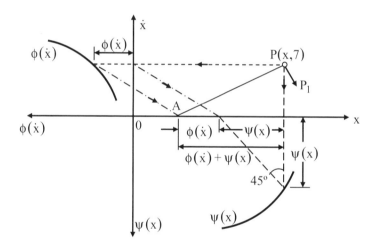

Figure 2.15 Pell's method for the construction of trajectory.

2.5.4 Multiple-Degrees-of-Freedom Systems

The general form of the equations of motion for a multiple-degrees-of-freedom system is governed by

$$[m]\{\ddot{x}\} + [c]\{\dot{x}\} + [k]\{x\} = \{F(t)\} \tag{2.173}$$

where the internal set of forces opposing the displacements are assumed to be nonlinear functions of $\{x\}$. For the linear case, these forces are equal to $[k]\{x\}$. To find the displacement vector $\{x\}$ that satisfies the nonlinear equilibrium in Eq. 2.173, it is necessary to perform an equilibrium iteration sequence in each time step.

Several numerical methods are available for the solution of the response of linear or nonlinear multiple-degrees-of-freedom systems. Several direct numerical integration methods also are available for the solution of Eq. 2.173. The term "direct" means that prior to the numerical integration, no transformation of the equations into a different form is carried out. However, if the integration must be carried out for many time steps, it may be more effective to first transform Eq. 2.199 into a form in which the step-by-step solution is faster.

2.6 Random Vibrations

In this section, we briefly discuss the theory of random vibrations by defining the random variable, the probability density function, the autocorrelation function, the power spectral density function, the Gaussian random process, the Fourier analysis-Fourier series and Fourier integral, the joint probability density function, and the cross-correlation function. The response of a single-degree-of-freedom vibrating system using the impulse response and frequency response approaches is presented. We also will discuss a method for calculating the response of a linear system subjected to stationary random excitations. The application of power spectral densities to vehicle dynamics is presented. The response of a single-degree-of-freedom system to random input and the response of a multiple-degrees-of-freedom system to random inputs are discussed.

A random variable is a variable whose value is determined by the outcome of a random experiment. Random variables are numerical values whose observed values are governed by the laws of probability. A random variable can be discrete or continuous. A random variable that has values that are countable is called a discrete random variable. Some examples of this are the number of students in a class, the toss of a coin, the number of customers visiting a bank during a given hour, the number of cars sold at a dealership in a given month, the color of a ball drawn from a collection of balls, and so forth. On the other hand, a random variable that can assume any value in one or more intervals is called a continuous random variable. Examples of this are the height of a person, the diameter of a manufactured shaft, the time to failure of a component, the repair time, the price of an automobile, the time taken to complete a medical examination, the duration of a snow storm, and so forth.

In the absence of any obvious pattern in a vibration record, the vibration is called a random vibration. In other words, if an identical experiment is performed several times and the records obtained are always alike, the process is called deterministic. On the contrary, when all conditions in the experiment remain unchanged but the records are changing continually, we then consider the process to be random, or nondeterministic. In this situation, a single record is not sufficient to provide a statistical description of the totality of possible records.

In a random process, instead of one time history, a whole family or an ensemble of possible time histories is described, as shown in Figure 2.16. Any single individual time history that belongs to the ensemble is referred to as a sample record.

Let $x^k(t_1)$ be the value of random variable $x(t)$ at time t_1, obtained from the kth record. Then the expected value (average or mean) $E[x(t_1)]$ of $x^k(t_1)$ for a fixed time t_1, obtained from all records (i.e., $k = 1, 2, 3, ..., n$) is

$$E[x(t_1)] = \lim_{n \to \infty} \frac{1}{n} \sum_{k=1}^{n} x^k(t_1) \qquad (2.174)$$

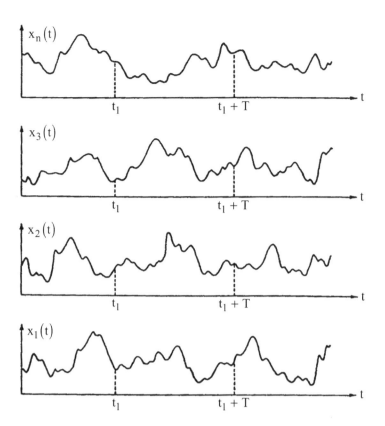

Figure 2.16 Ensemble of sample functions forming a random process.

If $E\left[x\left(t_1\right)\right]$ is independent of t (i.e., $E\left[x\left(t_1\right)\right] = E\left[x\left(t_1 + t\right)\right]$ for all t, then the random process is called stationary. If in addition to this property, each record is statistically equivalent to any other record such that $E\left[x\left(t_1\right)\right]$ in Eq. 2.174 can be replaced by a time average of a sample representative record $x\left(t\right)$ as

$$\bar{x}(t) = E\left(x\right) = \lim_{T\to\infty} \frac{1}{T}\int_0^T x\left(t\right)dt \tag{2.175}$$

then the stationary process is ergodic. For many applications, this assumption is fairly reasonable. The variance σ_x^2 of $x\left(t\right)$ is given by

$$E\left[x - E\left(x\right)\right]^2 = \lim_{T\to\infty} \frac{1}{T}\int_0^T \left[x - E\left(x\right)\right]^2 dt \tag{2.176}$$

For the special case with $E\left(x\right) = 0$, the variance σ_x^2 of x becomes its mean square value and is given by $\overline{x^2}\left(t\right)$, where

$$\overline{x^2}(t) = E\left[x^2\left(t\right)\right] = \lim_{T\to\infty} \frac{1}{T}\int_0^T x^2\left(t\right)dt \tag{2.177}$$

EXAMPLE E2.11

Find the temporal mean value and the mean square value of the function

$$x(t) = x_0 \sin\left(\frac{\pi t}{2}\right)$$

Solution:

The temporal mean value of the function is given by

$$\langle x(t) \rangle = \lim_{T \to \infty} \frac{1}{T} \int_{-T/2}^{T/2} x_0 \sin\left(\frac{\pi t}{2}\right) dt = \lim_{T \to \infty} \frac{1}{T} \left[\frac{-x_0 \cos\left(\frac{\pi t}{2}\right)}{\frac{\pi}{2}} \right]_{-T/2}^{T/2} = 0$$

The mean square value of the function is given by

$$\rangle x^2(t) \langle = \lim_{T \to \infty} \frac{1}{T} \int_{-T/2}^{T/2} x_0^2 \sin^2\left(\frac{\pi t}{2}\right) dt$$

$$= \lim_{T \to \infty} \frac{x_0^2}{T} \int_{-T/2}^{T/2} \left[\frac{1}{2} - \frac{1}{2}\cos \pi t\right] dt$$

$$= \lim_{T \to \infty} \frac{x_0^2}{T} \left[T - \frac{2}{\pi} \sin \frac{\pi T}{2} \right] = \frac{x_0^2}{2}$$

2.6.1 Probability Density Function

The probability density function of random data is the probability that the data will assume a value within some defined range at any instant of time. We consider a sample time history, as shown in Figure 2.17. The probability that $x(t)$ will occur within x and x + Δx can be obtained from the ratio $\frac{T_x}{T}$, where T_x indicates the total amount of time for which $x(t)$ falls within the range of x and x + Δx, that is, $T_x = \sum_{i=1}^{k} \Delta t_i$ and T is observed time. We define

$$\mathrm{Prob}\left[x < x(t) + \Delta x\right] = P(x) = \lim_{T \to \infty} \left(\frac{T_x}{T}\right)$$

$$\tag{2.178}$$

$$T_x = \sum_{i=1}^{k} \Delta t_i$$

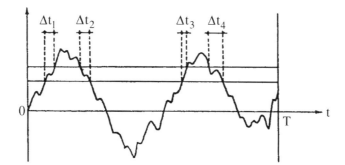

Figure 2.17 Probability measurement.

For small Δx, a probability density function $p(x)$ can be defined as

$$\text{Prob}\left[x < x(t) \leq x + \Delta x\right] = P(x) = p(x)\Delta x \tag{2.179}$$

More precisely,

$$\begin{aligned} p(x) &= \lim_{\Delta x \to 0} \frac{P(x)}{\Delta x} \\ &= \lim_{\Delta x \to 0} \frac{1}{\Delta x}\left[\lim_{T \to \infty} \frac{T_x}{T}\right] \end{aligned} \tag{2.180}$$

It is evident from Eq. 2.180 that $p(x)$ is the slope of the cumulative probability distribution $P(x)$. The area under the probability density curve between any two values of x represents the probability of the variable being in this interval. Also, the probability of $x(t)$ being between $x = \pm\infty$ is

$$P(\infty) = \int_{-\infty}^{\infty} p(x)dx = 1.0 \tag{2.181}$$

The mean value $\bar{x}(t)$ coincides with the centroid of the area under the probability density curve $p(x)$, as shown in Figure 2.18. Therefore, in terms of the probability density $p(x)$, the mean value is given by

$$\bar{x}(t) = \int_{-\infty}^{\infty} xp(x)dx \tag{2.182}$$

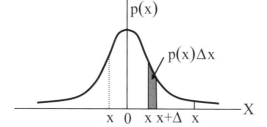

Figure 2.18 Probability density curve.

Likewise, the mean square value $(\bar{x})^2(t)$ is determined from the second moment to be

$$(\bar{x})^2(t) = \int_{-\infty}^{\infty} x^2 p(x) dx \qquad (2.183)$$

The variance σ_x^2, previously defined as the mean square value about the mean, is

$$\sigma_x^2 = \int_{-\infty}^{\infty} (x - \bar{x})^2 p(x) dx$$

$$= \int_{-\infty}^{\infty} x^2 p(x) dx - 2\bar{x} \int_{-\infty}^{\infty} x p(x) dx + (\bar{x})^2 \int_{-\infty}^{\infty} p(x) dx \qquad (2.184)$$

$$= (\bar{x})^2 - 2(\bar{x})^2 + (\bar{x})^2 = \bar{x}^2 - (\bar{x})^2$$

The standard deviation σ_x is the positive square root of σ_x^2.

EXAMPLE E2.12

The probability density function of a random variable x is given by

$$P(x) = \begin{cases} 0 & \text{for } x < 0 \\ 0.5 & \text{for } 0 \le x \le 2 \\ 0 & \text{for } x > 2 \end{cases}$$

Determine the mean value, the mean square value, and the standard deviation.

Solution:

The mean value is given by

$$\bar{x} = \int_{-\infty}^{\infty} x p(x) dx = \int_{0}^{2} 0.5x \, dx = 1.0$$

The mean square value is given by

$$(\bar{x})^2 = \int_{-\infty}^{\infty} x^2 p(x) dx = \int_{0}^{2} 0.5x^2 \, dx = 1.333$$

The variance σ_x^2 defined as the mean square value about the mean is

$$\sigma_x^2 = \int_{-\infty}^{\infty} (x - \bar{x})^2 p(x) dx$$

$$= \int_0^2 (x - 1)^2 (0.5) dx$$

$$= 0.5 \int_0^2 (x^2 - 2x + 1) dx$$

$$= 0.3333$$

Therefore, the standard deviation $\sigma_x = \sqrt{0.3333} = 0.5773$.

2.6.2 Autocorrelation Function

The autocorrelation function of a stationary random process is defined as the average value of the product $x(t)$ and $x(t + \tau)$. The process is sampled at time t and then again at time $t + \tau$, as shown in Figure 2.19, as

$$R_x(\tau) = \lim_{T \to \infty} \frac{1}{T} \int_0^T x(t) x(t + \tau) dt \tag{2.185}$$

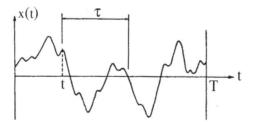

Figure 2.19 Autocorrelation measurement.

The quantity $R_x(\tau)$ is always a real-valued even function with a maximum occurring at $\tau = 0$, that is,

$$R_x(0) = \lim_{T \to \infty} \frac{1}{T} \int_0^T x^2(t) dt = (\bar{x})^2(t) \tag{2.186}$$

For large time intervals, with $\tau \to \infty$, the random process will be uncorrelated, and in this case,

$$R_x(\infty) = \left[\bar{x}(t)\right]^2$$

that is,

$$\bar{x}(t) = \sqrt{R_x(\infty)} \qquad (2.187)$$

The mean value of $x(t)$ is equal to the positive square root of the autocorrelation as the time displacement becomes very long.

2.7 Gaussian Random Process

The most widely used statistical distribution for modeling random processes is the Gaussian or normal random process. The probability density function $x(t)$ of a Gaussian random process is given by

$$p(x) = \frac{1}{\sqrt{2\pi}\,\sigma_x} e^{-1/2\left[\frac{x-\bar{x}}{\sigma_x}\right]^2} \qquad (2.188)$$

where $\bar{x}$ and σ_x are the mean value and standard deviation of x. By defining a standard normal variable, z is

$$z = \frac{x - \bar{x}}{\sigma_x}$$

Equation 2.188 becomes

$$p(x) = \frac{1}{\sqrt{2\pi}} e^{-1/2z^2} \qquad (2.189)$$

The probability of $x(t)$ in the interval $-k\sigma$ and $+k\sigma$ assuming $\bar{x} = 0$ is

$$\mathrm{Prob}\left[-k\sigma \le x(t) \le k\sigma\right] = \int_{-k\sigma}^{k\sigma} \frac{1}{\sqrt{2\pi}\,\sigma} e^{-1/2\frac{x^2}{\sigma^2}dx} \qquad (2.190)$$

where k is the positive number. Figure 2.20 shows the Gaussian probability density function, which is a bell-shaped curve symmetric about the mean value.

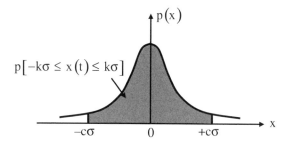

Figure 2.20 Gaussian probability distribution.

2.7.1 Fourier Analysis

2.7.1.1 Fourier Series

The periodic function $x(t)$ can be expressed in the form of a complex Fourier series as

$$x(t) = \sum_{n=-\infty}^{\infty} a_n e^{in\omega_0 t} \tag{2.191}$$

where τ is the period, $\omega_0 = \dfrac{2\pi}{\tau}$ is the fundamental frequency, and a_n are the complex Fourier coefficients. To determine a_n, we multiply both sides of Eq. 2.191 with $e^{-im\omega_0 t}$ and integrate over one time period. Thus,

$$\int_{-\pi/2}^{\pi/2} x(t)e^{-im\omega_0 t}dt = \sum_{n=-\infty}^{\infty} \int_{-\pi/2}^{\pi/2} a_n e^{i(n-m)\omega_0 t}dt$$

$$= \sum_{n=-\infty}^{\infty} a_n \int_{-\pi/2}^{\pi/2} \left[\cos(n-m)\omega_0 t + i\sin(n-m)\omega_0 t\right]dt \tag{2.192}$$

From Eq. 2.192, we obtain

$$a_n = \frac{1}{\tau} \int_{-\pi/2}^{\pi/2} x(t)e^{-in\omega_0 t}dt \tag{2.193}$$

Equation 2.191 implies that $x(t)$ of period τ can be written as a sum of an infinite number of harmonics that will have amplitudes given by Eq. 2.193 and frequencies that are multiples of ω_0. Hence, the difference between any two consecutive frequencies can be written as

$$\omega_{n+1} - \omega_n = (n+1)\omega_0 - \omega_0 n = \Delta\omega = \frac{2\pi}{\tau} = \omega_0 \tag{2.194}$$

The Fourier coefficients a_n in Eq. 2.193 will be real if $x(t)$ is a real and even function. Also,

$$a_n = a_{-n}^*$$

(2.195)

Hence, the mean square value, which is the time average of the square of the function $x(t)$, is given by

$$(\bar{x})^2(t) = \frac{1}{\tau} \int_{-\pi/2}^{\pi/2} x^2(t) dt = \frac{1}{\tau} \int_{-\pi/2}^{\pi/2} \left(\sum_{n=-\infty}^{\infty} a_n e^{in\omega_0 t} \right)^2 dt$$

$$= \frac{1}{\tau} \int_{-\pi/2}^{\pi/2} \left[\sum_{n=1}^{\infty} \left(a_n e^{in\omega_0 t} + a_n^* e^{-in\omega_0 t} \right) + a_0 \right]^2 dt$$

(2.196)

$$= \frac{1}{\tau} \int_{-\pi/2}^{\pi/2} \left[\sum_{n=1}^{\infty} 2 a_n a_n^* + a_0^2 \right]^2 dt$$

$$= a_0^2 + \sum_{n=1}^{\infty} 2|a_n|^2 = \sum_{n=-\infty}^{\infty} 2|a_n|^2$$

Equation 2.196 is called Parseval's theorem for periodic functions. Parseval's theorem is useful for converting time integration into frequency integration.

2.7.1.2 Fourier Integral

A nonperiodic function can be analyzed as a periodic function with an infinite period $(\tau \rightarrow \infty)$. Also, as $\tau \rightarrow \infty$, the frequency spectrum becomes continuous, and the fundamental frequency ω_0 becomes infinitesimal. We can denote ω_0 as $\Delta\omega$, and $n\omega_0$ as ω. Hence, Eq. 2.193 becomes

$$\lim_{\tau \rightarrow \infty} \tau a_n = \lim_{\tau \rightarrow \infty} \int_{-\pi/2}^{\pi/2} x(t) e^{-i\omega t} dt$$

$$= \int_{-\infty}^{\infty} x(t) e^{-i\omega t} dt$$

(2.197)

If we denote

$$X(\omega) = \lim_{\tau \rightarrow \infty} (\tau a_n) = \int_{-\infty}^{\infty} x(t) e^{-i\omega t} dt$$

(2.198)

then Eq. 2.191 becomes

$$x(t) = \lim_{\tau \to \infty} \sum_{n=-\infty}^{\infty} a_n e^{i\omega t} \frac{2\pi\tau}{2\pi\tau} = \lim_{\tau \to \infty} \sum_{n=-\infty}^{\infty} (a_n\tau) e^{i\omega t} \left(\frac{2\pi}{\tau}\right) \frac{1}{2\pi}$$

$$= \frac{1}{2\pi} \int_{-\infty}^{\infty} X(\omega) e^{i\omega t} d\omega$$

(2.199)

Equation 2.199 shows the frequency decompositions of the nonperiodic function $x(t)$ in a continuous frequency domain similar to Eq. 2.191. The mean square value of a nonperiodic function $x(t)$ can be obtained from Eq. 2.196. Thus,

$$\frac{1}{\tau} \int_{-\tau/2}^{\tau/2} x^2(x) dt = \sum_{n=-\infty}^{\infty} |a_n|^2$$

$$= \sum_{n=-\infty}^{\infty} a_n a_n^* \frac{\tau\omega_0}{\tau\omega_0} = \sum_{n=-\infty}^{\infty} a_n a_n^* \frac{\tau\omega_0}{\tau\left(\frac{2\pi}{\tau}\right)}$$

(2.200)

$$= \frac{1}{\tau} \sum_{n=-\infty}^{\infty} (\tau a_n)(a_n^*\tau) \frac{\omega_0}{2\pi}$$

Now as $\tau a_n \to X(\omega)$, $\left(a_n^*\right) \to X^*(\omega)$, and $\omega_0 \to d\omega$ as $\tau \to \infty$, Eq. 2.200 gives $\frac{2}{x(t)}$. Thus,

$$(\bar{x})^2(t) = \lim_{\tau \to \infty} \frac{1}{\tau} \int_{-\pi/2}^{\pi/2} x^2(t) dt = \int_{-\infty}^{\infty} \frac{|X(\omega)|^2}{2\pi\tau} d\omega$$

(2.201)

Equation 2.201 is known as Parseval's theorem for nonperiodic functions. Parseval's theorem is a useful tool for converting time integration into frequency integration.

EXAMPLE E2.13

Obtain the Fourier transform of the function shown in Figure E2.13(a), and plot the corresponding spectrum.

Solution:

$$x(t) = \begin{cases} A & -c \le t \le c \\ 0 & \text{elsewhere} \end{cases}$$

$$X(\omega) = \int_{-\infty}^{\infty} x(t)e^{i\omega t}dt = A\int_{-c}^{c} e^{-i\omega t}dt = \frac{A}{-i\omega}\left(e^{-i\omega t}\right)_{-c}^{c}$$

$$= \frac{Ai}{\omega}\left(e^{-i\omega c} + e^{+i\omega c}\right) = \frac{Ai}{\omega}\left(\cos\omega c - i\sin\omega c - \cos\omega c - i\sin\omega c\right) \qquad (E2.13.1)$$

$$= \frac{2Ac}{\omega c}\sin\omega c$$

Equation E2.13.1 shows that $X(\omega) = 2Ac$ as $\omega \to 0$ (Figure E2.13(b)).

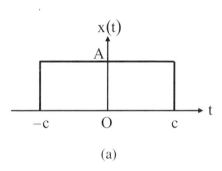

(a)

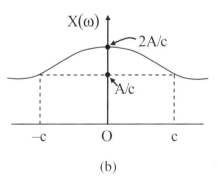

(b)

Figure E2.13 (a) Function $x(t)$, and (b) plot of the spectrum.

2.7.2 Response of a Single-Degree-of-Freedom Vibrating System

The equation of motion for a single-degree-of-freedom system shown in Figure 2.21 is given by

$$\ddot{y} + 2\zeta\omega_n\dot{y} + \omega_n^2 y = x(t) \qquad (2.202)$$

where

$$x(t) = \frac{F(t)}{m}$$

$$\omega_n = \sqrt{\frac{k}{m}}$$

$$\zeta = \frac{c}{C_c}$$

$$C_c = 2km$$

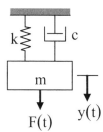

Figure 2.21 Single-degree-of-freedom vibrating system.

There are two ways to obtain the solution of Eq. 2.202: the impulse response method, or the frequency response method.

2.7.2.1 Impulse Response Method

In this method, we assume the forcing function $x(t)$ to be made up of a series of impulses of varying magnitude (Figure 2.22(a)), and the amplitude applied at time τ is $x(t)d\tau$. Also, if $y(t) = H(t - \tau)$ denotes the response of the unit impulse excitation $\delta(t - \tau)$, it is called the impulse response function. Note that the unit impulse applied at $t = \tau$ is denoted by

$$x(t) = \delta(t - \tau)$$

where $\delta(t - \tau)$ is called the Dirac delta function with

$$\delta(t - \tau) \to \infty \ \text{as} \ t \to \tau \qquad (2.203)$$

$\delta(t - \tau) = 0$ for all t except at $t = \tau$ and $\int\limits_{-\infty}^{\infty} \delta(t - \tau)dt = 1$ (Figure 2.22(b)).

The total response of the system can be obtained by superimposing the responses to the impulses of $x(t)d\tau$ applied at different values of $t = \tau$ (Figure 2.22(c)).

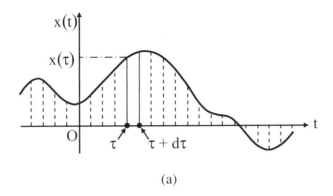

(a)

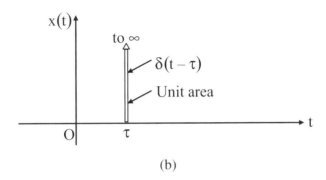

(b)

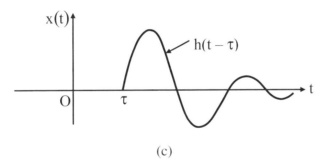

(c)

Figure 2.22 (a) Forcing function in the form of a series of impulses, (b) unit impulse excitation at $t = \tau$, and (c) impulse response function.

Hence, the response to the total excitation is given by the superposition or convolution integral

$$y(t) = \int_{-\infty}^{t} x(t)h(t - \tau)\,d\tau \qquad (2.204a)$$

where $x(\tau)h(t - \tau)$ is the response to the excitation $x(\tau)d\tau$.

Note here that because $h(t - \tau) = 0$ when $t < \tau$ or $t > \tau$, the upper limit of integration of Eq. 2.204 can be replaced by ∞. That is,

$$y(t) = \int_{-\infty}^{\infty} x(\tau)h(t - \tau)\,d\tau \qquad (2.204b)$$

Eq. 2.204b can be written as

$$y(t) = \int_{-\infty}^{\infty} x(\tau - \theta)h(\theta)\,d\theta \qquad (2.204c)$$

where we have changed the variable from τ to $\theta = t - \tau$ in Eq. 2.204b.

2.7.2.2 Frequency Response Method

The transient response is

$$x(t) = \frac{1}{2\pi} \int_{\omega=-\infty}^{\infty} X(\omega)e^{i\omega t}\,dt \qquad (2.205)$$

Eq. 2.205 indicates the superposition of components of different frequencies ω. If the forcing function of unit modulus is

$$\underset{\sim}{x}(t) = e^{i\omega t} \qquad (2.206)$$

then its response is given by

$$\underset{\sim}{y}(t) = H(\omega)e^{i\omega t} \qquad (2.207)$$

where $H(\omega)$ is the complex frequency response function.

The total response of the system by the superposition principle gives

$$y(t) = H(\omega)x(t) = \int_{-\infty}^{\infty} H(\omega)\frac{1}{2\pi}X(w)e^{i\omega t}\,d\omega$$

$$= \frac{1}{2\pi} \int_{-\infty}^{\infty} H(\omega)X(\omega)e^{i\omega t}\,d\omega \qquad (2.208)$$

or

$$y(t) = \frac{1}{2\pi} \int_{-\infty}^{\infty} Y(\omega)e^{i\omega t}d\omega \qquad (2.209)$$

where $Y(\omega)$ is the Fourier transform of the response function $y(t)$. We note from Eqs. 2.208 and 2.209 that

$$Y(\omega) = H(\omega)X(\omega) \qquad (2.210)$$

Equation 2.208 can be written as

$$y(t) = h(t) = \frac{1}{2\pi} \int_{-\infty}^{\infty} X(\omega)H(\omega)e^{i\omega t}d\omega \qquad (2.211)$$

where $X(\omega)$ is the Fourier transform of $x(t) = \delta(t)$

$$X(\omega) = \int_{-\infty}^{\infty} x(t)e^{-i\omega t}dt = \int_{-\infty}^{\infty} \delta(t)e^{-i\omega t}dt = 1 \qquad (2.212)$$

Now, because $\delta(t) = 0$ everywhere except at $t = 0$, where it has a unit area and $e^{-i\omega t} = 1$ at $t = 0$, Eqs. 2.211 and 2.212 give

$$h(t) = \frac{1}{2\pi} \int_{-\infty}^{\infty} H(\omega)e^{i\omega t}d\omega \qquad (2.213)$$

or

$$H(\omega) = \int_{-\infty}^{\infty} h(t)e^{-i\omega t}dt \qquad (2.214)$$

EXAMPLE E2.14

Figure E2.14 shows the simplified model of a two-wheeled road vehicle traveling over a rough guide way. It is assumed that the wheel is always in contact with the road surface. If the power spectral density of the road surface is S_0 and the constant speed of the vehicle is V, determine the mean square value of the vertical displacement of the vehicle mass m.

Solution:

Figure E2.14 can be considered as a single-degree-of-freedom system with random base excitation.

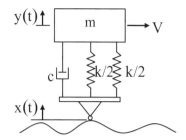

Figure E2.14 Two-wheeled road vehicle traveling over a rough guideway.

The equation of motion is

$$m\ddot{z} + c\dot{z} + kz = -m\ddot{x} \tag{E2.14.1}$$

where $z = y - x$.

The frequency response function of the system can be obtained as follows.

Let

$$x(t) = e^{i\omega t}$$
$$z(t) = H(\omega)e^{i\omega t} \tag{E2.14.2}$$

From Eqs. E2.14.1 and E2.14.2, we have

$$\left(-\omega^2 m + ic\omega + k\right)H(\omega)e^{i\omega t} = m\omega^2 e^{i\omega t}$$

or

$$H(\omega) = \frac{m\omega^2}{-\omega^2 m + ic\omega + k} = \frac{\omega^2}{\left(\omega_n^2 - \omega^2\right) + i2\zeta\omega\omega_n}$$

or

$$\left|H(\omega)\right|^2 = \frac{\omega^4}{\left(\omega_n^2 - \omega^2\right) + 4\zeta^2\omega^2\omega_n^2} \tag{E2.14.3}$$

The power spectral density of the response $z(t)$ is

$$S_z(\omega) = \left|H(\omega)\right|^2 S_x(\omega) \tag{E2.14.4}$$

Noting that $S_z(\omega) = S_0$, Eq. E2.14.4 becomes

$$S_z(\omega) = S_0 |H(\omega)|^2 \qquad (E2.14.5)$$

Hence, the mean square value of the relative displacement of the mass is given by

$$E(z^2) = \int_{-\infty}^{\infty} S_z(\omega)\, d\omega = S_0 \int_{-\infty}^{\infty} \left| \frac{\omega^2}{-\omega^2 + \omega_n^2 + \dfrac{i\omega c}{m}} \right| d\omega$$

$$(E2.14.6)$$

$$= S_0 \omega^4 \pi \left[\frac{\left(\dfrac{1}{\omega_n^2}\right)}{\left(\dfrac{c}{m}\right)} \right] = \frac{\pi \omega^4 S_0}{2\zeta \omega_n^3}$$

2.7.3 Power Spectral Density Function

The power spectral density (PSD) of a random process provides the frequency composition of the data in terms of the spectral density of its mean square value. The mean square value of a sample time-history record, in a frequency range ω and $\omega + \Delta\omega$, can be obtained by passing the sample record through a bandpass filter with sharp cutoff features and then computing the average of the squared output from the filter. The average square value will approach an exact mean square value as $T \to \infty$, that is,

$$\Psi_x^2(\omega, \Delta\omega) = \lim_{T\to\infty} \frac{1}{T} \int_0^T x^2(t, \omega, \Delta\omega)\, dt \qquad (2.215)$$

where $x(t, \omega, \Delta\omega)$ is the portion of $x(t)$ in the frequency range ω and $\omega + \Delta\omega$. For a small value of $\Delta\omega$, a power spectral density function $S_x(\omega)$ is defined as

$$\Psi_x^2(\omega, \Delta\omega) \cong S_x(\omega) \Delta\omega \qquad (2.216)$$

that is,

$$S_x(\omega) = \lim_{\Delta\omega\to 0} \frac{\Psi_x^2(\omega, \Delta\omega)}{\Delta\omega} = \lim_{\Delta\omega\to 0} \frac{1}{\Delta\omega} \lim_{T\to\infty} \frac{1}{T} \int_0^T x^2(t, \omega, \Delta\omega)\, dt \qquad (2.217)$$

The quantity $S_x(\omega)$ is always a real-valued nonnegative function.

In experimental work, a different unit of power spectral density often is used. The experimental spectral density is defined by $W(f)$, where f denotes the frequency in cycles per unit time. The relation between $S(\omega)$ and $W(f)$ is

$$W(f) = 4\pi S(\omega) \qquad \omega = 2\pi f \qquad (2.218)$$

For a stationary random process, the autocorrelation and power spectral density functions are related by a Fourier transform as

$$R(\tau) = \int_{-\infty}^{\infty} S(\omega)e^{i\omega\tau}d\omega \qquad (2.219)$$

$$S(\omega) = \frac{1}{2\pi}\int_{-\infty}^{\infty} R(\tau)e^{-i\omega\tau}d\tau \qquad (2.220)$$

In a limiting case, where $\tau = 0$,

$$R(0) = E\left[x^2(t)\right] = \int_{-\infty}^{\infty} S(\omega)d\omega \qquad (2.221)$$

that is, the mean square value is equal to the sum over all frequencies of $S(\omega)d\omega$; therefore, $S(\omega)$ may be interpreted as a mean square spectral density. Figure 2.23 shows probability density functions, autocorrelation functions, and power spectral density functions for four sample time-history records.

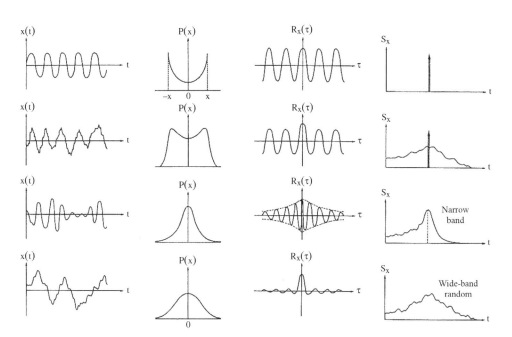

Figure 2.23 Probability density functions, autocorrelation functions, and power spectral density functions for four sample time-history records.

EXAMPLE E2.15

The spectral density of a random signal is given by $S(f) = 7 \times 10^{-5} \, m^2/cps$ between 10 and 800 cps and zero outside this frequency range. Find the standard deviation and the root mean square value of the signal if its mean value is 0.05 m.

Solution:

The mean square value of the signal is given by

$$(\overline{x})^2 = \int_0^\infty S(f)df = \int_{10}^{800} 7 \times 10^{-5} df = 0.0530 \, m^2$$

$$\sqrt{(\overline{x})^2} = \text{rms value} = 0.2302 \, m$$

The mean square value is

$$(\overline{x})^2 = (\overline{x})^2 + \sigma_x^2$$

Therefore,

$$\sigma_x^2 = (\overline{x})^2 - (\overline{x})^2 = 0.0530 - (0.05)^2 = 0.0505 \, m^2$$

or

$$\sigma_x = \sqrt{0.0505} = 0.2247 \, m$$

2.7.4 Joint Probability Density Function

The joint probability density $p(x,y)$ of two random variables is the probability that both variables assume values within some defined pair of ranges at any instant of time. If we consider two random variables $x(t)$ and $y(t)$, the joint probability density has this property: the fraction of ensemble members for which $x(t)$ lies between x and x + dx, and $y(t)$ lies between y and y + dy is $p(x,y)dx \, dy$. The joint probability densities are positive, and the probabilities of mutually exclusive events are additive. Also,

$$\int_{-\infty}^{\infty} p(x,y)dx \, dy = 1 \qquad (2.223)$$

When two variables are statistically independent, the joint probability density is given by

$$p(x,y) = p(x)p(y) \qquad (2.224)$$

2.7.5 Cross-Correlation Function

The cross-correlation function of two random variables indicates the general dependence of one variable on the other. The cross-correlation function of the time-history records $x(t)$ and $y(t)$ (Figure 2.24) is given as

$$R_{xy}(\tau) = \lim_{T \to \infty} \frac{1}{T} \int_0^T x(t) y(t + \tau) dt \qquad (2.225)$$

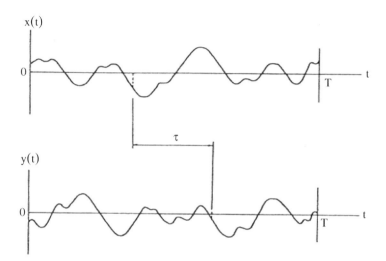

Figure 2.24 Cross-section measurements.

The function $R_{xy}(\tau)$ is always a real-valued function and can be either negative or positive. Also, $R_{xy}(\tau)$ does not necessarily have a maximum at $\tau = 0$, nor is $R_{xy}(\tau)$ an even function. However, $R_{xy}(\tau)$ is symmetric, that is,

$$R_{xy}(-\tau) = R_{xy}(\tau) \qquad (2.226)$$

when $R_{xy}(\tau) = 0$; then, $x(t)$ and $y(t)$ are said to be uncorrelated.

2.7.6 Application of Power Spectral Densities to Vehicle Dynamics

The guide way input power spectral density describes the frequency content of the guide way roughness. Because the guide way irregularity amplitude $Z(x)$ is a function of distance x measured along the guide way, we define the following spatial terms: A is the wavelength, $F = \left(\dfrac{1}{\lambda}\right)$ is the spatial frequency, and $\Omega = 2\pi F$ is the wave number (circular frequency). The mean square value of the amplitude can be obtained from the spatial power spectral density as

$$\overline{Z}^2 = \int_0^\infty S_x(F)dF = \int_0^\infty S_x(\Omega)d\Omega \qquad (2.227)$$

where $S_x(F)$ is the single-sided spatial frequency power spectral density, $S_x(\Omega)$ is the single-sided wave number power spectral density, and $S_x(F) = 2\pi S_x(\Omega)$.

With a vehicle traveling at a speed V,

$$\omega = V\Omega = \frac{2\pi V}{\lambda} \qquad (2.228)$$

and

$$S_x(\omega) = V^{-1}S_x(\Omega) = (2\pi V)^{-1}S_x(F) \qquad (2.229)$$

Thus, from the basic track spectrum, the temporal input power spectral density $S_x(F)$ is generated. Several track measurements have shown that $S_x(F)$ has the form

$$S_x(F) = \frac{C}{F^N}$$

or $\qquad (2.230)$

$$S_x(\lambda) = C\lambda^N$$

where C denotes the relative track roughness, and N is an exponent. Here, N lies between 1.5 and 4.0 for many track measurements and often is taken as 2 for many analytical studies.

We previously defined the complex frequency response function $H(\omega)$, which has a magnitude equal to the amplitude ratio and whose ratio of imaginary to real parts is equal to the tangent of the phase angle ϕ. The Fourier transforms of the response and excitation—that is, $X(\omega)$ and $F(\omega)$—are related through the frequency response function as

$$X(\omega) = H(\omega)F(\omega) \qquad (2.231)$$

This relation is valid for any arbitrary excitation $f(t)$. If the excitation is a stationary random process, then the response also will be a stationary random process. By using mathematical manipulations, it can be shown that for a linear system, the response mean square spectral density $S_x(\omega)$ and the mean square spectral density $S_f(\omega)$ of the excitation are related as

$$S_x(\omega) = |H(\omega)|^2 S_f(\omega) \qquad (2.232)$$

The mean square value of the response can be obtained as

$$R_x(0) = E\left[x^2(t)\right] = \frac{1}{2\pi}\int_{-\infty}^\infty |H(\omega)|^2 S_f(\omega)d\omega \qquad (2.233)$$

From Eqs. 2.231 and 2.232, it is evident that for a linear system, the response mean square spectral density and the mean square value can be calculated from the mean square spectral density of the excitation and the magnitude of the complete frequency response function $H(\omega)$, respectively.

If the excitation has a Gaussian probability distribution and the system is linear, then the response also will be Gaussian. This implies that for the stationary process, the probability distribution of the response is defined completely by the mean and the mean square values of the response.

2.7.7 Response of a Single-Degree-of-Freedom System to Random Inputs

As an example, we can consider a vehicle moving at constant speed V and can compute its response to various guide way surface irregularities. The vehicle is modeled as a single-degree-of-freedom system with linear suspension elements, as shown in Figure 2.25(a). The equations of motion for the vehicle are

$$m\ddot{y} + c(\dot{y} - \dot{z}) + k(y - z) = 0$$

$$m\ddot{y} + c\dot{y} + ky = c\dot{z} + kz \qquad (2.234)$$

$$m\ddot{y} + 2\xi\omega_n\dot{y} + \omega_n^2 y = \omega_n^2 f(t)$$

where $f(t) = (2\xi\omega_n)\dot{z} + z$ and $t = \dfrac{x}{V}$. The power spectral density of the vehicle response is given by

$$S_y(\omega) = |H(\omega)|^2 S_z(\omega) \qquad (2.235)$$

where

$$|H(\omega)|^2 = \left\{ \left[1 - \left(\frac{\omega}{\omega_n} \right)^2 \right]^2 + \left[2\xi \left(\frac{\omega}{\omega_n} \right) \right]^2 \right\}^{-1} \qquad (2.236)$$

If $S_z(\omega) = S_0$ (a constant), which represents the track input corresponding to white noise (Figure 2.25(b)), then from Eqs. 2.235 and 2.236, we obtain

$$S_y(\omega) = \frac{S_0}{\left[1 - \left(\dfrac{\omega}{\omega_n} \right)^2 \right]^2 + \left[2\xi \left(\dfrac{\omega}{\omega_n} \right) \right]^2} \qquad (2.237)$$

The response power spectral density $S_y(\omega)$ is plotted in Figure 2.25(c). The mean square value of the response will be

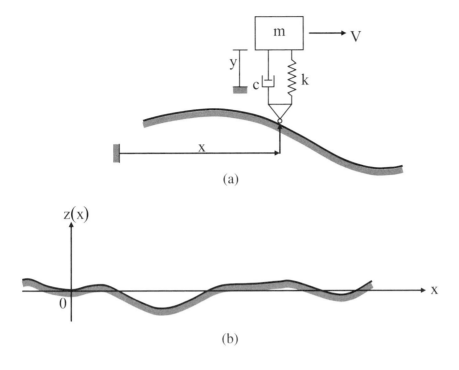

(a)

(b)

(c)

Figure 2.25 (a) Single-degree-of-freedom rail vehicle, (b) the power spectral density of the track input, and (c) the power spectral density of the vehicle response.

$$E\left[y^2(t)\right] = \frac{S_0}{2\pi} \int_{-\infty}^{\infty} \frac{d\omega}{\left[1 - \left(\frac{\omega}{\omega_n}\right)^2\right]^2 + \left[2\xi\left(\frac{\omega}{\omega_n}\right)\right]^2} \tag{2.238}$$

An integration of Eq. 2.238 can be performed by using the residue theorem of complex variables, which gives

$$E\left[y^2(t)\right] = S_0 \frac{\omega_n}{4\xi} \tag{2.239}$$

Because the random process is Gaussian with zero mean value, the mean square value equation (Eq. 2.239) is sufficient to determine the shape of the response probability density function. This makes it possible to evaluate the probability that the response $y(t)$ might exceed a given displacement.

2.7.8 Response of Multiple-Degrees-of-Freedom Systems to Random Inputs

The general equations of motion for a multiple-degrees-of-freedom discrete system with n degrees of freedom are written as

$$[m]\{\ddot{x}\} + [c]\{\dot{x}\} + [k]\{x\} = \{F(t)\} \qquad (2.240)$$

where $\{F(t)\}$ denotes the externally applied force(s). In Eq. 2.240, $[m]$, $[c]$, and $[k]$ are n × n mass, damping, and stiffness matrices, respectively. For linear systems, these matrices are constant, whereas for nonlinear systems, the elements of these matrices are functions of generalized displacements and velocities that are time dependent.

The response $\{x(t)\}$ of Eq. 2.240 consists of two parts. First, $\{x_h(t)\}$ is the transient response, that is, the homogeneous or complementary solution of the equations of motion, and, second, $\{x_p(t)\}$ is the steady-state or forced response, that is, the particular solution of the equations of motion.

We consider Eq. 2.240 and first solve the undamped free-vibration problem to obtain the eigenvalues and eigenvectors, which describe the normal modes of the system and the weighted modal matrix $[\bar{\phi}]$.

Let

$$\{x\} = [\bar{\phi}]\{y\} \qquad (2.241)$$

Substituting Eq. 2.241 into Eq. 2.240 yields

$$[m][\bar{\phi}]\{\ddot{y}\} + [c][\bar{\phi}]\{\dot{y}\} + [k][\bar{\phi}]\{y\} = \{F(t)\} \qquad (2.242)$$

Premultiply both sides of Eq. 2.242 by $[\bar{\phi}]^T$ to obtain

$$[\bar{\phi}]^T[m][\bar{\phi}]\{\ddot{y}\} + [\bar{\phi}]^T[c][\bar{\phi}]\{\dot{y}\} + [\bar{\phi}]^T[k][\bar{\phi}]\{y\} = [\bar{\phi}]^T\{F(t)\} \qquad (2.243)$$

Note that the matrices $[\bar{\phi}]^T[m][\bar{\phi}]$ and $[\bar{\phi}]^T[k][\bar{\phi}]$ on the left side of Eq. 2.243 are diagonal matrices that correspond to the matrices $[I]$ and $[\omega^2]$, respectively. However, the matrix $[\bar{\phi}]^T[c][\bar{\phi}]$ is not a diagonal matrix. If $[c]$ is proportional to $[m]$ or $[k]$ or both, the $[\bar{\phi}]^T[c][\bar{\phi}]$ becomes diagonal, in which case we can say that the system has proportional damping. The equations of motion then are completely uncoupled, and the ith equation will be

$$\ddot{y}_i + 2\xi_i\omega_i\dot{y}_i + \omega_i^2 y_i = \overline{f}_i(t) \qquad i = 1, 2, \ldots, n \qquad (2.244)$$

where $\overline{f}_i(t) = \left\{\overline{\phi}^{(i)}\right\}\{F(t)\}$. Thus, instead of n-coupled equations, we will have n-uncoupled equations.

We have already seen in Eq. 2.244 that the uncoupled equation of motion of the rth mode of a dynamic system that has proportional damping is given by

$$\ddot{y}_r + 2\xi_r\omega_r\dot{y}_r + \omega_r^2 y_r = \overline{f}_r(t) = \left\{\overline{\phi}^{(r)}\right\}^T \{F(t)\} \qquad (2.245)$$

where $\left\{\overline{\phi}^{(r)}\right\}$ represents the weighted rth modal vector of the undamped system. We introduce the following Fourier transforms of $y_r(t)$ and $\overline{f}_r(t)$, respectively, in the form

$$Y_r(\omega) = \int_{-\infty}^{\infty} y_r(t)e^{-i\omega t}dt$$

$$(2.246)$$

$$F_r(\omega) = \int_{-\infty}^{\infty} \overline{f}_r(t)e^{-i\omega t}dt = \sum_{j=1}^{n}\overline{\phi}_j(r)\int_{-\infty}^{\infty} F_j(t)e^{-i\omega t}dt$$

Then, we obtain Fourier transforms of both sides of Eq. 2.245 as

$$Y_r(\omega)\left[-\omega^2 + i2\xi_r\omega + \omega_r^2\right] = \omega_r^2 F_r(\omega) \qquad r = 1, 2, \ldots, n \qquad (2.247)$$

Equation 2.245 can be solved for $Y_r(\omega)$ as

$$Y_r(\omega) = H_r(\omega)F_r(\omega) \qquad r = 1, 2, \ldots, n \qquad (2.248)$$

where

$$H_r(\omega) = \left[1 - \left(\frac{\omega}{\omega_r}\right)^2 + i2\xi_r\left(\frac{\omega}{\omega_r}\right)\right]^{-1} \qquad r = 1, 2, \ldots, n \qquad (2.249)$$

The response correlation matrix $\left[R_x(\tau)\right]$ is given as

$$\left[R_x(\tau)\right] = \lim_{T\to\infty} \frac{1}{T}\int_{-T/2}^{T/2} \{x(t)\}\{x(t+\tau)\}^T dt \qquad (2.250)$$

Because the vector $\{x(t)\} = [\bar{\phi}]\{y(t)\}$, we can write Eq. 2.248 as

$$[R_x(\tau)] = \lim_{T\to\infty} \frac{1}{T} \int_{-T/2}^{T/2} [\bar{\phi}]\{x(t)\}\{x(t+\tau)\}^T [\bar{\phi}]^T dt$$

$$\quad (2.251)$$

$$= [\bar{\phi}][R_y(\tau)][\bar{\phi}]^T$$

where

$$[R_y(\tau)] = \lim_{T\to\infty} \frac{1}{T} \int_{-T/2}^{T/2} \{y(t)\}\{y(t+\tau)\}^T dt \quad (2.252)$$

is the response correlation matrix associated with generalized coordinates $y_r(t)$ (r = 1, 2, ..., n).

If $[H(\omega)]$ is the diagonal matrix of the frequency response function and $[H*(\omega)]$ is its conjugate, then the correlation matrix is

$$[R_y(\tau)] = \frac{1}{2\pi} \int_{-\infty}^{\infty} [H*(\omega)][S_f(\omega)][H(\omega)]e^{i\omega\tau}d\omega \quad (2.253)$$

where $[S_f(\omega)]$ is an n × n excitation matrix associated with the generalized forces $\bar{f}_r(t)$. Now $[S_f(\omega)]$ can be expressed in terms of the Fourier transform of the excitation correlation matrix $[R_f(\tau)]$ associated with $\bar{f}_r(t)$ as

$$[S_f(\omega)] = \int_{-\infty}^{\infty} [R_f(\omega)]e^{-i\omega\tau}d\tau \quad (2.254)$$

and $[R_f(\tau)]$ has the form

$$[R_f(\tau)] = \lim_{T\to\infty} \frac{1}{T} \int_{-T/2}^{T/2} \{\bar{f}(t)\}\{\bar{f}(t+\tau)\}^T dt \quad (2.255)$$

where $\{\bar{f}(t)\}$ is the vector of generalized forces $\bar{f}_r(t)$.

Therefore,

$$\{\bar{f}(t)\} = [\bar{\phi}]\{F(t)\} \quad (2.256)$$

and

$$\{\bar{f}(t+\tau)\}^T = \{F(t+\tau)\}^T [\bar{\phi}]^T$$

By substituting Eq. 2.256 into Eq. 2.255, we obtain

$$\left[R_f(\tau)\right] = \left[\bar\phi\right]\left[R_F(\tau)\right]\left[\bar\phi\right]^T \tag{2.257}$$

where

$$\left[R_F(\tau)\right] = \lim_{T\to\infty}\frac{1}{T}\int_{-T/2}^{T/2}\left\{F(t)\right\}\left\{F(t+\tau)\right\}^T dt \tag{2.258}$$

Introducing Eq. 2.257 into Eq. 2.254, we obtain

$$\left[S_f(\omega)\right] = \left[\bar\phi\right]\int_{-\infty}^{\infty}\left[R_F(\tau)\right]e^{-i\omega\tau}d\tau\left[\bar\phi\right]^T = \left[\bar\phi\right]\left[S_F(\omega)\right]\left[\bar\phi\right]^T \tag{2.259}$$

where

$$\left[S_F(\omega)\right] = \int_{-\infty}^{\infty}\left[R_F(\omega)\right]e^{-i\omega\tau}d\tau \tag{2.260}$$

is the excitation spectral matrix associated with the forces $F_i(t)$ ($i = 1, 2, ..., n$).

The response correlation matrix is obtained by substituting Eqs. 2.79 and 2.259 into Eq. 2.251

$$\left[R_x(\tau)\right] = \frac{1}{2\pi}\left[\bar\phi\right]\int_{-\infty}^{\infty}\left[H*(\omega)\right]\left[\bar\phi\right]\left[S_F(\omega)\right]\left[\bar\phi\right]^T\left[H(\omega)\right]e^{i\omega\tau}d\omega\left[\bar\phi\right]^T \tag{2.261}$$

and the autocorrelation function associated with the random response process $x_i(t)$ is

$$\left[R_{xi}(\tau)\right] = \frac{1}{2\pi}\left[\bar\phi_i\right]\int_{-\infty}^{\infty}\left[H*(\omega)\right]\left[\bar\phi\right]\left[S_F(\omega)\right]\left[\bar\phi\right]^T\left[H(\omega)\right]e^{i\omega\tau}d\omega\left[\bar\phi_i\right]^T \tag{2.262}$$

where $\left[\bar\phi_i\right]$ is the ith row matrix, that is, $\left[\bar\phi_i\right] = \left[\bar\phi_i^{(1)}, \bar\phi_i^{(2)}, ..., \bar\phi_i^{(n)}\right]$, which for $\tau = 0$ yields the mean square value

$$\left[R_{xi}(0)\right] = \frac{1}{2\pi}\left[\bar\phi_i\right]\int_{-\infty}^{\infty}\left[H*(\omega)\right]\left[\bar\phi\right]\left[S_F(\omega)\right]\left[\bar\phi\right]^T\left[H(\omega)\right]d\omega\left[\bar\phi_i\right]^T \tag{2.263}$$

This procedure can be employed to obtain the response of a multiple-degrees-of-freedom system, although the integral in Eq. 2.263 requires the use of the residue theorem.

2.8 Summary

In this chapter, we reviewed the analytical techniques for determining the response of dynamic systems. Both free and forced responses of linear single-degree-of-freedom systems and linear multiple-degrees-of-freedom systems are discussed by using the deterministic approach. We have reviewed the analytical techniques for determining the response of nonlinear dynamic systems. Both free and forced responses of single-degree-of-systems and multiple-degrees-of-freedom systems are discussed. Several methods are available for the solution of nonlinear single-degree-of-freedom systems. Some of the exact methods, approximate analytical techniques, and graphical procedures are discussed and presented.

A brief discussion of the theory of random vibrations was given by defining the random variable, probability density function, autocorrelation function, power spectral density function, Gaussian random process, Fourier analysis-Fourier series and Fourier integral, joint probability density function, and cross-correlation function. The response of a single-degree-of-freedom vibrating system using the impulse response and frequency response approaches was presented. A method for calculating the response of a linear system subjected to stationary random excitations was presented. The application of power spectral densities to vehicle dynamics is discussed briefly. The response of a single-degree-of-freedom system to random input and the response of a multiple-degrees-of-freedom system to random inputs were presented.

2.9 References

1. Anderson, G.L., "A Modified Perturbation Method for Treating Nonlinear Oscillation Problems," *Journal of Sound and Vibration*, Vol. 38, 1975, pp. 451–464.

2. Bendat, J.S., *Principles and Applications of Random Noise Theory*, Wiley, New York, 1958.

3. Bendat, J.S. and Piersol, A.G., *Engineering Applications of Correlation and Spectral Analysis*, Wiley, New York, 1980.

4. Bendat, J.S. and Piersol, A.G., *Random Data: Analysis and Measurement Procedures*, Wiley Interscience, New York, 1971.

5. Bendixson, T., "Sur les Courbes Defines par les Equations Differentielles," *Acta Mathematica*, Vol. 24, 1901, pp. 1–88.

6. Blackman, R.B. and Tukey, J.W., *The Measurement of Power Spectra*, Dover Publications, New York, 1958.

7. Brigham, E.O., *The Fast Fourier Transform*, Prentice-Hall, Englewood Cliffs, NJ, 1974.

8. Chetayev, N.G., *The Stability of Motion*, Pergamon Press, New York, 1961.

9. Clough, R.W. and Penzein, J., *Dynamics of Structures*, McGraw-Hill, New York, 1975.

10. Cramer, H., *The Elements of Probability Theory*, Wiley, New York, 1955.

11. Crandall, S.H., *Random Vibrations, Vol. 2*, The Technology Press of MIT, Cambridge, MA, 1963.

12. Crandall, S.H. and Mark, W.D., *Random Vibration in Mechanical Systems*, Academic Press, New York, 1963.

13. Dasarathy, B.V. and Srinivasan, P., "Study of a Class of Nonlinear Systems Reducible to Equivalent Linear Systems," *Journal of Sound and Vibration*, Vol. 7, 1968, pp. 27–30.

14. Den Hartog, J.P., *Mechanical Vibrations*, McGraw-Hill, New York, 1956.

15. Dukkipati, R.V. and Amyot, J.R., *Computer Aided Simulation in Railway Vehicle Dynamics*, Marcel-Dekker, New York, 1988.

16. Dukkipati, R.V., Anandarao, M., and Bhat, R.B., *Computer Aided Analysis and Design of Machine Elements*, New Age International Limited, New Delhi, 1999.

17. Garg, V.K. and Dukkipati, R.V., *Dynamics of Railway Vehicle Systems*, Academic Press, New York, 1984.

18. Hurley, W.C. and Rubinstein, M.F., *Dynamics of Structures*, Prentice-Hall, Englewood Cliffs, NJ, 1963.

19. Hutton, D.V., *Applied Mechanical Vibration*, McGraw-Hill, New York, 1981.

20. Jacobsen, L.S., "On a General Method of Solving Second-Order Differential Equations by Phase Plane Displacements," *J. Appl. Math.*, Vol. 19, 1952, pp. 543–553.

21. Lienard, A., "Etudes des Oscillatioins Entretenues," *Rev. Gen. De l'Elec.*, Vol. 23, 1928, pp. 901–912, 946–954.

22. Meirovitch, L., *Analytical Methods in Vibration*, Macmillan, New York, 1967.

23. Meirovitch, L., *Computational Methods in Structural Dynamics*, Sijhoff and Noordhoff International Publishers, The Netherlands, 1980.

24. Meirovitch, L., *Elements of Vibration Analysis*, McGraw-Hill, New York, 1986.

25. Meirovitch, L., *Method of Analytical Dynamics*, McGraw-Hill, New York, 1970.

26. Mickens, R.E., "Perturbation Solution of a Highly Nonlinear Oscillation Equation," *Journal of Sound and Vibration*, Vol. 68, 1980, pp. 153–155.

27. Newland, D.E., *Random Vibrations and Spectral Analysis*, Longman Group Limited, New York, NY, 1964.

28. Pell, W.H., "Graphical Solution of Single-Degree-of-Freedom Vibration Problems with Arbitrary Damping and Restoring Forces," *J. Appl. Mech.*, Vol. 24, 1957, pp. 311–312.

29. Popoulis, A., *Probability, Random Variables and Stochastic Processes*, McGraw-Hill, New York, 1965.

30. Rao, J.S., *Advanced Theory of Vibration*, Wiley, New Delhi, 1992.

31. Rao, S.S., *Mechanical Vibrations*, Addison-Wesley, Reading, MA, 1986.

32. Robson, J.D., *Random Vibrations*, Edinburgh University Press, Edinburgh, 1964.

33. Shabana, A.A., *Theory of Vibration—Volume I: An Introduction*, Springer-Verlag, New York, 1991.

34. Shabana, A.A., *Theory of Vibration—Volume II*, Springer-Verlag, New York, 1991.

35. Steidel, R.F., *An Introduction to Mechanical Vibration*, Wiley, New York, 1989.

36. Thomson, W.T., *Theory of Vibrations with Applications*, Prentice-Hall, Englewood Cliffs, NJ, 1981.

37. Timoshenko, S., Young, D.H., and Weaver, W., *Vibration Problems in Engineering*, Wiley, New York, 1974.

38. Tse, F.S., Morse, I.E., and Hinkle, R.T., *Mechanical Vibrations Theory and Applications, 2nd Edition*, Allyn and Bacon, London, 1978.

39. Wilkinson, J.H., *The Algebraic Eigenvalue Problem*, Clarendon Press, Oxford, 1965.

Chapter 3

Tire Dynamics

3.1 Introduction

As a significant part of the vehicle, pneumatic tires perform the following functions:

- Support the weight of the vehicle
- Cushion the vehicle over surface irregularities
- Provide sufficient traction for driving and braking
- Provide adequate steering control and directional stability

The primary forces acting on the vehicle are from pneumatic tires. The critical performances of the vehicle (e.g., driving, braking, stability, ride comfort, traveling) are related to pneumatic tires. Therefore, analyses of tire dynamics are the precondition and basis for the studies and analyses of vehicle dynamics.

The ground forces affecting the vehicle are applied through pneumatic tires. When the vehicle runs forward directly without side force, the forces exerted on the tires include the normal force F_z, rolling resistance moment M_y, tractive force F_x, and tractive moment $F_x \cdot R$. When the trailing wheel puts on the brake, the braking moment will increase the longitudinal resistance force. When the driving wheel moves under the driving moment, the direction of the ground longitudinal force is opposite to its direction under braking (Figures 3.1 and 3.2).

When the vehicle changes direction or if a side force is applied to the vehicle, a lateral force and moment will be developed on the tire (Figure 3.3). Because the lateral elasticity of the tire increases gradually as well as the lateral deformation at the tire-ground contact patch when the tire is rolling, there is a distance e that is called the pneumatic trail between the resultant of force of the lateral forces and the center of the contact patch. The product, $F_y \cdot e$, of the ground lateral forces applied to the tire and the pneumatic trail determines the self-aligning moment of the tire (Figure 3.4).

Figure 3.1 Forces exerted on the tire without lateral force under braking, where T$_b$ is the rolling resistance, R is the rolling radius, and a is the forward moving distance.

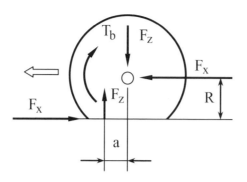

Figure 3.2 Forces exerted on the tire without lateral force under driving, where T$_b$ is the rolling resistance, R is the rolling radius, and a is the forward moving distance.

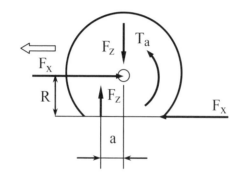

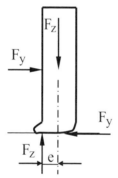

Figure 3.3 Lateral forces exerted on the tire.

Figure 3.4 Distribution of lateral forces at the contact patch.

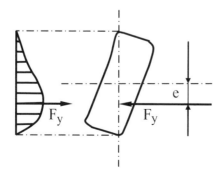

For a better analysis of the ground vehicle dynamics and the dynamics performance of the tire, we must analyze the characteristics of tire dynamics first and gain three tire forces and three tire moments.

To describe the characteristics of a tire and the forces and moments acting on it, it is necessary to define an axis system that serves as a reference for the definition of various parameters. Figure 3.5 shows one of the commonly used axis systems recommended by SAE International. The origin of the axis system is the center of tire contact. The x axis is the intersection of the wheel plane and the ground plane with a positive direction forward. The z axis is perpendicular to the ground plane with a positive direction downward. The y axis is in the ground plane, and its direction is chosen to make the axis system orthogonal and right-hand.

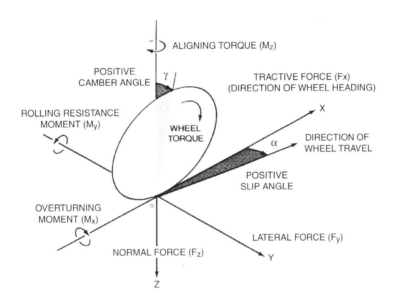

Figure 3.5 Commonly used axis system recommended by SAE International.

Three forces and three moments act on the tire from the ground. Tractive force (or longitudinal force) F_x is the component in the x direction of the resultant force exerted on the tire by the road. The lateral force F_y is the component in the y direction, and the normal force F_z is the component in the z direction. The overturning moment M_x is the moment about the x axis exerted on the tire by the road. The rolling resistance moment M_y is the moment about the y axis, and the aligning torque M_z is the moment about the z axis.

The tire is an elastic body that can be equivalent to a subsystem consisting of mass elements, elastic elements, and damping elements in terms of mechanics. The dynamics analysis and modeling of a tire are always the important part of vehicle dynamics analyses. This chapter presents mechanics characteristics of the tire chiefly at the directions vertical, longitudinal, and lateral, respectively, and introduces related dynamics models and concepts of tires.

3.2 Vertical Dynamics of Tires

The vertical dynamics of tires is the basis of research on lateral and longitudinal dynamics of tires. The vertical stiffness and damping characteristics, the classical vertical dynamic model of tires, and the advantages and disadvantages of tires models are discussed in the following sections.

3.2.1 Vertical Stiffness and Damping Characteristics of Tires

A pneumatic tire can cushion the vehicle over surface irregularities. The cushioning characteristics of tires have a direct relationship with the vertical stiffness and damping of the tires. The static stiffness of the tire, K_s, is determined by the slope of the static load-deflection curves. Basically, the static stiffness of the tire is a constant. The dynamic stiffness of the tire, K_d, which varies with the frequency of the dynamic load, is determined by the transfer characteristics of the dynamic load to the deflection when the tire is rolling. There is a distinct difference between the static stiffness and the dynamic stiffness of a tire.

In an experiment measuring tire stiffness, it was found that the force required to make the tire produce the same deflection is different from the force that makes the tire restore from the same deflection whether under static or dynamic load. Figure 3.6 shows the relationship between the tire load and its deflection. The close-up area shown in Figure 3.6 represents the dissipative power of the rolling tire, which manifests the damping characteristics of the tire. With an increase in the frequency of excitation, the damping characteristics decrease. This is opposite to the state of the decrease in tire inflation pressure, where damping characteristics increase.

Figure 3.6 Load-deflection relationship of a tire, where F_{z1} is the force required to make the tire produce the same deflection, and F_{z2} is the force that makes the tire restore from the same deflection under either static or dynamic load.

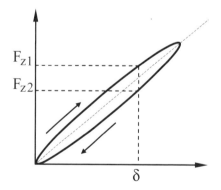

Figure 3.7 shows the variation of the dynamic stiffness of a tire with the excitation frequency: when the excitation frequency is zero, the corresponding stiffness is the static tire stiffness. Also, the dynamic stiffness of tires decreases sharply with the increase in excitation frequency. However, beyond a certain frequency, the influence of excitation becomes less important. Generally, the dynamic stiffness may be 10 to 15% less than the static value. Under the same tire size, the dynamic stiffness of radial ply tires is lower than the counterpart of bias ply tires. Inflation pressure has a noticeable influence on tire stiffness. Tire stiffness will decrease sharply as soon as the pressure varies in the same direction, which occurs because the compressed air supports 85% of the tire load while the remainder is exerted on the tire body.

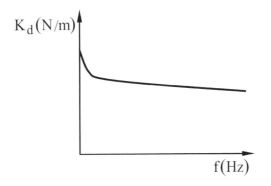

Figure 3.7 Dynamic stiffness-excitation frequency relationship of a tire.

3.2.2 Vertical Vibration Mechanics Models of Tires

Dynamic vertical forces exerting on the tires must be obtained exactly, which is the important content of analyses of the entire vehicle dynamics. The vertical dynamic model of tires is the basis for analyzing the vertical forces of tires. The mechanics characteristics of tire models not only embody the vertical elastic characteristic of tires, but also manifest the tire-ground contact relationship as well as the influence derived from surface irregularities.

When the influence of surface irregularities is less important with a constant velocity, the point contact tire model is suitable for the analysis of vehicle vibration dynamics under steady-state conditions. If the influence of surface irregularities is greater, the point contact model cannot represent the contact relationship between the tire and the road, even with a constant velocity. Thus, the fixed contact patch tire model is required. With the obvious variation of surface irregularities and nonsteady velocity, these models are incompetent for use in analyses of the vertical dynamics of vehicles. Here, the time-varying contact patch tire model is needed to exactly embody the vertical dynamics of the vehicle. These models will be introduced as follows.

3.2.2.1 Point Contact Model of Tires

An idealized tire can be equivalent to a spring-damping system, in which the elasticity characteristic of tires can be defined as the nominal stiffness and the resistance characteristic can be defined as the damping coefficient. In the point contact model of tires, it is assumed that the contact region between the tire and the ground is a point that is the projection of the center of the tire onto the ground (Figure 3.8). Therefore, the vertical dynamic force exerted on the tire can be defined as

$$F_z = K \cdot \left(h(X) - \zeta(X) \right) + C \cdot \left(\dot{h}(X) - \dot{\zeta}(X) \right) \tag{3.1}$$

where

$h(X)$ = surface irregularity just under the center of the tire when it is rolling to the site X

$\zeta(X)$ = displacement of the center of the axis arm

$$\dot{h}(X) \quad = \quad \text{slope of the surface irregularity just under the center of the tire}$$

$$\dot{\zeta}(X) \quad = \quad \text{velocity of the center of the axis arm}$$

$$K \quad = \quad \text{nominal stiffness of the tire}$$

$$C \quad = \quad \text{damping coefficient of the tire}$$

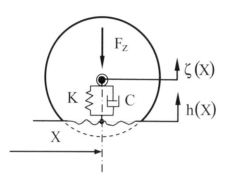

Figure 3.8 Point contact model of tires.

This tire model can be established and applied with the following conditions:

1. Tire stiffness is an equivalent concentrated stiffness coefficient that does not vary with the vertical dynamic load of the tire. As a result, the model cannot manifest nonlinear characteristics of tires with massive deformation under the nonsteady maneuvers such as sudden accelerating or severe braking.

2. $h(X) - \zeta(X)$ presents the equivalent deformation of the tire just under its center, which is related only to the displacement of the center of the axis arm and the surface irregularity under the contact point. When the influence of surface irregularities under the contact region is severe, there will be a massive error between the vertical deformation calculated and the practical tire deformation, which is reasonable only over a good road surface.

As these results 1 and 2 indicate, the point contact tire model is the most cursory and simplest one and is applied primarily to the analyses of vehicle dynamics in a state with little surface irregularities, constant velocity, and light variation of the tire load. The advantages of this model are simple mechanics characteristics and easy usage.

3.2.2.2 Fixed Contact Patch Model of Tires

The fixed contact patch model of tires (Figure 3.9) was created on the basis of the point contact tire model by converting the equivalent concentrated stiffness of the tires into the equivalent distributed stiffness within the fixed contact patch length of the tires

$$F_z = \int_l k_i \big(h(X - x) - \zeta(X) \big) dx + C \cdot \big(\dot{h}(X) - \dot{\zeta}(X) \big) \tag{3.2}$$

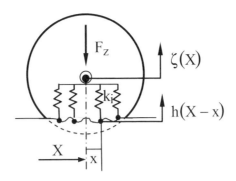

Figure 3.9 Fixed contact patch model of tires.

where

$$h(X - x) = \text{surface irregularity of the road with a distance of x apart from the contact point when the tire is rolling to the site X}$$

$$\zeta(X) = \text{displacement of the center of the axis arm}$$

$$\dot{h}(X) = \text{slope of the surface irregularity just under the center of the tire}$$

$$\dot{\zeta}(X) = \text{velocity of the center of the axis arm, } k_i, \text{ is the distributed stiffness coefficient of the tire corresponding to the site x}$$

$$C = \text{damping coefficient of the tire}$$

Compared to the point contact model of tires, the fixed contact patch model of tires embodies the comprehensive effect over the tire force from the surface irregularities within the contact patch, from which the dynamic tire force derived is closer to the practical value. Because it is assumed that the length of the contact patch does not vary, the equivalent stiffness of the tires is a constant value. Therefore, this model is not available for analyses of vehicle dynamics in maneuvers with strong variation of the contact patch, such as sudden accelerating or severe braking.

3.2.2.3 Time-Varying Contact Patch Model of Tires

The time-varying contact patch model of tires (Figures 3.10) is employed to offset the imperfections due to these models. Several preconditions are required for the model:

1. The distributions with respect to the elasticity and damping of the tire are uniform in the radial direction.

2. The influence of surface irregularities is neglected along the width of the tire.

3. The elastic deformations turn out only within the contact region of the tire.

Figure 3.10 illustrates a brief tire model according to the preceding conditions. Many stiffness and damping elements distribute equally in the radial direction within the contact patch. They deform in accordance with the variation of surface irregularities. Thus, the radial deformation and the corresponding velocity at the location that has an angle θ with the vertical direction can be expressed as

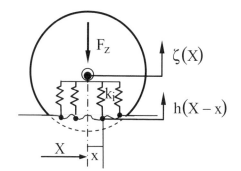

Figure 3.10 Time-varying contact patch model of tires.

$$\delta(X - x) = R - \frac{\left[h(X - x) - \zeta(X) \right]}{\cos\theta} \tag{3.3}$$

$$\dot{\delta}(X + x) = \frac{v \dfrac{dh(X + x)}{dx} - w}{\cos\theta} \tag{3.4}$$

where

x	=	distance apart from the contact point
R	=	free radius of the tire
v	=	horizontal velocity
w	=	vertical velocity
$h(X - x)$	=	surface irregularity of the location that has a distance of x from the projection of the tire center
$\dfrac{dh(X + x)}{dx}$	=	slope of the surface irregularity with a distance of x from the tire center

According to the geometric relationship shown in Figure 3.10, the angle θ is calculated as

$$\theta = \text{arctg}\left(\frac{x}{h(X + x) - \zeta(X)} \right) \tag{3.5}$$

Then, the radial force $F_c(x)$ acting on the spring and the damping elements can be written as

$$F_c(X + x) = \begin{cases} k\delta(X + x) + c\dot{\delta}(X + x) & \delta(X + x) \geq 0 \\ 0 & \delta(X + x) < 0 \end{cases} \tag{3.6}$$

When $\delta(X + x)$ is less than zero, it means that the tire does not deform at the contact point X + x, or that breakaway exists between the tire and the ground.

Thus, the vertical dynamic load acting on the tire is equal to the vector sum of the radial force derived from each element at the vertical direction, which is expressed as

$$F_z = \int \cos\theta \cdot F_c(X + x)dx \qquad (3.7)$$

Compared to the other tire models, the time-varying contact patch model of tires obtains various characteristics, as follows:

1. The dynamic vertical load acting on the tire is the comprehensive contribution of the surface irregularities on the contact patch rather than the single contact point just under the tire center, which complies with the practical filterability of the road surface. The load that is derived from the model is much closer to the experimental test data.

2. The equivalent concentrated stiffness coefficient of the tire varies with the contact patch. For instance, under driving conditions with sudden accelerating or severe braking, if the tire load rises dramatically, then the contact patch will vary in the same way. Thus, more elements will contribute to the deformation of the tire, and the stiffness will increase; in the contrary situation, the stiffness will decrease. Therefore, this model also can be called a nonlinear time-varying model.

3.2.3 Enveloping Characteristics of Tires

A tire obtains enveloping characteristics, which are defined as the tire being able to absorb the vertical influence of the surface irregularities when it is rolling on an irregular road. The surface irregularities and vertical elasticity of the tires are the fundamental reasons for the generation of enveloping characteristics of tires.

According to the preceding tire models, the second and the last models bear the enveloping characteristics, but the first does not. By the way, when the point contact model is applied to the analysis and calculation of vehicle dynamics, the road input to compute the tire vertical dynamic load is not the practical road input, but the equivalent enveloping road input. Thus, the formula for the force should be rewritten as

$$F_z = K \cdot \left(h_q(X) - \zeta(X) \right) + C \cdot \left(\dot{h}_q(X) - \dot{\zeta}(X) \right) \qquad (3.8)$$

where

$$h_q(X) = h(X + x) - \left\lfloor R - \sqrt{R^2 - x^2} \right\rfloor \qquad (3.9)$$

Adding a rigid wheel with a radius R to a point contact tire model to simulate the motion of the tire over irregular road, the practical contact point between the tire and the surface is not the projection of the tire center on the road, but it will offset a distance of x from the projection of the tire center on the road according to the geometric relationship related to the variation of surface irregularities and the rigid wheel. Thus, the trail of the rigid wheel center is equal to the effective surface irregularities, as shown in Figure 3.11.

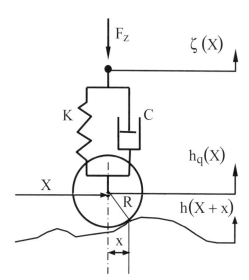

Figure 3.11 Point contact model with a rigid wheel.

The relationship between the effective surface irregularities and the practical surface irregularities again is defined from Eq. 3.9 as

$$h_q(X) = h(X + x) - \left[R - \sqrt{R^2 - x^2} \right] \tag{3.9}$$

Note that the point contact model with a rigid wheel can represent only the geometric enveloping characteristics of tires.

In fact, the tire also has elastic enveloping characteristics. For instance, when a tire is slowly rolling over a real road with a semi-sine wave heave or hollow, there will be a big difference from the trail of the tire center and the real surface irregularities due to the elasticity of the tires. Moreover, the amplitude of the former decreases with the increase in wavelength. Figure 3.12 shows the relationship of the trail of the tire center with the real surface irregularities when a time-varying contact tire is rolling over a surface with a semi-sine wave. Note that the shorter the wavelength, the more dramatic are the enveloping properties. The fixed contact patch model of tires has similar properties.

Figure 3.12 Effect of a surface with a semi-sine wave.

3.3 Tire Longitudinal Dynamics

Tire longitudinal dynamics mainly presents tire rolling resistance, the slip coefficient, and the relationship between the tire traction and brake forces and the road surface condition.

3.3.1 Tire Rolling Resistance

The rolling resistance of the tire, which is applied to the wheel, is due to the deformation of the tire at the tire/road interface. As is known from the tire experiment, tire deformation needs to consume energy, and the normal pressure distribution over the tire/road contact patch is not uniform because of the unequal force needed during the compression and elastic recovery, as shown in Figure 3.13. In the direction of travel, the normal force in the leading half of the contact patch is higher than that in the trailing half. The center point of the sum of the normal forces offsets a distance a to the centerline in the direction of rolling. The normal force produces a moment about the axis of rotation of the tire, which is the rolling resistance moment, expressed as $M_f = F_z \cdot a$. While the driving force F_{ax} applied to the wheel produces a moment to balance the rolling resistance moment, the equation can be expressed as

$$F_{ax} \cdot r = M_f \tag{3.10}$$

$$F_{ax} = \frac{M_f}{r} = F_z \cdot \frac{a}{r} \tag{3.11}$$

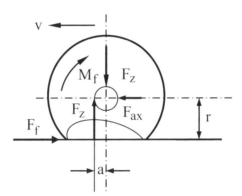

Figure 3.13 Distribution of the forces and moments of a rolling tire.

Set

$$f = \frac{a}{r}$$

Then

$$F_{ax} = F_z \cdot f \tag{3.12}$$

or

$$f = \frac{F_{ax}}{F_z} \tag{3.13}$$

The nondimensional coefficient f is defined as the rolling resistance coefficient. It is the ratio of the driving force and the normal force on the rolling wheel in certain conditions. Usually, tire rolling resistance changes linearly with the normal force of the wheel, that is,

$$F_f = f \cdot F_z \tag{3.14}$$

Experiments show that the rolling resistance coefficient is proportional to the tire deformation, while being inversely proportional to the radius of a loaded tire.

Table 3.1 lists the values of the rolling resistance coefficients of different road surfaces recorded in the literature. According to the U.S. standard, if v < 50 km/h, then f will be 0.0165. If v > 50 km/h, f may be expressed as

$$f = 0.0165\left[1 + 0.01(v - 50)\right] \tag{3.15}$$

TABLE 3.1
COEFFICIENT OF ROLLING RESISTANCE

Road Surface	Rolling Resistance Coefficient
Pneumatic tire of a car rolls	
on a coarse stone road surface	0.015
on a fine stone road surface	0.015
on a rolled asphalt mixed aggregate road surface	0.015
on an impacted coarse gravel road surface	0.012
on a mixed bituminous macadam road surface	0.025
on a soil road surface	0.05
on a field road surface	0.1~0.35
Pneumatic truck tire rolls on an asphalt mixed aggregate road surface	0.006~0.01
Steel wheel rolls on a field road surface	0.14~0.24
Steel wheel on a railroad surface	0.001~0.002

3.3.2 Rolling Resistance of the Tire with Toe-In

In actual vehicle structure, there is a toe-in angle of the front wheel. This produces a toe-in resistance acting on the front wheel in the opposite direction of traveling, as shown in Figure 3.14.

In this figure, δ_{v0} is the toe-in angle of the front wheel on one side. $F_{\delta v}$ is the side force due to the tire lateral deformation caused by the angle δ_{v0}. $F_{\delta v} = C_r \cdot \delta_{v0}$, where C_r is the cornering stiffness of the tire. When the angle δ_{v0} is small enough, the value of $\sin \delta_{v0}$ is approximately equal to δ_{v0}, that is, $\sin \delta_{v0} \approx \delta_{v0}$. Hence, the component of $F_{\delta v}$ of the wheel is the toe-in resistance

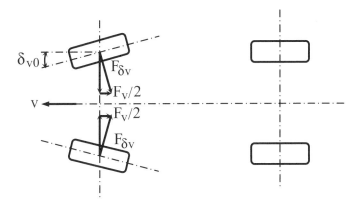

Figure 3.14 Toe-in resistance acting on the front wheel.

$$F_v = 2F_{\delta v} \cdot \sin \delta_{v0} = 2C_r \cdot \delta_{v0}^2 \qquad (3.16)$$

f_δ is defined as the toe-in resistance coefficient

$$f_\delta = \frac{C_r}{F_z} \cdot \delta_{v0}^2 \qquad (3.17)$$

Then, the toe-in resistance will be expressed as

$$F_v = 2f_\delta \cdot F_z \qquad (3.18)$$

3.3.3 Rolling Resistance of the Turning Wheel

The value of additional rolling resistance when the wheel is turning depends on the volecity of the vehicle, the turning radius, and the vehicle parameters. The rolling resistance coefficent f_R during a turn can be expressed as

$$f_R = f + \Delta f \qquad (3.19)$$

α_f and α_r are the slip angles of the front tire and rear tires, respectively, in the conditon of the steering angle δ_0. Correspondingly, F_{yf} and F_{yr} are the cornering forces to balance the centrifugal force of the vehicle when steering, as shown in Figure 3.15. According to the equilibrium of forces, F_{yf} and F_{yr} can be expressed as

$$F_{yf} = \frac{\cos \gamma \cdot F}{\sin \delta_0}$$

$$= m \frac{v^2}{R} \cdot \frac{\sin (\beta - \alpha_r)}{\sin \delta_0} \qquad (3.20)$$

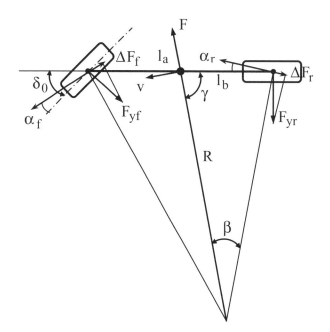

Figure 3.15 Rolling resistance of a turning wheel.

$$F_{yr} = \sin \gamma \cdot F - F_{yf} \cdot \cos \delta_0$$

$$= m \frac{v^2}{R} \cdot \left[\cos (\beta - \alpha_r) - \sin (\beta - \alpha_r) \cdot \mathrm{ctag} \delta_0 \right] \tag{3.21}$$

According to relationships of trigonometric function,

$$\beta = \arcsin \left(\frac{l_b \cos \alpha_r}{R} \right) \tag{3.22}$$

$$\gamma = \frac{\pi}{2} - \beta + \alpha_r \tag{3.23}$$

where m is the mass of the vehicle, v is the vehicle velocity when turning, and R is the turning radius.

Then the additional resistance applied on the front and rear wheels, respectively, can be obtained

$$\Delta F_f = \sin \alpha_f \cdot F_{yf}$$

$$= m \frac{v^2}{R} \cdot \frac{\sin (\beta - \alpha_r)}{\sin \delta_0} \cdot \sin \alpha_f \tag{3.24}$$

$$\Delta F_r = \sin \alpha_r \cdot F_{yr}$$

$$= m\frac{v^2}{R} \cdot \left[\cos(\beta - \alpha_r) - \sin(\beta - \alpha_r) \cdot \operatorname{ctag}\delta_0\right] \cdot \sin \alpha_r \tag{3.25}$$

The additional resistance coefficient under the condition of vehicle steering can be described as

$$\Delta f = \frac{\Delta F_f + \Delta F_r}{mg}$$

$$= \frac{v^2}{R \cdot g} \cdot \left\{ \frac{\sin(\beta - \alpha_r)}{\sin \delta_0} \cdot \sin \alpha_f + \left[\cos(\beta - \alpha_r) - \sin(\beta - \alpha_r) \cdot \operatorname{ctag}\delta_0\right]\sin \alpha_r \right\} \tag{3.26}$$

From Eq. 3.26, it can be seen that the additional rolling resistance coefficient increases with both the vehicle velocity and the steering angle increase, but decreases with the turning angle increase. It also depends on the vehicle parameters.

3.3.4 Longitudinal Adhesion Coefficient

When a driving (or braking) torque is applied to a pneumatic tire, a tractive force (or braking force), which is limited to the critical coefficient of the road adhesion, is developed at the tire-ground contact patch. The maximum force F_φ of a tire on hard surfaces is defined as adhesion force and is proportional to the normal force applied on the wheel during driving or braking as

$$F_\varphi = F_z \cdot \varphi \tag{3.27}$$

where φ is the adhesion coefficient, which varies with the state of the tire rolling or slipping.

When the driving torque of the wheel M_t is greater than the torque $F_\varphi \cdot r_d$ that is produced by the adhesion force around the wheel center, that is, $M_t > F_\varphi \cdot r_d$ (where r_d is the effective rolling radius of a tire, the vertical distance from the center of the wheel to the tire-ground contact point), the tire will be slipping. However, if the braking torque of the wheel M_b is more than the torque $F_\varphi \cdot r_d$, that is, $M_b > F_\varphi \cdot r_d$, then the tire will be skidding. Also, the tire will be in a state of rolling-slipping, with the variation of the driving torque or braking torque of the wheel. If ω is the angular speed of a rolling tire, then the longitudinal speed of the tire-ground contact point may be expresseed as $\omega \cdot r_d$, and if v_x is the linear speed of the tire relative to the ground, then the relationship of $\omega \cdot r_d$ and v_x can be used to describe the rolling or slipping state of the tire. When $\omega \cdot r_d = v_x$, there will be no relative motion at the tire-ground contact point. The tire then is in a state of pure rolling. When $\omega \cdot r_d > v_x$, there will be a negative linear velocity at the tire-ground contact point. Then the tire is rolling and slipping and will develop a longitudinal tractive force. When $\omega \cdot r_d < v_x$, there is positive linear velocity at the tire-ground contact point. The tire is rolling and sliding and will develop a longitudinal braking force.

To accurately describe tire slip in a braking maneuver longitudinal skid, s_b is defined as

$$s_b = \frac{v_x - \omega \cdot r_d}{v_x} \times 100\% \qquad (3.28)$$

Braking maneuver: If $s_b = 0$, then the tire is purely rolling. If $s_b = 100\%$, then the tire is purely skidding. If $0 < s_b < 100\%$, the tire is rolling and skidding, and the slip becomes serious as the longitudinal skid s_b increases.

To describe tire slip accurately in a tractive maneuver longitudinal slip, s_a is defined as

$$s_a = \frac{\omega \cdot r_d - v_x}{\omega \cdot r_d} \times 100\% \qquad (3.29)$$

Driving maneuver: If $s_a = 0$, then the tire is purely rolling. If $s_a = 100\%$, then the tire is purely spinning. If $0 < s_a < 100\%$, then the tire is rolling and slipping, and the slip becomes serious as the longitudinal slip s_a increases.

Both the longitudinal slip and skip describe the relationship of the rolling and sliding, with their value varying from 0 to 100%. Because driving and braking are opposite in the longitudinal direction, one single index, the slip ratio s, can be used to express both longitudinal slip and longitudinal skid. Zero is the division value. When $0 < s < 100\%$, it expresses a braking maneuver. When $s = 100\%$, the wheel will lock completely. When $-100\% < s < 0$, it expresses a driving maneuver. When $s = -100\%$, the wheels are spinning at a high angular speed, but the vehicle does not move forward.

Based on available experimental data, there is a closed relationship between the coefficient of road adhesion and the longitudinal slip, as shown in Figure 3.16, as follows:

1. If $|s| = 0 \sim 15\%$, the value of φ increases linearly with s.

2. If $|s| = 15\% \sim 30\%$, the value of φ reaches the maximum and is defined as the peak coefficient of road adhesion.

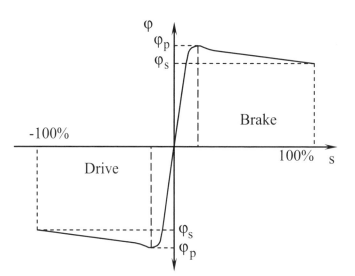

Figure 3.16 Variation of the coefficient of road adhesion with longitudinal slip.

3. If $|s| = 30\% \sim 100\%$, the value of φ gradually falls with the s increase and is defined as the sliding coefficient of road adhesion when $|s| = 100\%$.

The tractive effort or braking effort will reach the maximum when the tire is rolling and slipping. Correspondingly, the value of $|s|$ is between 15 and 30%. Therefore, it is important to avoid wheel lock-up during braking or wheel spinning during acceleration. This is one impetus to the development of anti-lock braking systems (ABS) and traction control systems for road vehicles. Further detailed discussion of braking and accelerating is provided in Chapters 7 and 8 of this book.

The peak value of the coefficient of road adhesion is approximately 1.2 times the sliding value on a dry surface and approximately 1.3 times on a wet surface. Table 3.2 gives the average peak and sliding values of the coefficient of road adhesion φ_p and φ_s on various surfaces.

TABLE 3.2
PEAK AND SLIDING VALUES OF THE
COEFFICIENT OF ROAD ADHESION

Surface	Peak Value φ_p	Sliding Value φ_s
Asphalt and concrete (dry)	0.8~0.9	0.75
Asphalt (wet)	0.5~0.7	0.45~0.6
Concrete (wet)	0.8	0.7
Gravel	0.6	0.55
Earth road (dry)	0.68	0.65
Earth road (wet)	0.55	0.4~0.5
Snow (hard-packed)	0.2	0.15
Ice	0.1	0.07

The coefficient of road adhesion depends mainly on the road texture and suface, tire structure, tread pattern, material, inflation pressure, normal loading on the wheel, travel speed of the vehicle, and so forth. Generally, on a given road, the larger the area of the tire-road contact patch, the larger the limited adhesion force will be produced, and then the higher the value of the adhesion coefficient. On asphalt or concrete, the tire adhesion coefficient is considerably higher than that on a field or ice. On wet surfaces, a lower tire adhesion coefficient than on dry surfaces usually is observed. The tire-road contact path of a tire with a wide tread is larger than that of a tire with a narrow tread; therefore, the adhesion coefficient of a wide-tread tire is better than that of a narrow-tread tire. The experiment data show that the adhesion coefficient of a radial tire is higher than that of bias tire, and that it is lower for a tire with high inflation pressure than that of a tire with low inflation pressure. In addition, the vehicle speed has an effect on the adhesion coefficient, in that a low travel speed can produce a relatively high adhesion coefficient of the tire.

3.3.5 Theoretical Model of Tire Longitudinal Force Under Driving and Braking

It is difficult to obtain an accurate longitudinal friction force of a tire through the equation $F_x = F_z \cdot \varphi$. In Julien's theory, it is assumed that the tire tread can be regarded as an elastic band, and that the contact patch is rectangular and the normal pressure is distributed uniformly. It is further assumed that the contact patch can be divided into an adhesion region and a sliding region. For a pure rolling tire, only an adhesion region exists. The adhesion region will become smaller and smaller when the tire slip becomes more and more serious. When the tire is sliding, the adhesion region will disappear; then, only a sliding region exists.

It is assumed that in the adhesion region, the longitudinal stresses depend on the longitudinal stiffness and the longitudinal elastic deformation of the tire, whereas in the sliding region, the longitudinal stresses depend on the adhesive properties of the tire-ground interface such as the contact pressure and the road/tire friction coefficient (Figure 3.17).

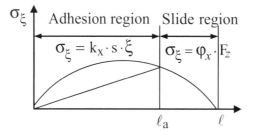

Figure 3.17 Variation of longitudinal stiffness with the tire contact patch length.

Then the road/tire friction coefficient can be approximately expressed linearly as

$$\varphi_x = \varphi_0\left(1 - A \cdot s\right) \tag{3.30}$$

$$A = \frac{\left(1 - \dfrac{\varphi_1}{\varphi_0}\right)}{s} \tag{3.31}$$

where

φ_0 = road/tire friction coefficient when no sliding takes place

φ_1 = road/tire friction coefficient when the slip ratio is S_1

A = parameter that can be obtained through the two points $(0, \varphi_0)$, (S_1, φ_1)

The normal force of the tire on the contact patch can be expressed as

$$F_z = k_z \cdot \delta \tag{3.32}$$

where

k_z = radial stiffness of the tire

δ = radial deformation of the tire

The longitudinal stress can be expressed as

$$\sigma_\xi = \begin{cases} k_x \cdot s \cdot \xi & 0 \leq \xi \leq \ell_a \\ \varphi_x \cdot p & \ell_a \leq \xi \leq \ell \end{cases} \tag{3.33}$$

where

σ_ξ = longitudinal stress of the tire tread

φ_x = tire/road longitudinal friction coefficient

ℓ = tire contact patch length

p = tire/road contact pressure

k_x = tire longitudinal stiffness per unit area in the contact patch

s = absolute value of the longitudinal slip ratio

ξ = tire longitudinal deformation

ℓ_a = length of the tire contact patch

The nondimensional tire contact patch length is expressed as

$$S_n = \frac{s}{S_c}$$

$$\ell_n = 1 - S_n$$

Then, the longitudinal stress of the tire tread is integrated in the longitudinal direction, and the longitudinal friction force can be expressed as

$$F_x = \begin{cases} K_x s \cdot \ell_n^2 + \varphi_x F_z \left(1 - 3\ell_n^2 + 2\ell_n^3\right) & s < S_c \\ \varphi_x F_z & s \geq S_c \end{cases} \tag{3.34}$$

where S_c is the critical value to divide the adhesion region and sliding region, expressed as

$$S_c = \frac{3\varphi_x F_z}{K_x} \tag{3.35}$$

where K_x is the total longitudinal stiffness.

3.4 Tire Lateral Dynamics

Tire lateral dynamics mainly discusses the cornering characteristics, roll characteristics, lateral forces at the tire-ground contact patch, and self-aligning torque caused by slipping. It also analyzes the relationships among slip angle, slip stiffness, cornering force, and self-aligning torque to obtain the precise lateral dynamics of the tire.

3.4.1 Tire Cornering Characteristics

Due to the road bank, lateral wind, or the centrifugal force when driving, a side force F_Y will be applied to the center of the tire. Correspondingly, a lateral reaction force also will be developed at the contact patch and is called the cornering force F_y. The phenomenon of side slip is due mainly to the tire lateral elasticity, which makes the driving direction deviate from the direction of the tire plane. When the tire begins to slip, the centerline that is connected by the points A_0, A_1, A_2, A_3, A_4 on the wheel tread will be distorted. Then it can be seen in Figure 3.18 that the points A_0, A_1, A_2, A_3, A_4 will be located on the points $A'_0, A'_1, A'_2, A'_3, A'_4$ during the wheel rolling. The angle between the path line a – a connected by those points and the line c – c of intersection of the wheel plane with the road surface will be formed. It usually is referenced as slip angle α, and the wheel rolls along the direction of a – a.

The value of the slip angle is related to that of the cornering forces. Experiments indicate that if the slip angle is not more than $4 \sim 5$, the lateral force increases linearly with the slip angle. When the vehicle is traveling normally, the lateral acceleration is not more than 0.4g, and the slip angle is less than $4 \sim 5°$. Subsequently, the relationship between the cornering force and the slip angle in practice can be shown as

$$F_y = K_y \cdot \alpha \tag{3.36}$$

K_y, which can be referenced to the slip stiffness of the tire, is the slope of the $F_y \sim \alpha$ curve at $\alpha = 0$. According to the right-hand rule, the positive cornering force generates the negative slip angle. The tire slip stiffness of the passenger car usually ranges from $-28,000$ to $-80,000$ N/rad, and it is an important parameter of automobile stability. The absolute value of the tire slip stiffness should be large to keep better manipulating stability. With the large cornering force, the slip angle increases at the higher rate. Partial sideslip has occurred on the tire-ground contact patch. When the cornering force comes up to the ultimate adhesion, the whole tire will begin to sideslip. Therefore, the maximum value of the cornering force is determined by the adhesion conditions. Generally, the larger the tire slip force is, the better the automobile limit performance.

Size, modal, and structure parameters of the tire have a great effect on the slip stiffness. A large-size and low-profile tire has high slip stiffness. Due to the wide contact patch of the radial ply tire, its slip stiffness also is very high. Moreover, the normal load on the tire strongly influences the cornering characteristics. The cornering force generally increases with an increase in the normal load. However, it will decrease if the perpendicular loads are too large, which makes the pressure of the ground-tire contact an inhomogeneous distribution. Inflation pressure also has a moderate effect on the cornering properties of a tire. In general, the cornering stiffness of a tire, which increases with increased inflation pressure, does not change when the tire pressure reaches a certain high level. However, the velocity of the vehicle has little influence on the slip stiffness.

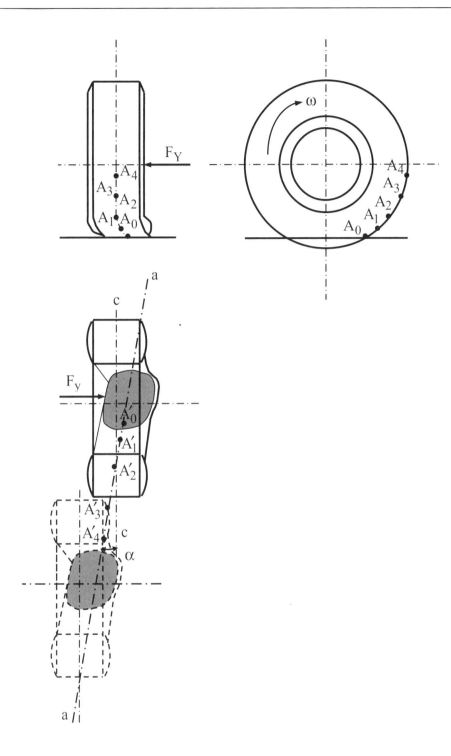

Figure 3.18 Cornering characteristics of a rolling tire.

When the tire is rolling, there is distance and a slip angle between the patch line and the wheel plane. The forepart of the patch is close to the wheel plane, which results in small lateral deformation; the rearward is far from it, which results in large lateral deformation. The patch distribution force is consistent with the deformation (Figure 3.19). When the tire is under cornering, the slip force is not added to the center of the patch. Also, the distance, which is called the pneumatic trail, is multiplied with the slip forces to form a torque (or couple), which tends to align the wheel plane with the direction

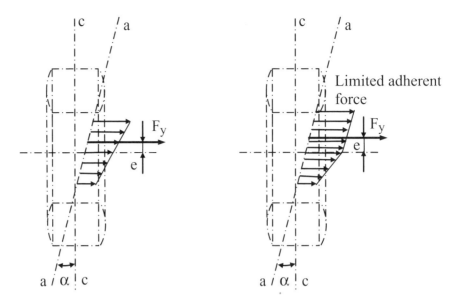

Figure 3.19 Distribution of the lateral ground force in the contact patch.

of motion. This torque is called the aligning or self-aligning torque and is one of the primary restoring moments that help the steered tire return to the original position after negotiating a turn.

The rearward of the contact patch will reach the ultimate adhesion if the cornering force is increased to a certain level. The ultimate adhesion area increases with the increased cornering force. A tire will slide if the area comes up to the whole contact patch area. The aligning torque first increases with an increase of the slip angle. It reaches a maximum at the slip angle of 4 ~ 6° and then decreases with a further increase. It reaches zero at the slip angle of 10 ~ 16° and decreases to a negative value with a further increase in the slip angle. The pneumatic trail has the same trend as the aligning torque. It first increases with an increase in the cornering forces and then decreases with a further increase.

At the same slip angle, a large tire always has high aligning torque. Also, the aligning torque of radial ply tires is higher than that of bias ply tires. Lower inflation pressure results in longer tire contact length and hence pneumatic trail. This causes an increase in the aligning torque.

Longitudinal force affects the aligning torque significantly. When the tire is exposed to a longitudinal force, it will spend some adhesion, which decreases the adhesion that the slide forces can use. Generally, the effect of a driving torque is to increase the aligning torque for a given slip angle. It reaches a maximum at a particular value and then decreases with a further increase. The braking torque has the opposite effect. Figure 3.20 shows the variation of aligning torque with the longitudinal force.

3.4.2 Mathematical Model of the Tire Cornering Characteristic

The tire cornering characteristic has a great effect on the maneuverability, stability, and safety of an automobile and is affected by many factors. Because it is difficult to conduct research in which the only factor on which to depend is tire performance, it is necessary to build a mathematical model to pursue a theoretical model. The purpose of

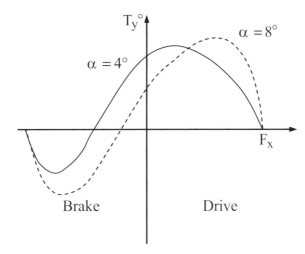

Figure 3.20 Variation of the aligning torque with the longitudinal force.

building the mathematical model is to acquire the mathematical relationship between the tire structure factor and the cornering characteristic. The second purpose is to provide a useful reference for the automobile design and automobile dynamics research.

3.4.2.1 Simplified Mathematical Model of the Tire Cornering Characteristic

The mathematical model is based on the following assumptions:

1.   The tire case is rigid, and the whole tire elasticity is applied on the layer of tread.

2.   The effect of the longitudinal force and skid is ignored. The tire can roll freely.

3.   The tire slip angle is zero.

4.   The friction coefficient between the patch and the road is constant.

According to these assumptions, we can draw a picture of the tire force situation as follows. When a tire rolls at the given slip angle α, we build a coordinate system on the front point of the patch. Thus, the whole patch length is 2a. With the condition of the side force, the tire lateral deformation increases with the increase of x and forms an angle α between the patch line and the tire center.

Distribution of the tire vertical load: Experiments indicate that the distribution of vertical load is not symmetrical at the patch length (Figure 3.21). It can be expressed as

$$q_z = \frac{F_z}{2a}\eta(\mu) \tag{3.37}$$

where $\mu = \frac{x}{a}$ and $\eta(\mu)$ is the load distribution function. It has two constraints:

$$\eta(\mu) = 0 \qquad \mu < 0 \ \text{ or } \ \mu > 2 \tag{3.38}$$

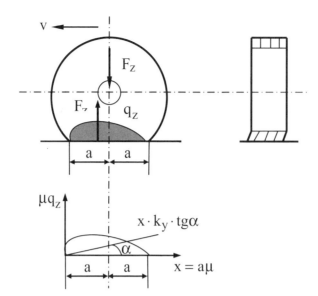

Figure 3.21 Vertical forces acting on the rolling tire within the contact patch.

According to the balance of force and moments in the z direction, we can get the equations as

$$
\begin{cases}
F_z = \displaystyle\int_0^{2a} q_z dx = \int_0^2 a q_z d\mu \\
F_z \cdot (a - \Delta) = \displaystyle\int_0^{2a} x q_z dx = \int_0^2 \mu a^2 q_z d\mu
\end{cases}
\tag{3.39}
$$

where Δ is defined as the distance between normal force F_z and the wheel centerline.

Substituting Eqs. 3.36 and 3.38,

$$
\begin{cases}
F_z = \displaystyle\int_0^2 \left(a \cdot \frac{F_z}{2a} \right) \cdot \eta(\mu) d\mu \\
F_z \cdot (a - \Delta) = \displaystyle\int_0^2 \left(a^2 \mu \cdot \frac{F_z}{2a} \right) \cdot \eta(\mu) d\mu
\end{cases}
\tag{3.40}
$$

After rearranging the terms, we can get

$$
\begin{cases}
\displaystyle\int_0^2 \eta(\mu) d\mu = 2 \\
\displaystyle\int_0^2 \mu \eta(\mu) d\mu = 2\left(1 - \frac{\Delta}{a} \right) = 2(1 - \theta) \qquad \theta = \frac{\Delta}{a}
\end{cases}
\tag{3.41}
$$

Tire cornering force and aligning torque: When the tire is rolling, the lateral deformation of the patch is proportional to the value of tgα. Therefore, the lateral load distribution is

$$q_y = k_y \cdot x \cdot tg\alpha \tag{3.42}$$

where k_y is the lateral distribution stiffness of the rubber layer, which is defined as a constant. Therefore, the equation of the cornering force and the aligning torque can be written as

$$\begin{aligned}
F_y &= \int_0^{2a} q_y dx \\
&= \int_0^2 a^2\mu \cdot k_y \cdot tg\alpha d\mu \\
&= 2a^2 k_y \cdot tg\alpha
\end{aligned} \tag{3.43}$$

$$\begin{aligned}
T_y &= \int_{-a}^a x \cdot q_y dx \\
&= \int_{-a}^a k_y \cdot tg\alpha \cdot x^2 dx \\
&= \frac{2}{3}a^3 k_y \cdot tg\alpha \\
&= \frac{a}{3}F_y
\end{aligned} \tag{3.44}$$

If we define $K_y = 2ak_y$ and substitute it in Eq. 3.43, we obtain the equation $F_y = K_y \cdot tg\alpha$. Compared with the experiment results, when the value of α is small, $tg\alpha \approx \alpha$, and $F_y = K_y \cdot \alpha$. Note that K_y is the slip stiffness of the tire.

If we define $e = \dfrac{a}{3}$ and substitute it in Eq. 3.44, we get the equation $T_y = F_y \cdot e$, where e is the tire trail.

Define the tire cornering force and the aligning torque when partial sideslip occurs on the tire-ground contact patch: With the large cornering force and the slip angle, experiments show that partial sideslip has occurred on the tire, because $q_y \geq \varphi \cdot q_z$ at the rearward patch. φ is the friction coefficient, and the origin position can be defined by $q_y = \varphi \cdot q_z$.

If we substitute Eqs. 3.37 and 3.43, we obtain

$$k_y \cdot x \cdot tg\alpha = \varphi \cdot \frac{F_z}{2a} \eta(\mu) \tag{3.45}$$

If we substitute the equations $x = \mu \cdot a$ and $K_y = 2ak_y$, we get

$$K_y \mu \cdot tg\alpha = \varphi \cdot F_z \eta(\mu) \tag{3.46}$$

The skid ratio can be defined as

$$\phi_y = \frac{K_y \cdot tg\alpha}{\varphi \cdot F_z} \tag{3.47}$$

Then

$$\phi_y = \frac{\eta(\mu')}{\mu'} \tag{3.48}$$

where μ' is referenced as the original position of skidding, and the value of it at the different slip angle or slip ratio can be acquired. The whole contact patch can be divided into two parts. First, $0 \leq \mu' \leq \mu$ is the part of adhesion, and its side force distribution is $q_y = k_y \cdot x \cdot tg\alpha$. Second, $\mu \leq \mu' \leq 2$ is the part of skidding, and its side force distribution is $q_z = \frac{F_z}{2a}\eta(\mu)$. The total side force can be expressed as

$$\begin{aligned}
F_y &= \int_0^{\mu'} a^2 \cdot k_y \cdot tg\alpha \cdot \mu \cdot d\mu + \int_{\mu'}^{2} a \cdot \varphi \cdot \frac{F_z}{2a} \cdot \eta(\mu) \cdot d\mu \\
&= \frac{\mu'^2}{4} \cdot K_y \cdot tg\alpha + \frac{\varphi \cdot F_z}{2} \int_{\mu'}^{2} \eta(\mu) d\mu \\
&= \frac{\mu'^2}{4} \phi_y \varphi \cdot F_z + \frac{\varphi \cdot F_z}{2} \left[\int_0^2 \eta(\mu) d\mu - \int_0^{\mu'} \eta(\mu) d\mu \right] \\
&= \left(\frac{\varphi \cdot F_z}{4} \right) \cdot \eta(\mu') \cdot \mu' + \frac{\varphi \cdot F_z}{2} \left[\int_0^2 \eta(\mu) d\mu - \int_0^{\mu'} \eta(\mu) d\mu \right]
\end{aligned} \tag{3.49}$$

The zero order and first order moment of $\eta(\mu)$ at the range of $0 \sim \mu$ can be defined as

$$m_0(\mu) = \int_0^{\mu} \eta(\mu) d\mu \tag{3.50}$$

$$m_1(\mu) = \int_0^\mu \mu \cdot \eta(\mu)d\mu \tag{3.51}$$

When $\mu = 2$,

$$m_0(2) = 2 \quad \text{and} \quad m_1(2) = 2(1-\theta) \tag{3.52}$$

The total cornering force with no dimension can be defined as

$$\frac{F_y}{\varphi \cdot F_z} = \begin{cases} \dfrac{\phi_y}{4} \times \mu'^2 + 1 - \dfrac{m_0(\mu')}{2} & 0 \leq \mu' < 2 \\ \phi_y & \mu' = 2 \end{cases} \tag{3.53}$$

According to Eq. 3.53, the nondimensional cornering force $\dfrac{F_y}{\varphi \cdot F_z}$ is a single variable function.

The total aligning torque is

$$\begin{aligned} T_z &= \int_0^{\mu'} a^3 \cdot tg\alpha \cdot k_y \cdot \mu^2 d\mu + \int_{\mu'}^2 a \cdot \mu \left[\varphi \cdot \frac{F_z}{2} \right] \cdot \eta(\mu)d\mu - F_y \cdot a \\ &= \frac{a}{6} K_y \cdot tg\alpha \cdot \mu'^3 + a \cdot \varphi \cdot F_z \cdot \left(1 - \theta - \frac{m_1(\mu')}{2} \right) - F_y \cdot a \end{aligned} \tag{3.54}$$

The nondimensional total aligning torque can be defined as

$$\frac{T_z}{\varphi \cdot F_z \cdot a} = \begin{cases} \dfrac{\mu'^2}{2}\left(\dfrac{\mu'}{3} - \dfrac{1}{2} \right) \times \phi_y - \theta + \dfrac{m_0(\mu') - m_1(\mu')}{2} & 0 \leq \mu' < 2 \\ \dfrac{\phi_y}{3} & \mu' = 2 \end{cases} \tag{3.55}$$

With the condition of vertical load distribution function, the procedure of acquiring nondimensional cornering characteristics is given as follows:

1. Given the nondimensional side skidding ratio, the original position of skidding can be obtained according to Eq. 3.48.

2. With the load distribution function, the value of $m_0(\mu)$ and $m_1(\mu)$ at $\mu = \mu'$ can be calculated by Eqs. 3.50 and 3.51. Then calculate the value of θ by Eq. 3.41.

3. Substitute the value of $\phi_y, \mu', m_0(\mu), m_1(\mu), \theta$ into Eqs. 3.53 and 3.55 to obtain the nondimensional cornering force $\dfrac{F_y}{\varphi \cdot F_z}$ and the aligning torque $\dfrac{T_z}{\varphi \cdot F_z \cdot a}$.

This kind of tire model is a rigid case and flexible tread. Its nondimensional cornering characteristic (including $\dfrac{F_y}{\varphi \cdot F_z} \sim \phi_y$ and $\dfrac{T_z}{\varphi \cdot F_z \cdot a} \sim \phi_y$) is related only to the vertical load distribution function. If the $\eta(\mu)$ function does not change when the vertical load F_z changes, the result of $\dfrac{F_y}{\varphi \cdot F_z} \sim \phi_y$ and $\dfrac{T_z}{\varphi \cdot F_z \cdot a} \sim \phi_y$ at different F_z can be drawn by only one curve.

The results of experiments indicate that the vertical load distribution function $\eta(\mu)$ always changes with the load F_z. When the load is low, the middle results of the $\eta(\mu)$ curve are large, and it trends to two sides of coordination with the increase in F_z (Figure 3.22).

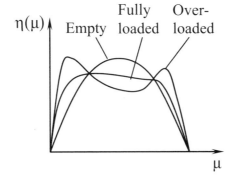

Figure 3.22 Variation of load distribution function with load and μ.

3.4.2.2 Cornering Characteristic with Lateral Bending Deformation of the Tire Case

The cornering characteristic with lateral bending deformation of the tire case is closer to reality. Consider the deformation of the tire case and tread with the function of cornering forces. Figure 3.23 shows the relationship between the contact patch and the wheel centric plane.

OX is the wheel centric plane with no cornering force. With the cornering force, the centric plane will move from line OX to line AA, and the tire case deforms to the line AEC. The contact patch rolls from point A along the direction of line APB. The angle between line APB and line AA is the slip angle α. T represents time , and point A rolls to point P. The distance between points P and E is the tread deformation y_r, and the distance between points E and X is the tire case deformation. It is composed of two parts—lateral deformation y_0 and lateral bending deformation y_b,—and they are related to the cornering forces

$$\begin{cases} y_0 = \dfrac{F_y}{k_{y0}} \\ y_b = F_y \cdot \dfrac{\xi(\mu)}{k_{yb}} \end{cases} \tag{3.56}$$

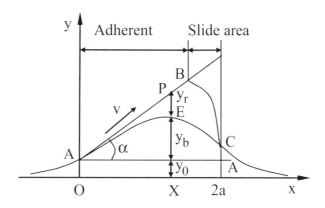

Figure 3.23 Relationship between the contact patch and the wheel centric plane.

where k_{y0} and k_{yb} are defined as two constants of the tire case deformation. $\mu = \dfrac{x}{a}$ is defined as the relative coordinate to point P, and $\xi(\mu)$ is the shape function of the tire case deformation.

We can easily obtain the following equation from Figure 3.23

$$y_r + y_b = x \cdot tg\alpha \qquad (3.57)$$

Respectively, we can define the slip ratio of the tire tread and case as

$$S_{yr} = \frac{y_r}{x}$$

and

$$S_{yb} = \frac{y_b}{x}$$

Then

$$y_r = a\mu \cdot tg\alpha - a\mu \cdot S_{yr} \qquad (3.58)$$

$$S_{yb} = F_y \cdot \frac{\xi(\mu)}{k_{yb} \cdot a\mu} \qquad (3.59)$$

Subsequently, the lateral shear stress distribution on the whole patch can be written as

$$q_y = k_{yr} \cdot y_r = k_{yr}\left(a \cdot \mu \cdot tg\alpha - q \cdot \mu \cdot S_{yb}\right) \qquad (3.60)$$

where k_{yr} is the lateral stiffness of the tire tread.

With a small slip angle α, there is no side skidding on the whole patch. Thus, the total cornering force can be expressed as

$$F_y = \int_0^{2a} q_r \cdot dx = \int_0^2 a^2 \mu \cdot k_{yr} \cdot tg\alpha \cdot d\mu - \int_0^2 a^2 \mu \cdot k_{yr} \cdot \frac{F_y}{k_{yb} \cdot a\mu} \xi(\mu) \cdot d\mu \quad (3.61)$$

After integrating

$$F_y = 2a^2 \cdot k_{yr} \cdot tg\alpha - a\frac{k_{yr}}{k_{yb}} F_y \int_0^2 \xi(\mu) d\mu \quad (3.62)$$

We can define the zero order moment of the relative deformation of the tire case as

$$D(\mu) = \left(a \cdot \frac{k_{yr}}{k_{yb}} \right) \int_0^u \xi(\mu) d\mu \quad (3.63)$$

When $\mu = 2$,

$$D(2) = 2a\frac{k_{yr}}{k_{yb}}$$

Not considering the deformation of the tire case, $K_{yr} = K_y = 2\alpha^2 k_{yr}$ leads to $\xi(\mu) = 0$, and the tire cornering force can be expressed as

$$F_y = 2a^2 k_{yr} \cdot tg\alpha$$
$$= K_{yr} \cdot tg\alpha \quad (3.64)$$

Considering the tire deformation, then

$$F_y = K_{yr} \cdot tg\alpha - F_y \cdot D(2)$$
$$= K_{yr} \cdot \frac{tg\alpha}{[1 + D(2)]} \quad (3.65)$$
$$= K_y \cdot tg\alpha$$

where $K_y = \frac{K_{yr}}{[1 + D(2)]}$, which includes the rigidity of the tire case and tread. Therefore, it is referenced as a general slip stiffness.

If we define $K_{yb} = a \cdot k_{yb}$, then we can obtain the following equation as

$$K_y = \frac{K_{yr} \cdot K_{yb}}{K_{yr} + K_{yb}} \quad (3.66)$$

General slip stiffness is similar to two springs in series connected by the slip stiffness of the tire case and tread. Due to the elasticity of the tire case, the whole tire slip stiffness decreases $\left[1 + D(2)\right]$ times.

With a small slip angle, the aligning torque is expressed as

$$
\begin{aligned}
T_z &= \int_0^2 a^2 \cdot q_y \cdot \mu d\mu - F_y \cdot a \\
&= \int_0^2 a^3 \cdot \mu^2 \cdot k_{yr} \left[tg\alpha - \frac{F_y}{(a \cdot k_{yb})} \cdot \frac{\xi(\mu)}{\mu} \right] d\mu - F_y \cdot a
\end{aligned}
\tag{3.67}
$$

Define the first order moment of the relative deformation function of the tire case as

$$
D_1(\mu) = \left(a \cdot \frac{k_{yr}}{k_{yb}} \right) \int_0^u \xi(\mu) \cdot \mu d\mu
\tag{3.68}
$$

Finally, the aligning torque is

$$
T_z = F_y \cdot a \left[\frac{1}{3} + \frac{4D(2)}{3} - D_1(2) \right]
\tag{3.69}
$$

according to the cornering properties of the tire model that considers lateral elasticity of the tire case. The main factors that have a great effect on the tire cornering properties are as follows:

1. **Tire load:** The slip stiffness of the tire tread is proportional to the load, while the slip stiffness of the tire case is opposite. Subsequently with the increase of load, K_{yr} increases and K_{yb} decreases, which results in a sharp increase of $D(2)$, and the general slip stiffness decreases rapidly.

2. **Tire abrasion:** When the tire begins to abrade, the shear stiffness of the tread increases, which increases the general slip stiffness. When the tire tread has abraded by half, the general slip stiffness will increase by 20%.

3. **Tire structure:** A passenger car tire usually has a few cord layers, and its inflation pressure is low, which results in the small stiffness of the passenger car tire case. Thus, the slip stiffness proportion $D(2) = \dfrac{K_{yr}}{K_{yb}}$ of the passenger car tire is larger than that of the truck. As a result, the tire slip stiffness of the passenger car obviously is affected by the variation of the load. When the rear axle is overloaded, it is easy for a passenger car to oversteer, with its tire trail also large. Hence, it is better for a passenger car to use a small kingpin caster angle, especially the negative angle. The slip stiffness of a truck tire case is high, and its general slip stiffness is not sensitive to the variation of load.

3.4.3 Rolling Properties of Tires

There is an angle between the wheel plane and the line perpendicular to the ground plane in the modern vehicle, which is called the camber angle of the tire. If the upper part of the wheel inclines outside the car body, the angle is positive. The front wheels of the car usually contain a tiny positive camber angle $\gamma = 5' \sim 10'$ with zero load, which makes the tire perpendicular to the ground surface with a little vaulting and abrades uniformly with the decrease in rolling resistance. The angle turns out to be slightly negative under full load.

A free-rolling tire with a camber angle would revolve about point O, as shown in Figure 3.24. However, the cambered tire in a vehicle is constrained to move in a straight line. There must be a side force F_{y0} acting on the tire center. Therefore, a lateral force in the direction of the camber is developed at the contact point, which is referred to as the camber thrust F_{yr}. The relationship of the camber thrust with the camber angle is linear and is defined as

$$F_{yr} = K_r \cdot \gamma$$

(3.70)

where K_r is defined as the camber stiffness. Note that a minus camber angle generates a positive camber thrust.

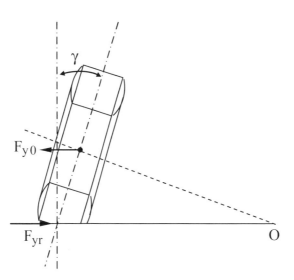

Figure 3.24 Side force of tires with a camber angle.

The relationship between camber thrust and camber angle (at zero slip angle) is illustrated in Figure 3.25. Three lines represent the cornering properties under a little slip angle corresponding to a camber angle that is positive, zero, and negative, respectively.

1. The ground lateral force with a zero slip angle is the camber thrust.

2. When there is a camber angle, the relationship between the cornering force and the camber angle as well as the cornering angle is given as

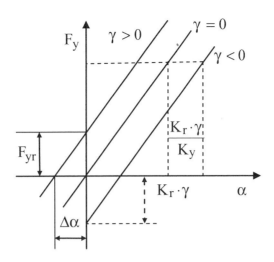

Figure 3.25 The relationship between camber thrust and camber angle (at zero slip angle).

$$F_y = K_y\alpha + K_r\gamma \qquad (3.71)$$

where K_y is the cornering stiffness of tires without inclination, K_r is the camber stiffness of the tires, α is the slip angle of the tires without inclination, and γ is the camber angle of the tires.

Note that with the increase in the camber angle, the contact state of the tire and the road surface would be worse, which would influence the maximum ground lateral force and degrade the ultimate performance of the vehicle. Therefore, ensure that the front wheel is perpendicular to the ground plane as much as possible under the steering driving condition.

According to the tire axis system, the positive slip angle corresponds to the negative cornering force and positive self-aligning moment. Also, positive camber angle corresponds to negative camber thrust and camber self-aligning moment.

3.4.3.1 Cambered Tire Models

In a simplified cambered tire model, three assumptions are as follows:

1. The whole elasticity of the tire is concentrated on the tire tread, without considering the elasticity of the tire carcass, and the tire is equivalent to a brush.

2. The width of the tire is neglected.

3. The form of the contact patch is a rectangle.

The broken line shown in Figure 3.26 represents the deformed part of the tire. A coordinate system Oxz is located at the left side of the contact patch, with a length of 2a. The vertical deformation of the tire is a function of x, which can be viewed as a quadratic function

$$\Delta z = \Delta z_m \mu (2 - \mu)$$

$$= \frac{a^2}{2R} \mu (2 - \mu) \qquad (3.72)$$

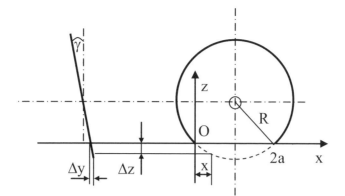

Figure 3.26 Vertical deformation of rolling tires with the camber angle in the contact patch.

where $\Delta z = \dfrac{a^2}{2R}$ represents the maximum deformation in the middle of the contact patch. Corresponding to x, the lateral deformation is given as

$$\Delta y = \Delta z \cdot tg\gamma$$

$$= \frac{a^2}{2R}\mu(2-\mu)\cdot tg\gamma \tag{3.73}$$

If it is assumed that the lateral distributed stiffness per length of tire tread is k_y, then the lateral stress at the site x can be drfined by

$$q_y = k_y \cdot \Delta y$$

$$= k_y a^2 \mu \cdot (2-\mu) \cdot \frac{tg\gamma}{2R} \tag{3.74}$$

Therefore, the lateral force with camber is written as

$$F_{yr} = \left(k_y \frac{a^2}{2R}\right) \cdot tg\gamma \int_0^{2a} \mu(2-\mu)dx$$

$$= \left(2k_y \frac{a^3}{3R}\right) \cdot tg\gamma \tag{3.75}$$

If we set the cornering stiffness of the tire as $K_y = 2a^2 k_y$, the lateral force with camber can be given by

$$F_{yr} = \left(\frac{a}{3R}\right) \cdot K_y \cdot tg\gamma = K_r \cdot tg\gamma \tag{3.76}$$

where $K_r = \left(\dfrac{a}{3R}\right)K_y$ represents the camber stiffness. The camber stiffness is proportional to the cornering stiffness, and the contact patch is inversely proportional to the radius of the tire.

3.4.3.2 Cambered Tire Model with Roll Elastic Deformation of the Tire Carcass

With the consideration of roll elastic deformation of the tire in the modeling of a cambered tire, there will be an angle, called the roll angle γ_c, that is formed between the tire carcass and felloe under the action of lateral force with the camber. In particular, for a radial ply tire, the influence of the roll of the tire cannot be neglected due to the large elasticity of the tire carcass.

With the roll force acting on the tire, the relationship between the roll force and the roll angle of the tire carcass and felloe is expressed as

$$tg\gamma_c = \frac{F_{yr}}{K_c} \tag{3.77}$$

where K_c is the camber stiffness of the tire carcass.

A part of the lateral deformation of the tire tread is absorbed by the roll of the tire carcass, and then the new deformation is given by

$$\Delta y' = \Delta y - \Delta z \cdot tg\gamma_c \tag{3.78}$$

According to Eq. 3.75, the roll force can be obtained by

$$F_{yr} = \left(\frac{a}{3R}\right)K_y \cdot \left(tg\gamma - \frac{F_{yr}}{K_c}\right) \tag{3.79}$$

Finally, the roll force of the tire owing to the roll of the tire can be gained as

$$F_{yr} = K_r' \cdot tg\gamma \tag{3.80}$$

where

$$K_r' = \frac{K_r \cdot K_c}{K_r + K_c} \tag{3.81}$$

Equation 3.81 indicates that the camber stiffness of the tire is formed exactly in series with the camber stiffness of the tire tread K_r and the camber stiffness of the tire carcass K_c. The camber stiffness of the tire carcass of the radial ply tire is less than the counterpart of the bias ply tire, so the former obtains a smaller camber stiffness. With the same camber angle, there should be a less toe-in angle for the former.

3.5 Tire Mechanics Model Considering Longitudinal Slip and Cornering Characteristics

Longitudinal force and lateral force that act on the wheel when the vehicle is in motion result from many factors, such as tire toe-in, camber, road surface roughness, lateral wind drag, and the steering of the wheel. Thus, in most cases, each tire moves under a combined slip condition. Thus, to establish a tire model under the condition of combined slip, it is very important to analyze the vehicle dynamics. Several main longitudinal slip and cornering characteristics of tire models are introduced as follows.

3.5.1 C.G. Gim Theoretical Model

In view of the difference between the cases of braking and driving, first we define the longitudinal and lateral slip ratio as

$$S_s = \begin{cases} \dfrac{v_x - v_c}{v_x} > 0 & \text{brake} \\ \dfrac{v_c - v_x}{v_c} < 0 & \text{drive} \end{cases} \tag{3.82}$$

$$S_\alpha = \begin{cases} |\tan\alpha| & \text{brake} \\ (1 - |S_s|) \cdot |\tan\alpha| & \text{drive} \end{cases} \tag{3.83}$$

where

v_c = circumference speed of the wheel

v_x = longitudinal velocity of the vehicle

v_y = lateral velocity of the vehicle

α = slip angle expressed as $\alpha = \text{arctg}\left(\dfrac{v_y}{v_x}\right)$

Due to the influence of longitudinal and lateral forces on the tire, the adhesion coefficient is defined as

$$\varphi = \varphi_0 \left(1 - A \cdot S_{sa}\right) \tag{3.84}$$

where

$$A = \dfrac{\left(1 - \dfrac{\varphi_1}{\varphi_0}\right)}{S_1} \tag{3.85}$$

$$S_{sa} = \sqrt{S_s^2 + S_a^2} \tag{3.86}$$

φ_0 is the static friction coefficient between the tire and the road surface, and S_{sa} is the associated parameter of the slip ratio.

The longitudinal, lateral, and comprehensive adhesion coefficients of the tire are given by

$$\varphi_x = \varphi \cdot \dfrac{S_s}{S_{sa}} = \varphi \cdot \cos\alpha \tag{3.87}$$

$$\varphi_y = \varphi \cdot \frac{S_a}{S_{sa}} = \varphi \cdot \sin\alpha \tag{3.88}$$

$$\varphi = \sqrt{\varphi_x^2 + \varphi_y^2} \tag{3.89}$$

Rolling and the slip critical point in the contact patch is

$$S_n = \frac{1}{3\varphi F_z}\sqrt{(K_s S_s)^2 + (K_a S_a)^2} \tag{3.90}$$

The slip critical point is

$$S_{sc} = \frac{3\varphi F_z}{K_s} \tag{3.91}$$

The cornering critical point is

$$S_{ac} = \frac{K_s}{K_a}\sqrt{S_{sc}^2 - S_s^2} \tag{3.92}$$

where

K_s = tire longitudinal stiffness

K_a = tire lateral stiffness

Set

$$\ell_n = 1 - S_n$$

Thus, the longitudinal force between the tire and road surface is expressed as

$$F_x = \begin{cases} K_s S_s \ell_n^2 + \varphi_x F_z\left(1 - 3\ell_n^2 + 2\ell_n^3\right) & S_s < S_{sc} \\ \varphi_x F_z & S_s \geq S_{sc} \end{cases} \tag{3.93}$$

The lateral force between the tire and the road surface is written as

$$F_y = \begin{cases} K_a S_a \ell_n^2 + \varphi_y F_z\left(1 - 3\ell_n^2 + 2\ell_n^3\right) & S_a < S_{ac} \\ \varphi_y F_z & S_a \geq S_{ac} \end{cases} \tag{3.94}$$

The self-aligning moment can be formulated as

$$T_Z = \begin{cases} \left[K_a S_a \left(-\dfrac{1}{2} + \dfrac{2}{3} \ell_n \right) + \dfrac{3}{2} \varphi_y F_z S_n^2 \right] \cdot 1 \cdot \ell_n^2 & S_a < S_{ac} \\ 0 & S_a \geq S_{ac} \end{cases} \qquad (3.95)$$

where ℓ is the contact length of the tire.

The parameters required by this model are specific and can be obtained by experiment, without fitting a large quantity of experimental data of the mechanical characteristics. The model is of high precision in the fitting of the cornering and longitudinal forces. However, due to using a simple physical mechanism in the tire model, there is a big difference between the theoretical calculation of the self-aligning moment and the testing data. The self-aligning moment derived from the model will be zero with a large slip angle, which cannot reflect the fact that there will be a negative self-aligning moment under a large slip angle.

3.5.2 K.H. Guo Tire Model

3.5.2.1 Steady-State Simplified Theoretical Tire Model

By assuming that elastic translation can take place in only the y direction on the carcass and angle about the z axis, and the displacement along the x axis can be neglected, the deformation of the tire contact patch under the condition of braking (or driving) and cornering is shown in Figure 3.27, where α is the slip angle, ABC is the tire centerline after cornering, and the total length of the tire contact patch is 2a. The translation deformation of the tire carcass resulting from the influence of the cornering force F_y is

$$y_b = \frac{F_y}{K_{by}} \qquad (3.96)$$

where

$$K_{by} = \text{lateral carcass stiffness}$$

$$R = \text{tire rolling radius}$$

$$\omega = \text{angular speed of the tire}$$

and the initial contact point A on the tire tread is rolling to point P through a time t. Thus, the counterpart on the tire plane turns from point O to point X.

$$x = \omega R t \qquad (3.97)$$

The deformations of the tire tread in the contact patch along the x and y directions can be written as

$$\begin{cases} \Delta x = vt \cos \alpha - x \\ \Delta y = vt \sin \alpha \end{cases} \qquad (3.98)$$

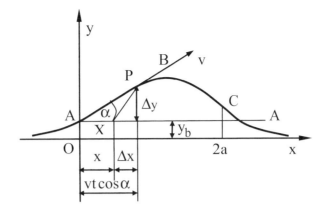

Figure 3.27 Deformation of the tire contact patch under the condition of braking (or driving) and cornering.

We can define the braking slip ratio S_b and the driving slip ratio S_d as

$$\begin{cases} S_b = \dfrac{\Delta x}{vt\cos\alpha} \\ S_d = -\dfrac{\Delta x}{x} \end{cases} \tag{3.99}$$

Hence,

$$\begin{cases} \Delta x = \dfrac{S_b x}{\left(1 - S_b\right)} = S_x x \\ \Delta y = \dfrac{x \cdot tg\alpha}{\left(1 - S_b\right)} = S_y x \end{cases} \tag{3.100}$$

We can define the longitudinal slip ratio and the cornering slip ratio as

$$\begin{cases} S_x = \dfrac{S_b}{\left(1 - S_b\right)} = -S_d \\ S_y = \dfrac{tg\alpha}{\left(1 - S_b\right)} = \left(1 + S_x\right) \cdot tg\alpha \end{cases} \tag{3.101}$$

Assuming that the stiffnesses of the tire tread in the x and y directions are k_x and k_y, respectively, the shear stresses in the x and y directions at point P in the contact patch correspondingly are expressed as, respectively,

$$\begin{cases} q_x = k_x \cdot \Delta x = k_x S_x \cdot x \\ q_y = k_y \cdot \Delta y = k_y S_y \cdot x \end{cases} \tag{3.102}$$

Tire without slip: The resultant shear stress in the contact patch is given by

$$\begin{aligned} q &= \sqrt{q_x^2 + q_y^2} \\ &= \sqrt{\left(k_x S_x\right)^2 + \left(k_y S_y\right)^2} \cdot x \end{aligned} \tag{3.103}$$

Define the direction of the resultant shear stress as

$$tg\theta = \frac{q_y}{q_x} = \frac{k_y S_y}{\left(k_x S_x\right)} \tag{3.104}$$

If we integrate the shear stress over the contact patch, we can obtain the total longitudinal force and cornering force as

$$\begin{cases} F_x = 2a^2 k_x \cdot S_x = K_x \cdot S_x \\ F_y = 2a^2 k_y \cdot S_y = K_y \cdot S_y \end{cases} \tag{3.105}$$

where the longitudinal slip stiffness is defined as $K_x = 2ak_x$, and the cornering stiffness is defined as $K_y = 2ak_y$. The resultant tangential force in the contact patch can be described as

$$\begin{aligned} F &= \sqrt{F_x^2 + F_y^2} \\ &= \sqrt{\left(K_x S_x\right)^2 + \left(K_y S_y\right)^2} \end{aligned} \tag{3.106}$$

The nondimensional longitudinal force $\overline{F}_x$, the lateral force $\overline{F}_y$, and the relative resultant tangential force $\overline{F}$ are defined, respectively, as

$$\begin{cases} \overline{F}_x = \dfrac{F_x}{\left(\varphi \cdot F_z\right)} \\[2mm] \overline{F}_y = \dfrac{F_y}{\left(\varphi \cdot F_z\right)} \\[2mm] \overline{F} = \dfrac{F}{\left(\varphi \cdot F_z\right)} \end{cases} \tag{3.107}$$

We can define the nondimensional longitudinal slip ratio ϕ_x, the lateral slip ratio ϕ_y, and the relative resultant slip ratio ϕ as

$$\begin{cases} \phi_x = \dfrac{K_x S_x}{\left(\varphi \cdot F_z\right)} \\[2mm] \phi_y = \dfrac{K_y S_y}{\left(\varphi \cdot F_z\right)} \\[2mm] \phi = \sqrt{\phi_x^2 + \phi_y^2} \end{cases} \tag{3.108}$$

where F_z is the tire vertical load, and φ is the friction coefficient between the tire and road surface. Then the nondimensional formulas of the tire tangential forces in respective directions under the condition without slip can be obtained as

$$\begin{cases} \overline{F}_x = \phi_x = \overline{F} \cdot \dfrac{\phi_x}{\phi} \\[2mm] \overline{F}_y = \phi_y = \overline{F} \cdot \dfrac{\phi_y}{\phi} \\[2mm] \overline{F} = \sqrt{\overline{F}_x^2 + \overline{F}_y^2} = \Phi \\[2mm] tg\theta = \dfrac{\phi_x}{\phi_y} \end{cases} \tag{3.109}$$

Assume that the nondimensional relative shear stress and the nondimensional coordinates are

$$\begin{cases} \overline{q} = \dfrac{2aq}{(\varphi \cdot F_z)} \\[3mm] u = \dfrac{x}{a} \end{cases} \tag{3.110}$$

Thus, under conditions without slip,

$$\begin{cases} q_x = \phi_x \cdot u \cdot \dfrac{\varphi \cdot F_z}{(2a)} \\[3mm] q_y = \phi_y \cdot u \cdot \dfrac{\varphi \cdot F_z}{(2a)} \\[3mm] q = \phi \cdot u \cdot \dfrac{\varphi \cdot F_z}{(2a)} \end{cases} \tag{3.111}$$

Therefore, the aligning moment can be obtained according to the shear stress in the contact patch as

$$T_z = -F_x \cdot \left(y_b + \frac{4a}{3} S_y \right) + F_y \cdot \frac{a}{3} = F_y \cdot D_x - F_x \cdot \left(D_y + y_b \right) \tag{3.112}$$

where the longitudinal trail is $D_x = \dfrac{a}{3}$, and the lateral trail is $D_y = \dfrac{4a}{3} \cdot S_y$.

The nondimensional aligning moment finally can be described as

$$\overline{T}_z = \frac{T_z}{(a \cdot \varphi \cdot F_z)} = \overline{F}_y \cdot D_x - \overline{F}_x \cdot \left(\overline{D}_y + \frac{y_b}{a} \right) \tag{3.113}$$

where the nondimensional trails are

$$\begin{cases} \overline{D}_x = \dfrac{1}{3} \\ \overline{D}_y = \dfrac{4}{3} \cdot S_y \end{cases} \qquad (3.114)$$

Tire under condition with slip: The resultant tangential stress in the slide region is $q(x) = \varphi \cdot q_z$, on the condition that the vertical pressure distribution q_z is known, and the coordinate of the slide starting point $x = x'$ or the relative coordinate $u = \dfrac{x}{a} = u' = \dfrac{x'}{a}$ can be obtained. Give the common formula of q_z as

$$q_z = \frac{F_z}{2a} \cdot \eta(u) \qquad (3.115)$$

where $\eta(u)$ is the nondimensional formula of the distribution form of the vertical load.

The tenable condition for the slip point is

$$x \cdot \sqrt{(k_x S_x)^2 + (k_y S_y)^2} = \varphi \cdot \frac{F_z}{2a} \cdot \eta(u) \qquad (3.116)$$

Then the nondimensional slipping conditions can be derived as

$$\begin{cases} \phi = \dfrac{\eta(u)}{u} \\ u \le 2 \end{cases} \qquad (3.117)$$

By assuming that the comprehensive relative slip ratio ϕ is known, the position of the slide starting point $u = u'$ can be derived from Eq. 3.117. When $u < u'$, in the region of adhesion,

$$\begin{cases} q_x = \phi_x \cdot u \cdot \dfrac{\varphi \cdot F_z}{(2a)} \\ q_y = \phi_y \cdot u \cdot \dfrac{\varphi \cdot F_z}{(2a)} \\ q = \phi \cdot u \cdot \dfrac{\varphi \cdot F_z}{(2a)} \end{cases} \qquad (3.118)$$

When $u \ge u'$, in the slide region,

$$q = \varphi \cdot q_z \qquad (3.119)$$

Therefore, the total tangential force can be formulated as

$$F = a\int_0^{u'} \varphi \cdot F_z \cdot \phi \cdot \frac{u}{(2a) \cdot du} + a\int_{u'}^{u} \varphi \cdot F_z \cdot \frac{\eta(u)}{(2a) \cdot du} \qquad (3.120)$$

The nondimensional formula of the total tangential force is

$$\overline{F} = \frac{1}{4}\phi \cdot u'^2 + 1 - \frac{m_0}{2} \tag{3.121}$$

where

$$m_0 = \int_0^{u'} \eta(u)du \tag{3.122}$$

With the assumption that the tangential stresses in the slide and adhesive regions are in the same direction, the nondimensional formula of the longitudinal and cornering forces can be expressed as

$$\begin{cases} \overline{F}_x = \overline{F} \cdot \dfrac{\phi_x}{\phi} = \left(\dfrac{\phi_x}{4}\right) \cdot u'^2 + \left(1 - \dfrac{m_0}{2}\right) \cdot \dfrac{\phi_x}{\phi} \\ \overline{F}_y = \overline{F} \cdot \dfrac{\phi_y}{\phi} = \left(\dfrac{\phi_y}{4}\right) \cdot u'^2 + \left(1 - \dfrac{m_0}{2}\right) \cdot \dfrac{\phi_y}{\phi} \end{cases} \tag{3.123}$$

Likewise, the aligning moment can be formulated as

$$\overline{T}_z = \phi_y \cdot \frac{u'^3}{6} + \frac{\phi_y}{\phi} \cdot \left(1 - \frac{\Delta}{a} - \frac{m_1}{2}\right) - \overline{F}_y - \frac{S_y}{6}$$
$$\cdot \phi_x \cdot u'^3 - \frac{S_y}{2} \cdot E \cdot \frac{\phi_x}{\phi^2} - \overline{F}_x \cdot \frac{y_b}{a} \tag{3.124}$$

where Δ is the eccentric moment generated by the vertical load

$$m_1 = \int_0^{u'} u \cdot \eta(u)du \tag{3.125}$$

$$E = \int_{u'}^{2} \eta^2(u)du \tag{3.126}$$

The generalized theoretical tire model gives a clear physical concept and is applied primarily to the description of steady-state tire dynamics when the vehicle is traveling over a good road surface under a normal driving condition. However, it is not suitable for the analysis of the nonsteady-state tire dynamics characteristics with the starting and emergency braking and severe steering conditions. Thus, the tire cornering and longitudinal slip characteristics of the nonsteady-state condition are available.

3.5.2.2 Nonsteady-State Semi-Empirical Tire Mechanics Model

Figure 3.28 describes the state of a tire in the nonsteady state of motion on the road, where XOY is the spatial fixed coordinate system, xoy is the tire coordinate system, o is the center of the tire contact patch with the coordinates (X,Y) in the spatial fixed coordinate system, and the angle formed between the intersection of the wheel plane and

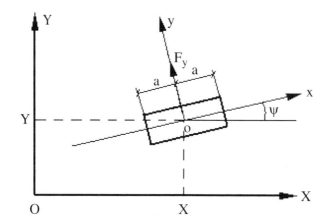

Figure 3.28 State of a tire in a nonsteady state of motion.

the ground plane and the OX axis of the spatial fixed coordinate system is the angle ψ, called the tire steering angle. The length of the tire contact patch is 2a.

We can define the tire nominal slip velocity of the contact patch center V_{syn} as

$$V_{syn} = V_r\left[\psi - \left(\frac{dY}{dX} + r\frac{d\gamma}{dX}\right) - a\left(\frac{d\psi}{dX} + \eta_\gamma\frac{\sin\gamma}{r}\right)\right] \qquad (3.127)$$

where

$$V_r = |\omega \cdot r|$$

and

ω = angular speed of the tire

r = effective rolling radius of the tire

γ = camber angle of the tire

η_γ = influence coefficient of the camber angle that represents the influence resulting from the deformation of the carcass on the tire effective camber angle

The nominal longitudinal slip velocity of the center of the tire contact patch is written as

$$V_{sxn} = V_x - V_r \qquad (3.128)$$

where V_x is the velocity of the tire center in the longitudinal direction, the component speed of the tire center relative to the ground in the x axis in the tire coordinate system.

With the assumption that the lateral displacement Y and the tire steering angle ψ are less, the longitudinal slip and cornering characteristics of the physical model with a nonsteady state can be described as shown in Figure 3.29. Thus, the effective longitudinal slip ratio

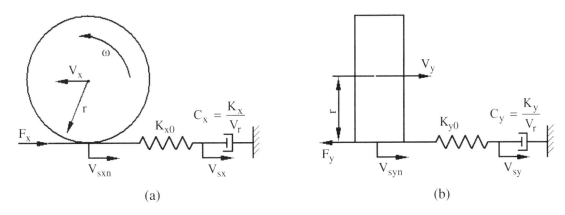

Figure 3.29 Longitudinal slip and cornering characteristics of a physical model with a nonsteady state: (a) nonsteady-state longitudinal slip characteristics physical model, and (b) nonsteady-state cornering characteristics physical model.

S_x and the effective lateral slip ratio S_y on the nonsteady state derived from this model can be given by

$$S_x = \frac{V_{sx}}{V_r} \tag{3.129}$$

$$S_y = \frac{V_{sy}}{V_r} \tag{3.130}$$

$$V_{sx} + \frac{\ell_x}{V_r} \dot{V}_{sx} = V_{sxn} \tag{3.131}$$

$$V_{sy} + \frac{\ell_y}{V_r} \dot{V}_{sy} = V_{syn} \tag{3.132}$$

where V_{sx} and V_{sy} are the effective longitudinal and lateral slip velocities, respectively.

The longitudinal relaxation length ℓ_x and the lateral relaxation length ℓ_y are defined, respectively, as

$$\ell_x = \frac{K_x}{K_{x0}} \tag{3.133}$$

$$\ell_y = \frac{K_y}{K_{y0}} \tag{3.134}$$

where K_{x0} and K_{y0} are the longitudinal and lateral carcass deformation stiffness, respectively, and K_x and K_y are the tire longitudinal and cornering stiffness corresponding to the effective slip ratio S_x and S_y

$$K_x = \frac{\partial F_x}{\partial s_x} \qquad (3.135)$$

$$K_y = \frac{\partial F_y}{\partial s_y} \qquad (3.136)$$

The relation obtained from experiments between the longitudinal friction coefficient φ_x as well as the lateral friction coefficient and the vertical load F_z can be formulated as

$$\begin{cases} \varphi_x = a_1 + a_2 F_z + a_3 F_z^2 \\ \varphi_y = b_1 + b_2 F_z + b_3 F_z^2 \end{cases} \qquad (3.137)$$

where a_1, a_2, a_3, b_1, b_2, and b_3 are experimental coefficients.

The relative longitudinal slip ratio ϕ_x, the lateral slip ratio ϕ_y, and the total slip ratio ϕ are

$$\begin{cases} \phi_x = \dfrac{K_x S_x}{\left(\varphi_x \cdot F_z\right)} \\ \phi_y = \dfrac{K_y S_y}{\left(\varphi_y \cdot F_z\right)} \\ \phi = \sqrt{\phi_x^2 + \phi_y^2} \end{cases} \qquad (3.138)$$

The nondimensional total tangential force $\overline{F}$ is written as

$$\overline{F} = 1 - \exp\left[-\phi - E_1 \phi^2 - \left(E_1^2 + \frac{1}{12}\right)\phi^3\right] \qquad (3.139)$$

where coefficient E_1 is

$$E_1 = \frac{0.5}{1 + \exp\left(-\dfrac{F_z - b_4}{b_5}\right)} \qquad (3.140)$$

where b_4 and b_5 are the experimental coefficients.

The tire longitudinal force F_x, lateral force F_y, and self-aligning moment T_z, respectively, are expressed as

$$\begin{cases} F_x = -\dfrac{\phi_x}{\phi}\varphi_x \cdot F_z \cdot \overline{F} \\[2mm] F_y = -\dfrac{\phi_y}{\phi}\varphi_y \cdot F_z \cdot \overline{F} \\[2mm] T_z = -F_y \cdot D_x - F_x \cdot y_0 \end{cases} \tag{3.141}$$

where D_x is the pneumatic trail formulated as

$$D_x = \left(D_{x0} + D_e\right) \cdot \exp\left(-D_1\phi - D_2\phi^2\right) - D_e \tag{3.142}$$

The relation between the initial value of the pneumatic trail D_{x0} and the vertical load F_z is determined by the experimental data formulated as

$$D_{x0} = c_1 + c_2 F_z + c_3 F_z^2 \tag{3.143}$$

The relation between the total value of the pneumatic trail D_e and the vertical load F_z is determined by the experimental data formulated as

$$D_e = c_4 + c_5 F_z + c_6 F_z^2 \tag{3.144}$$

D_1 is a coefficient as

$$D_1 = c_7 \exp\left(-\frac{F_z}{c_8}\right) \tag{3.145}$$

D_2 is a coefficient as

$$D_2 = c_9 \exp\left(\frac{F_z}{c_{10}}\right) \tag{3.146}$$

y_0 is the lateral translation deformation written by

$$y_0 = \frac{F_y}{K_{y0}} \tag{3.147}$$

K_{y0} is the lateral deformation stiffness expressed by

$$K_{y0} = d_1 F_z + d_2 F_z^2 \tag{3.148}$$

$c_1 \sim c_{10}$, $d_1 \sim d_2$ in these formulas are experimental coefficients.

3.5.3 H.B. Pacejka Magic Formula Model

The "magic formula" presented by Professor Pacejka in Holland Delft Industrial University has been shown to suitably match the experimental data of the dynamic tire characteristics by the combinational formula of the trigonometric function. Pacejka obtained a set of formulas of tire models that can describe the longitudinal and lateral forces and the self-aligning moment comprehensively with the same form. The formula is called the "magic formula."

Magic formula: In pure cornering and pure longitudinal slip conditions, the longitudinal and lateral force and self-aligning moment on the tire can be written as follows:

The lateral force is given by

$$F_x = D\sin\left(C\arctan\left\{B(1-E)(\sigma + S_h) + E\arctan\left[B(\sigma + S_h)\right]\right\}\right) + S_v \quad (3.149)$$

The longitudinal slip ratio is given by

$$\sigma = \frac{\omega}{\left(\dfrac{V}{R_e}\right)} - 1 \quad (3.150)$$

where ω is the angular speed of the tire, V is the linear speed of the tire center, R_e is the effective rolling radius of the tire, and

$$C = b_0 \quad (3.151)$$

$$D = \mu_p F_z \quad (3.152)$$

$$\mu_p = b_1 F_z + b_2 \quad (3.153)$$

$$B \cdot C \cdot D = \left(b_3 F_z^2 + b_4 F_z\right)e^{-b_5 F_z} \quad (3.154)$$

$$E = b_6 F_z^2 + b_7 F_z + b_8 \quad (3.155)$$

$$S_h = b_9 F_z + b_{10} \quad (3.156)$$

$$S_v = 0 \quad (3.157)$$

F_z is the vertical load on the tire (in kilonewtons [kN]).

The lateral force is given by

$$F_y = D \sin\left(C \arctan\left\{B(1-E)(\alpha + S_h) + E \arctan\left[B(\alpha + S_h)\right]\right\}\right) + S_v \quad (3.158)$$

where α is the slip angle (in degrees), γ is the camber angle (in degrees), and

$$C = a_0 \quad (3.159)$$

$$D = \mu_{yp} F_z \quad (3.160)$$

$$\mu_{yp} = a_1 F_z + a_2 \quad (3.161)$$

$$B \cdot C \cdot D = a_3 \sin\left[2 \arctan\left(\frac{F_z}{a_4}\right)\right](1 - a_5|\gamma|) \quad (3.162)$$

$$E = a_6 F_z + a_7 \quad (3.163)$$

$$S_h = a_8 \gamma + a_9 F_z + a_{10} \quad (3.164)$$

$$S_v = a_{11} \gamma F_z + a_{12} F_z + a_{13} \quad (3.165)$$

$$a_{11} = a_{111} F_z + a_{112} \quad (3.166)$$

F_z is the tire vertical load (in kilonewtons [kN]).

The self-aligning moment is formulated as

$$M_z = D \sin\left(C \arctan\left\{B(1-E)(\alpha + S_h) + E \arctan\left[B(\alpha + S_h)\right]\right\}\right) + S_v \quad (3.167)$$

where α is the slip angle (in degrees), γ is the camber angle (in degrees), and

$$C = c_0 \quad (3.168)$$

$$D = c_1 F_z^2 + c_2 F_z \quad (3.169)$$

$$B \cdot C \cdot D = \left(c_3 F_z^2 + c_4 F_z\right)(1 - c_6|\gamma|)e^{-c_5 F_z} \quad (3.170)$$

$$E = \left(c_7 F_z^2 + c_8 F_z + c_9\right)\left((1 - c_{10}|\gamma|)\right) \quad (3.171)$$

$$S_h = c_{11}\gamma + c_{12}F_z + c_{13} \tag{3.172}$$

$$S_v = \left(c_{14}F_z^2 + c_{15}F_z\right)\gamma + c_{16}F_z + c_{17} \tag{3.173}$$

F_z is the tire vertical load (in kilonewtons [kN]).

In the case of the combined condition of steering and braking, the vector diagram of the velocity is as shown in Figure 3.30.

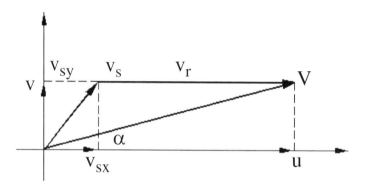

Figure 3.30 Vector diagram of velocity in the combined condition of steering and braking.

Set

$$\begin{cases} \sigma_x = \dfrac{v_{sx}}{V_r} \\[3mm] \sigma_y = \dfrac{v_{sy}}{V_r} \end{cases} \tag{3.174}$$

When $\alpha = 0$, define $\sigma = \dfrac{-v_{sx}}{u}$. Thus, the longitudinal slip ratio and the lateral slip ratio can be described as

$$\begin{cases} \sigma_x = \dfrac{v_{sx}}{u - v_{sx}} = \dfrac{-\sigma}{1 + |\sigma|} \\[3mm] \sigma_y = \dfrac{v_{sy}}{u - v_{sx}} = \dfrac{\tan(\alpha)}{1 + |\sigma|} \end{cases} \tag{3.175}$$

The correct values of the longitudinal and lateral slip ratios are given as

$$\begin{cases} \sigma_{xtot} = \dfrac{-\sigma}{1 + |\sigma|} - \delta\sigma \\[3mm] \sigma_y = \dfrac{\tan(\alpha)}{1 + |\sigma|} + \delta\alpha \end{cases} \tag{3.176}$$

where

$$\begin{cases} \delta\sigma = -S_h \\ \delta\alpha = -S_h - \dfrac{S_v}{B \cdot C \cdot D} \end{cases} \qquad (3.177)$$

Define

$$\sigma_{tot} = \sqrt{\sigma_{xtot}^2 + \sigma_{ytot}^2}$$

Thus, the tire mechanics characteristics are

$$\begin{cases} F_x = \dfrac{\sigma_{xtot}}{\sigma_{tot}} \cdot F_x\left(\sigma_{tot}\right) \\[2mm] F_y = \dfrac{\sigma_{ytot}}{\sigma_{tot}} \cdot F_y\left(\sigma_{tot}\right) \\[2mm] M_z = \dfrac{\sigma_{ytot}}{\sigma_{tot}} \cdot M_z\left(\sigma_{tot}\right) \end{cases} \qquad (3.178)$$

The magic formula presents the generalized tire mechanics property and high fitting precision, but it requires a massive quantity of calculation, which is suitable for the product design, vehicle dynamic simulation, and test comparison in the field of high-precision tire mechanics properties analysis.

Magic model: The magic model is empirical, requiring specification of a number of parameters determined from experimental measurements of the tire forces and moments. Such measurements require sophisticated test equipment, which makes the magic model impractical for organizations with modest tire-testing capabilities. Also, the magic model is not well suited to parametric studies of the impacts of snow- and ice-induced changes in the surface friction on vehicle performance. A static friction coefficient appears in the model, but it determines only the peak forces and moments. Its adjustment alone does not properly account for the variation in the surface friction commonly observed at different tire/road slip conditions. A full complement of force and moment measurements seems to be required instead for each road surface condition examined. Hence, an alternative tire model is needed that (1) requires only a limited set of easily measured tire input parameters; (2) is analytical in nature, with physically meaningful parameters that relate directly to tire characteristics and road surface conditions; (3) realistically accounts for changes in surface friction; and (4) is numerically simple and practical for use in complex vehicle simulations. A steady-state form of the solution is adequate because the problem being addressed is low-frequency vehicle maneuvering on smooth nondeformable road surfaces.

3.6 References

1. Allen, W.R., Rosenthal, T.J., and Chrstos, J.P., "Vehicle Dynamics Tire Model for Both Pavement and Off-Road Conditions," *Research into Vehicle Dynamics and Simulation*, SAE Special Publications, Vol. 1228, SAE Paper No. 970559, Society of Automotive Engineers, Warrendale, PA, 1997, pp. 27–38.

2. Ammon, D., "Vehicle Dynamics Analysis Tasks and Related Tire Simulation Challenges," *Vehicle System Dynamics*, Vol. 43, 2005, pp. 30–47.

3. Bohm, F., "Tire Models for Computational Car Dynamics in the Frequency Range up to 1000 Hz," *Vehicle System Dynamics*, Vol. 21, 1993, pp. 82–91.

4. Cabrera, J.A., Ortiz, A., Carabias, E., and Simon, A., "An Alternative Method to Determine the Magic Tire Model Parameters Using Genetic Algorithms," *Vehicle System Dynamics*, Vol. 41, No. 2, 2004, pp. 109–127.

5. Clover, C.L. and Bernard, J.E., "Longitudinal Tire Dynamics," *Vehicle System Dynamics*, Vol. 29, No. 4, 1998, pp. 231–259.

6. Cui, S. and Yu, Q., *Traveling Performance and Test of Vehicle Tire*, Mechanical Industry Press, China, 1995.

7. Deur, J., "Modeling and Analysis of Longitudinal Tire Dynamics Based on the LuGre Friction Model," *Advances in Automotive Control 2001, Proceedings Volume from the 3rd IFAC Workshop*, 2001, pp. 91–96.

8. Deur, J., Asgari, J., and Hrovat, D.A., "3D Brush-Type Dynamic Tire Friction Model," *Vehicle System Dynamics*, Vol. 42, No. 3, 2004, pp. 133–173.

9. Dong, P., Liu, Q., and Guo, K. (in Chinese), "Analysis of the E-Function for Non-Steady State Tire Cornering Mechanics," *Qinghua Daxue Xuebao/Journal of Tsinghua University*, Vol. 42, No. 2, 2002, pp. 251–253.

10. Gim, G., Choi, Y., and Kim, S., "A Semiphysical Tire Model for Vehicle Dynamics Analysis of Handling and Braking," *Vehicle System Dynamics*, Vol. 43, 2005, pp. 267–280.

11. Guan, D., Fan, C., and Xie, X., "A Dynamic Tyre Model of Vertical Performance Rolling over Cleats," *Vehicle System Dynamics*, Vol. 43, 2005, pp. 209–222.

12. Guo, K. and Liu, Q., "Generalized Theoretical Model of Tire Cornering Properties in Steady State Condition," *Heavy Vehicle and Highway Dynamics*, SAE Special Publications, Vol. 1308, SAE Paper No. 973191, Society of Automotive Engineers, Warrendale, PA, 1997, pp. 31–38.

13. Guo, K. and Liu, Q., "Model of Tire Enveloping Properties and Its Application on Modeling of Automobile Vibration Systems," *Developments in Tire, Wheel, Steering, and Suspension Technology*, SAE Special Publications, Vol. 1338, SAE Paper No. 980253, Society of Automotive Engineers, Warrendale, PA, 1998, pp. 21–27.

14. Guo, K. and Liu, Q., "Tire Models for Vehicle Dynamics Analysis in Steady State Condition," *Qiche Gongcheng/Automotive Engineering*, Vol. 20, No. 3, 1998, pp. 129–134.

15. Guo, K., Lu, D., Chen, S-K., Lin, W.C., and Lu, X-P., "The UniTire Model: A Nonlinear and Non-Steady-State Tire Model for Vehicle Dynamics Simulation," *Vehicle System Dynamics*, Vol. 43, 2005, pp. 341–358.

16. Guo, K-H., Yuan, Z-C., Lu, D., and Lin, B-Z., "Speed Prediction Ability of UniTire Steady-State Model," *Jilin Daxue Xuebao (Gongxueban)/Journal of Jilin University (Engineering and Technology Edition)*, Vol. 35, No. 5, 2005, pp. 457–461.

17. Holtschulze, J., Goertz, H., and Husemann, T., "A Simplified Tire Model for Intelligent Tires," *Vehicle System Dynamics*, Vol. 43, 2005, pp. 305–316.

18. Kao, B.G. and Warholic, T., "Tire Lateral Force Modeling and the Bushing Analogy Tire Model," *Tire Science and Technology*, Vol. 33, No. 1, 2005, pp. 18–37.

19. Konghui, Guo, *Automobile Handling Dynamics*, Jilin Press of Science and Technology, China, December 1991.

20. Liu, Q. and Guo, K. (in Chinese), "Numerical Simulation Algorithms of Tire Cornering Properties in Non-Steady State Conditions," *Qinghua Daxue Xuebao/Journal of Tsinghua University*, Vol. 39, No. 11, 1999, pp. 69–72.

21. Lugner, Peter, Pacejka, Hans, and Plochl, Manfred, "Recent Advances in Tire Models and Testing Procedures," *Vehicle System Dynamics*, Vol. 43, No. 6–7, 2005, pp. 413–436.

22. Miege, A.J.P. and Popov, A.A., "Truck Tire Modeling for Rolling Resistance Calculations under a Dynamic Vertical Load," *Proceedings of the Institution of Mechanical Engineers, Part D: Journal of Automobile Engineering*, Vol. 219, No. 4, 2005, pp. 441–454.

23. Miege, Arnaud J.P. and Popov, Atanas A., "The Rolling Resistance of Truck Tires under a Dynamic Vertical Load," *Vehicle System Dynamics*, Vol. 43, 2005, pp. 135–144.

24. Miyashita, N. and Kabe, K., "A Study of the Cornering Force by Use of the Analytical Tire Model," *Vehicle System Dynamics*, Vol. 43, 2005, pp. 123–134.

25. Pacejka, H.B., "Spin: Camber and Turning," *Vehicle System Dynamics*, Vol. 43, 2005, pp. 3–17.

26. Palkovics, L., El-Gindy, M., and Pacejka, H.B., "Modeling of the Cornering Characteristics of Tires on an Uneven Road Surface: A Dynamic Version of the 'Neuro-Tyre,'" *International Journal of Vehicle Design*, Vol. 15, No. 1–2, 1994, pp. 189–215.

27. Poliey, M. and Alleyne, A.G., "Dimensionless Analysis of Tire Characteristics for Vehicle Dynamics Studies," *Proceedings of the American Control Conference*, Vol. 4, 2004, pp. 3411–3416.

28. Schmeitz, A.J.C., Jansen, S.T.H., Pacejka, H.B., Davis, J.C., Kota, N.M., Liang, C.G., and Lodewijks, G., "Application of a Semi-Empirical Dynamic Tire Model for Rolling over Arbitrary Road Profiles," *International Journal of Vehicle Design*, Vol. 36, No. 2–3, 2004, pp. 194–215.

29. Shang, J., Guan, D., and Yam, L.H., "Study on Tire Dynamic Cornering Properties Using Experimental Modal Parameters," *Vehicle System Dynamics*, Vol. 37, No. 2, 2002, pp. 129–144.

30. Velenis, E., Tsiotras, P., Canudas, D-W.C., and Sorine, M., "Dynamic Tire Friction Models for Combined Longitudinal and Lateral Vehicle Motion," *Vehicle System Dynamics*, Vol. 41, No. 1, 2005, pp. 3–29.

31. Velumot, H-P., *Method and Simulation of Vehicle Dynamics*, Press of Peking University of Science and Technology, 1998.

32. Wille, R., Bohm, F., and Duda, A., "Calculation of the Rolling Contact Between a Tire and Deformable Ground," *Vehicle System Dynamics*, Vol. 43, 2005, pp. 483–492.

33. Yi, J., Alvarez, L., Claeys, X., Horowitz, R., and Canudas De Wit, C., "Emergency Braking Control with an Observer-Based Dynamic Tire/Road Friction Model and Wheel Angular Velocity Information," *Proceedings of the American Control Conference*, Vol. 1, 2001, pp. 19–24.

34. Yu, Z., *Automobile Theory*, Mechanical Industry Press, Third Ed., October 2000.

35. Zhang, H., *Automobile System Dynamics*, Tongji University Press, 1996.

36. Zuo, S., *Influence of Road Surface Roughness on Braking Performance of Tractor-Trailing*, Ph.D. dissertation submitted to Jilin University of Technology, June 1996.

37. Zuo, S., Liu, D., and Zhang J., "Tire Model Due to Braking Process for Vehicles on Uneven Road Surface," *Transactions of Nanjing University of Aeronautics & Astronautics*, 1997.

Chapter 4

Ride Dynamics

4.1 Introduction

Ride quality deals with passengers' comfort in the vehicle. The judgment of ride quality can vary from one customer to another according to their perception and experience. Several factors such as vibration, noise, heat, sitting posture, age, gender, and environment influence ride quality. Among these factors, vibration is considered the most important in determining the overall ride quality. Passengers in a moving vehicle are continuously exposed to a vibration environment caused by excitations from the road surface and from vehicle internal sources such as the powerplant, which is the combination of the engine and the transmission. Hence, this chapter discusses ride quality caused by vibration.

Ride comfort is one of the most important factors for a customer in determining whether or not to purchase a particular vehicle. Ride quality is an important attribute in modern car design and is a major contributor to customer satisfaction. For almost all vehicle OEMs (original equipment manufacturers) and suppliers, customer satisfaction has become the critical issue in vehicle product development and business strategy.

Passengers provide the ride quality evaluation when they ride in the vehicles. The judgment of ride quality by passengers is a subjective evaluation. Ride quality also can be evaluated by objective methods, including simulation and testing. Ride quality is directly related to the seat comfort design. The seat is mounted on the vehicle; thus, the vehicle design also influences ride quality. The vehicle is subjected to external and internal vibration sources that determine the magnitude of ride discomfort. Therefore, research on ride quality and ride dynamics includes studies on vibration sources and modeling of the vehicle, seat, and human body. Figure 4.1 lists the ride dynamics system.

The excitation to a passenger comes from both external and internal sources of the vehicle. The external sources include road input and wind excitation. Wind excitation has a significant impact on the vehicle body structure vibration that is transferred to interior sound. However, this contribution to passenger discomfort is less significant

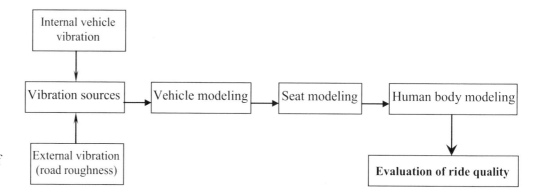

Figure 4.1 The system of ride dynamics.

compared with other vibration sources. The major external excitation is from an uneven road profile. The power spectral density (PSD) of passenger vibration is roughly proportional to the power spectral density of the road surface profile. Ride vibration also depends on the vehicle speed. The faster the vehicle is moving, the more severe the vibration will be.

The major internal excitation is from the engine. Engine vibration is transferred to the body through the engine mounts, exhaust hangers, and driveshaft bearings. The mount system design is a key element in reducing vibration. Engine vibration consists mainly of two sources. The first is the engine firing pulse, and the second is the inertia force from the powerplant. Usually, the firing pulse excitation is much higher than the inertia force.

The driveshaft and exhaust are mounted on the body through bearings, hangers, and isolators. The vibration caused by the driveshaft and exhaust will be transferred to the body. Sometimes, the vibration sources from the driveshaft and exhaust are severe and can even reach the same level as the engine vibration.

The modeling of the body is important in the analysis of ride dynamics. Two models are widely used in ride dynamics analysis. One is the quarter car model, and the other is the bounce-pitch model. The quarter car model includes the unsprung mass system, which consists of the tire and spindle, and the sprung mass system, which consists of the suspension and body. Only vertical vibration is considered in this model. This two-degrees-of-freedom model predicts two major vertical modes. The bounce-pitch model also is a two-degrees-of-freedom model, but it considers both vertical motion and pitch motion. In addition to the two commonly used models, many other complicated models are studied. In automotive engineering, finite element method (FEM) models are used by more and more engineers. FEM models provide not only accurate results but also can be integrated with other applications.

Excitation is transferred to the seat through the vehicle structure. The SEAT (seat effectiveness amplitude transmissibility) value is used to evaluate the effectiveness of the seat vibration isolation. Traditionally, linear seat cushion models have been used for the calculation of seat transmissibility. However, the seat cushion is a nonlinear structure according to recent research results (Pang *et al.* 2003, Ref. 21). This chapter describes both linear and nonlinear models.

Subjective and objective methods are used to evaluate ride quality. The VER (vehicle evaluation rating) is used for subjective evaluation. Passenger ride discomfort relates not only to the vibration amplitude but also to the vibration frequency. A weighting root-mean-square method and vibration dose value are used for objective evaluation. This chapter presents both linear and nonlinear seat-body models.

Traditional passive control to reduce the vibration from the engine and from road input is not effective for all frequency ranges. Active and semi-active controls have found more and more applications in modern vehicles. Active control possesses the capability to reduce vibration at any frequency, but its cost is high and its structure is complicated. Semi-active control can reach some level of active control effectiveness with lower cost and a simpler structure. This chapter introduces the controls of the engine isolation system and the suspension system.

4.2 Vibration Environment in Road Vehicles

This section deals with the vehicle vibration environment: outside excitation, and internal excitation. The uneven road profile is the major source of outside vibration. The vibration will be transferred to occupants through the suspension and seating systems. The major internal sources of vibration include engine excitation, driveline excitation, and exhaust vibration.

4.2.1 Vibration Sources from the Road

Vibration contributed by the road is due to the interaction of the tire tread with the profile of a rough or coarse road surface. The level of road excitation depends on the road profile, vehicle speed, and tire dynamic properties.

The road surface profile is one of the major excitation sources to ride dynamics. It has been found that the power spectral density of the interior vibration is roughly proportional to the power spectral density of the road surface profile.

Speed is another important factor in ride quality. Increasing the vehicle speed biases the vibration spectrum to higher frequencies and usually increases the overall vibration level.

The tire dynamic properties include the tire stiffness and pressure, the tire patch length and shape, and the width and size of the tire. These properties also influence the ride characteristics. The vibration levels increase as the tire pressure or stiffness increase. Shortening the tire patch length biases the vibration spectrum to higher frequencies. A wider tire typically involves a shorter tire patch, which will tend to bias frequencies upward.

4.2.1.1 Power Spectral Density in Spatial Frequency

Figure 4.2 shows a road profile. A vehicle travels on a road in the x direction, while the vehicle vibration is in the z direction. A road profile includes many waves. Spatial frequency, Ω, stands for the wavelength number per meter of the road, and the unit is m^{-1}. The spatial frequency is the inverse of the wavelength.

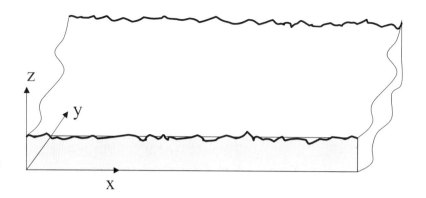

Figure 4.2 A road surface profile.

Roughness is defined as the road elevation profile and is used to evaluate the road smoothness. Usually, a road profile is a wide-band random function. The instantaneous values for a random function cannot be expressed by deterministic methods. Thus, statistical analysis is used to describe the road vibration input. Power spectral density (PSD) is a useful parameter to describe random function. The power spectral density of the road profile, $W_z(\Omega)$, is defined as the mean-square response of an ideal narrow-band filter divided by the frequency bandwidth of the filter $\Delta\Omega$ as $\Delta\Omega \to 0$ at frequency Ω

$$W_z(\Omega) = \lim_{\Delta\Omega \to 0} \frac{\overline{z_{\Delta\Omega}^2}}{\Delta\Omega} \tag{4.1}$$

where $\overline{z_{\Delta\Omega}^2}$ is the mean-square value within $\Delta\Omega$. Then, the mean-square value for the whole frequency range is

$$\overline{z^2} = \int_0^\infty W_z(\Omega) \, d\Omega \tag{4.2}$$

Or, the mean-square value during the wavelength L_i can be expressed as

$$\overline{z_{L_i}^2} = \frac{1}{L_i} \int_0^{L_i} z_i^2 dx \tag{4.3}$$

The power spectral density is a function of the spatial frequency. Based on road test data, a relation between the power spectral density of the road profile and the spatial frequency was proposed by ISO as

$$W_z(\Omega) = W_z(\Omega_0) \left(\frac{\Omega}{\Omega_0}\right)^{-N} \tag{4.4}$$

where Ω_0 is the reference spatial frequency and $\Omega_0 = \frac{1}{2\pi} \text{m}^{-1}$; $W_z(\Omega_0)$ is the profile spectral density corresponding to the reference spatial frequency. $W_z(\Omega_0)$ also is called the profile roughness coefficient. N is the frequency index and determines the frequency structure of the spectral density and the slope of the log-log curves.

The road roughness is divided into eight levels in the ISO standard from A to H. Table 4.1 lists the range and geometric mean values of the roughness coefficients.

TABLE 4.1
ROUGHNESS COEFFICIENTS OF ROAD PROFILES

	Roughness Coefficient $W_z(\Omega_0) \times 10^{-6}$	
Road Class	**Range**	**Geometric Mean**
A	8–32	16
B	32–128	64
C	128–512	256
D	512–2048	1024
E	2048–8192	4096
F	8192–32,768	16,384
G	32,768–131,072	65,536
H	131,072–524,288	262,144

The frequency index N can be chosen as several values for different road wavelengths. For most cases, N can be chosen as 2, that is, N = 2. For this case, Eq. 4.4 can be written as

$$W_z(\Omega) = \frac{W_z(\Omega_0)\Omega_0{}^2}{\Omega^2} \tag{4.5}$$

The power spectral density is inversely proportional to the square of the spatial frequency or is proportional to the square of the wavelength. Figure 4.3 plots the power spectral density of the road profile for the eight levels according to Eq. 4.5.

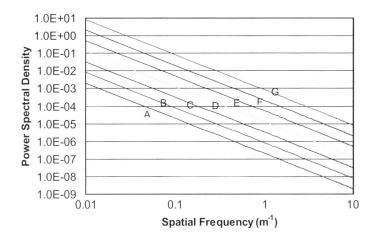

Figure 4.3 Power spectral density for different road profiles.

Usually, the elevation of the road surface along which a vehicle travels is used to describe the road roughness, which means that the vertical displacement input to the vehicle is the roughness. On the other hand, vibration is commonly expressed by acceleration and velocity. Hence, acceleration and velocity and their power spectral densities will be used as vibration input for vehicle ride dynamics analysis. The relation between displacement and acceleration can be obtained by harmonic analysis. Assume that the roughness displacement z of a road profile during a wavelength L can be expressed as

$$z(x) = A \sin\left(\frac{2\pi x}{L}\right) = A \sin(\Omega x) \tag{4.6}$$

where A is the amplitude of the road profile during the wavelength.

The corresponding velocity and acceleration are given by

$$\dot{z}(x) = \frac{2\pi A}{L} \cos\left(\frac{2\pi x}{L}\right) = A\Omega \cos(\Omega x) \tag{4.7}$$

$$\ddot{z}(x) = -\left(\frac{2\pi}{L}\right)^2 A \sin\left(\frac{2\pi x}{L}\right) = -A\Omega^2 \sin(\Omega x) \tag{4.8}$$

The mean square value of $\dot{z}(x)$ during the wavelength is obtained by

$$\overline{\dot{z}^2} = (2\pi\Omega)^2 \overline{z^2} \tag{4.9}$$

The mean square value of $\ddot{z}(x)$ is given by

$$\overline{\ddot{z}^2} = (2\pi\Omega)^4 \overline{z^2} \tag{4.10}$$

If Eqs. 4.9 and 4.10 are substituted into Eq. 4.1, the velocity and acceleration power spectral density can be obtained as

$$W_{\dot{z}}(\Omega) = (2\pi\Omega)^2 W_z(\Omega) \tag{4.11}$$

$$W_{\ddot{z}}(\Omega) = (2\pi\Omega)^4 W_z(\Omega) \tag{4.12}$$

For the special case where N = 2, substitute Eq. 4.5 into Eq. 4.12, and the velocity spectral density will be

$$W_{\dot{z}}(\Omega) = (2\pi\Omega)^2 \frac{W_z(\Omega_0)\Omega_0^2}{\Omega^2} = 4\pi^2 W_z(\Omega_0)\Omega_0^2 \tag{4.13}$$

The velocity spectral density is a constant and is independent of the spatial frequency. That is, the spectral density is "white noise."

4.2.1.2 Power Spectral Density in Temporal Frequency

The power spectral density in spatial frequency has been discussed. However, vehicle vibration usually is described by temporal frequency instead of spatial frequency. Thus, it is more convenient to express the spectral density of surface profiles using temporal frequency in hertz than using spatial frequency in meter^{-1} (m^{-1}). The temporal frequency relates to the vehicle speed and spatial frequency. The relation between the temporal frequency and spatial frequency can be expressed as

$$f = V * \Omega \qquad (4.14)$$

where f is the temporal frequency and V is the vehicle speed.

From Eq. 4.14, it can be seen that the temporal frequency is proportional to the vehicle speed for a given spatial frequency or a certain wavelength. The relation between the temporal frequency and spatial frequency can be plotted in Figure 4.4.

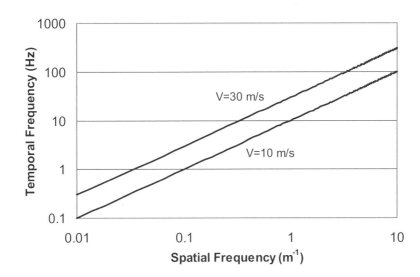

Figure 4.4 Relation between temporal frequency and spatial frequency.

The statistical data show that the spatial frequency for most road profiles is between 0.011 and 2.83 m^{-1}. The temporal frequency range is between 0.33 and 28.3 Hz for the corresponding spatial frequencies during the normal operation speed range of the vehicle (V = 10 ~ 30 m/s). Thus, the frequencies for the vehicle sprung mass (1 ~ 2 Hz) and the frequencies for the unsprung mass (10 ~ 15 Hz) fall into the road excitation frequency range, as shown in Figure 4.4.

Similar to Eq. 4.1, the power spectral density in temporal frequency can be expressed as

$$W_z(f) = \lim_{\Delta f \to 0} \frac{\overline{z_{\Delta f}^2}}{\Delta f} \qquad (4.15)$$

where Δf is the temporal frequency range. Substituting Eq. 4.14 into Eq. 4.15, the relation between the power spectral density in spatial frequency and that in temporal density can be obtained as

$$W_z(f) = \frac{1}{V} W_z(\Omega) \tag{4.16}$$

Substituting Eq. 4.4 into Eq. 4.16, the power spectral density in temporal frequency can be rewritten as

$$W_z(f) = \frac{W_z(\Omega_0)}{V}\left(\frac{\Omega}{\Omega_0}\right)^{-N} \tag{4.17}$$

For the special case where $N = 2$, the density is

$$W_z(f) = \frac{1}{V} W_z(\Omega_0)\left(\frac{\Omega}{\Omega_0}\right)^{-2}$$
$$= \frac{\Omega_0{}^2 W_z(\Omega_0)}{V\Omega^2} \tag{4.18}$$
$$= \frac{\Omega_0{}^2 W_z(\Omega_0) V}{f^2}$$

Performing similar manipulation on Eqs. 4.11 and 4.12, the velocity power spectral density and acceleration power spectral density are, respectively,

$$W_{\dot z}(f) = (2\pi f)^2 W_z(f)$$
$$= 4\pi^2 W_z(\Omega_0)\Omega_0{}^2 V \tag{4.19}$$

$$W_{\ddot z}(f) = (2\pi f)^4 W_z(f)$$
$$= 16\pi^4 W_z(\Omega_0)\Omega_0{}^2 V f^2 \tag{4.20}$$

Figure 4.5 shows the power density spectra of the road profile, velocity, and acceleration for road class A and a vehicle velocity of 10 m/s.

The velocity density spectrum is a constant for a given vehicle velocity. The road profile power spectral density decreases as the frequency increases. However, the acceleration power spectral density increases as the frequency increases. All three density spectra are proportional to the vehicle speed. The spectral densities increase as the speed increases. The acceleration density increases with the square of the temporal frequency, that is, the vibration transferred to the vehicle at high frequency is more severe than that transferred at low frequency.

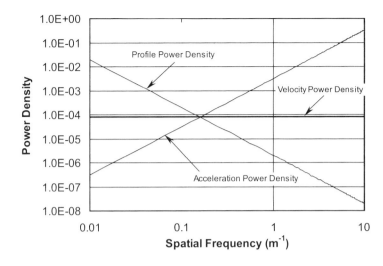

Figure 4.5 Power density spectra of the road profile, velocity, and acceleration.

4.2.2 Vehicle Internal Vibration Sources

Vehicle internal excitation sources come mainly from powertrain vibration. The power-train includes the engine, transmission, driveline system, exhaust system, and induction system. The powerplant is one of the major sources of vehicle vibration. The driveline, exhaust, and intake are connected with the powerplant. Vibration from the powerplant also is transferred to the vehicle body through these systems. Figure 4.6 shows the internal vibration sources. The induction system is connected with the engine using very soft tubing so that the engine vibration is almost isolated. This chapter describes in detail the excitation sources from the powerplant, driveline, and exhaust.

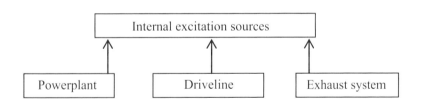

Figure 4.6 Vehicle internal vibration sources.

4.2.2.1 Vibration Sources from the Powerplant

4.2.2.1.1 Coordinates and Powerplant Modes

The two major vibration sources from the engine are the pressure variation in the cylinder and inertial imbalance forces. The pressure variation is developed during the combustion process. The inertial imbalance force is generated by the reciprocating components. The responses excited by the sources depend not only on the engine structure but also on other structures attached to the engine, such as the transmission. Figure 4.7 shows the two basic configurations of powerplants. The powerplant x direction is always defined along the engine-transmission direction.

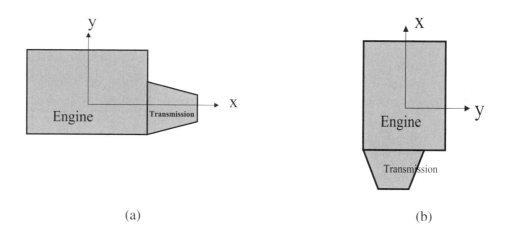

Figure 4.7 Powerplant coordinates system: (a) east-west direction, and (b) north-south direction.

Figure 4.8 is a vehicle drawing with its coordinates. The vehicle coordinates are defined as follows:

- x direction is defined from the vehicle front to the rear
- y direction is defined from the driver to the passenger
- z direction is defined from the bottom upward

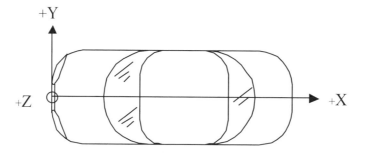

Figure 4.8 Vehicle coordinates system.

There are six motions for a rigid body: bounce motion, lateral motion, fore-aft motion, roll motion, pitch motion, and yaw motion. The linear motion along the x axis is the fore-aft motion, the linear motion along the y axis is the lateral motion, and the linear motion along the z axis is the bounce motion. The rotation around the x axis is the roll motion, the rotation around the y axis is the pitch motion, and the rotation around the z axis is the yaw motion. Figure 4.9 shows the six motions.

When packaged in a vehicle, the powerplant could have the same coordinates as the vehicle, as shown in Figure 4.10(a). This package of powerplant is called a north-south configuration. Most trucks and some large cars and sports cars have this kind of configuration. For this case, the six motions of the powerplant are the same as those for the vehicle body. However, the powerplant could be packaged in different coordinates, as shown in Figure 4.10(b). This package is called an east-west configuration. This configuration can be seen in most passenger cars and some small sport utility vehicles (SUVs).

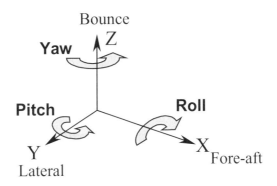

Figure 4.9 Six-motion rotation of a rigid body.

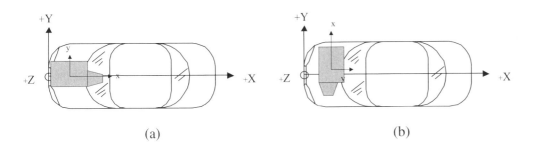

(a)   (b)

Figure 4.10 Powerplant configurations: (a) north-south configuration, and (b) east-west configuration.

The crankshaft of an engine is always along the x direction in the powerplant coordinates, so the most important engine excitations are around the x direction, that is, roll excitation. For a north-south configuration, the powerplant roll motion has the same direction as the vehicle roll motion. However, for an east-west configuration, the powerplant roll motion is different from the vehicle roll motion.

A powerplant usually is supported by several mounts and/or one or two roll restrictors. The configuration with three mounts and a roll restrictor is the most common case in passenger cars. The three mounts are installed on the front of the engine, on the rear of the engine, and on the transmission, respectively. These mounts have two major functions. The first is to support the powerplant weight, and the second is to reduce the powerplant vibration transferred to the vehicle body and to control the force transferred to the body from the road input. The roll restrictor usually is used to control the powerplant roll motion.

Figure 4.11 shows one possible configuration of these mounts, where a powerplant is supported on a subframe by three mounts and a roll restrictor. The subframe is a flexible structure, as is the powerplant. To primarily identify its modes and responses, the powerplant usually is assumed to be a rigid body, and the mounts are assumed to be fixed on the ground.

As a rigid body, the powerplant has six motions and six corresponding modes. The roll mode is the most critical because the engine excitation direction is the same as the mode. Two criteria guide the powerplant mount system design. One is that the roll mode frequency must be separated from the excitation frequency, and the powerplant vibration must be isolated effectively. The other is that the six modes must be decoupled.

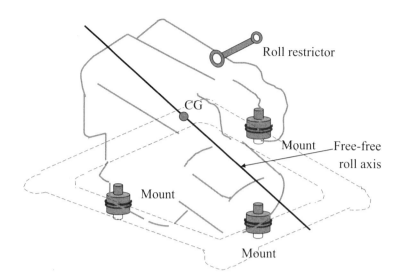

Figure 4.11 A power-plant mount system on a subframe.

4.2.2.1.2 Vibration Sources from Engine Firing Pulsation

The purpose of this section is to identify the vibration sources transferred to the human body. We also will study how to isolate the excitation sources that transfer to the vehicle body and how to minimize the coupling of disturbances to the vehicle occupants.

Engines used in most road vehicles are four-stroke engines. The engine finishes a working processing by four steps (strokes): air intake, air compression, combustion (explosion or firing), and exhaust. Figure 4.12 shows these steps. It takes 180° to process one step; therefore, the whole process takes 720° (i.e., two cycles).

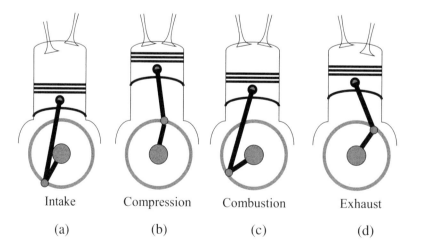

Figure 4.12 Four steps of the engine working processing: (a) intake, (b) compression, (c) combustion, and (d) exhaust.

Among the four steps, the combustion phase produces the force because the gas and air are combined and fired. The firing pulse drives the engine crankshaft to move and then is transferred to the vehicle wheels through the driveline.

For a single-cylinder engine, a firing pulse appears in every two cycles, or there is a "half" firing pulse for each cycle. The firing order, sometimes simply called the order, is defined as the firing number in each cycle. Thus, the basic firing order for a single-cylinder engine is the half order. For an engine with two cylinders, there is a half firing order for each cylinder in one cycle, that is, there exists a whole firing order in one cycle for the two cylinders. Hence, the basic firing order for a two-cylinder engine is first order. Similarly, there are two, three, and four firing pulses for four-, six-, and eight-cylinder engines, respectively, in one cycle. Thus, their corresponding firing orders are second order, third order, and fourth order, respectively.

The engine output provides the driving force to a vehicle, but it also causes vehicle vibration. The vibration usually is controlled by the engine mount isolation system. The design of an isolation system depends on the system natural frequency and the excitation frequency. To reduce the vibration excited by the firing pulse, it is necessary to understand the excitation frequency and the structure modes to be excited. The excitation frequency relates to the engine cylinders and speed. The engine speed range usually is from 600 to 6000 rpm. The idle speed range is from 600 to 1000 rpm, while the wide open throttle (WOT) speed range is from 1000 to 6000 rpm. The firing frequency f_{firing} for each engine speed can be calculated by

$$f_{firing} = \frac{\text{engine speed}}{60} * \text{firing order}$$

$$= \frac{\text{engine speed}}{60} * \frac{\text{number of cylinders}}{2}$$

(4.21)

The relation among the engine speed, frequency, and order can be expressed clearly in a plot. Figure 4.13 plots the first order and third order, but other orders also can be plotted.

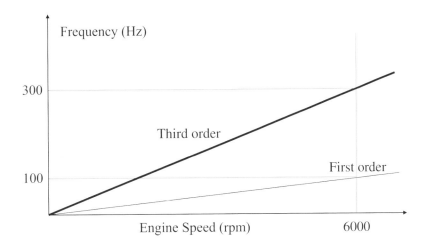

Figure 4.13 Relation between engine speed, order, and frequency.

The firing frequency varies with the engine speed. The lowest firing frequency is the most important because if vibration caused by this frequency is controlled, the higher excitation frequencies definitely will be controlled. Usually, the lowest engine working speed is 600 rpm. For a four-cylinder engine, second order is the dominant order; thus, its lowest firing frequency is 20 Hz. For a six-cylinder engine, third order is the dominant order; thus, the lowest firing frequency is 30 Hz. Table 4.2 lists the idle firing frequencies for different engines. The listed frequencies are the lowest excitation frequencies. These frequencies are important not only for powerplant isolation design but also for exhaust and driveline design.

TABLE 4.2
RELATION OF FIRING ORDER AND
FREQUENCY FOR DIFFERENT ENGINES

Engine Cylinder Number	Firing Order	Idle Firing Frequency (Assuming 600 rpm Idle Speed)
2	First	10 Hz
4	Second	20 Hz
6	Third	30 Hz
8	Fourth	40 Hz

4.2.2.1.3 Vibration Sources from Powerplant Inertia Forces and Moments

The second source of engine excitation is from the inertial unbalance (or imbalance) forces generated by the reciprocating components (piston and rods) or rotating (crankshaft and rod) masses within the engine. The directions of the inertia forces are both parallel to the piston axis and perpendicular to the crank and piston axes. The inertia torque acts about an axis, which is parallel to the crankshaft.

The inertia force is caused by system imbalance; thus, it increases with increasing engine speed because the force is proportional to the square of the operating speed. At high speeds, the inertia force could become the dominating excitation source.

For a multi-cylinder engine, the engine-unbalanced disturbance depends on the number and arrangement of cylinders in the engine. The engine inertia force can be canceled (or balanced) according to the engine cylinder number and crankshaft arrangement. For example, a four-cylinder engine has a vertical inertia force that acts on the engine block in addition to the oscillating torque about the crankshaft. Six- and eight-cylinder engines have no such inertia force, but they have torque oscillation. Generally, the forces become more balanced and are easier to handle with a greater number of cylinders. The inertia forces in six- and eight-cylinder engines are much smaller than those in two- and four-cylinder engines. Usually, the inertia imbalance force in a diesel engine is higher than that in a gasoline engine.

4.2.2.1.4 Powerplant Isolation Design

The most common method to reduce powerplant excitation is the use of an isolation system. The system analyzed with a six degrees-of-freedom model consists of the powerplant and supporting mounts. This is a complex spring-mass system. Many parameters, such as the powerplant moment of inertia, mass, center of gravity (CG), mount stiffness, and location, influence the effectiveness of isolation. The locations are always constrained by packaging limitations; thus, they must be chosen early in a vehicle development program and compete with other component package space.

The isolation design must perform the following functions:

- Location and support of the powertrain, which requires the mounts to be stiff

- Isolation of the powerplant vibration transferred to the body, which requires that the mounts are soft

- React to powerplant output torque and inertia unbalance

- React to dynamic acceleration and deceleration forces due to starting, stopping, cornering, road inputs, and truck and rail shipment

- Isolation of high-frequency vibration resulting from the powerplant structural elements

- Absorption of road and tire/wheel-induced vibration

The requirements usually conflict with each other. High mount stiffness is required for shock isolation at low frequency, but low stiffness is required for vibration isolation at high frequency. The dynamic stiffness of the mounts usually varies with the frequency. Well-designed hydraulic, semi-active, and active mounts can be closer to satisfying both conditions.

If an engine is chosen and an idle strategy is developed in a vehicle development, then the lowest firing frequency (f_{firing}) is known. The isolation design depends on the excitation frequency as well as the system natural frequencies. Thus, the first objective in vehicle design is to determine the powerplant rigid modes. In particular, the roll frequency is determined first because the roll motion is the most important one that must be isolated. Keep in mind that the firing pulses will cause torque to act on the engine block around the roll axle.

Vibration isolation theory can be used as a guide for the frequency choice. Let's consider a single-degree-of-freedom isolation system as shown in Figure 4.14. Applying Newton's law, the dynamic equation can be written as

$$M\ddot{z} = -C(\dot{z} - \dot{z}_0) - K(z - z_0) \qquad (4.22)$$

where M, C, and K are the mass, damping coefficient, and stiffness of the single-degree-of-freedom system, respectively.

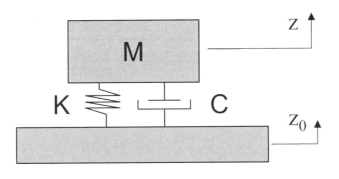

Figure 4.14 A single-degree-of-freedom isolation system.

The transmissibility, T, between the response z and excitation z_0 is given by

$$T = \frac{Z}{Z_0} = \frac{1 + j2\xi\lambda}{1 - \lambda^2 + j2\xi\lambda} \tag{4.23}$$

where $\lambda = \dfrac{\omega}{\omega_n}$ is the frequency ratio, ω is the excitation frequency, $\omega_n = \sqrt{\dfrac{K}{M}}$ is the natural frequency, and $\xi = \dfrac{C}{2\sqrt{KM}}$ is the damping ratio.

The magnitude of the transmissibility can be expressed as

$$|T| = \left|\frac{Z}{Z_0}\right| = \sqrt{\frac{1 + (2\xi\lambda)^2}{\left(1 - \lambda^2\right)^2 + (2\xi\lambda)^2}} \tag{4.24}$$

Figure 4.15 shows the transmissibility curves. The vibration can be reduced only in the region where the ratio between the excitation frequency and the natural frequency is greater than $\sqrt{2}$, that is, the isolation must meet the following condition of

$$\lambda = \frac{\omega}{\omega_n} = \frac{f_{firing}}{f_{roll}} > \sqrt{2} \tag{4.25}$$

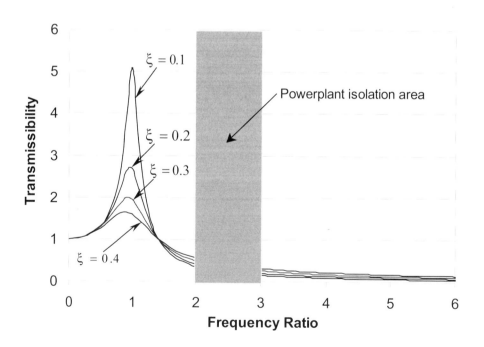

Figure 4.15 Transmissibility of a single-degree-of-freedom isolation system.

The isolation of powerplant roll motion is similar to that of a single-degree-of-freedom system. If the roll frequency f_{roll} satisfies $f_{roll} < \dfrac{f_{firing}}{\sqrt{2}}$, the roll motion will be reduced. It is desirable to design the roll mode frequency as low as possible. The lower the roll frequency, the lower the transmissibility needed. However, a soft mounting system is impractical because the mounting system will have to support the powerplant weight, react to output torque, and resist dynamic acceleration and deceleration. Thus, determining the mount rate becomes one of the key steps in the design of the mount system. To effectively isolate the powerplant roll excitation, in engineering practice the frequency ratio usually is chosen between 2 and 3, as shown in Figure 4.15, that is,

$$\frac{f_{firing}}{f_{roll}} = 2 \sim 3 \qquad (4.26)$$

If the system is effectively isolated at the minimal firing frequency, it will satisfy all excitation frequency ranges. It is desirable to minimize the roll mode frequency because a higher frequency mode will result in higher vibration and noise levels to the occupants inside the vehicle. In fact, an increase of 1 Hz in this frequency will cause an equivalent proportional 4 to 6 dB(A) vibration and noise increase at the occupants' touch points (Qatu *et al.* 2002).

Example E4.1

The idle speed of a six-cylinder engine is 600 rpm, and the firing order is third. The firing frequency is 30 Hz, that is, $f_{firing} = 30$ Hz. Determine the natural frequency of the powerplant roll mode that meets the isolation requirement.

Solution:

According to Eq. 4.26, if the firing frequency is twice the natural frequency, the roll frequency should be

$$f_{roll} = \frac{1}{2} f_{firing} = 15 \text{ Hz}$$

If the firing frequency is three times the natural frequency, the roll frequency should be

$$f_{roll} = \frac{1}{3} f_{firing} = 10 \text{ Hz}$$

Therefore, the roll mode frequency can be chosen between 10 and 15 Hz.

Note that in engineering practice, the roll mode frequency of a six-cylinder east-west engine usually is set between 10 and 12 Hz.

In addition to the roll mode frequency requirement, the acceleration reduction ratio η is used to evaluate the isolation effectiveness of the mounts. The acceleration attenuation is defined as

$$\eta = \frac{a_{active}}{a_{passive}} \qquad (4.27)$$

where a_{active} is the acceleration on the powerplant side (the active side), and $a_{passive}$ is the acceleration on the vehicle frame or subframe (the passive side). For a good isolation system, the ratio should be larger than 10, that is, the vibration transferred to the vehicle should be less than one-tenth of the engine vibration. The criterion can be expressed in decibels as

$$20 \lg \eta = 20 \lg \left(\frac{a_{active}}{a_{passive}} \right) > 20 \text{ dB} \qquad (4.28)$$

The roll mode frequency depends not only on the rate of its mounts but also on the locations of the mounts. Poor locations could introduce extra load to the system. To effectively design the isolation system, the mount locations should be chosen to allow the powerplant to move around its free-free torque roll axis. The free-free roll axis is the torque roll axis of the object when there are no external restraints to oppose its motion. This axis always goes through the center of gravity and depends only on the axis of the applied torque and the inertial properties of the object. If the mounts are located to allow the powerplant to move about the torque axis, no additional forces will be generated within the mounting system to cause it to move.

Figures 4.11 and 4.16 are two mounting systems. In Figure 4.11, the powerplant is supported on a subframe by three mounts, and the mounts are not located at the roll axis. Thus, extra load will be induced in addition to the roll torque. However, for the design shown in Figure 4.16, two mounts pass through the roll axis, and the extra load is avoided. The upper mount usually is fixed on the vehicle body, such as on the shock tower. This system is called a pendulum mounting system, which is the desirable system in mount design. However, it is not that easy to secure the desired pendulum mounting design because of package limitations and other requirements.

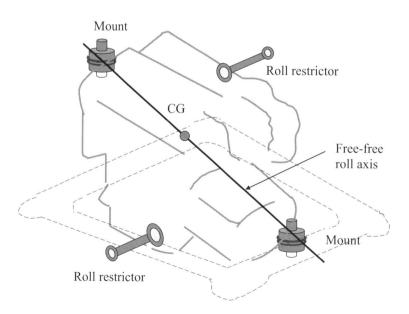

Figure 4.16 A power-plant pendulum mounting system.

Modal decoupling: An excitation is applied on a powerplant to cause a motion in one mode, but other modes are excited, too. The two modes are coupled with each other. Some of the six modes of the powerplant could be coupled. If the modes are coupled, it will be hard to isolate a single mode, which makes the isolation problem rather complicated. It is desired to decouple these six modes. Decoupling makes each mode move independently of the other modes. Complete decoupling may not be possible, and the objective then becomes to minimize the coupling. The decoupling of powerplant modes depends on its mount rates, locations, center of gravity, and moment of inertia. Decoupling is another critical issue in powerplant isolation (in addition to the roll mode frequency determination). The desired decoupling is for all modes to be independent, no matter what the excitation is. However, the desired decoupling situation may not be possible in actual cases. Mode energy is used to measure the strength of the coupling between the modes.

A powerplant is assumed to be a rigid body and supported by the mounts, as shown in Figure 4.17. The system has six degrees of freedom: three translation motions, and three rotation motions. The dynamic equations for a powerplant mounting system can be expressed as

$$\begin{cases} M\ddot{x}_c = F_x \\ M\ddot{y}_c = F_y \\ M\ddot{z}_c = F_z \\ I_{xx}\ddot{\alpha}_c - I_{xy}\ddot{\beta}_c - I_{xz}\ddot{\gamma}_c = M_x \\ I_{xy}\ddot{\alpha}_c - I_{yy}\ddot{\beta}_c - I_{yz}\ddot{\gamma}_c = M_y \\ I_{xz}\ddot{\alpha}_c - I_{yz}\ddot{\beta}_c - I_{zz}\ddot{\gamma}_c = M_z \end{cases} \tag{4.29}$$

where M is the powerplant mass; I_{xx}, I_{yy}, and I_{zz} are the moments of inertia; I_{xy}, I_{xz}, and I_{yz} are the products of inertia; x_c, y_c, and z_c are the coordinates of the powerplant center of gravity; α_c, β_c, and γ_c are the rotational angular acceleration around the x axis, y axis, and z axis, respectively; F_x, F_y, and F_z are the summation of all forces acting on the body in the x, y, and z directions, respectively; and M_x, M_y, and M_z are the moments acting around the x axis, y axis, and z axis, respectively.

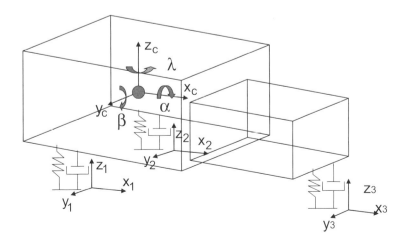

Figure 4.17 A six-degrees-of-freedom powerplant-mounting system.

Assuming harmonic free motion and applying Laplace transforms, the equations will become

$$\left([K] - \omega^2[M]\right)\Phi = \{0\} \tag{4.30}$$

where

$[K]$ = stiffness matrix of the mounting system

$[M]$ = mass matrix of the mounting system

Φ = eigenvector matrix of the six degrees of freedom

ω = natural frequency

Setting the determinant of Eq. 4.30 to zero, the natural frequency and modal shape of the mounting system will be obtained. The matrix of actual mode Φ can be expressed as ψ using the target modes Φ as a basis as

$$\Psi = \Phi^T M \hat{\Phi} \tag{4.31}$$

The modal energy matrix E is defined as the matrix of modal energies expressed as a percent as

$$E = 100\Psi\Psi \tag{4.32}$$

E has the property that the sum of each column and row totals 100%. The percentage of modal energy for each mode presents its coupling and decoupling with other modes. For example, if the roll mode energy at its roll frequency is 95%, then the roll mode is almost decoupled from the other modes. If a modal energy is only 50% of a particular frequency, then the other modes have the other 50% modal energy at this frequency; thus, this mode is not well decoupled.

EXAMPLE E4.2

A powerplant with a V6 engine is supported by three mounts and one roll restrictor. Table 4.3 lists the location and stiffness of the mounts. The powerplant weighs 270 kg. Its center of gravity is located at $x_c = 1700$ mm, $y_c = 0$ mm, and $z_c = 1200$ mm. The moments of inertia are $I_{xx} = 2.08 \text{ e}^7$ kg-mm^2, $I_{yy} = 1.05 \text{ e}^7$ kg-mm^2, and $I_{zz} = 1.95 \text{ e}^7$ kg-mm^2. The products of inertia are $I_{xy} = -1.6 \text{ e}^6$ kg-mm^2, $I_{xz} = 6.5 \text{ e}^5$ kg-mm^2, and $I_{yz} = 3.8 \text{ e}^6$ kg-mm^2. Determine the modal frequencies and the modal energy distribution.

TABLE 4.3
MOUNT LOCATION AND STIFFNESS OF
THE POWERPLANT MOUNTING SYSTEM

	Mount Location (mm)			Mount Stiffness (N/mm)		
	x	y	z	K_x	K_y	K_z
Mount 1	1500	200	1100	150	150	300
Mount 2	1900	200	1150	150	150	300
Mount 3	1750	−500	1050	100	100	300
Roll Restrictor	1750	−50	1600	60	1	1

Solution:

Substitute the parameters in this example into the modal equation and energy equation, and the natural frequency, modal shape, and modal energy can be obtained. Table 4.4 lists the calculation results.

TABLE 4.4
MODAL FREQUENCY AND MODAL ENERGY DISTRIBUTION

	Mode 1	Mode 2	Mode 3	Mode 4	Mode 5	Mode 6
Frequency (Hz)	6.50	5.97	9.00	10.51	7.69	11.42
Fore-aft	98.54	0.00	0.20	1.24	0.00	0.01
Lateral	0.00	97.06	0.40	0.00	0.85	1.69
Bounce	0.03	0.07	89.32	5.87	0.64	4.07
Roll	1.31	0.07	3.47	83.80	10.52	0.83
Pitch	0.00	2.08	4.86	0.42	0.09	92.55
Yaw	0.11	0.73	1.74	8.67	87.89	0.86

The results show that the modal energies for the fore-aft, lateral, and pitch modes are greater than 90%, which means that the modes are almost decoupled from the other modes. The bounce mode and yaw mode have 89.32 and 87.89% energy, respectively. The decoupling of these two modes from the other modes is acceptable. Usually, if the modal energy is greater than 85%, the decoupling of this mode is acceptable. Of course, the higher the modal energy, the better the decoupling. In this example, the roll modal energy is 83.80%. Therefore, the mounting system may need to be further optimized, although the roll mode frequency is as low as 10.51 Hz.

Bracket design: A mount is used to connect the powerplant and the body or subframe through two brackets, as shown in Figure 4.18(a). The function of the brackets is to solidly hold the mount to the powerplant and the body. However, if the stiffness of the brackets is low, they will work in the same way as springs, as shown in Figure 4.18(b).

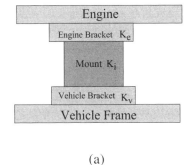

(a)

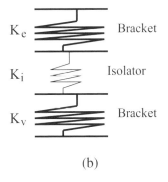

(b)

Figure 4.18 (a) Mount and bracket system, and (b) the stiffness series of the mount-bracket system.

The equivalent stiffness for the series connection of the three springs is

$$\frac{1}{K} = \frac{1}{K_e} + \frac{1}{K_i} + \frac{1}{K_v}$$

(4.33)

where K_e, K_i, and K_v are the stiffnesses of the engine bracket, the mount, and the body bracket, respectively.

If the stiffness of the brackets is too low, the equivalent stiffness of the assembly will be different from the designed stiffness of the mount K_i. Thus, the desirable roll mode frequency and mode decoupling cannot be met. The soft bracket in the mounting attachment also can result in resonance and the loss of isolation. Therefore, brackets on both mount sides must be sufficiently rigid. For rigid brackets, Eq. 4.33 becomes

$$K_e, K_i \rightarrow \infty \qquad K = K_i$$

(4.34)

Two rules are used to guide the bracket design. The first is that the stiffness of the brackets should be 6 to 10 times that of the stiffness of the mounts. The other is that the first natural frequency of the brackets should be above a certain value (500 Hz).

The mount location should make the natural frequency in the effective vibration isolation range and make the powerplant modes decoupled. The mount brackets must possess adequate stiffness. In addition to these requirements, the following conditions should be considered when the mount locations are determined:

- Locations should be chosen to allow the powerplant to move around an axis as close as possible to its free-free torque roll axis.

- Mounts should be located at the points of low powertrain vibration.

- Mounts should be located at points of low body sensitivity.

- Mounts should be located near suspension attachment points for great absorption of suspension disturbances.

- Powerplant vibration modes should be decoupled.

4.2.2.2 Vibration Sources from the Driveline

The main purpose of the driveline is to transmit torque from the powerplant to the wheels. For rear-wheel-drive (RWD) and all-wheel-drive (AWD) vehicles, the driveline includes a front driveshaft, a rear driveshaft (if more than one shaft is used), a front axle, a rear axle, half shafts, and other components. In front-wheel-drive (FWD) applications, half shafts define most of the driveline system. Figure 4.19 shows an all-wheel-drive driveline configuration. All-wheel drive is widely used in cars and sport utility vehicles (SUVs) today. Thus, the vibration sources caused by the driveline become more important to ride quality.

The driveline is supported on the vehicle body by the transmission mount, central bearings, and axle mounts. The driveline vibration will be transferred to the vehicle through the mounts and bearings. The mount and bearing locations are called the transmission

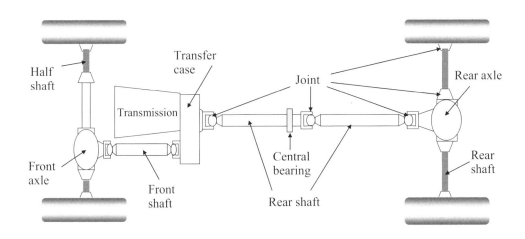

Figure 4.19 A driveline system for rear-wheel-drive-based all-wheel-drive vehicles.

plane, center bearing plane, and axle plane, respectively. The vibration transferred to the body will impact ride comfort. Most of the tactile vibration problems occur at frequencies below 100 Hz.

The driveline excitation sources can be described according to its components. The sources are the four generators: the transmission, driveshaft, front and rear axles, and universal joints. The mechanisms of these sources are divided into driveline imbalance, gear transmission error, and nonconstant velocity joint secondary couple excitation. Other excitations of constant velocity joints (CVJ) can appear similar to the plunging force of these joints. In addition to the component excitation sources, the driveshaft and the modes of the whole driveline and frequencies are important for driveline vibration.

4.2.2.2.1 Driveline Imbalance

The driveline imbalance forces are the results of individual imbalance effects and are related to the driveline rotation speed. The first order excitations are the results of driveline imbalance. The driveline will not produce disturbances on its bearing supports when the driveline is an axis-symmetric structure and its rotation axis coincides with the principal axis. However, this ideal case is not achieved (unless a balancing process is used). Thus, driveline imbalance happens. The driveshaft is the most common component for driveline imbalance. The following reasons can cause the imbalance:

- The driveshaft and other rotating components are not axis-symmetric. Figure 4.20(a) shows the center of gravity of a driveshaft that is offset from its geometric centerline.

- The driveshaft is supported off center on its supporting members.

- The driveshaft is not straight.

- The driveshaft itself deflects off its center (as a result of weight).

- Interface imbalance (runout) occurs. Figure 4.20(b) shows the centerlines of two connection shafts that are offset from each other.

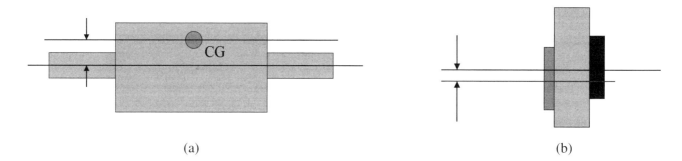

(a)   (b)

Figure 4.20 (a) The center of gravity is offset from its geometric center, and (b) the centerlines of two connection shafts are offset.

The imbalance produces centrifugal force that is proportional to the square of the rotational speed. The imbalance force F can be expressed as

$$F = mr\omega^2 \tag{4.35}$$

where

m = imbalance mass

r = distance from the imbalance mass center to the shaft rotation center

ω = shaft rotating angular speed

The imbalance force is proportional to both the imbalance mass and the offset distance; therefore, the mass and the distance are considered simultaneously as a contribution of the imbalance force, called static imbalance. The unit for the static imbalance is grams centimeter (g-cm). If the static imbalance is fixed, then the centrifugal force increases rapidly with the shaft rotating speed. The contribution by the rotation speed is dynamic imbalance. Figure 4.21 shows the centrifugal force varying with the shaft speed for different imbalance forces (g-cm).

Figure 4.21 The centrifugal force varying with the shaft speed for different static imbalance forces (g-cm).

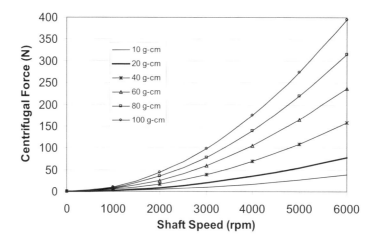

EXAMPLE E4.3

Calculate the imbalance force for a driveshaft with an imbalance of 50 g-cm at 1000 and 6000 rpm, respectively.

Solution:

According to Eq. 4.35, the imbalance force for the driveshaft at 1000 rpm is

$$F_{1000} = 50 \times 10^{-3} \times 10^{-2} \times \left(\frac{2\pi * 1000}{60} \right)^2 = 5.48 \text{ N}$$

The imbalance force at 6000 rpm is

$$F_{6000} = 50 \times 10^{-3} \times 10^{-2} \times \left(\frac{2\pi * 6000}{60} \right)^2 = 197.4 \text{ N}$$

This example shows that the dynamic force due to imbalance is significantly different for different rotation speeds. The imbalance force at 6000 rpm is much higher than that at 1000 rpm.

4.2.2.2.2 Gear Transmission Error

Figure 4.22 shows a pair of the two meshing gears. If a pair of gears is perfectly connected, the following conditions must be satisfied for both gears:

- The gear circles are geometrically perfect.
- The gears are perfectly aligned.
- The gears are infinitely stiff.

For a perfect gear pair, the relation between the two gears can be established as

$$R_1 \omega_1 = R_2 \omega_2 \qquad (4.36)$$

where R_1 and R_2 are the radii of the two gears, respectively, and ω_1 and ω_2 are the angular velocity of the two gears, respectively. In actual cases, the gears are not geometrically perfect, nor are they perfectly aligned or infinitely stiff. Thus, Eq. 4.36 is not always accurate. Instead, the actual meshing relation between the two gears is

$$R_1 \omega_1 \neq R_2 \omega_2 \qquad (4.37)$$

The difference between the actual meshing position and the ideal meshing position of the two gears is called the transmission error (TE) and is expressed as

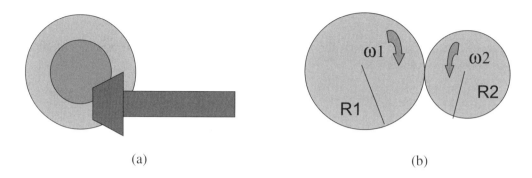

Figure 4.22 Two gear meshing pairs.

(a)

(b)

$$TE = \int_0^t \left(R_1\omega_1 - R_2\omega_2 \right) dt \qquad (4.38)$$

where T is the period of the gears meshing.

The transmission error can be found wherever gear meshing exists, such as in the transmission, the front and rear axles, and the transfer case. The transmission error causes the gear meshing excitation. The excitation will result in the driveshaft bending and in torsional vibration that is transferred to the body through the mounts and bearings. This first order of gear meshing excitation is at the first order of the driveline excitation times the number of teeth in the driving pinion of the axle.

To reduce the excitation caused by transmission error, the gear teeth should be optimized, and the gear meshing or interaction should be designed optimally. The modal separation between the bending and torsional modes of the driveshafts also can reduce the excitation.

4.2.2.2.3 Second Order Excitation

The first order excitation sources are mainly from the driveshaft imbalance. Higher orders are from the transmission error. The excitations caused by second order driveshaft speed are related to the driveline angles and the use of nonconstant velocity joints in the driveshaft, such as Cardan universal joints as shown in Figure 4.23. For a constant joint input speed, the output speeds go up or slow down twice per revolution. Thus, the system induces disturbances with frequencies that are twice the driveshaft rotation frequency. The magnitude of the disturbances increases with increase of joint angle

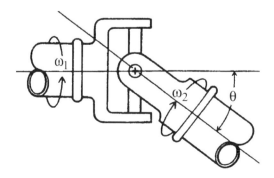

Figure 4.23 Cardan universal joint system.

and torque load. These disturbances usually cause vehicle takeoff shudder or drive pinion flutter at low speeds (10 mph). Universal joints also can induce torsional and inertia vibrations.

4.2.2.2.4 Driveshaft Modes and Driveline Modes

The vibration mechanism caused by components and component interactions (driveline imbalance, TE, second order excitation) is described. As a system, the design of the driveline modes and frequencies is another critical issue. The driveline includes several driveshafts and other components. The driveshaft modes are important in the whole driveline design. The driveshaft natural frequencies must be chosen above the driveline maximum excitation speed. If the driveshaft frequency is lower than the maximum excitation frequency, resonance of the driveline will occur. The driveshaft has a critical speed, which is defined as the shaft speed where the shaft rotating frequency matches the first bending mode/frequency of the shaft. Operation at or near critical speed can result in catastrophic failure of the driveshaft and/or the transmission or the transfer case. The critical speed can be avoided by the following suggestions:

- Increase the shaft stiffness (mostly by the use of a larger diameter) and reduce its weight.

- Use two- or three-piece shafts (reduce the length).

- Balance the shaft components.

- Separate the vertical and lateral P/T modes by at least 15%.

Figure 4.24 shows a simple driveline system. The driveline first bending frequency should be designed above the maximum excitation frequency.

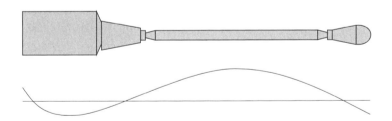

Figure 4.24 A simple driveline system and its first bending mode.

4.2.2.3 Vibration Sources from the Exhaust System

An exhaust system, defined from the exhaust manifold to the tailpipe, consists of a down pipe or Y pipe, catalyst, resonator, muffler, tailpipe, hangers, and other components, as shown in Figure 4.25. The fundamental purpose of the exhaust system is to route the exhaust gases from the engine and to keep the gases from entering the passenger compartment.

The exhaust system has interfaces with the engine and the vehicle body. The vibration is transferred from the powerplant to the exhaust system and then to the body structure through hangers, as shown in Figure 4.25. To reduce the vibration levels, the transfer

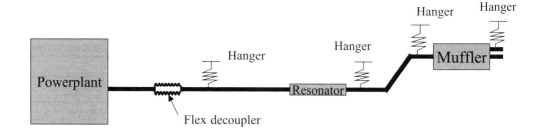

Figure 4.25 An exhaust system.

path is isolated. The flex decoupler is the key element to reducing the vibration transferred from the engine to the exhaust system. Hanger isolators are used to reduce the vibration transferred from the exhaust system to the body.

The decoupler works similarly to an isolator that possesses three linear stiffness values and three rotational ones. To effectively isolate the engine vibration, the decoupler should be placed as close to the powerplant roll axis as possible. The isolation effectiveness is determined by the stiffness of the decoupler and can be evaluated from the transfer function between the vibration responses of the two sides of the decoupler. The stiffness of the flex decoupler also is very sensitive to the hanger force transferred to the body structure. Therefore, the transfer function between the vibration of the two sides of the decoupler and the sensitivity of the hanger force over the decoupler stiffness are critical for the decoupler design.

It is important to understand the exhaust system dynamic characteristics in order to reduce the vibrations coming from the engine and transferring to the body. The exhaust system is a complicated system, and it is almost impossible to solve its vibration problem using traditional analysis. The structure is divided into many finite segments, and finite element analysis (FEA) is used to implement the structure vibration analysis. In finite element analysis, the beam, shell, and solid elements usually are used to mimic the real structure. In some cases, only beam elements are used to analyze the exhaust system for a simple and quick analysis to determine the locations of the hangers.

There are two major aspects in vibration analysis: modal analysis, and dynamic analysis. The exhaust system modes are critical for exhaust dynamic analysis. A normal modal analysis should be performed to ensure that the frequencies of the exhaust system do not line up with the frequencies of the powerplant and/or the body. Also, the same analysis is helpful in determining the nodal points on the exhaust system. These nodal points can be used as the locations where hanger rods can be attached to the exhaust system. This will minimize vibrational energy transferred to the body.

To mimic closely the real structure and engine excitation, we strongly suggest that a powerplant be included in the exhaust finite element model, as shown in Figure 4.26.

The powerplant in the model is represented by five rigid bar elements where the powerplant mass and the moments of inertia are added on the powerplant center of gravity. The powerplant is supported by engine mounts through three bar elements. The engine mounts are fixed (or grounded) to form the boundary condition. The exhaust is connected to the powerplant by two rigid bar elements. The engine load is applied on the location of the center of gravity. There are two advantages to modeling the exhaust with the powerplant: one is for load application, and the other is to check modal separation

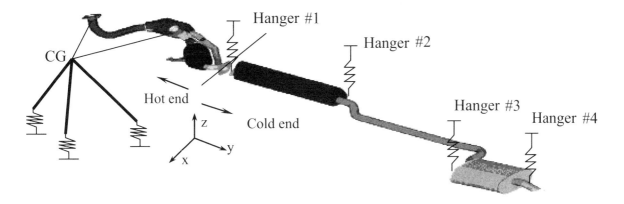

Hanger #1
Hanger #2
Hanger #3
Hanger #4
CG
Hot end
Cold end
z
y
x

Figure 4.26 An exhaust modeling and boundary condition.

between the powerplant and exhaust modes. The model could be called the exhaust-powerplant model. Although some engineers use the "big mass method" in vibration analysis, the powerplant information will be lost, and a real load cannot be applied to system.

The dynamic equation for the system can be expressed as

$$[M]\{\ddot{X}\} + [C]\{\dot{X}\} + [K]\{X\} = \{F\} \qquad (4.39)$$

where $[M]$, $[C]$, and $[K]$ are the mass matrix, damping matrix, and stiffness matrix of the exhaust-powerplant system, respectively. $\{X\}$ and $\{F\}$ are the displacement and force vectors, respectively.

In the modal analysis, the right side of Eq. 4.39 is set to zero. Most of the important modes to be considered are below 150 Hz. The exhaust first vertical bending frequency and first lateral bending frequency must be 10% away from engine idle firing frequency. For example, the idle speed for a six-cylinder engine is 600 rpm, and the corresponding firing frequency is 30 Hz. Thus, the exhaust bending frequencies must be avoided between 27 and 33 Hz. A straight exhaust system has less modal density than a curved exhaust system. Therefore, a straight exhaust system generally is recommended, from a modal analysis point of view.

The lineup of the exhaust modes with the body modes must be avoided because the exhaust system and the body structure are connected through hangers. The choice of hanger location is made by modal analysis. Theoretically, the hangers should be located at or near the nodes of the exhaust system and at points of low body sensitivity. Nodes typically occur near masses such as mufflers, making them good candidates for hangers. Package is another problem limiting hanger location, and a careful tradeoff analysis between noise, vibration, and harshness (NVH) and package must be considered.

The purpose of dynamic analysis is to determine the hanger force transferred to the body structure. The hanger forces depend on the engine excitation, as well as the number, stiffness, and locations of the hangers and the stiffness of the flex decoupler. The force should be as low as possible and within a certain target. The excitation is critical in the analysis. Several excitation methods of dynamic analysis are described next.

The first method is the large mass method. In this method, the powerplant is assumed to be a large mass, and only the roll rotation degree of freedom is set free, while the other five degrees of freedom remain fixed. A certain amount of acceleration is applied to the mass as system excitation. As mentioned, the method has a limitation because the powerplant information is lost, and its modes cannot be found.

The second method is that the excitation is put on the powerplant center of gravity. In this method, the exhaust system is modeled with the powerplant, and the powerplant is supported with the engine mount. Thus, the excitation resource information and powerplant modes are retained. A torque around the roll axis, M_x, is applied on the powerplant center of gravity as an excitation. The generic criterion is that the dynamic force in hangers, F_{hanger}, must be less than 10 N, that is,

$$F_{hanger} \leq 10 \text{ N} \quad \text{under condition} \quad M_x = 100,000 \text{ N} - \text{mm} \quad (4.40)$$

The third method is the same as the second method in the structure, but the load is different. In this method, a real load exciting the exhaust will be established. For a certain vehicle, a hanger force target usually is set by benchmarking other vehicles or from a general acoustical transfer function equation. To simulate a certain target force response, an actual powerplant excitation is needed. However, it usually is difficult to measure the actual powerplant excitation.

Force response usually is considered from 15 to 150 Hz. Frequencies below 15 Hz will not be considered. For an engine with the lowest idle speed of 600 rpm, the firing frequency is 20 Hz for four-cylinder engines and 30 Hz for six-cylinder engines, respectively. The exhaust modes below the firing frequency are hard to excite. Also, it is not easy to excite high-frequency modes. The most prominent modes of an exhaust system are between 20 and 50 Hz. Figure 4.27 shows a typical hanger force response of an exhaust system.

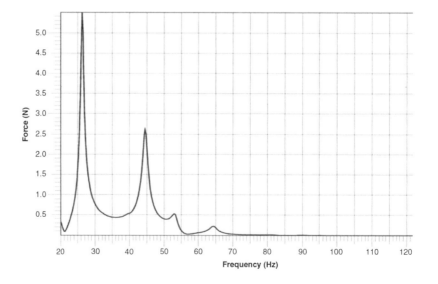

Figure 4.27 A typical hanger force response of an exhaust system.

Usually, there are two brackets (sometimes called rods) for a hanger set: a body side bracket, and an exhaust side bracket. The body side bracket is attached to the body and is used to support the isolator. The exhaust side bracket is attached to the exhaust system. The bracket designer must avoid cantilevered brackets because such a design will result in a lower bracket natural frequency and higher stress in the bracket and force in the hanger. The bracket also should be stiff enough to avoid resonance with connected structures. Currently, the common criterion used to guarantee natural frequency to reach a satisfied target can be expressed as

$$\text{Natural frequency} = \frac{\text{maximum engine rpm * number of cylinders}}{120} \qquad (4.41)$$

For example, for a six-cylinder engine with a maximum speed of 6000 rpm, the recommended minimum natural frequency is $6000 * \dfrac{6}{120} = 300 \text{ Hz}$. This principle applies to both the exhaust and body side brackets.

The exhaust excitation contributed to the body comes not only from the engine but also from the exhaust acoustic excitation. When an exhaust acoustic mode (resonance of gas inside the exhaust system) is excited with its structural mode, the exhaust system likewise will be excited. Usually, this excitation is not as critical as that of the engine, but in some sensitive frequencies, the excitation should be considered.

4.3 Vehicle Ride Models

A vehicle model is very important in analyzing ride dynamics. The vehicle structure is a complicated system with many degrees of freedom. The vehicle can be modeled in many ways according to the type of response. For example, in automobile engineering, a finite element model usually is used to calculate the vehicle body modes and dynamic response. However, building and running the finite element model is complicated, time consuming, and costly. In a ride quality analysis, only a few low frequencies are considered. Approximate solutions are enough to determine vehicle parameters, such as suspension damping and stiffness. Thus, simplified models will satisfy the requirements.

In ride dynamics analysis, a vehicle can be treated as a body installed on its suspension system that is supported by the tires and axles. The body and suspension constitute the sprung mass. The tires and axles constitute the unsprung mass. The two simplified models used in this chapter are the quarter car model and the bounce-pitch model. These models will be described in detail.

The purpose of ride dynamics is to find the relation between the response at the occupants' touch points and the vibration sources. In this section, vehicle body response will be the focus. The vehicle response will be covered in later sections. As discussed in the previous section, two major vibration sources are input to the body. One is from the uneven road profile, and the other is from the vehicle, such as the engine, driveline, and exhaust. Thus, the next section will analyze the relation between the body response and the road excitation, and the relation between the body response and the vehicle excitation.

4.3.1 Quarter Car Model

4.3.1.1 Modeling for the Quarter Car Model

A vehicle typically is a symmetric structure around the x axis of the vehicle. The vehicle body is supported by two front wheels and two rear wheels, and the body motions usually are coupled with the front and rear wheels. However, in many cases, the front and rear wheel motions are independent of each other. Thus, a quarter model, as shown in Figure 4.28, usually is used to analyze the ride characteristics. In this model, only the vertical vibration is analyzed.

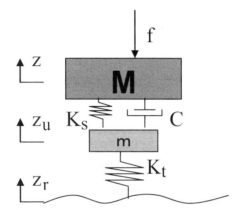

Figure 4.28 A quarter car model.

The quarter car model is a two-degrees-of-freedom model with two masses. The upper mass represents the vehicle body and the suspension mass, called sprung mass M. The lower mass is the weight of the tire and its axle, called the unsprung mass m. The quarter car model shows two frequencies associated with a vehicle suspension system.

Dynamic equations for the quarter car model can be obtained by applying Newton's second law as

$$M\ddot{z} + C\dot{z} + K_s z = K_s z_u + C\dot{z}_u + f \tag{4.42}$$

$$m\ddot{z}_u + C\dot{z}_u + (K_s + K_t)z_u = K_s z + C\dot{z} + K_t z_r \tag{4.43}$$

where

M = sprung mass

m = unsprung mass

C = suspension damping coefficient

K_s = suspension stiffness

K_t = tire stiffness

z = sprung mass displacement

z_u = unsprung mass displacement

z_r = road displacement input

f = vehicle internal excitation force

The displacement solutions and force excitation can be assumed as

$$z = Ze^{j\omega t}$$

$$z_u = Z_u e^{j\omega t}$$

$$z_r = Z_r e^{j\omega t}$$

$$f = Fe^{j\omega t}$$

(4.44)

Substituting Eq. 4.44 into Eqs. 4.42 and 4.43, the equations are transferred into the frequency domain as

$$\left(K_s - M\omega^2 + jC\omega\right)Z = \left(K_s + jC\omega\right)Z_u + F$$

(4.45)

$$\left(K_s + K_t - m\omega^2 + jC\omega\right)Z_u = \left(K_s + jC\omega\right)Z + K_t Z_r$$

(4.46)

For a passenger car, the sprung mass is approximately an order higher than the unsprung mass, whereas the sprung stiffness is approximately an order lower than the unsprung stiffness. The unsprung mass is neglected, and the two springs are connected in series. The approximate frequency ω_s for the sprung system can be obtained by

$$\omega_s = \sqrt{\frac{K_s K_t}{\left(K_s + K_t\right)M}}$$

(4.47)

Usually, the sprung mass is about ten times the unsprung mass (i.e., $M \approx 10\, m$), and the unsprung stiffness is about nine times the sprung stiffness (i.e., $K_t \approx 9\, K_s$). Thus, Eq. 4.47 can be simplified as

$$\omega_s = \sqrt{\frac{K_s K_t}{\left(K_s + K_t\right)M}} \approx \sqrt{\frac{K_s}{M}} = \omega_s{}'$$

(4.48)

where, $\omega_s{}'$ is the approximate value of ω_s.

The approximate frequency for the unsprung system can be calculated based on the assumption that the vehicle body is fixed and only the tire moves. The frequency is

$$\omega_u = \sqrt{\frac{K_s + K_t}{m}}$$

(4.49)

Usually, the sprung system frequency is around 1 Hz, whereas the unsprung system frequency is around 10 Hz. The two frequencies are so widely separated that the approximate frequencies are useful to understanding the system dynamic characteristics. The reason that the sprung system is designed for 1-Hz natural frequency is that this frequency is close to the walking pace frequency and generally does not excite resonances within the human body.

4.3.1.2 Modal Analysis for the Quarter Car Model

The damping ratio provided by the shock absorber is approximately 0.2 to 0.4. Thus, the damped natural frequency has little difference from the undamped natural frequency. To investigate the system natural frequencies and modal shapes, damping is neglected, and external excitation is not included. Substitute Eqs. 4.48 and 4.49 into Eqs. 4.45 and 4.46, and Eqs. 4.45 and 4.46 can be rewritten as

$$\left(\omega_s'^2 - \omega^2\right)Z - \omega_s'^2 Z_u = 0 \tag{4.50}$$

$$-\frac{K_s}{m}Z + \left(\omega_u^2 - \omega^2\right)Z_u = 0 \tag{4.51}$$

The characteristic equation of these equations can be written as

$$\omega^4 - \left(\omega_s'^2 + \omega_u^2\right)\omega^2 + \omega_s'^2\omega_u^2 - \omega_s'^2\frac{K_s}{m} = 0 \tag{4.52}$$

The two real characteristic values of the solutions of these equation are the frequencies of the system, which are

$$\omega_{1,2}^2 = \frac{\omega_s'^2 + \omega_u^2}{2} \pm \sqrt{\frac{1}{4}\left(\omega_s'^2 - \omega_u^2\right)^2 + \omega_s'^2\frac{K_s}{m}} \tag{4.53}$$

According to these relations between the sprung and unsprung masses, and between the sprung and unsprung stiffnesses, the approximate relation between the sprung and unsprung frequencies can be obtained as

$$\omega_u = \sqrt{\frac{K_s + K_t}{m}}$$

$$= \sqrt{\frac{K_s + 9K_s}{\frac{M}{10}}} \tag{4.54}$$

$$= 10\sqrt{\frac{K_s}{M}} = 10\omega_s'$$

If Eq. 4.54 is substituted into Eq. 4.53, the two frequencies of the system are

$$\omega_1 = 0.95\omega_s'$$

$$\omega_2 = 10\omega_s' \tag{4.55}$$

The results show that the first frequency of the system is close to the sprung system frequency, and the second frequency is close to the unsprung system frequency.

The mode shapes corresponding to these two natural frequencies can be obtained by substituting Eq. 4.55 into Eq. 4.50 or Eq. 4.51. The relation between Z and Z_u can be given by either Eq. 4.50 or Eq. 4.51. According to Eq. 4.50, the relation is

$$\frac{Z}{Z_u} = \frac{\omega_s'^2}{\omega_s'^2 - \omega^2} \tag{4.56}$$

Substitute Eq. 4.55 into Eq. 4.56, and the modal shapes of the two-degrees-of-freedom system will be obtained.

The first modal shape is

$$\left(\frac{Z}{Z_u}\right)_1 = \frac{\omega_s'^2}{\omega_s'^2 - \left(0.95\omega_s'\right)^2} = 10.26$$

The second modal shape is

$$\left(\frac{Z}{Z_u}\right)_2 = \frac{\omega_s'^2}{\omega_s'^2 - \left(10\omega_s'\right)^2} = -0.01$$

or

$$\left(\frac{Z_u}{Z}\right)_2 = -100$$

Figure 4.29 shows the two modal shapes for the two-degrees-of-freedom system. When the excitation frequency ω is close to the first natural frequency ω_1, the system will vibrate at frequency ω_1, and the mode shape is the first order mode shape. The body vibration is about ten times larger than that of the tires, as shown in Figure 4.29(b). The apparent motion comes from the body; thus, the vibration of the tires can be neglected. The first mode is called the body mode. On the other hand, if the excitation frequency is close to the second natural frequency ω_2, the amplitude of the tires is about 100 times larger than motion of the body, as shown in Figure 4.29(c). The body is almost at a static position, and the mode is called the tire mode. The separation of the two modes is significant for vibration isolation. One benefit is that the two modes can be controlled independently. The other benefit is that the body can be isolated effectively if the vehicle hits a bump and oscillates at the unsprung frequency.

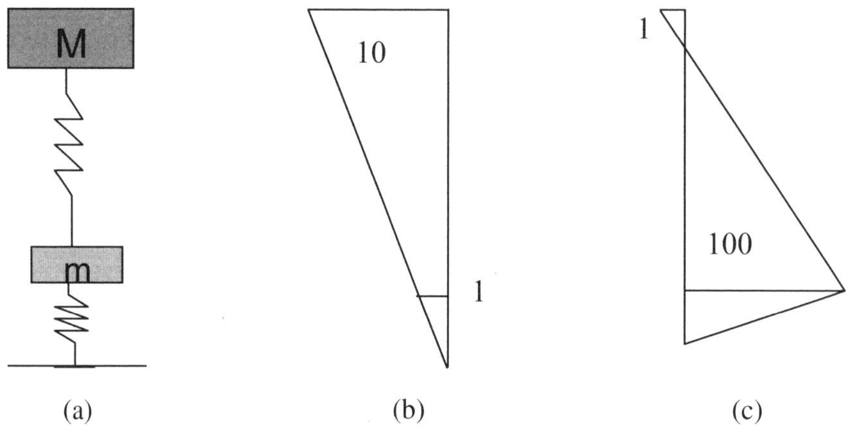

Figure 4.29 The quarter car model and its two modal shapes: (a) the quarter car model, (b) first mode, and (c) second mode.

(a)　　　　　　　　(b)　　　　　　　　(c)

4.3.1.3 Dynamic Analysis for the Quarter Car Model

When the vehicle travels over an undulating road, the excitation includes multi-frequency input. As mentioned, high frequency can be isolated effectively by the suspension, but low-frequency excitation will be transferred to the body and may even be amplified. Thus, the dynamic response must be analyzed over a wide frequency spectrum. The vibration sources are either from road excitation or from vehicle excitation. The body responses caused by the two sources will be analyzed. Transmissibility between the body response and road input, and the transmissibility between the body response and vehicle excitation, will be described.

4.3.1.3.1 Transmissibility Between the Body Response and Road Excitation

If the vehicle excitation is neglected and only the road excitation is considered (i.e., $F = 0$ in Eq. 4.45)), then the relation between the displacement at the sprung mass Z and that at the unsprung mass Z_u can be obtained as

$$\frac{Z_u}{Z} = \frac{\left(K_s - M\omega^2 + jC\omega\right)}{\left(K_s + jC\omega\right)} \tag{4.57}$$

Substituting Eq. 4.57 into Eq. 4.46, the transmissibility between the body response and road excitation Z_r is given as

$$\frac{Z}{Z_r} = \frac{K_t\left(K_s + jC\omega\right)}{\left(K_s + K_t - m\omega^2 + jC\omega\right)\left(K_s - M\omega^2 + jC\omega\right) - \left(K_s + jC\omega\right)^2} \tag{4.58}$$

The magnitude of the transmissibility can be expressed as

$$\left|\frac{Z}{Z_r}\right| = \frac{1}{P^2 + Q^2}\sqrt{\left(K_t K_s P + K_t CQ\omega\right)^2 + \left(K_t CP\omega - K_t K_s Q\right)^2} \tag{4.59}$$

where

$$P = mM\omega^2 - \left(K_s M + K_t M + K_s m\right)\omega^2 + K_s K_t$$

and

$$Q = K_t C\omega - C\left(m + M\right)\omega^3$$

The transmissibility must be controlled as low as possible, and the resonance should be avoided. Some factors, such as the damping of the shock absorber, the mass ratio, and the stiffness ratio between the sprung and unsprung systems, will influence the transmissibility. The influence of these parameters will be discussed.

Damping ratio: A system with the parameters of

$$M = 500 \text{ kg}$$

$$K_t = 180,000 \text{ N/m}$$

$$m = \frac{1}{10}M$$

$$K_s = \frac{1}{9}K_t$$

will be used as an example to analyze the influence of damping on transmissibility. The damping ratio

$$\xi = \frac{C}{2\sqrt{K_s M}}$$

is set to 0.1, 0.2, and 0.3, respectively. Figure 4.30 shows the transmissibility of the system with different damping.

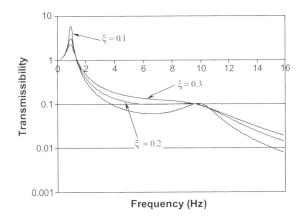

Figure 4.30 Transmissibility between the body response and road excitation with different damping.

Figure 4.30 shows two peaks of the transmissibility magnitude. One peak is at 1 Hz, and the other is at 10 Hz. The 1-Hz resonant frequency comes from the sprung system. The sprung system is designed deliberately to have the specific frequency that is in the range of the comfortable frequencies of the occupants. The 10-Hz frequency coming from the unsprung system also avoids the range of 4 to 8 Hz in which the human body is uncomfortable.

The transmissibility is reduced significantly with the damping increase at the resonant frequency of the sprung system. However, the resonant frequencies do not change with the damping variation, which indicates that the damping has little impact on the frequency. In the range between the sprung resonant frequency and the unsprung resonant frequency, the lower the damping, the better the effectiveness of isolation, which is the same conclusion as that made for a single-degree-of-freedom isolation system. At the unsprung resonant frequency, the peaks are almost the same for the three different dampings, which means that the damping does not impact the response at this frequency. However, above the unsprung resonant frequency, the higher the damping, the better the isolation.

Mass ratio: Based on the previous example, the system parameters are set to

$$M = 500 \, \text{kg}$$

$$K_t = 180,000 \, \text{N/m}$$

$$K_s = \frac{1}{9} K_t$$

$$\xi = 0.2$$

The effect of the mass ratio change on the transmissibility is studied. The mass ratio M/m is set to 5, 10, and 15, respectively.

Figure 4.31 shows the transmissibility changes with the mass ratio. The magnitude and resonant frequency for the sprung system are the same for different mass ratios. The change in the unsprung mass does not impact the performance of the sprung system. However, the response of the unsprung frequency shifts a lot. The lighter the mass, the higher the resonant frequency. Lighter mass also provides lower transmissibility magnitude between the sprung and unsprung frequencies, but the magnitude is increased above the unsprung frequency.

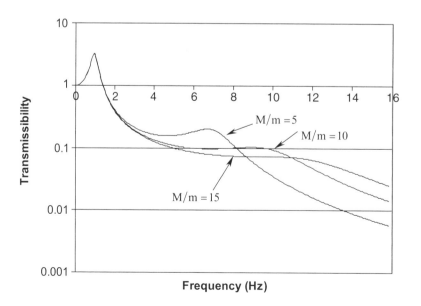

Figure 4.31 Transmissibility between the body response and road excitation with different mass ratios.

Stiffness ratio: The system parameters are reset as

$$M = 500 \text{ kg}$$

$$K_t = 180{,}000 \text{ N/m}$$

$$m = \frac{1}{10}M$$

$$\xi = 0.2$$

The influence of the stiffness ratio variation on transmissibility is studied. The stiffness ratio, $\dfrac{K_t}{K_s}$, is set to 7, 9, and 11, respectively.

Figure 4.32 shows the transmissibility change with the variation in stiffness ratio. At sprung frequencies, the higher the stiffness ratio, the lower the sprung frequency. This indicates that the suspension stiffness becomes relatively softer when the tire stiffness is increased. At the same time, the transmissibility magnitude decreases with the stiffness ratio increase. The higher stiffness ratio also has lower magnitude between the sprung and unsprung frequencies. Thus, the soft suspension will provide better isolation. However, at the unsprung frequency, the influence of the stiffness ratio on the magnitude and frequency is insignificant. Above the unsprung frequency, the magnitude decreases with the stiffness ratio increase.

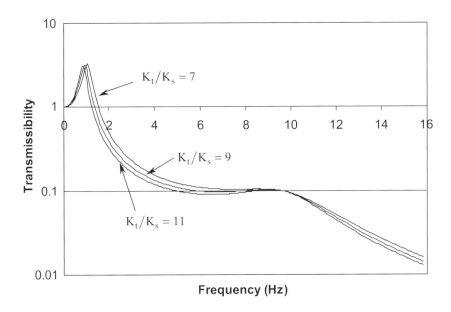

Figure 4.32 Transmissibility between the body response and road excitation with different stiffness ratios.

EXAMPLE E4.4

A quarter car model is used for a vehicle that has the following parameters:

$$M = 500 \text{ kg}$$

$$K_t = 180{,}000 \text{ N/m}$$

$$m = \frac{1}{10}M$$

$$K_s = \frac{1}{9}K_t$$

$$\xi = 0.3$$

A single-degree-of-freedom system, shown in Figure 4.14, has the same parameters as the sprung system of the quarter car model. Compare the transmissibility of the single-degree-of-freedom model with that of the quarter car model.

Solution:

The transmissibility of the single-degree-of-freedom model, as shown in Figure 4.14, is

$$\frac{Z_s}{Z_{r_s}} = \frac{K_s + j\omega C}{K_s - M\omega^2 + jC\omega} \tag{4.60}$$

where Z_s and Z_{r_s} are the displacement magnitudes of the body and road systems, respectively. The transmissibility magnitude of the system is

$$\left|\frac{Z_s}{Z_{r_s}}\right| = \sqrt{\frac{K_s^2 + \left(C\omega^2\right)^2}{\left(K_s - M\omega^2\right)^2 + \left(C\omega\right)^2}} \tag{4.61}$$

Figure 4.33 plots the transmissibility of the single-degree-of-freedom model and the quarter car model. The two curves have the same trend. At the sprung frequency, the transmissibility magnitude of the quarter car model is 2.29, whereas the number is 1.99 for the single-degree-of-freedom model. The quarter car model has a second resonant

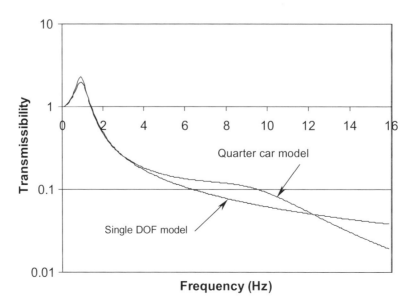

Figure 4.33 Transmissibility of the single-degree-of-freedom model and the quarter car model.

peak at 10 Hz, whereas the single-degree-of-freedom model does not. The difference between the two models shows the coupling effectiveness between the sprung system and the unsprung system in the quarter car model. The coupling makes isolation effectiveness worse than the single-degree-of-freedom system.

4.3.1.3.2 Transmissibility Between the Body Response and Vehicle Excitation

The transmissibility between the body response and road excitation has been analyzed. The powertrain is another important excitation source to the body response. It is essential to analyze the transmissibility between the body response and powertrain vibration. In this case, the road excitation is neglected (i.e., $Z_r = 0$). The relation between the displacement at the sprung mass Z and the displacement at the unsprung mass Z_u can be obtained from Eq. 4.46 as

$$\frac{Z_u}{Z} = \frac{K_s + jC\omega}{K_s + K_t - m\omega^2 + jC\omega} \tag{4.62}$$

Substitute Eq. 4.62 into Eq. 4.45, and the ratio between the body displacement and the vehicle excitation force as shown in Figure 4.28 is given as

$$\frac{Z}{F} = \frac{K_s + K_t - m\omega^2 + jC\omega}{\left(K_s - M\omega^2 + jC\omega\right)\left(K_s + K_t - m\omega^2 + jC\omega\right) - \left(K_s + jC\omega\right)^2} \tag{4.63}$$

The magnitude of the ratio can be expressed as

$$\left|\frac{Z}{F}\right| = \frac{1}{P^2 + Q^2}\sqrt{\left(\left(K_s + K_t - m\omega^2\right)P + QC\omega\right)^2 + \left(CP\omega - Q\left(K_s + K_t - m\omega^2\right)\right)^2} \tag{4.64}$$

Transmissibility usually is a non-unit value. However, the ratio of displacement over force in Eq. 4.64 has its unit. The ratio should be transferred to transmissibility to generally express the relation between response and excitation. The relation between the acceleration magnitude and the displacement magnitude can be expressed as

$$\left|\ddot{Z}\right| = \left|Z\right|\omega^2 \tag{4.65}$$

If the excitation force is divided by the mass, then the input acceleration can be obtained. Thus, the transmissibility can be written as

$$\left|\frac{\ddot{Z}}{\frac{F}{M}}\right| = \frac{M\omega^2}{P^2 + Q^2}\sqrt{\left(\left(K_s + K_t - m\omega^2\right)P + QC\omega\right)^2 + \left(CP\omega - Q\left(K_s + K_t - m\omega^2\right)\right)^2} \tag{4.66}$$

Some parameters, such as damping, stiffness ratio, and mass ratio, will influence the transmissibility. Three cases with different damping ratios, different mass ratios, and different stiffness ratios, respectively, will be discussed in the following material, similar to what was discussed in Section 4.3.1.3.1 "Transmissibility Between the Body Response and Road Excitation."

Damping ratio: Figure 4.34 shows the transmissibility for three damping ratios, where $\xi = 0.1$, 0.2, and 0.3. The results show that damping has significant influence on the transmissibility magnitude at the sprung frequency. The higher the damping, the lower the magnitude. The damping has almost no impact on the transmissibility beyond the sprung frequency range. Thus, the body response peak can be controlled independently by the suspension damping.

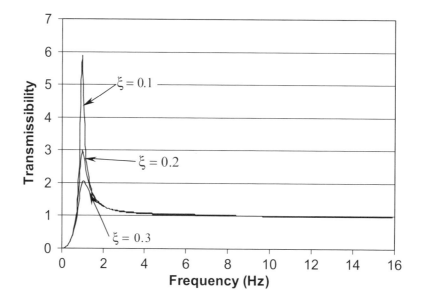

Figure 4.34 Transmissibility between the body response and vehicle excitation with three different damping ratios.

Mass ratio: Figure 4.35 shows the transmissibility for three mass ratios, where $M/m = 5$, 10, and 15. The variation of the mass ratio has little impact on the transmissibility at the unsprung frequency. For the rest of the frequency range, transmissibility remains the same for different mass ratios. The result shows that the weight design of the tire and axles is independent of the ride response.

Stiffness ratio: Figure 4.36 shows the transmissibility for three stiffness ratios, where $K_t/K_s = 7$, 9, and 11. The magnitude and sprung frequency become smaller with the increase in stiffness ratio, which indicates that softer suspensions are better for vibration isolation.

Analysis of the three cases shows that changing the damping, mass, and stiffness ratios influences only the body response at the sprung frequency. Vibration control for these cases is similar to that of the single-degree-of-freedom system.

4.3.1.3.3 Dynamic Response at Random Input

Transmissibility, or transfer function, is used to describe the relation between response and input. The transmissibility between the unsprung mass and the road is

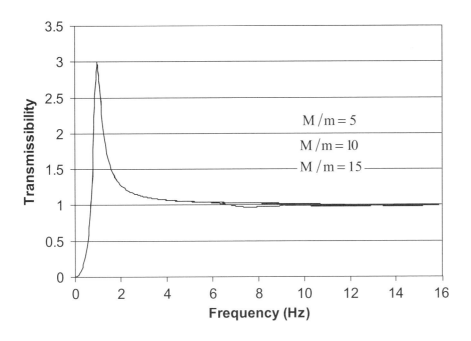

Figure 4.35 Transmissibility between the body response and vehicle excitation with three different mass ratios.

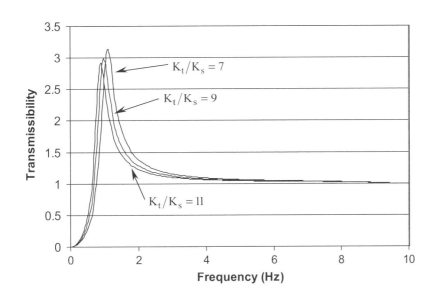

Figure 4.36 Transmissibility between the response and vehicle excitation with three different stiffness ratios.

$$H_1(f) = \frac{Z_u}{Z_r} \tag{4.67}$$

The transmissibility between the sprung and unsprung masses is

$$H_2(f) = \frac{Z}{Z_u} \tag{4.68}$$

Thus, the response on the sprung mass will be

$$z(t) = |H_1(f)||H_2(f)|z_r(t) \tag{4.69}$$

The mean square values of the vehicle response can be expressed by the transmissibility and the mean square values of road input

$$\bar{z}^2 = |H_1(f)|^2 |H_2(f)|^2 \bar{z}_r^2 \tag{4.70}$$

Equation 4.15, the power spectral densities relation between the vehicle response and road input, can be written as

$$W(f) = |H_1(f)|^2 |H_2(f)|^2 W_z(f) \tag{4.71}$$

This formula shows that if the road input density is given, the vehicle response will be obtained automatically as long as the vehicle ride model is provided.

4.3.2 Bounce-Pitch Model

A vehicle has both vertical and longitudinal dimensions, which means that the vehicle has a bounce motion and a pitch motion. As a vehicle travels on a road, the road excitations at the front and rear axles are not independent. The input to the rear axle is the same as that to the front axle but in a delayed time. The delayed time depends on the speed of the vehicle and the wheelbase. If the road wavelength is equal to or longer than the wheelbase, the vehicle will have only bounce motion. Otherwise, pitch motion does exist. Pitch motions are important because they are the primary sources of longitudinal vibrations at the location above the center of gravity. Any vertical and longitudinal vibrations on the vehicle could be a combination of bounce and pitch motions.

The quarter car model provides important information on the frequency and dynamic response at the vehicle body and axle that is useful for analysis of ride quality. However, only bounce motion is considered in the quarter car model. Thus, it is necessary to establish a model that includes both bounce and pitch modes. To comprehensively understand the pitch and bounce vibrations, a two-degrees-of-freedom model is used, as shown in Figure 4.37. In this model, the tires and suspension are supported by two separate springs. One spring is located at the front of the vehicle, and the other is located at the rear of the vehicle. The body is assumed to be rigid, and damping and the unsprung mass are neglected. The model is called the bounce-pitch model. By applying Newton's second law, the dynamic equations of the system can be expressed as

$$M\ddot{z} = -k_f(z + \ell_1\theta) - k_r(z - \ell_2\theta) \tag{4.72}$$

$$I_y\ddot{\theta} = -k_f(z + \ell_1\theta)\ell_1 + k_r(z - \ell_2\theta)\ell_2 \tag{4.73}$$

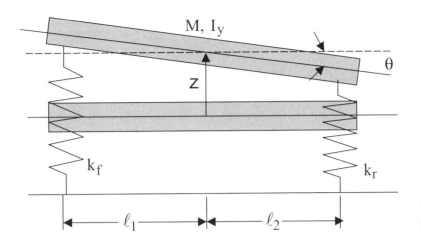

Figure 4.37 The bounce-pitch model.

where

k_f = stiffness of the front ride

k_r = stiffness of the rear ride

ℓ_1 = distance between the front axle and the center of gravity

ℓ_2 = distance between the rear axle and the center of gravity

I_y = pitch moment of inertia

Equations 4.72 and 4.73 can be rewritten as

$$\ddot{z} + \frac{k_f + k_r}{M}z + \frac{k_f\ell_1 - k_r\ell_2}{M}\theta = 0 \tag{4.74}$$

$$\ddot{\theta} + \frac{k_f\ell_1 - k_r\ell_2}{I_y}z + \frac{k_f\ell_1^2 + k_r\ell_2^2}{I_y}\theta = 0 \tag{4.75}$$

Assume

$$A_1 = \frac{k_f + k_r}{M} \tag{4.76}$$

$$A_2 = \frac{k_f\ell_1 - k_r\ell_2}{M} \tag{4.77}$$

$$A_3 = \frac{k_f\ell_1^2 + k_r\ell_2^2}{I_y}$$

$$= \frac{k_f\ell_1^2 + k_r\ell_2^2}{Mr_y^2} \tag{4.78}$$

where r_y^2 is the radius of gyration and $r_y^2 = \sqrt{\dfrac{I_y}{M}}$.

Substituting Eqs. 4.76, 4.77, and 4.78 into Eqs. 4.74 and 4.75, the two equations become

$$\ddot{z} + A_1 z + A_2 \theta = 0 \tag{4.79}$$

$$\ddot{\theta} + \frac{A_2}{r_y^2} z + A_3 \theta = 0 \tag{4.80}$$

In these equations, if θ is set to zero (i.e., $\theta = 0$), there will be no pitch motion. Only bounce motion exists. The equations become a pure bounce vibration equation as

$$\ddot{z} + A_1 z = 0 \tag{4.81}$$

The natural frequency of the pure bounce motion is $\omega_{nz} = \sqrt{A_1}$.

If z is set to zero (i.e., $z = 0$), the system will have only pitch motion. The equations become

$$\ddot{\theta} + A_3 \theta = 0 \tag{4.82}$$

The natural frequency of the pure pitch motion is $\omega_{n\theta} = \sqrt{A_3}$.

In most cases, neither θ nor z is zero. Thus, both bounce and pitch motions exist. The two motions are coupled. This coupling can be reflected on term A_2.

EXAMPLE E4.5

Calculate the pure bounce motion frequency and the pure pitch motion frequency for the system with following parameters:

Front ride rate: $k_f = 35{,}000$ N/m

Rear ride rate: $k_r = 40{,}000$ N/m

Sprung mass: $M = 1600$ kg

Radius of gyration: $r_y = 1.2$ m

Distance between the front axle and the vehicle center of gravity: $\ell_1 = 1.5$ m

Distance between the rear axle and the vehicle center of gravity: $\ell_2 = 1.8$ m

Solution:

According to the provided parameters, A_1 and A_3 can be calculated as

$$A_1 = \frac{k_f + k_r}{M}$$

$$= \frac{35,000 + 40,000}{1600}$$

$$= 46.9 \text{ s}^{-2}$$

$$A_3 = \frac{k_f \ell_1^2 + k_r \ell_2^2}{I_y} = \frac{k_f \ell_1^2 + k_r \ell_2^2}{M r_y^2}$$

$$= \frac{35000 * 1.5^2 + 40000 * 1.8^2}{1600 * 1.2^2}$$

$$= 90.4 \text{ s}^{-2}$$

Thus, the pure bounce motion frequency is

$$f_{nz} = \frac{\sqrt{A_1}}{2\pi}$$

$$= \frac{\sqrt{46.9}}{2\pi}$$

$$= 1.09 \text{ Hz}$$

The pure pitch motion frequency is

$$f_{n\theta} = \frac{\sqrt{A_3}}{2\pi}$$

$$= \frac{\sqrt{90.4}}{2\pi}$$

$$= 1.51 \text{ Hz}$$

To obtain the coupled natural frequencies of the system, Eqs. 4.79 and 4.80 must be solved. Assume

$$z = Z \sin \omega t \qquad (4.83)$$

$$\theta = \Theta \sin \omega t \qquad (4.84)$$

Substitute Eqs.4.83 and 4.84 into Eqs. 4.79 and 4.80 to obtain

$$\left(A_1 - \omega^2\right)Z + A_2\Theta = 0 \tag{4.85}$$

$$\frac{A_2}{r_y^2}Z + \left(A_3 - \omega^2\right)\Theta = 0 \tag{4.86}$$

The eigenvalues for Eqs. 4.85 and 4.86 can be solved by

$$\omega^4 - \left(A_1 + A_3\right)\omega^2 + \left(A_1A_3 - \frac{A_2^2}{r_y^2}\right) = 0 \tag{4.87}$$

The eigenvalues of Eq. 4.87 or the natural frequencies of the coupled system are

$$\omega_{n1}^2 = \frac{A_1 + A_3}{2} - \sqrt{\frac{1}{4}\left(A_1 - A_3\right)^2 + \frac{A_2^2}{r_y^2}} \tag{4.88}$$

$$\omega_{n2}^2 = \frac{A_1 + A_3}{2} + \sqrt{\frac{1}{4}\left(A_1 - A_3\right)^2 + \frac{A_2^2}{r_y^2}} \tag{4.89}$$

The vibration modes, or the amplitude ratio between the bounce and pitch modes, can be obtained from Eq. 4.85 or Eq. 4.86. Using Eq. 4.85, the amplitude ratios corresponding to ω_{n1} and ω_{n2} are

$$\left(\frac{Z}{\Theta}\right)_{\omega_{n1}} = \frac{A_2}{\omega_{n1}^2 - A_1} \tag{4.90}$$

$$\left(\frac{Z}{\Theta}\right)_{\omega_{n2}} = \frac{A_2}{\omega_{n2}^2 - A_1} \tag{4.91}$$

Among the two amplitude ratios, one is negative and the other is positive. To further understand the bounce and pitch modes, the concept of oscillation center is introduced. Oscillation center is a motion center for pure bounce motion or for pure pitch motion. The center depends on the two mode amplitude ratios.

If the amplitude ratio is positive, the oscillation center will be behind the vehicle body center of gravity, that is, the center is located on the right of the center of gravity, as shown in Figure 4.38. The center location can be determined by

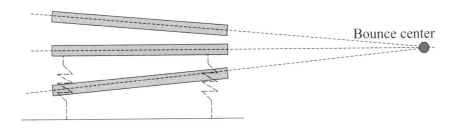

Figure 4.38 The bounce center of the bounce-pitch model.

$$L_{c1} = \frac{A_2}{\omega_{n1}^2 - A_1} \qquad (4.92)$$

The center usually is large enough to fall outside the wheelbase. For this case, the body motion is dominated by the bounce mode, and the associated frequency is the bounce frequency ω_{n1}. The location is called the bounce center.

On the other hand, if the amplitude ratio is negative, the oscillation center will be ahead (left) of the center of gravity, as shown in Figure 4.39. The center location can be determined by

$$L_{c2} = \frac{A_2}{\omega_{n2}^2 - A_1} \qquad (4.93)$$

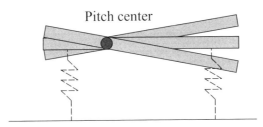

Figure 4.39 The pitch center of the bounce-pitch model.

The center usually is small enough to fall within the wheelbase. For this case, the body motion is dominated by the pitch mode, and the associated frequency is the pitch frequency ω_{n1}. The location is called the pitch center.

Usually, the vehicle body motion includes both bounce and pitch modes. Only when $A_2 = 0$ (i.e., $k_r \ell_2 = k_f \ell_1$) will the body have the pure bounce mode and the pure pitch mode. For this special case, the bounce center will be at infinite distance from the center of gravity, and the pitch center will be at the center of gravity. The bounce and pitch oscillations are totally independent. Ride quality is poor for this case because the motions will be irregular.

The other special case happens at $r_y^2 = \ell_1 \ell_2$. If the condition is substituted into Eqs. 4.92 and 4.93, then the special locations are found, $L_{c1} = \ell_1$ and $L_{c2} = \ell_2$. In this case, one oscillation center will be at the front spring attached point, and the other center is located at the rear spring attached point. The vehicle model can be simplified as two independent models—one is located at the front wheel, and the other is located at the rear wheel. The front wheel motion will have no influence on the rear wheel motion, and vice versa. This is a desirable condition for good ride quality.

EXAMPLE E4.6

Determine the bounce and pitch centers and their frequencies of the vehicle in Example E4.5.

Solution:

The three values A_1, A_2, and A_3 used for calculating the frequencies are determined first. A_1 and A_3 were calculated in Example E4.5. A_2 can be calculated from Eq. 4.77 as

$$
\begin{aligned}
A_2 &= \frac{k_f \ell_1 - k_r \ell_2}{M} \\
&= \frac{35,000 * 1.5 - 40,000 * 1.8}{1600} \\
&= -12.2
\end{aligned}
$$

Substituting these values into Eqs.4.88 and 4.89, the frequencies will be given as

$$
\omega_{n1}^2 = \frac{A_1 + A_3}{2} - \sqrt{\frac{1}{4}(A_1 - A_3)^2 + \frac{A_2^2}{r_y^2}} = 44.65
$$

$$
\omega_{n2}^2 = \frac{A_1 + A_3}{2} + \sqrt{\frac{1}{4}(A_1 - A_3)^2 + \frac{A_2^2}{r_y^2}} = 92.65
$$

Thus,

$$
\omega_{n1} = 6.68, \quad \text{or} \quad f_{n1} = \frac{\omega_{n1}}{(2\pi)} = 1.06 \text{ Hz}
$$

$$
\omega_{n1} = 9.63, \quad \text{or} \quad f_{n12} = \frac{\omega_{n2}}{(2\pi)} = 1.53 \text{ Hz}
$$

Then, the locations of the bounce and pitch oscillation centers can be determined by Eqs. 4.92 and 4.93, respectively, as

$$L_{c1} = \frac{A_2}{\omega_{n1}^2 - A_1}$$

$$= \frac{-12.2}{44.65 - 46.9}$$

$$= 5.4 \text{ m}$$

$$L_{c2} = \frac{A_2}{\omega_{n2}^2 - A_1}$$

$$= \frac{-12.2}{92.65 - 46.9}$$

$$= -0.27 \text{ m}$$

The bounce center is on the rear side and is 5.4 meters from the center of gravity. The pitch center is within the wheelbase and is 0.27 meter in front of the center of gravity. For a passenger car, the bounce frequency usually is between 1.0 and 1.5 Hz. The pitch mode frequency is a little higher than the bounce frequency.

4.3.3 Other Modeling

The quarter car and the bounce-pitch models are introduced in this chapter. Only bounce motion is considered in the quarter car model. The bounce-pitch model includes both bounce and pitch motions. The results reflect the basic characteristics of ride quality in vehicle dynamics.

If more accuracy is needed, more complicated vehicle models could be used. Figure 4.40 shows a complicated plane model. In the model, the front and rear suspensions are separated, and the vehicle center mass is considered. If the roll mode must be considered in ride dynamic analysis, a three-dimensional model is introduced. Figure 4.41 shows a three-dimensional model. The bounce, pitch, and roll motions are simulated simultaneously.

With the development of engineering software, more accurate analysis has been widely used in automobile engineering, such as finite element analysis, modal synthesis analysis, and others. In advanced modeling, the vehicle body, suspension, tires, and other components are divided into fine elements. Dynamic responses at the seat and steering wheel are calculated by finite element analysis. In modern finite element models, a car usually is divided into more than one million elements. It takes considerable time and

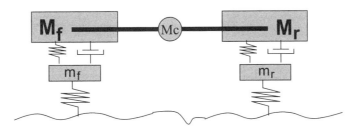

Figure 4.40 A complicated plane model.

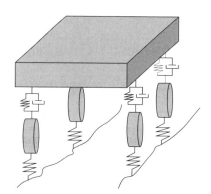

Figure 4.41 A three-dimensional vehicle model.

is costly to perform calculation on such a model. Thus, the vehicle can be divided into several subsystems, and finite element analysis is applied to them. The results from these subsystems then are integrated to predict the whole vehicle responses by the modal syntheses method. Readers interested in this technology can refer to other materials.

4.4 Seat Evaluation and Modeling

4.4.1 Introduction

A seat is attached to the vehicle floor, and the occupants sit on the seat, as shown in Figure 4.42(a). Vibration sources from the road and powerplant are transmitted to the occupants through the seat. The seat is a critical vibration transfer path. Ride comfort is determined by both vibration magnitude and frequency. Occupants sitting on car seats have different levels of discomfort for different frequencies. A constant vibration magnitude does not produce the same level of discomfort at all frequency ranges. Low-frequency vibrations at 1 to 2 Hz have little impact on the human body because the frequency is close to those induced by people's normal behavior (e.g., people walking). However, in the frequency range of 4 to 8 Hz, the occupants are sensitive to vertical vibrations. Usually, vibrations are amplified at this frequency range. At higher frequencies above 10 Hz, the sensitivity to vibrations is less.

Ride quality is an important factor for customer satisfaction. The seat is one of the most important loops in determining ride quality, as shown in Figure 4.1. One major function of the seat is to isolate the vibration from the vehicle body to the human body. Thus,

Figure 4.42 (a) Seat response-excitation system, and (b) seat isolation system.

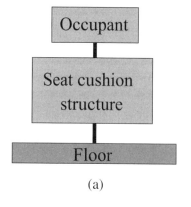

(a)

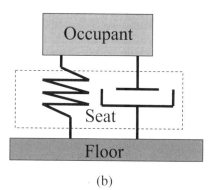

(b)

seat quality evaluation is a critical task. Traditionally, subjective tests by occupants are used for the evaluation. Customers frequently rate different ergonomic aspects of automotive seats according to their senses. However, a subjective evaluation cannot be reflected directly on the seat design.

In recent years, scientists and engineers have been looking for objective evaluation methods. Seating comfort is related directly to the dynamic characteristics of the seat. An automotive seat has inherent vibration characteristics, based on its materials of construction and design. Thus, understanding of and modeling of seat dynamics are important in the analysis of ride comfort.

The vibration transmissibility of seat foam cushion is a key element in identifying ride comfort. A relation between ride comfort and cushion physical properties does exist. From the vibration isolation viewpoint, the function of the seat is similar to that of an isolator, as shown in Figure 4.42(b). The dynamic responses either can be amplified at some frequencies or can be attenuated at others. One objective of seat design is to achieve an overall vibration reduction compared with a rigid seat such as a wooden seat. Based on this assumption, various models have been established to better understand seat structure and to simulate occupants' responses.

In this chapter, several objective evaluation methods are introduced. The first one is the seat effective amplitude transmissibility (SEAT) value method. In this method, the seat comfort level is evaluated by only one calculated value. The second method is that in which the seat comfort level is evaluated by the root-mean-square (RMS) velocity. The third method is the transmissibility comparison method based on seat modeling. In addition to seat modeling, the seat modal decoupling from the vehicle and suspension modes should be considered because these systems are connected.

4.4.2 SEAT Value

An evaluation of seat comfort should include three factors: an input vibration spectrum from the vehicle floor, an output spectrum at the vehicle occupant, and transmissibility of the seat structure. A satisfactory evaluation method must take into account all three factors at all significant frequencies. Usually, the evaluation data are expressed as a spectrum, but the three factors can be combined into a single value to quantify seat comfort. The seat effective amplitude transmissibility (SEAT) value suggested by Griffin is the most common. The SEAT value is defined as

$$\text{SEAT\%} = \left[\frac{\int W_{ss}(f) E_i^2(f) df}{\int W_{ff}(f) E_i^2(f) df} \right]^{1/2} \times 100 \qquad (4.94)$$

where $W_{ss}(f)$ and $W_{ff}(f)$ are acceleration spectra for the seat and floor, respectively, and $E_i(f)$ is the frequency weighting function for the human response in the frequency range of interest. Table 4.5 lists the weighting functions in the vertical direction. The use of frequency weighting can make the value more suitable for the evaluation of human body comfort. The weighting function is plotted in Figure 4.43.

TABLE 4.5
SEAT FREQUENCY WEIGHTING FUNCTIONS

Frequency Range (Hz)	Weighting Functions
$0.5 \sim 2.0$	$E_i(f) = 0.4$
$2.0 \sim 5.0$	$E_i(f) = \dfrac{f}{5.0}$
$5.0 \sim 16.0$	$E_i(f) = 1.0$
$16.0 \sim 80$	$E_i(f) = \dfrac{16.0}{f}$

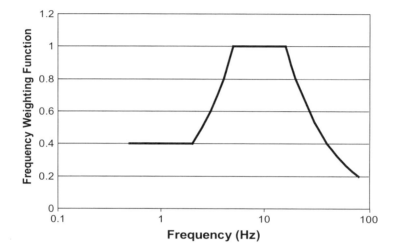

Figure 4.43 Vertical weighting values for seat cushion structure.

If the seat transmissibility $H_s(f)$ is given by measurement or calculation, the SEAT value is

$$\text{SEAT}\% = \left[\frac{\int W_{ff}(f) H_s(f) E_i^2(f) df}{\int W_{ff}(f) E_i^2(f) df} \right]^{1/2} \times 100 \qquad (4.95)$$

The SEAT value provides a simple numerical assessment of the seat isolation efficiency. The integrals cover all the interested floor excitation frequencies ranging from 0.5 to 80 Hz. The SEAT value may be interpreted as a percentage of the overall ride value on the evaluated seat to the value on a rigid seat. A SEAT value of 100% means that the seat vibration is the same as the floor vibration, that is, the seat is a rigid one. If the SEAT value is less than 100%, the seat vibration is less than the floor vibration, that is, some vibration from the floor is isolated. A value greater than 100% indicates that the vibration transferred to the seat is amplified; an occupant feels more discomfort sitting on the seat than on the floor. The seat degrades the ride comfort. A seat with a SEAT value of 50% means that the seat produces half as much discomfort as a seat with a SEAT value of 100%.

The SEAT value is important for evaluating seat design because the value includes the excitation, seat response, and human body sensitivity. The value implies whether vibration on the floor is reduced or amplified. Usually, vertical vibration is much more important than the vibration from the other two directions. However, if the vehicle has significant roll and pitch vibration with a low center of rotation, the motions in the lateral and longitudinal directions will increase with height. The SEAT values in these directions may be greater than 100%.

The SEAT value expressed in Eqs. 4.94 and Eq. 4.95 is suitable for excitation and response with a low crest value. The crest value is defined as the modulus of the ratio of the maximum instantaneous peak value of the frequency-weighted acceleration over its RMS value. For cases where the floor input and the seat response possess high crest values, the SEAT value is calculated by vibration dose values (VDV) as

$$SEAT\% = \frac{VDV \text{ on the seat}}{VDV \text{ on the floor}} \times 100 \qquad (4.96)$$

Vibration dose values will be introduced in detail in Section 4.5 "Discomfort Evaluation and Human Body Model." Equation 4.96 is a more suitable expression for transient vibration and shock input.

4.4.3 Seat Velocity

The SEAT value provides a simple method for evaluating seat comfort. The value is useful and convenient in comparing the overall ride comfort of different seats. However, the seat response spectrum may be important because it includes both frequency and amplitude data.

Figure 4.44 shows human body acceleration sensitivity over frequency. The most sensitive frequency for the human body is between 4 and 8 Hz. Acceleration is a horizontal

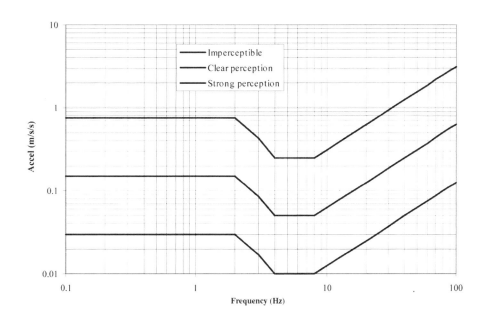

Figure 4.44 Human body acceleration sensitivity curve.

line within the frequency range. It increases linearly with the frequency increase above 8 Hz. If the acceleration above 8 Hz is integrated, the obtained velocity is a horizontal line. Usually, frequencies higher than 8 Hz are considered when a seat is evaluated. Thus, velocity is more suitable to represent the seat response than acceleration in a wide frequency range because of its constant value. During vehicle development, a constant velocity usually is set up as a target for seat comfort.

Figure 4.45 shows velocity spectra for two seats and a target varying with engine speed (revolutions per minute [rpm]) at idle. Suppose the seat velocity target is 40 mm/s. It can be seen that both seat responses are above the target, but Seat 2 is better than Seat 1.

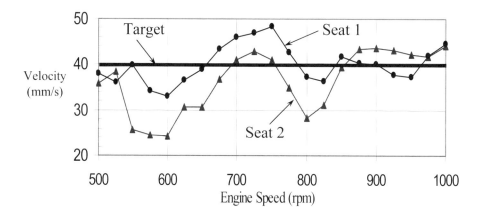

Figure 4.45 Two seat velocity responses and a target.

4.4.4 Linear Seat Modeling and Transmissibility

The velocity spectrum as already described here is one way to evaluate seat ride quality over frequency. Another common way is transmissibility. The purpose of a seat design is to minimize the transmissibility of vibration through the seat to the occupant, especially at some sensitive frequencies. The transmissibility includes both magnitude and frequency. Frequency information is part of the SEAT and VDV values. Thus, the transmissibility between the seat response and the floor input is useful in understanding the seat structure characteristics.

A seat can be assumed to be simply a linear structure, including linear stiffness and linear damping, as shown in Figure 4.42(b). The magnitude of the transmissibility is expressed in Eq. 4.61 and is plotted in Figure 4.33. In Eq. 4.61, the mass includes the mass of the body, head, torso, and skeleton. The linear seat model provides reasonable accuracy for certain constant floor excitation. However, if the floor excitation magnitude changes, the stiffness and damping will change accordingly, which means that the seat is a nonlinear structure.

4.4.5 Nonlinear Seat Modeling and Transmissibility

Many researchers and engineers have investigated the dynamic mechanism of seat cushion structure. In most theoretical analyses, the seat cushion is regarded as a linear structure, and linear models were used to simulate seat response, human body response, and even the SEAT values.

Some tests had shown that the seat cushion structure has nonlinear characteristics. Fairley and Griffin (1989) conducted an experiment that indicated that the compliance of foam-padded spring seats changes with frequency. Casati *et al.* (1999) measured seat cushion dynamic stiffness that increases with the load and frequency increase. Kinkekaar and Neal (1998) and Yu (1999) tested seat foam structure to evaluate a subjective dynamic comfort study and to provide similar results. Pang *et al.* (2003) (Ref. 21) tested a series of seats and revealed the nonlinearity of the seat cushion structure. The experiment is described as follows.

An electrohydraulic motion simulator (vibration platform), as shown in Figure 4.46, was used to conduct the seat testing. One accelerator was mounted on the platform surface (floor), and another was located on top of the seat cushion (seat surface). A sandbag weighing 356 N (80 lb) was placed on the seat cushion. The platform excitation used in the test ranged from 0.05 to 0.45g RMS in increments of 0.05g RMS. The transmissibility between the seat surface and the floor were measured. Figure 4.47 depicts the collection of measured transmissibility between the seat surface and the floor of a sports car seat. Figure 4.48 depicts the measured transmissibility collection for a luxury car seat. Two interesting characteristics were found from the tested transmissibility: (1) The peak frequencies decrease with the excitation increment increase, and (2) The peak amplitudes decrease as the excitation increment increases. For a linear structure, its transmissibility is independent of input and output to the system, and superposition law is applicable. The test results revealed that the seat cushion structure is a nonlinear system. The system has both nonlinear stiffness and nonlinear damping.

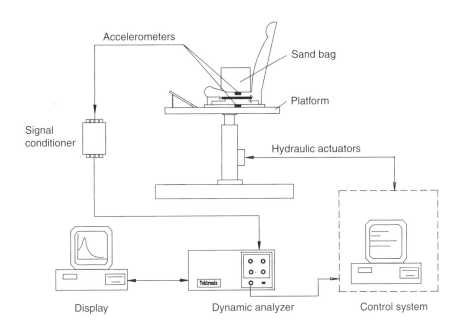

Figure 4.46 An experimental setup for a seat transmissibility test.

It could be difficult to reveal the nonlinear dynamic model of a seat cushion structure due to its complexity. The vibration transmitted to the driver through the seat depends on the seat structure as well as the human body sitting on the seat. To investigate the seat dynamic characteristics and to build a nonlinear seat model, a sandbag is put on top of the seat. The seat and the sandbag form a single degree-of-degree system, as shown

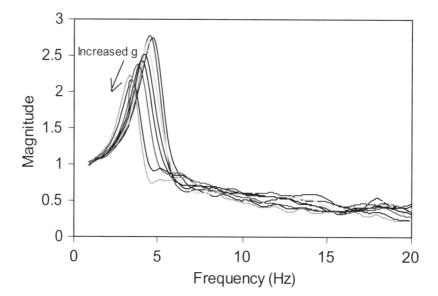

Figure 4.47 Transfer function between the seat butt acceleration and the seat track acceleration for an 80-lb sandbag placed on a sports car seat. The seat track inputs are 0.05 to 0.45g RMS (smoothed data).

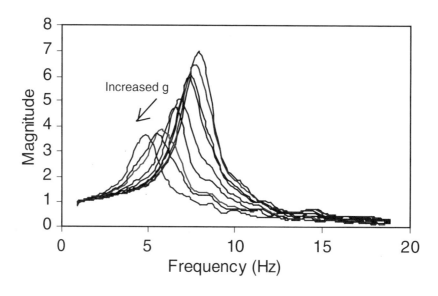

Figure 4.48 Transfer function between the seat butt acceleration and the seat track acceleration for an 80-lb sandbag on a luxury car seat. The seat track inputs were 0.05 to 0.45g RMS (smoothed data).

in Figure 4.49. Based on the nonlinear phenomenon from the tests, a nonlinear model was established by the authors (Pang *et al.* 2003, Ref. 21).

The nonlinear stiffness force of the seat cushion can be expressed as

$$F_k(z_0, z) = \frac{k_1}{1 + k_2|z - z_0|} = \frac{k_1}{1 + k_2|\delta|} \tag{4.97}$$

The nonlinear damping force of the seat cushion is

$$F_d(z_0, z) = c_1(\dot{z} - \dot{z}_0) + c_2|\dot{z} - \dot{z}_0|(\dot{z} - \dot{z}_0) = c_1\dot{\delta} + c_2|\dot{\delta}|\dot{\delta} \tag{4.98}$$

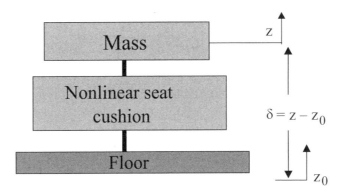

Figure 4.49 The seat modeling.

Therefore, the dynamic equation of the seat model can be written as

$$m\ddot{\delta} = -\frac{k_1}{1+k_2|\delta|}\delta - c_1\dot{\delta} - c_2|\dot{\delta}|\dot{\delta} - m\ddot{z}_0 \qquad (4.99)$$

where k_1 and k_2 are the nonlinear stiffness, and c_1 and c_2 are the linear and nonlinear damping coefficients, respectively. The coefficient can be identified by the least squares method. Table 4.6 lists the stiffness and damping coefficients for both a sports car and a luxury car.

TABLE 4.6
STIFFNESS AND DAMPING COEFFICIENTS
FOR A SPORTS CAR AND A LUXURY CAR

	k_1 (N/m)	k_2 (N/m)	c_1 (Ns/m)	c_2 (Ns^2/m^2)
Sports car	120,000	1,400	230	3000
Luxury car	35,000	625	380	600

The comparison between Figures 4.47 and 4.48 indicates that the transmissibility magnitude of the sports car seat is higher than that of the luxury car seat over all frequency ranges. The magnitudes for the sports car seat are approximately three times higher than those of the luxury car seat for the same excitation. The distribution of peak frequencies for both seats also may provide an important clue on how to construct an automotive seat. The human body experiences significant discomfort when vibrated vertically between 4 and 8 Hz. The transmissibility peaks of the luxury car are below 1 after 5.8 Hz for the entire excitation range. On the other hand, the sports car seat exhibits amplification of the input at a frequency range from 4 to 8 Hz for all excitation levels. This investigation shows whether a seat is likely to be perceived as comfortable when tested subjectively. The models are useful for other vehicle seats, but parameters for each seat should be identified according to the corresponding test data.

4.5 Discomfort Evaluation and Human Body Model

4.5.1 Discomfort and Subjective Evaluation

Ride quality is related to whole-body vibration that concerns an occupant's total feeling in a vibration environment, and the effect of interest is not local to any particular point of contact. Local vibration occurs when one or more limbs are in contact with a vibrating surface.

Ride quality can be described by the degree of discomfort or comfort. Discomfort depends on many factors, including the vibration magnitude, frequency, direction, and duration; the occupant's posture position, body size, age, and gender; and noise and other factors. For example, posture can have a large influence on the amount of vibration transmitted to a seated person and can determine the extent of any detrimental effects. Occupants feel more uncomfortable riding in a car for one hour than they do when they ride in a car for 10 minutes. Vibration input axes also influence the degree of discomfort. Most of the discussion here will focus on the vertical direction. The study of ride quality specifically exposed to a vibration environment is intended to establish the relation between the occupant response and vibration sources. Figure 4.50 shows a human body sitting posture.

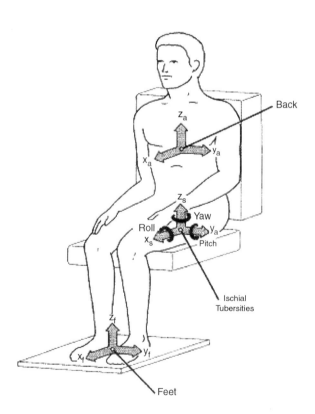

Figure 4.50 A human body sitting posture.

Ride quality is inherently a subjective measure. Different occupants who are exposed to the same vibration level may perceive different comfort levels. Human body responses to vibration vary greatly over time and from one person to another. The occupants often modify their subjective evaluation according to their expectations. Thus, a subjective and relative evaluation method, sometimes called a vehicle evaluation rating (VER), is used for the subjective evaluation of discomfort. Table 4.7 lists the relation between VER values and the subjective evaluation rating. Usually, a group of people is organized to drive a vehicle and then is asked to provide their subjective ratings on the ride quality of that vehicle. The rating value is used as a reference in vehicle dynamics.

TABLE 4.7
RELATION BETWEEN VER VALUES AND
THE SUBJECTIVE EVALUATION RATING

Unacceptable				Borderline	Acceptable				
1	2	3	4	5	6	7	8	9	10
Not acceptable		Objectionable		Requires improvement	Medium	Light	Very light	Trace	Not noticeable

ISO 2631 and 2631-1 also provide the equivalent comfort contours for subjective rating, as shown in Figure 4.51. The equivalent levels of perceived comfort vary with vibration magnitude, frequency, axis, and duration. ISO 2631 (Frequency Weighting to Vibration) is a guideline for the comfort evaluation of human exposure to whole-body vibration. The standard is concerned with translational and rotational vibration in the frequency range of 0.5 to 80 Hz and the duration range of 1 minute to 24 hours.

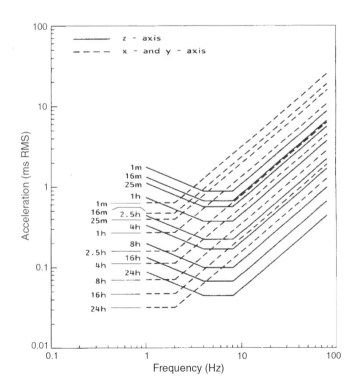

Figure 4.51 Human discomfort contour for different vibration levels and durations (ISO).

ISO 2631 and 2631-1 also provide the relation between the subjective evaluation of ride discomfort and the objective acceleration, as listed in Table 4.8. The vibration should be controlled below 0.5 m/s^2 for the body to be comfortable.

TABLE 4.8
RELATION BETWEEN RIDE DISCOMFORT
AND VIBRATION ACCELERATION

Vibration (m/s^2)	Reaction
< 0.315 m/s^2	Not uncomfortable
0.315 to 0.63 m/s^2	A little uncomfortable
0.5 to 1 m/s^2	Fairly uncomfortable
0.8 to 1.6 m/s^2	Uncomfortable
1.25 to 2.5 m/s^2	Very uncomfortable
>2 m/s^2	Extremely uncomfortable

4.5.2 Objective Evaluation of Ride Discomfort

In addition to the subjective methods, objective evaluation is critical because the subjective evaluation varies greatly from person to person. Some objective evaluation methods to represent the quantitative ride quality value are described here.

Vibration magnitude and duration are two major factors determining the level of discomfort. Table 4.8 provides the relation between discomfort and the vibration magnitude. It may seem intuitively obvious that occupants feel more uncomfortable for one hour of vibration exposure than for 10 minutes of it. Clearly, that exposure time is an important factor in evaluating discomfort, as shown in Figure 4.51. Griffin 1986 suggested a simple average method that includes both vibration acceleration and duration

$$a^4 t = \text{constant} \tag{4.100}$$

The expression shows that the discomfort experienced by an occupant is proportional to the duration, t, and the fourth power of the acceleration, a. For the same vibration magnitude, the longer the time the occupant is exposed to vibration, the more discomfort he or she experiences. However, the acceleration magnitude is more critical than the duration. This expression in Eq. 4.100 will be used as the basis of the vibration dose value (VDV).

4.5.2.1 Weighted Root-Mean-Square Method

The root-mean-square analysis method is used for the subjective evaluation of ride quality. In this method, both the vibration magnitude and the duration are included. As mentioned, the human body is very sensitive to excitation frequency; thus, acceleration must be weighted over the whole frequency range of interest. A weighted RMS acceleration, $\bar{a}_w$, is used for the objective evaluation and is defined as

$$\bar{a}_w = \left[\frac{1}{T} \int_0^T a_w^2(t) dt \right]^{1/2} \tag{4.101}$$

where $a_w(t)$ is the weighted acceleration in time domain, and T is the measured duration. Here, $a_w(t)$ is determined by the measured acceleration $a(t)$ and the frequency weighting function of Table 4.5.

The frequency weighted RMS acceleration shall be determined by the weighting function and the appropriate addition of narrow band or one-third octave band data. The overall weighted acceleration a_w is obtained from the expression

$$a_w = \left[\sum_i (E_i a_i)^2 \right]^{1/2} \qquad (4.102)$$

where E_i is the weighting factor for ith one-third octave band if the one-third band data are used. a_i is the RMS acceleration for the ith one-third octave band.

The concept of weighted RMS acceleration integrates the effect of vibration level over time. The method provides a single value to judge ride comfort. However, the statistical method is good only for smooth vibration where there are no transient peaks. Thus, its application has some limitation. For vibration with a crest factor below or equal to 9, the weighted RMS acceleration is useful to evaluate human body discomfort.

4.5.2.2 Objective Evaluation by the Vibration Dose Value

For some cases in which the crest factors are higher than 9, the weighted RMS acceleration will underestimate the instantaneous effect of vibration, such as occasional shock, transient vibration, and so forth. For these cases, the vibration dose value (VDV) is used to evaluate the discomfort index. The VDV is defined as

$$VDV = \left[\int_{t=0}^{t=T} a_w^4(t) dt \right]^{1/4} \qquad (4.103)$$

where $a_w(t)$ is the instantaneous frequency weighted acceleration.

The VDV method is more sensitive to peak values than the weighted RMS acceleration because the fourth power instead of the second power of acceleration time history is used. Similar to the weighted RMS acceleration, in VDV we also integrate the effect of the vibration level over the whole duration. This provides a single value with which to judge ride comfort.

When the vibration exposure consists of two or more periods with different magnitude, the VDV for total exposure should be calculated from the fourth root of the sum of the fourth power of the individual VDV as

$$VDV_{total} = \left[\sum_i VDV_i^4 \right]^{1/4} \qquad (4.104)$$

4.5.3 Linear Human Body Modeling

Both the weighted RMS acceleration and the vibration dose value integrate the vibration magnitude and duration and provide a single evaluation value of ride quality. However, sometimes, not only overall value but also spectra are needed. The body responses (acceleration and/or velocity) are used to evaluate ride discomfort. The spectra also are important in identifying excitation sources and vibration transfer paths. Either measurement or calculation can be used to obtain the spectra.

A human body model is used to obtain the spectra for the subjective evaluation of ride quality. The human body can be assumed as a series of lumped masses connected by springs and dampers to the skeleton. Many tests show that transmissibility between the body response and seat excitation has two peaks. Thus, a two-degrees-of-freedom model is suitable to represent the human body. Figure 4.52 shows a two-degrees-of-freedom human body model supported on a frame. In this model, m_1 is the mass of the head and upper torso, m_2 is the mass of the main body, and m_3 is the supported mass of the skeleton. Here, m_3 is much smaller than m_1 and m_2.

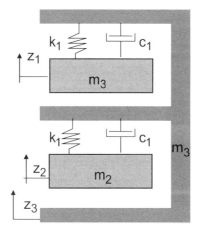

Figure 4.52 A two-degrees-of-freedom human body model supported on a skeleton frame.

The human body is a proven linear system as documented in the ISO standard shown in Figure 4.52. The dynamic equations of the model can be written as

$$m_1\ddot{z}_1 = -k_1(z_1 - z_3) - c_1(\dot{z}_1 - \dot{z}_3) \tag{4.105}$$

$$m_2\ddot{z}_2 = -k_2(z_2 - z_3) - c_2(\dot{z}_2 - \dot{z}_3) \tag{4.106}$$

$$m_3\ddot{z}_3 = -k_1(z_3 - z_1) - c_1(\dot{z}_3 - \dot{z}_1) - k_2(z_3 - z_2) - c_2(\dot{z}_3 - \dot{z}_2) \tag{4.107}$$

where z_1, z_2, and z_3 are the displacement of the head, body, and skeleton, respectively. The skeleton moves together with the seat surface.

Applying Laplace transforms and substituting $j\omega$ for s, Eqs. 4.105 and 4.106 become

$$\left[\left(k_1 - m_1\omega^2\right) + jc_1\omega\right]Z_1 = \left(k_1 + jc_1\omega\right)Z_3 \tag{4.108}$$

$$\left[\left(k_2 - m_2\omega^2\right) + jc_2\omega\right]Z_2 = \left(k_2 + jc_2\omega\right)Z_3 \tag{4.109}$$

Thus, the transmissibility between the head and the seat surface is

$$\left|\frac{Z_1}{Z_3}\right| = \sqrt{\frac{k_1^2 + \left(c_1\omega\right)^2}{\left(k_1 - m_1\omega^2\right)^2 + \left(c_1\omega\right)^2}} \tag{4.110}$$

The transmissibility between the body and the seat surface is

$$\left|\frac{Z_2}{Z_3}\right| = \sqrt{\frac{k_2^2 + \left(c_2\omega\right)^2}{\left(k_2 - m_2\omega^2\right)^2 + \left(c_2\omega\right)^2}} \tag{4.111}$$

If the vibration spectrum at the seat surface is given, then the responses at the head and body will be obtained by Eqs. 4.110 and 4.111, respectively. Clearly, the expressions have the same transmissibility as that of a single-degree-of-freedom isolation system.

4.5.4 Objective Evaluation by Nonlinear Seat–Human Body Modeling

According to Eqs. 4.105 and 4.106, the motions of the head and body are independent of each other. Actually, when an occupant sits in a vehicle, his or her head motion and body motion will interact with each other. After the seat model is integrated with the body model, the interaction will occur. Thus, the human body and seat models must be considered simultaneously and subjected to floor excitations. Figure 4.53 shows the human body–seat–floor series plot.

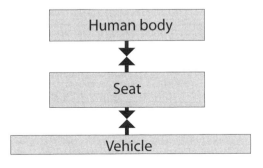

Figure 4.53 The human body–seat–floor connection system.

Figure 4.54 shows an occupant sitting on a seat supported by a vibration platform that simulates the vehicle floor vibration. The floor vibration will be transferred to the occupant through the seat. As mentioned in Section 4.4, "Seat Evaluation and Modeling," the seat is a nonlinear structure; thus, the body–seat combined model also is a nonlinear system because it consists of a nonlinear seat model and a linear human body model, as shown in Figure 4.55.

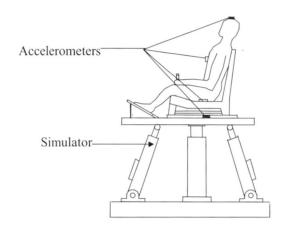

Figure 4.54 An occupant sitting on a seat supported by a vibration platform.

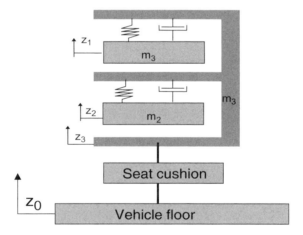

Figure 5.55 A seat–human body nonlinear model.

The dynamic equations of the nonlinear body–seat system can be expressed as

$$m_1\ddot{z}_1 = -k_1(z_1 - z_3) - c_1(\dot{z}_1 - \dot{z}_3) \tag{4.112}$$

$$m_2\ddot{x}_2 = -k_2(x_2 - x_3) - c_2(\dot{x}_2 - \dot{x}_3) \tag{4.113}$$

$$m_3\ddot{z}_3 = -k_1(z_3 - z_1) - c_1(\dot{z}_3 - \dot{z}_1) - k_2(z_3 - z_2) - c_2(\dot{z}_3 - \dot{z}_2)$$

$$-\frac{K_1}{1 + K_2|z_3 - z_0|}(z_3 - z_0) - C_0(\dot{z}_3 - \dot{z}_0) - C_1|(\dot{z}_3 - \dot{z}_0)|(\dot{z}_3 - \dot{z}_0) \tag{4.114}$$

The responses at the head and body are used for ride quality evaluation. However, the analytical transmissibility or transfer function cannot be obtained because the system is nonlinear. Instead, the transmissibility can be calculated by numerical simulation. Figure 4.56 shows the transfer function between the head response and the floor excitation for a sports car with 0.1, 0.2, and 0.3g RMS white noise acceleration excitation to the floor. The magnitudes and peak frequencies of the transfer functions shift to smaller values with the increased excitation. Figure 4.57 shows the transfer function between the body response and the floor excitation for a sports car with 0.1, 0.2, and 0.3g RMS white noise acceleration excitation to the floor. The results are similar to the head response. Both the head and body responses over the floor excitation have the same trend as the seat response. Thus, the nonlinearity of the system is due to the nonlinear seat structure.

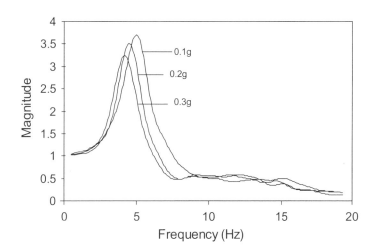

Figure 4.56 Analytical transfer function between the head and seat track for a sports car with 0.1, 0.2, and 0.3g RMS white noise floor acceleration input.

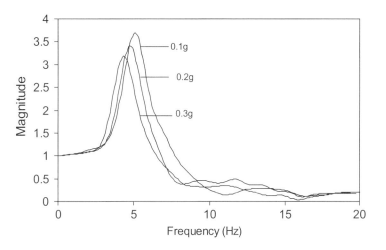

Figure 4.57 Analytical transfer function between the body and seat track for a sports car with 0.1, 0.2, and 0.3g RMS white noise floor acceleration input.

The human body is a complicated structure. To reflect responses at any point on the complex body, a more detailed model, such as a finite element model, should be established.

4.6 Active and Semi-Active Control

4.6.1 Introduction

As described in Section 4.2, road input and powerplant excitation are two major vibration sources to ride quality. The isolators are used to reduce the vibration transferred to the human body. To achieve low transmissibility, the isolator stiffness should be as low as possible. Lighter damping also is desirable for lowering transmissibility at higher frequencies. However, the low stiffness of an isolator will cause larger transient response of the powerplant under shock excitation caused by sudden acceleration, deceleration, braking, and riding on uneven roads. Thus, high stiffness and high damping of the isolators are desired to minimize the shock input. The conflicting requirements between vibration isolation and shock isolation are hard to satisfy by traditional isolators. Relative low stiffness at high frequency for an elastomeric isolator is good for vibration isolation, but its low damping at low speed is bad for shock isolation. Most hydraulic isolators possess high stiffness and high damping at low speeds and are good for reducing shock input. However, their high stiffness at high frequency is a disadvantage for vibration isolation. Some hydraulic isolators can provide a better tradeoff of these requirements.

To resolve these conflicting requirements, some semi-active or active controls have been applied in vehicles, such as suspension control and engine control. However, many aspects of this field are still limited to academic research instead of automotive industrial applications.

4.6.2 Basic Control Concepts

Figure 4.58 shows an open-loop control system. In this system, the process is completely identified (i.e., the output depends totally on the input). One particular input (say, $\bar{u}$) corresponds to a specified and desired output (say, $\bar{y}$). To control the process and achieve the desired output, one needs to supply only an input $\bar{u}$ to the system. Figure 4.59 shows closed-loop control. Instead of controlling the output of the process by picking the control signal $\bar{u}$ that produces the desired $\bar{y}$, the control u is generated as a function of the system error, defined as the difference between the desired output $\bar{y}$ and the actual output y, where $e = \bar{y} - y$. The error is amplified or multiplied by a gain K and then is provided as input to the process.

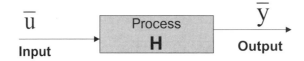

Figure 4.58 Open-loop control.

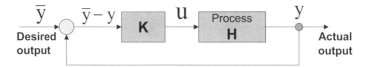

Figure 4.59 Closed-loop control.

Both active and semi-active controls have been used to control vehicle suspension and powerplant systems. All the active controls are closed-loop controls, whereas the semi-active controls can be either closed- or open-loop controls. In control analysis, dynamic equations usually are provided first and then are transferred to state-space equations. Then control algorithms are applied on the state-space equations. A control system must be controllable, observable, and robust.

4.6.3 Active Control

In active control, a counteracting dynamic force is created by one or more actuators to suppress the disturbance force transferred to the system, as shown in Figure 4.60. In other words, an active energy source should be supplied continuously to counteract the disturbance forces. Figure 4.61 shows an active control system for an engine. The control system consists of these components:

* Passive isolator (elastomeric or hydraulic mount)
* Mechanical actuator
* Sensor
* Electronic controller
* Extra energy supplier

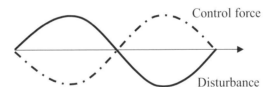

Figure 4.60 Disturbance and control force.

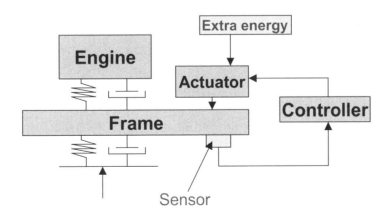

Figure 4.61 A control system for an engine mount.

The passive mounts are used to support the engine. The actuator provides dynamic force that is at the same amplitude and at the 180° phase difference when compared with the disturbance forces. The sensor picks up the disturbance signal and feedback to the controller. The controller controls the actuator to provide the desired force to the system. The controller can be either a feedback or a feedforward type. The active control is implemented with the closed-loop architecture. Many control methods have been studied in academic fields, such as optimal control, bi-state control, fuzzy control, neural control, and others. However, the control methods used in the automotive industry usually are simple control mechanisms, such as PID (proportional, integral, and derivative) controls including proportional control, integration control, and differentiation control. The extra energy supplier continuously provides energy to the actuator to suppress the disturbance force.

Figure 4.62 is an active control system for a half vehicle model. The vehicle body is assumed as a rigid body with mass M_b and moment of inertia I_P. The vehicle has both bounce and pitch motions. The body is supported by the front and rear wheels, which are represented by two unsprung masses, respectively. Thus, it is a four degrees-of-freedom system. A constant displacement hydraulic pump is used to supply hydraulic force to control motion of the front suspension and the rear suspension. The state-space equation of the system can be written as

$$\dot{x}(t) = Ax(t) + Bu(t) + B_1 x_0(t) \tag{4.115}$$

where A, B, and B_1 are constant matrices. The output equation is

$$y(t) = Cx(t) + Du(t) \tag{4.116}$$

where C and D are constant matrices.

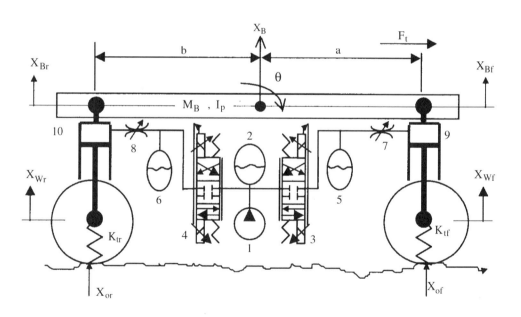

Figure 4.62 A four-degrees-of-freedom half car model with a control system. (El-Demerdash et al., 1999.)

The proportional, integral, and derivative controller (PID) is designed as

$$y_L = K_{LP} y_f + K_{LI} \int y_f dt + K_{LD} \frac{dy_f}{dt} \tag{4.117}$$

where K_{LP}, K_{LI}, and K_{LD} are control gains.

The objective of the control design is to minimize the factors that influence ride quality, such as vertical body acceleration at the center of gravity, pitch acceleration, front and rear suspension working spaces, suspension deflection, and so forth. Linear quadratic regular control (LQR) and the composite controller consisting of the PID controller and the LQR controller are applied on the full-state feedback control. Figure 4.63 shows the front suspension deflection with and without control. The deflection with the control system is much lower than that without control. Ride quality is significantly improved by the controller.

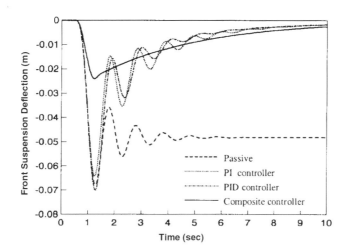

Figure 4.63 Front suspension deflection for the half car model, with and without the controller. (El-Demerdash *et al.*, 1999.)

Figure 4.64 shows an example of the engine mount control system. There are two conventional mounts, one active controlled mount (ACM), and one roll restrictor. The two active mounts are installed on the front of the engine to reduce idle vibration. Figure 4.65 is the vehicle seat track vibration acceleration comparison with ACM and conventional mounts. The result shows that the ACM mount reduced the floor vibration by approximately 10 dB. Interior sound measurement also shows that the vehicle is quieter with the ACM when compared with conventional mounts.

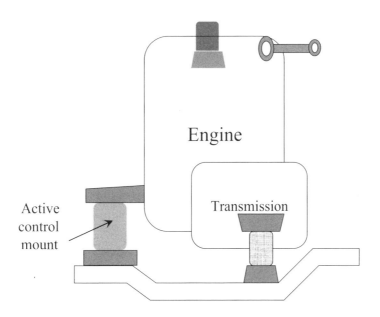

Figure 4.64 A diesel engine and isolation system with active mounts.

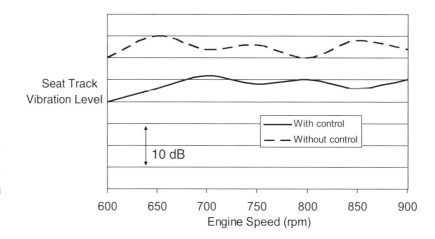

Figure 4.65 Vehicle floor vibration comparison with active and conventional mounts.

Active isolators overcome the limitation of passive ones. They can be controlled to make the mounts stiff at low frequencies and soft at high frequencies (i.e., frequency distribution can be controlled). Damping also can be tuned by the isolators. Thus, the active system provides both superior vibration and shock isolation effectiveness. However, an extra energy supplier, sensor, and controller are needed in the system. The active system increases cost, energy, and weight. In addition, its reliability can be degraded, and it must be maintained carefully.

4.6.4 Semi-Active Control

Active control can automatically adjust control devices to provide the desired force to reduce the disturbance input. However, as mentioned, the restrictive features that make an active system impractical are its high cost, extra energy, increased weight, and reliability concerns. Thus, semi-active control has been more attractive in automobile engineering in recent years. The basic idea of semi-active control usually is to dissipate the vibration energy by changing the dynamic properties of the isolator. In most cases, damping control is preferred because of its simplicity of implementation. Semi-active control can be either a closed-loop or open-loop mechanism.

Figure 4.66(a) shows a closed semi-active loop for suspension control. A semi-active vibration actuator (SAVA) as shown in Figure 4.66(b) replaces the traditional damper. There are two chambers of the hydraulic cylinder. A pipe is connected with the two chambers, while a control is installed on the pipe. The valve controls the fluid flow between the chambers. Vibration energy will be stored and then released to produce the desired control force and to reduce the disturbance. A simple "bi-state" control algorithm is used in this system to open or close the valve. When the valve is closed, the fluid in the actuator acts similarly to a stiff spring. The fluid "spring," once compressed by the deflection of the structure, becomes an energy storage device. The microcontroller is programmed to optimally regulate the valve to both store and release the energy in the fluid and to maximize dissipation when required. When the valve is fully open, the actuator acts similarly to a shock absorber. The resistance to the fluid flow through the valve orifice dissipates the vibration energy. Applying continuity, conservation of

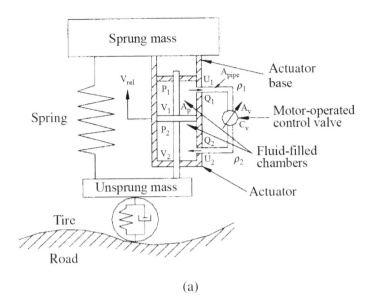

(a)

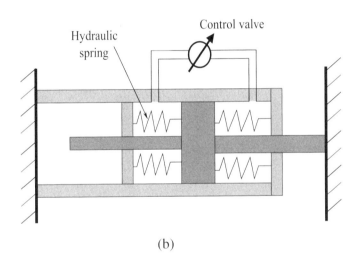

(b)

Figure 4.66 (a) A closed-loop semi-active control system for a quarter car model, and (b) a semi-active vibration actuator (SAVA).

volume, and energy balance at the control valve, the full order fluid dynamics of the SAVA will be obtained.

Figure 4.67 depicts an acceleration comparison of the sprung mass for the system with and without control under a constant velocity road excitation. A noticeable reduction in the sprung mass acceleration is achieved.

Most semi-active controls in automotive engineering use the application of isolation mechanisms filled with electro-rheological (ER) fluid or magneto-rheological (MR) fluid. Electro-rheological fluid is a special fluid with suspensions of highly polarizable tiny particles. Furthermore, ER fluids are capable of changing their fluid viscosity under an electric field. The large change in apparent viscosity coupled with the swift response of ER fluids makes them interesting and advantageous materials to be applied to a variety of electrically controlled systems in the vehicle, such as the engine mount and shock

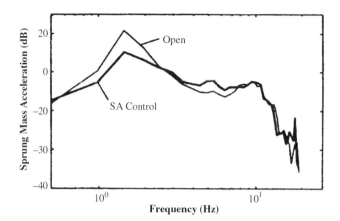

Figure 4.67 The acceleration of sprung mass with and without control. (Mo *et al.*, 1999)

absorbers. Figure 4.68(a) shows an engine mount filled with ER fluids. Figure 4.68(b) shows a transmissibility comparison between the vehicle body (frame) and the engine with and without electrical current applied. The transmissibility under an electric field is reduced compared to that without an electric field.

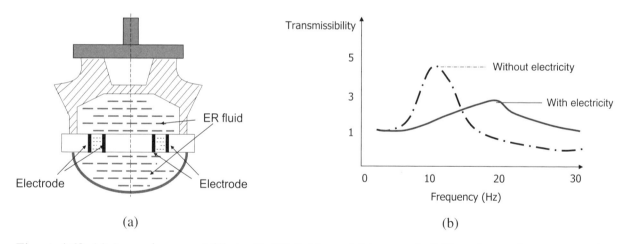

Figure 4.68 (a) An engine mount filled with ER fluids, and (b) transmissibility of a passive mount and semi-active mount with ER fluids.

The seat is one of the major systems that improves ride quality. Seats usually are designed to isolate the driver from most vibration conditions that he or she experiences. However, the seat could run out of its travel and hit the end stop under severe shock conditions. Thus, the driver will be exposed to harmful levels of vibration and shock. One solution to protect drivers from routine vibration and end-stop collision simultaneously is the semi-active damper. A semi-active damper is filled with MR fluid as shown in Figure 4.69. This damper resembles a conventional one, but there are major differences between semi-active dampers and conventional dampers. Semi-active dampers

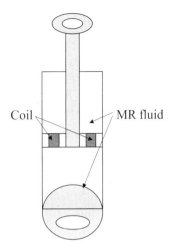

Figure 4.69 A semi-active damper filled with MR fluid.

have electromagnetic coils located with the piston around the orifice, and they have MR fluid.

Magneto-rheological fluid has special characteristics. The MR fluid forms a chain in the direction perpendicular to the fluid flow through the orifice when a magnetic field exists. Without the magnetic field, the fluid is at a random and free-flowing condition. Figure 4.70 shows the two conditions. The MR fluid is capable of changing from a free-flowing liquid to a near-solid state in fewer than 10 ms, enabling real-time control of damping force. A controller is used together with the damper and generates the command current to the electromagnet to automatically adjust the damping characteristics.

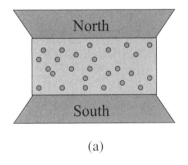

(a)

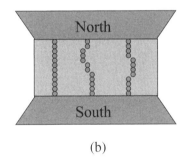

(b)

Figure 4.70 Magneto-rheological particles in fluid: (a) without a magnetic field, and (b) with a magnetic field.

Table 4.9 lists the test results of the maximum weighted acceleration on the seat cushion for a passive damper and a semi-active damper. The semi-active damper achieved dramatic reductions in the maximum weighted acceleration, even close to the middle ride position. Specially, the system reduced the maximum acceleration measured on the seat by 46% at the mid-ride position, 35% at 25 mm above the mid-ride position, 31% at 25 mm below mid-ride position, 49% at 50 mm above the mid-ride position, and 36% at 50 mm below the mid-ride position.

In recent years, some researchers (Foumani and Khajepour, 2002) studied a semi-active engine mount where shape memory alloy (SMA) wires were embedded in rubber as

TABLE 4.9
COMPARISON OF MAXIMUM WEIGHTED ACCELERATION (M/S²)
FOR A PASSIVE DAMPER AND A SEMI-ACTIVE DAMPER

Leveled Seat Height	Standard Passive Damper	Semi-Active Damper	Percent Reduction
+50 mm	22.71	11.66	48.66
+25 mm	14.11	9.21	34.73
Mid-ride	9.83	5.32	45.88
−25 mm	15.64	10.74	31.33
−50 mm	23.3	14.86	36.22

part of the hydraulic engine mount, as shown in Figure 4.71. These SMAs are known as the materials that can memorize their original shapes because their crystalline structure changes with temperature. In low temperatures, they are in the martensite phase with a low Young's modulus; at high temperatures, their Young's modulus increases and becomes in the austenite phase. The SMAs easily can be plastically deformed by approximately 5% in low temperatures; once heated, they can recover the deformation. Due to their internal electrical resistance, temperature control can be achieved easily by controlling the electrical current passing through the wire. Thus, the dynamic properties of the mount are changed. A simple on-off control is sufficient for the semi-active mount.

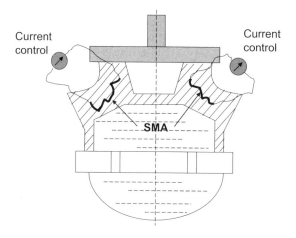

Figure 4.71 A shape memory alloy (SMA) semi-active mount.

The semi-active control system is much simpler than the architecture of an active control system. In some applications, the semi-active control can achieve almost the same control effect as the active control system but at less cost. The simple structure also makes the semi-active control system possess higher reliability than the active control system. However, almost all reported work shows that the semi-active control system is used mainly to improve the performance at the low-frequency range. To improve the high-frequency vibration isolation, active control techniques must be employed.

4.7 Summary

In this chapter, ride dynamics was divided into excitation sources, vehicle ride modeling, seating modeling, and human body modeling. We reviewed the vehicle road surface uneven profile and powertrain excitation sources. The quarter car model and the bounce-pitch model were described in detail. This chapter also introduced subjective and objective methods for evaluating seat isolation effectiveness and the ride comfort of passengers. Active and semi-active control methods for engine isolation and suspension systems also were described.

4.8 References

1. Aoki, K., Shikata, T., Hyoudou, Y., Hirade, T., and Aihara, T., "Application of an Active Control Mount (ACM) for Improved Diesel Engine Vehicle Quietness," SAE Paper No. 1999-01-0832, Society of Automotive Engineers, Warrendale, PA, 1999.

2. Brunning, A.D., "The Relationship Between Objective Measurement and Perception of Low Levels of Sensory Inputs Experienced in a Motor Vehicle," M.Sc. thesis, University of Bradford, Bradford, UK, 1996.

3. Casati, F.M., Berthevas, P.R., Herrington, R.M., and Miyazaki, Y., "The Contribution of Molded Polyurethane Foam Characteristics to Comfort and Durability of Car Seats," SAE Paper No. 1999-01-0585, Society of Automotive Engineers, Warrendale, PA, 1999.

4. Crede, C.E., *Vibration and Shock Isolation*, John Wiley & Sons, New York, NY, 1952.

5. EI-Demerdash, S.M., Selim, A.M., and Corlla, D.A., "Vehicle Body Attitude Using an Electronically Controlled Active Suspension," SAE Paper No. 1999-01-0724, Society of Automotive Engineers, Warrendale, PA, 1999.

6. Fairley, T.E. and Griffin, M.J., "The Apparent Mass of the Seated Human Body: Vertical Vibration," *Journal of Biomechanics*, Vol. 22, No. 2, 1989, pp. 81–94.

7. Foumani, M.S. and Khajepour, A., "Application of Shape Memory Alloys to a New Adaptive Hydraulic Mount," SAE Paper No. 2002-01-2163, Society of Automotive Engineers, Warrendale, PA, 2002.

8. Gillespie, T.D., *Fundamental of Vehicle Dynamics*, Society of Automotive Engineers, Warrendale, PA, 1991.

9. Griffin, M.J., "Evaluation of Vibration with Respect to Human Response," SAE Paper No. 860047, Society of Automotive Engineers, Warrendale, PA, 1986.

10. Griffin, M.J., *Handbook of Human Vibration*, Academic Press Limited, London, 1996.

11. Harris, C.M., *Shock and Vibration Handbook*, 4th Ed., McGraw-Hill, New York, NY, 1996.

12. ISO 2631-1, "Mechanical Vibration and Shock—Evaluation of Human Exposure to Whole-Body Vibration, Part I: General Requirement," 1997.

13. Kim, T., Cho, Y., Yoon, Y., and Park, S., "Dynamic Ride Quality Investigating and DB of Ride Values for Passenger and RV Cars," SAE Paper No. 2001-01-0384, Society of Automotive Engineers, Warrendale, PA, 2001.

14. Kinkekaar, M.R. and Neal, B.L., "The Influence of Polyurethane Foam Dynamics on the Vibration Isolation Character of Full Foam Seats," SAE Paper No. 980657, Society of Automotive Engineers, Warrendale, PA, 1998.

15. Kitazaki, S. and Griffin, M.J., "A Modal Analysis of Whole-Body Vertical Vibration, Using a Finite Element Model of the Human Body," *J. Sound and Vibration*, 200, No. 1, 1997, pp. 83–103.

16. Lewitzke, C. and Lee, P., "Application of Elastomeric Components for Noise and Vibration Isolation in the Automotive Industry," SAE Paper No. 2001-01-1447, Society of Automotive Engineers, Warrendale, PA, 2001.

17. Matsumoto, Y. and Griffin, M.J., "Movement of the Upper-Body of Seated Subjects Exposed to Vertical Whole-Body Vibration at the Principal Reasonance Frequency, *J. Sound and Vibration*, Vol. 215, No. 4, 1998.

18. McManus, S.J. and St. Clair, K.A., "Vibration and Shock Isolation Performance of a Commercial Semi-Active Vehicle Seat Damper," SAE Paper No. 2000-01-3408, Society of Automotive Engineers, Warrendale, PA, 2000.

19. Mo, C., Sunwoo, M., and Patten, W.N., "Bistate Control of a Semiactive Suspension," SAE Paper No. 1999-01-0725, Society of Automotive Engineers, Warrendale, PA, 1999.

20. Pang, J., Kurrle, P., Qatu, M., and Rebandt, R., "Attribute Analysis and Criteria for Automotive Exhaust Systems," SAE 2003 World Congress, Detroit, MI, March 2003.

21. Pang, J., Qatu, M., Dukkipati, R., Sheng, G., and Patten, W.N., "Model Identification for Nonlinear Automotive Seat Cushion Structure," *Intl. J. Vehicle Noise and Vibration*, Vol. 1, Nos. 1 and 2, 2003.

22. Park, S., Cheung, W., Cho, Y., and Yoon, Y., "Dynamic Ride Quality Investigation for Passenger Car," SAE Paper No. 980660, Society of Automotive Engineers, Warrendale, PA, 1998.

23. Pielemeier, W.J., *et al.*, "The Estimation of Seat Values from Transmissibility Data," SAE Paper No. 2001-01-0392, Society of Automotive Engineers, Warrendale, PA, 2001.

24. Qatu, M., Sarifi, M., and John F., "Robustness of Powertrain Mount System for Noise, Vibration and Harshness at Idle," *J. Automobile Engineering*, Vol. 216, 2002.

25. Tchernychouk, V., Rakheja, S., Stiharu, I., and Boileau, P.E., "Study of Occupant-Seat Models for Vibration Comfort Analysis of Automotive Seats," SAE Paper No. 2000-01-2688, Society of Automotive Engineers, Warrendale, PA, 2000.

26. Ushijima T., Takano, K., and Kojima, H., "High Performance Hydraulic Mount for Improving Vehicle Noise and Vibration," SAE Paper No. 880073, Society of Automotive Engineers, Warrendale, PA, 1988.

27. Wei, L. and Griffin, J., "The Prediction of Seat Transmissibility from Measures of Seat Impedance," *J. Sound and Vibration*, Vol. 214, No. 1, 1998.

28. Wong, J.Y., *Theory of Ground Vehicles*, Wiley-Interscience Publication, John Wiley & Sons, New York, NY, 1978.

29. Yu, L.C. and Khameneh, K.N., "Automotive Seating Foam: Subjective Dynamic Comfort Study," SAE Paper No. 1999-01-0588, Society of Automotive Engineers, Warrendale, PA, 1999.

30. Yu, Y., Peelamedu, S.M., Naganathan, N.G., and Dukkipati, R.V., "Automotive Vehicle Engine Mounting System: A Survey," *J. Dynamic Systems, Measurement, and Control*, Vol. 123, June 2001.

Chapter 5

Vehicle Rollover Analysis

5.1 Introduction

Vehicle vulnerability to rollover is one of the most important vehicle safety attributes. In this chapter, we will introduce the rollover mechanism and present scenarios in which rollover could happen.

5.1.1 Rollover Scenario

Much attention has been paid to automotive rollover accidents during recent decades. Rollover accidents are a serious threat to the life of automobile occupants. Rollover is defined as a vehicle rotation of 90° or more around its longitudinal axis. The vehicle body will contact the ground, and the occupants will be injured. Figure 5.1 shows a vehicle rollover scenario.

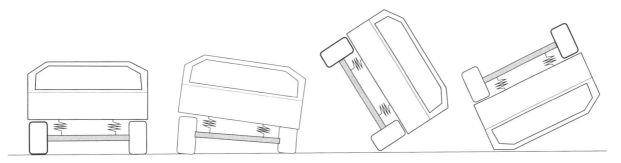

Figure 5.1 A vehicle rollover scenario.

Several factors will affect the vulnerability of a vehicle to rollover. Among these factors are tire and vehicle characteristics, environmental conditions, and drivers. A combination of vehicle and tire design is among the reasons for rollover. For example, vehicles with a higher center of gravity (CG) such as trucks are more vulnerable to rollover when a simple but severe steering input is given. Environmental conditions also may cause rollover. Such conditions are related primarily to road conditions such as ice or bumpy roads. Drivers with dangerous driving habits likewise may cause rollover. Rollover accidents may result from one or a combination of these factors.

Rollover can happen on a flat road, on a cross-slope road, or off road. Rollover can be divided into two categories: tripped rollover, and untripped rollover. Tripped rollover is caused by a vehicle hitting an obstacle. This rollover is caused by vehicle inclination where the center of gravity of the vehicle exceeds the stable point. National Highway and Traffic Safety Administration (NHTSA) reports show that only 10% or less of all rollover crashes occur under untripped conditions for the on-road environment. However, major rollovers occur in an off-road environment where the road conditions are unpredictable. Rollover for on-road accidents usually is caused by dangerous maneuvers that are induced either by high lateral acceleration or by yaw instability. The high lateral acceleration produces sufficient centrifugal force to pull the vehicle and cause it to rotate around its outside tire. During yaw motion, the tires produce saturated forces that cause tire sliding and rollover. Most rollover occurs under the following conditions:

Traveling at high speed on a curved road: When a vehicle travels on a curved road, lateral centrifugal force will pull it in an outboard motion, as shown in Figure 5.2.

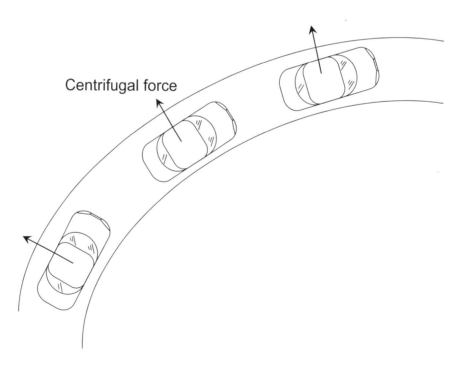

Centrifugal force

Figure 5.2 A vehicle traveling on a curved road.

Severe cornering maneuver: This case is similar to the preceding case where the centrifugal force pulls the vehicle to rollover. For example, a driver avoiding an accident and steering rapidly can cause a yaw disturbance.

Traveling on a collapsing road and suddenly providing steering input for a vehicle with a low level of roll stability: This kind of rollover also is caused by a yaw disturbance. Figure 5.3 shows a vehicle rollover scenario caused by severe steering input.

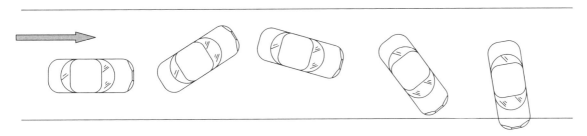

Figure 5.3 A vehicle rollover scenario caused by severe steering input.

Losing control due to a rapid decrease of friction, such as driving on an icy road: Steering can cause yaw motion because forces on the tires in the lateral direction are strong enough to roll the vehicle. The forces also produce lateral acceleration on the vehicle center of gravity. The forces are produced by the friction between the tires and the road; therefore, these rollover scenarios are called friction rollovers. Figure 5.4 shows a vehicle losing control because of low friction on the road.

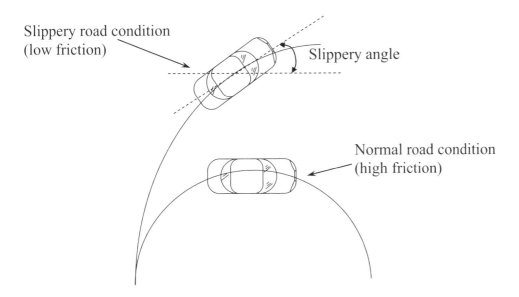

Figure 5.4 A vehicle losing control because of low friction on the road.

Laterally sliding off the road: An example of this case would be a vehicle that can be decelerated by a barrier.

Sliding from a cliff: Figure 5.5 shows this type of scenario, in which a vehicle is sliding from a cliff.

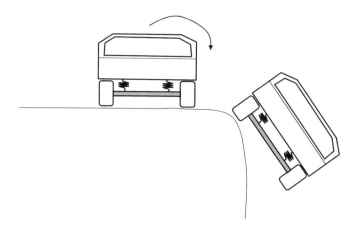

Figure 5.5 A vehicle sliding from a cliff.

5.1.2 Importance of Rollover

Rollover has been an important safety issue of vehicles. Rollover is a major reason for severe and fatal injuries, especially for trucks and sport utility vehicles (SUVs). National Automotive Sampling System (NASS) data (Hinch *et al.*, 1992) show that approximately 85 to 90% of rollover occurred in single-vehicle accidents. The vehicles were out of control prior to overturning. Of these single-vehicle crashes, more than 50% resulted in fatalities with the vehicles rolling over. Rollover crashes are the most dangerous type of collision for all classes of light vehicles. The rate of rollover accidents for SUVs and small trucks is higher than that of passenger cars.

Rollover accidents account for a small percentage of all vehicle accidents, but they account for a large percentage of fatal accidents. In 1997, NASS (James *et al.*, 1997) reported that only 8% of car accidents resulted in rollover, but rollover accounted for 17% of all fatal injuries. In 1996, rollovers were 1.8% of all crashes in the United States but represented 9.9% of all fatal crashes (SAE Paper No. 1999-01-0122). Some reports showed that approximately 50% of single-vehicle accident fatalities resulted from rollover, and approximately 10% of multi-vehicle accident fatalities were caused by rollover (SAE Paper No. 950315). Almost 10,000 people die each year in rollover accidents in the United States (Hinch *et al.*, 1992).

More materials on rollover safety and rollover control will be introduced in Section 5.7 of this chapter.

5.1.3 Research on Rollover

Because rollover is the major safety concern, it is important to develop testing and analytical methods to predict rollover and to improve design to control rollover. Rollover studies include field accident data, statistical analysis, occupant kinematics, injury mechanisms, rollover model analysis and testing, vehicle stability factors, rollover accident reconstruction, rollover control, and others.

The study of vehicle rollover is a complicated topic. The phenomenon of rollover is difficult to predict. Rollover involves many vehicle factors and some nonvehicle factors, such as the road friction coefficient, obstacles on the road, the type of road shoulder, road and shoulder inclination angles, driver steering patterns, and others. To simplify the complicated overall analysis, a steady-state rigid model and a steady-state suspended model are most commonly used. Some simple metrics have been developed to evaluate the static and dynamic stability of vehicles, such as the static stability factor (SSF), the tilt table ratio (TTR), the critical sliding velocity, the dynamic stability index, and others. Among the indicators, the static stability factor and the tilt table ratio are most commonly used.

A vehicle dynamic model involves many systems and parameters. Of particular interest are the steering and tire systems. Some simple dynamic models include only several degrees of freedom. The models can be used for quick review of rollover characteristics and parameter tuning. More complicated dynamic models include many components, and systems can be built using specific software such as ADAMS, NADSdyna, VDANL, and SIMPACK. Vehicle rollover scenarios can be animated on the computer screen.

The taller and narrower a vehicle is, the easier it tends to roll over. The most effective way to keep the vehicle from rolling over is to make the center of gravity of the vehicle as low as possible and/or to make the vehicle wider (i.e., increase the track width of the vehicle). The use of tires with less lateral force will help minimize the potential for rollover. Active control is one of the effective methods to prevent rollover. Some control algorithms are integrated with a dynamic model to simulate rollover and its control. Active control in rollover prevention is an active field of research. Research on rollover also includes parameter sensitivity analyses, dynamic stability studies, and correlation between vehicle design parameters and rollover statistical data.

Many studies have been done by the NHTSA on vehicle rollover since the 1970s. In 1973, NHSTA proposed an Advance Notice of Proposed Rule Making (ANPRM) to evaluate vehicle rollover tendencies on smooth and dry roads. Since then, several procedures have been proposed, but no standard procedures to evaluate vehicle rollover stability nor a suitable rollover stability metric have been adopted by NHSTA. Other organizations, such as ISO, SAE, and CU (Customers Union), have developed some test procedures. The severe lane-change maneuver procedure to evaluate vehicle road-holding ability was developed by ISO. SAE proposed a procedure to evaluate both steady-state and transient behavior. CU developed an obstacle avoidance maneuver procedure, and ISO and CU developed full-scale vehicle handling tests to evaluate the dynamic response of a vehicle in an emergency situation.

5.1.4 Scope of This Chapter

This chapter introduces the concept of and possible scenarios for rollover. Static rollover models, including rigid and suspended models, are analyzed. In evaluating the static rollover threshold, the tilt table ratio and slide pull ratio are the most commonly used metrics. Many parameters influence the rollover threshold, such as the vehicle center-of-gravity height, track width, suspension stiffness, tire stiffness, tire lateral force and friction, and others. The influence of these parameters on the static rollover threshold is analyzed. The dynamic model is developed by extending the quasi-static model, which includes rotational and lateral types of motion. This chapter briefly introduces the rigid model, dependent suspension model, and independent suspension model. Dynamic rollover thresholds are based on the energy that is needed to cause the vehicle to roll over. The dynamic rollover thresholds include the dynamic stability index, rollover prevention energy reserve, rollover prevention metric, and critical sliding velocity.

This chapter emphasizes the importance of rollover analysis and control, and it introduces the relation between rollover accidents and occupants. The rollover sensing system provides a warning when the vehicle is close to the rollover condition, and the control system activates an anti-rollover mechanism. Active anti-roll control has been widely used in modern vehicles, such as the active suspension, active roll-bar, active steering, anti-rollover braking system, and others.

5.2 Rigid Vehicle Rollover Model

Vulnerability to rollover using simple vehicle models is the subject of this section. The models are rigid body models and typically are used in the early stages of product development.

5.2.1 Rigid Vehicle Model

A vehicle is a complicated structure. In dynamic analysis, the vehicle usually is simplified as a combination of many rigid bodies. Spring, damping, and other connection components are used to connect these rigid bodies. To quickly understand basic rollover characteristics, assume that a vehicle is a rigid body model, as shown in Figure 5.6. The road can either be flat or have a slope angle of θ. The deflection of the tires and suspension is neglected. When the vehicle undergoes a turn, centrifugal force caused by its body pulls it outward from the turning center. The centrifugal force equals the lateral acceleration of the vehicle multiplied by the weight of the vehicle, or ma_y.

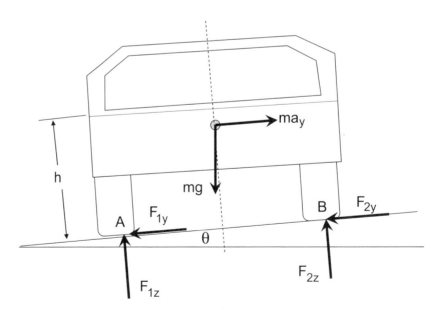

Figure 5.6 A rigid vehicle model.

Taking the moment at point B, the vehicle dynamic equation of Figure 5.6 can be written as

$$I_b\ddot{\varphi} = F_{1z}d - (mg\sin\theta)h - (mg\cos\theta)\frac{d}{2} + ma_yh \qquad (5.1)$$

where I_b is the vehicle moment of inertia around point B, h and d are the vehicle center-of-gravity height and track width, respectively, and F_{1z} is the vertical force acting on the inner tires.

Usually, the slope angle θ is very small; therefore, $\sin\theta \approx \theta$ and $\cos\theta \approx 1$. Equation 5.1 can be simplified as

$$I_b\ddot{\phi} = F_{1z}d - mgh\theta - mg\frac{d}{2} + ma_yh \tag{5.2}$$

At the beginning moment when rollover occurs, the rotation velocity and acceleration are neglected (i.e., $\ddot{\phi} = 0$). This situation is called the quasi-static state (steady state). The discussions in this section and the next section are based on an assumption of the quasi-static state. Equation 5.2 can be simplified as

$$F_{1z}d - mgh\theta - mg\frac{d}{2} + ma_yh = 0 \tag{5.3}$$

5.2.2 Steady-State Rollover on a Flat Road

Rollover begins if the load on the inner tires is zero (i.e., $F_{1z} = 0$). Equation 5.3 can be simplified as

$$\frac{a_y}{g} = \frac{d}{2h} + \theta \tag{5.4}$$

The lateral acceleration depends on the slope angle, the vehicle center-of-gravity height, and the vehicle track width. The larger the slope angle, the higher the rollover lateral acceleration. Usually, the slope angle θ is much smaller than $\frac{d}{(2h)}$. The first term, $\frac{d}{(2h)}$, often is called the first-order estimate of the static rollover threshold.

EXAMPLE E5.1

The center-of-gravity height of a truck is 1.8 m, and its tread width is 2 m. The truck travels on a road with a slope of 5°. Calculate the lateral acceleration of the vehicle at the moment rollover begins.

Solution:

According to Eq. 5.4, the lateral acceleration is

$$a_y = \left(\frac{d}{2h} + \theta\right)g$$

$$= \left(\frac{1.8}{2*2} + \frac{5}{180}*\pi\right)g$$

$$= 0.45g + 0.087g$$

$$= 0.537g$$

The result shows that the lateral acceleration caused by the road angle is much smaller than that of the vehicle parameters h and d.

On a flat road surface, $\theta = 0$. Equation 5.4 can be simplified as

$$a_y = \frac{d}{2h}g \qquad (5.5)$$

Equation 5.5 represents the relation between the lateral acceleration and the two basic dimensions of the vehicle: the center-of-gravity height (h), and the tread width (d). If $a_y > \frac{d}{2h}g$, rollover will occur. The wider and the lower the center of gravity of the vehicle is, the more difficult it is for rollover to occur. Thus, Eq. 5.5 indicates the static roll stability and is called the rollover threshold for a flat road surface. The vehicle static rollover threshold is the maximum lateral acceleration that the vehicle can sustain in a steady-state turn on a level surface without rolling over.

Equation 5.5 also can be written as a nondimensional expression as

$$\frac{a_y}{g} = \frac{d}{2h} \qquad (5.6)$$

Equation 5.6 is a ratio of half of the track width to the height of the center of gravity of the vehicle. This ratio is called the static stability factor (SSF). The static stability factor is a useful and simple metric for studying rollover propensity because only two basic properties of the vehicle are used for calculating the static stability factor. The static stability factor reflects the most fundamental dynamic relation of the vehicle.

The static stability factor is obtained under the assumption that the vehicle is a rigid body, and deflections of the suspension and tire are neglected. When a vehicle is turning, the lateral force pulls the center of gravity of the vehicle to shift toward the outside of the tire and causes the outside tire and suspension deformation, which will significantly change the vehicle center-of-gravity movement and track width. All of these influencing factors reduce the vehicle rollover stability and the static stability factor. The estimation by Eq. 5.6 is not conservative, in that it virtually overestimates the actual static rollover threshold of real vehicles.

Thus, the static stability factor overestimates the rollover threshold. Although the vehicle is not a rigid body and the static stability factor is not an accurate metric to predict the rollover threshold, it provides the first guideline for a vehicle design and is useful for doing benchmark comparisons for different vehicles.

The inverse term of the roll threshold is called the rollover propensity (i.e., $\frac{2h}{d}$). A higher number of the propensity means a lower rollover threshold.

5.2.3 Tilt Table Ratio

Figure 5.7 is a simple tilt test table, with a vehicle placed on its surface. The table is tilted from the horizontal by an angle ϕ. The vehicle is in steady state. The vehicle weight, in milligrams (mg), can be divided in two directions: one perpendicular to the tilt table surface, and the other parallel to the tilt table surface. The weight perpendicular to the tilt table surface is the actual weight acting on the vehicle, called the simulated weight, and the corresponding force can be expressed as

$$F_1 = m_s g = mg \cos \phi \tag{5.7}$$

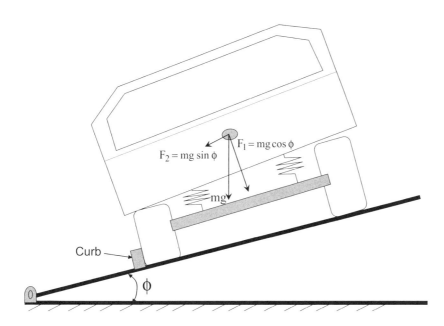

Figure 5.7 Tilt test table with a vehicle placed on it.

The weight parallel to the tilt table surface models the simulated centrifugal force. Assume that a simulated lateral acceleration, a_s, is acted on the vehicle, and a centrifugal force pulls the vehicle into rollover. The centrifugal force can be expressed as

$$F_2 = m_s a_s = mg \sin \phi \tag{5.8}$$

During the test, the tilt of the table is increased slowly until the tire at the high side loses contact with the surface of the table. The angle at which this occurs is called the tip-over angle. Dividing Eq. 5.7 by Eq. 5.8, the ratio of the simulated lateral acceleration over g is obtained and is called the tilt table ratio (TTR), which is expressed as

$$TTR = \frac{a_s}{g} = \tan \phi \tag{5.9}$$

The tilt table ratio is an estimator of the vehicle static rollover threshold. It is a simple test to estimate the vehicle rollover threshold because it is determined only by the tilt angle and does not need any additional measurement, such as the vehicle center of gravity or weight. Furthermore, the test is safe and nondestructive to the vehicle.

The tilt table ratio is better used to estimate the rollover of unstable vehicles, where there is a small tilt angle, than stable vehicles. The tilt table ratio is more accurate in simulating a vehicle steady turn at high lateral acceleration or a vehicle on an embankment or side slope than the static stability factor, where a rigid vehicle model is used. The tilt table ratio replicates the physical road-vehicle condition, but it has some disadvantages. Because the real weight acting on the vehicle (mg cos φ) is less than the vehicle weight, the lateral centrifugal force is less than the actual force. For most light-duty vehicles, the centrifugal force on a vehicle on the tilt table is approximately 60 to 80% of that while the vehicle turns on real roads. Thus, the vehicle rollover stability will be overestimated. On the other hand, the reduced vertical load acting on the compliant tires and suspension of the vehicle may rise relative to the normal position, which results in a higher center-of-gravity position. Thus, the static rollover stability is underestimated. These two influencing factors tend to cancel each other.

Some vehicles could slide due to the limited surface-tire friction before rollover occurs. To prevent that, a small curb usually is placed next to the lower tires, as shown in Figure 5.7. However, the presence of a curb can raise the effective side force of the center of the tire. That is, the effective center-of-gravity height is reduced, and the static rollover stability is overestimated.

5.2.4 Side Pull Ratio

The tilt table ratio is not a good estimator of rollover for more stable vehicles. To overcome this disadvantage, another static rollover estimator, the side pull ratio (SPR), is induced. Figure 5.8 shows the side pull ratio test setup. A vehicle is placed on a high-friction surface. Then a pull force through a belt is applied to the sprung mass at the vehicle center of gravity and slowly pulls laterally. The test is in a quasi-static state, and the belt is maintained in a horizontal line as the vehicle moves during the test.

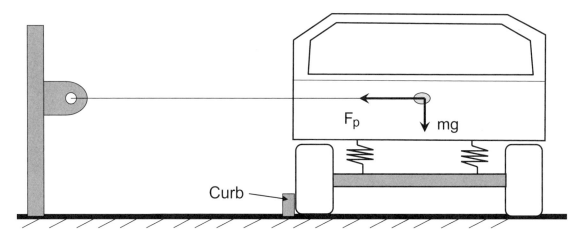

Figure 5.8 A side pull ratio (SPR) test setup.

The side pull ratio is defined as the ratio of the minimum side pull force (F_p) required to lift all tires on one side of the vehicle from the ground to the vehicle weight and is expressed as

$$SPR = \frac{F_p}{mg} \tag{5.10}$$

The test attempts to simulate lateral acceleration that acts on the vehicle in steady-state turns and reaches the rollover threshold. The side pull ratio uses the proper magnitude of vertical and lateral loads applied to the vehicle. The forces acting on the vehicle at the start of rollover are more accurate than those in the static stability factor and the tilt table ratio. Thus, the side pull ratio is a good estimator of the static rollover threshold.

During the side pull test, the pulling force is acted upon the sprung mass at the vehicle center of gravity. However, the vehicle consists of many "rigid bodies," and each has its own center of gravity. The pull force represents the inertia forces of all of these rigid bodies; thus, the force transmitted from the sprung mass to the unsprung mass is higher than that in a real maneuver. The higher force causes a higher deflection of the suspension, which will lead to higher lateral displacement of the vehicle center of gravity. That is, the center-of-gravity height is reduced, and this causes the side pull ratio to be lower than the actual lateral acceleration at rollover.

The other disadvantage is that the belt could damage the vehicle body when a force equal to the vehicle weight is applied. More belts or wider belts can reduce potential damage but add extra weight. The belt load is applied on a single point on the vehicle. This is different from the real distributed load. The single contact point is assumed as the center-of-gravity point. During the pulling procedure, the center-of-gravity height changes.

Similar to the tilt table test, a side curb also is used in the side pull ratio test because friction between the road surface and the tires is not enough to hold the vehicle before rollover occurs. This will influence the vehicle center-of-gravity height.

The side pull test cannot give the absolute lateral acceleration when rollover occurs. It usually is used to study the static rollover stability of vehicles.

5.3 Suspended Vehicle Rollover Model

In the rigid model, the deflections of the tires and suspension are neglected. However, these deflections will impact vehicle rollover stability. When rollover occurs, lateral force will pull the vehicle body outward. The outside suspension and tires will be compressed, which will cause the vehicle center of gravity to shift to the outside, accelerating the possibility of rollover. This section discusses the influence of suspension stiffness and tire stiffness on the rollover threshold.

5.3.1 Steady-State Rollover Model for a Suspended Vehicle

Figure 5.9 shows a suspended vehicle rollover model. The outside tire is compressed, and the inside tire is released, so there exists a deflection angle between the two tires, donated by θ_1. The vehicle body is supported on the suspension. The centrifugal force pulls the body to the outside and compresses the right suspension spring. At the same time, the left suspension spring is extended. Another deflection angle exists between the two suspension springs and is donated by θ_2.

The vehicle center of gravity rotates around an imaginary center, called the rollover center, which is point C in Figure 5.9. When the vehicle is stationary on a flat road, a line connecting the center of gravity and the rollover center, called the rollover line, is in the vertical direction. During turning or cornering, the rollover line rotates an

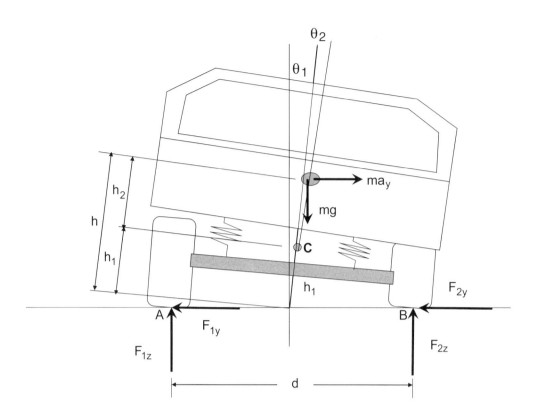

Figure 5.9 Suspended rollover model.

angle θ_1 due to the deflection of the tires and another angle θ_2 due to the suspension deflection. Thus, the total angle at which the rollover line rotates is $\theta_1 + \theta_2$, as shown in Figure 5.9.

Taking the moment about the contact point B between the outside wheel and the ground, the equilibrium equation is

$$F_{1z}d + ma_y\left[h_1\cos\theta_1 + \left(h - h_1\right)\cos\left(\theta_1 + \theta_2\right)\right]$$

$$- mg\left[\frac{d}{2} - h_1\sin\theta_1 - \left(h - h_1\right)\sin\left(\theta_1 + \theta_2\right)\right] = 0 \tag{5.11}$$

where h_1 is the distance from the rollover center to the ground at the vehicle stationary condition, h_2 is the distance from the vehicle center of gravity to the rollover center, h is the vehicle center-of-gravity height, and $h = h_1 + h_2$.

Equation 5.11 can be rewritten as

$$F_{1z}d + ma_y\left[h_1\cos\theta_1 + h_2\cos\left(\theta_1 + \theta_2\right)\right]$$

$$- mg\left[\frac{d}{2} - h_1\sin\theta_1 - h_2\sin\left(\theta_1 + \theta_2\right)\right] = 0 \tag{5.12}$$

Because θ_1 and θ_2 are small numbers, it is assumed that $\cos\theta_1 \approx 1$, $\cos\left(\theta_1 + \theta_2\right) \approx 1$, $\sin\theta_1 \approx \theta_1$, and $\sin\left(\theta_1 + \theta_2\right) \approx \theta_1 + \theta_2$. When rollover occurs, the vertical force on the inside tire is zero (i.e., $F_{1z} = 0$). Inserting these assumptions into Eq. 5.12, we obtain

$$\frac{a_y}{g} = \frac{d}{2h} - \theta_1 - \frac{h_2}{h}\theta_2 \qquad (5.13)$$

Equation 5.13 is the rollover threshold for the suspended vehicle model. Compare Eq. 5.13 with Eq. 5.6, and it is clear that this rollover threshold is lower than that of the rigid vehicle model. The first item on the right side of Eq. 5.13 represents the rollover threshold for the rigid vehicle model. The second and third terms are the contributions from the tire deflection and suspension deflection, respectively. The contributions from the tire deflection and suspension deflection on the rollover threshold will be discussed separately.

5.3.2 Contribution from the Tire Deflection

In this section, the suspension is assumed to be rigid, and only the tire deflection is considered, as shown in Figure 5.10. The outside tire is compressed Δr from the original height r. At the same time, the inside tire is released and is not subjected to any load. When the vehicle is at a straight stationary condition, each of the two tires takes half of the vehicle weight. However, when rollover occurs, the whole vehicle weight is supported by the outside tire. The increased weight supported by the tire is

$$\Delta F_{2z} = \frac{mg}{2} \qquad (5.14)$$

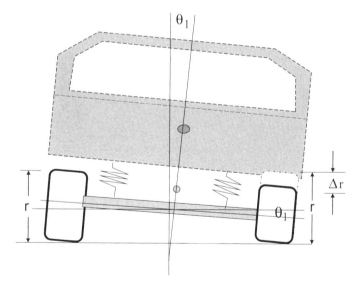

Figure 5.10 Deflected tire and rigid body model.

The tire deflection, Δr, compared with the stationary condition, is

$$\Delta r = \frac{\frac{mg}{2}}{K_{tire}} \qquad (5.15)$$

where K_{tire} is the tire stiffness.

Thus, the rotation angle at the rollover incipient caused by tire deflection is

$$\theta_1^R = \frac{\Delta r}{\dfrac{d}{2}} = \frac{mg}{dK_{tire}} \tag{5.16}$$

The rotation angle θ_1 is determined by the track width d, the tire stiffness k_{tire}, and the vehicle weight. If the tire is rigid (i.e., $k_{tire} \to \infty$), the tire deflection and angle will go to zero (i.e., $\theta_1 \to 0$). The tire stiffness is the key element in controlling the deflection angle.

5.3.3 Contribution from the Suspension Deflection

In this section, the suspension deflection together with the tire deflection is considered, as shown in Figure 5.11.

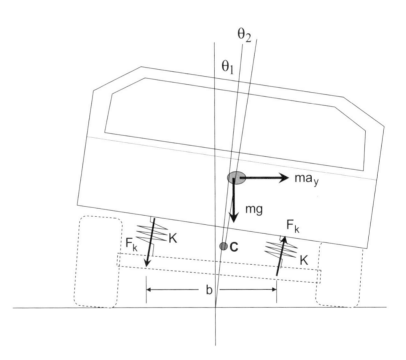

Figure 5.11 Deflected suspension model.

During turning or cornering, the lateral force pulls the vehicle body toward the outside direction, and the suspension outside spring is compressed. The outside spring is compressed by Δz, and the inside spring is extended by Δz, which can be expressed as

$$\Delta z = \frac{b}{2} tg\theta_2 \approx \frac{b}{2}\theta_2 \tag{5.17}$$

where b is the distance between the two suspension springs, as shown in Figure 5.11.

Taking the moment around the rollover center C, we obtain

$$ma_yh_2\cos(\theta_1+\theta_2)+mgh_2\sin(\theta_1+\theta_2)-F_Kb=0 \qquad (5.18)$$

where F_K is the suspension spring force acting on the vehicle body, and $F_K=K\Delta z$. K is the suspension spring stiffness at one side.

For a small deflection angle, $\cos(\theta_1+\theta_2)\approx1$ and $\sin(\theta_1+\theta_2)\approx\theta_1+\theta_2$, Eq. 5.18 can be simplified as

$$ma_yh_2+mgh_2(\theta_1+\theta_2)-bK\Delta z=0 \qquad (5.19)$$

Substituting Eq. 5.17 into Eq. 5.19 gives

$$\theta_2=\frac{a_y}{g}\frac{mgh_2}{\dfrac{b^2}{2}K-mgh_2}+\frac{mgh_2}{\dfrac{b^2}{2}K-mgh_2}\theta_1 \qquad (5.20)$$

The rotation angle is determined by the suspension stiffness K, the distance between two suspension springs, the lateral centrifugal acceleration, the distance between the center of gravity and the rollover center, and the vehicle weight. If the suspension is rigid (i.e., the stiffness $K\to\infty$), then the deflection angle caused by the suspension will be zero (i.e., $\theta_2\to0$). The higher the location of the center of gravity is, the larger the deflection angle becomes, and the vehicle body becomes more unstable. During turning, the lateral force always pulls the vehicle body toward the outside, that is, the deflection angle is clockwise. Hence, $\theta_2\geq0$, so we obtain the relation

$$b\geq\sqrt{\frac{2mgh_2}{K}} \qquad (5.21)$$

The distance between the two suspension springs should be long enough to satisfy Eq. 5.21; otherwise, the body will be unstable.

If Eqs. 5.16 and 5.20 are compared, the rotation angle θ_1 caused by the tire deflection is determined only by the vehicle parameters, but the rotation angle θ_2 depends not only on the vehicle parameter but also on the lateral acceleration and θ_1.

EXAMPLE E5.2

A truck weighs 12,000 kg. Its roll center height is 0.65 m, and the vehicle body center-of-gravity height is 1.7 m. The suspension stiffness is 80,000 N/m. Design a suspension springs' distance that makes the system stable.

Solution:

The height from the body center of gravity to the rollover center, h_2, is

$$h_2 = h - h_1$$

$$= 1.7 - 0.65$$

$$= 1.05 \text{ m}$$

According to Eq. 5.21, the minimal distance between the two springs is

$$b = \sqrt{\frac{2mgh_2}{K}}$$

$$= \sqrt{\frac{2 \times 12,000 \times 9.8 \times 1.05}{80,000}}$$

$$= 1.76 \text{ m}$$

5.3.4 Parameters Influencing the Suspended Rollover Model

Substituting Eqs. 5.16 and 5.20 into Eq. 5.13, the normalized lateral acceleration can be expressed as

$$\frac{a_y}{g} = \frac{d}{2h} - \frac{mg}{dK_{tire}} - \frac{a_y}{g} \frac{mgh_2^2}{h\left(\dfrac{b^2}{2}K - mgh_2\right)} - \frac{mgh_2^2}{h\left(\dfrac{b^2}{2}K - mgh_2\right)} \frac{mg}{dK_{tire}} \quad (5.22)$$

Equation 5.22 can be rewritten as

$$\frac{a_y}{g} = \frac{d}{2h} \frac{h\left(\dfrac{b^2}{2}K - mgh_2\right)}{\dfrac{b^2}{2}hK - mghh_2 + mgh_2^2} - \frac{mg}{dK_{tire}} \quad (5.23)$$

To reflect the roll moment influence on the lateral acceleration, the roll stiffness of the suspension, K_{θ_2}, is introduced. The two suspension springs shown in Figure 5.11 provide the torque to the vehicle body, which causes the body roll moment. The torque T can be expressed as

$$T = F_k b = K\Delta z b = K\frac{b}{2}\theta_2 b = K_{\theta_2}\theta_2 \quad (5.24)$$

Subsequently, the roll stiffness is

$$K_{\theta_2} = \frac{b^2}{2}K \quad (5.25)$$

Substituting Eq. 5.25 into Eq. 5.23 gives

$$\frac{a_y}{g} = \frac{d}{2h} \frac{h\left(K_{\theta_2} - mgh_2\right)}{hK_{\theta_2} - mghh_2 + mgh_2^2} - \frac{mg}{dK_{tire}} \tag{5.26}$$

The right side of Eq. 5.26 can be extended into three independent terms as

$$\frac{a_y}{g} = \frac{d}{2h} - \frac{d}{2h} \frac{mgh_2^2}{hK_{\theta_2} - mghh_2 + mgh_2^2} - \frac{mg}{dK_{tire}} \tag{5.27}$$

To clearly express Eq. 5.27, the dimensionless stiffness is introduced for the tire stiffness and suspension stiffness.

Thus, the dimensionless stiffness for the tire is

$$K_{tire}^* = \frac{K_{tire}}{\dfrac{mg}{d}}$$

and the dimensionless stiffness for the suspension is

$$K_{\theta_2}^* = \frac{K_{\theta_2}}{mgh_2}$$

Substitute these dimensionless stiffness values into Eq. 5.27, and the normalized lateral acceleration can be expressed as

$$\frac{a_y}{g} = \frac{d}{2h} - \frac{d}{2h} \frac{1}{K_{tire}^*} - \frac{d}{2h} \frac{1}{1 + \dfrac{h}{h_2}\left(K_{\theta_2}^* - 1\right)} \tag{5.28}$$

In the right side of Eq. 5.28, the three terms have their own physical meanings. The first depends only on the basic vehicle dimensions. The second has a relation with the tire dimensionless stiffness. The third term depends on the suspension dimensionless stiffness and the basic dimensions of the vehicle. The influences of the tire and suspension stiffness will be discussed in detail in the following case studies.

Case A: Assume the tire and suspension are infinitely stiff (i.e., $K_{tire} \to \infty$ and $K_{\theta_2} \to \infty$). Equation 5.28 becomes

$$\frac{a_y}{g} = \frac{d}{2h} \tag{5.29}$$

Equation 5.29 is the same expression as Eq. 5.6. The suspended model becomes the rigid model. Or, we can say that the rigid rollover model is a special case of the suspension rollover model when the tire stiffness and suspension stiffness become infinitely high. In this case, the rollover lateral acceleration is determined by only two vehicle dimensions: the center-of-gravity height, and the track width.

Case B: Assume the suspension is rigid (i.e., $K_{\theta_2} \rightarrow \infty$), and the only flexibility considered is that of the tire. Equation 5.28 becomes

$$\frac{a_y}{g} = \frac{d}{2h} - \frac{1}{K^*_{tire}} \tag{5.30}$$

The rotation angle at the rollover incipient caused by the tire deflection, θ_1^R, can be expressed by the dimensionless stiffness as

$$\theta_1^R = \frac{1}{K^*_{tire}} \tag{5.31}$$

Substitute Eq. 5.31 into Eq. 5.30 to obtain the expression

$$\frac{a_y}{g} = \frac{d}{2h} - \theta_1^R \tag{5.32}$$

The normalized lateral acceleration of rollover is controlled by the basic dimensions of the vehicle as well as the tire dimensionless stiffness. In this case, the lateral acceleration for rollover to occur is reduced by $\theta_1^R g$ (i.e., the rollover threshold is reduced). The higher the tire dimensionless stiffness parameter is, the smaller the rotational angle caused by the tire deflection. If the stiffness reaches infinity high, the second item in Eq. 5.32 becomes zero. When the tire stiffness is high enough, the lateral acceleration caused by the tire deflection becomes negligible.

EXAMPLE 5.3

A loaded truck weighs 15,000 kg, and its tread width is 2 m. The roll center height is $h_1 = 0.7$ m, and the body center-of-gravity height is $h = 1.8$ m. The suspension is assumed to be rigid. Calculate the normalized lateral acceleration.

Solution:

According to Eq. 5.31, the normalized lateral acceleration is

$$\frac{a_y}{g} = \frac{2}{2 \times 1.8} - \frac{1}{K^*_{tire}}$$

$$= 0.556 - \frac{1}{K^*_{tire}}$$

The tire stiffness should be high enough to support the whole vehicle. Because the lateral acceleration is larger than zero when the rollover occurs, the tire dimensional stiffness satisfies the expression

$$K^*_{tire} > 1.8$$

The relation between the normalized lateral acceleration and the tire dimensionless stiffness is plotted in Figure 5.12.

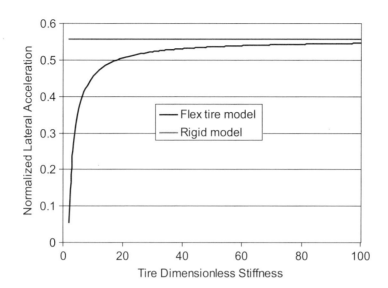

Figure 5.12 The relation between the normalized lateral acceleration and the tire dimensionless stiffness.

When the tire dimensionless stiffness reaches 18, the lateral acceleration is 0.5g (i.e., $a_y = 0.5g$), which reaches 90% of its maximum value of 0.556g. When the dimensionless stiffness reaches a certain value (18 in this example), the increase rate of the lateral acceleration gradually reduces. However, if the dimensional stiffness is below a certain value, the lateral acceleration is significantly influenced by the stiffness. If the stiffness is smaller than some value, rollover occurs easily at a very low lateral acceleration. Hence, to prevent vulnerability to rollover, it is necessary to increase the stiffness to a certain value.

If rollover occurs at 0.5g lateral acceleration, the rollover angle of the vehicle becomes

$$\theta_1^R = \frac{1}{K_{tire}^*} = \frac{1}{18} * \frac{180}{3.14} = 3.18°$$

The relation between the rollover angle and the tire dimensionless stiffness, $\theta_1^R = \frac{1}{K_{tire}^*}$, is plotted in Figure 5.13.

Case C: Assume a tire is rigid (i.e., $K_{tire} \to \infty$), but the suspension deflection is finite. Equation 5.28 will be simplified as

$$\frac{a_y}{g} = \frac{d}{2h} - \frac{d}{2h} \frac{1}{1 + \frac{h}{h_2}\left(K_{\theta_2}^* - 1\right)} \tag{5.33}$$

Equation 5.33 is similar to Eq. 5.32. The second term on the right side of Eq. 5.32 is the rotational angle caused by the tire deflection. In Eq. 5.33, the second term on the right side is another rotational angle, θ_2^R, caused by the suspension roll moment. Define this angle as

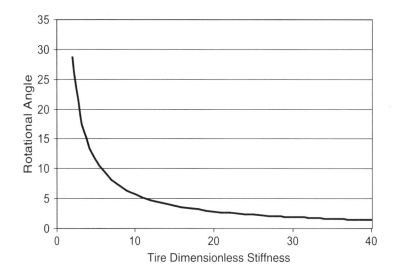

Figure 5.13 The relation between the rollover angle and the tire dimensionless stiffness.

$$\theta_2^R = \frac{d}{2h} \frac{1}{1 + \dfrac{h}{h_2}\left(K_{\theta_2}^* - 1\right)} \tag{5.34}$$

Thus, Eq. 5.33 can be rewritten as

$$\frac{a_y}{g} = \frac{d}{2h} - \theta_2^R \tag{5.35}$$

The normalized lateral acceleration of rollover is controlled by the vehicle basic dimensions as well as the suspension dimensionless stiffness. In this case, the lateral acceleration needed for rollover to occur is reduced by $\theta_2^R g$ than that of the rigid vehicle model (i.e., the rollover threshold is reduced). The higher the suspension dimensionless stiffness is, the smaller the rotational angle θ_2^R. If the stiffness becomes infinitely high, the second item in Eq. 5.35 goes to zero. When the suspension stiffness is high enough, the lateral acceleration caused by the suspension moment is negligible.

EXAMPLE E5.4

Assume that the tire is rigid, but the suspension stiffness is flexible. The remainder of the vehicle parameters are the same as those in example E5.3. Determine the required suspension roll stiffness for the lateral acceleration to be 85% of that of the rigid vehicle model. Also, calculate the normalized lateral acceleration.

Solution:

According to Eq. 5.33, the normalized lateral acceleration is

$$\frac{a_y}{g} = 0.556 - \frac{0.556}{1.636 K^*_{\theta_2} - 0.636}$$

The suspension stiffness should be high enough for the lateral acceleration to be larger than zero when rollover occurs. Thus,

$$0.556 - \frac{0.556}{1.636 K^*_{\theta_2} - 0.636} > 0$$

that is,

$$K^*_{\theta_2} > 1$$

The relation between the normalized lateral acceleration and the suspension dimensionless stiffness is plotted in Figure 5.14. The plot is similar to the one shown in Figure 5.12.

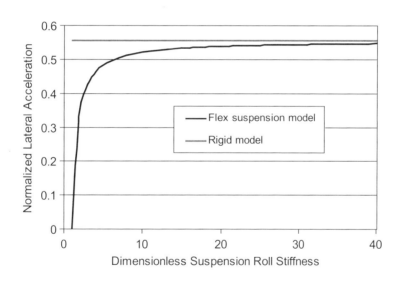

Figure 5.14 The relation between the normalized lateral acceleration and the suspension dimensionless stiffness.

The lateral acceleration needs to be 85% that of the rigid model, that is,

$$\frac{a_y}{g} = 0.556 - \frac{0.556}{1.636 K^*_{\theta_2} - 0.636}$$

$$= 0.85 * 0.556$$

The dimensionless stiffness is $K_{\theta_2}^* = 4.46$. Therefore, the roll stiffness is

$$K_{\theta_2} = mgh_2 K_{\theta_2}^*$$

$$= 15,000 * 9.8 * (1.8 - 0.7) * 4.46$$

$$\approx 721,000 \ \text{Nm/rad}$$

The lateral acceleration corresponding to the stiffness is

$$a_y = 0.85 * 0.556g = 0.473g$$

Similar to Figure 5.13, the relation between the rotational angle and the suspension roll dimensionless stiffness can be plotted as well. The rotational angle corresponding to the lateral acceleration 0.473g is 0.0834 rad or 4.78°.

Case D: Both the tire stiffness and the suspension stiffness are flexible.

The normalized lateral acceleration for this case can be written as

$$\frac{a_y}{g} = \frac{d}{2h} - \theta_1^R - \theta_2^R \tag{5.36}$$

The first term is the rollover threshold for the rigid vehicle model. The second and third terms are the corrected parameters of the rollover threshold for the tire deflection and suspension roll moment, respectively.

EXAMPLE E5.5

A loaded truck weighs 15,000 kg, and the tread width is 2 m. The roll center height is $h_1 = 0.7$ m, and the body center-of-gravity height is h = 1.8 m. Both the dimensionless stiffness for the tire and the suspension are 12. Calculate the rotational angle and the lateral acceleration at the moment rollover begins. Also compare the contribution of the tire and the suspension to the rotational angle.

Solution:

The rotational angle due to the tire deflection is

$$\theta_1^R = \frac{1}{K_{tire}^*}$$

$$= \frac{1}{12}$$

$$= 0.083 \ \text{rad, or } 4.76°$$

The rotational angle due to the suspension roll moment is

$$\theta_2^R = \frac{d}{2h} \frac{1}{1 + \dfrac{h}{h_2}\left(K_{\theta_2}^* - 1\right)}$$

$$= \frac{2}{2*1.8} \frac{1}{1 + \dfrac{1.8}{1.1}(12 - 1)}$$

$$= 0.029 \text{ rad, or } 1.68°$$

For the same dimensionless stiffness, the rotational angle caused by the tire deflection is larger than that of the suspension roll moment.

The normalized lateral acceleration is

$$\frac{a_y}{g} = \frac{d}{2h} - \theta_1^R - \theta_2^R$$

$$= \frac{2}{2*1.8} - 0.083 - 0.029$$

$$= 0.444$$

If only the tire deflection is considered, then the normalized acceleration is $0.556 - 0.083 = 0.473$. On the other hand, if only the suspension roll is considered, then the normalized acceleration is $0.556 - 0.029 = 0.527$.

5.4 Dynamic Rollover Model

In static rollover analysis, the lateral acceleration is considered, and a vehicle is in a quasi-static situation. In addition to the lateral acceleration, the roll angular velocity and the roll angular acceleration should be considered in the vehicle rollover dynamic model. Dynamic models can be simple ones such as the rigid model, or more complicated ones such as the body-suspension model. These models can be treated analytically. In industry, dynamic models usually are built by some specific tools, including many degrees of freedom. The whole scenario of rollover motion can be simulated and animated.

5.4.1 Rigid Dynamic Model

Figure 5.15 shows a simple rigid dynamic model. Suppose the center of gravity is coincident with the roll axis. The dynamic equations can be written as

$$I_s\ddot{\varphi} = \left(F_{2z} - F_{1z}\right)\frac{d}{2} + \left(F_{2y} + F_{1y}\right)h \qquad (5.37a)$$

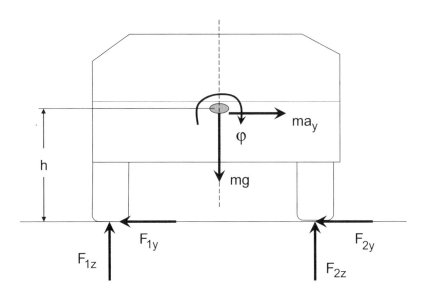

Figure 5.15 A simple and rigid dynamic rollover model.

$$ma_y = -F_{1y} - F_{2y} \tag{5.37b}$$

$$F_{1z} + F_{2z} = mg \tag{5.37c}$$

When rollover occurs, the left tire will leave the ground, so $F_{1y} = F_{1z} = 0$. Substitute Eqs. 5.37b and 5.37c into Eq. 5.37a to obtain

$$I_s\ddot{\varphi} = mg\frac{d}{2} - ma_yh \tag{5.38}$$

5.4.2 Dynamic Rollover Model for a Dependent Suspension Vehicle

Figure 5.9 is a dependent suspension model. The model can be divided into two parts: the body (sprung mass), and the suspension (unsprung mass). Force diagrams can be drawn separately, as shown in Figures 5.16 and 5.17.

The body is regarded as a rigid body, and its rotation center (C) and center of gravity are not coincident. The rollover dynamic equation for the body can be written as

$$m_s\ddot{z}_s = -mg + F_{k2}\cos\left(\theta_1 + \theta_2 + \varphi\right) - F_{k1}\cos\left(\theta_1 + \theta_2 + \varphi\right) \tag{5.39a}$$

$$m_s\ddot{y}_s = F_{k2}\sin\left(\theta_1 + \theta_2 + \varphi\right) - F_{k1}\sin\left(\theta_1 + \theta_2 + \varphi\right) \tag{5.39b}$$

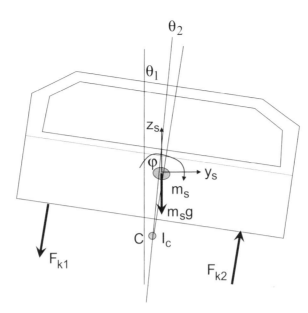

Figure 5.16 Body (sprung mass) diagram of the suspended vehicle model.

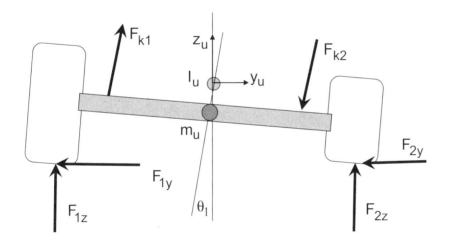

Figure 5.17 Suspension (unsprung mass) diagram of the suspended vehicle model.

$$I_c\ddot{\varphi} = \left(m_s g - m_s \ddot{z}_s\right)h_2 \sin\left(\theta_2 + \varphi\right) + m_s \ddot{y}_s h_2 \cos\left(\theta_2 + \varphi\right) - \left(F_{k1} + F_{k2}\right)\frac{b}{2} \quad (5.39c)$$

where I_c is the moment of inertia of the body around the roll center, φ is the rotational angle, and F_{k1} and F_{k2} are the spring forces.

The suspension model is regarded as a rigid body, and its center of gravity is at point U. Its rotation center is at point C. When rollover starts, the forces between the inner tire and surface vanish (i.e., $F_{1y} = F_{1z} = 0$). The dynamic equations for the dependent suspension are

$$m_u\ddot{z}_u = -m_u g + F_{k1}\cos(\theta_1 + \theta_2 + \varphi) - F_{k2}\cos(\theta_1 + \theta_2 + \varphi) + F_{2z} \qquad (5.39d)$$

$$m_u\ddot{y}_u = F_{k1}\sin(\theta_1 + \theta_2 + \varphi) - F_{k2}\sin(\theta_1 + \theta_2 + \varphi) - F_{2y} \qquad (5.39e)$$

$$I_u\ddot{\varphi} = (F_{k1} + F_{k2})\frac{b}{2}\cos(\theta_1 + \theta_2 + \varphi) + F_{2y}\left\{h_1 - \frac{d}{2}\cos(\theta_1 + \varphi)\right\} - F_{2z}\frac{d}{2}\cos(\theta_1 + \varphi) \quad (5.39f)$$

where I_u is the moment of inertia of the unsprung mass around the roll center C.

If Eqs. 5.38a through 5.38d are combined, dynamic parameters will be obtained.

5.4.3 Dynamic Rollover Model for an Independent Suspension Vehicle

Figure 5.18 shows a dynamic rollover model for a vehicle with independent suspension. It is a half vehicle model, and the body is regarded as a rigid body. The tires are regarded as concentrated masses, and their moment of inertia is neglected. The model has seven degrees of freedom: body lateral motion, vertical motion and rotation, and vertical and lateral motion for each tire. Newton's equilibrium equations or the Lagrange method can be used to build the dynamic model.

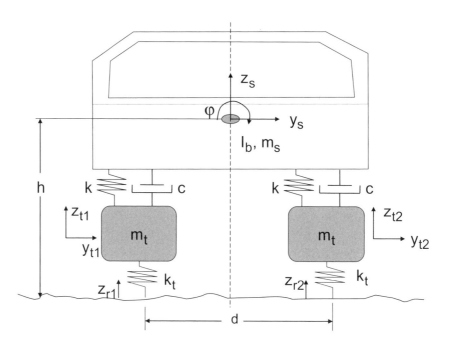

Figure 5.18 A dynamic rollover model for a vehicle with independent suspension.

5.4.4 Rollover Simulation Tools

In this section, three dynamic models are introduced. These simple models have several degrees of freedom, which are useful for the analysis of base rollover characteristics. When more detailed information is needed, these models have their limitations, and more complicated rollover models must be developed. In industry, rollover models usually are built using specific software such as ADAMS, Carsim, and NADSdyna.

With these tools, the whole virtual vehicle (Figure 5.19) or system (a suspension as shown in Figure 5.20) can be built. Rollover can be simulated by these virtual vehicle models. In addition, the models can be used for other dynamic attributes analysis, such as lane change, steering maneuver, accelerating, braking, cornering, handling, active vehicle dynamics, and others.

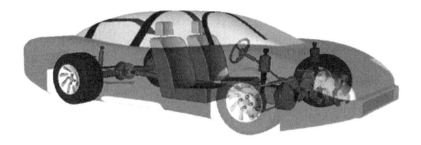

Figure 5.19 A virtual vehicle dynamic model. (www.carsim.com.)

Figure 5.20 A virtual suspension dynamic model. (www.carsim.com.)

The virtual vehicle models provide many benefits for vehicle dynamics analysts. The models are especially valuable for parameter studies and optimization. The models can simulate different road terrains as well as forces and torques applied on the tires. Dynamic performance can be simulated, and the results can be animated on a computer screen. Figure 5.21 shows a truck ADAMS model on a designed road. Thus, the vehicle development time and cycle can be shortened, and the cost can be reduced. The risk caused by rollover testing also can be reduced significantly.

The virtual vehicle models are extremely useful for situations where testing is dangerous (e.g., rollover testing) and parameters are difficult to obtain experimentally. Rollover is sensitive to road conditions and transient disturbances. The tools are proven to simulate vehicle rollover well. Figure 5.22 is an example of a rollover scenario called a fishhook maneuver. Fishhook is a difficult rollover test that requires a robotic programmable steering controller in place of the driver. However, with the help of the simulation tools, the fishhook procedure and performance can be simulated and animated on screen.

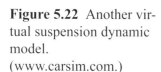

Figure 5.21 A truck ADAMS model on a designed road. (Elliott and Wheeler, 2001.)

Figure 5.22 Another virtual suspension dynamic model. (www.carsim.com.)

5.5 Dynamic Rollover Threshold

In the evaluation of the static rollover threshold, the vehicle is considered in a quasi-static state. Static stability factor, tilt table ratio, and side pull ratio are introduced. In addition to the definition of static rollover threshold, there is interest in studying the roll stability under transient and dynamic conditions. Several dynamic rollover thresholds are used, and some of them will be introduced in this section as follows:

- Dynamic stability index (DSI)
- Rollover prevention energy reserve (RPER)
- Rollover prevention metric (RPM)
- Critical sliding velocity (CSV)

These metrics are related to the vehicle lateral and rotational accelerations and/or velocities. The dynamic stability index includes rotational acceleration and velocity in addition to lateral acceleration. The last three thresholds are based on the concept of estimating the amount of energy needed to cause a vehicle to roll over. In these thresholds, the kinetic energy and potential energy are compared with the energy needed to raise the vehicle center of gravity and cause rollover.

5.5.1 Dynamic Stability Index

Equation 5.36 can be rewritten as

$$\frac{a_y}{g} + \frac{I\ddot{\phi}}{mgh} = \frac{d}{2h} \tag{5.40}$$

Compared with Eq. 5.6, Eq. 5.40 has one more term on the left side, $\dfrac{I\ddot{\varphi}}{mgh}$. Because the rotational energy is considered, the left side of the equation is called the dynamic stability index (DSI), that is,

$$\text{DSI} = \frac{a_y}{g} + \frac{I\ddot{\varphi}}{mgh} \tag{5.41}$$

If the dynamic stability index is larger than the static stability factor, rollover will occur. In the analysis of the static rollover threshold, only lateral acceleration is considered as a cause of rollover. However, the roll velocity and acceleration also contribute to rollover. The dynamic stability index includes both. It provides a closer view of roll instability during dynamic testing, and it represents a practical metric to describe dynamic rollover propensity.

5.5.2 Rollover Prevention Energy Reserve

The rollover prevention energy reserve (RPER) function is defined as the difference between the energy required to bring a vehicle to its tip-over position, E_v, and the sum of the instantaneous rotational kinetic energy and potential energy, E_k. The rollover prevention energy reserve is expressed as

$$\text{RPER} = E_v - E_k \tag{5.42}$$

Prior to rollover of a vehicle, the motion of the vehicle includes both lateral and rotational motions. The vehicle kinetic energy of rollover motion consists of two parts: the kinetic energy caused by the lateral motion, and that caused by rotation, expressed as

$$E_k = \frac{1}{2}mV^2 + \frac{1}{2}I_B\dot{\varphi}^2 \tag{5.43}$$

where, m, V, and $\dot{\varphi}$ are the vehicle mass, lateral velocity, and rotational velocity, respectively. It is the moment of inertia at the hinge point B, as shown in Figure 5.23. I_B can be expressed in terms of the moment of inertia at the center of gravity I by the expression

$$I_B = I + m\left(h^2 + \frac{d^2}{4}\right) \tag{5.44}$$

If the rollover prevention energy reserve is positive, the kinetic energy is not enough to roll over the vehicle. If the rollover prevention energy reserve is negative, the kinetic energy exceeds the required energy to roll over the vehicle; thus, rollover will occur. During the rollover procedure, the rollover prevention energy reserve is changed from positive to negative. How soon the rollover prevention energy reserve is changed to negative and how rapidly the rollover prevention energy reserve is reduced below zero indicates the severity of the rollover. The rollover prevention energy reserve is a dynamic function of the vehicle motion, so it can be used to describe the dynamic rollover threshold. However, the measurement of energy is time consuming and expensive. Additionally, it is difficult to predict the storage of the tire and the dissipative energy.

5.5.3 Rollover Prevention Metric

The rollover prevention metric (RPM) is defined as the percentage difference between the lateral kinetic energy and the initial rotational energy as

$$RPM = 100 \frac{T_0 - T_1}{T_0} \% \tag{5.45}$$

where T_0 and T_1 are the vehicle lateral kinetic energy and rotational energy, respectively. These terms are expressed as

$$T_0 = \frac{1}{2} m V^2 \tag{5.46}$$

$$T_1 = \frac{1}{2} I_B \dot{\varphi}_0^2 \tag{5.47}$$

where φ_0 is the initial rotational velocity.

According to the conservation of momentum, the momentum at impact is

$$m V h = I_B \dot{\varphi}_0 \tag{5.48}$$

Substituting Eq. 5.48 into Eq. 5.47, the initial rotational energy becomes

$$T_1 = \frac{1}{2} \frac{m^2 V^2 h^2}{I} \tag{5.49}$$

Substitute Eqs. 5.46 and 5.49 into Eq. 5.45 to obtain

$$RPM = 100 \left(1 - \frac{m h^2}{I} \right) \% \tag{5.50}$$

The rollover prevention metric is not a function of lateral speed, and it depends only on the vehicle mass, the center-of-gravity height, and the moment of inertia.

5.5.4 Critical Sliding Velocity

When a vehicle slides on a road at a lateral velocity V and hits an object, it pivots about a hinge with an initial angular velocity $\dot{\varphi}_0$. The lateral kinetic energy depends on the lateral velocity. The lateral kinetic energy is transferred to the rotational kinetic energy. The rotational kinetic energy must be large enough to raise the vehicle center of gravity to the point where it is just above the pivot point about which the vehicle is rolling. Subsequently, we have

$$\frac{1}{2} I_B \dot{\varphi}_0^2 \geq m g \left(\sqrt{h^2 + \left(\frac{d}{2} \right)^2} - h \right) \tag{5.51}$$

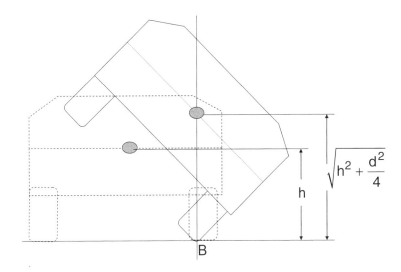

Figure 5.23 The center of gravity of a vehicle is raised to the point where rollover occurs.

The critical sliding velocity (CSV) is defined as the minimum lateral velocity required to initiate rollover over an obstacle. Substitute Eq. 5.47 into Eq. 5.50 to obtain

$$V^2 \geq \frac{2gI_B}{mh}\left(\sqrt{1 + \left(\frac{d}{2h}\right)^2} - 1\right)$$

(5.52)

The critical sliding velocity can be expressed as

$$V = \sqrt{\frac{2gI_B}{mh}\left(\sqrt{1 + \left(\frac{d}{2h}\right)^2} - 1\right)}$$

(5.53)

The critical sliding velocity depends on only the vehicle parameters, including its mass.

5.6 Occupant in Rollover

5.6.1 Overview of the Occupant and Rollover

Rollover includes three phases: the tripping phase, the airborne phase, and the ground-impact phase. In the tripping phase, occupants tend to tilt in the direction of the vehicle moment. In the airborne phase, the occupants tend to move upward and outward, away from the vehicle center of gravity. In the ground-impact phase, the occupants tend to continue moving until they impact the vehicle interior.

There are two types of injuries in rollover accidents. One is that the occupants are ejected from the vehicle when they do not wear seatbelts. The other is that occupants using seatbelts are injured by contacting with the vehicle body during airborne movement. This section focuses on the second case. The major purpose of the analysis of occupants' motion in rollover is to investigate the kinematics and forces experienced by them. The head and neck are the most vulnerable parts of the human body. Injury to these parts can be a result of contact with the roof and the upper vehicle interior during the ground-impact phase. The occupants' motion during the airborne phase is determined by restraint and centripetal forces that tend to project the occupant upward and outward relative to the compartment. Thus, it is important to understand the nature and extent of head and neck motion with respect to the interior vehicle structures.

Occupant kinematics in rollover accidents have been investigated experimentally and analytically using a variety of tools. Usually, anthropomorphic test dummies (ATDs), such as the Hybrid III ATD, are used for testing. The commonly used occupant models are the MADYMO ATD and the human facet computational model. The occupants' compartment used to investigate the motion of the occupants includes the human facet model and the vehicle interior model.

5.6.2 Testing of an Occupant Model

An occupant rollover test can be processed using test fixtures in a laboratory setting. The occupants used in rollover testing include ATDs and human volunteers. Human volunteers can be tested in a ground-based controlled rollover impact system (GB-CRIS). However, realistic rollover crash tests with people are forbidden because of the high risk of injury. Only dummies are used for frontal, side, or rear impacts in rollover crash tests. The 50% Hybrid III dummy probably is the most widely used surrogate for evaluating automotive safety in frontal crash testing.

An occupant's upward and outward motion and head excursion can be tested statically and dynamically. The static test is useful for estimating the magnitude of head excursion in a controlled acceleration (1g) environment but cannot quantify head excursion or head trajectory during a dynamic roll event. The dynamic test with a ground-based fixture can quantify head excursion in a dynamic environment.

The purpose of the rollover test is to investigate the occupants' kinematics in rollover crash scenarios, including head excursion and time histories. Linear accelerations in three directions (x, y, and z) and three rotational velocities (roll, pitch, and yaw) of the vehicle are measured. The vehicle kinematics can be obtained from these signals. Motion can be recorded by a camera, and the occupant excursion can be observed.

Figure 5.24 is a ground-based controlled rollover impact system experimental setup. A vehicle is mounted on a controlled rollover impact system and can be rotated at various rates about a principle axis that passes through its mass center. The vehicle is equipped with a continuous loop, a three-point seatbelt system configured with a sliding latch plate, an adjustable upper anchorage, and an emergency locking retractor mounted in the lower B-pillar. A Hybrid III ATD was positioned in the driver's seat. The kinematics of the occupant was measured by rotating the ground-based controlled rollover impact system fixture clockwise and counterclockwise. The fixture is rotated from the rest condition up to a steady-state roll rate. During the roll-up procedure, the vehicle is held for a period of several seconds at nominal constant rates to facilitate the steady-state evaluation of head excursion. Head excursion in both the outboard direction (y direction) and the upward direction (z direction) is recorded by cameras mounted inside the vehicle and is quantified by two-dimensional digitization of the Hybrid III ATD head position.

Figure 5.24 A ground-based controlled rollover impact system experimental setup. (Newberry *et al.*, 2005.)

5.6.3 *Simulation of Occupant Rollover*

Simulation of occupant rollover is a quick and cost-effective way to understand different crash configurations and design variants, especially during the development of new vehicles. The simulated models of the vehicle and dummies are used to calculate the kinematics and dynamics of the crash.

Many occupant models have been developed in past years based on cadaver data. MADYMO is the most widely used one in automotive engineering. MADYMO is a mathematical dynamic modeling software package developed by TNO Automotive. Unlike dummy models that are developed for a particular loading direction, the MADYMO occupant model is a multidirectional model; it is not designed for a specific impact direction. Therefore, it is suitable for complex vehicle accident scenarios such as rollovers. MADYMO has two types of occupant models: a finite element (FE) model, and a multi-body facet model. The finite element model includes both rigid bodies and deformable bodies. This model can be used to simulate global deformations and kinematics as well as local deformations of components and flesh or skin. The multi-body facet models use a rigid surface finite element method. Massless shell elements are used to model the outer surfaces. The multi-body skeleton model is constructed by connecting the outer surface and deformable bodies. The skeleton is made up of chains of rigid bodies connected by kinematic joints. The joint characteristics, joint motion, and inertial properties are based on published biomechanical data. Kinematic joints combined with dynamic restraint models are used to model the deformation of flexible components. The geometry of the model is represented by ellipsoids. Contact force characteristics for the ellipsoids are used to describe the interaction between the model and its environment and to represent the deformation of soft materials such as skin.

A multi-body facet model combined with a vehicle model forms an occupant compartment model, as shown in Figures 5.25 and 5.26. This model is used to predict the occupants' kinematics and contact forces. The finite element model for the vehicle

Figure 5.25 An occupant and compartment model. (Praxl *et al.*, 2003.)

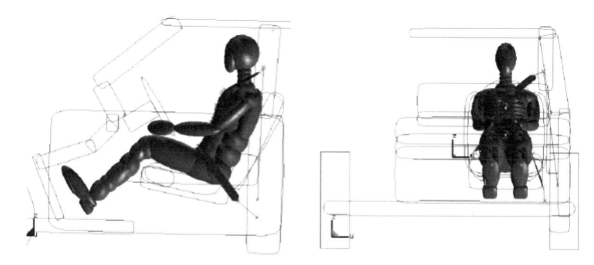

Figure 5.26 A MADYMO model with the vehicle model. (Newberry *et al.*, 2005.)

includes the vehicle bottom, roof, doors, windows, seats, belts, instrument panel, and other components. The model can be used to calculate the kinematics and dynamics of linked bodies.

The occupant must be placed in the correct position in the vehicle to obtain accurate simulation results. Two major constraints of the occupant are the contact between the occupant and the seat and the contact with the seatbelt. The occupant is placed in the seat in such a way that an equilibrium between the occupant's weight and the seat cushion force occurs. The seatbelt is modeled by a finite element method, as shown in Figure 5.27. The finite element belt is placed near the outer geometry of the occupant model. The model is run until the end points of the belt move backward and the belt is wrapped around the facet.

The occupant kinematics is simulated, including the overall kinematics and the time histories of the head to contact with the side window. Figure 5.27 shows an occupant model position 30 ms after head-to-side-window contact.

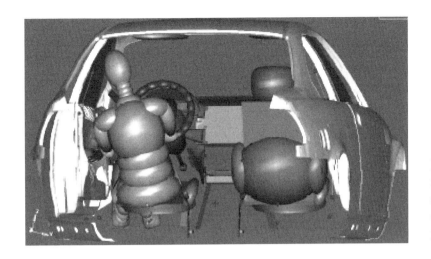

Figure 5.27 An occupant model position 30 ms after head-to-side-window contact. (Praxl *et al.*, 2003.)

5.7 Safety and Rollover Control

5.7.1 Overview of Rollover Safety

According to either fatality statistics or incapacitate statistics, rollover is one of the most dangerous types of vehicle accidents. Rollover is second only to frontal crashes in the level of severity in terms of fatalities per registered vehicle. There are two types of injuries in rollover. One is the ejection of the vehicle occupants, and the other is the occupants' impact with the interior structure of the vehicle.

NASS data show that there were approximately 200,000 rollovers in 1991. Among these accidents, about 56,000 cause serious incapacitating injuries, and about 9,000 are fatalities. The NHTSA (1999) data indicate that rollovers account for less than 5% of all vehicle crashes, but they account for about 15% of serious injuries and 25% of fatalities. Table 5.1 lists the accident distribution rate for different types of accidents and the corresponding harm rates and harm per occupant using 1988–1994 NASS and CDS (Crashworthiness Data System) data. Harm is a method developed by the NHTSA for weighting injuries according to their severity and frequency. Rollover accounts for only 8% of all accidents, but the injury or harm is 17%. The major reason for such a high injury rate is the higher number of incidents of ejection in rollover accidents.

Single-vehicle rollover is the most frequent rollover accident of all vehicles. The NASS data show that there were 137,600 rollover accidents of passenger cars in 1989. Among these, 124,800 were single-vehicle rollover accidents. The NASS-CDS database indicates that from 1995 to 1999, 81% of rollovers were single-vehicle crashes.

Single-vehicle rollover accidents are one of the major reasons for fatalities in vehicle accidents. According to FARS (Fatal Accident Reporting System) data, there were 15,901 fatalities in single-vehicle accidents in 1990. Among the accidents, 8,088 were caused by rollover, which occupies 51% of all fatalities. In 1999, 10,140 people were fatally injured in light-vehicle rollovers. Of that number, 8,348 were in single-vehicle rollover crashes. In 2001, 8,842 people lost their lives in single-vehicle rollover crashes. The number is 50% of all fatalities in single-vehicle crashes and 21% of all traffic fatalities.

TABLE 5.1
DISTRIBUTION RATE FOR DIFFERENT TYPES OF ACCIDENTS AND
CORRESPONDING HARM RATES AND HARM PER OCCUPANT
(1988–1994 NASS AND CDS, AND SAE PAPER NO. 970396*)

	Occupants	Harm	Harm per Occupant
Front	43%	37%	2.2
Side	23%	30%	3.4
Rear	8%	2%	0.8
Rollover	8%	17%	5.9
Other	5%	10%	5.5
Unknown	13%	4%	0.9
Total	100%	100%	

* James *et al.*, 1997.

From 1985 to 2000, the total number of single-vehicle crashes in the United States was reduced by 10%. However, the number of fatalities in rollover did not change much, as shown in Figure 5.28.

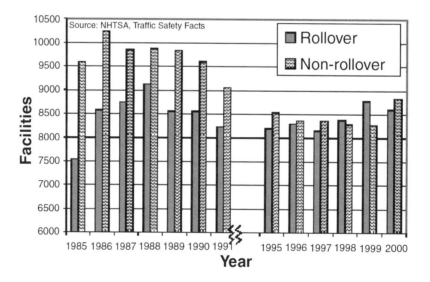

Figure 5.28 Fatalities in single-vehicle crashes in the United States from 1985 to 2000. (Baumann and Eckstein, 2004.)

Some types of vehicles have more frequent rollover accidents than others. Figure 5.29 shows the number of rollover fatalities per million registered vehicles for different types of vehicles. Sport utility vehicles (SUVs) have the highest rates of rollover fatalities. Rollover fatalities for cars are much lower than those for SUVs and are lower than those for pickup trucks and vans.

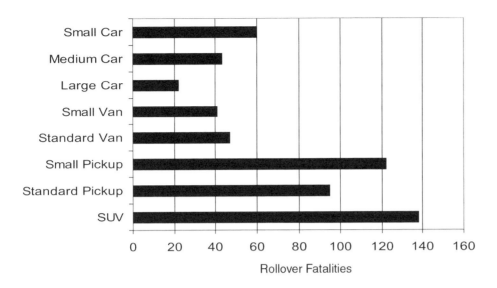

Figure 5.29 The number of rollover fatalities per million registered vehicles for different types of vehicles, averaged from 1985 to 1990. (Hinch *et al.*, 1992.)

Table 5.2 lists rollover accident data collected by NASS in 1989. For light trucks and vans (LTVs), the number of single-vehicle rollover accidents was 65,800 from the overall 75,600 rollover accidents. Single-vehicle accidents account for about 90% of overall rollover accidents. Figure 5.30 shows statistics of fatal rollover crashes for different types of vehicles from 1988 to 2002 in the United States. This figure shows that the SUV rollover fatality has increased rapidly in recent years.

TABLE 5.2
ROLLOVER ACCIDENTS AND
SINGLE-VEHICLE ROLLOVER ACCIDENTS
(Hinch, 1992)

Type of Vehicle	Overall Number of Accidents	Number of Single-Vehicle Accidents	Rate of Single-Vehicle Accidents
Passenger cars	137,600	124,800	90%
LTVs	75,600	65,800	87%

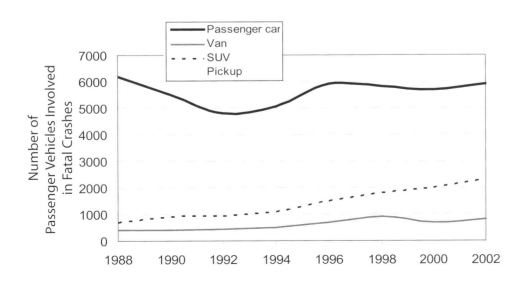

Figure 5.30 Passenger vehicles involved in fatal crashes, by vehicle body type and year (Baumann and Eckstein, 2004).

Most serious and fatal injuries in rollover accidents result from the ejection of occupants. The seatbelt is an effective tool to prevent such occupant ejection. The two functions of seatbelts are to prevent the ejection of the occupants from the vehicle and to balance the position of the occupants inside the vehicle. Unbelted occupants have a higher risk of ejection than those who are belted. The use of restraint reduces the risk of serious injury by more than 80% and fatal injuries by more than 65%. NASS and CDS data (Hinch, 1992) from 1982 to 1989 show that unrestrained occupants were 4.7 times more likely to be seriously injured than restrained occupants.

During the rollover phase, in addition to preventing ejection, seatbelt restraints provide substantial benefits to non-ejected occupants. Unrestrained and non-ejected occupants receive serious injuries six times more often than restrained occupants. Rollover fatalities can be reduced by as much as 75% if all occupants use seatbelts.

Table 5.3 shows that the injuries and fatalities from accidents can be reduced significantly by the use of seatbelts, especially for rollover accidents. Table 5.4 shows the distribution of occupants, harm percentage, and harm value per occupant for a vehicle that is involved in a rollover accident. Among the 39% of the occupants who did not use seatbelts, the harm is 73%, whereas the harm is only 27% for the 61% of occupants who were restrained.

TABLE 5.3
SEATBELT EFFECTIVENESS IN FATALITY REDUCTION,
(JAMES *ET AL.*, 1997)

	Overall Fatality Reduction	Reduction from Ejection Prevention	Reduction from Impacts with Vehicle Interior
All accidents	41%	18%	23%
Rollover accidents	68%	50%	18%
Rollover as first event	80%	64%	16%

TABLE 5.4
DISTRIBUTION OF OCCUPANTS, HARM PERCENTAGE,
AND HARM VALUE PER OCCUPANT
(JAMES *ET AL.*, 1997)

	Occupants	Harm	Harm per Occupant
Unrestrained	39%	73%	11.1
Restrained	61%	27%	2.6

For restrained occupants, the bodies move inside the vehicle and impact the interior structure. The head is the most frequently injured body region and is 4.5 times more likely to be injured than the neck during rollover.

5.7.2 Sensing of Rollover

Rollover is a complicated event. Many factors influence the initiation of rollover and the subsequent motion of the vehicle. Thus, it is important to develop a rollover detection and prevent system. Active rollover control has focused mainly on two areas: rollover detection, and anti-rollover systems. A rollover detection system or sensing system consists of sensor architecture and the algorithm to activate the control system. The two major functions of a rollover sensing system are as follows:

- Accurate estimation of the vehicle roll rate, roll angle, lateral and vertical acceleration, and so forth

- Timely and accurate activation of the rollover restraint system

A tripped rollover is caused by a vehicle hitting against an obstacle or by lateral slipping. The vehicle is rotated around its longitudinal direction, as shown in Figure 5.31(a). An untripped rollover is caused by the vehicle inclination where the vehicle center of gravity exceeds the stable point. The vehicle will be under vertical or lateral translation, as shown in Figure 5.31(b). If rollover is caused by both of these cases, the rollover is a combination of a rotation moment and a translation force, as shown in Figure 5.31(c).

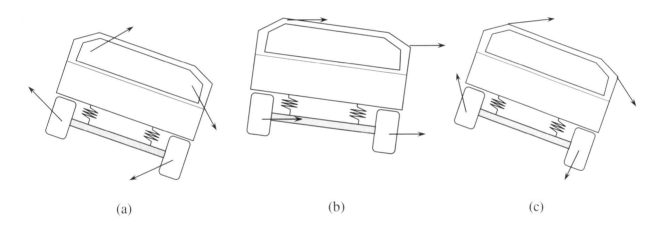

(a)	(b)	(c)

Figure 5.31 Rollover motion: (a) rotation, (b) lateral translation, and (c) rotation and translation.

Rollover includes the rotation and translation of the vehicle. Subsequently, signals that can be used as input to a rollover detector include the roll angle, roll angle rate, lateral and vertical accelerations, suspension deflection, and others. The roll angle is used as a signal because it directly correlates to the current roll stability of the vehicle. The roll angle rate is used because it is related to the rotation energy. Lateral and vertical acceleration are used because they are related to the dynamic energy. These signals are integrated in an algorithm as shown in Figure 5.32 and then are summarized and calculated. The calculated results are compared with the rollover threshold, and then the control command is sent to deploy the rollover restraint system.

The detector identifies signals using different sensors. Traditionally, there are two major sensors. One is a gravity-based sensor such as an inclinometer or tilt sensor. Gravity-based sensors respond slowly and can produce errors when exposed to linear acceleration.

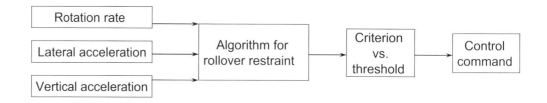

Figure 5.32 Rollover detection system.

Likewise, they do not perform correctly during an airborne phase, on rough road, and during aggressive driving conditions. The other type of sensor is an angular rate sensor based on the Coriolis effect. The sensor is good for rollover detection but is not accurate for many other applications such as collision. Today, some special sensors have been designed, such as a satellite rollover sensor. Rollover sensors and algorithms are designed to be robust to environmental inputs such as vibration and shock.

5.7.3 Rollover Safety Control

The following methods are used to control vehicle rollover and to protect occupants:

* Seatbelts, including seatbelt pretensioners and electromechanical seatbelt retractor locks

* Inflatable head restraints/window curtains

* Active rollover controls

* Roll-bars

* Curtain airbags

In this section, only active rollover control will be introduced. It is a new field, and advanced technology is employed. In recent years with the development of advanced control technology and the cost reduction of electronic and control equipment, active control has been widely used in the automotive industry. Active rollover control is used to improve both comfort and handling. The following are the major anti-rollover control applications:

* Active suspensions

* Active roll-bars

* Active steering

* Anti-rollover braking systems

Active suspensions and active roll-bars directly control the vehicle roll motion. The active steering reduces vehicle oversteering and the vehicle yaw moment.

Active suspension: The suspension supports the vehicle body, controls handling, and provides ride comfort. In recent years, customers (especially SUV customers) demanded more comfortable ride and more stable handling. To satisfy their demands, active suspension usually is used. Active suspension is a regular suspension with a control system, as shown in Figure 5.33. The control system uses electrohydraulic equipment to generate controlled forces, $F_{control}$, to react to rollover moments.

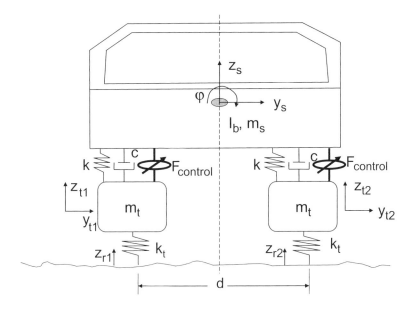

Figure 5.33 An active suspension.

Figure 5.33 is a simplified half-vehicle model. The rollover index first must be defined. The signals, such as roll angle, roll angle rate, and lateral acceleration, are integrated in the algorithm. In critical situations where the rollover index indicates that the vehicle is under rollover danger, an emergency rollover prevention control in the suspension is made active.

Active anti-roll-bar: An active anti-roll-bar system consists of stabilized bars, a hydraulic actuator, and an electronic control unit (ECU), as shown in Figure 5.34. Each stabilized bar connects to a hydraulic actuator at one end. The ECU provides command to the hydraulic actuator according to the obtained lateral acceleration signal.

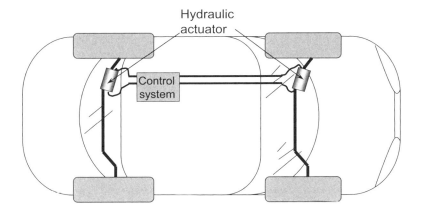

Figure 5.34 An active anti-roll-bar system.

An active anti-roll-bar hydraulically determines the variation of the equivalent stiffness of the anti-roll-bars. The stiffness of the front and rear roll-bars also can be balanced. The vehicle load distribution is influenced by the roll-bar stiffness distribution. An active anti-roll-bar can effectively reduce the roll angle, control the roll motion, and improve ride comfort and handling by optimizing the ratio between the front and rear roll stiffness.

Active steering: Steering input significantly influences lateral vehicle dynamics. Excessive steering command may result in unstable vehicle motion. When a vehicle is at a severe left turn, its body rolls to the right in a clockwise direction, as shown in Figure 5.35. To reduce or reverse the unstable roll movement, the steer angle must be reduced or reversed by an active steering system. Active steering with individual wheel torque control has the capability to recover a vehicle from an unstable roll state.

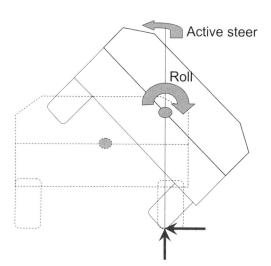

Figure 5.35 Rollover and steering.

Anti-rollover braking system: An anti-rollover braking (ARB) system monitors vehicle motion. When a rollover is about to start, the ARB controls the front brakes. The brakes reduce the cornering capability of the front tires, which causes the vehicle to turn less sharply and reduces its speed. When the vehicle passes the rollover warning threshold, the ARB stops control, and the brakes are released.

5.8 Summary

In this chapter, we reviewed the rollover scenario and its importance. Rigid models and the corresponding steady-static rollover thresholds were studied. This chapter also introduced suspension vehicle rollover models, and the influence of tires and the suspension stiffness were analyzed. In dynamic rollover models, roll acceleration/velocity in addition to lateral acceleration were considered. This chapter also provided dynamic rollover thresholds. The occupants' safety during rollover likewise was described. Finally, this chapter treated rollover sensing and anti-rollover systems.

5.9 References

1. Baumann, F.W. and Eckstein, L., "Effects Causing Untripped Rollover of Light Passenger Vehicles in Evasive Maneuvers," SAE Paper No. 2004-01-1057, Society of Automotive Engineers, Warrendale, PA, 2004.

2. Chrstos, J.P. and Guenther, D.A., "The Measurement of Static Rollover Metrics," SAE Paper No. 920582, Society of Automotive Engineers, Warrendale, PA, 1992.

3. Dahlberg, E., "A Method Determining the Dynamic Rollover Threshold of Commercial Vehicles," SAE Paper No. 2000-01-3492, Society of Automotive Engineers, Warrendale, PA, 2000.

4. Danesin, D., Krief, P., Sorniotti, A., and Velardocchia, M., "Active Roll Control to Increase Handling and Comfort," SAE Paper No. 2003-01-0962, Society of Automotive Engineers, Warrendale, PA, 2003.

5. Das, N.S., Suresh, B.A., and Wambold, J.C., "Estimation of Dynamic Rollover Threshold of Commercial Vehicles Using Low Speed Experimental Data," SAE Paper No. 932949, Society of Automotive Engineers, Warrendale, PA, 1993.

6. Davis, M., Marting, P., Eger, R., Majjad, R., and Nasr, N., "Rollover Simulation Based on a Nonlinear Model," SAE Paper No. 980208, Society of Automotive Engineers, Warrendale, PA, 1998.

7. d'Entremont, K.L., "The Effect of Light-Vehicle Design Parameters in Tripped-Rollover Measurement—A Statistical Analysis Using an Experimentally Validated Computer Model," SAE Paper No. 950315, Society of Automotive Engineers, Warrendale, PA 1995.

8. Elliott, A.S. and Wheeler, G., "Validation of ADAMS Models of Two USMC Heavy Logistic Vehicle Design Variants," SAE Paper No. 2001-01-2734, Society of Automotive Engineers, Warrendale, PA, 2001.

9. Ervin, R., Winkler C., and Karamihas, S., "A Centrifugal Concept for Measuring the Rollover Threshold of Light-Duty Vehicles," SAE Paper No. 2002-01-1603, Society of Automotive Engineers, Warrendale, PA, 2002.

10. Frimberger, M., Wolf, F., Scholpp, G., and Schmidt, J., "Influences of Parameters at Vehicle Rollover," SAE Paper No. 2000-01-2669, Society of Automotive Engineers, Warrendale, PA, 2000.

11. Gertsch, J. and Eichelhard, O., "Simulation of Dynamic Rollover Threshold for Heavy Trucks," SAE Paper No. 2003-01-3385, Society of Automotive Engineers, Warrendale, PA, 2003.

12. Gillespie, T.D., *Fundamentals of Vehicle Dynamics*, Society of Automotive Engineers, Warrendale, PA, 1992.

13. Hac, A., "Influence of Chassis Characteristics on Sustained Roll, Heave, and Yaw Oscillations in Dynamic Rollover Testing," SAE Paper No. 2005-01-0398, Society of Automotive Engineers, Warrendale, PA, 2005.

14. Hecker, F., *et al.*, "Vehicle Dynamics Control for Commercial Vehicles," SAE Paper No. 973284, Society of Automotive Engineers, Warrendale, PA, 1997.

15. Heydinger, G.J. and Howe, J.G., "Analysis of Vehicle Response Data Measured During Severe Maneuvers," SAE Paper No. 2000-01-1644, Society of Automotive Engineers, Warrendale, PA, 2000.

16. Hinch, J., Shadle, S., and Klein, T.M., "NHTSA's Rollover Rulemaking Program— Results of Testing and Analysis," SAE Paper No. 920581, Society of Automotive Engineers, Warrendale, PA, 1992.

17. James, M.B., Allsop, D.L., Nordhagen, R.P., and Decker, R., "Injury Mechanisms and Field Accident Data Analysis in Rollover Accidents," SAE Paper No. 970396, Society of Automotive Engineers, Warrendale, PA, 1997.

18. Jialiang Le, Jerry, McCoy, Robert W., and Chou, Clifford C., "Early Detection of Rollovers with Associated Test Development," SAE Paper No. 2005-01-0737, Society of Automotive Engineers, Warrendale, PA, 2005.

19. Kazemi, R. and Soltani, K., "The Effects of Important Parameters on Vehicle Rollover with Sensitivity Analysis," SAE Paper No. 2003-01-0170, Society of Automotive Engineers, Warrendale, PA, 2003.

20. Lai, W., "Evaluation of Human Surrogate Models for Rollover," SAE Paper No. 2005-01-0941, Society of Automotive Engineers, Warrendale, PA, 2005.

21. Lund, Y.I. and Bernard, J.E., "Analysis of Simple Rollover Metrics," SAE Paper No. 950306, Society of Automotive Engineers, Warrendale, PA, 1995.

22. Marine, M.C., Wirth, J.L., and Thomas, T.M., "Characteristics of Road Rollover," SAE Paper No. 1999-01-0122, Society of Automotive Engineers, Warrendale, PA, 1999.

23. Nalecz, A.G., Lu, Z., and d'Entermont, K.L., "An Investigation into Dynamic Measures of Vehicle Rollover Propensity," SAE Paper No. 930831, Society of Automotive Engineers, Warrendale, PA, 1993.

24. Newberry, W., Carhart, M., Lai, W., Corrigan, C.F., and Croteau, J., "A Computational Analysis of the Airborne Phase of Vehicle Rollover: Occupant Head Excursion and Head-Neck Posture," SAE Paper No. 2005-01-0943, Society of Automotive Engineers, Warrendale, PA, 2005.

25. Padmanaban, Jeya and Davis, Martin Stone, "Examination of Rollover Accident Characteristics Using Field Performance Data," SAE Paper No. 2002-01-2056, Society of Automotive Engineers, Warrendale, PA, 2002.

26. Praxl, N., Schönpflug, M., and Adamec, J., "Application of Human Models in Vehicle Rollover," SAE Paper No. 2003-01-2188, Society of Automotive Engineers, Warrendale, PA, 2003.

27. Ruhl, R.L. and Ruhl, R.A., "Prediction of Steady State Roll Threshold for Loaded Flat Bed Trailers—Theory and Calculation," SAE Paper No. 973261, Society of Automotive Engineers, Warrendale, PA, 1997.

28. Salaani, M.K., Heydinger, G.J., and Grygier, P.A., "Heavy Tractor-Trailer Vehicle Dynamics Modeling for the National Advanced Driving Simulator," SAE Paper No. 2003-01-0965, Society of Automotive Engineers, Warrendale, PA, 2003.

29. Shim, T. and Toomey, D., "Investigation of Active Steering/Wheel Torque Control at the Rollover Limit Maneuver," SAE Paper No. 2004-01-2097, Society of Automotive Engineers, Warrendale, PA, 2004.

30. Slaates, P.M. and Coo, P.D., "Safety Restraint Systems in Heavy Truck Rollover Scenarios," SAE Paper No. 2003-01-3424, Society of Automotive Engineers, Warrendale, PA, 2003.

31. Viano, D.C. and Parenteau, C.S., "Rollover Crash Sensing and Safety Overview," SAE Paper No. 2004-01-0342, Society of Automotive Engineers, Warrendale, PA, 2004.

32. Waller, E. and Schiffmann, J.K., "Development of an Automotive Rollover Sensor," SAE Paper No. 2000-01-1651, Society of Automotive Engineers, Warrendale, PA, 2000.

33. Wielenga, T.J. and Chace, M.A., "A Study in Rollover Prevention Using Anti-Rollover Braking," SAE Paper No. 2000-01-1642, Society of Automotive Engineers, Warrendale, PA, 2000.

34. www.adams.com, the ADAMS website.

35. www.carsim.com, the CarSim website.

36. www.madymo.com, the MADYMO website.

37. Yang, H. and Liu, Y., "A Robust Active Suspension Controller with Rollover Prevention," SAE Paper No. 2003-01-0959, Society of Automotive Engineers, Warrendale, PA, 2003.

Chapter 6

Handling Dynamics

6.1 Introduction

Vehicle handling behavior is becoming increasingly important for today's discerning customers. The handling characteristics refer to the dynamic properties to steering input or environmental disturbance. Therefore, two essential issues associated with handling characteristics are direction control and disturbance stabilization. Vehicle dynamics plays an important role in the responsiveness of a vehicle. The performance of a vehicle depends on its handling characteristics. Handling is a measure of the directional response of a vehicle and is one of the most important characteristics from a vehicle dynamics point of view (Refs. 1 to 18).

This chapter addresses the basic handling features of vehicles. A simplified model for vehicle dynamics will be presented. Based on the solutions of the model, steady-state characteristics of handling and the transient response of the vehicle as well as a frequency domain analysis will be discussed. The effects of suspension and the steering system on handling characteristics also are addressed. Overturning limit handling characteristics then are presented. Nonlinear models are introduced, and the simulation and testing of handling characteristics are presented. Traditional vehicle dynamics techniques view the vehicle as a system, which typically is simplified to a planar model in one of three planes: yaw, pitch, and roll. These planes are defined relative to the vehicle chassis at static equilibrium conditions. Figure 6.1 shows examples of these types of models. Several of the yaw plane models have been developed, with the most widely known being the bicycle model, as shown on the right side of Figure 6.1. Yaw plane models often are used to predict vehicle handling response within the linear range because any additional degrees of freedom for the vehicle sprung mass have little effect. However, outside the linear handling range, additional degrees of freedom must be accounted for, because tire traction force saturation and load transfer effects play predominant roles.

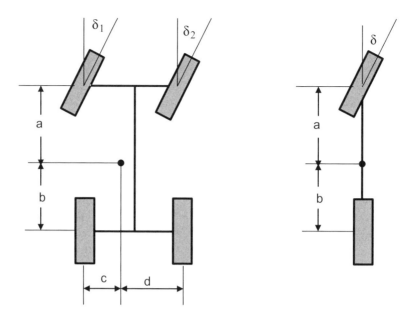

Figure 6.1 Simplified yaw plane models.

6.1.1 Tire Cornering Forces

In addition to gravitational forces and aerodynamic disturbance, the dominant forces and moments influencing a vehicle are the tire forces. In the cornering process, the tire develops a lateral force and will experience lateral slip associated with a slip angle.

The slip angle is the angle between the tire's direction of heading and its direction of travel, as shown in Figure 6.2. The lateral force, denoted by F_y, is named as the cornering force if the tire camber angle is ignored. Experiments show that cornering forces increase with the slip angle for a given tire load. Figure 6.3 shows the cornering force as a function of the slip angle. At a low slip angle, such as 5° or less, the relationship is linear and can be expressed as

$$F_y = C_\alpha \alpha \qquad (6.1)$$

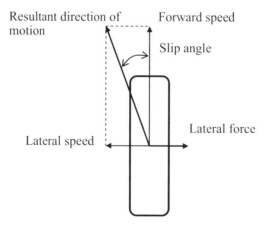

Figure 6.2 The tire force.

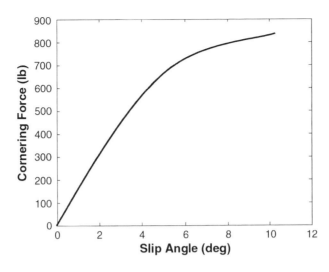

Figure 6.3 Cornering force as a function of slip angle.

The proportional constant C_α is known as the cornering stiffness, which is the slope of the curve of the cornering force versus the slip angle α at $\alpha = 0$.

EXAMPLE E6.1

A typical car has C_α in the range of 30 to 80 kN/rad. Many factors affect corning stiffness, including the tire size, type, number of plies, cord angles, and wheel width, as well as the tread. For a given tire, the inflation pressure and load are the main factors. Figures 6.4 and 6.5 show the effects of these factors.

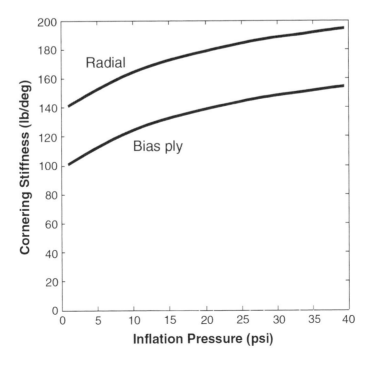

Figure 6.4 Cornering stiffness as a function of inflation pressure.

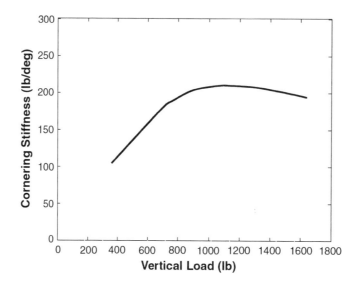

Figure 6.5 Cornering stiffness as a function of vertical load.

6.1.2 Forces and Torques in the Tire Contact Area

When a tire slips, at any point in the contact area between the tire and track, normal and friction forces are developed, and some moment is associated with it. According to the profile design of the tire, the contact area forms an area that is not necessarily coherent. The effect of the contact forces can be described fully by a vector of force and a torque in reference to a point in the contact area. The vectors are described in a track-fixed coordinate system. The z axis is normal to the track, and the x axis is perpendicular to the z axis and perpendicular to the wheel rotation axis y. In Figure 6.6, F_x is the longitudinal or circumferential force, F_y is the lateral force, F_z is the vertical force or wheel load, M_x is the tilting torque, M_y is the rolling resistance torque, and M_z is the self-aligning and bore torque. Nonsymmetric distributions of force in the contact patch cause torques around the x and y axes. The tilting torque M_x occurs when the tire is cambered. M_y also contains the rolling resistance of the tire. In particular, the torque around the z axis is relevant in vehicle dynamics. It consists of two parts:

$$M_z = M_B + M_s$$

(6.2)

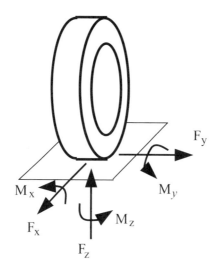

Figure 6.6 Contact forces and torques.

Rotation of the tire around the z axis causes the bore torque M_B. The self-aligning torque M_S reflects the fact that, in general, the resulting lateral force is not applied in the center of the contact patch.

6.2 The Simplest Handling Models—Two-Degrees-of-Freedom Yaw Plane Model

To study the basic handling dynamic characteristics, we make use of a two-degrees-of-freedom vehicle model, the bicycle model, which is the simplest one to describe the lateral motion of the vehicle.

The model considers the lateral and yaw dynamics of the vehicle, as shown in Figure 6.7. At high speeds, the radius of turn is much larger than the wheelbase of the vehicle. The difference between the steer angles on the outside and inside front wheels is negligible. The two front wheels can be represented by one wheel at a steer angle δ, with a cornering force equivalent to that of both wheels. The same assumption is made for the rear wheels. The basic handling model has the following assumptions:

- The road surface is flat and level.

- The vehicle structure is rigid, including the suspension system.

- The steering system is either ignored or assumed to be rigid.

- The forward speed of the vehicle is constant.

- The vehicle undergoes only small perturbations.

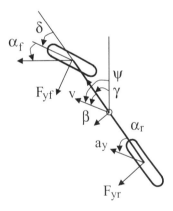

Figure 6.7 Bicycle model.

The slip angle is less than $5°$, and the cornering stiffness is constant. The front and rear cornering forces are represented by $F_{yf} = C_f \alpha_f$ and $F_{yr} = C_r \alpha_r$, where C_f and C_r are the cornering stiffness of the front and rear tires, respectively. Assume the sideslip angle β at the center of gravity (CG), which is defined as the angle between the longitudinal axis and the local direction of travel. The sideslip angle is defined by the difference between the vehicle heading angle γ and the direction angle of velocity ψ as

$$\beta = \psi - \gamma \tag{6.3}$$

In general, the sideslip angle will be different at every point on a car during cornering. Define the yaw rate or yaw velocity as $r = \dot{\psi}$. Denote that F_{xf} is the longitudinal force at the front axle, and F_{xr} is the longitudinal force at the rear axle, v is the forward velocity, R is the radius of the turn, δ is the steer angle at the front wheels (deg), L is the wheelbase (m), a_y is the lateral acceleration (g), W_f is the load on the front axle (N), W_r is the load on the rear axle (N), F_w is the wind gust excitation force, M_w is the wind gust excitation moment, and Vx and Vy are the longitudinal and lateral components of the center-of-gravity velocity, respectively. Assuming a constant longitudinal velocity of Vx = v, I is the moment of inertia of the vehicle about its yaw axis, m is the vehicle mass, and a and b are the distance of the front and rear axles from the center of gravity, respectively, where L = a + b. Apply Newton's second law at the center of gravity along the longitudinal and lateral directions, respectively, to obtain

$$F_{xr} + F_{xf}\cos\delta - F_{yf}\sin\delta = mv^2\frac{\sin\beta}{R} + m\dot{v}\cos\beta \tag{6.4}$$

$$F_{yr} + F_{yf}\cos\delta + F_{xf}\sin\delta - F_w = mv^2\frac{\cos\beta}{r} - m\dot{v}\sin\beta \tag{6.5}$$

Considering the moment equilibrium at the center of gravity, we obtain

$$F_{yf}a\cos\delta - F_{xf}a\sin\delta = I\dot{r} + M_w + F_{yr}b \tag{6.6}$$

and consider the relationship

$$F_{yf} = C_f\alpha_f$$
$$F_{yr} = C_r\alpha_r \tag{6.7}$$

For small angles, we can write

$$\sin\beta = \beta$$
$$\sin\delta = \delta$$
$$\cos\alpha = 1$$
$$\cos\delta = 1 \tag{6.8}$$
$$\alpha_f \approx \beta + \delta - a\frac{r}{v}$$
$$\alpha_r \approx \beta + b\frac{r}{v}$$

Substituting these equations in Eq. 6.7, we get

$$F_{yf} = C_f\left(\beta + \delta - a\frac{r}{v}\right) \tag{6.9}$$

$$F_{yr} = C_r \left(\beta + b \frac{r}{v} \right) \tag{6.10}$$

Noting $\dot{\gamma} = \dfrac{v}{R}$, we obtain the following relationship among yaw angle, forward velocity, and sideslip angle as

$$\psi = \beta + \gamma$$
$$r = \dot{\beta} + \dot{\gamma} \tag{6.11}$$

Substituting Eqs. 6.9 through 6.11 into Eqs. 6.4 through 6.6, and ignoring the wind gust excitation, we obtain

$$F_{xr} + F_{xf} + \delta C_r \left(-\beta - \delta + a\frac{r}{v} \right) + mv\beta \left(\dot{\beta} - r \right) - m\dot{v} = 0 \tag{6.12a}$$

$$C_r \left(\beta + b\frac{r}{v} \right) + C_f \left(\beta + \delta - a\frac{r}{v} \right) + \delta F_{xr} + mv \left(\dot{\beta} - r \right) + m\dot{v}\beta = 0 \tag{6.12b}$$

$$C_f a \left(\beta + \delta - a\frac{r}{v} \right) + \delta F_{xr} a - I\dot{r} - C_r b \left(\beta + b\frac{r}{v} \right) = 0 \tag{6.12c}$$

Equation 6.12 is a set of nonlinear equations. By assuming forward velocity to be constant (namely, $\dot{v} = 0$), we can rearrange these equations as

$$r \left(\delta C_f \frac{a}{v} - mv\beta \right) + \dot{\beta}mv\beta - \delta C_f \beta - \delta^2 C_f + F_{xf} + F_{xr} = 0 \tag{6.13a}$$

$$r \left(C_r \frac{b}{v} - C_f \frac{a}{v} - mv \right) + \dot{\beta}mv + \beta \left(C_f + C_r \right) + \delta \left(C_f + F_{xf} \right) = 0 \tag{6.13b}$$

$$\dot{r}I + r \left(C_f \frac{a^2}{v} + C_r \frac{b^2}{v} \right) + \beta \left(C_r b - C_f a \right) + \delta a \left(-C_r - F_{xf} \right) = 0 \tag{6.13c}$$

If the forward velocity v, the cornering stiffness C_f and C_r, and the parameters a, b, m, and I are known, with a given steer angle δ, the sideslip angle β, yaw or vehicle heading ψ, and the direction angle of velocity γ can be calculated. Generally, $C_f \gg F_{xf}$ because the longitudinal force will not affect the lateral motion. For the case of $F_{xf} = 0$, by ignoring Eq. 6.13a, we have

$$\dot{\beta} = -\left(\frac{C_r + C_f}{mv} \right)\beta + \left(\frac{C_f b - C_r a}{mv^2} - 1 \right)r + \left(\frac{C_f + F_{xf}}{mv} \right)\delta \tag{6.14a}$$

$$\dot{r} = -\left(\frac{C_r b - C_f a}{I} \right)\beta - \left(\frac{C_r b^2 + C_f a^2}{I} \right)r + \frac{a\left(C_f + F_{xf} \right)}{I}\delta \tag{6.14b}$$

Or as a state equation, we have

$$P\dot{X} + QX = RU \tag{6.15}$$

where X is a state vector and U is the input vector

$$X = \begin{bmatrix} \beta \\ r \end{bmatrix}$$

$$U = [\delta]$$

$$P = \begin{bmatrix} m & \\ & I \end{bmatrix}$$

$$Q = \begin{bmatrix} \dfrac{C_r + C_f}{v} & \dfrac{-C_r b + C_f a}{v} + v \\ \dfrac{-C_r b + C_f a}{v}\beta & \dfrac{C_r b^2 + C_f a^2}{v} \end{bmatrix} \tag{6.16}$$

$$R = \begin{bmatrix} \dfrac{C_f}{v} \\ \dfrac{C_f a}{I} \end{bmatrix}$$

By introducing the following parameters

$$a_{11} = -\frac{C_r + C_f}{mv}$$

$$a_{12} = -\left(\frac{C_r b - C_f a}{mv^2} - 1\right)$$

$$a_{21} = -\frac{C_r b - C_f a}{I}$$

$$a_{22} = -\frac{C_r b^2 + C_f a^2}{Iv} \tag{6.17}$$

$$b_1 = \frac{C_f + F_{xf}}{mv}$$

$$b_2 = \frac{a\left(C_f + F_{xf}\right)}{I}$$

Equation 6.14 can be expressed as

$$\dot{\beta} = a_{11}\beta + a_{12}r + b_1\delta$$

$$\dot{r} = a_{21}\beta + a_{22}r + b_2\delta \tag{6.18}$$

By eliminating one variable β, we obtain

$$\ddot{r} - (a_{11} + a_{22})\dot{r} + (a_{11}a_{22} - a_{12}a_{21})r = b_2\dot{\delta} + (a_{21}b_1 - a_{11}b_2)\delta \qquad (6.19)$$

Similarly, we can get another as

$$\ddot{\beta} - (a_{11} + a_{22})\dot{\beta} + (a_{11}a_{22} - a_{12}a_{21})\beta = b_2\dot{\delta} + (a_{12}b_2 - a_{22}b_1)\delta \qquad (6.20)$$

Note that both equations are similar to the equations for a single-degree-of-freedom system as we discussed in Chapter 1. The term $a_{11}a_{22} - a_{12}a_{21}$ corresponds to the square of the natural frequency, and the term $-a_{11} -a_{22}$ corresponds to the damping. The tire actually behaves similarly to a spring and damper, and the characteristics of the tire are similar to the stiffness and damping. The yaw rate and the sideslip angle as well as the directional angle have identical natural frequencies and damping. If the excitation is from the wind gust disturbance instead of the input of the steer angle, then the following equation governs the motion of the vehicle handling:

$$\dot{\beta} = a_{11}\beta + a_{12}r + \frac{F_w}{mv}$$
$$\dot{r} = a_{21}\beta + a_{22}r + \frac{M_w}{I} \qquad (6.21)$$

6.3 Steady-State Handling Characteristics

Cornering is defined as the ability of the vehicle to travel a curved path. Many factors affect the cornering ability of a vehicle, such as the tire construction, tire tread, road surface, alignment, and tire loading. As a vehicle turns a corner, centrifugal force pushes outward on the center of gravity of the vehicle. Centrifugal force is resisted by the traction of the tires. The interaction of these two forces moves weight from the side of the vehicle on the inside of the turn to the outside of the vehicle, and the vehicle leans. As this occurs, weight leaves the springs on the inside, and that side of the vehicle rises. This weight goes to the springs on the outside, and that side of the vehicle lowers. This is what is known as body roll. When the cornering requirement of a particular maneuver is less than the traction that can be provided by the tires, the car will go where it is pointed and steered. However, if the cornering force exceeds the available traction from the tires, the tires will slip across the road surface and will skid. In this chapter, we will discuss the steady-state handling characteristics.

6.3.1 Yaw Velocity Gain and Understeer Gradient

Consider steady cornering, when a vehicle drives through a curve at low lateral acceleration and low lateral forces are needed. At the wheels, then hardly any lateral slip occurs. In the ideal case, with vanishing lateral slip, the wheels move only in the circumferential direction. Therefore, $\ddot{\beta} = 0$, $\dot{\beta} = 0$, $\dot{\delta} = 0$, $\delta = $ constant, and $F_{xf} = 0$, and Eq. 6.20 becomes

$$(a_{11}a_{22} - a_{12}a_{21})\beta = (a_{12}b_2 - a_{22}b_1)\delta \qquad (6.22)$$

Then we have

$$\frac{\beta}{\delta} = \frac{a_{12}b_2 - a_{22}b_1}{a_{11}a_{22} - a_{12}a_{21}} = \frac{-C_f C_r b(a + b) + C_f a m v^2}{C_f C_r (a + b)^2 + (C_r b - C_f a) m v^2} \tag{6.23}$$

Similarly,

$$\frac{r}{\delta} = \frac{a_{21}b_1 - a_{11}b_2}{a_{11}a_{22} - a_{12}a_{21}} = \frac{v(a + b)C_f C_r}{(a + b)^2 C_f C_r + m v^2 (b C_r - a C_f)} \tag{6.24}$$

This is defined as yaw velocity gain. We define the understeer gradient as

$$K = \frac{m(b C_r - a C_f)}{L^2 C_f C_r} \tag{6.25}$$

with the unit as deg/g or rad/m/s^2.

Example E6.2

A typical understeer gradient for a car is K = 0.0024 s^2/m at 2.94 m/s^2, and K = 0.0026 s^2/m at 4.9 m/s^2.

Example E6.3

A typical North American car has an average K = 0.45 deg/m/s^2, and a typical European car has an average K = 0.265 deg/m/s^2.

From Eqs. 6.24 and 6.25, we obtain the yaw velocity gain as

$$\frac{r}{\delta} = \frac{\dfrac{v}{L}}{1 + K v^2} \tag{6.26}$$

This ratio represents a gain that is proportional to velocity in the case of a neutral steer vehicle.

Example E6.4

For v = 22.35 m/s, a_y = 3.92 m/s^2, and $\frac{r}{\delta}$ is in the range of 0.16 to 0.33 s^{-1}.

Larger lateral acceleration leads to nonlinearity of the tire slip angle. The steady-state response of the vehicle can be classified into three categories in terms of the value of K, as shown in Figure 6.8.

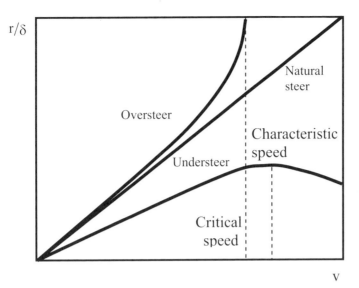

Figure 6.8 Yaw velocity gain as a function of speed.

6.3.1.1 Neutral Steer

When $K = 0$, $\dfrac{r}{\delta} = \dfrac{v}{L}$, and the yaw velocity gain is proportional to the velocity of the vehicle, with a slope of $\dfrac{1}{L}$. It is named as neutral steer and is a pure rolling case with radius of $R \approx \dfrac{L}{\delta}$. Note that this relationship is the cornering characteristics of the vehicle when it travels with very low velocity and without sideslip angle. When the sideslip angle is $\beta = 0$, the steering angle $\delta \approx \dfrac{L}{R}$ is named as the Ackerman angle. The radius of turn is $R = \dfrac{L}{\delta}$, the yaw velocity is $r = \left(\dfrac{v}{L}\right)\delta$, and the yaw velocity gain is $\left.\dfrac{r}{\delta}\right|_{s} = \dfrac{v}{L}$.

On a constant-radius turn, no change in the steer angle will be required as the speed is varied.

6.3.1.2 Understeer

When $K > 0$, the denominator of Eq. 6.20 is larger than 1, and $\dfrac{r}{\delta}$ is not linearly proportional to the velocity of the vehicle. It is a downward bending curve. The larger the value of K, the lower the yaw velocity gain and, accordingly, the larger the understeering. When the vehicle speed reaches $v_{ch} = \sqrt{\dfrac{1}{K}}$, the yaw velocity gain reaches its maximum value. Its value is exactly half of that of the neutral steer vehicle. The vehicle speed v_{ch} is defined as the characteristic speed, which is used to quantify the magnitude of understeering. When the value of the understeer increases, K increases, and v_{ch} decreases. The response is always stable and decreases with the increase in the speed of the vehicle.

6.3.1.3 Oversteer

When K < 0, the denominator of Eq. 6.26 is smaller than 1, and $\frac{r}{\delta}$ is not linearly proportional to the velocity of the vehicle. It is an upward bending curve. The smaller the value of K, the higher the yaw velocity gain and, accordingly, the larger the oversteering. When the vehicle speed reaches $v_{cr} = \sqrt{\frac{-1}{K}} = \sqrt{\frac{l^2 C_f C_r}{m(aC_f - bC_r)}}$, the yaw velocity gain reaches an infinite value. This speed characterizes a critical speed where the vehicle starts to lose stability. As $\frac{r}{\delta}$ approaches infinity, any trivial steering angle input results in an infinite yaw velocity, which means that the radius of turn is infinitely small; the vehicle has radical cornering and sideslip, or even rollover. Because the oversteer design of a vehicle has the trend of instability, vehicles generally are designed with appropriate understeer properties. For the understeer case, the yaw velocity increases with the speed of the vehicle until it reaches the characteristic speed. Then the yaw velocity begins to decrease thereafter. The characteristic speed is of significance as the speed at which the vehicle is most responsive in yaw. Figure 6.9 shows a typical curve of yaw rate gain as a function of vehicle speed.

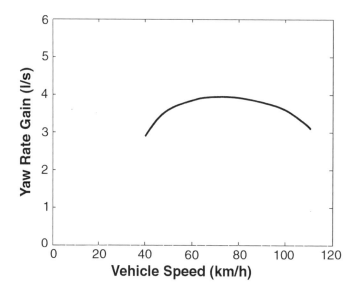

Figure 6.9 Yaw rate gain as a function of vehicle speed.

6.3.2 Difference Between Slip Angles of the Front and Rear Wheels

The difference between the slip angles of the front and rear wheels on a constant-radius turn is another useful index. For a vehicle traveling forward with a speed of v, the sum of forces in the lateral direction from the tires must equal the mass times the centripetal acceleration

$$F_{yf} + F_{yr} = ma_y = \frac{mv^2}{R} \tag{6.27}$$

where R is the radius of the turn. Consider a vehicle with a moment equilibrium about the center of gravity. The sum of the moment from the front and rear lateral forces results in

$$F_{yf}a - F_{yr}b = 0 \tag{6.28}$$

From Eqs. 6.27 and 6.28, we can write

$$F_{yr} = \frac{mav^2}{(LR)}$$

$$F_{yf} = \frac{mbv^2}{(LR)} \tag{6.29}$$

By denoting

$$\alpha_f = \frac{mav^2}{(C_f LR)}$$

and

$$\alpha_r = \frac{mav^2}{(C_r LR)}$$

we obtain

$$K = \frac{1}{a_y L}(\alpha_f - \alpha_r) \tag{6.30}$$

$$(\alpha_f - \alpha_r) = Ka_y L \tag{6.31}$$

This relationship is plotted in Figure 6.10. To measure the steady-state response of the vehicle, we apply a fixed value of the steering angle, which enables the vehicle to travel with a constant-radius turn of different speed, measure the front and rear tire slip angles, and apply the difference between the front and rear tire slip angles versus lateral acceleration a_y curve to evaluate the steady response. When the lateral acceleration is larger than 3 to 4 m/s/s, $(\alpha_f - \alpha_r)$ is no longer linearly proportional to the lateral acceleration. This is due to the cornering stiffness, which is no longer in the linear range. From Eq. 6.31, we know that if $(\alpha_f - \alpha_r) > 0$, $K > 0$ and the vehicle is in understeer. Likewise, when $(\alpha_f - \alpha_r) = 0$, $K = 0$ and the vehicle is in neutral steer. When $(\alpha_f - \alpha_r) < 0$, $K < 0$ and the vehicle is in oversteer. Consider Eqs. 6.26 and 6.30. We have $\delta = \frac{L}{R} + LKa_y$. Therefore, we have $\delta = \frac{L}{R} + (\alpha_f - \alpha_r)$. This equation can be viewed from the model plot and can be rearranged as

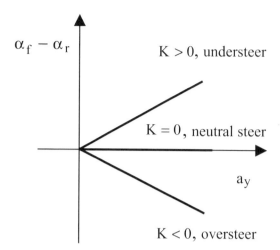

Figure 6.10 $(\alpha_f - \alpha_r)$ as a function of acceleration.

$$R = \frac{L}{\delta - (\alpha_f - \alpha_r)} \qquad (6.32)$$

If the slip angle can be ignored at very low speed, we get the radius of turn corresponding to the Ackerman angle as

$$R_0 = \frac{L}{\delta} \qquad (6.33)$$

When the vehicle speed is raised and there is proper lateral acceleration, there is sufficient slip angle at the front and rear tires. If $(\alpha_f - \alpha_r) > 0$, then $R > R_0$ and it is understeer. If $(\alpha_f - \alpha_r) < 0$, then $R < R_0$.

6.3.3 Ratio of Radius of Turn

If the slip angle can be ignored at very low speed, we get the radius of turn R_0. When the vehicle speed is raised and there is proper lateral acceleration, there are enough slip angles at the front and rear tires, and then a different R, the ratio of the radius, can be applied. To characterize the steady-state response of vehicle, we derive

$$R = \frac{v}{r} = \frac{(1 + Kv^2)L}{\delta} = (1 + Kv^2)R_0 \qquad (6.34)$$

Then

$$\frac{R}{R_0} = (1 + Kv^2) \qquad (6.35)$$

When $K = 0$, $\dfrac{R}{R_0} = 1$, which means that the radius of turn of the vehicle with neutral steer does not vary with the vehicle speed and remains as R_0. When $K > 0$, $\dfrac{R}{R_0} > 1$, which means the turn radius of the understeer vehicle is always larger than R_0, and moreover, the radius of turn increases with the increase in the speed of the vehicle. When $K < 0$, $\frac{R}{R_0} < 1$, which means the turn radius of the understeer vehicle is always smaller than R_0, and moreover, the radius of turn decreases with the increase in the speed of the vehicle.

Static margin: From Eq. 6.25, the sign of the understeer gradient depends on the sign of $bC_r - aC_f$, which is referred to as the static margin. The static margin is another measure of the steady-state handling behavior. It is determined by the point on the vehicle where a lateral force will not generate any steady-state yaw velocity (i.e., the neutral steer point). We define a neutral steer line as shown in Figure 6.11. The neutral steer line is the locus of points in the x-y plane along which external forces produce no steady-state yaw velocity. The static margin is defined as the distance the neutral steer point falls behind the center of gravity, normalized by the wheelbase, that is,

$$\text{Static margin} = \frac{e}{L}$$

When the point is behind the center of gravity, the static margin is positive and the vehicle is in understeer. At the center of gravity, the static margin is zero and the vehicle is in neutral steer. When the point is ahead of the center of gravity, the vehicle is in oversteer.

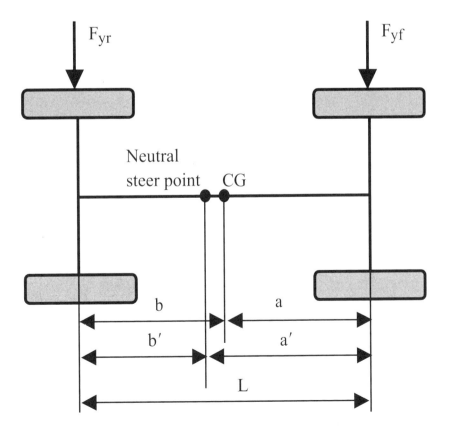

Figure 6.11 Schematic of the static margin.

EXAMPLE **E6.5**

The static margin of a typical vehicle falls in the range of 0.05 to 0.07 behind the center of gravity. We can determine the location of the neutral steer point by the moment of equilibrium.

When lateral forces are applied at the neutral steer point, the front and rear tires yield the same slip angle α, together with the lateral tire force $F_{yf} = C_f\alpha$ and $F_{yr} = C_r\alpha$. The distance between the neutral steer point and the front axis is

$$a' = \frac{F_{yr}L}{F_{yf} + F_{yr}} = \frac{C_r}{C_f + C_r}L \qquad (6.36)$$

The static margin is the difference between the distance of the neutral steer point to the front axis and the distance of the center of gravity to the front axis

$$\text{Static margin} = \frac{a' - a}{L} = \frac{C_r}{C_f + C_r} - \frac{a}{L} = \frac{bC_r - aC_f}{C_f + C_r} \qquad (6.37)$$

When the neutral steer point coincides with the center of gravity, the static margin is zero. If the lateral force applied at the center of gravity causes the front and rear tires to have the same slip angle, the vehicle possesses neutral steer properties. When the center of gravity is in front of the neutral steer point, the static margin is negative. If the lateral force applied at the center of gravity causes the slip angle of the rear tire to be larger than that of the slip angle of the front tire, the vehicle has oversteer properties. When the center of gravity is behind the neutral steer point, the static margin is positive. Then the lateral force applied at the center of gravity causes the slip angle of the front tire to be larger than the slip angle of the rear tire, and the vehicle has understeer properties.

6.4 Dynamic Characteristics of Handling

In this section, we will discuss transient analysis and frequency domain analysis. We also will describe the stability analysis associated with the system eigenvalue.

6.4.1 Handling Damping and Natural Frequency

Next we discuss the free response characteristics of steering a vehicle. Consider Eqs. 6.19 and 6.20, where the system can be treated as a single-degree-of-freedom system as discussed in Chapter 2. For Eqs. 6.19 and 6.20, we consider the free response case, namely,

$$b_2\dot{\delta} + (a_{21}b_1 - a_{11}b_2)\delta = 0$$

and

$$b_2\dot{\delta} + (a_{12}b_2 - a_{22}b_1)\delta = 0$$

By assuming $\omega_n^2 = a_{11}a_{22} - a_{12}a_{21}$ and $2\xi\omega_n = -(a_{11} + a_{22})$, we define ω_n as the natural circular frequency of the yaw velocity, and ξ as the damping factor of the yaw velocity. Hence, we obtain

$$\omega_n = \sqrt{\frac{mv(aC_f - bC_r) + C_fC_r\frac{L^2}{v}}{mvI}}$$

$$= \frac{L}{v}\sqrt{\frac{C_fC_r(1 + Kv^2)}{mI}}$$

(6.38)

where ω_n is an important factor in assessing the transient response. It is desired to have a higher ω_n for vehicles.

The damping factor or ratio can be derived as

$$\xi = \frac{-\left[m(a^2C_f + b^2C_r) + I(C_f + C_r)\right]}{2mIL\sqrt{\dfrac{C_fC_r(1 + Kv^2)}{mI}}}$$

$$= \frac{-\left[m(a^2C_f + b^2C_r) + I(C_f + C_r)\right]}{2L\sqrt{mIC_fC_r(1 + Kv^2)}}$$

(6.39)

EXAMPLE E6.6

A typical car has $f_n = \dfrac{\omega_n}{2\pi}$ about 1 Hz, and the damping is in the range of $\xi = 0.5 - 0.8$.

Figures 6.12 and 6.13, respectively, show the damping ratio and the natural frequency as functions of the vehicle speed under different cornering stiffness differences.

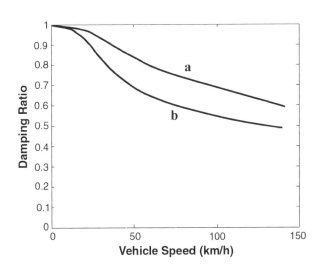

Figure 6.12 Damping as a function of speed for different cornering stiffness differences, where $\Delta C = C_r - C_f$: (a) ΔC_1, and (b) ΔC_2, $\Delta C_2 > \Delta C_1$.

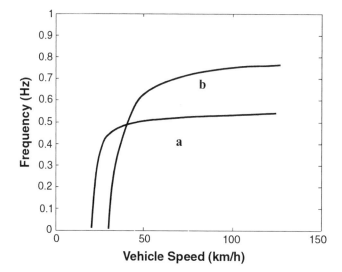

Figure 6.13 Natural frequency as a function of speed for different cornering stiffness differences, where $\Delta C = C_r - C_f$: (a) ΔC_1, and (b) ΔC_2, $\Delta C_2 > \Delta C_1$.

To find the transient response, Eq. 6.19 can be rewritten as

$$\ddot{r}(t) + 2\xi\omega_n\dot{r}(t) + \omega_n^2 r(t) = 0 \tag{6.40}$$

To solve Eq. 6.40, we assume that

$$r(t) = A\,e^{st} \tag{6.41}$$

where A is a constant and s is a parameter that remains to be determined. By substituting Eq. 6.41 into Eq. 6.40, one obtains

$$\left(s^2 + 2\xi\omega_n s + \omega_n^2\right)A\,e^{st} = 0 \tag{6.42}$$

Because $A\,e^{st} \neq 0$, then

$$s^2 + 2\xi\omega_n s + \omega_n^2 = 0 \tag{6.43}$$

Equation 6.43 is known as the characteristic equation of the system. This equation has the following two roots

$$s_1, s_2 \left(-\xi \pm \sqrt{\xi^2 - 1}\right)\omega_n \tag{6.44}$$

Solution a:

$\xi < 1$ (underdamped condition):

$$s_1, s_2 = \left(-\xi \pm i\sqrt{1 - \xi^2}\right)\omega_n$$

Then the yaw velocity can be solved as

$$r(t) = A\exp(-\xi\omega_n t)\cos\left(\omega_n\sqrt{1-\xi^2}\,t - \phi\right) \qquad (6.45)$$

or

$$r(t) = A\exp(-\xi\omega_n t)\cos(\omega_d t - \phi) \qquad (6.46)$$

where $\omega_d = \omega_n\sqrt{1-\xi^2}$ is the damped frequency of the system. Constants A and ϕ are determined from the initial conditions.

Solution b:

$\xi > 1$ (overdamped condition):

$$s_1, s_2 = \left(-\xi \pm \sqrt{\xi^2 - 1}\right)\omega_n$$

Then the yaw velocity can be solved as

$$r(t) = A_1\exp\left(-\xi + \sqrt{\xi^2 - 1}\right)\omega_n t$$
$$+ A_2\exp\left(-\xi - \sqrt{\xi^2 - 1}\right)\omega_n t \qquad (6.47)$$

The motion is aperiodic and decays exponentially with time. Constants A_1 and A_2 are determined from the initial conditions.

Solution c:

$\xi = 1$ (critically damped condition):

$$s_1 = s_2 = -\omega_n$$

Then the yaw velocity can be solved as

$$r(t) = (A_1 + A_2)\exp(-\omega_n t) \qquad (6.48)$$

Equation 6.48 represents an exponentially decaying response. The constants A_1 and A_2 depend on the initial conditions.

6.4.2 Step Steer Input Response

Yaw velocity properties can be characterized by the transient response to step function input or ramp function input from the steering angle. The frequency response of the yaw velocity can be obtained from the transient response test of the impulse input of the steering angle. Next we will derive the step input response by considering the step steer input to the following equation from Eq. 6.19

$$\ddot{r}(t) + 2\xi\omega_n\dot{r}(t) + \omega_n^2 r(t) = B_1\dot{\delta} + B_0\delta \tag{6.49}$$

where $B_1 = b_2$, $B_0 = (a_{21}b_1 - a_{11}b_2)$. Consider the following step input of the front tire steering angle as

$$\delta(t) = \begin{cases} 0, t < 0 \\ \delta_0, t \geq 0 \end{cases}$$

$$\dot{\delta}(t) = \begin{cases} 0, t < 0 \\ 0, t \geq 0 \end{cases} \tag{6.50}$$

Then for $t \geq 0$, Eq. 6.49 can be simplified as

$$\ddot{r}(t) + 2\xi\omega_n\dot{r}(t) + \omega_n^2 r(t) = B_0\delta_0 \tag{6.51}$$

The nonhomogeneous solution of Eq. 6.51 is

$$r = \frac{B_0\delta_0}{\omega_n^2} = \frac{\dfrac{v}{L}}{1 + Kv^2}\delta_0 = \frac{r}{\delta}\bigg|_s \delta_0 = r_0 \tag{6.52}$$

This is defined as the steady-state yaw velocity. Combining the homogenous solution given by Eq. 6.45 and the nonhomogeneous solution given by Eq. 6.52 yields

$$r(t) = \frac{B_0\delta_0}{\omega_n^2} + A\exp(-\xi\omega_n t)\cos(\omega_d t - \phi) \tag{6.53}$$

or

$$r(t) = \frac{B_0\delta_0}{\omega_n^2} + A_1\exp(-\xi\omega_n t)\cos\omega_d t + A_2\exp(-\xi\omega_n t)\sin\omega_d t \tag{6.54}$$

It can be shown that

$$A_1 = -\frac{B_0\delta_0}{\omega_n^2}$$

$$A_2 = \frac{B_0\delta_0}{\omega_n^2}\left(\frac{B_1\omega_n^2}{B_0} - \xi\omega_n\right)\frac{1}{\omega_d}$$

$$= r_0\left(\frac{-mva\omega_n}{LC_r} - \xi\right)\frac{1}{\sqrt{1 - \xi^2}}$$

$$A = \sqrt{A_1^2 + A_2^2}$$

$$= r_0 \sqrt{\left[\left(\frac{-mva}{LC_r}\right)^2 \omega_n^2 + \frac{2mva\xi\omega_n}{LC_r} + 1\right]\frac{1}{1-\xi^2}} \tag{6.55}$$

$$\phi = \arctan\frac{A_1}{A_2}$$

$$= \arctan\left[\frac{-\sqrt{1-\xi^2}}{\dfrac{-mva\omega_n}{LC_r} - \xi}\right] \tag{6.56}$$

Then

$$r(t) = r_0\sqrt{[(\frac{-mva}{LC_r})^2\omega_n^2 + \frac{2mva\xi\omega_n}{LC_r} + 1]\frac{1}{1-\xi^2}}\exp(-\xi\omega_n t)\cos\left(\omega_d t - \arctan\left[\frac{-\sqrt{1-\xi^2}}{\dfrac{-mva\omega_n}{LC_r} - \xi}\right]\right) \tag{6.57}$$

This is the yaw velocity transient response to step input of the front tire steer angle. From Eq. 6.57, we know that when $t = \infty$, $r(\infty) = \left.\frac{r}{\delta}\right|_s \delta_0 = r_0$, which means that the yaw velocity approaches a steady-state yaw velocity. For the transition process, the response is a damped oscillation, as shown in Figure 6.14.

EXAMPLE E6.7

A typical overshoot ranges $\frac{r_{max}}{r_0} = 1.12 - 1.65$, with a testing condition 31.3 m/s and $a_y = 0.4g$.

To further quantify the response, more dynamic factors can be defined as follows. Response time τ is defined as the time needed for yaw velocity to attain a steady value r_0 for the first time, after the step input is applied. Typically, it is desired to have smaller τ.

To derive τ, from Eq. 6.53 when $t = \tau$, we have $A\exp(-\xi\omega_n\tau)\cos(\omega_d\tau - \phi) = 0$. Therefore, $\cos(\omega_d\tau - \phi) = 0$. Then

$$\tau = -\frac{\Phi}{\omega_d} = \frac{\arctan\left[\dfrac{\sqrt{1-\xi^2}}{\dfrac{-mva\omega_n}{LC_r} - \xi}\right]}{\omega_n\sqrt{1-\xi^2}} \tag{6.58}$$

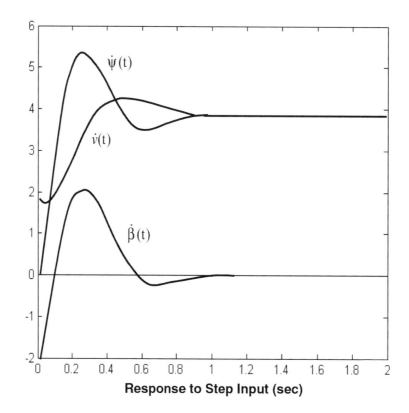

Figure 6.14 Yaw rate, sideslip velocity, and directional velocity response.

Another dynamic factor is ε, the time needed to attain the first peak which is the peak response time. This is used to assess the transient response. Taking the first derivative of Eq. 6.53, we obtain

$$\frac{dr(t)}{dt} = A\exp(-\xi\omega_n t)\big[-\xi\omega_n \sin(\omega_d t + \phi) + \omega_d \cos(\omega_d t + \phi)\big] \qquad (6.59)$$

Consider

$$\left.\frac{dr(t)}{dt}\right|_{t=\varepsilon} = 0$$

We have

$$-\xi\omega_n \sin(\omega_d t + \phi) + \omega_d \cos(\omega_d t + \phi) = 0$$

$$\tan(\varepsilon\omega_d + \phi) = \frac{\omega_d}{\xi\omega_n}$$

Therefore,

$$\varepsilon = -\frac{\arctan\dfrac{\omega_d}{\xi\omega_n} - \Phi}{\omega_d} = \frac{\arctan\left[\dfrac{\sqrt{1-\xi^2}}{\xi}\right]}{\omega_n\sqrt{1-\xi^2}} + \tau \qquad (6.60)$$

Another dynamic factor is the vehicle factor T.B., which is defined as the product of the peak response time and the slip angle of the center of gravity, where T.B. = $\varepsilon\beta$.

EXAMPLE E6.8

A typical car has the range of $\varepsilon = 0.23$ to 0.59 sec and T.B. = 0.25 to 1.45 sec, for a tested condition of $v = 31.3$ m/s and $a_y = 0.4g$.

6.4.3 Ramp Steer Input Response

Next we give the ramp input response by considering the ramp steer input of

$$\delta(t) = \begin{cases} 0, & t < 0 \\ kt, & t > 0 \end{cases} \qquad (6.61)$$

$$\ddot{r} - (a_{11} + a_{22})\dot{r} + (a_{11}a_{22} - a_{12}a_{21})r = b_2\dot{\delta} + (a_{21}b_1 - a_{11}b_2)\delta$$

If the damping can be ignored, the solution can be derived as

$$r = \frac{k}{\omega_n^2}\left[b_2 + (a_{21}b_1 - a_{11}b_2)t - \frac{1}{\omega_n}(a_{21}b_1 - a_{12}b_2)\sin\omega_n t\right] \qquad (6.62)$$

6.4.4 Impulse Input Excitation Response

Wind gust excitation can be represented as impulse input excitation. Next we will derive the impulse step response by considering the impulse steer input

$$\delta = \delta_0\delta(t)$$

where $\delta(t)$ is the Dirac delta function, which is defined as

$$\delta(t) = 0 \qquad t \neq 0 \qquad (6.63a)$$

and

$$\int_{-\infty}^{\infty} \delta(t)dt = 1 \qquad (6.63b)$$

From the system equation

$$\ddot{r} - (a_{11} + a_{22})\dot{r} + (a_{11}a_{22} - a_{12}a_{21})r = b_2\dot{\delta} + (a_{21}b_1 - a_{11}b_2)\delta$$

the solution is derived as

$$r(t) = \frac{(a_{21}b_1 - a_{11}b_2)\delta_0}{\omega_d}\exp(-\xi\omega_n t)\sin\omega_d t \qquad (6.64)$$

If we consider the wind gust impulse excitation, the system equation can be derived from Eq. 6.21 as

$$\ddot{r} - (a_{11} + a_{22})\dot{r} + (a_{11}a_{22} - a_{12}a_{21})r = \frac{a_{21}F_w}{(mv)} - a_{11}M_w \qquad (6.65)$$

The solution is

$$r(t) = \frac{a_{21}F_w - a_{11}M_w mv}{\omega_d mv}\exp(-\xi\omega_n t)\sin\omega_d t \qquad (6.66)$$

Figure 6.15 shows the yaw, sideslip, and directional angle response to impulse excitation.

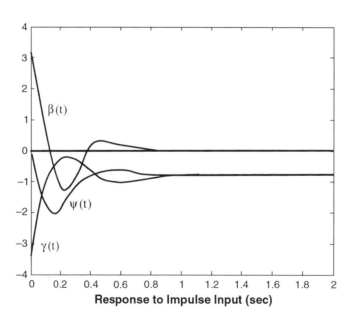

Figure 6.15 Yaw, sideslip, and directional angle response to impulse excitation.

6.4.5 Frequency Response of Yaw Velocity

We now consider the response of a bicycle model to a harmonic excitation, based on the equation of system motion, which is Eq. 6.19. The approach uses a frequency domain analysis and is based on the concept of transfer and harmonic response functions. By taking the Laplace transformation of Eq. 6.19 or Eq. 6.49, we obtain

$$\left[s^2 \bar{r}(s) - r(0)s - \dot{r}(0) \right] + 2\xi\omega_n \left[s\bar{r}(s) - r(0) \right] + \omega_n^2 \bar{r}(s) = B_1 s\delta(s) + B_0 \delta(s) \quad (6.67)$$

It follows that

$$\bar{r}(s) = \frac{B_1 s\delta(s) + B_0 \delta(s)}{s^2 + 2\xi\omega_n s + \omega_n^2} + \frac{(s + 2\xi\omega_n)r(0) + \dot{r}(0)}{s^2 + 2\xi\omega_n s + \omega_n^2} \quad (6.68)$$

By ignoring the homogeneous solution, Eq. 6.68 can be written as

$$\bar{r}(s) = \frac{A(s)}{B(s)} \quad (6.69)$$

where $A(s)$ and $B(s)$ are polynomials, with $B(s)$ of a higher order. The response $r(t)$ is found by taking the inverse Laplace transformation of Eq. 2.48. If only the forced solution is considered, we can define impedance transform as

$$Z(s) = \left[\frac{\bar{\delta}(s)}{\bar{r}(s)} \right] = \frac{\left(s^2 + 2\xi\omega_n s + \omega_n^2 \right)}{\left(B_1 s + B_0 \right)} \quad (6.70)$$

A transfer function $H(s)$ is defined by

$$H(s) = \frac{1}{Z(s)} = \left(B_1 s + B_0 \right)\left(s^2 + 2\xi\omega_n s + \omega_n^2 \right)^{-1} \quad (6.71)$$

The yaw rate and sideslip velocity transfer function can be solved, respectively, as

$$H_{r\delta}(s) = \frac{b_2 s + a_{21}b_1 - a_{11}b_2}{s^2 - (a_{22} + a_{11})s + (a_{11}a_{22} - a_{12}a_{21})} \quad (6.72)$$

$$H_{\beta\delta}(s) = \frac{b_1 s + a_{12}b_2 - a_{22}b_1}{s^2 - (a_{22} + a_{11})s + (a_{11}a_{22} - a_{12}a_{21})} \quad (6.73)$$

By using the definition of

$$B_1 = b_2$$

$$B_1' = b_1$$

$$B_0 = \left(a_{21}b_2 - a_{11}b_1 \right)$$

$$B_0' = a_{12}b_2 - a_{22}b_1$$

we have

$$H_{r\delta}(i\omega) = \frac{B_1 i\omega + B_0}{-\omega^2 + 2i\omega\xi\omega_n + \omega_n^2}$$

$$= \frac{B_1 i\omega + B_0}{-\omega^2 + 2i\omega\xi\omega_n + \omega_n^2} \frac{-\omega^2 - 2i\omega\xi\omega_n + \omega_n^2}{-\omega^2 - 2i\omega\xi\omega_n + \omega_n^2}$$

$$= \frac{2B_1\xi\omega_n\omega^2 + B_0\left(\omega_n^2 - \omega^2\right)}{\left(\omega_n^2 - \omega^2\right)^2 + \left(2\xi\omega\omega_n\right)^2} + \frac{B_1\omega\left(\omega_n^2 - \omega^2\right) - 2B_0\xi\omega_n\omega}{\left(\omega_n^2 - \omega^2\right)^2 + \left(2\xi\omega\omega_n\right)^2}i$$

$$= A(\omega) + B(\omega)i \qquad (6.74)$$

The amplitude characteristics are

$$D(\omega) = \sqrt{\left[A(\omega)\right]^2 + \left[B(\omega)\right]^2} \qquad (6.75)$$

The phase characteristics are

$$\Phi(\omega) = \arctan\frac{B(\omega)}{A(\omega)} \qquad (6.76)$$

Similarly,

$$H_{\beta\delta}(i\omega) = \frac{B_1' i\omega + B_0'}{-\omega^2 + 2i\omega\xi\omega_n + \omega_n^2}$$

$$= \frac{B_1' i\omega + B_0'}{-\omega^2 + 2i\omega\xi\omega_n + \omega_n^2} \frac{-\omega^2 - 2i\omega\xi\omega_n + \omega_n^2}{-\omega^2 - 2i\omega\xi\omega_n + \omega_n^2}$$

$$= \frac{2B_1'i\xi\omega_n\omega^2 + B_0'\left(\omega_0^2 - \omega^2\right)}{\left(\omega_n^2 - \omega^2\right)^2 + \left(2\xi\omega\omega_n\right)^2} + \frac{B_1'\omega\left(\omega_0^2 - \omega^2\right) - 2B_0'\xi\omega_n\omega}{\left(\omega_n^2 - \omega^2\right)^2 + \left(2\xi\omega\omega_n\right)^2}i$$

$$= A'(\omega) + B'(\omega)i \qquad (6.77)$$

The amplitude characteristics are

$$D'(\omega) = \sqrt{\left[A'(\omega)\right]^2 + \left[B'(\omega)\right]^2} \qquad (6.78)$$

The phase characteristics are

$$\left|\Phi'(\omega) = \arctan\frac{B'(\omega)}{A'(\omega)} \qquad (6.79)\right.$$

Figure 6.16 shows the typical amplitude and phase frequency response of the yaw velocity. The x axis is the logarithmic of the input frequency. The gain is represented by level (unit decibels [dB]).

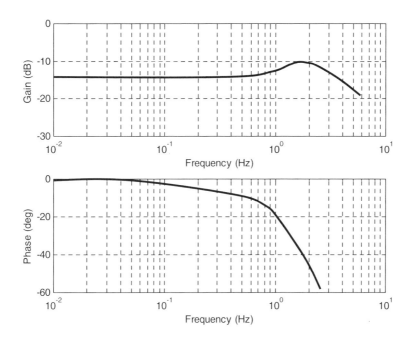

Figure 6.16 Amplitude and phase frequency response of the yaw velocity.

The amplitude frequency response plot reflects the capability of the vehicle to follow steering orders with different frequencies. The amplitude frequency response curve levels off at the low frequency range, and the amplitude increases with the increase in frequency. The system tends toward resonance when a specific frequency is attained. If the frequency increases further, the amplitude reduces again and drops rapidly. From the viewpoint of steer stability, it is desirable to have a flatter amplitude frequency curve with a higher resonant frequency. Also, it is desirable to have a smaller phase difference to allow the vehicle to have a rapid response. Several parameters related to the frequency response curve have been used to quantify the steer stability:

- Steady gain, $H_{r\delta}(0)$, the amplitude ratio at zero frequency

- Resonant frequency, f_r, the higher the f_r, the better the steer stability

- Amplitude ratio at the resonant point $\dfrac{H_{r\delta}(f_r)}{H_{r\delta}(0)}$: it is desirable to have a smaller $\dfrac{H_{r\delta}(f_r)}{H_{r\delta}(0)}$

- Phase delay angle at a frequency of 0.1 Hz : it represents the response speed at low speed steering and is desirable for it to be zero

- Phase delay angle at a frequency of 0.6 Hz : it represents the response speed at fast speed steering and is desirable to be smaller

EXAMPLE **E6.9**

A typical car has $f_r = 1.2$ Hz, $H_{r\delta}(f_r) - H_{r\delta}(0) = 3.38$ dB and a phase delay of $26°$.

6.4.6 Stability Analysis

If the transient response $r(t)$ converges to a steady-state value, then the vehicle handling will be stable. When $r(t)$ approaches infinity, the vehicle handling exhibits instability. It is well known that the system natural properties will determine its stability. From Eq. 6.46, when $\xi \leq 1$, as long as $\xi\omega_n > 0$, the solution converges; otherwise, it is divergent or unstable. It can be shown that

$$\xi\omega_n = \frac{-\left[m\left(a^2 C_f + b^2 C_r\right) + I\left(C_f + C_r\right)\right]}{2mvI} \tag{6.80}$$

Because C_f and C_r are negative, $\xi\omega_n > 0$. If $\xi < 1$, the solution will converge to zero. When $\xi > 1$, the characteristic roots should be negative for the solution to converge to zero, namely, $-\xi\omega_n \pm \sqrt{\left(\xi\omega_n\right)^2 - \omega_n^2}$ should be negative. In other words, ω_n^2 should be positive for the yaw velocity to converge. It can be shown that

$$\omega_n^2 = \frac{\left(aC_f - bC_r\right)}{I} + \frac{C_f C_r L}{mv^2 I} \tag{6.81}$$

The sign of the first term depends on the vehicle steady-state response. Because the understeer gradient is defined as

$$K = \frac{m\left(bC_r - aC_f\right)}{L^2 C_f C_r} \tag{6.82}$$

$$bC_r - aC_f = \frac{KL^2 C_f C_r}{m} \tag{6.83}$$

When the vehicle has an understeer feature, $K > 0$ and $bC_r - aC_f > 0$. The first term of Eq. 6.81 is positive when the vehicle has an understeer feature, $K < 0$ and $bC_r - aC_f < 0$. The first term of Eq. 6.81 is negative. The second term of Eq. 6.81 is positive, and when the vehicle speed is slow, it is a very large value; therefore, ω_n^2 tends to be positive despite the sign of the first term. Then the yaw velocity $r(t)$ converges, and the vehicle is stable. With an increase in the vehicle speed, the second term becomes smaller and smaller. When $bC_r - aC_f < 0$ due to oversteer, ω_n^2 is likely to be negative, $r(t)$ diverges, and the vehicle is not stable. The velocity when $\omega_n^2 = 0$ with oversteer is defined as the critical speed v_{cr}. When the speed is larger than v_{cr} and $\omega_n^2 < 0$, the vehicle is unstable. By allowing ω_n^2 to equal zero, the critical speed can be derived as $v_{cr} = \sqrt{-\frac{1}{k}}$, which was given in Eq. 6.31.

6.4.7 Curvature Response

Curvature gain is another useful factor in some analyses. Consider

$$v = R\dot{\gamma} = \frac{\dot{\gamma}}{\rho} = \frac{(\dot{\psi} - \dot{\beta})}{\rho}$$

where

$$\frac{\rho}{\delta}(i\omega) = \frac{\dot{\gamma}}{\delta v} = \frac{1}{v}\left(\frac{\dot{\psi}}{\delta} - \frac{\dot{\beta}}{\delta}\right) = \frac{1}{v}\left(H_{\dot{\gamma}\delta} - H_{\dot{\beta}\delta}\right) \qquad (6.84)$$

Substituting the yaw velocity and sideslip transfer functions from Eqs. 6.74 and 6.77 into Eq. 6.84, we obtain

$$\frac{\rho}{\delta}(i\omega) = \frac{b_2(a_{11} - i\omega + a_{12}i\omega) - b_1(\omega^2 + a_{22}i\omega + a_{21})}{a_{12}v(\omega^2 + a_{22}i\omega + a_{21}) - (a_{11} - i\omega + a_{12}i\omega)(a_{22}v - vi\omega)} \qquad (6.85)$$

Let us consider the following two cases.

In the first case, when $\omega = 0$ or δ is constant, Eq. 6.85 leads to

$$\frac{\rho}{\delta}(i\omega) = \frac{b_2 a_{11} - b_1 a_{21}}{a_{12}a_{21}v - a_{11}a_{22}v} = \frac{C_f C_r(a + b)}{C_f C_r(a + b)^2 - mv^2(C_f a - C_r b)} \qquad (6.86)$$

When $v = 0$, the curvature gain becomes

$$\frac{\rho}{\delta}(i\omega) = \frac{1}{a + b} \qquad (6.87)$$

If $C_f a = C_r b$, even $v \neq 0$, Eq. 6.87 still holds.

In the second case, if ω is very big or $\omega^2 \gg \omega$, the curvature gain is given by

$$\frac{\rho}{\delta}(i\omega) = \frac{C_f a_{11}}{mv^2} = \frac{1}{v}b_1 \qquad (6.88)$$

The curvature gain is inversely proportional to the speed. Figure 6.17 shows both cases.

6.5 Chassis System Effects on Handling Characteristics

The previous models and the motion equations employed many assumptions and simplifications to the real vehicle system. Numerous secondary effects to vehicle handling are caused by the kinematics and compliances in the chassis systems. The effects include roll steer, roll camber, suspension compliance steer, suspension compliance camber,

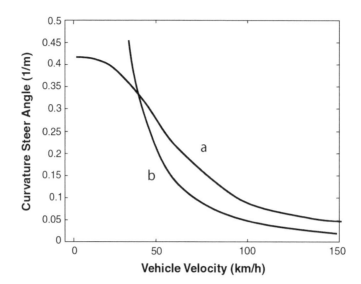

Figure 6.17 Curvature gain as a function of vehicle velocity for two cases.

steering kinematics, and steering compliance steer. These have not been taken into account by the previous models and the motion equations. For simplification, we next will discuss their respective effects on understeer gradients.

For all of these influences, second-order approximations of the actual responses are used. Separate values for left versus right are used where appropriate. As indicated, many factors in a vehicle may influence the cornering force developed in the presence of a lateral acceleration, in addition to the cornering stiffness. Any factor influencing the cornering force developed at a wheel will have a direct effect on the directional response. The suspensions and steering system are among the primary sources of these influences. In this section, we will discuss the suspension and steer system affecting handling. We will concentrate on steady-state cornering and touch on some of the dynamic characteristic effects. Because the steady-state cornering characteristics of a vehicle usually are quantified by the understeer gradient, we derive the understeer coefficient for different factors.

6.5.1 Lateral Force Transfer Effects on Cornering

A previous study showed that handling behavior is dependent on the ratios of the axial load/cornering stiffness ($\frac{ma}{C_fL}$, $\frac{mb}{C_rL}$) on the front and rear axles. The ratios are defined as cornering compliance, which indicates the number of degrees of slip angle at an axle per g of lateral force imposed at that point. If the front axle is more compliant than the rear axle (understeer vehicle), a lateral disturbance results in more sideslip at the front axle. Hence, the vehicle turns away from the disturbance, as illustrated in Figure 6.18.

If the rear axle has more cornering compliance (oversteer), the rear of the vehicle drifts outward, and this turns into disturbance. The lateral acceleration acting at the center of gravity adds to the disturbance force, further increasing the turning response and precipitating instability.

The cornering forces of pneumatic tires depend on load. Figure 6.19 shows a typical example of the lateral force as a function of vertical load for different slip angles. For

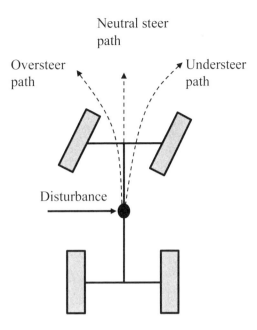

Figure 6.18 Schematic of the effect of a center-of-gravity disturbance on the steering handling.

a vehicle with some load on each wheel, appropriate lateral forces will be developed by each wheel at some slip angle. In hard cornering, the loads typically might change to lower values on the inside wheel and higher values on the outside wheel. Then the average lateral force from both tires will be reduced to an appropriate value. Consequently, the tires will have to assume a greater slip angle to maintain the lateral force necessary for turn. If these are the front tires, the front will plough outward, and the vehicle will understeer. If they are on the rear tires, the rear will slip outward, and the vehicle will oversteer. In fact, this mechanism is at work on both axles of all vehicles, but whether it contributes to understeer or oversteer depends on the balance of the roll moments distributed on the front and rear axles. More roll moment on the front axle contributes to understeer, whereas more roll moment on the rear axle contributes to oversteer. Figure 6.20 shows the schematics of the roll moment applied to an axle.

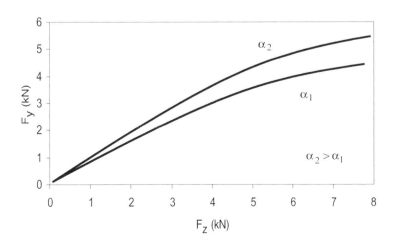

Figure 6.19 Lateral force F_y over wheel load F_z at different slip angles.

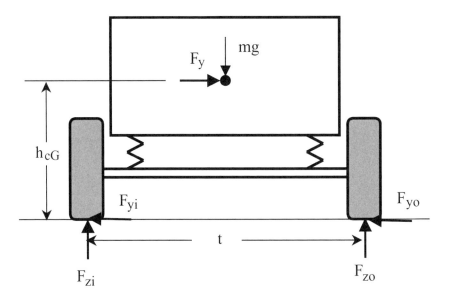

Figure 6.20 Schematic of the forces applied on a vehicle.

The suspensions can be modeled as two spring components. The separation of the springs could cause them to develop a roll-resisting moment proportional to the difference in the roll angle between the body and axle. Assume the center of gravity is identical to the roll center, which can be considered as the point on the body at which a lateral force application will produce no roll angle and that it is the point around which the axle rolls when a pure roll moment is applied. By applying Newton's second law, the relationship between wheel loads and the lateral force and roll angle can be established. In addition to the vertical forces, there is a net lateral force imposed at the tire, F_y, which is the sum of the lateral forces on the inside and outside wheels

$$F_{zo} - F_{zi} = \frac{2F_y h_{cG}}{t} = 2\Delta F_z \qquad (6.89)$$

in which F_{zo} and F_{zi} are the loads on the outside and inside wheels in the turn, respectively. $F_y = F_{yi} + F_{yo}$ is the lateral force, h_{cG} is the roll center height, and t is the track width. ΔF_z is the load change on each wheel. Note that the lateral load transfer has two different mechanisms. The first is $\dfrac{2F_y h_{cG}}{t}$, which is the lateral load transfer due to cornering forces. This mechanism originates from the lateral force imposed on the axle and is an instantaneous effect. It is independent of the roll angle of the body and the roll moment distribution. Second are the lateral load transfers due to vehicle roll, which we ignored for simplification in the preceding analysis. This effect depends on the roll characteristics and may lag the changes in the cornering conditions. It depends on the front/rear roll moment distribution. Generally, the roll moment distribution on vehicles is likely to be biased toward the front wheels due to a number of factors.

With the expression of roll moments for the front and rear axles, the difference of the load between the left and right wheels on the axles can be calculated. To translate the lateral load transfer into an effect on understeer gradient, we need to use data that link the tire cornering force to the slip angle and load. At the given conditions, the slip angle on each axle will change when the load transfer is taken into account. The difference between the change in the front and rear represents the understeer effect. In the preceding analysis, the cornering characteristics of the tires simply were modeled linearly by a constant named as the cornering stiffness.

To quantify the load nonlinear sensitivity effect as shown in Figure 6.21, the inside and outside tires must be treated separately. For simplification, the stiffness nonlinearity depending on the load is characterized by using a polynomial. The cornering stiffness of each tire can be represented by a second-order polynomial, and each lateral force developed will be given by

$$F_y' = C'\alpha = \left(a_1 F_z - b_1 F_z^2\right)\alpha \tag{6.90}$$

where F_y' is the lateral force of one tire, C' is the nonlinear cornering stiffness of one tire, a_1 and b_1 are the coefficients of polynomials, and F_z is the load on one tire. Therefore, for both tire cases,

$$F_y = \left(a_1 F_{zo} - b_1 F_{zo}^2 + a_1 F_{zi} - b_1 F_{zi}^2\right)\alpha \tag{6.91}$$

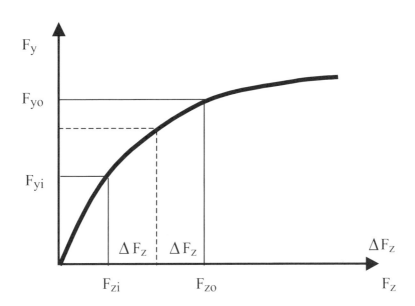

Figure 6.21 The effect of lateral force transfer to the lateral forces acted on the inside and outside wheel.

Assume the load change on each wheel to be ΔF_z

$$\begin{aligned} F_{z0} &= F_z + \Delta F_z \\ F_{zi} &= F_z - \Delta F_z \end{aligned} \tag{6.92}$$

Then we have

$$F_y = \left[a_1\left(F_z + \Delta F_z\right) - b_1\left(F_z + \Delta F_z\right)^2 + a_1\left(F_z - \Delta F_z\right) - b_1\left(F_z - \Delta F_z\right)^2\right]\alpha \tag{6.93}$$

Equation 6.93 reduces to

$$F_y = \left[2a_1F_z - 2b_1F_z^2 - 2b_1\Delta F_z^2 \right]\alpha \tag{6.94}$$

or

$$F_y = \left[C_\alpha - 2b_1\Delta F_z^2 \right]\alpha \tag{6.95}$$

Then for two tires on the front and the rear, we have, respectively,

$$F_{yf} = \left[C_{\alpha f} - 2b_1\Delta F_{zf}^2 \right]\alpha_f = \frac{W_f v^2}{(Rg)} \tag{6.96}$$

$$F_{yr} = \left[C_{\alpha r} - 2b_1\Delta F_{zr}^2 \right]\alpha_r = \frac{W_r v^2}{(Rg)} \tag{6.97}$$

Note that the steer angle necessary to maintain a turn is given by

$$\delta = \frac{L}{R} + \alpha_f - \alpha_r \tag{6.98}$$

By eliminating the slip angle from Eq. 6.98, we obtain

$$\delta = \frac{L}{R} + \frac{\dfrac{W_f v^2}{(Rg)}}{C_{\alpha f} - 2b_1\Delta F_{zf}^2} - \frac{\dfrac{W_r v^2}{(Rg)}}{C_{\alpha r} - 2b_1\Delta F_{zr}^2} \tag{6.99}$$

When $C_\alpha \gg 2b_1\Delta F_z^2$, Eq. 6.99 can be approximated as

$$\delta = \frac{L}{R} + \left[\left(\frac{W_f}{C_{\alpha f}} - \frac{W_r}{C_{\alpha r}} \right) + \left(\frac{W_f}{C_{\alpha f}}\frac{2b_1\Delta F_{zf}^2}{C_{\alpha f}} - \frac{W_r}{C_{\alpha r}}\frac{2b_1\Delta F_{zr}^2}{C_{\alpha r}} \right) \right]\frac{v^2}{(Rg)}$$
$$= \frac{L}{R} + \left[K_{tires} + K_{ll} \right]\frac{v^2}{(Rg)} \tag{6.100}$$

where the understeer gradient is

$$K_{tires} = \frac{W_f}{C_{\alpha f}} - \frac{W_r}{C_{\alpha r}}$$

and the understeer contributed by the lateral load transfer on the tires is

$$K_{llt} = \frac{W_f}{C_{\alpha f}}\frac{2b_1\Delta F_{zf}^2}{C_{\alpha f}} - \frac{W_r}{C_{\alpha r}}\frac{2b_1\Delta F_{zr}^2}{C_{\alpha r}} \tag{6.101}$$

The values for ΔF_{zf} and ΔF_{zr} can be obtained from Eqs. 6.96 and 6.97 as a function of the lateral acceleration. Because all the variables in Eq. 6.101 are positive, the contribution from the front axle is always understeer, and that from the rear axle is always negative. This means that it is an oversteer effect.

6.5.2 Steering System

The angle at the steering wheel may not be the decisive factor for the driving behavior. Actually, the steer angle at the road wheels is the dominant factor, which can differ from the steering wheel angle because of elasticity, friction influences, and a servo-support. At fast steering movements, the dynamic rise of the tire forces also plays an important role. Because the understeer gradient usually is measured at the steering wheel, the compliance in the steering system that allows the road wheel to deviate from the steering wheel input has an important effect on the understeer gradient and other handling properties. Figure 6.22 shows a schematic of the steering system and the system stiffness effects. Assuming that K_{ss} is the steer stiffness between the road wheel and the steering wheel, we can derive

$$K_{ss} = \frac{K_{sc}K_{sl}}{S^2 K_{sc} + K_{sl}} \tag{6.102}$$

in which K_{sc} and K_{sl} are the steering column stiffness and the road wheel axle stiffness, respectively. The steering ratio S is defined as the ratio of the steering wheel rotation angle to the steer angle at the road wheels.

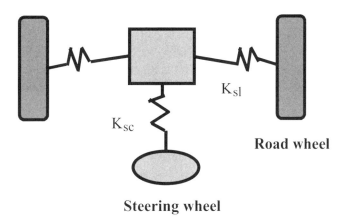

K_{sl}

K_{sc}

Road wheel

Steering wheel

Figure 6.22 Schematic of the steering system and system stiffness.

The contribution of the steering system to understeer depends on the front wheel load and the caster angle. Based on a simple analysis for understeer influences in which the lateral forces and the aligning torques are dominant, the understeer gradient can be derived as

$$K_{strg} = \frac{W_f \left(r v_c + p \right)}{K_{ss}} \tag{6.103}$$

where

K_{strg} = understeer gradient (deg/g) due to the steering system

W_f = front wheel load

r = wheel radius

p = pneumatic trail associated with the align torque

v_c = characteristic angle called the Cater angle

It can be seen that the Cater angle and the aligning torque effects add to understeer in the presence of a compliant steering system. In addition to experimentally determining K_{ss} statically, it is important to characterize the dynamic characteristics of a steering system in the frequency domain.

Rack and pinion is the most common steering system on passenger cars. The rack may be located either in front of or behind the axle. The rotations of the steering wheel first are transformed by the steering box to the rack travel and then are transmitted via the drag links to the wheel rotations. Hence, the overall steering ratio depends on the ratio of the steer box and on the kinematics of the steer linkage. Typically, it ranges from 15 to 20 to 1 on passenger cars and 20 to 36 to 1 on trucks. The directional response of a vehicle depends on the dynamics of the steering system. A good steering control provides accurate feedback about how the vehicle reacts to the road. The frequency response analysis can be used to design and analyze the dynamics of a manual rack-and-pinion steering system. We can obtain the transfer function between the angle of rotation of the front tire and the angle of rotation of the steering wheel.

Figure 6.23 shows the dynamic steer and camber measurement system. An autocollimator is a noncontact optical device that measure the dynamic steer and camber angles at the wheels. Scaffolding made of aluminum pipes is attached to the body with suction cups and clamps. The dynamic steer and camber measurement equipment and the steering transducer give us the input to the vehicle (i.e., steering wheel angle) and the output at the wheels (i.e., steer angle). With this information, the amount of compliance in the system during any steady-state or transient turning maneuver can be calculated. Other valuable data also can be measured, such as the steer angle change during braking and acceleration, the slip angle at the wheels, and the steering ratio of the vehicle. The frequency response of the steering system is analyzed in the steady state. The steady-state response of the steering system to a sinusoidal steering torque input is carried out. A stable linear time-invariant system subjected to a sinusoidal steering input will have, at steady state, a sinusoidal output of the same frequency as the input. However, the amplitude and phase will differ from the input, depending on the transfer function, as shown in Figure 6.24. We obtain the transfer function between the angle of rotation of the front tire and the angle of rotation of the steering wheel. The characteristics of the system then are derived to understand the damping frequency and the damping ratio.

6.5.3 Camber Change Effect

Camber angle is defined as the inclination of a wheel outward from the body. Camber on a wheel will produce a lateral force known as camber thrust. Typically, camber angle generates much less lateral force than slip angle. Approximately 4 to 6° of camber produce only the same lateral force as 1° of slip angle on a bias ply tire. The camber

Figure 6.23 Experimental testing system for a tire and steering system. (Gorder *et al.*, 2000.)

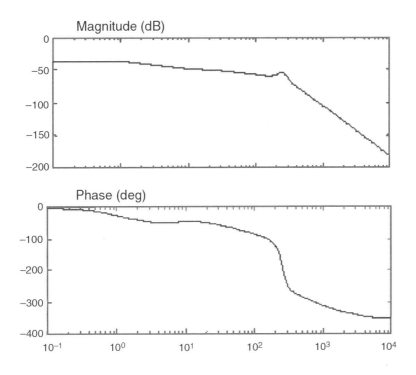

Figure 6.24 Frequency analysis of the steering system.

stiffness of radial tires generally is lower than that for bias ply tires; hence, as much as 10 to 15° are required on a radial tire. However, camber thrust still contributes to the cornering force in addition to the slip angle. Thus, it affects understeer. The camber thrust of bias ply tires is strongly affected by inflation pressure, although not so for radial tires, and it is relatively insensitive to load and speed for both radial and bias ply tires. Camber angles are small on solid axles, and it is possible to change the lateral forces by 10% or less on an independent wheel suspension. However, camber can play an important role in cornering, and camber changes both as a result of body roll and the normal camber change in jounce or rebound.

Consider that the lateral force depends not only on the slip angle of a tire, but also on the camber angle. Assume a linear relationship of

$$F_y = C_\alpha \alpha + C_\gamma \gamma \tag{6.104}$$

Thus,

$$\alpha = \frac{F_y}{C_\alpha} - \frac{C_\gamma}{C_\alpha} \gamma \tag{6.105}$$

Both F_y and γ are related to the lateral acceleration.

On the other hand, the total camber angle during cornering is

$$\gamma_g = \gamma_b + \phi \tag{6.106}$$

in which γ_g, γ_b, and ϕ are the camber angle with respect to the ground, the camber angle of the wheel with respect to the body, and the roll angle of the vehicle, respectively. The camber angle arising from the suspension is a function of the roll angle. Thus, we obtain the derivative of the camber angle with the roll angle from the analysis of the suspension characteristics. The relationship depends on the geometry of the suspension, For different suspensions, a camber gradient can be derived as

$$\frac{\partial \gamma}{\partial \phi} = f_\gamma \left(W_t, \Omega, \phi \right) \tag{6.107}$$

in which W_t is the track width, Ω is the suspension geometry, and ϕ is the roll angle. The roll angle is related to the lateral acceleration through previous equations. Finally, from Eq. 105, we obtained the equations for α_f and α_r as

$$\alpha_f = \frac{W_f}{C_\alpha} a_y - \frac{C_\gamma}{C_\alpha} \frac{\partial \gamma_f}{\partial \phi} \frac{\partial \phi}{\partial a_y} a_y$$

$$\alpha_r = \frac{W_r}{C_\alpha} a_y - \frac{C_\gamma}{C_\alpha} \frac{\partial \gamma_r}{\partial \phi} \frac{\partial \phi}{\partial a_y} a_y \tag{6.108}$$

When these are substituted into the turning equation, it takes the form of

$$\delta = \frac{L}{R} + \left[\left(\frac{W_f}{C_{\alpha f}} - \frac{W_r}{C_{\alpha r}} \right) + \left(\frac{C_{\gamma f}}{C_{\alpha f}} \frac{\partial \gamma_f}{\partial \phi} - \frac{C_{\gamma r}}{C_{\alpha r}} \frac{\partial \gamma_r}{\partial \phi} \right) \frac{\partial \phi}{\partial a_y} \right] \frac{v^2}{(Rg)} \tag{6.109}$$

Hence, the understeer deriving from the camber angles on each axle is expressed as

$$K_{camber} = \left(\frac{C_{\gamma f}}{C_{\alpha f}} \frac{\partial \gamma_f}{\partial \phi} - \frac{C_{\gamma r}}{C_{\alpha r}} \frac{\partial \gamma_r}{\partial \phi} \right) \frac{\partial \phi}{\partial a_y} \tag{6.110}$$

The tire lateral dynamic performance actually is measured using the sinusoidal lateral slip angle input.

6.5.4 Roll Steer Effect

When a vehicle rolls in the cornering process, the suspension system functionally leads to the wheels steering. Roll steer is referred to as the steering motion of the wheels with respect to the sprung mass that is due to the rolling motion of the sprung mass. Roll steer could lag the steer input of the vehicle in handling. The steer angle can affect handling as it changes the angle of the wheels with respect to the direction of travel. Assuming ε to be the roll steer coefficient on an axle (degrees steer/degree roll) and ϕ as the roll angle of the vehicle, we can derive the understeer gradient contribution from roll steer as

$$K_{\text{Roll steer}} = (\varepsilon_f - \varepsilon_r)\frac{\partial \phi}{\partial a_y} \tag{6.111}$$

A positive roll steer coefficient causes the wheels to steer to the right in a right-hand roll. When the vehicle is turning to one side, a positive roll steer on the front axle steers out of the turn and is understeer. Conversely, a positive roll steer on the rear axle is oversteer. As such, the suspension will allow the axle to roll about an imaginary axis that may be inclined with respect to the longitudinal axis of the vehicle. The kinematics of the suspension could be considered as functionally equivalent to a leading or trailing cantilever beam, and the roll axis inclination is equal to that of the beam. Given an initial inclination angle β on the beam, when the vehicle body rolls, the beam on the inside wheel rotates downward while the beam on the outside wheel rotates upward. If the initial orientation of a rear axle trailing beam is angled downward, the effect of the trailing beam angle change is to pull the inside wheel forward while pushing the outside wheel rearward. This results in roll steer of the solid axle contributing to oversteer. The roll steer coefficient is equal to the inclination angle ($\varepsilon = \beta$) of the trailing beam. On a rear trailing beam system, roll understeer is achieved by keeping the transverse pivots of the trailing beam below the wheel center.

6.5.5 Lateral Force Compliance Steer

Because many flexible parts such as soft bushings are used in suspension linkages, lateral compliance in the suspension contributes to steer. The compliance steer can be represented as rotation about a yaw center. For a forward yaw center on a rear axle, the compliance allows the axle to steer toward the outside of the turn and causes oversteer. Conversely, a rearward yaw center results in understeer. The opposite is true on a front axle: a rearward yaw center is oversteer, and a forward yaw center is understeer. The effects of lateral force compliance steer on handling can be characterized by defining an appropriate coefficient as

$$A = \frac{\delta_c}{F_y} \tag{6.112}$$

where δ_c is the steer angle and F_y is the lateral force. Through the analysis of the kinematics of linkages, the coefficients on independent wheel suspensions and on steered wheels can be determined.

The lateral force experienced on an axle is represented by the load on the axle times the lateral acceleration. Thus, on the front axle,

$$\delta_{cf} = A_f W_f a_y \tag{6.113}$$

Because the understeer effect is related directly to the steer angles produced on the front and rear axles, the understeer arising from lateral force compliance steer is

$$K_{lfcs} = A_f W_f - A_r W_r \tag{6.114}$$

6.5.6 Aligning Torque Effects

The aligning torque on the tires always resists the attempted turn of the vehicle; therefore, it is the source of an understeer effect. Aligning torque is attributed to the lateral forces developed by a tire at a point behind the tire center. The distance is known as the pneumatic trail. Its effects on handling features can be determined by assuming that the lateral forces are developed at a distance p behind each wheel. The understeer term can be expressed as

$$K_{at} = w\frac{d}{L}\frac{C_{\alpha f} + C_{\alpha r}}{C_{\alpha f}C_{\alpha r}} \tag{6.115}$$

Because the C_α values are positive, the aligning torque effect is positive (understeer) and cannot ever be negative (oversteer). The understeer due to this mechanism normally is less than 0.5 deg/g. However, aligning torque is indirectly responsible for an additional understeer mechanism through its influence on the steering system.

6.5.7 Effect of Tractive Forces on Cornering

The preceding analysis does not consider the effects of the drive forces at the wheels. Next, we will look at the effect of the drive forces present at the front and rear wheels. By considering the drive forces, the application of Newton's second law to the bicycle model in the lateral direction yields

$$\frac{W_f v^2}{(Rg)} = F_{yf}\cos(\alpha_f + \delta) + F_{xf}\sin(\alpha_f + \delta)$$

$$\frac{W_r v^2}{(Rg)} = F_{yf}\cos\alpha_r + F_{xr}\sin\alpha_r \tag{6.116}$$

in which F_{xf} and F_{xr} are the tractive forces on the front and rear axles, respectively. We can derive

$$\delta = \frac{\dfrac{L}{R}}{1 + \dfrac{F_{xf}}{C_{\alpha f}}} + \frac{\dfrac{W_f}{C_{\alpha f}}\dfrac{v^2}{(Rg)}}{1 + \dfrac{F_{xf}}{C_{\alpha f}}} + \frac{\dfrac{W_r}{C_{\alpha r}}\dfrac{v^2}{(Rg)}}{1 + \dfrac{F_{xr}}{C_{\alpha r}}} \tag{6.117}$$

Considering that $\dfrac{F_{xf}}{C_{\alpha f}}$ and $\dfrac{F_{xr}}{C_{\alpha r}}$ are much less than 1, Eq. 6.117 can be simplified as

$$\delta = \frac{\dfrac{L}{R}}{1 + \dfrac{F_{xf}}{C_{\alpha f}}} + \left[\left(\frac{W_f}{C_{\alpha f}} - \frac{W_r}{C_{\alpha r}} \right) - \left(\frac{W_f}{C_{\alpha f}} \frac{F_{xf}}{C_{\alpha f}} - \frac{W_r}{C_{\alpha r}} \frac{F_{xr}}{C_{\alpha r}} \right) \right] \frac{v^2}{(Rg)} \qquad (6.118)$$

This final turning formula takes into account the effect of tractive forces. The three terms on the right side of the equation are interpreted, respectively, as the Ackerman steer angle modified by the tractive force on the front axle, which is

$$\frac{\dfrac{L}{R}}{1 + \dfrac{F_{xf}}{C_{\alpha f}}}$$

the understeer gradient, unchanged from its earlier form, which is

$$\frac{W_f}{C_{\alpha f}} - \frac{W_r}{C_{\alpha r}}$$

and the effect of tractive forces on the understeer behavior of the vehicle, which is

$$\frac{W_f}{C_{\alpha f}} \frac{F_{xf}}{C_{\alpha f}} - \frac{W_r}{C_{\alpha r}} \frac{F_{xr}}{C_{\alpha r}}$$

Generally, the understeer coefficient K is the result of the tire, vehicle, suspension, and steering system parameters. Its total value consists of the effects of K_{tires}, K_{camber}, $K_{rollsteer}$, K_{lfcs}, K_{at}, K_{llt}, and K_{strg}, which are affected, respectively, by the tire cornering stiffness, camber thrust, roll steer, lateral force compliance steer, aligning torque, lateral load transfer, and the steering system.

6.6 Handling Safety—Overturning Limit Handling Characteristics

When a vehicle gets cornering with a 90° or larger angle, there exists an overturning hazard. There are two kinds of overturning hazards. One is the maneuver-induced rollover, and the other is the tripped rollover. This chapter limits our discussion to only the maneuver-induced rollover in the context of handling dynamics. The overturning hazard of a vehicle is determined primarily by the width of the vehicle and the height of the center of gravity.

In Chapter 5, we discussed the steady-state characteristics of the overturning limit, which is applicable to the case of small lateral acceleration. For larger lateral acceleration cases, we need to study the overturning response. The modeling of the vehicle will not be discussed here. In the following, we briefly cover the basic features derived from the dynamic model of an overturning vehicle.

Figure 6.25 shows a typical overturning response of a vehicle under step input steering. It can be seen that under the step input steering, the vehicle roll angle has an overshoot after it reaches and crosses the steady value. This suggests that the inside wheel of the vehicle could leave the ground even when the vehicle is under a smaller lateral acceleration than the steady-state rollover threshold. In other words, the dynamic rollover threshold is smaller than the steady-state rollover threshold. For a car or van, the rollover threshold of the step input steering is smaller than $\dfrac{t}{(2h_{CG})}$ by 30%, whereas the difference is 50% for a truck. The overshoot depends on the roll damping. Figure 6.26 shows the rollover thresholds as functions of the critical damping ratio. The zero damping gives the smallest rollover threshold. The rollover threshold increases with the increase of damping ratio but levels off at the higher damping ratio range. When the vehicle is provided with sinusoidal acceleration input, the roll response depends on the input response. Figure 6.27 shows the roll response as a function of input frequency. The rollover threshold approaches the steady-state rollover threshold. The threshold decreases with the increase of the input frequency until it reaches the minimum at the roll natural frequency, and then it increases again rapidly.

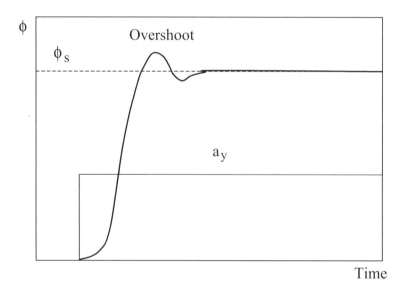

Figure 6.25 The rollover response under step input.

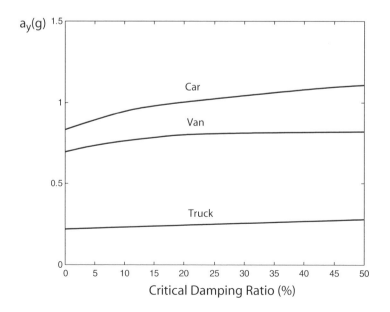

Figure 6.26 The rollover threshold as a function of the critical damping ratio for different vehicles.

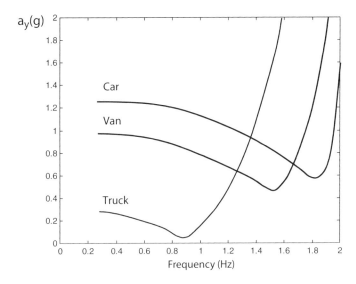

Figure 6.27 The rollover response as a function of input frequency.

6.7 Nonlinear Models of Handling Dynamics

6.7.1 Multiple-Degrees-of-Freedom System Models

In Section 6.2, we used a two-degrees-of-freedom model with sideslip and yaw rate to capture the basic characteristics of handling. Many existing models can simulate the motion of an actual automobile. Most of these fall at the two opposite ends of the spectrum: very simplified models for a specific use (the most common being the bicycle model), or high-order models covering the full degrees of freedom for a vehicle.

The simplified models do not have enough complexity to be used for many complicated applications. Some are targeted at specific scenarios and are good at simulating those circumstances but have poor results for general use. These models make many simplifying assumptions. If the circumstances to be studied are not within the scope of these assumptions, then the results may be inaccurate. Many of these models are linear, so they give poor results for maneuvers where the effects of nonlinearity in the vehicle become significant. For many studies with wide-ranging and/or severe inputs, these models are inadequate.

The other end of the spectrum involves the high-order vehicle models. These models certainly are more complete, have good general applicability, and can provide excellent results. The downside is that they are much more complicated and require a significant amount of information about the vehicle being modeled. In particular, to model several of the degrees of freedom requires detailed information about the suspension design, which may not be readily available. In addition, depending on the scenarios being studied, many of the degrees of freedom may not be necessary to predict vehicle handling response within the linear range, because any additional degrees of freedom for the vehicle sprung mass have little effect. Outside the linear handling range, additional degrees of freedom must be accounted for, because tire tractive force saturation and load transfer effects play a predominant role, as discussed briefly in the previous section.

The bicycle model is simplified from a complicated three-degrees-of-freedom system described by Eq. 6.12, which includes yaw, sideslip, and longitudinal degree. A more applicable three-degrees-of-freedom system model is the one including the lateral, yaw, and roll degree-of-freedom system of the car, and the vehicle consists of a sprung mass and a front and rear unsprung mass that are interconnected by a fixed roll axis. Several roll-induced effects such as roll steer and roll camber (due to suspension geometry), which have important influences on handling responses, are included in this model. The effects of longitudinal and lateral weight also are considered in this model. The vertical motions of sprung mass and the vertical and roll motions of the unsprung mass were not considered in this model. Finally, the roll plane model has been applied widely to characterize the lateral load transfer effects during handling maneuvers. This model is based on the concept of an axle roll center. The roll center is the point where the moments acting on the sprung mass of the vehicle sum to zero and thus is effectively the pivot point for the chassis rotation with respect to the ground.

For a more complicated analysis, we can use a four-degrees-of-freedom system model, which includes the longitudinal, lateral, roll, and yaw degrees. Or we can use an eight-degrees-of-freedom system model, which includes the longitudinal, lateral, roll, and yaw degrees. Rotation of four wheels with the chassis (i.e., the sprung mass) is modeled as a rigid six-degrees-of-freedom body, with a known mass and inertia properties. It comprises the mass of the powertrain, too.

A more complicated system is a fifteen-degrees-of-freedom system that is indispensable to research on the vehicle stability control system, with the brake system simply modeled as the torque actuator, three degrees of freedom for translation of the body, three degrees of freedom for rotation of the body, four degrees of freedom for the vehicle travel of the wheel, four degrees of freedom for rotation of the wheels, and one degree of freedom for the steering. This system can be used as standalone for vehicle handling performance prediction, integrated with control system algorithms for their synthesis and offline precalibration.

Nowadays, systems with seventeen, nineteen, or even larger numbers of degrees of freedom are becoming more popular in practical simulation. With large programs, multiple body dynamics models have been used. Also, an effort to predict vehicle handling characteristics upfront in the design process has been assuming an increasingly important role, with torque management to the wheels and other important developments. With vehicle dynamics refinement taking center stage, it has become increasingly accepted that the use of well-developed computer-aided-engineering (CAE) models presents the best approach for upfront prediction of vehicle behavior. Meaningful results can be derived, and projections made, from the CAE model only if the CAE results are correlated against real-world tests.

6.7.2 An Eight-Degrees-of-Freedom System Model

Next we introduce an eight degrees-of-freedom system model to illustrate the nonlinear model. Before generating the equations of motion for the vehicle, the assumptions about the model must be defined clearly. For this model, the single most significant assumption is that the vehicle is traveling on a smooth road such that there are no vertical motions of the wheels. With this assumption, the degrees of freedom associated directly with vertical motion can be neglected. Because a simplified model that still provides adequate results for vehicle handling was desired, pitch was excluded. Given these assumptions, the vehicle model is assumed to have the following eight degrees of freedom, which are shown by the thick arrows in Figure 6.28:

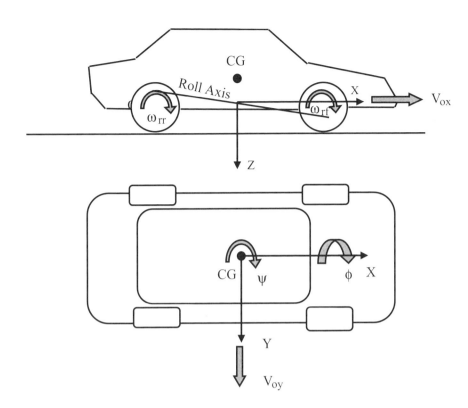

Figure 6.28 Eight-degrees-of-freedom and chassis coordinate system origin and axes.

1. Translation in the longitudinal direction

2. Translation in the lateral direction

3. Body roll relative to the chassis about the roll axis

4. Yaw about the vertical axis

5. Rotation of the left and right front wheels, the left rear wheel, and the right rear wheel

To generate the equations of motion, three major coordinate systems are used. The first is the inertial reference frame, the second is fixed to the vehicle chassis, and the third is fixed to the vehicle body. All equations are generated from the vehicle chassis reference frame. In addition, wheel-fixed coordinate frames are used to calculate the tire forces and moments. This means a positive x axis forward, a positive y axis to the right, and a positive z axis downward. Positive rotations are determined by the right-hand rule for these axes. The variables used for each of the degrees of freedom are as follows:

V_{ox} = longitudinal velocity of the chassis origin

V_{oy} = lateral velocity of the chassis origin

ϕ = body roll angle from the vertical

ψ = the vehicle heading angle

ω = angular velocity of the wheel

By integrating the longitudinal and lateral velocities and using the heading angle to convert into global coordinates, the global position of the vehicle also can be calculated. Using the preceding eight degrees of freedom and the assumptions described, the eight equations of motion for the vehicle are

$$\dot{V}_{ox} = \frac{\sum F_x + M_s\left(2h_s\dot{\phi}\dot{\psi}\cos\phi + h_s\ddot{\psi}\sin\phi\right)}{m} + V_{oy}\dot{\psi} \qquad (6.119a)$$

$$\dot{V}_{oy} = \frac{\sum F_y - M_s\left(h_s\ddot{\phi}\cos\phi - h_s\dot{\phi}^2\sin\phi - h_s\dot{\psi}^2\sin\phi\right)}{m} - V_{ox}\dot{\psi} \qquad (6.119b)$$

$$\ddot{\phi} = \frac{1}{I_{xxso}}\left\{\sum T_{xs} - I_{xzso}\ddot{\psi}\cos\phi - M_s h_s a_{oy}\cos\phi \right.$$
$$\left. + \left(I_{yys} - I_{zzs} + M_s h_s^2\right)\dot{\psi}^2\sin\phi\cos\phi\right\} \qquad (6.119c)$$

$$\ddot{\psi} = \frac{1}{I_{zzo}}\left\{\sum T_z - I_{xzso}\ddot{\phi}\cos\phi + I_{xzso}\dot{\phi}^2\sin\phi + M_s h_s a_{ox}\sin\phi \right.$$
$$\left. - 2\left(I_{yys} - I_{zzs} + M_s h_s^2\right)\dot{\phi}\dot{\psi}\sin\phi\cos\phi\right\} \qquad (6.119d)$$

$$\dot{\omega}_{lf} = \frac{F_{xwlf}R_{wlf} - M_{blf}}{I_{wlf}} \qquad (6.119e)$$

$$\dot{\omega}_{rf} = \frac{F_{xwrf}R_{wrf} - M_{brf}}{I_{wrf}} \qquad (6.119f)$$

$$\dot{\omega}_{lr} = \frac{F_{xwlr}R_{wlr} - M_{blr}}{I_{wlr}} \qquad (6.119g)$$

$$\dot{\omega}_{rr} = \frac{F_{xwrr}R_{wrr} - M_{blrr}}{I_{wrr}} \qquad (6.119h)$$

The forces acting in Eq. 6.119a are given by

$$\sum F_x = F_{xlf} + F_{xrf} + F_{xlr} + F_{xrr} + F_{xw} \qquad (6.119i)$$

which is the sum of the longitudinal forces in the chassis coordinates from all four tires plus the longitudinal aerodynamic forces. Because the tire forces are generated in a local wheel reference frame, the lateral and longitudinal tire forces must be multiplied by the appropriate sine and cosine of the steer angle for each wheel to convert to chassis coordinates. Any other x direction forces of interest could be added (rolling resistance of tires, for example). The forces acting in Eq. 6.119b are similar and are given by

$$\sum F_y = F_{ylf} + F_{yrf} + F_{ylr} + F_{yrr} + F_{yw} \tag{6.119j}$$

In Eq. 6.119c, the sum of the moments acting on the sprung mass about the roll axis (assumed the same as the x axis) is given by

$$\sum T_{xs} = T_{\phi f} + T_{\phi r} + M_s g h_s \sin \phi + T_{\phi w} \tag{6.119k}$$

where the last term is the aerodynamic roll moment. The first two terms are given by

$$T_{\phi f} + T_{\phi r} = -\left(K_{\phi f} + K_{\phi r}\right)\phi - \left(B_{\phi f} + B_{\phi r}\right)\dot{\phi} \tag{6.119l}$$

where K_{ϕ} and B_{ϕ} are the roll stiffness and roll damping, respectively. For Eq. 6.119d, the sum of the moments acting on the vehicle about the z axis is given by

$$\sum T_z = \left(F_{ylf} + F_{yrf}\right)l_f - \left(F_{ylr} + F_{yrr}\right)l_r + \frac{t_f}{2}\left(F_{xlf} - F_{xrf}\right) + \frac{t_r}{2}\left(F_{xlr} - F_{xrr}\right) \tag{6.119m}$$
$$+ M_{zlf} + M_{zrf} + M_{zlr} + M_{zrr} + M_{zw}$$

where the first four terms are the influence of tire forces (again in chassis coordinates), the second four terms are the tire aligning moments, and the final term is the aerodynamic yaw moment.

In addition, the accelerations of the chassis origin in the longitudinal and lateral directions are given by

$$a_{ox} = \dot{V}_{ox} - V_{oy}\dot{\psi} \tag{6.119n}$$

$$a_{oy} = \dot{V}_{oy} - V_{ox}\dot{\psi} \tag{6.119o}$$

These form the primary equations for the model. The accelerations at any other point in the vehicle, such as the center of gravity, can be calculated by using the results of these equations and knowing the location of the point in question relative to the origin. Then taking into account the tire model and the effects of the suspension and steering effects, we can use the model to conduct the simulations.

6.8 Testing of Handling Characteristics

Vehicle field testing was designed to minimize the effects of driver variability and to provide repeatable open-loop experiments. There are various testing approaches, such as static alignment, which determines the kinematic properties of the vehicle (i.e., steer and camber curves, steer ratio); constant radius cornering, which determines understeer, sideslip at the wheels, and overall compliance in the steering and suspension system during a steady-state maneuver; swept steer, which determines steering sensitivity, sideslip at the wheels, and overall compliance in the steering and suspension system during a steady-state maneuver; and frequency response, which defines the vehicle response to a range of different frequency inputs at the steering wheel. In addition, many nonstandard tests are conducted when special concerns or issues arise.

Because the steady-state cornering characteristics of a vehicle usually are quantified by the understeer gradient, we discuss how to determine the understeer gradient and then extend our discussion to transient response analysis and frequency analysis.

Methods for experimental measurement of understeer gradient are based on the definition of the gradient defined by

$$\delta = \frac{L}{R} + Ka_y \qquad (6.120)$$

Because the derivation of this equation assumes the vehicle to be in a steady-state operating condition, understeer is defined as a steady-state property. For experimental measurement, the vehicle must be placed into a steady-state turn with appropriate measures of the quantities in Eq. 6.120 so that the value of K can be determined. Several test methods have been suggested as the means to measure this property: constant radius, constant speed, and constant steer angle.

6.8.1 Constant Radius Turn

The constant radius turn is an important event in fingerprinting. In this test, the vehicle drives along a curve with a constant radius at various speeds. The steer angle or the steering wheel angle required to maintain the vehicle on course at various forward speeds and the corresponding lateral acceleration are measured. The curve of the steer angle versus the lateral acceleration is plotted, and the slope of the curve can be determined. For a constant radius turn, the slope is the understeer coefficient

$$\frac{d\delta}{d\left(\dfrac{a_y}{g}\right)} = K \qquad (6.121)$$

If the steer angle required to maintain the vehicle on a constant radius turn is the same for all forward speeds, the vehicle is neutral steer. If the slope is positive, the vehicle is considered to be understeer. If the slope is negative, the vehicle is considered to be oversteer. For real vehicles, the value of the understeer coefficient varies with the operation condition due to the nonlinear behavior of the tires, suspension, and load transfer. The vehicle usually has understeer characteristics at low lateral acceleration and oversteer characteristics at high lateral acceleration. The steering tendency of a real vehicle is determined by the driving maneuver of the steady-state cornering. The maneuver is performed quasi-static. The driver tries to keep the vehicle on a circle with the given radius R. He or she slowly increases the driving speed v and with this, because of $a_y = \dfrac{v^2}{R}$, the lateral acceleration until reaching the limit. Figure 6.29 displays typical results.

The constant radius method has the advantage that minimal instrumentation is required, but it has the disadvantage that it is difficult to execute in an objective fashion. Determination of a precise steering wheel angle is difficult because of the deviations necessary to keep the vehicle on the selected radius of turn. The minimum radius of turn for this test procedure normally is 30 m. For a two-axle vehicle, the understeer gradient is not affected by the radius of the circle.

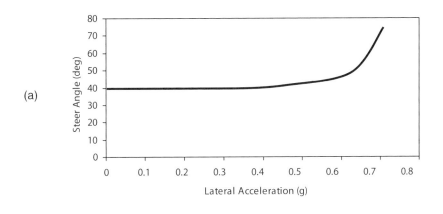

(a)

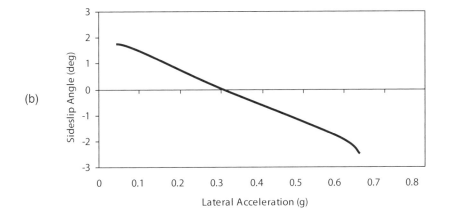

(b)

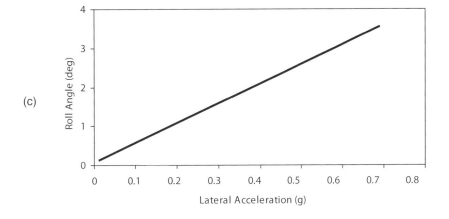

(c)

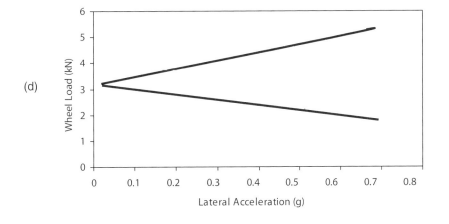

(d)

Figure 6.29 Steady-state cornering: a rear-wheel-driven car on R = 100 m. (a) Steer angle versus lateral acceleration. (b) Sideslip angle versus lateral acceleration. (c) Roll angle versus lateral acceleration. (d) Wheel load versus lateral acceleration.

6.8.2 Constant Speed Test

The constant speed test is another test that allows the vehicle to drive at a constant speed at various turning radii. In the test, the steer angle and the lateral acceleration are measured and plotted. The slope of the curve is given by

$$\frac{d\delta}{d\left(\dfrac{a_y}{g}\right)} = \frac{gL}{v^2} + K \tag{6.122}$$

If the slope equals the value $\dfrac{gL}{v^2}$, the vehicle is neutral steer. If the slope is larger than $\dfrac{gL}{v^2}$, the vehicle is considered to be understeer. If the slope is smaller than $\dfrac{gL}{v^2}$, the vehicle is considered to be oversteer, particularly when the slope is zero. This indicates that the vehicle is oversteer and that the vehicle is at the threshold of instability. The steer angle and yaw velocity also can be measured during the test to evaluate the handling characteristics.

6.8.3 Constant Steer Angle Test

In this test, the vehicle drives with a specific steer wheel angle at various forward speeds. The lateral accelerations at various speeds are measured. The curvature can be calculated from the lateral acceleration and vehicle speed and then plotted as a function of acceleration. The slope of the curve is given by

$$\frac{d\rho}{d\left(a_y\right)} = -\frac{K}{L} \tag{6.123}$$

If the slope is zero, the vehicle is considered to be neutral steer. If the slope is negative, the vehicle is considered to be understeer. If the slope is positive, the vehicle is considered to be oversteer.

6.8.3.1 Dynamic Testing

The dynamic response of a vehicle often is tested with many maneuvers, including step steer, slowly increasing steer, pulse steer, and double lane change input. In practice, a step steer input usually is used only to judge vehicles subjectively. Figure 6.30 represents a step steer input. Figure 6.31 shows the lateral acceleration, yaw velocity, roll angle,

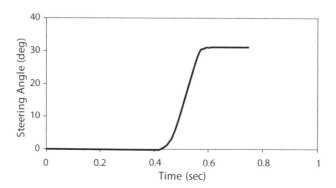

Figure 6.30 Step steer input.

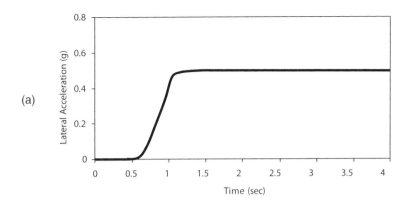

(a)

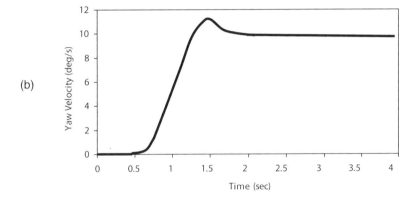

(b)

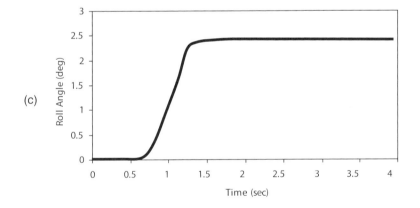

(c)

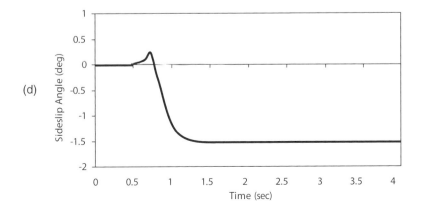

(d)

Figure 6.31 The responses of step steer input. (a) Lateral acceleration versus time. (b) Yaw velocity versus time. (c) Roll angle versus time. (d) Sideslip angle versus time.

and sideslip response to the step steer input. Excesses in yaw velocity, roll angle, and especially sideslip angle are felt as annoying. The vehicle behaves dynamically very well, with almost no excess at the roll angle and lateral acceleration, and with small excess at the yaw velocity and sideslip angle.

6.8.3.2 Simulations and Testing Validation

The trend toward multibody models in the vehicle industry is increasing due to commercially available, simpler, and easier-to-use interfaces, CAD software, and multibody dynamics software.

A popular model is NAVDyn (Non-Linear Analysis of Vehicle Dynamics) that addresses these issues. It includes the dynamics relevant to vehicle handling, with very few simplifying assumptions. The model includes relevant nonlinearity in the dynamic equations, as well as in the kinematics and compliances of the steering and suspension systems. In addition, the model allows for four-wheel independent braking and steering inputs.

In the following examples, the vehicle dynamics code employed is NADSdyna. This software was developed specifically for real-time vehicle simulation and is intended for use in the U.S. Department of Transportation (DOT)–sponsored National Advanced Driving Simulator (NADS). The core of NADSdyna is a numerical multibody code, RTRD, that has been developed using a recursive form of the Newton–Euler dynamic equations, based on joint–coordinate reference frames. NADSdyna is the vehicle dynamics program of NADS and includes vehicle subsystems (i.e., steering, brake, powertrain, engine, aerodynamic, and tire forces) that interact with the RTRD.

The RTRD was selected on a competitive basis by the government to be the core program for NADS for its generality and ability to systematically model a wide range of mechanical systems. The RTRD is used to precisely model chassis/suspension/steering linkages to capture geometric nonlinearity. It offers great capabilities in terms of rigid-body mechanism modeling, and it accounts for all nonlinear kinematics effects, such as roll steer, roll centers, and squat/lift forces. The RTRD can model either open- or closed-loop chains.

To validate the model, a test maneuver relevant to vehicle handling is applied because the model is intended only to represent performance on a smooth road. In particular, the same tests that were used to compare simulations were used for this validation. The following maneuvers are compared with simulation: slowly increasing steer, step steer, pulse steer, and double lane change. For limit performance maneuvers, only threshold or relatively close results should be expected, because the mechanics of these extreme conditions (especially tires) are very difficult to model. The yaw rate can be measured by a rate-gyro or calculated from the measured lateral acceleration divided by the vehicle speed. The lateral acceleration can be measured by an accelerometer.

For each maneuver, the driver control inputs were repeated, and the mean values and 95% confidence intervals were computed. The simulation is driven using the mean values. The exceptions to this statistical analysis are the slowly increasing steer and lane-change maneuvers. Because of the variability of the driver inputs, these maneuvers differ from one run to the next. Thus, several runs were performed, and a representative run is compared with the simulation using the same driver input.

EXAMPLE E6.10

A car has a weight of 8.449 kN on the front axle and 6.898 kN on the rear, with a wheelbase of 2.55 m. The tires have the following cornering stiffness values:

Load (N)	Cornering Stiffness (N/deg)	Cornering Coefficient (N/N/deg)
999	297.5	0.298
1998	537	0.269
2997	759	0.253
3996	999	0.250
4995	1141	0.228
5994	1332	0.222

Determine the following cornering properties for the vehicle: the Ackerman steer angle for 500-, 200-, 100-, and 15.36-m turn radius; the understeer gradient; the characteristic speed; the lateral acceleration gain at 96 km/h; the yaw velocity gain at 96 km/h; the sideslip angle at the center of gravity on a 246-m radius turn at 96 km/h; and the static margin.

Solution:

(a) The Ackerman steer angles can be solved as

$$\delta_{500} = \frac{L}{R} = 0.96 \text{ deg}$$

$$\delta_{200} = 2.4 \text{ deg}$$

$$\delta_{100} = 4.8 \text{ deg}$$

$$\delta_{50} = 9.6 \text{ deg}$$

(b) To find the understeer gradients, we need to know the cornering stiffness of the tires at the prevailing loads. On the front axle, the tire load is 4218 N per tire. Interpolating the cornering stiffness data between the loads of 3996 and 4995 N leads to a stiffness of 1030 N/deg at 4218 N. On the rear axle, the load is 3445 N per tire. Again interpolating between the appropriate loads in the tire data, we obtain a cornering stiffness of 8662 N/deg. Finally, we obtain

$$K = \frac{W_f}{C_{\alpha f}} - \frac{W_r}{C_{\alpha r}}$$

$$= \frac{950 \text{ lb}}{232 \text{ lb/deg}} - \frac{776 \text{ lb}}{195 \text{ lb/deg}}$$

$$= (4.09 - 3.98) \text{ deg}$$

$$= 0.11 \text{ deg}$$

(c) The characteristic speed is determined by

$$V_{char} = \sqrt{\frac{Lg}{K}} = 428.8 \text{ km/h}$$

(d) The lateral acceleration gain is

$$\frac{a_y}{\delta} = \frac{\dfrac{V^2}{57.3 \, Lg}}{1 + \dfrac{KV^2}{57.3 \, Lg}} = 7.448 \text{ m/s}^2/\text{deg}$$

(e) The yaw rate gain is given by

$$\frac{r}{\delta} = \frac{\dfrac{V}{L}}{1 + \dfrac{KV^2}{Lg}} = 9.95 \, \frac{\text{deg/s}}{\text{deg}}$$

(f) To obtain the sideslip angle, we need to find the value of c, which is the distance from the center of gravity to the rear axle. This is obtained from a simple moment balance about the front axle

$$c = 1.419 \text{ m}$$

$$\beta = \frac{c}{R} - \frac{W_r V^2}{C_{\alpha r} R_g} = -0.865 \text{ deg}$$

(g) To find the static margin, it is necessary to find the neutral steer point. The neutral steer point (nsp) is the point on the side of the vehicle where one can push laterally and produce the same slip angle at both the front and rear tires. From a moment balance in the plan view, we can show that the distance from the neutral steer point to the rear axle (c′) must be

$$c' = WB \frac{C_{af}}{C_{af} + C_{ar}} = 139.75 \text{ m}$$

The neutral steer point is 0.0215 m (0.8% of the wheelbase) behind the center of gravity.

EXAMPLE E6.11

A passenger car has an equal arm (parallel) independent front suspension and a conventional solid rear axle with a leaf spring suspension. The front suspension has a roll stiffness $K_{\phi f}$ of 168.5 mN/deg. The leaf springs have a rate of 20,240 N/m and a lateral separation of 1.012 m.

(a) What is the rear suspension roll stiffness?

(b) If the sprung mass is 12,210 N at a center-of-gravity height of 0.2024 m above the roll axis, what is the roll rate?

(c) Assuming a camber stiffness that is 10% of the cornering stiffness, estimate the understeer gradient due to camber effects.

(d) The rear leaf springs have an effective trailing arm angle of $-7°$. (The negative sign means the pivot of the arms is below the wheel center.) What is the understeer gradient due to rear roll steer?

Solution:

(a) The rear suspension roll stiffness can be computed as

$$K_\phi = 0.5 \, K_s s^2 = 20.3 \text{ mN/deg}$$

(b) The roll rate can be calculated as

$$\frac{d\phi}{da_y} = \frac{Wh_1}{K_{\phi f} + K_{\phi r} - Wh_1} = 1.07 \text{ deg/m/s}^2$$

(c) The understeer gradient due to camber effects can be estimated as

$$K_{camber} = \left(\frac{C_{\gamma f}}{C_{\alpha f}} \frac{\partial \gamma_f}{\partial \phi} - \frac{C_{\gamma r}}{C_{\alpha r}} \frac{\partial \gamma_r}{\partial \phi} \right) \frac{\partial \phi}{\partial a_y}$$

For an independent front suspension with parallel equal arms, the wheel does not incline with jounce and rebound. Therefore, the camber angle will change exactly with the roll angle, and the gradient for the front axle is 1. The rear axle is a solid axle that does not roll significantly. Therefore, its gradient is 0. As such,

$$K_{camber} = 0.107 \text{ deg/m/s}^2$$

(d) The understeer gradient due to roll steer on the rear axle is

$$K_{roll\,steer} = (\varepsilon_f - \varepsilon_r) \frac{d\phi}{da_y}$$

Because we are concerned with only the rear axle, only the second term must be determined. A solid axle will exhibit roll steer dependent on the effective angle of the imaginary trailing arms, which is $-7°$. Then,

$$K_{roll\,steer(rear)} = 0.13 \text{ deg/m/s}^2$$

EXAMPLE E6.12

A car has the following parameters:

Parameter	Symbol	Units	Type A
Mass	M	kg	1000
Inertia	I	kg m^2	1500
Distance from front axle to center of gravity	a	m	1.25
Distance from rear axle to center of gravity	b	m	1.25
Front cornering stiffness	C_f	kN/rad	53
Rear cornering stiffness	C_r	kN/rad	53
Velocity	U	m/s	20

This system has neutral steer, and the eigenvalues are –5.30 and –5.52. When the centers of gravity are changed to yield a = 1.15 m, b = 1.35 m, and a = 1.35 m, b = 1.15 m, respectively, derive the handling parameters.

Solution:

Parameter	Symbol	Unit	Understeer	Neutral Steer	Oversteer
Distance from front axle to center of gravity	a	m	1.15	1.25	1.35
Distance from rear axle to center of gravity	b	m	1.35	1.25	1.15
Stability margin	$bC_r - aC_f$	kNm/rad	10.6	0	–10.6
Understeer gradient	K	deg/g	0.85	0	–0.85
Critical velocity	U_{crit}	m/s	–	∞	41

EXAMPLE E6.13

Calculate the system eigenvalues as a function of velocity when the velocity changes from 10 to 50 m/s.

Solution:

The system eigenvalues as a function of velocity are plotted in Figure E6.13.

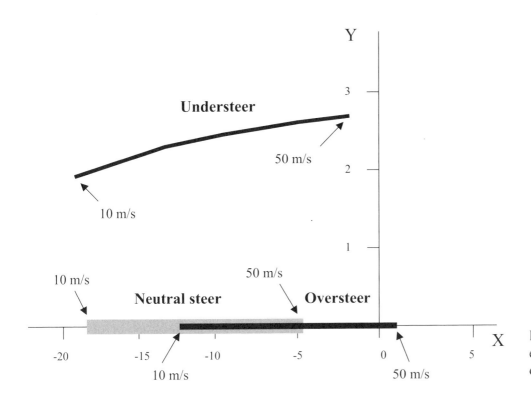

Figure E6.13 System eigenvalues as a function of velocity.

6.9 Summary

In this chapter, we presented the simplest handling model and investigated the steady-state handling characteristics, which included the yaw rate gain and understeer gradient, the difference of the slip angles of the front and rear wheels, the ratio of the radius of turn and lateral acceleration gain, and the curvature gain. We also described the dynamic characteristics by using the simplest handling model, which included the damping and natural frequency, time response, and frequency response properties, as well as stability features. The effect of the chassis system on handling characteristics then was discussed, and some of the overturning limit handling characteristics also were described briefly. Finally, the nonlinear modeling and testing of handling characteristics were presented.

6.10 References

1. Crolla, D. and Yu, V., *Vehicle Dynamics and Control*, China Communication Press, 2004.

2. Demerly, J.D. and Youcef-Toumi, Kamel, "Non-Linear Analysis of Vehicle Dynamics (NAVDyn): A Reduced Order Model for Vehicle Handling Analysis," SAE Paper No. 2000-01-1621, Society of Automotive Engineers, Warrendale, PA, 2000.

3. Elliott, A.S. and Wheeler, G., "Validation of ADAMS Models of Two USMC Heavy Logistic Vehicle Design Variants," SAE Paper No. 2001-01-2734, Society of Automotive Engineers, Warrendale, PA, 2001.

4. Gillespie, T.D., *Fundamentals of Vehicle Dynamics*, Society of Automotive Engineers, Warrendale, PA, 1992.

5. Gorder, K.V., Thomson, D., and Basas, J., "Steering and Suspension Test and Analysis," SAE Paper No. 2000-01-1626, Society of Automotive Engineers, Warrendale, PA, 2000.

6. Heydinger, G.J., Bixel, R.A., Durisek, N.J., Yu, E., and Guenther, D.A., "Effects of Loading on Vehicle Handling," SAE Paper No. 980228, Society of Automotive Engineers, Warrendale, PA, 1998.

7. Kamel, S.M., Guenther, D.A., and Heydinger, Gary J., "Vehicle Dynamics Modeling for the National Advanced Driving Simulator of a 1997 Jeep Cherokee," SAE Paper No. 1999-01-0121, Society of Automotive Engineers, Warrendale, PA, 1999.

8. Kamel, S.M. and Heydinger, G.J., "Model Validation of the 1997 Jeep Cherokee for the National Advanced Driving Simulator," SAE Paper No. 2000-01-0700, Society of Automotive Engineers, Warrendale, PA, 2000.

9. Kazemi, R., Hamedi, B., and Javadi, B., "Improving the Ride and Handling Qualities of a Passenger Car via Modification of Its Rear Suspension Mechanism," SAE Paper No. 2000-01-1630, Society of Automotive Engineers, Warrendale, PA, 2000.

10. Salaani, M.K., Christos, J.P., and Guenther, D.A., "Parameter Measurement and Development of a NADSdyna Validation Data Set for a 1994 Ford Taurus," SAE Paper No. 970564, Society of Automotive Engineers, Warrendale, PA, 1997.

11. Salaani, M.K., Grygier, P.A., and Heydinger, G.J., "Model Validation of the 1998 Chevrolet Malibu for the National Advanced Driving Simulator," SAE Paper No. 2001-01-0141, Society of Automotive Engineers, Warrendale, PA, 2001.

12. Sheng, G., Jiang, C.H., and Yang, S.Z., "Dynamic Modeling of Pass-By Vehicle on Bridge," *Journal of Vibration, Measurement & Diagnosis*, Vol. 12, No. 2, 1992, pp. 20–26.

13. Shi, H.M, Sheng, G., and Wu, Y., *Mechanical Vibration System: Analysis, Measurement, Modeling, and Control*, HUST Press, Wuhan, 1991.

14. Silvio, C., Data, Pascali, L., and Santi, C., "Handling Objective Evaluation Using a Parametric Driver Model for ISO Lane Change Simulation," SAE Paper No. 2002-01-1569, Society of Automotive Engineers, Warrendale, PA, 2002.

15. Vijayakumar, S. and Barak, P., "Application of Bond Graph Technique and Computer Simulation to the Design of Passenger Car Steering System," SAE Paper No. 2002-01-0617, Society of Automotive Engineers, Warrendale, PA, 2002.

16. Wong, J.Y., *Theory of Ground Vehicle*, 3rd Ed., John Wiley and Sons, New York, 2001.

17. Yang, S.Z., Shi, H.M., Yang, K., Wu, Y., and Sheng, G., *Modeling of Dynamic Mechanical Systems*, MEATU Press, Beijing, 1990.

18. Yu, Z.S., *Auto Theory*, Machine Press, Beijing, 1989.

Chapter 7

Braking

7.1 Introduction

The braking performance of a vehicle is one of its most important safety features. This performance is judged by drivers under various conditions. One condition closely related to vehicle safety is the emergency stop. Drivers expect certain behavior from the vehicle when under emergency. The direct expectation of the braking system under emergency conditions is simply avoiding an accident.

Performance of the braking system also is judged when under constant deceleration (i.e., normal stop) condition. Likewise, brakes can be used under normal driving conditions to adjust the speed of the vehicle in traffic.

Pure braking is simple to understand. The input to the braking system is a direct force at the braking pedal by the driver with the direct response of deceleration in the longitudinal motion. The braking mechanism transfers the braking force applied by the driver at the braking pedal into a torque applied on the wheel, preventing the wheels from rotation, either on a continuous basis or in a slip regulation fashion. This will result in large friction forces between the tires and the road, which reduces the speed of the vehicle.

Analyses of braking systems can be performed on various types of vehicles, including on-road and off-road vehicles. The emphasis of this book is on-road vehicles. Road vehicles can have one axle or multiple axles. They also can have a trailer. The fundamental theory presented and equations derived are focused on single-axle vehicles without a trailer. The same methods implemented here can be used to analyze multiple-axle vehicles, and they can be used for vehicles with trailers. Such analysis is considered beyond the scope of this book.

Although much of the treatment made here is intended for passenger cars, the analysis and theory can be used for light trucks, sport utility vehicles (SUVs), and other similar vehicles.

7.1.1 Types of Automotive Brakes

Automotive brakes used by standard road vehicles can be one of two types: drum brakes, or disk brakes. Figure 7.1 shows both types of brakes.

(a)   (b)

Figure 7.1 (a) Drum brakes, and (b) disk brakes. (Courtesy of Bosch Corporation.)

Each of the two types of brakes has advantages and disadvantages over the other type. In relation to braking performance, drum brakes require less actuation force (because of a higher brake factor and better mechanical advantage) on the brake pedal when compared with disk brakes. Also, installation of parking brake features is easier with drum brakes. On the other hand, disk brakes offer better (i.e., more consistent) distribution of forces on the wheel when compared with drum brakes.

Other motion control devises exist and frequently are used as auxiliary devices to help the braking action of the drum and/or disk brakes. Engine brakes are one version of such devices. In these brakes, the torque transmission can be interrupted by disengaging the clutch or selecting a neutral gear position. If the throttle is closed for a vehicle in motion, the engine will put a retarding force on the vehicle due to losses experienced to balance the kinetic energy of the vehicle. These forces are limited, and their effectiveness is increased if the compressor action of the engine is increased by shutting the exhaust. This type of retardation device is referred to as exhaust brakes, which consist of a throttle in the exhaust duct system that can be controlled mechanically or electronically. The brake action that results from this will depend on the gearing and vehicle speed. The brake torque generated by shutting the exhaust can reach up to 70% of the drive torque. Other improvements can be made to the camshaft timing to increase the compression and subsequently the retardation. This leads to improvement of the engine brake torque to be up to 100% of the drive torque. Brake torque devices have been used more on diesel engines. The use of engine brakes has been shown to reduce the use of foundation brakes by about 20% in normal driving conditions. Recent research on this auxiliary braking system includes the work of Moklegaard *et al.* (2000).

Another type of retardation device is the hydraulic retarders, which use viscous damping as the mechanism to produce the retarding torque. This retarder is similar to the hydraulic clutch. The retarding torque is produced by the rotor that pumps a fluid to a stator, which reflects back against the rotor, and a continuous internal pumping cycle is developed. The reaction forces and hence the retarding torque are absorbed by the rotor, which is connected to the wheels of the vehicle. The torque produced by hydraulic retarders becomes higher at high speeds, which is a clear advantage. Also, these retarders are independent of the vehicle engine, drivetrain, or electronic system; thus, they are robust in their action.

A third type of retardation is the electric retarders, which are based on producing eddy currents with a metal disc rotating between two electromagnets. This develops a retarding torque on the rotating disc. The eddy current results in heating the disc. The cooling of the disc is accomplished by means of convection heat transfer with ventilated rotors. These retarders need high temperatures to have effective cooling capacity. However, high temperatures reduce the retardation effectiveness.

These retarding devices are installed in conjunction with foundation disk and drum brakes to help maximize the braking force. They also are used to reduce the use of foundation braking devices, especially when light braking is needed.

7.1.2 Braking Distance and Deceleration

Among the attributes in which braking performance is measured are the braking distance and deceleration. If the vehicle is moving as a result of a consistent and constant braking force and is changing its velocity from V_i to V_f (which would be zero in the case of a complete stop), the acceleration a can be calculated as

$$a = \frac{\left(V_f - V_i\right)}{t} \tag{7.1}$$

where

$$t \quad = \text{ time}$$

$$V_f \ = \text{ final vehicle velocity}$$

$$V_i \ = \text{ initial vehicle velocity}$$

Note that this acceleration will be a negative number when the vehicle is decelerating. If the total braking force on both axles is F_b, then from Newton's second law of motion, the deceleration is $\dfrac{F_b}{M}$, where M is the mass of the vehicle.

The distance traveled during braking (X_b) can be calculated by

$$X_b = \frac{\left(V_f^2 - V_i^2\right)}{2 \times a} \tag{7.2}$$

EXAMPLE E7.1

A 2000-kg vehicle traveling at 100 km/h is stopped under a constant braking force of 10,000 N. Find both the time traveled during the deceleration and the stopping distance.

Solution:

The velocity can be written as

$$V_i = \frac{100 \text{ km}}{\text{hr}} \times \frac{1000 \text{ m}}{\text{km}} \times \frac{\text{hr}}{3600 \text{ sec}} = 27.78 \text{ m/sec}$$

The acceleration of the vehicle is calculated first as

$$a = -\frac{F_b}{M} = -\frac{10,000}{2000} = -5 \text{ m/sec}^2$$

The time needed to reach a complete stop is

$$t = \frac{\left(V_f - V_i\right)}{a} = \frac{(0 - 27.78)}{-5} = 5.56 \text{ sec}$$

The stopping distance is

$$X_b = \frac{\left(V_f^2 - V_i^2\right)}{2 \times a} = \frac{0 - (27.78)^2}{2 \times (-5)} = 77.16 \text{ m}$$

Requirements of stopping distance for various driving conditions have been established by regulatory institutions (e.g., FMVSS 121). More regulations can be found in the references used by Shaffer and Radlinski (2003). According to the latter reference, the early regulations of 1936/1937 included only five brake-related requirements, paraphrased by Shaffer and Radlinski (2003) as follows:

1. Adequate brakes are required on all vehicles, with two separate, independent means of application. If the application means are connected, failure of one part shall not leave the vehicle without adequate brakes.

2. Brakes on one or more units of a combination vehicle must be adequate to stop and hold the combination.

3. Adequate brakes defined by stopping distances and decelerations are listed as:

   Vehicles with brakes on all wheels must be capable of stopping in 30 ft from 20 mph or sustained deceleration of 14 ft/sec^2. Vehicles with brakes on some wheels must be capable of stopping in 45 ft from 20 mph or sustained deceleration of 9.5 ft/sec^2.

4. For a combination vehicle, the trailer brakes must come on first or at the same time as those on the towing vehicle.

5. The parking brake system must be able to hold the vehicle on any hill on which it is operated.

Later, in 1962, new regulations were established (e.g., Shaffer and Radlinski, [2003]). In these new regulations, the deceleration was no longer specified as sustained, but as "not less than," meaning peak, and the requirements for the new column, the percentage of the braking force (BF) with respect to the gross vehicle weight (GVW), also were "not less than." The stopping distance, which was listed as not more than the numbers listed in the Table, is now clarified as measured from the point of first movement of the brake pedal. There were seven categories of commercial vehicles (CVs) listed. Vehicles must meet three requirements: braking force, required deceleration, and stopping distance. The Federal Motor Carrier Safety Regulation (FMCSR) modified these later in 1971 to take the form of Table 7.1. Note that commercial vehicles are reduced to only six, three of which are classified as passenger-carrying vehicles and the remaining three as property-carrying vehicles.

7.2 Brake Torque Distribution

One of the most important characteristics of the brake system is the distribution of torque on the rim of the wheel to ensure certain deceleration behavior. The torque on the rim is a direct result of the actuation force. We will discuss this for both drum brake and disk brake systems.

7.2.1 Drum Brakes

Figure 7.2 shows the mechanical forces acting on the wheel as a result of applying a typical braking force in a drum brake.

Note that as the forces are applied on the brake pedal, they transfer directly through the brake system to the brake cylinders and translate into a force F applied on the brake shoes. Soon after the application of the brake force, the shoes are pushed against the rim. This generates the friction forces between the moving drum and the stationary shoe.

7.2.1.1 Mechanical Advantage

In reality, friction will occur along an arc of contact between the shoes of the drum brake and the wheel rim. For simplicity, let us assume that the friction actually occurs at points A and B during braking. We then can take the moment about the pivots (for shoe B) to obtain

$$\sum M_{Pivot} = eF + e_2\mu N_B - e_1 N_B = 0 \qquad (7.3)$$

where

M_{Pivot} = moment around the pivot

e = vertical distance between the pivot and the contact with the brake cylinder

e_1 = vertical distance between the pivot and the center of application of the friction force

e_2 = horizontal distance between the pivot and the center of application of the friction force

μ = coefficient of friction

N_B = normal force

TABLE 7.1

SAFETY REQUIREMENTS FOR DIFFERENT TYPES OF VEHICLES

Type of Motor Vehicle	Service Brake Systems			Emergency Brake Systems
	Braking Force as a Prcentage of Gross Vehicle or Combination Weight	Deceleration in Feet per Second	Application and Braking Distance in Feet from Initial Speed of 20 mph	Application and Braking Distance in Feet from Initial Speed of 20 mph
Passenger-Carrying Vehicles:				
(1) Vehicles with a seating capacity of 10 or fewer persons, including driver, and built on a passenger chassis	65.2	21	20	54
(2) Vehicles with a seating capacity of more than 10 persons, including driver, and built on a passenger car chassis; vehicles built on a truck or bus chassis and having a manufacturer's GVWR of 10,000 lb or less	52.8	17	25	66
(3) All other passenger-carrying vehicles	43.5	14	35	85
Property-Carrying Vehicles:				
(1) Single unit vehicles having a manufacturer's GVWR of 10,000 lb or less	52.8	17	25	66
(2) Single unit vehicles having a manufacturer's GVWR of more than 10,000 lb, except truck tractors; combinations of a two-axle towing vehicle and trailer having a GVWR of 3000 lb or less; all combinations of two or fewer vehicles in driveaway or towaway operation.	43.5	14	35	85
(3) All other property-carrying vehicles and combinations of property-carrying vehicles	43.5	14	40	90

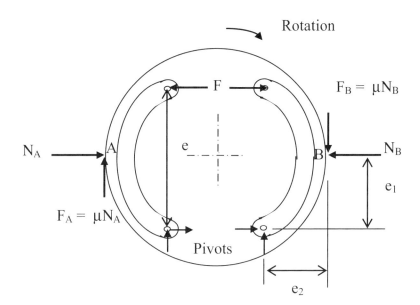

Figure 7.2 Forces acting on the shoes of a drum brake.

Note that the friction force F_B at that point of contact is

$$F_B = \mu N_B \tag{7.4}$$

Mathematical treatment of these equations gives the ratio of the friction force to the applied force as

$$\frac{F_B}{F} = \frac{\mu e}{e_1 - \mu e_2} \tag{7.5}$$

Similarly for the other shoe, we can find that

$$\frac{F_A}{F} = \frac{\mu e}{e_1 + \mu e_2} \tag{7.6}$$

Equations 7.5 and 7.6 define the mechanical advantage obtained by using the drum shoe mechanism.

EXAMPLE E7.2

In a drum shoe system where $e = 2e_1$ and $e_1 = e_2$, determine the mechanical advantage of both shoes. Assume a coefficient of friction equal to 0.5.

Solution:

$$\frac{F_B}{F} = \frac{\mu e}{e_1 - \mu e_2} = \frac{0.5 \times 2 \times e_1}{0.5 \times e_1} = 2$$

$$\frac{F_A}{F} = \frac{\mu e}{e_1 + \mu e_2} = \frac{0.5 \times 2 \times e_1}{1.5 \times e_1} = 0.67$$

This example shows that shoe B experiences higher mechanical advantage than shoe A. Also, if the coefficient of friction gets close to 1, the denominator gets closer to 0 and a high advantage is obtained. This will result in brake lock-up and wheel skid, which has a smaller friction coefficient and thus a smaller braking force.

One important observation should be made. Assume a clockwise rotation, and let us observe the force generation on the right-hand shoe, which frequently is referred to as the leading shoe. As the force F is applied, this will result in the reaction force N_B, which will result in the friction force ($F_B = \mu N_B$). This friction force will add to the total moment done around the pivot by the actuation force itself. To balance that moment, a higher value of N_B will be needed. This will create a higher friction force, a higher moment, and a higher N_B again. This system often is referred to as self energizing, and soon the maximum pressure of the brake shoe is obtained.

7.2.1.2 Torque Calculations

A more detailed analysis of forces on the shoes will require treatment of the arc of contact between the shoe and the wheel rim. We will conduct this analysis for the leading shoe (the right shoe in a clockwise rotation). Assume a shoe width of b and that the pressure is linearly proportional from the distance to the pivot. In other words, the pressure at the pivot will remain at zero, and the maximum pressure is achieved at an angle that is perpendicular to the line connecting the pivot with the center point (Figure 7.3). Equation 7.7 then applies for the pressure distribution on the shoe as

$$p = p_{max} \sin \theta \tag{7.7}$$

where p is the pressure at any angle θ on the lining, and p_{max} is the maximum pressure over the range of contact.

At an angle θ, the normal force is

$$dN = pbrd\theta \tag{7.8}$$

where b is the width of the shoe and r is the radius to the lining. Proper mathematical manipulation will yield the moment resulting from the frictional forces (M_F) over the arc length of the shoe contact with the drum as

$$M_F = \mu p_{max} br \int_{\theta_1}^{\theta_2} \sin \theta \; (r - a_1 \cos \theta) d\theta \tag{7.9}$$

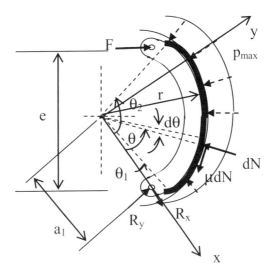

Figure 7.3 Pressure distribution along the brake shoe.

Note that the arm of the frictional force μdN is the expression $(r - a_1 \cos\theta)$. The moment produced by the normal forces (pressure) about the pivot is

$$M_N = p_{max} br a_1 \int_{\theta_1}^{\theta_2} \sin^2\theta \, d\theta \qquad (7.10)$$

Applying the summation of moments equation around the pivot will yield the actuation force equation

$$F = \frac{M_N - M_F}{e} \qquad (7.11)$$

Here, note that a brake system can be designed where it will be self-locked (i.e., it does not need an actuation force). This is achieved by setting M_N equal to M_f. The brake must be designed such that $M_N > M_f$ to prevent self-locking.

The torque T applied to the drum as a result of the frictional forces can be obtained by the integration product of the friction force and the radius to the drum of the friction area. This will yield

$$T = \mu p_{max} b \, r^2 \left(\cos\theta_1 - \cos\theta_2\right) \qquad (7.12)$$

The reaction forces at the pivot (hinge) can be obtained by summing the forces in the a and y directions as

$$R_x = p_{max} b \, r \left(\int_{\theta_1}^{\theta_2} \sin\theta \, \cos\theta \, d\theta - \mu \int_{\theta_1}^{\theta_2} \sin^2\theta \, d\theta \right) - F_x \qquad (7.13)$$

$$R_y = p_{max}b \quad r\left(\mu \int_{\theta_1}^{\theta_2} \sin\theta \, \cos\theta \, d\theta + \int_{\theta_1}^{\theta_2} \sin^2\theta \, d\theta \right) - F_y \qquad (7.14)$$

Derivation can be made for the left shoe. The equations of M_f, M_N, and the torque T will remain the same as those of the right shoe. (Note that the maximum pressure on the left shoe usually is less than that on the right shoe in a clockwise rotation.) The following are the resulting equations for the actuation force and the hinge reactions. The actuation force is

$$F = \frac{M_N + M_F}{e} \qquad (7.15)$$

Note here that the left shoe is not self-energizing and needs an actuation force to balance the moments. The hinge (i.e., pivot) reactions are

$$R_x = p_{max}b \quad r\left(\int_{\theta_1}^{\theta_2} \sin\theta \, \cos\theta \, d\theta + \mu \int_{\theta_1}^{\theta_2} \sin^2\theta \, d\theta \right) - F_x \qquad (7.16)$$

$$R_y = p_{max}b \quad r\left(-\mu \int_{\theta_1}^{\theta_2} \sin\theta \, \cos\theta \, d\theta + \int_{\theta_1}^{\theta_2} \sin^2\theta \, d\theta \right) - F_y \qquad (7.17)$$

Note here that the actuation force on both shoes is the same. However, the maximum pressure in the left shoe is not the same as that of the right shoe. The limiting maximum pressure is always in the right shoe (for clockwise rotation).

EXAMPLE E7.3

Consider the brake system shown in Figure E7.3. The width of the shoe system is 30 mm, and the coefficient of friction is 0.3. The friction material used permits a maximum pressure value of 1000 kPa. Find the actuation force, the total torque applied by the brake system, and the reaction force at the pivots of both shoes.

Solution:

First we will analyze the right-hand shoe. The distance a_1 from the pivot to the center is

$$a_1 = \sqrt{(100)^2 + (50)^2} = 111.80 \text{ mm}$$

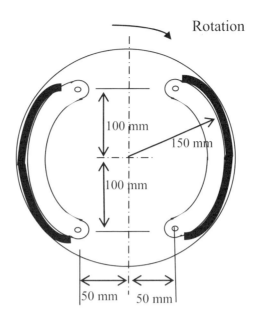

Figure E7.3 Braking system described in Example E7.3.

The angles $\theta_1 = 0$, and θ_2 can be found as

$$\theta_2 = 180 - 2\tan^{-1}\left(\frac{50}{100}\right) = 180 - 2 \times 26.5 = 127 \text{ deg}$$

Next, we need to find the moment from the friction forces

$$M_F = \mu p_{max} br\left[\left(-r\cos\theta\right)_0^{\theta_2} - a_1\left(\frac{\sin^2\theta}{2}\right)_0^{\theta_2}\right] = \mu p_{max} br\left[r - r\cos\theta_2 - \frac{a_1}{2}\sin^2\theta_2\right]$$

$$= 0.3 \times 10^6 \times 0.03 \times 0.15\left[0.15 - 0.15\cos 127° - \frac{0.118}{2}\sin^2 127°\right]$$

$$= 275.7 \text{ Nm}$$

The moment from the normal forces is

$$M_N = p_{max} bra_1\left[\frac{\theta}{2} - \frac{1}{4}\sin 2\theta\right]_0^{\theta_2} = p_{max} bra_1\left[\frac{\theta_2}{2} - \frac{1}{4}\sin 2\theta_2\right]$$

$$= 10^6 \times 0.03 \times 0.15 \times 0.1118\left[\frac{1}{2}\frac{\pi \times 127}{180} - \frac{1}{4}\sin\left(2 \times 127°\right)\right]$$

$$= 677.8 \text{ Nm}$$

The applied or actuation force is

$$F = \frac{M_N - M_f}{e} = \frac{677.8 - 275.7}{0.2} = 2010.4 \text{ N}$$

Note that this force is the resultant force, which can be broken into x and y components (i.e., F_x and F_y). The reactions at the pivot of the right-hand shoe are

$$R_x = p_{max} b \ r \left(\int_{\theta_1}^{\theta_2} \sin\theta \ \cos\theta \ d\theta - \mu \int_{\theta_1}^{\theta_2} \sin^2\theta \ d\theta \right) - F_x$$

$$= 1000 \times 0.03 \times 0.15 \left(\frac{1}{2} \sin^2 127 - 0.3 \left(\frac{\pi \times 127}{2 \times 180} - \frac{1}{4} \sin(2 \times 127) \right) \right) - 2.010 \sin 26.5$$

$$= -1.2778 \text{ kN}$$

$$R_y = p_{max} b \ r \left(\mu \int_{\theta_1}^{\theta_2} \sin\theta \ \cos\theta \ d\theta + \int_{\theta_1}^{\theta_2} \sin^2\theta \ d\theta \right) - F_y$$

$$= 1000 \times 0.03 \times 0.15 \left(0.3 \left(\frac{1}{2} \sin^2 127 \right) + \left(\frac{\pi \times 127}{2 \times 180} - \frac{1}{4} \sin(2 \times 127) \right) \right) - 2.010 \cos 26.5$$

$$= 4.6961 \text{kN}$$

The resultant force at the right pivot is

$$R = \sqrt{R_x^2 + R_y^2}$$

$$= \sqrt{(-1.2778)^2 + (4.6961)^2}$$

$$= 4.8668 \text{ kN}$$

Next, we will find the torque of the right-hand shoe T_R as

$$T_R = \mu p_{max} b r^2 \left(\cos\theta_1 - \cos\theta_2 \right)$$

$$= 0.3 \times 1000 \times 0.032 \times (0.15)^2 \left(\cos(0) - \cos(127) \right)$$

$$= 0.3240 \text{ kN} \cdot \text{m}$$

Before finding any of the values of the left-hand shoe, we will need to find the maximum pressure on that shoe. We will use the fact that the actuation force on both shoes is the same to determine the maximum pressure on the left shoe as

$$M_F = \mu p_{max} br \left[\left(-r\cos\theta \right)_0^{\theta_2} - a_1 \left(\frac{\sin^2\theta}{2} \right)_0^{\theta_2} \right]$$

$$= \mu p_{max} br \left[r - r\cos\theta_2 - \frac{a_1}{2}\sin^2\theta_2 \right]$$

$$= 0.3 \times p_{max} \times 0.03 \times 0.15 \left[0.15 - 0.15\cos 127° - \frac{0.1118}{2}\sin^2 127° \right]$$

$$= 2.757 \times 10^{-4} p_{max}$$

The moment from the normal forces is

$$M_N = p_{max} b\, r\, a_1 \left[\frac{\theta}{2} - \frac{1}{4}\sin 2\theta \right]_0^{\theta_2}$$

$$= p_{max} b\, r\, a_1 \left[\frac{\theta_2}{2} - \frac{1}{4}\sin 2\theta_2 \right]$$

$$= p_{max} \times 0.03 \times 0.15 \times 0.1118 \left[\frac{1}{2}\frac{\pi \times 127}{180} - \frac{1}{4}\sin\left(2 \times 127°\right) \right]$$

$$= 6.778 \times 10^{-4} p_{max}$$

The applied or actuation force is

$$F = \frac{M_N + M_f}{e} = \frac{(6.778 + 2.757) \times 10^{-4} p_{max}}{0.2} = 2010.4 \text{ N}$$

or

$$p_{max} = \frac{2010.4 \times 0.2}{(6.778 + 2.757) \times 10^{-4}} = 4.21691 \times 10^5 = 421.7 \text{ kPa}$$

Thus, the torque on the left-hand shoe is

$$T_L = \mu p_{max} br^2 \left(\cos\theta_1 - \cos\theta_2 \right)$$

$$= 0.3 \times 421 \times 0.032 \times (0.15)^2 \left(\cos(0) - \cos(127) \right)$$

$$= 0.1366 \text{ kN} \cdot \text{m}$$

The total torque on the shoes is

$$T = T_L + T_R = 0.4606 \text{ kN} \cdot \text{m}$$

Finally, we need to find the reaction forces on the pivot of the left-hand shoe as

$$R_x = p_{max} b\, r \left(\int_{\theta_1}^{\theta_2} \sin\theta\, \cos\theta\, d\theta + \mu \int_{\theta_1}^{\theta_2} \sin^2\theta\, d\theta \right) - F_x$$

$$= 421.7 \times 0.03 \times 0.15 \left(\frac{1}{2}\sin^2 127 + 0.3\left(\frac{\pi \times 127}{2 \times 180} - \frac{1}{4}\sin(2 \times 127) \right) \right) - 2.0104 \sin 26.5$$

$$= 0.4751\ kN$$

$$R_y = p_{max} b\, r \left(\mu \int_{\theta_1}^{\theta_2} \sin\theta\, \cos\theta\, d\theta - \int_{\theta_1}^{\theta_2} \sin^2\theta\, d\theta \right) - F_y$$

$$= 421.7 \times 0.03 \times 0.15 \left(-0.3\left(\frac{1}{2}\sin^2 127 \right) + \left(\frac{\pi \times 127}{2 \times 180} - \frac{1}{4}\sin(2 \times 127) \right) \right) - 2.0104 \cos 26.5$$

$$= 0.5761\ kN$$

The resultant force at the right pivot is

$$R = \sqrt{R_x^2 + R_y^2}$$

$$= \sqrt{(0.4751)^2 + (0.5761)^2}$$

$$= 0.7467\ kN$$

Note that the left-hand shoe experiences less pressure and less reaction forces on its pivot than the right-hand shoe.

Let us use the parameters from the preceding example to find the total torque as a function of the coefficient of friction. Table 7.2 shows the variations in both the application force (needed to achieve a maximum pressure of 1000 kPa in the right-hand shoe) and the total torque.

The most important observation is to note that the actuation force is related linearly to the coefficient of friction. However, the total torque has a nonlinear relation with the coefficient of friction.

A customer's "brake feel" is related mostly to the brake force needed to generate a braking action. The brake designer can choose two shoes for the B and one for the A, if a higher advantage (usually referred to as a brake factor) is desired. Brake designers should be careful because a higher advantage may result in a less robust design, particularly when the heat generated in the system slightly alters the coefficient of friction. Also, this may result in the undesired noise of brake squeal.

7.2.2 Disk Brakes

A more robust design in terms of the actuation force needed is that of a disk brake. In this type of brake system, the mechanical advantage is practically 1, and slight

TABLE 7.2
RELATION AMONG COEFFICIENT OF FRICTION,
ACTUATION FORCE, AND TOTAL TORQUE

Coefficient of Friction	Actuation Force (F)	Total Torque (T)
0	3388.9	0
0.05	3159.1	101.14
0.1	2929.4	190.21
0.15	2699.6	269.24
0.2	2469.9	339.84
0.25	2240.0	403.29
0.3	2010.4	460.63
0.35	1780.6	512.69
0.4	1550.8	560.18
0.45	1321.1	603.66
0.5	1091.3	643.63

variation of the coefficient of friction will not result in magnified application force. For disk brakes, however, the application force is relatively high.

Disk brakes can be implemented by applying a frictional disk over the whole exposed area of the disk of the driven spindle. This is not common in automotive engineering mainly because of the high energy that needs to be absorbed. For that reason, the disk brake shoes are applied only to a limited surface of the disk of the driven spindle. This will expose the remaining surface to air to dissipate heat and cool the surface. Caliper disk brakes are used extensively in the automotive industry, particularly for the front wheels. If the friction surface area of contact between the brake shoes and the driven disk is A and the applied force is F, then the pressure over the shoe pads is

$$ p = \frac{F}{A} \tag{7.18a} $$

The torque applied at the wheel is

$$ t = \mu F \, r_{ave} \tag{7.18b} $$

where r_{ave} is the average radius from the center of the moving element to the contacting surface. Note that the torque applied is linearly dependent on the coefficient of friction; thus, variations in this coefficient due to heat will not be magnified as was observed for drum brakes due to the nonlinearity.

7.2.3 Consideration of Temperature

One important consideration that should be made when designing a braking system is that of heat generation and temperature rise. This is important for two reasons. The first is that temperature rise affects the coefficient of friction considerably and thus affects the actuation force needed for a certain braking action. The second is that the generated heat is the main reason for cumulative damage in the braking system.

The energy that needs to be absorbed by the braking systems of the vehicle is the kinetic energy (KE) of the vehicle

$$KE = \frac{1}{2} M \left(V_i^2 - V_f^2 \right) \tag{7.19}$$

where

M = mass of the vehicle

V_f = zero if the vehicle is to reach a complete stop

This energy is to be absorbed by the braking system over the braking period and will be transferred into heat energy at the braking system, resulting in a temperature rise.

The power absorption is a function of the braking force and the speed. If the braking force is constant over the braking action, the power absorption is highest at the beginning of the braking action, when the speed is maximum. The power at that point is the direct product of the speed times the braking force. Over the time interval of the braking system, the average power can be calculated as the kinetic energy divided by the time interval.

EXAMPLE E7.4

A large sedan of 1779.29 N (4000 lb) traveling at 128.75 km/hr (80 mph) is to reach a complete stop in 10 seconds. Find the energy and average power.

Solution:

$$M = \frac{4000 \text{ lb}}{32.2 \text{ ft/sec}^2} = 124.22 \text{ slugs}$$

$$V_i = \frac{80 \text{ mile}}{hr} \times \frac{5280 \text{ ft}}{mile} \times \frac{hr}{3600 \text{ sec}} = 117.33 \text{ ft/sec}$$

$$KE = \frac{1}{2} M \left(V_i^2 - V_f^2 \right)$$

$$= \frac{1}{2} \times 124.22 \frac{\text{lb} \cdot \text{sec}^2}{\text{ft}} \times \left(117.33 \frac{\text{ft}}{\text{sec}} \right)^2$$

$$= 855,075 \text{ ft} \cdot \text{lb}$$

$$Power = \frac{855,075 \text{ ft} \cdot \text{lb}}{10 \text{ sec}} \frac{HP}{550 \frac{\text{ft} \cdot \text{lb}}{\text{sec}}}$$

$$= 155 \text{ HP}$$

The temperature rise at the brake can be estimated using Eq. 7.20 as

$$\Delta \text{Temp} = \frac{KE}{Sm}$$ (7.20)

where

S = specific heat (500 J/kg C° for steel)

m = mass of the braking system

EXAMPLE E7.5

Estimate the temperature rise of a vehicle with a mass of 2000 kg and traveling at 100 km/hr, to reach a complete stop in 10 seconds. The braking system has a mass of 10 kg. Let S = 500 J/(kg C°).

Solution:

$$V_i = \frac{100 \text{ km}}{\text{hr}} \times \frac{1000 \text{ m}}{\text{km}} \times \frac{\text{hr}}{3600 \text{ sec}} = 27.78 \text{ m/sec}$$

$$KE = \frac{1}{2} M \left(V_i^2 - V_f^2 \right)$$

$$= \frac{1}{2} \times 2000 \text{ kg} \times \left(27.78 \frac{\text{m}}{\text{sec}} \right)^2$$

$$= 771,605 \text{ J}$$

$$\Delta \text{Temp} = \frac{771,605}{500 \times 10} = 154.3 \text{ C}°$$

The coefficient of friction is a function of various environmental factors, most notably the velocity of the vehicle and the temperature of the shoes. The challenge to the application engineer is to try to find the value of the brake coefficient. Unfortunately, the application of a high brake force will result in high temperatures (exceeding 600°F) and permanent changes to the braking torque. Recently, advanced finite element software packages have been used to predict the temperature and the resulting changes of the brake torque performance of the system.

7.3 Load Transfer During Braking

The fundamental forces acting on a vehicle when undergoing braking were discussed in Section 3.3.5 of Chapter 3. Some of these forces will be reviewed briefly here. First we will consider braking of a vehicle on a horizontal road (i.e., no grade).

7.3.1 Simple Braking on a Horizontal Road

Figure 7.4 gives the free body diagram of the vehicle that shows the forces acting on the vehicle under normal driving conditions. Note that we have neglected aerodynamic forces, rolling resistance, and powerplant resistance in this simple vehicle model. Applying Newton's second law in the horizontal direction yields

$$F_{x1} + F_{x2} = Ma \qquad (7.21)$$

where

$$F_{x1} = \text{friction force on the front wheels}$$

$$F_{x2} = \text{friction force on the rear wheels}$$

$$a = \text{acceleration}$$

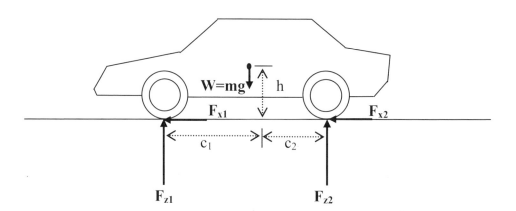

Figure 7.4 A simple vehicle model on a horizontal plane.

Applying Newton's second law in the vertical direction yields

$$F_{z1} + F_{z2} = Mg \qquad (7.22)$$

where

$$F_{z1} = \text{vertical force on the front wheels}$$

$$F_{z2} = \text{vertical force on the rear wheels}$$

$$g = \text{acceleration of gravity}$$

Summing the moment about the center of gravity point yields

$$F_{z1}c_1 - F_{z2}c_2 + \left(F_{x1} + F_{x2}\right)h = 0 \qquad (7.23)$$

where

$$c_1 = \text{horizontal (parallel to the road surface) distance between the contact point of the front wheel and the center of gravity}$$

c_2 = horizontal distance between the contact point of the rear wheel and the center of gravity

h = vertical (perpendicular to the road surface) distance between the contact points of the contact of the wheels with the ground and the center of gravity

Note that Eqs. 7.21 through 7.23 involve four unknowns: F_{x1}, F_{x2}, F_{z1}, and F_{z2}. This is an indeterminate system, where one additional equation is needed to solve for all the unknowns. Fortunately, two of the unknowns can be found by substituting the first equation into the third equation. This yields

$$F_{z1}c_1 - F_{z2}c_2 + Mah = 0 \qquad (7.24)$$

Equations 7.22 and 7.24 can be used directly to solve for the vertical forces (axle loads) independently as

$$F_{z1} = Mg\left(\frac{c_2}{c_1 + c_2}\right) - \frac{h}{c_1 + c_2}Ma$$

$$\qquad (7.25)$$

$$F_{z2} = Mg\left(\frac{c_1}{c_1 + c_2}\right) + \frac{h}{c_1 + c_2}Ma$$

Note that we are assuming that the vehicle is accelerating. Deceleration will be treated simply by using a negative value for a.

Also note that when the vehicle is braking and negative a is used, the axle force on the front wheels will be larger than those forces experienced under cruising driving conditions (i.e., no acceleration or deceleration). This is felt by drivers, particularly when the vehicle reaches a complete stop.

EXAMPLE E7.6

Consider a midsize sedan that weighs 14,234.31 (3200 lb N) under a certain load. The wheelbase of the vehicle is 9 ft or 108 in. (2.74 m), and the center of gravity is 1.067 m (3.5 ft) from the front axle and 0.508 m (20 in.) above the ground. Determine the load on each axle if the brake is under no acceleration, decelerating 1.524 m/s² (5 ft/sec²), 3.048 m/s² (10 ft/sec² 3), and 4.572 m/ s² (15 ft/sec²).

Solution:

The following are given in the problem:

$\quad$ W $\quad$ = 3200 lb

$\quad$ $c_1 + c_2$ = 108 in.

$\quad$ h $\quad$ = 20 in.

$\quad$ c_1 $\quad$ = 42 in.

First we will find the mass and c_2

$$M = \frac{3200}{32.2} = 99.38 \text{ slugs}$$

$$c_2 = 108 - 42 = 66 \text{ in.}$$

Apply Eq. 7.25 to obtain the loading under no acceleration (i.e., a = 0) as

$$F_{z1} = Mg\left(\frac{c_2}{c_1 + c_2}\right) - \frac{h}{c_1 + c_2}Ma$$

$$= 99.38 \times 32.2\left(\frac{66}{108}\right) - \frac{20}{108} \times 99.38 \times 0$$

$$= 1955.56 \text{ lb}$$

$$F_{z2} = Mg\left(\frac{c_1}{c_1 + c_2}\right) + \frac{h}{c_1 + c_2}Ma$$

$$= 99.38 \times 32.2\left(\frac{42}{108}\right) + \frac{20}{108} \times 99.38 \times 0$$

$$= 1244.44 \text{ lb}$$

Similarly for a = –5 ft/sec^2,

$$F_{z1} = 99.38 \times 32.2\left(\frac{66}{108}\right) - \frac{20}{108} \times 99.38 \times (-5)$$

$$= 2047.57 \text{ lb}$$

$$F_{z2} = 99.38 \times 32.2\left(\frac{42}{108}\right) + \frac{20}{108} \times 99.38 \times (-5)$$

$$= 1152.43 \text{ lb}$$

For a = –10 ft/sec^2,

$$F_{z1} = 99.38 \times 32.2\left(\frac{66}{108}\right) - \frac{20}{108} \times 99.38 \times (-10)$$

$$= 2139.59 \text{ lb}$$

$$F_{z2} = 99.38 \times 32.2\left(\frac{42}{108}\right) + \frac{20}{108} \times 99.38 \times (-10)$$

$$= 1060.41 \text{ lb}$$

Finally, for a = –15 ft/sec^2,

$$F_{z1} = 99.38 \times 32.2 \left(\frac{66}{108}\right) - \frac{20}{108} \times 99.38 \times (-15)$$

$$= 2231.61 \text{ lb}$$

$$F_{z2} = 99.38 \times 32.2 \left(\frac{42}{108}\right) + \frac{20}{108} \times 99.38 \times (-15)$$

$$= 968.39 \text{ lb}$$

Note that in the previous example, the total load on the axles remains at 3200 lb, which is the weight of the vehicle. In reality, what has happened is that load transferred from the rear axle to the front axle as the vehicle was decelerating. Figure 7.5 shows the load distribution as a function of the deceleration.

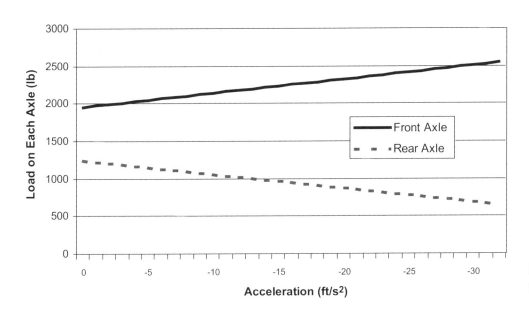

Figure 7.5 Load distribution under braking.

7.3.2 Effect of Aerodynamic and Other Forces

A typical driver notices that speed reduction can be achieved without braking action. In other words, if the vehicle is not cruising and the driver takes his or her foot off the gas pedal, deceleration is observed. This leads us to conclude that other resistance forces cause the vehicle to decelerate.

In the previous section, we studied a simple vehicle model where the vehicle is on a horizontal plane and under its own weight and braking forces without other loading conditions. Other forces should be accounted for in actual driving conditions. Even on a horizontal surface, the rolling resistance of the tires, the aerodynamic resistance, and the driveline drag should be accounted for.

7.3.2.1 Rolling Resistance

The rolling resistance is between the tires and the drive axle. If the rolling coefficient is distinguished by μ_r, then the rolling resistance can be estimated by

$$R_r = \mu_r \, Mg \tag{7.26}$$

The rolling resistance is caused by the hysteresis in the tire materials that results from the deflection of the tire while rolling. Friction between the tire and the road when sliding and resistance due to air circulating inside the tire contribute minimally to the rolling resistance.

The tire structure, including its construction and material and the operating conditions, is found to be the major factor that affects the rolling resistance. Operating conditions include the road surface, temperature, pressure, and speed. The coefficient of rolling resistance varies between 0.012 and 0.02 on concrete surfaces, with radial tires offering lower coefficients than bias tires. This coefficient reaches 0.1 when the vehicle is driven on hard soil, and it reaches 0.2 to 0.3 when the vehicle is driven on sand. The rolling resistance usually is less for tires with large diameters.

7.3.2.2 Aerodynamic Drag

Aerodynamic forces (drag) also should be considered when studying the braking of a vehicle. The aerodynamic resistance is generated mainly by air flow around the vehicle and air flow through the vehicle. Air flow around the vehicle is mainly the flow of air over the exterior of the vehicle. Air flow through the vehicle is basically the flow of air through the engine radiator for the purposes of heating, cooling, and ventilation. The first of the two accounts for the main aerodynamic resistance. While skin friction forces are important, the normal pressure of the air on the vehicle acts against the motion of the vehicle and contributes the most to the aerodynamic drag.

Aerodynamic resistance is estimated as

$$R_a = \frac{\rho}{2} C_D A_F V_r^2 \tag{7.27}$$

where

$$\rho \quad = \quad \text{density of the air}$$

$$C_D = \quad \text{drag coefficient}$$

$$A_F = \quad \text{projected area of the front of the vehicle}$$

$$V_r \quad = \quad \text{relative velocity of the vehicle with respect to the wind}$$

The air density is a function of the altitude and temperature. Normal operating temperatures can lead to approximately 20% variation in the density and similarly for altitude. Low temperatures at low altitudes give rise to higher density. The coefficient of drag historically has been a subject of many studies. An aerodynamically designed vehicle can reach a coefficient of drag that is as low as 0.3. It also can go as high as 0.5 if the car is not designed properly for aerodynamic resistance. This coefficient reaches the value of 0.8 for vans. The frontal area of the vehicle is the area that a person sitting in front of the vehicle sees. It can go from 1.7 m^2 for a mini car to a value of 2.2 m^2 for a full-size luxury car. The most important observation from Eq. 7.27 is to note that the aerodynamic resistance is a function of the square of the speed.

Note that equations for aerodynamic lift can be written similarly to that of Eq. 7.27, that is,

$$R_L = \frac{\rho}{2} C_L A_F V_r^2 \qquad (7.28)$$

The aerodynamic lift acts upward on the vehicle. The coefficient of aerodynamic lift (C_L) usually is determined in wind tunnel testing.

Similarly, the aerodynamic pitching can be calculated as

$$M_a = \frac{\rho}{2} C_M L_C V_r^2 \qquad (7.29)$$

where

L_C = characteristic length of the vehicle

C_M = coefficient of aerodynamic pitching moment obtained from wind tunnel testing

7.3.2.3 Powertrain Resistance

The powertrain of the vehicle consists of its powerplant (engine and transmission) and the driveline. These are the rotating components of the vehicle. Naturally, internal friction is present in these elements, mostly in the gearing actions of the transmission or power takeoff or axle units in a 4 × 4 drive system.

The resistance of the powertrain usually is negligible in the consideration of the braking system of a typical vehicle.

7.3.2.4 Load Transfer on a Horizontal Plane

Now we will consider the aerodynamic drag and rolling resistance in developing the load transfer equations. Figure 7.6 shows the vehicle model on a horizontal plane, with consideration of the rolling resistance and aerodynamic drag.

Note that the resultant of the aerodynamic drag is represented by the force R_a acting at some height h_a. These are found either from wind tunnel testing or, more recently, from computational fluid dynamics (CFD) software packages. The vertical forces (axle loads) can be found in a manner similar to that used in Section 7.3.1 as

$$F_{z1} = Mg\left(\frac{c_2}{c_1 + c_2}\right) - \frac{h_a}{c_1 + c_2} R_a - \frac{h}{c_1 + c_2} Ma$$
$$F_{z2} = Mg\left(\frac{c_1}{c_1 + c_2}\right) + \frac{h_a}{c_1 + c_2} R_a + \frac{h}{c_1 + c_2} Ma \qquad (7.30)$$

Deceleration will be treated simply by using a negative value for a. Often, h_a is assumed to be the same as h.

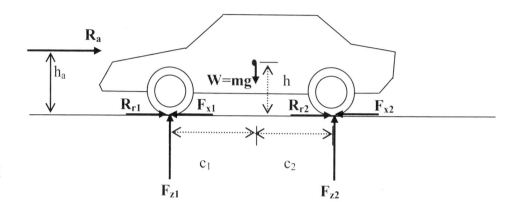

Figure 7.6 Vehicle model on a horizontal plane.

7.3.3 Effect of Grade

Now we will consider the effect of grade as well as aerodynamic drag and rolling resistance in developing the load transfer equations. Figure 7.7 shows the vehicle model on an inclined surface with consideration of rolling resistance and aerodynamic drag.

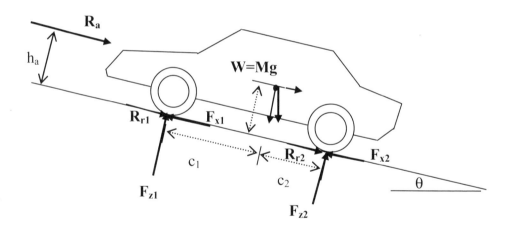

Figure 7.7 Vehicle model on an incline (grade).

The vertical forces (axle loads) can be found in a manner similar to that used in Section 7.3.1 as

$$-F_{z1}(c_1 + c_2) + Mgc_2 \cos\theta - h_a R_a - hMg\sin\theta = hMa$$

$$F_{z2}(c_1 + c_2) - Mgc_2 \cos\theta - h_a R_a - hMg\sin\theta = hMa$$

(7.31a)

or

$$F_{z1} = \frac{1}{c_1 + c_2}\left(Mgc_2 \cos\theta - h_a R_a - hMa - hMg\sin\theta\right)$$

$$F_{z2} = \frac{1}{c_1 + c_2}\left(Mgc_1 \cos\theta + h_a R_a + hMa + hMg\sin\theta\right)$$

(7.31b)

where

θ = grade or slope angle

R_a = aerodynamic drag

h_a = vertical distance (perpendicular to the road surface) between the road and the center of the aerodynamic force

Note that when the vehicle is going uphill, the negative sign should be used for the last term. Deceleration will be treated simply by using a negative value for a.

EXAMPLE E7.7

Consider a midsize sedan that weighs 14,234.31 N (3200 lb) under a certain load. The vehicle is going up a hill with a 5° slope at 26.82 m/s (60mph, 88 ft/sec). The frontal area of the vehicle is 2.044 m² (22.0 ft²), and the coefficient of drag is 0.4. The aerodynamic resistance is acting on the center of gravity. The wheelbase of the vehicle is 2.74 m (9 ft), and the center of gravity is 1.372 m (4.5 ft) from the front axle and 0.508 m (20 in.) above the ground. Determine the load on each axle if the vehicle is under no acceleration. What would the loads be if the vehicle is decelerating at 3.048 m/s² (10 ft/sec²). The density of air is 1.226 kg/m³ (0.002378 slug/ft³). Calculate the forces when going downhill at a 5° slope with the same speed.

Solution:

The aerodynamic resistance first is calculated as

$$R_a = \frac{\rho}{2} C_D A_F V_r^2$$

$$= \frac{0.002378}{2} \times 0.4 \times 22 \times (88)^2$$

$$= 81.03 \text{ lb}$$

The forces on the axles when traveling on a horizontal road are

$$F_{z1} = \frac{1}{c_1 + c_2} \left(Mgc_2 \cos\theta - h_a R_a - hMa - hMg\sin\theta \right)$$

$$= \frac{1}{c_1 + c_2} \left(Mgc_2 - h_a R_a - hMa \right)$$

$$= \frac{1}{108} \left(3200 \times 66 - 20 \times 81 - 20 \times 99.38 \times (-10) \right) = 2124.6 \text{ lb}$$

$$F_{z2} = \frac{1}{c_1 + c_2}\left(Mgc_1\cos\theta + h_aR_a + hMa + hMg\sin\theta\right)$$

$$= \frac{1}{c_1 + c_2}\left(Mgc_1 + h_aR_a + hMa\right)$$

$$= \frac{1}{108}\left(3200 \times 42 + 20 \times 81 + 20 \times 99.38 \times (-10)\right) = 1075.4 \text{ lb}$$

The forces on the axles when traveling uphill are

$$F_{z1} = \frac{1}{c_1 + c_2}\left(Mgc_2\cos\theta - h_aR_a - hMa - hMg\sin\theta\right)$$

$$= \frac{1}{108}\left(3200 \times 66 \times \cos(5) - 20 \times 81 - 20 \times 99.38 \times (-10) - 20 \times 3200 \times \sin(5)\right)$$

$$= 2065.5 \text{ lb}$$

$$F_{z2} = \frac{1}{c_1 + c_2}\left(Mgc_1\cos\theta + h_aR_a + hMa + hMg\sin\theta\right)$$

$$= \frac{1}{108}\left(3200 \times 42 \times \cos(5) + 20 \times 81 + 20 \times 99.38 \times (-10) + 20 \times 3200 \times \sin(5)\right)$$

$$= 1122.3 \text{ lb}$$

When the vehicle travels downhill, the forces are

$$F_{z1} = \frac{1}{c_1 + c_2}\left(Mgc_2\cos\theta - h_aR_a - hMa + hMg\sin\theta\right)$$

$$= \frac{1}{108}\left(3200 \times 66 \times \cos(5) - 20 \times 81 - 20 \times 99.38 \times (-10) + 20 \times 3200 \times \sin(5)\right)$$

$$= 2168.8 \text{ lb}$$

$$F_{z2} = \frac{1}{c_1 + c_2}\left(Mgc_1\cos\theta + h_aR_a + hMa - hMg\sin\theta\right)$$

$$= \frac{1}{108}\left(3200 \times 42 \times \cos(5) + 20 \times 81 + 20 \times 99.38 \times (-10) - 20 \times 3200 \times \sin(5)\right)$$

$$= 1019.0 \text{ lb}$$

The load described here represents the weight distribution at the initial braking. Aerodynamic forces will reduce as the speed starts decreasing and then will vanish at the full stop.

Comparing the results with those obtained in Example 7.5 where the aerodynamic loads were neglected, it is observed that the addition of aerodynamic forces has reduced the load on the front axle but has increased the load on rear axle. Also note that the this represents normal operating conditions and not severe ones.

7.4 Optimal Braking Performance

Design consideration of the automotive brake system, where four brakes exist on the four wheels, requires considering the vehicle safety and a brake lock-up mechanism. In particular, vehicle stability and the ability to steer are of major concern, particularly when braking is conducted on surfaces with a low coefficient of friction. Indeed, a braking strategy needs to be developed to achieve the highest level of safety. We will start by considering the braking of a single axle but will move later to consider the braking of both axles.

7.4.1 Braking of a Single Axle

First we will consider the braking of a single axle, either front or rear. To gain some insight into the behavior of the vehicle, we will treat only the braking force and the weight of the vehicle on a horizontal plane.

7.4.1.1 Braking of the Front Axle

Let us assume the brakes are applied on the front axle. In this case, the braking force at the front wheels will be equal to the coefficient of friction multiplied by the normal force at those wheels as

$$F_{x1} = -\mu F_{z1} \tag{7.32}$$

The braking force at the rear wheels will be zero. The maximum deceleration then can be found from Eq. 7.21 by utilizing Eq. 7.25 as

$$Ma_F = -\mu Mg\left(\frac{c_2}{c_1 + c_2}\right) + \mu\left(\frac{h}{c_1 + c_2}\right)Ma_F \tag{7.33}$$

Note that we replaced a by a_F to indicate that the equation is valid only when the front brakes alone are applied. The equation can be solved directly for the maximum deceleration as

$$a_F = -g\frac{\mu\left(\dfrac{c_2}{c_1 + c_2}\right)}{1 - \mu\left(\dfrac{h}{c_1 + c_2}\right)} \tag{7.34}$$

7.4.1.2 Braking of the Rear Axle

Similar treatment can be made if the brakes are applied only to the rear wheels. In this case, the braking force on the front wheels is zero, and the braking force on the rear wheels is

$$F_{x2} = -\mu F_{z2} \tag{7.35}$$

We can develop an equation for the maximum deceleration when under rear wheel braking as

$$a_R = -g \frac{\mu \left(\dfrac{c_1}{c_1 + c_2} \right)}{1 + \mu \left(\dfrac{h}{c_1 + c_2} \right)} \tag{7.36}$$

Figure 7.8 shows the deceleration of a typical compact vehicle.

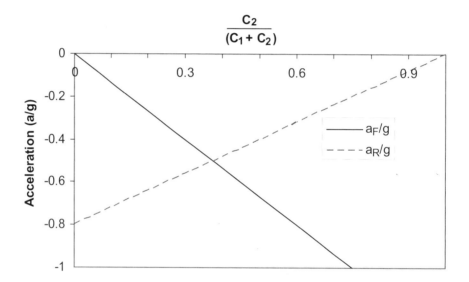

Figure 7.8 Deceleration under front or rear braking, where $\mu = 1$, $h = 0.5$ m, and $c_1 + c_2 = 2.0$.

Note that for typical vehicle designs and loading conditions in the vehicle, the value of $\dfrac{c_2}{(c_1 + c_2)}$ ranges from 0.4 to 0.6. Within this range of application, the maximum deceleration that can be obtained when only the rear brakes are activated is between the values –0.48 and –0.32g. This deceleration is between –0.53 and –0.8g when the vehicle is under front braking alone. This shows that for normal braking conditions in a single drive braking, better deceleration can be achieved with front axle braking.

Figure 7.9 shows a similar graph when the coefficient of friction is 0.5 (instead of 1). Within the range of normal application, the maximum deceleration that can be obtained when only the rear brakes are activated is between the values –0.27 and –0.18g. This deceleration is between –0.23 and –0.34g when the vehicle is under front braking alone. Note also that, in general, front axle braking gives higher deceleration except in the vicinity of $\dfrac{c_2}{(c_1 + c_2)}$ of 0.4, where the higher deceleration is obtained with rear axle braking.

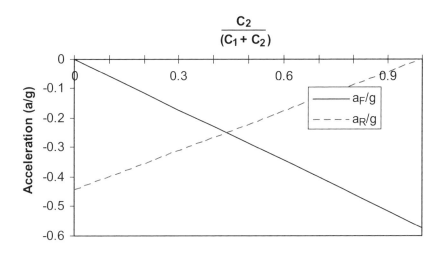

Figure 7.9 Deceleration under front or rear braking, where $\mu = 0.5$, $h = 0.5$ m, and $c_1 + c_2 = 2.0$.

In real application, brakes are supplied at all four wheels to achieve the maximum braking force before lock-up occurs. The weight is distributed differently on both axles, depending on the design and loading conditions as well as environmental factors (e.g., aerodynamics and grade). Assuming that the same coefficient of friction exists between all four wheels and the surface, the maximum braking force that can be achieved at each axle is directly proportional to the load on the axle. To achieve the maximum braking force for the whole vehicle, the braking force distribution between the front and rear wheels should be proportional to that of the axle load. In other words, if the front axle carries 60% of the load and the rear axle carries the remaining 40% of the load, achieving a maximum braking force will require that 60% of the total braking force should be provided by the front wheels and 40% provided by the rear wheels. This distribution of braking forces, which can be designed into the system, ensures that lock-up occurs at the same time for both axles. However, lock-up of any or both axles has significant consequences that should be discussed.

7.4.1.3 Safety Considerations

If the rear tires lock up first, the vehicle will lose stability of direction. Indeed, when driving on surfaces with a low coefficient of friction (e.g., ice), rear lock-up of the brake system results in the vehicle rotating by an angle of 180°. At rear lock-up, the rear axles cannot resist any lateral movement. Slight forces, because of side winds or other factors, will cause the rear wheels to move laterally. This will result in a yawing moment of the inertia force about the yaw center to be developed, causing further lateral motion of the rear wheels. In turn, this will result in a larger arm of the inertia forces and a larger moment, which causes again further lateral motion. This self-energizing moment will not stop until the vehicle turns 180°. Thus, the directional stability of the vehicle is completely lost, and serious safety consequences are possible.

On the other hand, if the lock-up starts at the front axle, directional control of the vehicle is lost. The driver will not be able to control the steering of the vehicle. Front tire lock-up does not cause the kind of directional instability noticed with rear wheel lock-up.

Note that when directional stability is lost, the driver cannot regain it. In other words, the vehicle will continue with the self-energizing yawing moment until the vehicle reaches a complete stop with the 180° turn. However, the driver can regain the ability to steer if the brakes are released slightly to prevent the front lock-up. Because of these considerations, braking requirements often call for front lock-up to occur ahead of rear lock-up. Although this will not allow us to achieve maximum braking force with both lock-ups happening at the same time, it is worthwhile to design the braking system such that front lock-up occurs ahead of rear lock-up for the previously mentioned safety reasons. The reader is referred to the book by Limpert (1992) and the paper by Dahlberg (1999) for more detailed analyses of the problem.

7.4.2 Braking at Both Axles

In today's vehicles, both axles have a braking system. Disk brakes commonly are used for the front axle and drum brakes for the rear axle. As mentioned, the brake designer needs the maximum braking force that can be obtained from the braking system at hand and needs to ensure that front lock-up occurs first under various loading and environmental conditions. Let us assume that based on certain emergency conditions, the brake designer chooses to distribute the braking load such that a certain ratio (K_f) of the total braking force goes to the front brakes. The remainder that goes to the rear brakes is $1 - K_f$ of the total braking force

$$F_{x1} = K_f F_x$$
$$F_{x2} = (1 - K_f) F_x$$
(7.37a)

$$F_{x1} = \frac{K_f}{1 - K_f} F_{x2}$$
$$F_{x2} = \frac{1 - K_f}{K_f} F_{x1}$$
(7.37b)

where F_x is the total friction force. Consider the free body diagram of the vehicle under direct braking forces $-F_{x1}$ for the front axle and $-F_{x2}$ for the rear axle. The vehicle is still under the effects of aerodynamic forces as well as grade and rolling resistance. The total braking force is $-F_x = -F_{x1} - F_{x2}$. Note that the negative signs indicate that the direction of the force is opposite to that shown in Figure 7.7. The total rolling resistance force is $R_r = R_{r1} + R_{r2}$. The total weight is distributed as

$$F_{z1} = \frac{1}{c_1 + c_2} \left(Mgc_2 \cos\theta - h_a R_a - hMa - hMg\sin\theta \right)$$
$$F_{z2} = \frac{1}{c_1 + c_2} \left(Mgc_1 \cos\theta + h_a R_a + hMa + hMg\sin\theta \right)$$
(7.31b)

Note that $F_{z1} + F_{z2}$ will always be

$$F_{z1} + F_{z2} = \frac{1}{c_1 + c_2} \left(Mgc_2 \cos\theta \right) + \frac{1}{c_1 + c_2} \left(Mgc_1 \cos\theta \right)$$
$$= Mg\cos\theta$$
(7.38)

Considering Figure 7.7 again and summing the forces in the direction of motion yields

$$F_{x1} + F_{x2} - R_a - R_r - Mg\sin(\theta) = Ma \qquad (7.39)$$

7.4.2.1 Front Lock-Up

Under front lock-up, $F_{x1} = -\mu F_{z1}$. Substituting this and Eq. 7.31 into Eq. 7.39 yields

$$-\mu\left(\frac{1}{c_1 + c_2}\left(Mgc_2\cos\theta - h_aR_a - hMa_F - hMg\sin\theta\right)\right)\left(1 + \frac{1 - K_f}{K_f}\right)$$
$$- R_a - R_r - Mg\sin(\theta) = Ma_F \qquad (7.40)$$

Rearranging for a_F gives

$$-\mu\left(\frac{1}{c_1 + c_2}\left(Mgc_2\cos\theta - h_aR_a - hMg\sin\theta\right)\right)\left(1 + \frac{1 - K_f}{K_f}\right)$$
$$- R_a - R_r - Mg\sin(\theta) \qquad (7.41)$$
$$= \left(-\frac{\mu h}{c_1 + c_2}\left(1 + \frac{1 - K_f}{K_f}\right) + 1\right)Ma_F$$

or simply

$$a_F = \frac{-\mu\left(\dfrac{1}{c_1 + c_2}\left(Mgc_2\cos\theta - h_aR_a - hMg\sin\theta\right)\right)\left(1 + \dfrac{1 - K_f}{K_f}\right) - R_a - R_r - Mg\sin(\theta)}{\left(-\dfrac{\mu h}{c_1 + c_2}\left(1 + \dfrac{1 - K_f}{K_f}\right) + 1\right)M} \qquad (7.42)$$

Note that the acceleration a has a subscript F to indicate that the acceleration obtained by using the equation is valid only when the front tires are locked up.

Ignoring aerodynamic drag, grade, and rolling resistance and using the following simplification,

$$1 + \frac{K}{1 - K_f} = \frac{1 - K_f}{1 - K_f} + \frac{K_f}{1 - K_f} = \frac{1}{1 - K_f}$$

and

$$1 + \frac{1 - K_f}{K_f} = \frac{K_f}{K_f} + \frac{1 - K_f}{K_f} = \frac{1}{K_f}$$

The deceleration becomes

$$a_F = -\frac{\mu\left(\dfrac{c_2}{c_1+c_2}\right)\left(1+\dfrac{1-K_f}{K_f}\right)}{\left(-\dfrac{\mu h}{c_1+c_2}\left(1+\dfrac{1-K_f}{K_f}\right)+1\right)}g = \frac{\dfrac{\mu}{K_f}\left(\dfrac{c_2}{c_1+c_2}\right)}{\left(1-\dfrac{\mu h}{K_f(c_1+c_2)}\right)}g \qquad (7.43)$$

Note that this expression may yield a denominator of zero. This will yield a very high number for a_F, which will mean that rear lock-up must happen first in these circumstances. Equation 7.43 will reduce to that of Eq. 7.34 when a front axle alone is under breaking ($K_f = 1$).

7.4.2.2 Rear Lock-up

Equation 7.38 can be rewritten as

$$F_{x2}\left(1+\frac{K_f}{1-K_f}\right) - R_a - R_r - Mg\sin(\theta) = Ma \qquad (7.44)$$

When under rear lock-up, $F_{x2} = -\mu F_{z2}$. Substituting this and Eq. 7.31 into Eq. 7.44 yields

$$-\mu\left(\frac{1}{c_1+c_2}\left(Mgc_1\cos\theta + h_a R_a + hMa_R + Mg\sin\theta\right)\right)\left(1+\frac{K_f}{1-K_f}\right)$$
$$- R_a - R_r - Mg\sin(\theta) = Ma_R \qquad (7.45)$$

Rearranging for a_R gives

$$-\mu\left(\frac{1}{c_1+c_2}\left(Mgc_1\cos\theta + h_a R_a + Mg\sin\theta\right)\right)\left(1+\frac{K_f}{1-K_f}\right)$$
$$- R_a - R_r - Mg\sin(\theta) \qquad (7.46)$$
$$= \left(+\frac{\mu h}{c_1+c_2}\left(1+\frac{K_f}{1-K_f}\right)+1\right)Ma_R$$

Simplifying, we obtain

$$a_R = \frac{-\mu\left(\dfrac{1}{c_1+c_2}\left(Mgc_1 + h_a R_a + Mg\sin\theta\right)\right)\left(1+\dfrac{K_f}{1-K_f}\right) - R_a - R_r - Mg\sin(\theta)}{\left(\dfrac{\mu h}{c_1+c_2}\left(1+\dfrac{K_f}{1-K_f}\right)+1\right)M} \qquad (7.47)$$

Ignoring aerodynamic drag, grade, and rolling resistance, the deceleration is

$$
a_R = -\frac{\mu\left(\dfrac{c_1}{c_1 + c_2}\right)\left(1 + \dfrac{K_f}{1 - K_f}\right)}{\left(\dfrac{\mu h}{c_1 + c_2}\left(1 + \dfrac{K_f}{1 - K_f}\right) + 1\right)}g
\qquad (7.48)
$$

Note that Eq. 7.48 will reduce to that of Eq. 7.36 when a rear axle alone is under braking ($K_f = 0$).

The front tires will lock up first if $|a_F| < |a_R|$; otherwise, the rear tires will lock up first.

Example E7.8

A 21,351.46-N (4800-lb) passenger car has a 2844.80-mm (112-in.) wheelbase and a center of gravity of 1270.00 mm (50 in.) behind the front axle and 508 mm (20 in.) above the ground. The braking effort distribution gives the front axle 60% of the total braking force. The car is moving on a horizontal plane. Determine which tires will lock up first if the car is moving on a surface with first a 0.75 coefficient of friction, and second, a 0.25 coefficient of friction. Ignore drag and rolling resistance.

Solution:

M	=	4800 lb = 149.07 slugs
$c_1 + c_2$	=	112 in.
c_1	=	50 in.
c_2	=	62 in.
K_f	=	0.6
h	=	20
θ	=	0
R_a	=	0
R_r	=	0

First surface: $\mu = 0.75$

Acceleration under front lock-up is

$$a_F = -\frac{\mu\left(\dfrac{c_2}{c_1+c_2}\right)\left(1+\dfrac{1-K_f}{K_f}\right)}{\left(-\dfrac{\mu h}{c_1+c_2}\left(1+\dfrac{1-K_f}{K_f}\right)+1\right)}g$$

$$= -\frac{0.75\left(\dfrac{62}{112}\right)\left(1+\dfrac{0.4}{0.6}\right)}{\left(-\dfrac{0.75\times 20}{112}\left(1+\dfrac{0.4}{0.6}\right)+1\right)}g = -0.8908g$$

Acceleration under rear lock-up is

$$a_R = -\frac{\mu\left(\dfrac{c_1}{c_1+c_2}\right)\left(1+\dfrac{K_f}{1-K_f}\right)}{\left(\dfrac{\mu h}{c_1+c_2}\left(1+\dfrac{K_f}{1-K_f}\right)+1\right)}g$$

$$= -\frac{0.75\left(\dfrac{50}{112}\right)\left(1+\dfrac{0.6}{0.4}\right)}{\left(\dfrac{0.75\times 20}{112}\left(1+\dfrac{0.6}{0.4}\right)+1\right)}g = -0.2510g$$

Because $|a_F| > |a_R|$, the rear tires will lock up first.

Second surface: $\mu = 0.25$

Acceleration under front lock-up is

$$a_F = -\frac{\mu\left(\dfrac{c_2}{c_1+c_2}\right)\left(1+\dfrac{1-K_f}{K_f}\right)}{\left(-\dfrac{\mu h}{c_1+c_2}\left(1+\dfrac{1-K_f}{K_f}\right)+1\right)}g$$

$$= -\frac{0.25\left(\dfrac{62}{112}\right)\left(1+\dfrac{0.4}{0.6}\right)}{\left(-\dfrac{0.25\times 20}{112}\left(1+\dfrac{0.4}{0.6}\right)+1\right)}g = -0.2492g$$

Acceleration under rear lock-up is

$$a_R = -\frac{\mu\left(\dfrac{c_1}{c_1 + c_2}\right)\left(1 + \dfrac{K_f}{1 - K_f}\right)}{\left(\dfrac{\mu h}{c_1 + c_2}\left(1 + \dfrac{K_f}{1 - K_f}\right) + 1\right)}g$$

$$= -\frac{0.25\left(\dfrac{50}{112}\right)\left(1 + \dfrac{0.6}{0.4}\right)}{\left(\dfrac{0.25 \times 20}{112}\left(1 + \dfrac{0.6}{0.4}\right) + 1\right)}g = -0.2510g$$

Because $|a_F| < |a_R|$, the front tires will lock up first.

The preceding example shows the sensitivity of tire lock-up to the coefficient of friction (whether the front or rear tires lock up first). The vehicle is to be driven on various types of surfaces with a high coefficient of friction (e.g., concrete surfaces) and a low coefficient of friction (e.g., icy or slippery roads). For safety reasons, the brakes should be designed such that front lock-up occurs first under a wide range of surfaces. Also, the brakes should be designed such that the maximum braking force will be obtained.

Figure 7.10 shows how the acceleration under front and rear lock-ups changes with the coefficient of friction for the parameters described in Example 7.8. It clearly shows that rear lock-up is quite possible at surfaces with higher coefficients of friction. Figure 7.11 shows that only with changing the braking effort distribution such that 80% goes to the front brake will front brake lock-up occur for all types of surfaces. However, this is done only at the expense of maximum braking force. In other words, a higher braking force can be achieved with a braking effort factor of 60% than that of an 80% factor. This must be sacrificed to ensure front lock-up occurs first for all surfaces.

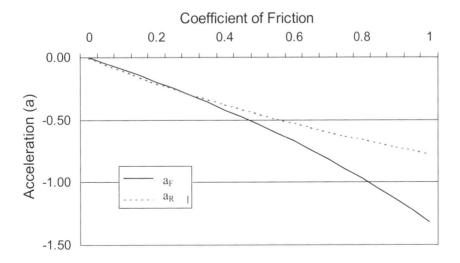

Figure 7.10 Acceleration under front and rear lock-up conditions for $c_1 = 50$ in., $c_2 = 62$ in., $h = 20$, and $K_f = 0.5$.

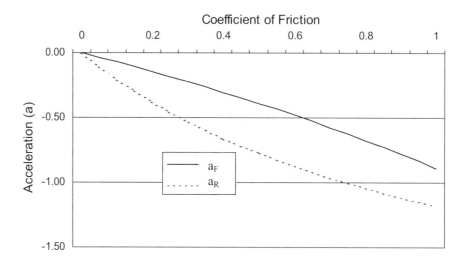

Figure 7.11 Acceleration under front and rear lock-up conditions for $c_1 = 50$ in., $c_2 = 62$ in., $h = 20$, and $K_f = 0.4$.

7.4.3 Achieving Optimal Braking Performance

Maximum braking performance is achieved when a maximum braking force is obtained. It is becoming clear from the previous examples that the maximum braking force is achieved when both front and rear approach lock-up at the same time. To achieve that, the braking force at each axle should be proportional to the axle load. In other words, if the front axle experiences 60% of the load, then the maximum braking force is achieved when the braking force is distributed such that 60% of the braking force is directed toward the front axle. Three major factors affect the braking force distribution. The first is that the loading conditions of the vehicle (fully loaded versus lightly loaded) will cause movement of the center of gravity of the vehicle. This actually means that a certain optimal distribution for a certain loading condition may not be the same as that for another loading condition. The second factor is the coefficient of friction. Load distribution on the front and rear axles is directly related to the coefficient of friction. The coefficient of friction varies considerably between dry and wet surfaces and among various types of surface constructions and finishing. The third factor is safety considerations that require the front brakes to lock up first. In today's automobiles, the braking system is designed with a feedback control system. Sensors are attached to both axles to determine the axle load and to distribute the braking force hydraulically to both axles according to the axle load distribution, with some bias toward achieving front lock-up first.

Thus, a robust braking design will require the involvement of the braking engineers early in the vehicle development to determine the center of gravity of the vehicle under various loads. As shown by the previous examples in this chapter, this is a key input to the braking design and its optimal performance.

In addition, various countries have established different criteria and requirements for the braking system to ensure its safety and acceptable performance. Such requirements take into consideration the initial speed at which braking is to occur, the condition of the braking system (i.e., new versus worn), and the loading of the vehicle (fully loaded versus lightly loaded). Most requirements are written for dry surfaces with a relatively high coefficient of friction. On such surfaces, the requirements call for an average deceleration in the range of 0.5 to 0.65g, depending on the vehicle initial speed, the condition of the brake system, and loading conditions. Optimal (i.e., maximum) average braking deceleration is noticed to be achieved when the starting speed is around 60 mph (96 km/hr).

Consider a vehicle that weighs 4800 lb (21,351.46 N) and has a wheelbase of 112 in. (2844.80 mm). Its center of gravity is 50 in. (1270.00 mm) behind the front axle and 20 in. (508.00 mm) above the ground, and the coefficient of friction is 0.75. The acceleration found for the assumptions of the front and rear lockups are given in Table 7.3 for various values of brake force distribution factors. The bold numbers show which deceleration is smaller. The lock-up associated with the smaller deceleration happens first.

TABLE 7.3
ACCELERATION UNDER FRONT AND REAR LOCK-UP ASSUMPTIONS FOR VARIOUS FRONT BRAKE FORCE DISTRIBUTION FACTORS
(c_2 = 62 IN., c_1 + c_2 = 112 IN., μ = 0.75, AND h= 20 IN.)

K_f	a_F/g	A_r/g
0.0001	3.102	**–0.295**
0.05	4.947	**–0.309**
0.1	12.237	**–0.324**
0.15	–25.833	**–0.340**
0.2	–6.284	**–0.359**
0.25	–3.577	**–0.379**
0.3	–2.500	**–0.401**
0.35	–1.921	**–0.427**
0.4	–1.560	**–0.456**
0.45	–1.314	**–0.490**
0.5	–1.134	**–0.528**
0.55	–0.998	**–0.573**
0.6	–0.891	**–0.627**
0.65	–0.804	**–0.692**
0.6875	**–0.750**	**–0.750**
0.7	**–0.733**	–0.772
0.75	**–0.674**	–0.872
0.8	**–0.623**	–1.003
0.85	**–0.580**	–1.179
0.9	**–0.542**	–1.431
0.95	**–0.509**	–1.820
1	**–0.479**	–2.500

The observation is made that by controlling only the brake force distribution factor, we can obtain the maximum braking force. This is achieved at a point where a_R and a_F are close to each other. Note that at K_f = 0.6875, both acceleration values a_R and a_F are the same (–0.75g), and maximum friction is achieved.

The maximum braking force occurs when both F_{x1} and F_{x2} approach their lock-up (or maximum values) simultaneously. This can be achieved only if we set $F_{x1} = -\mu F_{z1}$ and $F_{x2} = -\mu F_{z2}$. The total friction force is

$$
\begin{aligned}
F_x &= F_{x1} + F_{x2} \\
&= -\mu F_{z1} - \mu F_{z2} \\
&= -\mu \left(F_{z1} + F_{z2} \right) \\
&= -\mu Mg \cos(\theta)
\end{aligned}
\tag{7.49}
$$

The maximum deceleration is

$$
\begin{aligned}
a &= \frac{F_{x1} + F_{x2} - R_a - R_r - Mg\sin(\theta)}{M} \\
&= \frac{-\mu Mg\cos(\theta) - R_a - R_r - Mg\sin(\theta)}{M}
\end{aligned}
\tag{7.50}
$$

The only way both front and rear brakes would lock up at the same time is if the front brake distribution factor is equal to the ratio of the front axle load to the total load, or

$$
\begin{aligned}
K_{f\max} = \frac{F_{x1}}{F_x} = \frac{-\mu F_{z1}}{-\mu Mg\cos(\theta)} &= \frac{-\dfrac{\mu}{c_1 + c_2}\left(Mgc_2\cos\theta - h_a R_a - hMa - hMg\sin\theta\right)}{-\mu Mg\cos(\theta)} \\
&= \frac{\left(Mgc_2\cos\theta - h_a R_a - hMa - hMg\sin\theta\right)}{\left(c_1 + c_2\right)Mg\cos(\theta)}
\end{aligned}
\tag{7.51}
$$

$$
K_{f\max} = \frac{\left(Mgc_2\cos\theta - h_a R_a - h\left(-\mu Mg\cos(\theta) - R_a - R_r - Mg\sin(\theta)\right) - hMg\sin\theta\right)}{\left(c_1 + c_2\right)Mg\cos(\theta)}
\tag{7.52}
$$

Ignoring aerodynamic forces, rolling resistance, and grade, the front brake distribution factor becomes

$$
K_{f\max} = \frac{\left(c_2 + \mu h\right)}{\left(c_1 + c_2\right)}
\tag{7.53}
$$

EXAMPLE E7.9

Consider Example 7.8, where the 21,351.46-N (4800-lb) passenger car has a 2844.80-mm (112-in.) wheelbase and a center of gravity of 1270.00 mm (50 in.) behind the front axle and 508 mm (20 in.) above the ground. The car is moving on a horizontal plane, and the drag and rolling resistance are negligible. Determine the maximum braking force, maximum deceleration, and distribution factor to achieve them. Consider first a 0.75 coefficient of friction, and second, a 0.25 coefficient of friction.

Solution:

First surface with a 0.75 coefficient of friction:

The front brake distribution factor to achieve the maximum braking force is

$$K_f = \frac{(c_2 + \mu h)}{(c_1 + c_2)} = \frac{60 + 0.75 \times 20}{112} = 0.6875$$

The maximum braking force F_x is

$$F_x = -\mu Mg \cos(\theta) = -0.75 \times 4800 = -3600 \text{ lb}$$

The maximum deceleration is

$$a = \frac{-\mu Mg \cos(\theta) - R_a - R_r - Mg \sin(\theta)}{M} = -24.15 \text{ ft/sec}^2$$

Note that the maximum deceleration is obtained when both a_R and a_F are equal. At that point, they both equal the coefficient of friction times the acceleration of gravity.

Second surface with a 0.25 coefficient of friction:

The front brake distribution factor to achieve the maximum braking force is

$$K_f = \frac{60 + 0.25 \times 20}{112} = 0.5804$$

The maximum braking force F_x is

$$F_x = -\mu Mg \cos(\theta) = -0.25 \times 4800 = 1200 \text{ lb}$$

The maximum deceleration is

$$a = -\mu g = -8.05 \text{ ft/sec}^2$$

To calculate the forces on both axles (including both friction and load forces), we need to develop a strategy to tackle the problem. The strategy is based on the following:

1. For the given problem, determine a_F and a_R from Eqs. 7.42 and 7.47.

2. If $|a_F| < |a_R|$, then front lock-up occurs first. The following equations should be used:

$$F_{z1} = \frac{1}{c_1 + c_2}\left(Mgc_2 \cos\theta - h_a R_a - hMa_F - hMg \sin\theta\right)$$

$$F_{z2} = \frac{1}{c_1 + c_2}\left(Mgc_1 \cos\theta + h_a R_a + hMa_F + hMg \sin\theta\right)$$

Note that

$$F_{z1} + F_{z2} = Mg\cos\theta$$

$$F_{x1} = -\mu F_{z1}$$

$$F_{x2} = \frac{1 - K_f}{K_f} F_{x1}$$

or

$$F_{x2} = Ma_F - F_{x1}$$

The total friction force is

$$F_x = F_{x1} + F_{x2}$$

3. If $|a_F| > |a_R|$, then rear lock-up occurs first. The following equations should be used:

$$F_{z1} = \frac{1}{c_1 + c_2}\left(Mgc_2\cos\theta - h_a R_a - hMa_R - hMg\sin\theta\right)$$

$$F_{z2} = \frac{1}{c_1 + c_2}\left(Mgc_1\cos\theta + h_a R_a + hMa_R + hMg\sin\theta\right)$$

Note that

$$F_{z1} + F_{z2} = Mg\cos\theta$$

$$F_{x2} = -\mu F_{z2}$$

$$F_{x1} = \frac{K_f}{1 - K_f} F_{x2}$$

or

$$F_{x1} = Ma_R - F_{x2}$$

The total friction force is

$$F_x = F_{x1} + F_{x2}$$

EXAMPLE E7.10

Consider Example 7.9. To ensure front axle lock-up on a dry surface, the front brake distribution factor is chosen at 0.7. The vehicle is moving on a low-friction surface with a 0.25 coefficient of friction. Determine the maximum braking force and which axle brakes would lock up first.

Solution:

Because the drag, rolling resistance, and grade are ignored, acceleration under front-lock-up is

$$a_F = -\frac{\mu\left(\dfrac{c_2}{c_1 + c_2}\right)\left(1 + \dfrac{1 - K_f}{K_f}\right)}{\left(-\dfrac{\mu h}{c_1 + c_2}\left(1 + \dfrac{1 - K_f}{K_f}\right) + 1\right)}g$$

$$= -\frac{0.25\left(\dfrac{62}{112}\right)\left(1 + \dfrac{0.3}{0.7}\right)}{\left(-\dfrac{0.25 \times 20}{112}\left(1 + \dfrac{0.3}{0.7}\right) + 1\right)}g = -0.2112g$$

Acceleration under rear lock-up is

$$a_R = -\frac{\mu\left(\dfrac{c_1}{c_1 + c_2}\right)\left(1 + \dfrac{K_f}{1 - K_f}\right)}{\left(\dfrac{\mu h}{c_1 + c_2}\left(1 + \dfrac{K_f}{1 - K_f}\right) + 1\right)}g$$

$$= -\frac{0.25\left(\dfrac{50}{112}\right)\left(1 + \dfrac{0.7}{0.3}\right)}{\left(\dfrac{0.25 \times 20}{112}\left(1 + \dfrac{0.7}{0.3}\right) + 1\right)}g = -0.3238g$$

Because $|a_F| < |a_R|$, front lock-up occurs first. The normal forces are

$$F_{z1} = \frac{1}{c_1 + c_2}\left(Mgc_2 - hMa_F\right)$$

$$= \frac{1}{112}\left(149 \times 32.2 \times 62 - 20 \times 149 \times (-0.2112)\right)$$

$$= 2838 \text{ lb}$$

$$F_{z2} = \frac{1}{c_1 + c_2}\left(Mgc_1 + hMa_F\right)$$

$$= \frac{1}{112}\left(149 \times 32.2 \times 50 + 20 \times 149 \times (-0.2112)\right)$$

$$= 1962 \text{ lb}$$

Note that

$$F_{z1} + F_{z2} = 4800 \text{ lb}$$

$$F_{x1} = -\mu F_{z1} = -0.25 \times 2838 = -710 \text{ lb}$$

$$F_{x2} = \frac{1 - K_f}{K_f} F_{x1} = \frac{1 - 0.7}{0.7} \times (-710) = -304 \text{ lb}$$

The total friction force is

$$F_x = F_{x1} + F_{x2} = -1014 \text{ lb}$$

Table 7.4 shows the acceleration and force distribution on each axle for the case of Example 7.9 and a coefficient of friction of 0.75. Note that the maximum braking forces occur when the K_{fmax} of 0.6875 found in Example 7.9 is used. Also note that at that value, a_F and a_R are equal.

The maximum braking force is plotted in Figure 7.12 as a function of the front brake distribution factor.

In a typical design problem, one needs to consider various surfaces and loading conditions. For example, let us consider the following problem where a vehicle is under different loading conditions and on various surfaces.

EXAMPLE E7.11

Consider a vehicle that weighs 3200 lb (14,235 N) when lightly loaded (driver only), with a center of gravity at 40 in. (1016.0 mm) behind the front axle and 18 in. (457.2 mm) above the road, and 4000 lb (17,793 N) when loaded with a center of gravity of 51 in. (1295.4 mm) behind the front axle and 20 in. (508.0 mm) above the road. The wheelbase is 108 in. (2743.2 mm). The vehicle is to achieve maximum possible braking force and front lock-up under the following severe conditions:

1. The vehicle is loaded and on a dry surface of $\mu = 0.8$.

2. The vehicle is loaded and on a slippery surface of $\mu = 0.2$.

3. The vehicle is lightly loaded and on a dry surface of $\mu = 0.8$.

4. The vehicle is lightly loaded and on a slippery surface of $\mu = 0.2$.

Determine the value of K_f that you recommend, knowing that the braking system your company installs has a constant value (i.e., no feedback control system). Determine the axle loads and the maximum braking force.

TABLE 7.4
FORCE DISTRIBUTION FOR DIFFERENT VALUES OF K_f UNDER FRONT AND REAR LOCK-UP ASSUMPTIONS
(c_2 = 62 IN., $c_1 + c_2$ = 112 IN., μ = 0.75, h = 20 IN.)

K_f	a_F/g	F_{z1}	F_{z2}	F_{x1}	F_{x2}	F_x	a_F/g	F_{z1}	F_{z2}	F_{x1}	F_{x2}	F_x
0.0001	3.102						-0.295	2910	1890	0	-1417	-1471
0.05	4.947						-0.309	2922	1878	-74	-1409	-1483
0.1	12.237						-0.324	2935	1865	-155	-1399	-1554
0.15	-25.833						-0.340	2949	1851	-245	-1388	-1633
0.2	-6.284						-0.359	2964	1836	-344	-1377	-1721
0.25	-3.577						-0.379	2982	1818	-455	-1364	-1818
0.3	-2.500						-0.401	3001	1799	-578	-1349	-1927
0.35	-1.921						-0.427	3023	1777	-718	-1333	-2050
0.4	-1.560						-0.456	3048	1752	-876	-1314	-2190
0.45	-1.314						-0.490	3077	1723	-1057	-1292	-2350
0.5	-1.314						-0.528	3110	1690	-1268	-1268	-2535
0.55	-0.998						-0.573	3149	1651	-1514	-1239	-2752
0.6	-0.891						-0.627	3195	1605	-1806	-1204	-3010
0.65	-0.804						-0.692	3250	1550	-2159	-1162	-3321
0.6875	-0.750	3300	1500	-2475	-1125	-3600	-0.750	3300	1500	-2475	-1125	-3600
0.7	-0.733	3286	1514	-2464	-1056	-3521	-0.772					
0.75	-0.674	3235	1565	-2426	-809	-3235	-0.872					
0.8	-0.623	3191	1609	-2394	-598	-2992	-1.003					
0.85	-0.580	3154	1646	-2366	-417	-2783	-1.179					
0.9	-0.542	3122	1678	-2341	-260	-2601	-1.431					
0.95	-0.509	3093	1707	-2320	-122	-2442	-1.820					
1	-0.479	3068	1732	-2301	0	-2301	-2.500					

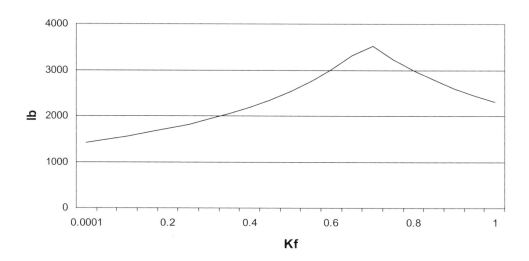

Figure 7.12 Braking force as a function of the front brake force distribution factor.

Solution:

This problem is best solved by writing a spreadsheet with all possible tests. Table E7.11 shows the optimal value of K_f (Table E7.11A) and the associated acceleration and axle loads (Table E7.11B), as well as the maximum brake force for each of the cases.

To avoid rear lock-up under all cases, the K_f must be chosen above the maximum value of 0.763 determined for case 3. Thus, $K_f = 0.8$ is chosen. For this value, the acceleration and loads are obtained for each of the cases mentioned (Table E7.11B).

Note that with the choice made for K_f, front lock-up occurs first in all cases. To achieve this, major sacrifice is made for maximum braking force, especially for case 2, where the maximum braking force is decreased from 800 to 553 lb.

As already stated, the front brake force distribution factor K_f in today's automotive brakes can be altered, depending on the particular application at hand. This can be done straightforwardly to account for variations in the axle loads, either because of different decelerations or because of different vehicle loading conditions (lightly loaded versus loaded conditions). In the preceding example, a factor of 0.8 can be used for the lightly loaded cases and 0.7 for the loaded cases. This will enable us to achieve higher braking force at the surfaces with a low coefficient of friction. The K_f factor cannot be altered straightforwardly to account for the different coefficients of friction that various surfaces may offer. In that case, an advanced anti-lock braking system (ABS) can be used.

European specifications call for the following regulations for vehicles that do not have an anti-lock braking system (Buschman *et al.,* 1992). First, the acceleration should always remain higher than 0.3g, where g is the acceleration of gravity. Second, the braking force in the front should be lower than that of the rear when the vehicle acceleration is between 0.5 and 0.8g.

Modern braking systems have electronic braking force distribution (EBD) systems. These are proven to be more robust because the input signal describes current vehicle and driving conditions. An electronic braking force distribution system takes into account friction variations of the lining and road surfaces (Buschman *et al.,* 1992).

TABLE E7.11A
OPTIMAL VALUE FOR K_f FOR EXAMPLE E7.11

Case	c_1	c_2	h	μ	Mg	K_{fmax}	a_F/g	F_{z1}	F_{z2}	F_{x1}	F_{x2}	F_x
1	51	57	20	0.8	4000	0.6759	−0.800	2704	1296	−2163	−1037	−3200
2	51	57	20	0.2	4000	0.5648	−0.200	2259	1741	−452	−348	−800
3	40	68	18	0.8	3200	0.763	−0.800	2241	759	−1953	−607	−2560
4	40	68	18	0.2	3200	0.663	−0.200	2121	1079	−424	−216	−640

TABLE E7.11B
LOAD DISTRIBUTION FOR $K_f = 0.8$ FOR EXAMPLE E7.11

Case	c_1	c_2	h	μ	Mg	K_f	a_F/g	F_{z1}	F_{z2}	F_{x1}	F_{x2}	F_{x2}	F_x	a_R/g
1	51	57	20	0.8	4000	0.8	−0.6477	2591	1409	−2073	−518	−518	−2591	−1.0851
2	51	57	20	0.2	4000	0.8	−0.1383	2214	1786	−443	−111	−111	−553	−0.3984
3	40	68	18	0.8	3200	0.8	−0.7556	2418	782	−1934	−484	−484	−2418	−0.8889
4	40	68	18	0.2	3200	0.8	−0.1643	2102	1098	−420	−105	−105	−526	−0.3175

7.5 Considerations of Vehicle Safety

Brakes are fundamentally a safety feature to enhance motion control of the vehicle in a manner that will enable the driver to avoid a potential accident. The ability of brakes to deliver the function of motion control is dependent on the traction between the tires and the road. If this traction is lost, this safety feature is compromised. Hence, the main objective of additional safety features of brakes is to maximize the traction effort between the tires and the road.

As already stated, lock-up conditions will change the rolling action of the tires with a sliding action. This considerably reduces the coefficient of friction. It also will cause the vehicle to lose the ability to steer (under front lock-up conditions) or a loss of direction stability (under rear lock-up conditions). In addition, the distance traveled during braking will be much larger than that without lock-up conditions. This will cause the vehicle to behave in an unpredictable manner to the automotive customer, reducing his or her control over the vehicle and, more importantly, not give the required performance to avoid a potential accident. Thus, preventing lock-up was the challenge facing engineers over time. Engineers had the objective of achieving the maximum force from the braking system and minimizing the possibility of lock-up conditions. With advances in electronic controls and sensors, a system has been used effectively on vehicles to prevent lock-up conditions from occurring. The key to this is for sensors planted for this purpose to assess skid, or slipping, conditions of the wheels. Brakes then are electronically and momentarily released to bring the wheel to the non-skid condition. Brakes are applied again to reach the skid condition, gaining higher speed reduction over a short period of time. This process is repeated until a complete stop, or the desired speed, is achieved. This system, referred to as an anti-lock braking system (ABS), is becoming a standard safety feature on many vehicles and is optional on most others.

The anti-lock braking system offers better treatment of the complex surfaces frequently encountered in cold regions. In areas where the temperature reaches the subfreezing point at night, pockets of water will freeze. This causes variations in the coefficient of friction between the two wheels of the same axle. Thus, a condition of the wheels of the same axle having one side on a low coefficient of friction and the other wheel on a higher coefficient of friction can be encountered. Under such a condition, the maximum braking force will be sacrificed if the same braking force is given to both wheels on the same axis.

There are three major components of the anti-lock braking system: the sensors to monitor wheel speeds and compare those speeds with the vehicle speed, the electronic control unit (ECU), and the solenoid, which is used to release and reapply the pressure to the brakes. It is important to understand the relationship between the skid (or slip) condition of the tires and the braking force.

7.5.1 Skid (Slip) Condition and Braking

When forces (driving and braking) are applied to the tire, the tire experiences a condition of skid (slippage) over the road surface. The speed of the vehicle under free-rolling condition (i.e., no skid) can easily be related to the rotational speed of the wheels as

$$V = r_t \, \omega \tag{7.54}$$

where

V = vehicle speed

r_t = radius of the tire

ω = angular speed of the tire

In real applications and when a braking force is applied, variation is observed between V and $r_t\omega$. This can be observed when the wheels are under lock-up conditions. Under a lock-up condition, the angular speed of the tire is zero ($\omega = 0$) while the vehicle is still moving ($V \neq 0$). Total skid and zero rolling are observed at that point. Some skid is observed between the two cases of free rolling and the wheel lock-up. If we define the first as the case of 0% skid and the last as a 100% skid condition, the following equation can be written to describe skid at any point between the two extreme cases:

$$i = \left(1 - \frac{V}{r_t\omega}\right) \times 100\% \tag{7.55}$$

Note that different definitions can be used for the skid of the tire. In fact, SAE International offers the following definition (SAE J670e):

$$i_{SAE} = \left(\frac{r_t\omega}{V} - 1\right) \times 100\% \tag{7.56}$$

Note that i_{SAE} gives the value of 0% skid when the vehicle is under free-rolling condition and the value of –100% when the tires are locked up.

The phenomenon of tire-road interaction is complex. The coefficient of friction between the tires and the road is a function of many environmental parameters, including the surface structure, the type of tire, the tire pressure, dry versus wet, wet versus icy, and so forth. The coefficient of friction under given environmental conditions is found to be a function of the skid between the tire and the ground, achieving a peak value only when a certain percentage of skid is achieved (Figure 7.13).

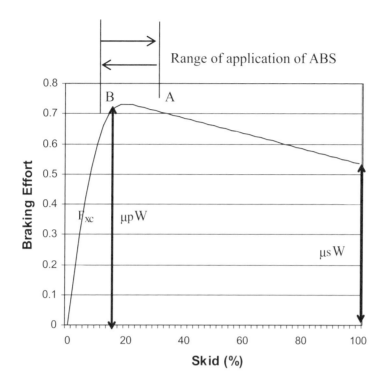

Figure 7.13 Braking coefficient versus skid.

As the braking torque is applied to the wheel, a proportional braking force is developed between the tire and the ground. The skid condition is a function of the braking force applied. The peak value of the braking force usually is achieved with a skid of 15 to 20% occurring at the tire. Any additional skid after that (as is the case when the tires are locked up) will reduce the friction to the skid value (almost null on icy roads) and approximately 70 to 80% of the peak value on dry roads. This will reduce the braking efficiency (in the sense that longer distance is needed when wheels lock up than when they do not).

Note that a portion of the curve shown in Figure 7.13 is linear. This portion starts at the origin and ends at a critical value F_{xc} and has a slope of C. For that portion, the braking force is

$$F_x = iC \qquad (7.57)$$

where $F_x < F_{xc}$. The critical value is

$$F_{xc} = \frac{\mu_p W}{2} \qquad (7.58)$$

where W is the total weight. Note that when F_x is below F_{xc}, the relationship between the braking effort and the skid is linear, indicating that the apparent skid is due only to the longitudinal stiffness of the tire. When F_x exceeds F_{xc}, actual slipping between the tread of the tire and the ground takes place. The braking effort is (Wong, 2001)

$$F_x = \mu_p W \left(1 - \frac{\mu_p W}{4iC} \right)$$ (7.59)

Note the nonlinearity between the braking effort and the skid. Equation 7.59 applies until the maximum peak braking effort value is obtained. At that point, lock-up occurs, and skid moves rapidly from the 20% or so skid to 100% almost instantly with a reduced coefficient of friction.

7.5.2 Anti-Lock Braking System

Returning to Figure 7.13, it is clear that maximum braking can be achieved if the brake efforts remain within the vicinity of the peak value. The danger in conventional braking systems is that as soon as the braking force reaches that peak value, excessive skid occurs almost instantly. At that point, pressure is released off the brake in an anti-lock braking system with excessive skid, which returns the braking effort to a value below that of the peak. Effort then is applied again to reach the maximum, and so forth.

This phenomenon is best described with the help of Figure 7.14. Under normal braking with the braking effort below its peak value, speed reduction is noticed both for the vehicle and the wheels. When the braking force reaches its maximum value and excessive skid is observed (by noticing that the vehicle speed is reducing at a constant rate while the wheel speeds are reducing rapidly), the brake pressure is released. The wheel then will gain speed, and an optimal value of skid is retained. The pressure then is applied again to return the braking force to its maximum value, and so forth.

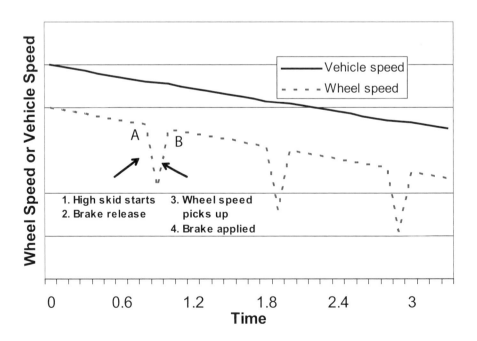

Figure 7.14 Wheel speed and vehicle speed under activation of an anti-lock braking system.

Comparing Figure 7.13 with Figure 7.14, point A on Figure 7.14 is the point at which skid starts to occur excessively (i > 25% or so). This is associated with point A on Figure 7.13. At that point, the anti-lock braking system recognizes that the tire is locked up, and the pressure is released off the brakes. This will bring the system to point B, where the wheel is gaining angular speed and the lock-up condition is freed. The pressure then is applied to maximize the braking action. This system will ensure that the braking effort will remain at its maximum range (Figure 7.13), giving the best braking performance and avoiding lock-up and the associated safety concerns of losing directional stability of the vehicle or the ability to steer the vehicle.

The angular speed of the tires can be measured easily with electromagnetic pulse pickups mounted on the tires or the driveline system. One needs to mount such pickups on each wheel to give more control over each of the wheels and to observe their skid status. The angular speed first is found as a direct measurement, and the angular deceleration is derived from that. A challenging issue of the anti-lock braking system is to determine the vehicle speed. Instead, a linear accelerometer is mounted on the vehicle to find its linear deceleration. The comparison between vehicle deceleration and angular deceleration will help determine if the tires are locked up or not.

The control system of the anti-lock braking system consists of three main components. The first is a signal processing unit that measures the angular deceleration of the tire and the linear deceleration of the vehicle. The second is a module that calculates skid based on the signals processed and determines if skid can be avoided when certain criteria are met. The third is a module for sending the needed signal for the pressure control system of the brakes.

Variations exist in different anti-lock braking system on how to determine lock-up conditions. One of these is the direct comparison between the linear acceleration of the vehicle and the angular deceleration of the tire. In the second scenario, the angular deceleration of the vehicle is monitored, and if it exceeds a certain value, it is held in memory for a certain time. During this time, both the linear deceleration of the vehicle and the angular deceleration of the tire are monitored closely. If a certain criterion is met, the anti-lock braking system will be activated. In another control system, skid of the vehicle is estimated, and the angular speed of the tire is measured. If both meet a certain criterion, the anti-lock braking system is activated. At low speeds, the anti-lock braking system is found to run into errors in determining skid conditions and linear deceleration of the vehicle. The threshold of skid for determining lock-up is increased at low speeds. The reapplication of pressure is modulated either by continuous monitoring of the angular tire deceleration and vehicle linear deceleration or by a fixed time-delay mechanism.

Different layouts of anti-lock braking systems are used in different applications. They all depend on the number of channels for the application and/or the release of the pressure and sensors used to determine the tire angular deceleration. For the front tires, each tire is monitored by its independent sensor, and an independent channel is used for pressure release and application. For the rear tires, some anti-lock braking systems use one sensor and one pressure regulator, two sensors and one pressure regulator, or two sensors and two pressure regulators.

7.6 Pitch Plane Models

Vehicle braking may induce some of the fundamental vehicle modes, which will result in an uncomfortable ride. Automotive customers experience the bounce and/or pitch types of motion when driving. Bounce basically is a mode in which the vehicle acts as a one-degree-of-freedom system, where its mass is simply setting on the "spring" tires. This motion is experienced particularly if some road input is adding to the excitation. Pitch mode is the other mode that is experienced by customers when the vehicle is undergoing acceleration or deceleration. Deceleration is due mostly to a braking action.

A representative model of the vehicle would have to account for both bounce and pitch modes. If the whole vehicle is represented as a mass, setting on two different springs representing front tires and rear tires, it is possible to obtain a reasonable representation of the vehicle. Figure 7.15 shows a two-degrees-of freedom model in which both bounce and pitch modes can be accounted for.

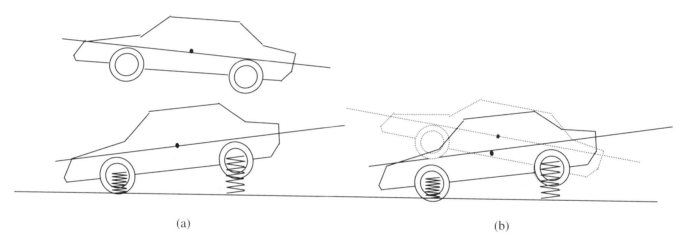

(a)   (b)

Figure 7.15 Bounce and pitch modes: (a) mainly bounce mode, and (b) mainly pitch mode.

Pitch and bounce modes generally are coupled. The simple vehicle model shown in Figure 7.16 is a two-degrees-of-freedom model. Such a model can be used for the assessment of vehicle bounce and pitch modes. Note that normal loads are affected by the load transfer that occurs during braking. Chapter 4 gave a more detailed analysis.

7.7 Recent Advances in Automotive Braking

The major attributes that drive today's braking technology are their safety and their noise and vibration, particularly brake squeal. Other aspects of new evolving braking technology relate to brake by wire, reuse of braking energy, and heat (for fuel economy improvement).

Among recent technologies related to brakes is the use of brake assist, also referred to as "power brakes." Among the recent research on brake assist is the work of Tamura *et al.* (2001), who presented work on a brake assist system with a preview function.

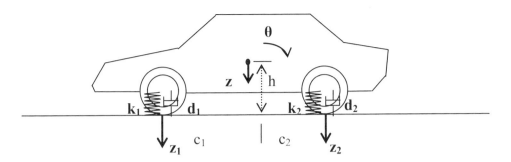

Figure 7.16 Simple two-degrees-of-freedom (pitch plane) vehicle model.

Such assist can be electrical or mechanical. The work of Feigel and Schonlau (1999) also should be mentioned. Figure 7.17 shows a cross section for a mechanical power assist.

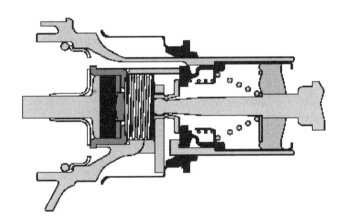

Figure 7.17 Mechanical power assist. (Feigel and Schonlau, 1999.)

Utilizing vehicle braking force to achieve a higher level of vehicle stability was the subject of various research. Estimation of the actual coefficient of friction between the tire and the road and adjusting the brake force distribution to achieve the desired dynamics were the subjects of the study by Yoon *et al.* (2004). Vehicle yaw stability during braking was studied by Seluga *et al.* (2004). Such studies are important to the discussion of accident reconstruction (Chapter 10).

Noise and vibrations induced by braking action is another corner of research receiving attention. In particular, brake squeal has been the subject of considerable research. Among this research is the work of Ouyang *et al.* (2005) on a review of numerical analysis of automotive disk brake squeal. Test and evaluation of automotive disk brake squeal also was discussed by Chen *et al.* (2003). Brake judder, another noise phenomenon induced by brake action, was the subject of several studies (e.g., Abdelhamid, 1997).

The possibility of recovering vehicle kinetic energy is one advantage of electric and hybrid electric vehicles. When a vehicle drives in heavy traffic, more than half of the total energy is dissipated in the brakes. Therefore, recovering braking energy is an effective approach for improving the driving range of electric vehicles and the energy efficiency or fuel economy of hybrid vehicles. Gao *et al.* (1999) investigated the effectiveness of regenerative braking.

There has been good progress recently in using computational tools to estimate braking performance. The use of advanced finite element methods (FEM) and codes as well as those of computational fluid dynamics (CFD) has become standard engineering practice in the design and analysis of these tools. Qian (2002) studied aerodynamic shape optimization using a CFD parametric model. The application was intended for brake cooling. Other applications can be found in various SAE International transactions and the open literature on using computer-aided engineering (CAE) tools. Figures 7.18 and 7.19 show the results obtained in some recent publications using finite element method and computational fluid dynamics tools.

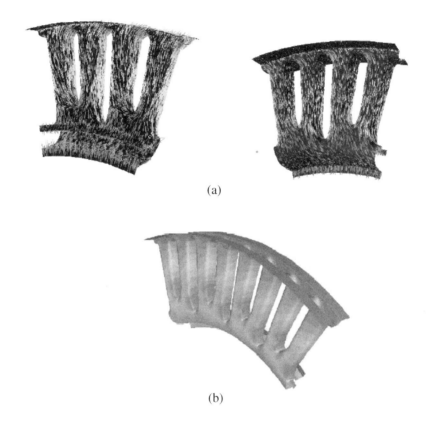

(a)

(b)

Figure 7.18 Use of computer-aided engineering (CAE) tools to predict brake performance: (a) velocity vector distribution, and (b) static pressure contour. (Qian, 2002.)

Brake by wire is another technology that is evolving rapidly, particularly with the recent popularity of hybrid vehicles and electric cars, where higher voltage is now available. Among recent research on brake by wire are those of Yoon *et al.* (2004) and Langenwalter and Kelly (2003).

7.8 Summary

The brake system constitutes an important safety feature in the vehicle that allows the driver to reduce speed either to allow for safe steering or to avoid accidents. The two

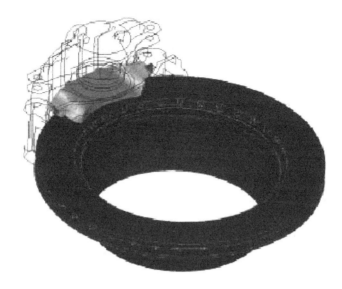

Figure 7.19 The use of finite element methods (FEM) in vibration analysis of brake systems. (Shi *et al.*, 2001.)

types of brake systems used in automotive applications are disk and drum brakes. Disk brakes usually are used for the front wheels, and drum brakes usually are used for the rear brakes. As the brakes are applied, a force will move the brake system closer to the wheels, allowing touching of the wheels and applying pressure. This results in a torque at each wheel against the angular motion of the wheel and thus a reduction in speed. In some cases where excessive brake force is applied, the brakes will prevent the wheels from rotating (i.e., a lock-up condition is encountered). In these cases, the wheels will slide on the road with less coefficient of friction than that experienced with the wheels rolling (with some skid).

Some of the weight of the vehicle transfers to the front wheels during braking, allowing higher friction forces for these wheels. The brake force should be distributed proportionally to the vertical load at each wheel to allow for maximum braking efforts. Also, when maximum braking effort is applied, the brake design should allow for front wheel lock-up to occur ahead of the rear wheel lock-up. This will allow the loss of directional control (i.e., steerability of the vehicle). Although this is a safety hazard, it is of less danger that the directional instability encountered with rear wheel lock-up.

New vehicles are equipped with anti-lock braking systems that allow braking of each wheel independently. This enables greater control of wheel motion and recovering of the wheel spinning after a lock-up condition is encountered. The anti-lock braking systems further allow for maximum braking force, even when different coefficients of friction are encountered by each wheel.

Finally, an overview of recent advances in brake technology is provided. In particular, discussions about technology-related brake safety, brake noise and vibration, brake by wire, brake energy reuse, and brake assist are included.

7.9 References

1. Abdelhamid, M., "Brake Judder Analysis: Case Studies," *Society of Automotive Engineers Transactions,* SAE Paper No. 973018, Society of Automotive Engineers, Warrendale, PA, 1997.

2. Buschmann, G., Ebner, H., and Kuhn, W., "Electronic Brake Force Distribution Control, A Sophisticated Addition to ABS," *Society of Automotive Engineers Transactions,* SAE Paper No. 920646, Society of Automotive Engineers, Warrendale, PA, 1992.

3. Chen, F., Abdelhamid, M., Blaschke, P., and Swayzel, J., "On Automotive Disc Brake Squeal Part III: Test and Evaluation," *Society of Automotive Engineers Transactions,* SAE Paper No. 2003-01-1622, Society of Automotive Engineers, Warrendale, PA, 2003.

4. Dahlberg, E., "Yaw Instability Due to Longitudinal Load Transfer During Braking in a Curve," *Society of Automotive Engineers Transactions,* SAE Paper No. 1999-01-2952, Society of Automotive Engineers, Warrendale, PA, 1999.

5. FMVSS 121, "Stability and Control of Medium and Heavy Duty Vehicles During Braking," 49 CFR Part 571, *Federal Register Final Rule,* Vol. 60, No. 47, March 10, 1995, pp. 13216–13309.

6. Feigel, H.J. and Schonlau, J., "Mechanical Brake Assist—A Potential New Standard Safety Feature," *Society of Automotive Engineers Transactions,* SAE Paper No. 1999-01-0480, Society of Automotive Engineers, Warrendale, PA, 1999.

7. Gao, Y., Chen, L., and Ehsani, M., "Investigation of the Effectiveness of Regenerative Braking for EV and HEV," *Society of Automotive Engineers Transactions,* SAE Paper No. 1999-01-2910, Society of Automotive Engineers, Warrendale, PA, 1999.

8. Gillespie, T.D., *Fundamentals of Vehicle Dynamics,* Society of Automotive Engineers, Warrendale, PA, 1992.

9. Gillespie, T.D., "Heavy Truck Ride," *Society of Automotive Engineers Transactions,* SAE Paper No. 850001, Society of Automotive Engineers, Warrendale, PA, 1985.

10. Langenwalter, J. and Kelly, B., "Virtual Design of a 42V Brake-by-Wire System," *Society of Automotive Engineers Transactions,* SAE Paper No. 2003-01-0305, Society of Automotive Engineers, Warrendale, PA, 2003.

11. Limpert, R., *Brake Design and Safety,* Society of Automotive Engineers, Warrendale, PA, 1992.

12. Moklegaard, L, Stafanopoulou, A., and Schmidt, J., "Transition from Combustion to Variable Compression Braking," *Society of Automotive Engineers Transactions,* SAE Paper No. 2000-01-1228, Society of Automotive Engineers, Warrendale, PA, 2000.

13. Ouyang, H., Nack, W., Yuan, Y., and Chen, F., "Numerical Analysis of Automotive AISC Brake Squeal: A Review," *International Journal of Vehicle Noise and Vibration,* Vol. 1, Nos. 3/4, 2005, pp. 207–231.

14. Qian, C., "Aerodynamic Shape Optimization Using CFD Parametric Model with Brake Cooling Application," *Society of Automotive Engineers Transactions*, SAE Paper No. 2002-01-0599, Society of Automotive Engineers, Warrendale, PA, 2002.

15. SAE J670e, "Vehicle Dynamics Terminology," Society of Automotive Engineers, Warrendale, PA, 1978.

16. Seluga, K., Obert, R., and Ojalvo, I., "Articulated Vehicle Yaw Stability During Braking—A Parametric Study," *Society of Automotive Engineers Transactions*, SAE Paper No. 2004-01-2630, Society of Automotive Engineers, Warrendale, PA, 2004.

17. Shaffer, S. and Radlinski, R., "Braking Capability Requirements for In-Use Commercial Vehicles—A Chronology," *Society of Automotive Engineers Transactions,* SAE Paper No. 2003-01-3397, Society of Automotive Engineers, Warrendale, PA, 2003.

18. Shi, T., Dessouki, O., Warzecha, T., and Chang, W., "Advances in Complex Eigenvalue Analysis for Brake Noise," *Society of Automotive Engineers Transactions*, SAE Paper No. 2001-01-1603, Society of Automotive Engineers, Warrendale, PA, 2001.

19. Shigley, J.E. and Mischke, C.R., *Mechanical Engineering Design*, 5th Ed., McGraw-Hill, New York, 1989.

20. Tamura, T., Inoue, H., Watanabe, T., and Maruko, N., "Research on a Brake Assist System with a Preview Function," *Society of Automotive Engineers Transactions,* SAE Paper No. 2001-01-0357, Society of Automotive Engineers, Warrendale, PA, 2001.

21. Wong, J.Y., *Theory of Ground Vehicles,* 3rd Ed., John Wiley and Sons, New York, 2001.

22. Yoon, P., Kang, H.-J., and Hwang, J., "Braking Status Monitoring for Brake-By-Wire Systems," *Society of Automotive Engineers Transactions,* SAE Paper No. 2004-01-0259, Society of Automotive Engineers, Warrendale, PA, 2004.

Chapter 8

Acceleration

8.1 Introduction

Among the many attributes customers look for when purchasing a vehicle is its power. Power ranks as the top attribute with customers seeking sport vehicles, and it ranks among the top ten attributes for most other customers. Vehicle power affects the load-carrying capacity and acceleration of the vehicle. In this chapter, we will study the acceleration of vehicles under various driving and loading conditions.

Vehicle acceleration is limited by two main factors. The first is its power. In power-limited acceleration, the maximum acceleration is obtained when the vehicle is driven at its maximum power, provided that no slippage occurs at the tires. The second factor is traction. In traction-limited acceleration, the maximum acceleration is obtained when the tires start slipping. At low speeds and in low gears, acceleration typically is limited by traction. At high speeds and in high gears, acceleration is limited by the power generated by the powerplant. The term "powerplant" will refer to the engine and transmission.

If the vehicle is moving as a result of consistent and constant tractive forces at the tires, changing its velocity from an initial velocity V_i (which would be zero if one starts from an idle status) to a final velocity V_f, the acceleration can be calculated as

$$a = \frac{\left(V_f - V_i\right)}{t} \tag{8.1}$$

where t is time. Note that Eq. 8.1 is the same as that used in Chapter 7 (Eq. 7.1) for braking and is repeated here for convenience. The only difference is that the acceleration will be a positive number when the vehicle is accelerating. If the total tractive force on both axles is F_b, then from Newton's second law of motion, the acceleration is $\frac{F_b}{M}$, where M is the mass of the vehicle and its load.

The distance traveled during acceleration (X_b) can be calculated by (repeated from Eq. 7.2 for convenience)

$$X_b = \frac{\left(V_f^2 - V_i^2\right)}{2 \times a} \tag{8.2}$$

EXAMPLE E8.1

A 2000-kg vehicle is to achieve a maximum speed of 100 km/hr in 6 seconds. Determine the tractive force needed between the tires and the road to achieve this speed and the distance it takes to reach that speed from an idle condition (i.e., complete stop).

Solution:

The final velocity can be written as

$$V_f = \frac{100 \text{ km}}{\text{hr}} \times \frac{1000 \text{ m}}{\text{km}} \times \frac{\text{hr}}{3600 \text{ sec}} = 27.78 \text{ m/sec}$$

The acceleration can be found from

$$a = \frac{V_f - V_i}{t} = \frac{27.78 - 0}{6} = 4.63 \text{ m/sec}^2$$

The force needed to achieve this acceleration is

$$F = M \times a$$

$$= 2000 \text{ kg} \times 4.63 \text{ m/sec}^2$$

$$= 9260 \text{ kg m/sec}^2$$

$$= 9260 \text{ N}$$

The distance needed to reach the maximum speed is

$$X_b = \frac{(V_f^2 - V_i^2)}{2 \times a}$$

$$= \frac{(27.78)^2 - 0}{2 \times (4.63)}$$

$$= 83.34 \text{ m}$$

In our quest for detailed understanding of vehicle acceleration, we need to understand the load transfer characteristics of an accelerating vehicle.

8.2 Load Transfer During Acceleration

The fundamental forces acting on an accelerating vehicle were discussed in Chapter 3. These equations and the treatment that followed are similar to that used in Chapter 7. Some of these forces will be reviewed briefly here. First we will consider acceleration of a vehicle on a horizontal road (i.e., no grade).

8.2.1 Simple Acceleration on a Horizontal Road

Figure 8.1 gives the free body diagram of the vehicle that shows the forces acting on it under normal driving conditions. Note that we have neglected aerodynamic forces, rolling resistance, and powerplant resistance in this simple vehicle model. The mathematical manipulation of Newton's second law yields the equations derived in Chapter 7 and will not be repeated here. The vertical forces at the wheels are

$$F_{z1} = Mg\left(\frac{c_2}{c_1 + c_2}\right) - \frac{h}{c_1 + c_2}Ma$$

$$F_{z2} = Mg\left(\frac{c_1}{c_1 + c_2}\right) + \frac{h}{c_1 + c_2}Ma$$

(8.3)

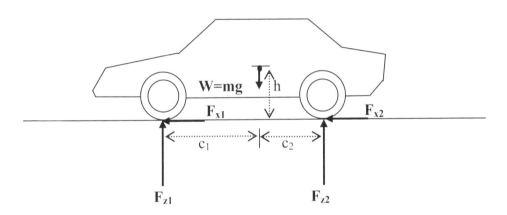

Figure 8.1 Simple vehicle model on a horizontal plane.

These equations are the same as Eq. 7.25 from Chapter 7, repeated here for convenience. Note that we are assuming that the vehicle is accelerating; thus, a positive value for the acceleration (a) is applied.

When the vehicle is accelerating, the axle force on the front wheels will be smaller than those forces experienced under cruising driving conditions (i.e., no acceleration or deceleration). As the vehicle accelerates, the load is transferred from the front axle to the rear axle. The total axle load of both axles remains equal to the weight of the vehicle.

EXAMPLE E8.2

Consider a midsize sedan that weighs 14,234.31 N (3200 lb) under a certain load. The wheelbase of the vehicle is 2.74 m (9 ft, or 108 in.), and the center of gravity is 1.067 m (3.5 ft) from the front axle and 0.508 m (20 in.) above the ground. Determine the load on each axle if the vehicle is under no acceleration, accelerating at 0.610 m/s^2 (2 ft/sec^2) and 1.829 m/s^2 (6 ft/sec^2).

Solution:

The following are given in the problem:

$$W = 3200 \text{ lb}$$

$$c_1 + c_2 = 108 \text{ in.}$$

$$h = 20 \text{ in.}$$

$$c_1 = 42 \text{ in.}$$

First we will find the mass and c_2 as

$$M = \frac{3200}{32.2} = 99.38 \text{ slugs}$$

$$c_2 = 108 - 42 = 66 \text{ in.}$$

Apply Eq. 8.3 (or Eq. 7.25 from Chapter 7) to obtain the loading under no acceleration (i.e., a = 0) as

$$F_{z1} = Mg\left(\frac{c_2}{c_1 + c_2}\right) - \frac{h}{c_1 + c_2}Ma$$

$$= 99.38 \times 32.2\left(\frac{66}{108}\right) - \frac{20}{108} \times 99.38 \times 0$$

$$= 1955.56 \text{ lb}$$

$$F_{z2} = Mg\left(\frac{c_1}{c_1 + c_2}\right) + \frac{h}{c_1 + c_2}Ma$$

$$= 99.38 \times 32.2\left(\frac{42}{108}\right) + \frac{20}{108} \times 99.38 \times 0$$

$$= 1244.44 \text{ lb}$$

Similarly, for a = 2 ft/sec^2,

$$F_{z1} = 99.38 \times 32.2\left(\frac{66}{108}\right) - \frac{20}{108} \times 99.38 \times (2)$$

$$= 1918.76 \text{ lb}$$

$$F_{z2} = 99.38 \times 32.2 \left(\frac{42}{108} \right) + \frac{20}{108} \times 99.38 \times (2)$$

$$= 1281.24 \text{ lb}$$

For a = 6 ft/sec²,

$$F_{z1} = 99.38 \times 32.2 \left(\frac{66}{108} \right) - \frac{20}{108} \times 99.38 \times (6)$$

$$= 1845.14 \text{ lb}$$

$$F_{z2} = 99.38 \times 32.2 \left(\frac{42}{108} \right) + \frac{20}{108} \times 99.38 \times (6)$$

$$= 1354.86 \text{ lb}$$

The total load on the axles remains at 3200 lb, which is the weight of the vehicle. Load transfers from the front axle to the rear axle as the vehicle accelerates. Figure 8.2 shows the load distribution as a function of acceleration, using the parameters of Example E8.2. Of course, maximum acceleration is limited by the power of the vehicle.

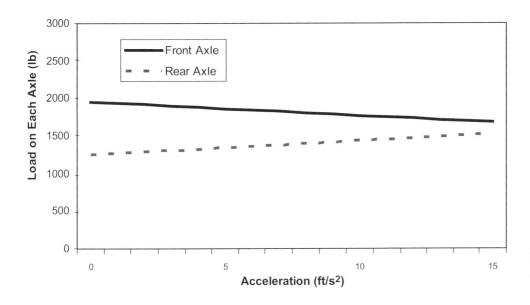

Figure 8.2 Load transfer during acceleration.

8.2.2 Effect of Aerodynamic and Other Forces

In this section, other forces are considered in the accelerating vehicle model. Similar to Chapter 7 on braking, a more accurate accelerating vehicle model must take other factors into consideration. In particular, the rolling resistance of the tires, the aerodynamic drag resistance, and the driveline drag should be accounted for.

The rolling resistance is between the tires and the drive axle. If the rolling coefficient is distinguished by μ_r, then the rolling resistance between the tire and the drive axle can be written as in Eq. 7.26 of Chapter 7, or

$$R_r = \mu_r Mg$$

The coefficient of rolling resistance varies between 0.012 and 0.02 on concrete surfaces, with radial tires offering lower coefficients than bias ply tires. This coefficient reaches 0.1 when the vehicle is driven on hard soil, and it reaches 0.2 to 0.3 when the vehicle is driven on sand. The rolling resistance usually is less for tires with large diameters.

The aerodynamic resistance is generated mainly by air flow around the vehicle and air flow through the vehicle. The aerodynamic resistance is estimated by Eq. 7.27 in Chapter 7 as

$$R_a = \frac{\rho}{2}C_D A_F V_r^2$$

where

$$\rho \quad = \quad \text{density of the air}$$

$$C_D \quad = \quad \text{drag coefficient}$$

$$A_F \quad = \quad \text{projected area of the front of the vehicle}$$

$$V_r \quad = \quad \text{relative velocity of the vehicle with respect to the wind}$$

The coefficient of drag ranges between 0.3 and 0.5 for passenger cars and can reach a value of 0.8 for vans. Similarly, aerodynamic lift and pitching can be treated (see Eqs. 7.28 and 7.29 in Chapter 7).

Now we will consider the aerodynamic drag and rolling resistance in developing the load transfer equations. Figure 8.3 (which is the same as Figure 7.6 in Chapter 7) shows the vehicle model on a horizontal plane, including the forces of rolling resistance and aerodynamic drag.

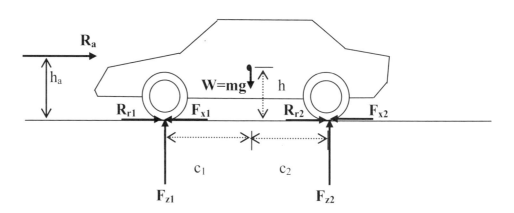

Figure 8.3 Vehicle model on a horizontal plane.

Note that the resultant of the aerodynamic drag is represented here by the force R_a acting at some height h_a. These are found either from wind tunnel testing or, more recently, from advanced computer models using computational fluid dynamics (CFD) software. The vertical forces (axle loads) can be written as (Eq. 7.30 from Chapter 7)

$$F_{z1} = Mg\left(\frac{c_2}{c_1 + c_2}\right) - \frac{h_a}{c_1 + c_2}R_a - \frac{h}{c_1 + c_2}Ma$$

$$F_{z2} = Mg\left(\frac{c_1}{c_1 + c_2}\right) + \frac{h_a}{c_1 + c_2}R_a + \frac{h}{c_1 + c_2}Ma \qquad (8.4)$$

Acceleration will be treated simply by using a positive value for a. Often, h_a is assumed to be the same as h.

8.2.3 Effect of Grade

Now we will consider the effect of grade as well as aerodynamic drag and rolling resistance forces in developing the load transfer equations. Figure 8.4 shows the vehicle model on an inclined surface, with consideration of rolling resistance and aerodynamic drag.

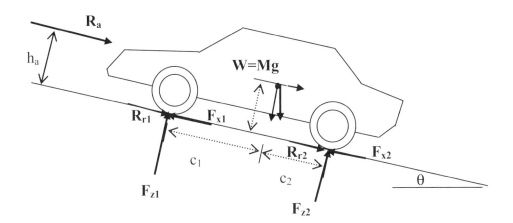

Figure 8.4 Vehicle model on an incline (grade): (a) front-wheel drive, and (b) all-wheel drive or four-by-four (front-wheel-drive based).

The vertical forces (axle loads) for a vehicle going uphill are (Eq. 7.31b from Chapter 7 is repeated here for convenience)

$$F_{z1} = \frac{1}{c_1 + c_2}\left(Mgc_2 \cos\theta - h_aR_a - hMa - hMg\sin\theta\right) = hMa$$

$$F_{z2} = \frac{1}{c_1 + c_2}\left(Mgc_1 \cos\theta + h_aR_a + hMa + hMg\sin\theta\right) = hMa \qquad (8.5)$$

Note that the sign of the last term inside the bracket should be reversed when the vehicle is going downhill. Acceleration will be treated simply by using a positive value for a. The total tractive force F_x ($F_{x1} + F_{x2}$) can be found as

$$F_x = R_a + R_r + Ma + Mg\sin\theta \qquad (8.6)$$

or

$$F_x - R_r = R_a + Ma + Mg\sin\theta \qquad (8.7)$$

where R_r is the total rolling resistance force of both axles ($R_r = R_{r1} + R_{r2} = Mg\,\mu_r$). Assuming $h_a = h$ and substituting Eq. 8.7 into Eq. 8.5 yields the load on the front axle as

$$
\begin{aligned}
F_{z1} &= \frac{1}{c_1 + c_2}\Big(Mgc_2\cos\theta - h\big(R_a + Ma + Mg\sin\theta\big)\Big) \\[2mm]
&= \frac{1}{c_1 + c_2}\Big(Mgc_2\cos\theta - h\big(F_x - R_r\big)\Big)
\end{aligned}
\qquad (8.8)
$$

The load on the rear axle is

$$
\begin{aligned}
F_{z2} &= \frac{1}{c_1 + c_2}\Big(Mgc_1\cos\theta + h\big(R_a + Ma + Mg\sin\theta\big)\Big) \\[2mm]
&= \frac{1}{c_1 + c_2}\Big(Mgc_1\cos\theta + h\big(F_x - R_r\big)\Big)
\end{aligned}
\qquad (8.9)
$$

EXAMPLE E8.3

Consider a midsize sedan that weighs 14,234.3 N (3200 lb) under a certain load. The vehicle is going uphill at a 5° slope at a speed of 26.82 m/s (60 mph). The frontal area of the vehicle is 2.044 m^2 (22.0 ft^2), and the coefficient of drag is 0.4. The aerodynamic resistance is acting on the center of gravity. The wheelbase of the vehicle is 2.74 m (9 ft), and the center of gravity is 1.372 m (4.5 ft, or 66 in.) from the front axle and 0.508 m (20 in.) above the ground. Determine the load on each axle if the vehicle is under no acceleration. What would the loads be if the vehicle is accelerating at 1.524 m/s^2 (5 ft/sec^2)? The density of air is 1.226 kg/m^3 (0.002378 slug/ft^3). Calculate the axle forces when going downhill at a 5° slope with the same speed.

Solution:

The aerodynamic resistance first is calculated as

$$
\begin{aligned}
R_a &= \frac{\rho}{2}C_D A_F V_r^2 \\[2mm]
&= \frac{0.002378}{2} \times 0.4 \times 22 \times (88)^2 \\[2mm]
&= 81.03 \text{ lb}
\end{aligned}
$$

When traveling on a horizontal road, the forces on the axles are

$$F_{z1} = \frac{1}{c_1 + c_2}\left(Mgc_2\cos\theta - h_aR_a - hMa - hMg\sin\theta\right)$$

$$= \frac{1}{c_1 + c_2}\left(Mgc_2 - h_aR_a - hMa\right)$$

$$= \frac{1}{108}\left(3200 \times 66 - 20 \times 81 - 20 \times 99.38 \times 5\right)$$

$$= 1848.5 \text{ lb}$$

$$F_{z2} = \frac{1}{c_1 + c_2}\left(Mgc_1\cos\theta + h_aR_a + hMa + hMg\sin\theta\right)$$

$$= \frac{1}{c_1 + c_2}\left(Mgc_1 + h_aR_a + hMa\right)$$

$$= \frac{1}{108}\left(3200 \times 42 + 20 \times 81 + 20 \times 99.38 \times 5\right)$$

$$= 1351.5 \text{ lb}$$

When traveling uphill, the forces on the axles are

$$F_{z1} = \frac{1}{c_1 + c_2}\left(Mgc_2\cos\theta - h_aR_a - hMa - hMg\sin\theta\right)$$

$$= \frac{1}{108}\left(3200 \times 66 \times \cos(5) - 20 \times 81 - 20 \times 99.38 \times 5 - 20 \times 3200 \times \sin(5)\right)$$

$$= 1789.4 \text{ lb}$$

$$F_{z2} = \frac{1}{c_1 + c_2}\left(Mgc_1\cos\theta + h_aR_a + hMa + hMg\sin\theta\right)$$

$$= \frac{1}{108}\left(3200 \times 42 \times \cos(5) + 20 \times 81 + 20 \times 99.38 \times 5 + 20 \times 3200 \times \sin(5)\right)$$

$$= 1398.4 \text{ lb}$$

When the vehicle is traveling downhill, the forces are

$$F_{z1} = \frac{1}{c_1 + c_2}\left(Mgc_2\cos\theta - h_aR_a - hMa + hMg\sin\theta\right)$$

$$= \frac{1}{108}\left(3200 \times 66 \times \cos(5) - 20 \times 81 - 20 \times 99.38 \times 5 + 20 \times 3200 \times \sin(5)\right)$$

$$= 1892.7 \text{ lb}$$

$$F_{z2} = \frac{1}{c_1 + c_2}\left(Mgc_1\cos\theta + h_a R_a + hMa - hMg\sin\theta\right)$$

$$= \frac{1}{108}\left(3200 \times 42 \times \cos(5) + 20 \times 81 + 20 \times 99.38 \times 5 - 20 \times 3200 \times \sin(5)\right)$$

$$= 1295.1\ \text{lb}$$

Observe that uphill travel increased the forces on the rear wheels. On the other hand, downhill travel increased the forces on the front wheels. As will be shown in the next section, this will add to the advantages of rear-wheel-drive (RWD) vehicles in accelerating on uphill slopes.

The load described here represents the weight distribution at the initial acceleration from a startup speed of 60 mph. Aerodynamic forces will increase as the speed increases during acceleration. As was found earlier for vehicles under braking, the aerodynamic forces reduce the load on the front axle and increase it on the rear axle. Accurate analysis of the force distribution as the vehicle accelerates should take into consideration aerodynamic drag.

8.3 Traction-Limited Acceleration

One factor that limits acceleration is traction between the tires and the road. This is particularly felt by drivers of front-wheel-drive (FWD) vehicles under startup conditions and when accelerating on smooth surfaces. We will study the maximum attainable acceleration of various vehicles having different drive mechanisms.

8.3.1 Drivetrain Configurations

Many drive mechanisms are available in the marketplace for road vehicles. In particular, there are front-wheel-drive (FWD) systems, rear-wheel-drive (RWD) systems, and all-wheel-drive (AWD) or four-by-four (4×4) systems. In the latter system, different mechanisms exist for torque distribution between the front and rear axles.

A front-wheel-drive system usually consists of the engine configured such that the crankshaft direction is parallel to the axle direction (line connecting the front wheels). Such a configuration is called an east-west configuration. Figure 8.5 shows a typical front-wheel-drive configuration. The figure also shows how a front-wheel-drive system can be configured or adapted for an all-wheel-drive application. This can be done by adding several components. The first is a power takeoff unit (PTU) that consists of bevel gears to produce a line of motion perpendicular to that given at the output of the transmission. The second is the driveline, which may consist of a single-, two-, or three-piece driveshaft. Large cars and trucks tend to have a single-piece (relatively large in diameter) shaft. Small cars tend to have a two-piece or three-piece (smaller in diameter) driveline. The driveline also includes one or more center bearing(s) to mount the shaft(s) to the vehicle body. The third major unit is the coupler. In all-wheel-drive systems, the coupler is a unit of a clutch mechanism that engages the rear wheels (axle) only when a slip condition at the front wheels is observed. At this point, we will establish a difference between the all-wheel-drive and four-by-four systems. The all-wheel-drive systems are those systems in which the other axle (i.e., the rear axle in this case) is engaged as needed, without input from the driver. In other words, the engagement happens when a

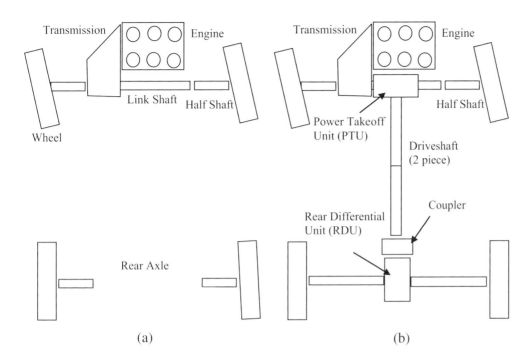

Transmission / Engine — Link Shaft — Half Shaft — Wheel — Rear Axle

(a)

Transmission / Engine — Power Takeoff Unit (PTU) — Half Shaft — Driveshaft (2 piece) — Coupler — Rear Differential Unit (RDU)

(b)

Figure 8.5 (a) Front-wheel-drive (FWD) and (b) front-wheel-drive-based all-wheel-drive (AWD) systems.

slip condition of the front axle is met. The four-by-four system is a system in which the drive system is either front-wheel drive all the time or can engage the other axle (rear) with an on/off switch activated by the driver. The four-by-four system may not need a coupler. The last major unit in the front-wheel-drive-based all-wheel-drive system is the rear differential unit (RDU). This unit consists of a set of bevel gears to distribute the torque and motion from the driveshaft(s) to the rear wheels.

Front-wheel-drive vehicles offer a major package advantage when compared with rear-wheel-drive ones. The power pack is compact, allowing more space to be used for the cabinet. Initially, front-wheel drive was used for light vehicles; however, one recently can find vehicles with V6 and V8 engines that are front-wheel drive delivering more than 200 hp. Because the engine is mounted on the front chassis systems, more load will be given to the front, or driving, wheels. This offers better traction capability and road handling, especially on icy and wet roads. Furthermore, front-wheel-drive vehicles are less sensitive to wind loads. This arrangement offers a shorter power flow from the engine to the driving wheels, thus minimizing power losses in the system.

Among the disadvantages of the front-wheel-drive system is its compact engine design, forcing shorter engines to be used. The gains made by offering higher loads to the driving front wheels are compromised with these wheels wearing off early. Also, this configuration presents challenges to the engineers designing the mount system. Frequently, poor mount design would lead to inferior noise and vibration performance. Because the transmission is packaged closer to the left (or right) wheels, it needs longer shafts to be connected with the right wheel than it does with the left wheel. This will allow the longer shaft to twist more under torque, causing the vehicle to steer unintentionally. This phenomenon is referred to as torque steer.

Torque distribution among front and rear axles is engineered to give maximum traction, as will be seen later. In certain applications, a distribution is determined *a priori* without an active control mechanism. Among common distributions are the 50/50 distribution, in which half the torques goes to each axle, and the 40/60 distribution for rear-wheel-drive-based all-wheel-drive and 60/40 for front-wheel-drive-based all-wheel-drive systems.

A rear-wheel-drive system consists of the engine configured such that the crankshaft direction is perpendicular to the axle direction. This configuration is called a north-south configuration. Figure 8.6 shows a typical rear-wheel-drive configuration. Similar to the front-wheel-drive configuration, the rear-wheel-drive system can be configured for all-wheel-drive and four-by-four applications. Several components must be added. The first is the transfer case (TC) that is added at the output end of the transmission. This unit consists of a chain and a set of spur gears. The purpose of the transfer case is to bring some torque back to the front wheels. The second major unit is the front driveshaft, and the third major unit is the coupler. The last major unit in the rear-wheel-drive-based all-wheel-drive system is the front differential unit (FDU).

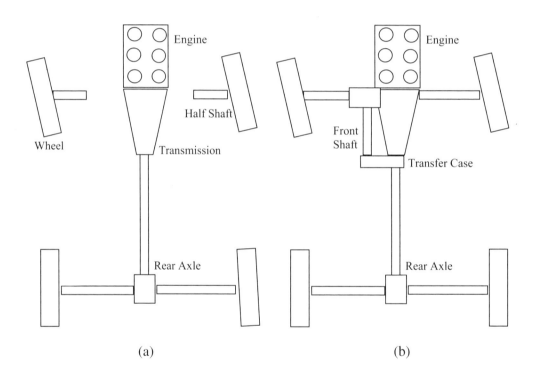

Figure 8.6 (a) Rear-wheel-drive and (b) rear-wheel-drive-based all-wheel-drive or four-by-four systems.

Other changes take place as a result of making an all-wheel-drive (or four-by-four) version of a front-wheel-drive or rear-wheel-drive system. The major obstacle typically facing engineers in modern vehicles is the issue of package. All-wheel-drive and four-by-four systems are more challenging to package.

The rear-wheel-drive vehicle design offers certain important advantages. This design sets virtually no limit on the engine length. Thus, larger engines can be used (e.g., V8, V10, and V12). Insulation of the engine noise with this mount system is relatively easy and more robust when compared with the front-wheel-drive system. In addition, this system is more robust to torque steer. Rear-wheel-drive offers balanced weight distribution between the front and rear wheels, allowing longer life for the tires before they wear out.

Other configurations exist in the marketplace. These configurations include rear-engine vehicles. Such a configuration often is incorporated into sport vehicles used for racing purposes. In these vehicles, the engine is placed in the rear of the vehicle. Other configurations place the engine beneath the body of the vehicle. For more information on drive configurations, the reader is referred to the book by Reimpell and Stoll (1996).

We will start by considering single-axle drive but will move later to consider all-wheel-drive systems.

8.3.2 Front-Wheel Drive

First let us consider a horizontal plane with no resistance or drag forces. For front-wheel-drive systems, the tractive forces are applied to the front wheels. These forces will equal the coefficient of friction multiplied by the normal force at those wheels

$$F_{x1} = \mu F_{z1} \tag{8.10}$$

The tractive force at the rear wheels will be zero. The maximum acceleration can be found from Newton's second law

$$F_{x1} + F_{x2} = Ma$$

and by utilizing Eq. 8.3 as

$$Ma_F = \mu Mg \left(\frac{c_2}{c_1 + c_2} \right) - \mu \left(\frac{h}{c_1 + c_2} \right) Ma_F \tag{8.11}$$

These equations are similar to Eqs. 7.32 and 7.33 from Chapter 7. Note the sign changes due to tractive force being opposite in direction to the braking force. We replaced the acceleration symbol a by a_F to indicate that this equation is used to calculate the maximum acceleration when the tractive effort is applied to the front axle. This equation can be solved directly for the vehicle acceleration as

$$a_F = g \frac{\mu \left(\dfrac{c_2}{c_1 + c_2} \right)}{1 + \mu \left(\dfrac{h}{c_1 + c_2} \right)} \tag{8.12}$$

The maximum tractive force applied in a front-wheel-drive system is

$$F_{x1} = M \times a_F$$

When the grade, rolling resistance, and aerodynamic drag are considered, Eq. 8.10 is substituted into Eq. 8.8. This yields the maximum tractive forces as

$$F_{z1} = \frac{1}{c_1 + c_2}\left(Mgc_2\cos\theta - h\left(R_a + Ma + Mg\sin\theta\right)\right)$$

$$= \frac{1}{c_1 + c_2}\left(Mgc_2\cos\theta - h\left(F_{x1} - R_r\right)\right) \tag{8.13}$$

$$F_{x1} = \frac{\mu}{c_1 + c_2}\left(Mgc_2\cos\theta - h\left(F_{x1} - R_r\right)\right)$$

Substituting $R_r = \mu_r Mg$ into Eq. 8.13 and rewriting that equation for F_x gives

$$F_{x1} + \frac{\mu h}{c_1 + c_2}F_{x1} = \frac{\mu Mg}{c_1 + c_2}\left(c_2\cos\theta + h\mu_r\right)$$

or

$$F_{x1} = \frac{\dfrac{\mu Mg}{c_1 + c_2}\left(c_2\cos\theta + h\mu_r\right)}{1 + \dfrac{\mu h}{c_1 + c_2}} \tag{8.14}$$

Note that F_{x2} is zero for a front-wheel-drive system. Equation 8.14 gives the maximum tractive force that can be applied to a front-wheel-drive system. The maximum acceleration can be obtained from Eq. 8.6.

8.3.3 Rear-Wheel Drive

Similar treatment can be made if the tractive forces are applied to only the rear wheels. In this case, the tractive force on the front wheels is zero. The maximum tractive force on the rear wheels is

$$F_{x2} = \mu F_{z2} \tag{8.15}$$

The maximum acceleration for rear-wheel-drive vehicles becomes

$$a_R = g\frac{\mu\left(\dfrac{c_1}{c_1 + c_2}\right)}{1 - \mu\left(\dfrac{h}{c_1 + c_2}\right)} \tag{8.16}$$

Figures 8.7 and 8.8 show the acceleration of a typical compact vehicle and the impact of the location of the center of gravity on the vehicle maximum traction-limited acceleration for both front-wheel-drive and rear-wheel-drive systems. In these figures, aerodynamic drag, rolling forces, and grade are not included.

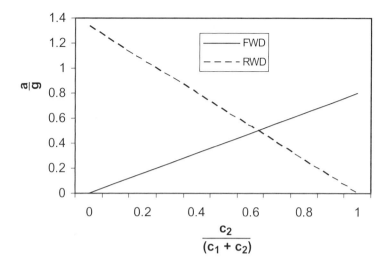

Figure 8.7 Acceleration for front-wheel-drive and rear-wheel-drive vehicles, where $\mu = 1$, $h = 0.5$ m, and $c_1 + c_2 = 2.0$.

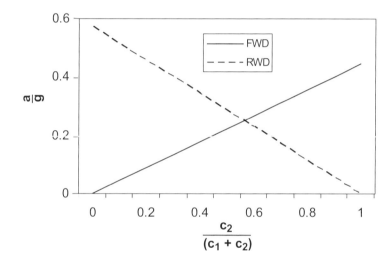

Figure 8.8 Acceleration for front-wheel-drive and rear-wheel-drive vehicles, where $\mu = 0.5$, $h = 0.5$ m, and $c_1 + c_2 = 2.0$.

Note that for typical vehicle designs and loading conditions, the value of $\dfrac{c_2}{c_1 + c_2}$ ranges from 0.4 to 0.6. Within this range of application, the maximum acceleration that can be obtained for rear-wheel-drive systems is between the values of 0.22 and 0.34g when $\mu = 0.5$, and between 0.53 and 0.8g when $\mu = 1$. For front-wheel-drive systems, the maximum acceleration that can be achieved is between 0.18 and 0.27g when $\mu = 0.5$, and between 0.32 and 0.48g when $\mu = 1$. This shows that for normal acceleration conditions in a single drive, higher levels of maximum acceleration can be achieved with rear-wheel-drive. Thus, sport vehicles are almost always rear-wheel drive.

In a rear-wheel-drive vehicle, grade, rolling resistance, and aerodynamic drag can be included in a fashion similar to that done for a front-wheel-drive system. This yields Eq. 8.17 for maximum tractive forces applied to the rear wheels as

$$F_{x2} = \dfrac{\dfrac{\mu Mg}{c_1 + c_2}\left(c_1 \cos\theta - h\mu_r\right)}{1 - \dfrac{\mu h}{c_1 + c_2}} \tag{8.17}$$

Note that F_{x1} is zero for a rear-wheel-drive system. Equation 8.17 gives the maximum tractive force that can be applied to a rear-wheel-drive vehicle. The maximum acceleration can be obtained from Eq. 8.6.

8.3.4 All-Wheel-Drive and Four-by-Four Systems

In recent years, all-wheel-drive vehicles and four-by-four ones (also called four-wheel-drive) have received considerable attention. In four-by-four systems, the tractive forces of the vehicle are distributed on all its wheels under all driving conditions. The tractive forces may be distributed equally or have some bias toward a certain axis (front or rear). In all-wheel-drive systems, the vehicle normally is driven with one axis (either rear or front) until wheel spinning is detected. Only when one or more of the wheels is detected to be spinning is a certain portion of the torque distributed to the other axle.

Let us assume that the tractive torque is distributed such that a certain ratio K_f of the total torque goes to the front axle. The remainder that goes to the rear axle is $1 - K_f$ of the total torque. We then can generate Eq. 8.18 in a manner similar to that used in Eq. 7.37b in Chapter 7

$$F_{x1} = \frac{K_f}{1 - K_f} F_{x2}$$

$$F_{x2} = \frac{1 - K_f}{K_f} F_{x1} \tag{8.18}$$

Consider the free body diagram of the vehicle under direct tractive forces F_{x1} for the front axle and F_{x2} for the rear axle. The vehicle is still under the effects of aerodynamic forces as well as grade and rolling resistance. The total tractive force is F_x ($F_{x1} + F_{x2}$). The total rolling resistance force is R_r ($R_{r1} + R_{r2}$). The total weight is distributed as in Eq. 8.5. Equation 8.6 then can be written as

$$F_{x1} + F_{x2} - R_a - R_r - Mg\sin(\theta) = Ma \tag{8.19}$$

8.3.4.1 Front Skid

Under a front skid condition, $F_{x1} = \mu F_{z1}$. Substituting this and Eq. 8.5 into Eq. 8.19 yields

$$\mu\left(\frac{1}{c_1 + c_2}\left(Mgc_2\cos\theta - h_a R_a - hMa_F - hMg\sin\theta\right)\right)\left(1 + \frac{1 - K_f}{K_f}\right)$$
$$- R_a - R_r - Mg\sin\theta = Ma_F \tag{8.20}$$

Rearranging for a_F gives

$$\mu\left(\frac{1}{c_1 + c_2}\left(Mgc_2\cos\theta - h_aR_a - hMg\sin\theta\right)\right)\left(1 + \frac{1 - K_f}{K_f}\right) - R_a - R_r - Mg\sin(\theta)$$

$$= \left(\frac{\mu h}{c_1 + c_2}\left(1 + \frac{1 - K_f}{K_f}\right) + 1\right)Ma_F \tag{8.21}$$

or simply

$$a_F = \frac{\mu\left(\frac{1}{c_1 + c_2}\left(Mgc_2\cos\theta - h_aR_a - hMg\sin\theta\right)\right)\left(1 + \frac{1 - K_f}{K_f}\right) - R_a - R_r - Mg\sin(\theta)}{\left(\frac{\mu h}{c_1 + c_2}\left(1 + \frac{1 - K_f}{K_f}\right) + 1\right)M} \tag{8.22}$$

Note that the acceleration symbol a has a subscript F to indicate that the acceleration obtained is the maximum acceleration achievable when the front tires are under the skid condition.

Ignoring aerodynamic drag, grade, and rolling resistance, the acceleration is

$$a_F = \frac{\mu\left(\frac{c_2}{c_1 + c_2}\right)\left(1 + \frac{1 - K_f}{K_f}\right)}{\left(\frac{\mu h}{c_1 + c_2}\left(1 + \frac{1 - K_f}{K_f}\right) + 1\right)}g \tag{8.23}$$

Note that Eqs. 8.20 through 8.23 are similar to those of Eqs. 7.40 through 7.43 from Chapter 7 used earlier in braking. Equation 8.23 will reduce to that of Eq. 8.12 when only a front axle is under loading in a front-wheel-drive system ($K_f = 1$).

8.3.4.2 Rear Skid

Equation 8.19 can be rewritten as

$$F_{x2}\left(1 + \frac{K_f}{1 - K_f}\right) - R_a - R_r - Mg\sin(\theta) = Ma \tag{8.24}$$

When under rear skid, $F_{x2} = \mu F_{z2}$. Substituting this and Eq. 8.5 into Eq. 8.24 yields

$$\mu\left(\frac{1}{c_1 + c_2}\left(Mgc_1\cos\theta + h_aR_a + hMa_R + Mg\sin\theta\right)\right)\left(1 + \frac{K_f}{1 - K_f}\right)$$

$$- R_a - R_r - Mg\sin(\theta) = Ma_R \tag{8.25}$$

Rearranging for a_R gives

$$\mu\left(\frac{1}{c_1 + c_2}(Mgc_1\cos\theta + h_aR_a + Mg\sin\theta)\right)\left(1 + \frac{K_f}{1 - K_f}\right) - R_a - R_r - Mg\sin(\theta)$$

$$= \left(-\frac{\mu h}{c_1 + c_2}\left(1 + \frac{K_f}{1 - K_f}\right) + 1\right)Ma_R \qquad (8.26)$$

Simplify as

$$a_R = \frac{\mu\left(\frac{1}{c_1 + c_2}(Mgc_1 + h_aR_a + Mg\sin\theta)\right)\left(1 + \frac{K_f}{1 - K_f}\right) - R_a - R_r - Mg\sin(\theta)}{\left(-\frac{\mu h}{c_1 + c_2}\left(1 + \frac{K_f}{1 - K_f}\right) + 1\right)M} \qquad (8.27)$$

Equations 8.25 through 8.28 are similar to those of Eqs. 8.45 to 7.28. Ignoring aerodynamic drag, grade, and rolling resistance, the acceleration is

$$a_R = \frac{\mu\left(\frac{c_1}{c_1 + c_2}\right)\left(1 + \frac{K_f}{1 - K_f}\right)}{\left(-\frac{\mu h}{c_1 + c_2}\left(1 + \frac{K_f}{1 - K_f}\right) + 1\right)}g \qquad (8.28)$$

Note that Eq. 8.28 will reduce to that of Eq. 8.16 in a rear-wheel-drive vehicle where $K_f = 0$. Also note that in Eq. 8.28, the denominator comes closer to zero, which yields a higher value for a_R. This indicates the increased possibility that a front skid will happen first.

The front tires will skid first if $a_F < a_R$; otherwise, the rear tires will skid first.

Example E8.4

A 21,351.46-N (4800-lb) passenger car has a wheelbase of 2844.80 mm (112 in.) and a center of gravity 1270.00 mm (50 in.) behind the front axle and 508 mm (20 in.) above the ground. The tractive effort distribution gives the front axle 50% of the total tractive force. The car is moving on a horizontal plane. Determine which tires will skid first if the car is moving on a surface with first a 0.75 coefficient of friction, and second a 0.25 coefficient of friction. Ignore drag and rolling resistance.

Solution:

$$M = 4800 \text{ lb} = 149.07 \text{ slugs}$$

$$c_1 + c_2 = 112 \text{ in.}$$

$$c_1 = 50 \text{ in.}$$

$$c_2 = 62 \text{ in.}$$

$$K_f = 0.5$$

$$h = 20$$

$$\theta = 0$$

$$R_a = 0$$

$$R_r = 0$$

First surface: $\mu = 0.75$

Acceleration under front skid is

$$a_F = \frac{\mu\left(\dfrac{c_2}{c_1 + c_2}\right)\left(1 + \dfrac{1 - K_f}{K_f}\right)}{\left(\dfrac{\mu h}{c_1 + c_2}\left(1 + \dfrac{1 - K_f}{K_f}\right) + 1\right)}g$$

$$= \frac{0.75\left(\dfrac{62}{112}\right)\left(1 + \dfrac{0.5}{0.5}\right)}{\left(\dfrac{0.75 \times 20}{112}\left(1 + \dfrac{0.5}{0.5}\right) + 1\right)}g$$

$$= 0.6549g$$

Acceleration under rear skid is

$$a_R = \frac{\mu\left(\dfrac{c_1}{c_1 + c_2}\right)\left(1 + \dfrac{K_f}{1 - K_f}\right)}{\left(-\dfrac{\mu h}{c_1 + c_2}\left(1 + \dfrac{K_f}{1 - K_f}\right) + 1\right)}g$$

$$= \frac{0.75\left(\dfrac{50}{112}\right)\left(1 + \dfrac{0.5}{0.5}\right)}{\left(-\dfrac{0.75 \times 20}{112}\left(1 + \dfrac{0.5}{0.5}\right) + 1\right)}g$$

$$= 0.9146g$$

Because $a_F < a_R$, the front tires will skid first.

Second surface: $\mu = 0.25$

Acceleration under front skid is

$$a_F = \frac{\mu\left(\dfrac{c_2}{c_1+c_2}\right)\left(1+\dfrac{1-K_f}{K_f}\right)}{\left(\dfrac{\mu h}{c_1+c_2}\left(1+\dfrac{1-K_f}{K_f}\right)+1\right)}g$$

$$= \frac{0.25\left(\dfrac{62}{112}\right)\left(1+\dfrac{0.5}{0.5}\right)}{\left(\dfrac{0.25\times 20}{112}\left(1+\dfrac{0.5}{0.5}\right)+1\right)}g$$

$$= 0.2541g$$

Acceleration under rear skid is

$$a_R = \frac{\mu\left(\dfrac{c_1}{c_1+c_2}\right)\left(1+\dfrac{K_f}{1-K_f}\right)}{\left(-\dfrac{\mu h}{c_1+c_2}\left(1+\dfrac{K_f}{1-K_f}\right)+1\right)}g$$

$$= \frac{0.25\left(\dfrac{50}{112}\right)\left(1+\dfrac{0.5}{0.5}\right)}{\left(-\dfrac{0.25\times 20}{112}\left(1+\dfrac{0.5}{0.5}\right)+1\right)}g$$

$$= 0.2451g$$

Because $a_F > a_R$, the rear tires will skid first.

Similar to our finding under braking conditions, the preceding example shows the sensitivity of the tire skid (whether the front or rear tires skid first) to the coefficient of friction. The vehicle is to be driven on various types of surfaces with a high coefficient of friction (e.g., concrete surfaces) and a low coefficient of friction (e.g., icy or slippery roads). The drive system should be designed such that the maximum tractive force is obtained.

Figure 8.9 shows how the acceleration, under front and rear skid conditions, changes with the coefficient of friction for the parameters described in Example E8.4. This figure clearly shows that rear skid is quite possible at surfaces with a lower coefficient of friction. Figure 8.10 shows that by changing only the tractive effort distribution such that 60% goes to the front axle, front axle skid will occur for all types of surfaces. Any factor higher than 55% will guarantee front skid conditions to occur first for all surfaces.

8.3.5 Optimal Tractive Effort

Maximum tractive effort is obtained when all wheels skid at the same time. To achieve this, torque distribution between the front and rear axle must follow the load on each

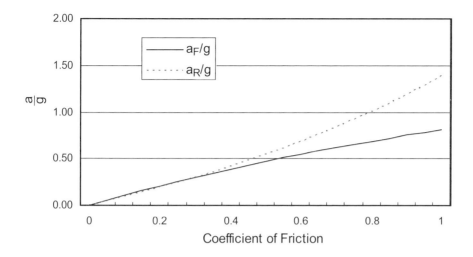

Figure 8.9 Acceleration under front and rear skid conditions for $c_1 = 50$ in., $c_2 = 62$ in., h = 20, and $K_f = 0.5$.

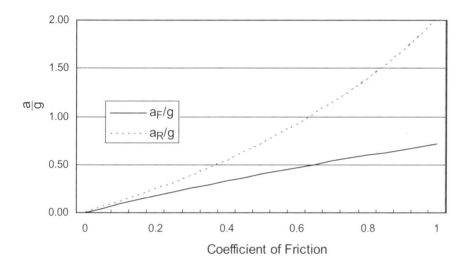

Figure 8.10 Acceleration under front and rear skid conditions for $c_1 = 50$ in., $c_2 = 62$ in., h = 20, and $K_f = 0.6$.

axle, assuming that the wheels of both axles have the same coefficient of friction. In other words, if the front axle is experiencing 40% of the vehicle load under certain driving conditions, the maximum tractive effort is achieved if the motor torque is distributed such that 40% of it goes to the front axle. Similar to braking systems, three major factors affect the torque distribution between the front and rear axles. The first is that the loading conditions of the vehicle (fully loaded versus lightly loaded) will cause movement of the center of gravity of the vehicle. This actually means that a certain optimal distribution for a certain loading condition may not be the same as that for another loading condition. The second factor is the coefficient of friction. The coefficient of friction varies considerably between dry and wet surfaces and between various types of surface constructions and finishing. The third factor is safety considerations and/or vehicle handling objectives. This may dictate that one axle will experience skid conditions first (usually the front axle). Also similar to brakes, the drive system can be designed with a feedback control algorithm. Sensors can be attached to both axles to determine the axle load and to distribute motor torque (mechanically or hydraulically) to both axles according to the axle load distribution, with some bias toward achieving front skid conditions first (if desired).

Consider the vehicle treated in Example E8.4. The vehicle weighs 4800 lb (21,326 N) and has a wheelbase of 112 in. (2844.2 mm). Its center of gravity is 50 in. (1270.0 mm) behind the front axle and 20 in. (508 mm) above the ground and has a coefficient of friction of 0.75 with the road surface. The acceleration found for the assumptions of the front and rear lock-ups are given in Table 8.1 for various values of the torque distribution factor. The bold numbers are the smaller acceleration values. The skid condition that is associated with the smaller acceleration happens first.

The maximum tractive force occurs when both F_{x1} and F_{x2} approach their skid (or maximum values) simultaneously. This can be achieved only if we set $F_{x1} = \mu F_{z1}$ and $F_{x2} = \mu F_{z2}$. The total friction force is

$$
\begin{aligned}
F_x &= F_{x1} + F_{x2} \\
&= \mu F_{z1} + \mu F_{z2} \\
&= \mu\left(F_{z1} + F_{z2}\right) \\
&= \mu Mg\cos\theta
\end{aligned}
\tag{8.29}
$$

The maximum tractive force is achieved at a point where a_R and a_F are close to each other. Note that at $K_f = 0.4196$, both acceleration values a_R and a_F are the same (0.75g), and maximum friction is achieved. From Eq. 8.19, the acceleration is

$$
\begin{aligned}
a &= \frac{F_{x1} + F_{x2} - R_a - R_r - Mg\sin(\theta)}{M} \\
&= \frac{\mu Mg\cos(\theta) - R_a - R_r - Mg\sin(\theta)}{M}
\end{aligned}
\tag{8.30}
$$

The only way both the front and rear drives would skid at the same time is if the front torque distribution factor is equal to the ratio of the front axle load to the total load, or

$$
\begin{aligned}
K_{f\,max} &= \frac{F_{x1}}{F_x} = \frac{\mu F_{z1}}{\mu Mg\cos(\theta)} \\
&= \frac{\dfrac{\mu}{c_1 + c_2}\left(Mgc_2\cos\theta - h_a R_a - hMa - hMg\sin\theta\right)}{\mu Mg\cos(\theta)} \\
&= \frac{\left(Mgc_2\cos\theta - h_a R_a - hMa - hMg\sin\theta\right)}{\left(c_1 + c_2\right)Mg\cos(\theta)}
\end{aligned}
\tag{8.31}
$$

or, by further manipulation,

$$
K_{f\,max} = \frac{\left(Mgc_2\cos\theta - h_a R_a - h\left(\mu Mg\cos(\theta) - R_a - R_r - Mg\sin(\theta)\right) - hMg\sin\theta\right)}{\left(c_1 + c_2\right)Mg\cos(\theta)}
\tag{8.32}
$$

TABLE 8.1
ACCELERATION AND FORCE DISTRIBUTION UNDER FRONT AND REAR SKID ASSUMPTIONS FOR VARIOUS FRONT TORQUE DISTRIBUTION FACTORS
($c_2 = 62$ IN., $c_1 + c_2 = 112$ IN., $\mu = 0.75$, h = 20 IN.)

K_f	a_F/g	F_{z1}	F_{z2}	F_{x1}	F_{x2}	F_x	a_R/g	F_{z1}	F_{z2}	F_{x1}	F_{x2}	F_x
0.0001	3.098						0.387	2326	2474	0	1856	1856
0.05	2.257						0.410	2305	2495	98	1871	1969
0.1	1.775						0.437	2283	2517	210	1888	2098
0.15	1.462						0.468	2256	2544	337	1908	2244
0.2	1.243						0.503	2226	2574	483	1930	2413
0.25	1.081						0.543	2191	2609	652	1957	2609
0.3	0.957						0.591	2150	2650	852	1987	2839
0.35	0.858						0.649	2101	2699	1090	2024	3114
0.4	0.778						0.718	2041	2759	1379	2069	3448
0.4196	**0.750**	2014	2786	1511	2090	3600	0.750	2014	2786	1510	2089	3600
0.45	0.711	2048	2752	1536	1877	3413	0.805					
0.5	0.655	2096	2704	1572	1572	3144	0.915					
0.55	0.607	2137	2663	1603	1311	2914	1.059					
0.6	0.566	2172	2628	1629	1086	2715	1.258					
0.65	0.530	2203	2597	1652	890	2542	1.550					
0.7	0.498	2230	2570	1673	717	2390	2.016					
0.75	0.470	2255	2545	1691	564	2255	2.885					
0.8	0.445	2276	2524	1707	427	2134	5.068					
0.85	0.422	2295	2505	1722	304	2025	20.833					
0.9	0.402	2313	2487	1735	193	1927	−9.868					
0.95	0.383	2329	2471	1747	92	1839	−3.989					
1	0.366	2343	2457	1757	0	1757	−2.500					

Ignoring aerodynamic forces, rolling resistance, and grade, the front torque distribution factor becomes

$$K_{f\,max} = \frac{(c_2 - \mu h)}{(c_1 + c_2)} \qquad (8.33)$$

Table 8.1 also displays the tractive forces on each axle for various front torque distribution factors. Figure 8.11 shows the total maximum tractive force.

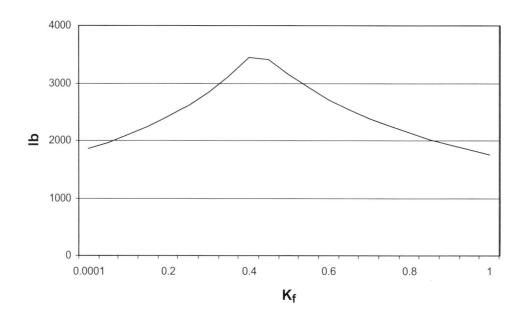

Figure 8.11 Maximum tractive force and the front torque distribution factor.

EXAMPLE E8.5

Consider Example E8.4, where the 21,351.46-N (4800-lb) passenger car has a wheelbase of 2844.80 mm (112 in.) and a center of gravity 1270.00 mm (50 in.) behind the front axle and 508 mm (20 in.) above the ground. The car is moving on a horizontal plane, and drag and rolling resistance are negligible. Determine the maximum tractive force, maximum acceleration, and the distribution factor to achieve them. The coefficient of friction is 0.25.

Solution:

The front brake distribution factor to achieve the maximum braking force is

$$K_f = \frac{(c_2 + \mu h)}{(c_1 + c_2)}$$

$$= \frac{62 - 0.25 \times 20}{112}$$

$$= 5089$$

The maximum braking force F_x is

$$F_x = \mu Mg \cos(\theta)$$
$$= 0.25 \times 4800$$
$$= 1200 \text{ lb}$$

The maximum acceleration is

$$a = \frac{\mu Mg \cos(\theta) - R_a - R_r - Mg \sin(\theta)}{M}$$
$$= 8.05 \text{ ft/sec}^2$$

Note that the maximum acceleration is obtained when both a_R and a_F are equal. At that point, they both are proportional to the coefficient of friction times the axle load.

To calculate the forces on both axles (including both friction and load forces), we need to develop a strategy to tackle the problem. The strategy is based on the following.

Determine for the given problem a_F and a_R from Eqs. 8.23 and 8.28.

If $|a_F| < |a_R|$, front skid occurs first. Use the equations

$$F_{z1} = \frac{1}{c_1 + c_2}\left(Mgc_2 \cos\theta - h_a R_a - hMa_F - hMg\sin\theta\right)$$

$$F_{z2} = \frac{1}{c_1 + c_2}\left(Mgc_1 \cos\theta + h_a R_a + hMa_F + hMg\sin\theta\right)$$

Note that

$$F_{z1} + F_{z2} = Mg\cos\theta$$

$$F_{x1} = \mu F_{z1}$$

$$F_{x2} = \frac{1 - K_f}{K_f}F_{x1}$$

or

$$F_{x2} = Ma_F - F_{x1}$$

The total friction (tractive) force is

$$F_x = F_{x1} + F_{x2}$$

If $|a_F| > |a_R|$, rear skid occurs first. Use the equations

$$F_{z1} = \frac{1}{c_1 + c_2}\left(Mgc_2\cos\theta - h_aR_a - hMa_R - hMg\sin\theta\right)$$

$$F_{z2} = \frac{1}{c_1 + c_2}\left(Mgc_1\cos\theta + h_aR_a + hMa_R + hMg\sin\theta\right)$$

Note that

$$F_{z1} + F_{z2} = Mg\cos\theta$$

$$F_{x2} = \mu F_{z2}$$

$$F_{x1} = \frac{K_f}{1 - K_f}F_{x2}$$

or

$$F_{x1} = Ma_R - F_{x2}$$

The total friction (tractive) force is

$$F_x = F_{x1} + F_{x2}$$

Note the similarity between this strategy and the strategy developed for brakes (Chapter 7).

EXAMPLE E8.6

Consider Example E8.5. The vehicle strategy requires a front axle skid on a dry surface, and the front torque distribution factor is chosen at 0.7. The vehicle is moving on a low friction surface with a 0.25 coefficient of friction. Determine the maximum tractive force and which axle would skid first.

Solution:

Because drag, rolling resistance, and grade are ignored, the acceleration under front skid is

$$a_F = \frac{\mu\left(\dfrac{c_2}{c_1 + c_2}\right)\left(1 + \dfrac{1 - K_f}{K_f}\right)}{\left(\dfrac{\mu h}{c_1 + c_2}\left(1 + \dfrac{1 - K_f}{K_f}\right) + 1\right)}g$$

$$= \frac{0.25\left(\dfrac{62}{112}\right)\left(1 + \dfrac{0.3}{0.7}\right)}{\left(\dfrac{0.25 \times 20}{112}\left(1 + \dfrac{0.3}{0.7}\right) + 1\right)}g$$

$$= 0.1368g$$

The acceleration under rear skid is

$$a_R = \frac{\mu\left(\dfrac{c_1}{c_1 + c_2}\right)\left(1 + \dfrac{K_f}{1 - K_f}\right)}{\left(-\dfrac{\mu h}{c_1 + c_2}\left(1 + \dfrac{K_f}{1 - K_f}\right) + 1\right)}g$$

$$= \frac{0.25\left(\dfrac{50}{112}\right)\left(1 + \dfrac{0.7}{0.3}\right)}{\left(-\dfrac{0.25 \times 20}{112}\left(1 + \dfrac{0.7}{0.3}\right) + 1\right)}g$$

$$= 0.7184g$$

Because $a_F < a_R$, the front skid occurs first. The normal forces are

$$F_{z1} = \frac{1}{c_1 + c_2}\left(Mgc_2 - hMa_F\right)$$

$$= \frac{1}{112}\left(149 \times 32.2 \times 62 - 20 \times 149 \times (0.1368)\right)$$

$$= 2517 \text{ lb}$$

$$F_{z2} = \frac{1}{c_1 + c_2}\left(Mgc_1 + hMa_F\right)$$

$$= \frac{1}{112}\left(149 \times 32.2 \times 50 + 20 \times 149 \times (0.1368)\right)$$

$$= 2283 \text{ lb}$$

Note that

$$F_{z1} + F_{z2} = 4800 \text{ lb}$$

$$F_{x1} = \mu F_{z1}$$
$$= 0.25 \times 2517$$
$$= 629 \text{ lb}$$

$$F_{x2} = \frac{1 - K_f}{K_f} F_{x1}$$
$$= \frac{1 - 0.7}{0.7} \times (629)$$
$$= 157 \text{ lb}$$

The total friction force is

$$F_x = F_{x1} + F_{x2} = 786 \text{ lb}$$

In a typical design problem, one needs to consider various surfaces and loading conditions. For example, let us consider the following problem where a vehicle is under different loading conditions and on various surfaces.

EXAMPLE E8.7

Consider a vehicle that weighs 14,235 N (3200 lb) when lightly loaded (driver only), with a center of gravity 1016.0 mm (40 in.) behind the front axle and 457.2 mm (18 in.) above the road, and 17,793 N (4000 lb) when loaded, with a center of gravity at 1295.4 mm (51 in.) behind the front axle and 508.0 mm (20 in.) above the road. The wheelbase is 2743.2 mm (108 in.). The vehicle is to achieve maximum possible tractive force and front skid under the following severe conditions:

Vehicle is loaded and on a dry surface of $\mu = 0.8$.

Vehicle is loaded and on a wet surface of $\mu = 0.2$.

Vehicle is lightly loaded and on a dry surface of $\mu = 0.8$.

Vehicle is lightly loaded and on a wet surface of $\mu = 0.2$.

Determine the value of K_f that you recommend, knowing that the drive system your company installs has a constant torque distribution value (no feedback control system). Determine the axle loads and the maximum tractive force.

Solution:

This problem is best solved by writing a spreadsheet with all possible tests. Table E8.7 shows the optimal value of K_f (Table E8.7A) and the associated acceleration and axle loads (Table E8.7B), as well as the maximum brake force for each case.

TABLE E8.7A
OPTIMAL VALUE FOR K_f FOR EXAMPLE E8.7

Case	c_1	c_2	h	μ	Mg	K_{fmax}	a_F/g	F_{z1}	F_{z2}	F_{x1}	F_{x2}	F_x
1	51	57	20	0.8	4000	0.3796	0.800	1519	2481	1215	1985	3200
2	51	57	20	0.2	4000	0.4907	0.200	1963	2037	393	407	800
3	40	68	18	0.8	3200	0.4963	0.800	1588	1612	1271	1289	2560
4	40	68	18	0.2	3200	0.5963	0.200	1908	1292	382	258	640

TABLE E8.7B
LOAD DISTRIBUTION FOR $K_f = 0.65$ FOR EXAMPLE E8.7

Case	c_1	c_2	h	μ	Mg	K_f	a_F/g	F_{z1}	F_{z2}	F_{x1}	F_{x2}	F_x
1	51	57	20	0.8	4000	0.65	0.5290	**1719**	**2281**	**1375**	**741**	**2116**
2	51	57	20	0.2	4000	0.65	0.1536	1997	2003	399	215	615
3	40	68	18	0.8	3200	0.65	0.6430	1672	1528	1337	720	2058
4	40	68	18	0.2	3200	0.65	0.1843	1917	1283	383	206	590

To ensure front skid under all cases, the K_f must be chosen above the maximum value of 0.5963 determined for the fourth case. Thus, $K_f = 0.65$ is chosen. For this value, Table E8.7B shows the acceleration and loads obtained for each case.

Note that with the choice made for K_f, front skid will occur first in all cases. To achieve this, major sacrifice is made for the maximum tractive force, especially for the first case, where the maximum tractive force is decreased from 3200 to 2116 lb.

In today's automotive drive systems, the front torque distribution factor K_f can be altered in a feedback control system. This will account for variations in the axle loads because of either different accelerations or different vehicle loading conditions (e.g., lightly loaded versus loaded conditions). This will enable the engineers to achieve higher tractive force at the surfaces with a low coefficient of friction. The K_f factor cannot be altered straightforwardly to account for the different coefficients of friction that various surfaces may offer. In that case, an advanced traction control system can be used.

8.4 Power-Limited Acceleration

The engine is the source of the power and energy of a vehicle. It transforms the energy from a source (typically chemical) into a kinetic energy, resulting in rotation of the crankshaft.

The ideal powerplant puts forth constant power at all speeds. This power is defined as force times linear velocity, or torque times angular velocity, as

$$P = F \times V$$

or   (8.34)

$$P = T \times \omega$$

where ω is the angular velocity usually measured in radians per second (rad/s). In actual applications, the angular velocity is measured in engine revolutions per minute (rpm).

The torque and power relation in an ideal powerplant can be determined based on this constant power assumption, as shown in Figure 8.12. Note that for such an ideal powerplant, at low speeds, torque is needed more than power.

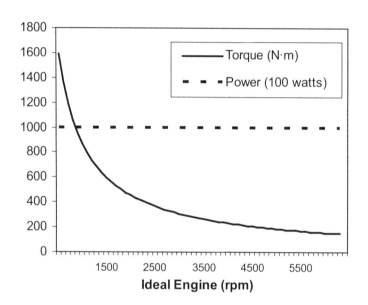

Figure 8.12 Power and torque characteristics in an ideal engine.

Some electric motors can deliver this ideal output of the powerplant. On the other hand, the internal combustion engine performs differently. Despite this major shortcoming, it is still the most frequently used engine in automotive applications. Perhaps the abundance and affordability of the fuel are among the main reasons for that.

The powertrain consists of many components and systems. These systems include the engine cooling system, the drivetrain, the air conditioning, and others. In addition, some systems connect the powerplant to other systems. An example of these is the steering system, which has the power steering pump on the engine and the steering gear on the chassis of the vehicle. In addition to these systems (or subsystems), the engine carries other components such as the alternator. The arrangement of these subsystems and components will depend heavily on the powerplant configuration. Figure 8.13 shows a typical north-south rear-wheel-drive engine configuration. The key elements of the powerplant include its engine and transmission. Both of these will be discussed in detail in the next sections.

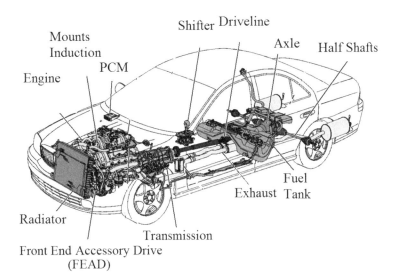

Shifter Driveline
Mounts
Induction   Axle   Half Shafts
PCM
Engine

Radiator   Fuel
                         Exhaust Tank
Transmission
Front End Accessory Drive
(FEAD)

Figure 18.13 Components and layout of a typical north-south rear-wheel-drive powertrain.

8.4.1 The Engine

Performance of an engine in the sense of torque and power versus engine revolutions per minute is only one of a few attributes that are critical in selecting an engine for a particular application. Automotive customers demand high torque at low speeds and high power in general. They also demand that the source of energy be economical and easy to replenish when consumed. To achieve high efficiency, the power-to-weight ratio is a factor considered by engineers and automotive customers. In addition, government regulations require fewer harmful emissions to the environment.

Among the many engines considered or used for automotive applications are the following:

Gas turbine engine. This engine has a good power-to-weight ratio and lower emissions when compared with internal combustion diesel or gasoline engines. Tilagone *et al.* (2005) demonstrated the environmental and emission advantages of natural gas as a fuel source of engines for urban vehicles. However, this engine has low efficiency under no or little load conditions. This engine is not as popular as gasoline or diesel engines. One of the reasons could be the lack of an infrastructure that supports the delivery of this fuel to consumers (Unich, 1993).

Electric motor. This motor can be driven by batteries or fuel cells. Battery-driven road vehicles face the challenge of providing the customer demanded power and speed as well as driving range. The advancement of electric vehicles is completely dependent on that of battery technology. Fuel-cell-driven vehicles and fuel cell technology have received increased attention lately (e.g., Cantemir *et al.*, 2004). Fuel cells release the chemical energy and convert it directly to electrical energy. The fuel cells are filled with hydrogen that is combined with air to produce vapor, with little or no harm to the environment. Figure 8.14c shows a fuel-cell powertrain layout.

Stirling-cycle engine. In this engine, compressed helium and hydrogen are used to produce mechanical work. This engine was invented in 1816 by Robert Stirling, and its original use was to pump water out of mines. Small engines were built for fans and even were developed for an artificial heart that could be powered with an alpha-particle-emitting plutonium radioisotope (Oman, 1999). In the 1930s, a Dutch firm developed

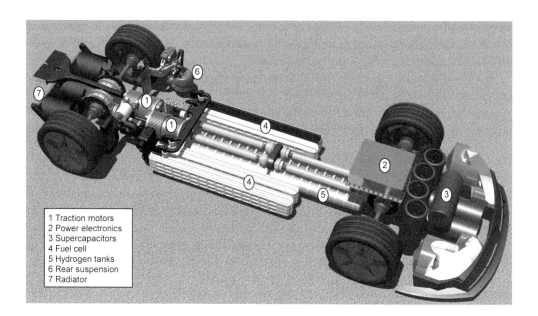

1 Traction motors
2 Power electronics
3 Supercapacitors
4 Fuel cell
5 Hydrogen tanks
6 Rear suspension
7 Radiator

Figure 8.14 Fuel cell powertrain layout (Cantemir *et al.*, 2004).

a small Stirling engine for driving generators that powered radios in remote regions. Transistor radios ended this use. General Motors subsequently developed this "Meijer" Stirling engine version for one application in the U.S. Army. The Stirling engine theoretically can convert the heat from burning hydrogen with Carnot-cycle efficiency. No other heat engine, including those based on cycles such as the diesel and Rankine engines, can achieve Carnot-cycle efficiency. A Stirling engine once was tested in a car. Although efficient, it was unsatisfactory because it did not give instant response to a driver's demand for acceleration power.

Rankine vapor cycle engine. This engine has many of the favorable characteristics of the ideal engine, including close-to-ideal torque delivery and low emissions. However, its low power-to-weight ratio is a major disadvantage.

Internal combustion engine. Gasoline or diesel is used in internal combustion engines as the source of energy. These fuels enter the combustion chamber and are given an electric charge to ignite the combustion process and release this energy. The high power-to-weight ratio of these engines makes them among the most frequently used engines in the automotive industry. Figure 8.13 shows a typical rear-wheel-drive powertrain layout with an internal combustion engine.

The main difference between a gasoline (petrol) and diesel engine is in how the ignition takes place. A gasoline engine draws a mixture of air and gasoline into the cylinder, compresses it, and ignites it with a spark to provide mechanical energy. On the other hand, a diesel brings in only air and compresses it, raising the temperature to the point where the fuel will ignite spontaneously when it is injected. This feature gives the diesel superior efficiency. The best car diesels have a thermal efficiency of approximately 43% compared with approximately 30% for gasoline engines. Because the process in a diesel engine needs air to be compressed to a higher degree than a gasoline engine, it causes more vibration. Thus, diesel engines must be stiffer and heavier to withstand those vibrations. This also results in the higher durability of diesel engines than that of gasoline engines. The higher compression ratio of diesel engines produces a higher torque output than that of equivalent gasoline engines. On the other hand, diesel power output usually is lower than that of an equivalent gasoline engine. This is the case because diesel engine speeds are lower than gasoline engine speeds. The diesel engine favored by the 1930s diesel pioneers, Citroën and Mercedes-Benz, was a similar type to that used in European passenger cars until the 1990s. The indirect injection (IDI) or pre-chamber design suited passenger cars best. Even now, indirect injection engines are still used in some European cars and light commercial vehicles.

Noise is the main problem of diesel engines. The fuel explodes on contact with the hot air in the engine, and the force of that explosion gives rise to the traditional diesel tick or rattle noise. To control that explosion, fuel is injected into a small chamber in the cylinder head, known as the pre-chamber, which is connected to the main combustion chamber directly over the piston. Combustion spreads rapidly into the main combustion chamber where the explosive force acts on the piston and forces it downward. The drawback of this controlled combustion is lower efficiency.

Efficiency could be further improved by injecting the fuel directly into the main combustion chamber and eliminating the pre-chamber. The introduction of control electronics made this type of engine appropriate to cars. Fiat introduced the first direct injection (DI) car engine in the Croma in 1988. The drawback to direct injection engines was greater combustion noise than indirect injection diesels. However, direct injection engines used up to 15% less fuel than indirect injection engines. This is one reason why premium brands such as Mercedes-Benz and BMW were late adopters of direct injection engines.

Electronic control systems meant that injection timing could be tightly controlled as required. Better diesel engines need high injection pressures to burn fuel more efficiently. The higher the fuel pressure, the better the fuel mixes with the air, completely burning more of the injected fuel in the process. That means more efficiency and cleaner emissions.

Diesel engines are believed to increase in popularity, especially when fuel economy becomes one of the more important vehicle attributes to consumers.

Hybrid engines use an electric, relatively high-voltage battery to drive the vehicle when the demand for power is low, while an internal combustion engine is available when the demand for power is high. This type of powerplant is achieving favorable characteristics, particularly in the area of fuel economy. These engines now are installed on many small and medium-sized vehicles, but their initial cost remains a disadvantage. Figure 8.15 shows an exploded view of a Volkswagen hybrid drive (Kalberlah, 1991)

Nakamura *et al.* (1991) studied the potential of various engines in the future. The four-stroke engine of basically conventional structure will remain the mainstream powerplant up to the future. The engine must be improved in thermal efficiency and with significantly less tailpipe emissions. Future engines will turn toward downsizing, lighter weight, higher speed, and design diversity to meet diverse requirements from the standpoints of global energy saving, preservation of the environment and natural resources, and higher performance of the vehicle. Engine electronics and sensor technology must be improved for reliability and service-free operation. Progress is expected in catalytic technology to meet increasingly stringent emissions control requirements. Electric vehicles could be an alternative means of transportation for a limited application from the viewpoint of preserving urban environment and of utilizing possible excess electric

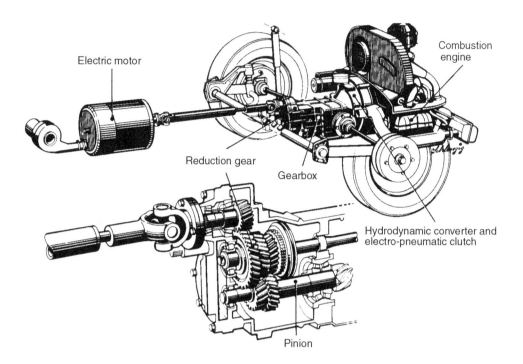

Figure 8.15 Exploded view of a Volkswagen hybrid drive (Kalberlah, 1991).

energy supply in the future. Finally, various fuels that are alternatives to petroleum will continue to be investigated.

Next, we will spend more time reviewing the characteristics of the internal combustion engine because it is the most common engine used in automotive applications.

8.4.2 Internal Combustion Engines

The description of the combustion process and internal engine design are beyond the scope of this book. The focus here is on the engine output in terms of both torque and power that relate to vehicle dynamics.

The torque and power relations with the engine speed of typical gasoline and diesel engines are given in Figures 8.16 and 8.17, respectively. In these engines, the torque and power are delivered at a lower speed (e.g., idle speed) but reach maximum values at an intermediate speed.

The performance of these engines is not ideal. They do not produce constant power and hyperbolically reducing torque as a function of the engine revolutions per minute. Instead, the torque reaches a maximum value around 3000 rpm for gasoline engines and at a lower speed for diesel engines. After it reaches its maximum torque, losses within the engine reduce its output torque. Power reaches a maximum value at an engine speed higher than the one at which the maximum torque is obtained. The maximum engine speed is controlled to be slightly higher than that at which maximum power is achieved.

It should be mentioned that these torque and power characteristics are gross outputs. What the engine delivers in reality to the transmission is less than that. In particular, losses happen in the air handling systems (induction and exhaust), cooling systems (water

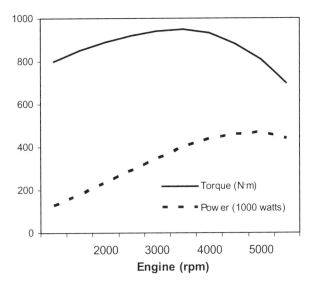

Figure 8.16 Power and torque characteristics in a gasoline engine.

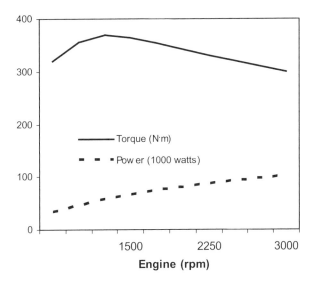

Figure 8.17 Power and torque characteristics in a diesel engine.

pump and fan), front end accessory drive (FEAD) components (i.e., power steering pump and alternator), and climate control systems (e.g., air conditioning). Air handling systems including exhaust and induction alone can consume approximately 10% of the total power. At approximately 3000 rpm, air conditioning consumes approximately 4 kW of power, the cooling system of the water pump and fan consume approximately 3 kW, and the alternator and power steering pump consume another 1.5 kW for each. This leads to approximately 10 kW of losses to the engine power before the power is delivered to the transmission input shaft. In total, these losses may add up to more than 20% of the engine power.

In addition to these factors, the output torque and power of the powerplant are functions of the ambient temperature and pressure. In general, low ambient pressure and/or high ambient temperature reduce the output power of the engine.

SAE International has published standards for finding the engine performance at various ambient pressure and temperature values (SAE J1995). In these standards are used a reference absolute pressure of $B_o = 100$ kPa or 14.5 psi, an absolute air dry pressure of $B_{do} = 99$ kPa or 14.36 psi, and an inlet temperature of $T_o = 25°C$. The power P under given temperature T in Celsius, dry pressure B_d in kiloPascals (kPa), as compared with that found under the standard temperature and pressure values, are given using the formula

$$\frac{P}{P_o} = \frac{1}{1.18\left(\left(\frac{99}{B_d}\right)\left(\frac{T+273}{298}\right)^{1/2}\right) - 0.18}$$ (8.35)

Equation 8.35 is applicable to gasoline engines. Figure 8.18 gives the variation of the power ratio ($\frac{P}{P_o}$) as a function of temperature and pressure. This shows that within reasonable ambient variations, the output power of the engine may increase or decrease by approximately 10%.

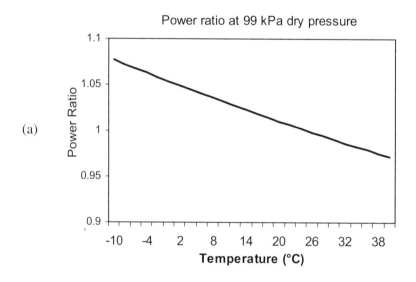

(a)

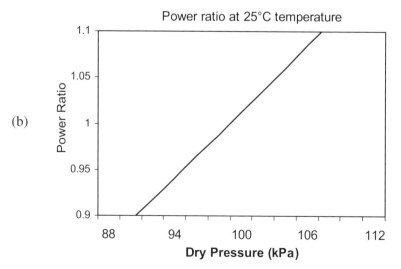

(b)

Figure 8.18 Power variation as a function of (a) temperature and (b) dry pressure variations (SAE J1995).

The power prediction under various temperature and pressure values for diesel engines is given in the SAE J1995 standard, and readers are advised to refer to this standard. However, note that for diesel engines, the power ratio varies with $\frac{1}{T}$ rather than $\frac{1}{\sqrt{T}}$ as in gasoline engines.

One major difference between gasoline and diesel engines is the energy-to-weight ratio. At their highest efficiency, diesel engines seem to output twice the energy obtained by gasoline engines for the same weight.

Some of the useful relations in converting power among different units are

$$P\left(\frac{ft \times lb}{sec}\right) = \text{Torque}\left(ft \times lb\right) \times \text{speed}\left(\frac{rad}{sec}\right)$$

$$P\,(hp) = \frac{\text{Torque}\left(ft \times lb\right) \times \text{speed}\left(\dfrac{rad}{sec}\right)}{550}$$

$$= \frac{\text{Torque}\left(ft \times lb\right) \times \text{speed}\left(rpm\right)}{5252}$$

$$\text{Power (kW)} = 0.746\ \text{power (hp)}$$

$$1\ hp = 550\frac{ft \times lb}{sec}$$

(8.36)

Before we study the transmission, let us consider an example on power calculations of a typical engine.

EXAMPLE E8.8

A V6 gasoline engine is outputting a maximum of 220 hp of gross power at 5000 rpm. Air induction losses are found to be 2.5% at that engine speed. The hot and cold end exhaust losses are found to be 7%. In addition, losses due to air conditioning, power steering, the alternator, and other accessories add up to 10 hp under specific driving conditions. The ambient temperature is 5°C, and the dry pressure is 95 kPa. Determine the power available to the transmission.

Solution:

The total losses of the induction and exhaust systems are

$$2.5\% + 7\% = 9.5\%$$

The power available to the accessories and transmission is

$$90.5\% \times 220\ hp = 199.1\ hp$$

The power available to the transmission under standard conditions is

$$199.1 - 10 = 190.1 \text{ hp}$$

The power ratio under the given conditions is

$$\frac{P}{P_o} = \frac{1}{1.18\left(\left(\frac{99}{B_d}\right)\left(\frac{T + 273}{298}\right)^{1/2}\right) - 0.18}$$

$$= \frac{1}{1.18\left(\left(\frac{99}{95}\right)\left(\frac{5 + 273}{298}\right)^{1/2}\right) - 0.18} = 0.9924$$

The horsepower available to the transmission is

$$0.9924 \times 190.1 \text{ hp} = 188.7 \text{ hp}$$

8.4.3 The Transmission

As mentioned, an ideal powerplant delivers constant power throughout the engine speed range. However, the internal combustion engine delivers output torque that peaks at a certain speed, and power that peaks at a higher speed. Generally, the vehicle needs high torque at low speeds (for acceleration) and low torque at high speeds. To compensate for that shortcoming of internal combustion engines, a gearing mechanism is needed to deliver the required torque at the given speed. This gearing mechanism is the transmission.

Many types of transmissions are available in the marketplace. The automatic transmission is common in North America, whereas the manual transmission is widespread in Europe and the rest of the world. Recently, the continuously variable transmission (CVT) is finding applications in small vehicles.

In general, the transmission provides the required startup conditions, in forward and reverse, for a fully loaded vehicle on a relatively steep gradient (at approximately 30%). Under these conditions, the vehicle needs to accelerate, and high tractive forces at the tires are required. To produce these forces, higher torque is needed than that typically produced by the engine. Thus, the output engine speed should be reduced, and a lower gear (with a high gear reduction ratio) is needed. On the other hand, the higher gear (with the lower gear reduction ratio) is determined based on the desired maximum vehicle speed. To ensure smooth transition between these limits, other gear ratios are used in the transmission. For that purpose, typical transmissions for passenger road vehicles have three to six gear ratios.

At maximum vehicle speed, acceleration is not needed any further, and the engine should deliver enough power to maintain that speed and overcome the internal resistance of the transmission, rolling resistance of the tires, aerodynamic resistance, and reasonable grade (3 to 6%). Based on these factors, the gear ratio of the highest gear can be selected.

Thus, the transmission gear ratios are to be selected such that they bring the torque output of a typical engine (Figure 8.16) to that of the ideal engine (Figure 8.12). Figure 8.19 shows how the transmission output torque (and thus the tractive forces at the tires) changes with different gears.

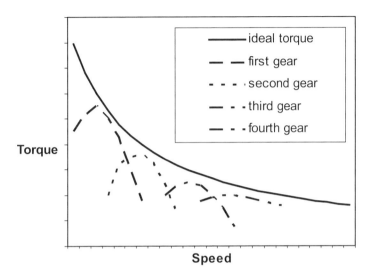

Figure 8.19 Transmission torque output as a result of different gear engagement.

8.4.3.1 Manual Transmissions

A manual transmission consists of a clutch for engagement with the drive axle, a gear box that consists of the different gear sets that yield the desired gear ratios, a propeller shaft, and a drive axle. If there is no overdrive gear and as a general practice, the highest gear delivers no gear reduction. In other words, the speed of the engine is the same as that of the gear output shaft.

Common practice has been to select the maximum engine speed to be slightly higher (approximately 10% higher) than the speed at which the maximum power is obtained. The gear ratio of the highest gear (with the lowest gear reduction ratio) is found using the formula

$$\xi_n = \frac{n_e r (1 - i)}{V_{max} \xi_x} \qquad (8.37)$$

where

n_e = engine speed corresponding to the maximum speed of the vehicle

r = radius of the tire

i = tire slip

V_{max} = maximum speed of the vehicle

ξ_x = gear ratio in the drive axle

Note that if the lowest gear reduction ratio in the highest gear is selected to be 1 (i.e., $\xi_n = 1$), then Eq. 8.37 can be used to find the gear ratio of the drive axle (i.e., ξ_x).

The gear reduction ratio of the lowest gear is to be determined based on the ability of the vehicle to climb a certain grade (say, 30%) when fully loaded. Other criteria also may be used. One of these criteria states that the tires of the vehicle must not spin when climbing the maximum possible gradient.

Consider first rear-wheel-drive systems. At startup, drag can be neglected. Thus, Eq. 8.37 can be written as

$$
\begin{aligned}
F_x &= F_{x2} \\
&= R_r + Ma + Mg\sin\theta \\
&= \mu_r Mg + Ma + Mg\sin\theta
\end{aligned}
\tag{8.38}
$$

The maximum slope (θ_{max}) at which the vehicle can withstand the load (without acceleration) can be calculated by equating Eq. 8.38 (with a = 0) with Eq. 8.17. This yields

$$
Mg\sin(\theta_{max}) + \mu_r Mg = \frac{\dfrac{\mu Mg}{c_1 + c_2}\left(c_1\cos\theta_{max} - h\mu_r\right)}{1 - \dfrac{\mu h}{c_1 + c_2}}
\tag{8.39}
$$

Simplifying further yields

$$
\sin\left(\theta_{max}\right) = \frac{\dfrac{\mu}{c_1 + c_2}\left(c_1\cos\theta_{max} - h\mu_r\right)}{1 - \dfrac{\mu h}{c_1 + c_2}} - \mu_r
\tag{8.40}
$$

Equation 8.40 is a nonlinear equation that can be solved for θ_{max}. At small angles of θ_{max}, the term $\cos\left(\theta_{max}\right)$ can be approximated to 1, and the equation can be solved directly for θ_{max}. The value then can be substituted back into Eq. 8.40 to obtain a more accurate estimate. The process can be repeated until the desired accuracy is achieved.

Similar mathematical manipulation can be made for front-wheel-drive systems. This yields

$$
\sin\left(\theta_{max}\right) = \frac{\dfrac{\mu}{c_1 + c_2}\left(c_2\cos\theta_{max} + h\mu_r\right)}{1 + \dfrac{\mu h}{c_1 + c_2}} + \mu_r
\tag{8.41}
$$

EXAMPLE E8.9

Consider a vehicle that weighs 15,000 N when lightly loaded (driver only), with a center of gravity at 1.000 m behind the front axle and 0.460 m above the road. Consider a rolling resistance of 0.02. The wheelbase is 2.750 m. Find the maximum angle that can be withstood without losing traction for both front-wheel drive and rear-wheel drive when the coefficient of friction is 0.2.

Solution:

Given

$$c_1 + c_2 = 2.750 \text{ m}$$

$$c = 1.000 \text{ m}$$

$$\mu_r = 0.02$$

$$h = 0.46 \text{ m}$$

$$\mu = 0.2$$

For rear-wheel-drive vehicles:

First trial, $\theta_{max} = 0$

$$\sin(\theta_{max}) = \frac{\dfrac{\mu}{c_1 + c_2}(c_2 \cos\theta_{max} - h\mu_r)}{1 - \dfrac{\mu h}{c_1 + c_2}} - \mu_r$$

$$= \frac{\dfrac{0.2}{2.75}(1.75 \times \cos(0) - 0.46 \times 0.02)}{1 - \dfrac{0.2 \times 0.46}{2.75}} - 0.02$$

$$= 0.1110$$

Second trial, $\theta_{max} = \sin^{-1}(0.1110) = 0.1111 \text{ rad}$

$$\sin(\theta_{max}) = \frac{\dfrac{0.2}{2.75}(1.75 \times \cos(0.1111 \text{ rad}) - 0.46 \times 0.02)}{1 - \dfrac{0.2 \times 0.46}{2.75}} - 0.02$$

$$= 0.1101$$

Third trial, $\theta_{max} = \sin^{-1}(0.1101) = 0.1104 \text{ rad}$

$$\sin(\theta_{max}) = \frac{\dfrac{0.2}{2.75}(1.75 \times \cos(0.1101 \text{ rad}) - 0.46 \times 0.02)}{1 - \dfrac{0.2 \times 0.46}{2.75}} - 0.02$$

$$= 0.1102$$

Fourth trial, $\theta_{max} = \sin^{-1}(0.1102) = 0.1104$ rad = 6.325 deg

For front-wheel-drive vehicles:

First trial, $\theta_{max} = 0$

$$\sin(\theta_{max}) = \frac{\dfrac{\mu}{c_1 + c_2}(c_2 \cos\theta_{max} + h\mu_r)}{1 + \dfrac{\mu h}{c_1 + c_2}} + \mu_r$$

$$= \frac{\dfrac{0.2}{2.75}(1.75 \times \cos(0) + 0.46 \times 0.02)}{1 + \dfrac{0.2 \times 0.46}{2.75}} + 0.02$$

$$= 0.1263$$

Second trial, $\theta_{max} = \sin^{-1}(0.1263) = 0.127$ rad

$$\sin(\theta_{max}) = \frac{\dfrac{0.2}{2.75}(1.75 \times \cos(0.1267 \text{ rad}) + 0.46 \times 0.02)}{1 + \dfrac{0.2 \times 0.46}{2.75}} + 0.02$$

$$= 0.1428$$

Third trial, $\theta_{max} = \sin^{-1}(0.1428) = 0.1433$ rad

$$\sin(\theta_{max}) = \frac{\dfrac{0.2}{2.75}(1.75 \times \cos(0.1433 \text{ rad}) + 0.46 \times 0.02)}{1 + \dfrac{0.2 \times 0.46}{2.75}} + 0.02$$

$$= 0.1425$$

Fourth trial, $\theta_{max} = \sin^{-1}(0.1425) = 0.1430$ rad = 8.193 deg

Note that front-wheel-drive vehicles are more capable of achieving this low friction. For high friction, rear-wheel-drive vehicles have better capability for climbing slopes.

Assume

T_{emax} = maximum engine torque

η_t = transmission efficiency

ξ_1 = gear ratio of the lowest gear

The total output torque available for the wheels is

$$T_{max} = T_{emax} \times \eta_t \times \xi_x \times \xi_1 \tag{8.42}$$

This must be equal to the total torque produced by the tractive force. This torque is calculated by using Eq. 8.7 for the tractive force and multiplying it by the rolling radius r. This gives (note a = 0)

$$T_{max} = (\mu_r \, Mg + Mg \sin \theta) r \qquad (8.43)$$

Equating Eqs. 8.42 and 8.43 yields the gear ratio of the lowest gear as

$$\xi_1 = \frac{Mg(\sin(\theta_{max}) + \mu_r)r}{T_{emax} \, \xi_x \, \eta_t} \qquad (8.44)$$

With the gear ratios of the high gear ξ_n and low gear ξ_1 determined from Eqs. 8.6 and 8.13, respectively, we need to establish the gear ratios of the intermediate gears.

For commercial vehicles, it is desired that the engine speed remains in the range that achieves the highest fuel economy. Thus, the gear ratios are set in a manner that follows geometric progression. To find these ratios, a general transmission ratio is to be found, based on the number of gear ratios (n), the low gear ratio ξ_1, and the high gear ratio ξ_n. This factor is

$$K_g = \sqrt[n-1]{\frac{\xi_n}{\xi_1}} \qquad (8.45)$$

The gear ratios of the intermediate gears then are found by using

$$\xi_n = K_g \xi_{n-1} \qquad (8.46)$$

For passenger vehicles, the gear ratios are found in a manner that is biased toward achieving the maximum vehicle speed in the shortest time. Also, higher gears are given closer ratios because shifting happens more frequently in these gears. Thus, the transmission factor K_g is not constant for passenger vehicles. Note that the lowest gear ratios ($\frac{\xi_1}{\xi_2}$) are approximately 10 to 15% higher than the value of K_g calculated by Eq. 8.45, and for the highest gear ratios ($\frac{\xi_{n-1}}{\xi_n}$) are 10 to 15% lower than the calculated value of K_g.

The average of all these ratios comes very close to that found using Eq. 8.45. Table 8.2 shows the gear ratios found in three small vehicles.

TABLE 8.2
GEAR RATIOS IN PASSENGER VEHICLES

	Gear Ratios						$\frac{\xi_1}{\xi_2}$	$\frac{\xi_2}{\xi_3}$	$\frac{\xi_3}{\xi_4}$	$\frac{\xi_4}{\xi_5}$	
	1st	2nd	3rd	4th	5th	K_g					Average
Audi A4 1.8	3.50	2.12	1.43	1.03	0.84	1.429	1.651	1.483	1.388	1.226	1.437
BMW 525i	4.20	2.49	1.67	1.24	1.00	1.432	1.687	1.491	1.347	1.240	1.441
Ford Escort 1.8	3.42	2.14	1.45	1.03	0.77	1.452	1.598	1.476	1.408	1.338	1.455

The mechanical efficiency of a transmission describes the losses due to friction between the engaging gears as well as friction within the bearings. This efficiency is a function of the speed and torque transmitted. The mechanical efficiency usually is between 90 and 98%, but it is lower for higher torque values. The transmission efficiency also is lower for lower gears and is higher for higher gears.

For a manual transmission at any engine speed, the tractive force available at the tires is calculated by

$$F_x = \frac{T_e \xi_o \eta_t}{r} \qquad (8.47)$$

where

T_e = engine output torque

ξ_o = overall reduction ratio

η_t = overall efficiency of the transmission

The engine speed is related to the vehicle speed as

$$V = \frac{n_e r (1-i)}{\xi_o} \qquad (8.48)$$

where i is the slip (taken between 2 and 5%), and n_e is the engine speed.

Recently, automated manual transmissions (AMTs) have been receiving attention. These are different from manual transmissions because the driver does not actuate the power of the clutch and gear shifting. Clutch actuation and gear shifting occur by means of a hydraulic system or an electric motor and are operated electronically. The clutch pedal is obsolete; the clutch control is connected to the position and movement of the selector lever. The mechanical connection between the selector lever and the transmission is omitted; the transmission is controlled electronically (shift by wire). This offers more possibilities regarding the design of selector units as with the usual mechanical shifting systems. The driver usually activates the gearshift. With the implementation of a shift program in the transmission control unit, an automatic mode is easily established so that an AMT can execute gearshifts automatically. The automatic operation mode is selected by a separate switch.

8.4.3.2 Automatic Transmissions

Automatic transmissions are quite common in North America and less common in the rest of the world. Compared with the manual transmission, the automatic transmission offers more convenience to the driver because the driver does not need to manually shift at various speeds when the automatic transmission is used. However, the automatic transmission generally is more expensive and is not as economical with fuel efficiency. Figure 8.20 shows a typical automatic transmission used in a north-south powerplant configuration.

The main components of an automatic transmission are the torque converter (Figure 8.21) and the gear box. In automotive applications, the torque converter is a three-element hydrodynamic drive unit that links the engine to the automatic transmission. The torque

Figure 8.20 A typical automatic transmission used in a north-south powerplant configuration. ZF six-speed automatic transmission. (Wagner, 2001.)

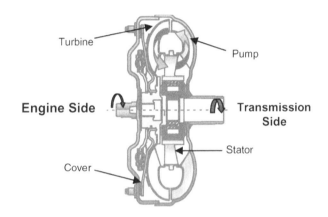

Figure 8.21 Cross-sectional view of a torque converter (Luk Inc., Lee and Lee, 2004).

converter decouples the engine from the transmission at idle conditions. It also increases the engine torque for input to the transmission during a launch condition. The first element of the torque converter is the impeller. It is a centrifugal pump that converts mechanical power from the engine to fluid power. The fluid power from the impeller is directed to the second element, a radial turbine that converts it back to mechanical power which then is transmitted to the transmission. The third element of the torque converter, the reactor or stator, redirects the working fluid back to the impeller.

The power output of the torque converter (P_{tcout}) is equal to the output torque (T_{tcout}) times the output speed (ω_{tcout}), and the power input (P_{tcin}) is the input torque (T_{tcin}) times the input speed (ω_{tcin}). The efficiency of the torque converter η_{tc} is defined as

$$\eta_{tc} = \frac{P_{out}}{P_{in}} = \frac{T_{tcout}\omega_{tcout}}{T_{tcin}\omega_{tcin}} \tag{8.49}$$

Note that the subscript tc stands for the torque converter.

The efficiency of the torque converter and the torque ratio are displayed in Figure 8.22 as functions of the speed ratio. Note that at a speed ratio of zero (idle condition), the torque is the highest and the efficiency is the lowest.

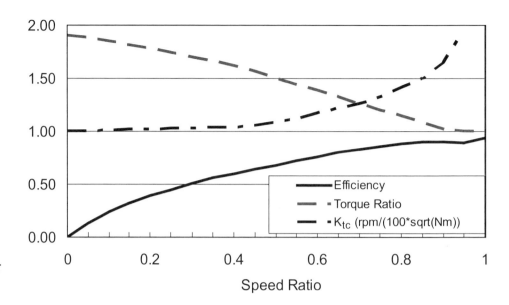

Figure 8.22 Efficiency and torque ratios as functions of the speed ratio for a torque converter.

At a maximum speed, the torque converter acts as a fluid coupling device, giving the highest efficiency and a torque ratio of 1.

Figure 8.23 shows the tractive force as a function of the vehicle speed and the various gear ratios. At the lowest gears, the vehicle speed is low, and acceleration can be high (under wide throttle). Hence, the tractive forces are high. At higher speeds, the need is mainly to maintain speed rather than the need for acceleration. This figure also shows how grade affects the gear ratio used. At no grade levels, the vehicle speed can be maintained with the highest gear, whereas at a certain grade, a lower gear must be chosen to maintain the vehicle speed. If the vehicle condition (speed and gear) are below the grade line given, the vehicle speed cannot be maintained (not enough torque or tractive forces).

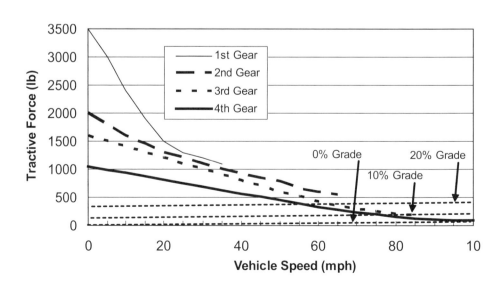

Figure 8.23 Relation between tractive forces at various gears and vehicle speeds.

The introduction of new technology such as torque converter lock-up clutches and electronic transmission control units and an increase in the number of speeds and optimized components and parts all have contributed to the fact that the power losses of automatic transmissions were reduced drastically during the last decade. The fuel consumption of an automatic transmission today is approaching that of a vehicle with a manual transmission. Good shift quality and electronically controlled shift programs that adapt to driving conditions and driving behavior guarantee high driving comfort.

The introduction of five-speed transmissions since 1989, six-speed transmissions in general applications later, and the broadening numerous functions such as the manual shift mode using the lever to shift the gears (Tiptronic) have added strongly to the image of improvement in automatic transmissions. The following are the advantages of high-speed (i.e., five- or six-speed) automatic transmissions:

- Fuel consumption savings by approximately 5%

- Reduced emissions

- Improved acceleration by approximately 5%

- Reduced weight by approximately 10%

To determine the working output speed and torque of the torque converter, we need to specify a parameter called the capacity factor, defined as

$$K_{tc} = \frac{\omega_{tc}}{\sqrt{T_{tc}}} \tag{8.50}$$

This also is given in Figure 8.22. The steps to be taken in finding the output speed and the torque of the torque converter are as follows:

- Determine the engine speed and torque (given).

- Find the engine capacity factor. The capacity factor of the torque converter is the same as that of the engine.

- Use a graph similar to Figure 8.22 to determine the speed ratio, efficiency, and torque ratio.

- Calculate the output speed and torque as well as the efficiency.

EXAMPLE E8.10

An engine delivers 300 Nm torque at an engine speed of 2500 rpm. It is coupled to a torque converter and an automatic speed transmission. Determine the output speed and the output torque of the converter of the characteristics given in Figure 8.22. Determine the efficiency of the torque converter.

Solution:

The engine speed and torque are given as

$$T_e = 300 \text{ Nm}$$

$$n_e = 2500 \text{ rpm}$$

Find the engine capacity factor. The capacity factor of the torque converter is the same as that of the engine

$$K_{tc} = K_e$$

$$= \frac{\omega_e}{\sqrt{T_e}}$$

$$= \frac{2500}{\sqrt{300}}$$

$$= 144.3 \frac{\text{rpm}}{\sqrt{\text{Nm}}}$$

Use a graph similar to Figure 8.22 to determine the speed and torque ratio.

The speed ratio is 0.81, and the torque ratio is 1.13. Then,

$$n_{tc} = 0.81 \times 2500 = 2025 \text{ rpm}$$

$$T_{tc} = 1.13 \times 300 = 339 \text{ Nm}$$

Using Eq. 8.49, the efficiency is

$$\eta_{tc} = \frac{P_{out}}{P_{in}}$$

$$= \frac{T_{tcout}\omega_{tcout}}{T_{tcin}\omega_{tcin}}$$

$$= \frac{2025 \times 339}{2500 \times 300} = 0.91$$

This also could be found from Figure 8.22. Note that higher torque was given at a lower speed. This will help give a higher level of acceleration in startup condition (as desired by typical drivers).

For an automatic transmission, we need to modify Eqs. 8.47 and 8.48 to reflect the implementation of the torque converter. Instead of having the engine speed (n_e) and the engine torque (T_e) for a manual transmission, we will have the output speed of the torque converter (n_{tc}) and the output torque of the torque converter (T_{tc}). Hence, for the automatic transmission, the tractive force available at the tires is calculated by

$$F_x = \frac{T_{tc}\xi_o\eta_t}{r} \tag{8.51}$$

where ξ_o is the overall reduction ratio, and η_t is the overall efficiency of the transmission.

The vehicle speed is related to the engine speed (similar to Eq. 8.48) as

$$V = \frac{n_{tc}r(1 - i)}{\xi_o} \qquad (8.52)$$

where i is the slip (taken between 2 and 5%).

8.4.3.3 Continuously Variable Transmissions (CVTs)

Continuously variable transmissions (CVTs) are transmissions where the creation of the ratio occurs by means of a variator with axial repositioning of a conical-shaped pair of discs. A chain or a belt between the discs transfers torque with friction contact. Figure 8.24 shows a typical continuously variable transmission used in an east-west powerplant configuration (Ford's Five Hundred and Freestyle).

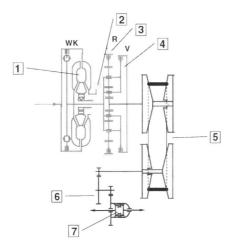

1 Torque Converter

2 Pump

3 Reverse Gear Set

4 Forward Clutch

5 Variator

6 Final Drive

7 Differential

Figure 8.24 Typical continuously variable transmission used in an east-west power-plant configuration for a ZF CFT 30 continuously variable transmission for front-wheel drive. (Wagner *et al.*, 2004.)

Continuously variable transmissions are receiving interest for use in small and medium vehicles with low torque (300 Nm or less). Applications where high torque is needed (greater than 350 Nm) generally are not suitable for continuously variable transmissions.

The system shown in Figure 8.24 is based on the Van Doorne belt concept. This system has two conically faced pulleys, with the effective radius of the pulleys (and hence the reduction ratio) changing as the distance x (Figure 8.24) changes.

The other potential mechanism is the use of a toroidal (or Perby) system. Figure 8.25 shows the basic mechanism.

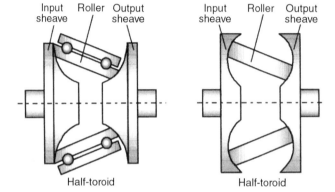

Figure 8.25 The basic mechanism of a toroidal continuously variable transmission. (Wagner, 2001.)

Power is transferred through the first torus disc by friction on rollers and flows over a second sheave to the transmission output shaft. Twisting of the rollers changes the ratio. The fundamental advantage is the high torque capacity, where it is higher than with a Van Doorne continuously variable transmission. The toroidal continuously variable transmission also offers good driving comfort and the possibility of rapid ratio change. Disadvantages are the package requirements, the weight, and manufacturing costs. Furthermore, no real solution for traction oil has been found. Oils that currently are employed in development have not fulfilled satisfying functions over the whole temperature range of –40 and 150°C (Wagner, 2001).

The first toroidal transmissions went into mass production with General Motors in the 1930s, but the concept did not really catch on. Nissan undertook the second application in 1999 with a toroidal transmission for 350 Nm, a gear ratio of 4.36, and a weight of 105 kg for use in a standard driveline. These numbers are not competitive in comparison to a six-speed automatic transmission.

The three types of transmissions described here remain the ones used in the vast majority of vehicles today. In particular, the automated manual transmission (AMT) and the continuously variable transmission (CVT) are expected to receive increased interest, mainly due to the advantages they offer with fuel economy and shift quality. Conventional automatic transmissions are expected to gain higher efficiency levels, particularly with the introduction of six-speed transmissions.

8.4.4 Vehicle Acceleration

As stated, the tractive forces at the tires must be calculated to determine the vehicle acceleration and travel distance. In Section 8.3, we found the maximum possible acceleration based on the traction limitation. On the other hand, finite engine torque is the main factor that limits acceleration at off-idle conditions and on dry surfaces.

The ultimate objective of the engine power and torque is to deliver tractive forces at the driving axles. These tractive forces result in the acceleration of the vehicle.

First we must account for the losses in the front end accessory drive, induction, and exhaust. These power losses are direct losses in the torque. Keep in mind that these losses are calculated at a specific engine speed (i.e., one that corresponds to maximum power). These losses directly impact the torque delivered by the engine as the speed remains the same. Once the net engine power and torque are determined, we need to find the torque output of the transmission, taking into account the gear ratio of the transmission, the impact of the torque converter, and the system inefficiencies. Once the output torque of the transmission is determined, the axle gear ratio(s), if available, as well as driveline system inefficiencies must be considered to find the torque available for the driving axle(s) to accelerate the vehicle.

In addition to these losses, keep in mind that the resulting torque does more than simply accelerate the vehicle. This torque is used to rotate all rotating elements in the vehicle, including the wheels. To take into consideration the effect of inertia of the vehicle rotating elements, we will introduce the mass factor (γ_m). This is needed for application of Newton's second law to calculate vehicle acceleration as

$$F_{net} = F - \sum R = \gamma_m ma \qquad (8.53)$$

where

$$\sum R = R_r + R_a + ...$$

F_{net} is the net tractive force causing the vehicle to accelerate. The mass factor is

$$\gamma_m = 1 + \frac{1}{Mr^2}\left\{\sum I_w + \sum I_1\xi_1^2 + \sum I_2\xi_2^2 + ... + \sum I_n\xi_n^2\right\} \qquad (8.54)$$

where

$\quad$ M $\quad=\quad$ mass of the vehicle

$\quad$ r $\quad=\quad$ rolling radius of the wheels

$\quad$ I_w $\quad=\quad$ mass moment of inertia for a wheel

$\quad$ I_i $\quad=\quad$ moments of inertia of the rotating components attached to the drivetrain (with I = 1, 2, ...)

The following formula is suggested as an approximation for passenger cars (Taborek, 1957)

$$\gamma_m = 1.04 + 0.0025\xi_0^2 \qquad (8.55)$$

where ξ_0 is the overall transmission ratio. Knowing that the acceleration is the first derivative of velocity with respect to time $\left(a = \dfrac{dV}{dt}\right)$, Eq. 8.53 can be written as

$$F_{net} = F_x - \sum R = \gamma_m M\left(\frac{dV}{dt}\right) \tag{8.56}$$

Knowing that the net force (F_{net}) is a function of velocity (because of resistance forces, which are functions of velocity), this expression can be written to calculate time traveled between velocity V_i and V_f as

$$t = \gamma_m M \int_{V_i}^{V_f} \frac{dV}{F_{net}} \tag{8.57}$$

Similarly, the distance traveled (S) during this change in velocity would be

$$S = \gamma_m M \int_{V_i}^{V_f} \frac{VdV}{F_{net}} \tag{8.58}$$

Keep in mind during acceleration the shifting time that takes place when going from a certain gear to another. Each shift is approximately 1 second, with the automatic transmission offering less shifting time than the manual transmission.

In general, Eqs. 8.57 and 8.58 should be used to calculate the key vehicle attributes of time and distance traveled to achieve a certain level of velocity. This is of particular and key interest to vehicles with a "sporty" image. The difficulty in these equations is in the integration that must be carried numerically. Realize that the engine speed (and torque) during acceleration changes. This results in a variable tractive force F_x. Also, the driver may shift the transmission gear during this acceleration, leading to a second source of variation in the tractive force. In addition, as the vehicle gains speed, the drag changes, which adds another reason for variation in the net force (F_{net}). All of these factors lead to a higher level of complexity when the acceleration time and/or distance are calculated.

EXAMPLE E8.11

A vehicle (including wheels) weighs 22,215 N (5000 lb). Each wheel weighs 266.6 N (60 lb), and the rolling radius of the wheels is 355.6 mm (14 in.). The radius of gyration is 279.4 mm (11 in.). The engine is driven at 3000 rpm and is delivering 406.27 Nm (300 ft-lb). The total drivetrain efficiency is 86%, and the transmission is in fourth gear with a ratio of 4.5 to 1. The components rotating at the engine speed have an inertia of 0.6 slug ft². Assume a slip of the running gear at 2% and a coefficient of rolling resistance of 2%. The front area of the vehicle is 1.858 m² (20 ft²), the aerodynamic drag coefficient is 0.4, and the speed of the wind is zero. Determine the acceleration at that point.

Solution:

Given

$$M = \frac{5000}{32.2} = 155.2800 \text{ slugs}$$

$$m_w = \frac{60}{32.2} = 1.8634 \text{ slugs}$$

$$r = 14 \text{ in.}$$

$$r_g = 11 \text{ in.}$$

$$n_e = 3000 \text{ rpm}$$

$$T_e = 300 \text{ ft-lb}$$

$$\eta = 0.86$$

$$\xi_t = 4.5$$

$$\Sigma I = 0.6 \text{ slug ft}^2 = 86.4 \text{ slug in.}^2$$

$$i = 0.02$$

$$\mu_r = 0.02$$

$$C_D = 0.4$$

$$A_F = 20 \text{ ft}^2$$

1. Calculate the mass factor (γ_m) as

$$I_w = m\left(r_g\right)^2$$

$$= 1.8634 \left(11\right)^2$$

$$= 225.47 \text{ slug in.}^2$$

$$\gamma_m = 1 + \frac{1}{Mr^2}\left\{\sum I_w + \sum I_1\xi_1^2 + \sum I_2\xi_2^2 + \dots + \sum I_n\xi_n^2\right\}$$

$$= 1 + \frac{1}{155.28 \times (14)^2}\left\{4 \times 225.47 + 86.4 \times 4.5^2\right\}$$

$$= 1.0871$$

2. Calculate the tractive force (Eq. 8.47)

$$F_x = \frac{T_e\xi_o\eta_t}{r} = \frac{36{,}000 \times 4.5 \times 0.86}{14} = 995.14 \text{ lb}$$

3. Calculate the vehicle speed (Eq. 8.48)

$$V = \frac{n_e r(1-i)}{\xi_o}$$

$$= \frac{3000 \times (2\pi) \times 14 \times (1 - 0.02)}{4.5}$$

$$= 57{,}470.20 \text{ in./min}$$

$$= 54.4225 \text{ mph}$$

Because the wind is at zero speed,

$$V_r = V = 54.4225 \text{ mph}$$

4. Calculate the rolling resistance force (Eq. 7.26)

$$R_r = \mu_r \, mg$$

$$= 0.02 \times 5000 = 100 \text{ lb}$$

5. Calculate the aerodynamic drag force (Eq. 7.27)

$$R_a = \frac{\rho}{2} C_D A_F V_r^2$$

$$= \frac{0.002378}{2} \times 0.4 \times 20 \times \frac{(57{,}470)^2}{(12)^2 \times (60)^2}$$

$$= 60.60 \, \frac{\text{slug.ft}}{\text{s}^2}$$

$$= 60.60 \text{ lb}$$

Note the unit conversion factors.

6. Calculate the acceleration (Eq. 8.53)

$$a = \frac{F_{net}}{\gamma_m m}$$

$$= \frac{F - \sum R}{\gamma_m m}$$

$$= \frac{995.14 - 100.0 - 60.60}{1.0871 \times 155.28}$$

$$= 4.944 \text{ ft/s}^2$$

EXAMPLE E8.12

Consider Example E8.11. Assume that the engine torque remained constant during a takeoff acceleration. Also, acceleration is happening while the transmission is engaged in the same gear as in the previous example. Determine how long it would take the vehicle to reach a constant speed of 60 mph.

Solution:

1. The unit conversion is

$$V_f = 60 \frac{mile}{hr}$$

$$= 60 \times \frac{5280.0 \text{ ft}}{3600 \text{ s}}$$

$$= 88.0 \text{ ft/s}$$

2. Determine the drag (Eq. 7.27) as

$$R_a = \frac{\rho}{2} C_D A_F V_r^2$$

$$= \frac{0.002378}{2} \frac{\text{slug}}{\text{ft}^3} \times 20 \text{ ft}^2$$

$$= 0.02378 \frac{\text{slug}}{\text{ft}} V_r^2$$

3. Determine F_{net} as

$$F_{net} = F - \sum R$$

$$= 995.14 - 100.0 - 0.02378 \text{ V}_r^2$$

$$= 895.14 - 0.02378 \text{ V}_r^2$$

4. Determine the time (Eq. 8.57) as

$$t = \gamma_m m \int_{V_i}^{V_f} \frac{dV}{F_{net}}$$

$$= 1.0871 \times 155.28 \times \int_0^{88} \frac{dV}{895.14 - 0.02378 \text{ V}_r^2}$$

$$= 18.11 \text{ s}$$

The last example is a simple example that shows achievable acceleration with nonconstant net force (a function of velocity because of drag). Real problems would involve varying the engine revolutions per minute and varying the torque and gear ratio, as well as other complexities.

8.5 Safety Features

The most important safety concern that is encountered by customers trying to accelerate a vehicle is the loss of traction or slip condition. Although drivers generally possess more control while accelerating than braking (where drivers may lose control), traction is important to move the vehicle away from a possible accident or hazardous situation. Two features usually are used for that purpose. These are the use of a limited slip axle and traction control. The first is an axle that uses a clutch or locking device to make torque bias to a wheel that still has traction when the other wheel on the axle slips. Traction control applies brakes to slipping wheels.

8.5.1 Limited Slip Axle

While driving around a corner, the wheel on the outside of a given axle rotates more than the wheel on the inside. For this to occur, a differential gear set is incorporated in the axles. This differential unit receives torque from an input (the driveshaft) and distributes it between the axle shafts (or half shafts). It allows those outputs to rotate at different speeds.

A normal (or open) differential unit behaves such that if one output loses traction, little torque will be sent to the other output. In other words, neither of the wheels can receive more torque than the amount required to slip either one of them. If a front-wheel-drive vehicle has ice under the right front wheel, then that wheel will spin when torque is applied. If 25 ft-lb (33.87 Nm) caused that right front wheel to slip, the left front wheel will receive only 25 ft-lb (33.87 Nm).

Some axles incorporate a limited slip device to govern the differential to prevent this condition from occurring. This device directs some torque to the wheel that still has traction. Some of the devices that can be used as axle limited slip devices include the following:

1. **Plate clutch.** In this device, a series of plates is coated with friction material that provides resistance when the wheels move at different speeds. These have limited torque capacity. Figure 8.26 shows one of these devices.

2. **Viscous coupling.** This device causes silicone fluid to be sheared between a series of clutch plates when the wheels move at different speeds. Cost is a drawback of these devices.

3. **Gerodisc/Visco-Lok devices.** In these devices, a difference in speed between the two wheels activates a pump that compresses a clutch. This, in turn, controls the slip. More information is available in a paper by Gassmann and Barlage, 1996.

4. **Torque multiplying differential.** This differential sends a multiple of the traction available at the slipping wheel to the wheel that still has traction. For example, if one wheel on an axle slips at 33.87 Nm (25 ft-lb) of torque, the other axle may receive 101.57 Nm (75 ft-lb).

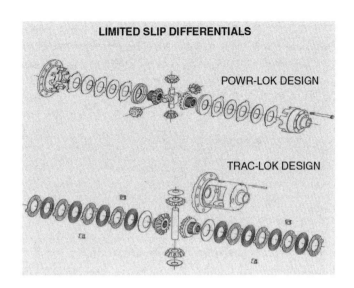

Figure 8.26 Dana's limited slip axle design (Vettel and Bielik, 2003).

These devices are among those available on the market to control the amount of torque distributed at each wheel.

8.5.2 Traction Control

The traction control system works by applying the brake to the spinning wheel. This allows torque to be sent to the wheel that still has traction. For example, if the right wheel slips after 33.87 Nm (25 ft-lb) of torque, then a certain amount of brake torque (say, 75 ft-lb, or 101.57 Nm) is applied to that wheel. There is effectively a total of 135.4 Nm (100 ft-lb) of resistance at that right wheel—25 ft-lb from the limited traction, and 75 ft-lb from the brake. Hence, the left wheel will be able to receive 135.4 Nm (100 ft-lb). This torque is applied to the ground where traction is still present. Some traction control systems may reduce torque delivered by the engine to help control slip. One mechanism to perform that is to start in second gear when slippery roads are experienced.

Engineers have the objective of achieving the maximum tractive force. With advances in electronic controls and sensors, sensors are planted to assess skid, or slipping, conditions of the wheels. Traction control systems work in a manner opposite that of an anti-lock braking system (ABS). At slip during acceleration, the brakes are applied electronically and momentarily to bring the wheel to the nonskid condition. Throttle is applied again to reach the skid condition, gaining higher speed over a short period of time. This process is repeated until full traction at the desired speed is achieved. Traction control is now a standard safety feature on many vehicles and is available as an optional feature on most others. Similar to the advantages of anti-lock braking systems, traction control systems offer better treatment of complex surfaces frequently encountered in cold regions or in off-road maneuvers. Variations in the coefficient of friction between the two wheels of the same axle are possible. Thus, a condition of the wheels of the same axle having one side on a low coefficient of friction and the other wheel on a higher coefficient of friction can be encountered. Under such a condition, the maximum tractive force will be sacrificed if the same force is given to both wheels on the same axis.

When driving forces are applied to the tire, the tire experiences a condition of skid (slippage) over the road surface. The speed of the vehicle under a free-rolling condition (i.e., no skid) can easily be related to the rotational speed of the wheels

$$V = r\omega$$

as given in Chapter 7. Here,

V = vehicle speed

r = rolling radius of the tire

ω = angular speed of the tire

Variation is observed between V and $r\omega$ under a slipping condition, where the angular speed of the tire is nonzero, while the vehicle is not moving ($V = 0$). Some skid is observed between the two cases of free rolling and the complete slippage. If we define the first as the case of 0% skid and the last as a 100% skid condition, Eq. 8.59 can be written to describe skid at any point between the two extreme cases as

$$i = \left(1 - \frac{V}{r\omega}\right) \times 100\% \tag{8.59}$$

As discussed, the coefficient of friction between the tires and the road is a function of many environmental parameters. This coefficient under given environmental conditions is a function of the skid between the tire and the ground, achieving a peak value only when a certain percentage of skid is achieved. The reader is encouraged to review the material on anti-lock braking systems to understand how traction control works.

8.6 Summary

Vehicle acceleration is one of the important attributes of an automobile. It can be limited by tractive forces at the tires or engine power and torque capacity.

This chapter showed how maximum acceleration can be obtained under certain driving conditions, assuming an engine that can deliver the necessary torque. Different driveline configurations (front-wheel drive, rear-wheel drive, and all-wheel drive) were treated. In particular, we showed how maximum tractive forces can be achieved for vehicles with conventional traction and those with traction control systems that distributed torque to the front and rear axles according to the axle vertical load.

The various types of engines available in the automotive field were overviewed. Of particular interest was the internal combustion engine. The performance of such an engine was described in detail.

In addition to the performance of the engine, the performance of the transmission and the drivetrain elements was discussed at length. The maximum torque delivered by the engine was discussed. This torque is compromised by losses due to exhaust, induction, front end accessory drive, and rotating components in the engine and transmission, as well as the driveline and drive axle. This yields a reduced torque level at the driving axle to accelerate the vehicle.

Various drive possibilities were discussed, and equations were derived for front-wheel-drive, rear-wheel-drive, and all-wheel-drive vehicles. Safety features related to traction also were overviewed.

8.7 References

1. Cantemir, C., Hubert, C., Rizzoni, G., and Demetrescu, B., "High Performance Fuel Cell Sedan," SAE Paper No. 2004-01-1003, Society of Automotive Engineers, Warrendale, PA, 2004.

2. Gassmann, T. and Barlage, J., "VISCO-LOK: A Speed-Sensing Limited Slip Device with High Torque Progressive Engagement," *SAE Transactions*, SAE Paper No. 960718, Society of Automotive Engineers, Warrendale, PA, 1996.

3. Genta, G., *Motor Vehicle Dynamics,* World Scientific Publishing, River Edge, NJ, 1997.

4. Gillespe, T.D., *Fundamentals of Vehicle Dynamics*, Society of Automotive Engineers, Warrendale, PA, 1992.

5. Kalberlah, A., "Electric Hybrid Drive Systems for Passanger Cars and Taxis," *SAE Transactions*, SAE Paper No. 910247, Society of Automotive Engineers, Warrendale, PA, 1991.

6. Kluger, M.A. and Long, D.M., "An Overview of Current Automatic, Manual and Continuously Variable Transmission Efficiencies and Their Projected Future Improvements," *SAE Transactions*, SAE Paper No. 1999-01-1259, Society of Automotive Engineers, Warrendale, PA, 1999.

7. Lee, J. and Hoon Lee, H., "Dynamic Simulation of Nonlinear Model-Based Observer for Hydrodynamic Torque Converter System," *SAE Transactions*, SAE Paper No. 2004-01-1228, Society of Automotive Engineers, Warrendale, PA, 2004.

8. Milliken, W.F. and Milliken, D.L., *Race Car Vehicle Dynamics,* Society of Automotive Engineers, Warrendale, PA, 1995.

9. Nakamura, H., Motoyama, H., and Kiyota, Y., "Passenger Car Engines for the 21st Century," *SAE Transactions*, SAE Paper No. 911908, Society of Automotive Engineers, Warrendale, PA, 1991.

10. Oman, H., "New Energy Management Technology Gives Hybrid Cars Long Battery Life," *SAE Transactions*, SAE Paper No. 1999-01-2468, Society of Automotive Engineers, Warrendale, PA, 1999.

11. Reimpell, J. and Stoll, H., *The Automotive Chassis: Engineering Principles,* Society of Automotive Engineers, Warrendale, PA, 1996.

12. SAE J670e, "Vehicle Dynamics Terminology," Society of Automotive Engineers, Warrendale, PA, 1978.

13. SAE J1995, "Engine Power Test Code-Spark Ignition and Compression Ignition—Gross Power Rating," Society of Automotive Engineers, Warrendale, PA, 1995.

14 Taborek, J.J., "Mechanics of Vehicles," *Machine Design,* May 30–December 26, 1957.

15. Tilagone, R., Venturi, S., and Monnier, G., "Natural Gas—An Environmentally Friendly Fuel for Urban Vehicles: The SMART Demonstrator Approach," *SAE Transactions*, SAE Paper No. 2005-01-2186, Society of Automotive Engineers, Warrendale, PA, 2005.

16. Unich, A., Bata, R., and Lyons, D., "Natural Gas: A Promising Fuel for IC Engines," *SAE Transactions*, SAE Paper No. 930929, Society of Automotive Engineers, Warrendale, PA, 1993.

17. Vettel, P. and Bielik, P., "Advances in Additive Technology for Limited Slip Axles—Part I Friction Test Development," *SAE Transactions*, SAE Paper No. 2003-01-3234, Society of Automotive Engineers, Warrendale, PA, 2003.

18. Wagner, G., "Application of Transmission Systems for Different Driveline Configurations in Passenger Cars," *SAE Transactions*, SAE Paper No. 2001-01-0882, Society of Automotive Engineers, Warrendale, PA, 2001.

19. Wagner, G., Remmlinger, U., and Fischer, M., "CFT30—A Chain Driven CVT for FWD 6 Cylinder Application," *SAE Transactions*, SAE Paper No. 2004-01-0648, Society of Automotive Engineers, Warrendale, PA, 2004.

Chapter 9

Total Vehicle Dynamics

9.1 Introduction

In the previous chapters, the dynamics of various vehicle maneuvers was analyzed. These maneuvers or driving conditions included steering, handling, braking, and acceleration. Each of these conditions was discussed mostly as a stand-alone maneuver. In actual driving, one maneuver is experienced at a time for most drivers and/or driving conditions. However, for experienced drivers competing in car racing and for special events, more than one maneuver can be experienced at a time. Steering and braking to avoid an accident is one combination among these possible events. Steering and acceleration (while taking off in the parking lot or going around a curve in car racing) is another possible driving condition.

The purpose of this chapter is to bring to the reader the various driving conditions that can be experienced by a driver when driving a vehicle. In vehicle design and development, several extreme driving conditions are tested, and measurements are taken to express the driver's experience in technical terms.

Each driver has certain priorities in mind for vehicle attributes. For example, a young driver tends to prefer a small car with some power. He or she pays less attention to comfort and/or size of the car. On the other hand, an older driver tends to prefer a more comfortable and bigger vehicle. He or she pays less attention to power. Although ride, comfort, and safety play high on an older driver's list of attributes, style, power, and acceleration are the more important attributes to young drivers. There also is evidence that the customer's attitude toward the vehicle and its attributes is race and gender dependent, in addition to being age dependent.

9.1.1 Subjective and Objective Evaluations

The dynamics of a vehicle usually is expressed by the driver in various general subjective terms. Terms and phrases such as "fun to drive," "comfortable," "responsive," "sporty," and others frequently are used to describe the driver's experience with the vehicle. In fact, these phrases and terms can be used as metrics and can be rated subjectively, with 10 being the best to 0 being the worst. Both industry and customer advocacy groups as well as research centers perform what are called "customer clinics," where subjective assessments by customers are gathered. In most cases, these are actual customers, and the data are gathered and analyzed statistically. In other cases, experts who don the customer's hat deliver their assessments.

The engineering challenge is to develop objective metrics that can describe, or even predict, the terms used by the customer. For example, when the customer describes a vehicle as responsive, what does he or she exactly mean? Does he or she mean the vehicle accelerates with the desired or expected acceleration when it is given some throttle? Or does he or she mean that the vehicle responds instantly to steering?

The first challenge faced by engineers in vehicle design is to establish the target customer and then to recognize the general subjective vehicle dynamics terms that describe the customer's wants and needs. This task requires a genuine multi-team effort, where marketing works hand in hand with engineering. The second challenge for the engineer is to describe the set of maneuvers and driving conditions that relate to the target customer's experience with the vehicle. In addition, the third challenge for the engineer is to find the actual objective metrics that should be used (e.g., acceleration of the vehicle at a certain throttle, vibration level at a point such as seat track) and the actual values of these metrics that best describe the level desired by the customer. For example, at a certain steering torque and/or position, the customer expects the vehicle to steer at a certain radius. If the vehicle steers at a smaller radius, then it oversteers; if the steering radius is found to be more than the expected, the vehicle understeers. Needless to say, both oversteering and understeering are error states to the customer. In summary, engineers start their vehicle design by defining the target customer, describing his or her wants and needs in subjective terms, defining the testing condition or driving maneuvers that describe these wants and needs, deriving metrics that best relate to these wants and needs, and then determining the values that describe the customer's level of comfort with these metrics at these maneuvers.

9.1.2 Target Setting

Design targets can be set based only on the customer's wants and needs. The business case of the vehicle, which precedes the engineering case, is that the vehicle is built to be sold for a profit. Once the customer's wants and needs are established, the question of competitiveness in the marketplace is raised. In particular, questions should be asked about how competitors or potential competitors are addressing these wants and needs, how much success they have achieved, and where they would be by the time the engineered vehicles hit the market. The last question is of great importance because engineers need to know not only the status of the industry today but where it will be a number of years from now, when the product for which the targets are being set hits the market.

Knowing the customer provides the vision of the engineers and where the industry should be going. On the other hand, knowing the industry provides the strategy that should be followed to be competitive or move ahead of the competition. In addition, engineers should know what they can deliver, based on the current state of knowledge and available hardware.

Thus, targets should best describe the customer's wants and needs by taking into consideration the current market status. Targets should be measurable objectively. Subsequently, the hardware design process can be started. Using advanced engineering simulation techniques, the behavior of the vehicle with specific hardware can be assessed. It is understood that not all of the customer's wants and needs can be achieved with a certain vehicle package. At that point, "trade-off" studies are performed. These studies will compromise certain attributes in favor of others, depending on the customer's list of priorities.

Trade-off studies describe the level of attributes that can be achieved when a certain package, design limitations, or constraints are imposed on the engineering problem. It is understood that the intention is to give customers the best value for their wants and needs. Consideration of the target customer and/or group of customers and their priorities in the vehicle selection process is of great importance when conducting these trade-off studies. For example, a trade-off between "responsive" and "robust" may be needed. A vehicle that responds well to the driver's steering torque input may not be "forgiving" for a mother driving while talking to her children on their way to school. Also, trade-off studies among various attributes on one hand and between these attributes and cost in general will be performed. Indeed, most of the engineering task is to optimize the vehicle package such that the best value of these attributes is given to the customer.

9.1.3 Vehicle Dynamics Tests and Evaluations

During the target-setting process, knowledge of where the competition is with the vehicle dynamics attribute constitutes one of the basic pillars of vehicle design. To achieve such knowledge, competitors' products must be benchmarked or evaluated. This dictates the establishment of certain maneuvers that simulate the customer's behavior. In addition, engineers need to establish the measurable metrics that describe the customer's attitude toward the vehicle and the range of these metrics where a certain customer's comfort level is achieved.

Benchmarking occurs early in the design or product development process. In addition to benchmarking, engineers need to assess their own design at a later stage of the product development process, particularly when prototype hardware is available. Both of these actions (benchmarking and design assessment) necessitate establishing standardized tests for vehicle dynamics.

The challenge of coming up with these tests and procedures to describe the driver's interaction with the vehicle and its dynamics, as well as the metrics and the metric levels, constitutes the biggest challenge that the engineer faces. Based on such assessment of the customer, target values for certain metrics that can be measured objectively are established. The following is a general description of the many tests and evaluations that can be performed on the vehicle to describe the driver's interaction with it.

9.1.3.1 Ride

Ride was discussed in Chapter 4. Among the objectives of ride tests is to find the low-frequency movements of the vehicle. This includes assessment of vehicle movement on various roads, including smooth and rough roads. Assessments include the behavior of the vehicle under discrete events such as impact. These events cause substantial suspension movement. These tests typically describe the magnitude and characteristics of the vehicle motions relative to the road as a result of road inputs. They address the question of how the vehicle body movement relates to the road profile. Does it follow the profile exactly, or does it absorb it? They also address the delay of the vehicle response to road input, the nature of damping (overdamped or critically damped), and the relative motion of the front and rear (pitching) of the vehicle.

Disturbances to the driver and occupants as a result of the vehicle body motions also are evaluated. In particular, abruptness to occupants in the vertical direction, the lateral motions of the head and body, and impact of the suspension motions under various stop conditions are among the many tests and studies made on the vehicle.

Assessment of the vehicle ride includes its vibrations caused by road disturbances and felt by the occupants through the seat, steering wheel, floor pan, controls, and other customer touch points. Road-induced pitch and bounce motions of the body are among the examples. Another example is shake, which refers to the low-frequency (5- to 30-Hz) oscillations of the body, frame, suspension, powertrain, and seat that are felt by the driver or passengers.

9.1.3.2 Steering

Steering can be tested under different maneuvers. Some of these maneuvers are summarized as follows:

- **Parking**. Parking refers to very-low-speed driving and parking situations. Among the attributes considered are the steering effort, the uniformity of the effort, and the ability of the steering wheel to return to its original straight-ahead position. In addition, the ability to maneuver the vehicle into and out of restricted spaces is evaluated. This includes the steering wheel angle required for parking maneuvers or "angle demand."

- **Straight-ahead controllability**. This refers to the characteristics of steering around the straight-ahead position to allow the driver confident control of the steering. Tests cover the vehicle response to small steering corrections required to keep the vehicle in a straight path or lane changes. The acceleration in various directions due to small inputs on steering wheels around the straight-ahead position is evaluated. The symmetry of the vehicle behavior for steering inputs to the right and left also is evaluated. Tests also include control of the vehicle motion around the longitudinal axis, as well as the relationship between small steering wheel torque or angle and the response of the vehicle (torque feedback).

- **Cornering controllability**. These tests are intended to describe driver confidence in the vehicle control while cornering, particularly at moderate and high speeds. The tests include the reaction of the vehicle to steering inputs, both when turning the vehicle into a corner and when keeping the vehicle on an intended curvature. They also include steering to return to the straight-ahead position.

- **Steering disturbances**. These refer to unwanted vehicle reactions, steering wheel motions, or torque feedback that can result from sources other than the driver's steering inputs. Such sources include insufficient steering system performance. An example is torque steer. This relates to deviations from the course during power takeoff, pedal tip-ins, gear changes, or others. The disturbance of these maneuvers to the stability of the straight-ahead is controllability tested. Nibble, or the vibration of the steering wheel around its rotational axis on smooth surfaces due to tire imbalance or force variations, also is studied.

9.1.3.3 Handling

Handling includes studies similar to those done under steering. Among these are the following:

- **Straight-ahead stability**. This is the ability of the vehicle to travel in a straight line without driver input. Tests include assessment for the directional stability of the vehicle when disturbed by road or other input. Also included is the sensitivity to side wind, passing, or being passed by other traffic (e.g., large trucks). Vehicle behavior including its roll, yaw, and deviation from straightness as well as corrective actions should be assessed.

- **Cornering stability**. Among these tests performed is whether the vehicle understeers or oversteers under constant throttle and varying steering input. The predictability and comfort level of steering are studied. Tests include studying whether braking on various types of surfaces changes the understeer or oversteer characteristics.

- **Transitional stability**. This is the control of the vehicle during lane changes or similar maneuvers. Stability and controllability of the vehicle under different steering inputs and levels are tested. Steering inputs can be continuous and smooth or abrupt.

9.1.3.4 Braking

Three types of braking maneuvers are considered. These are light, moderate, and maximum braking.

Light and moderate braking cover the low-to-moderate deceleration range of braking. Vehicle reaction to operation of the brakes, including the pedal effort, is tested. Measurement of pedal travel and pedal force before noticing deceleration is made. Controllability of the vehicle when under light braking is considered. The effect of the speed at which the braking force is applied on the vehicle behavior also is included. The smoothness and progression of the transition from light to moderate braking also is considered.

Maximum braking is the operating point at which wheel lock-up is induced during straight-ahead driving. It should be evaluated on consistent friction surfaces. The time delay between the vehicle response and the pedal force engagement is evaluated. The travel of the braking pedal and its relation with the force applied are tested.

Braking disturbances also are evaluated. These refer to any disturbance within or initiated by the braking system of the vehicle.

Finally, parking brake operation is evaluated. This refers to the overall function of the parking brake. This includes the force required to engage and hold the vehicle on a grade, the travel required to engage and hold it, the parking brake location and ease of use, and other attributes.

Other studies also are made on the relationship between the pedal effort and travel, the lateral free play of the brake pedal, pedal ergonomics, and others.

9.1.3.5 Performance

Performance relates to the powertrain and acceleration characteristics (basic powertrain power and torque, including gearing). Performance feel is the customer's perception of performance. In addition to power and acceleration, it includes the effects of shift response and ratio selection, sound quality, traction control, transmission, all-wheel-drive (AWD) performance, and accelerator control characteristics.

- **Standing start.** This includes part throttle and wide open throttle. Performance feel is tested from a standing start throughout the driving range into top gear.

- **Rolling start.** This includes testing or rating performance feel from a rolling 2- to 8-km/hr (1- to 5-mph) start, at steady part throttle (or wide open throttle), throughout the driving range into top gear. The time or distance to accelerate to various speeds also is considered, with an automatic transmission in overdrive and a manual transmission in low gear.

Other evaluations include city performance, traveling in the range of 30 to 80 km/hr (20 to 50 mph), and highway performance with speeds in the range of 65 to 110 km/hr (40 to 70 mph). Tests also can be performed on drivability and engine performance. Drivability relates to the calibration of the engine and the quality of the fuel charging system. It should be evaluated at low and high altitudes and during cold and hot operation. Tests include starting and stability of the engine speed at both idle and while driving. In addition, the performance of the transmission can be evaluated.

Tests on the automatic transmission include shift quality. This is the customer's perception of how quickly, smoothly, quietly, consistently, and predictably the transmission events occur. It applies to torque converter clutch events as well as to gear ratio changes. Generally, the quicker, smoother, quieter, more consistent, and more predictable the transmission event is to the customer, the better. Evaluation of an automatic transmission includes testing upshifts, downshifts with or without braking, and shift timing (i.e., the speed and vehicle conditions where shifting occurs). Assessment also includes the ergonomics of shift and static engagement. Static engagement refers to moving the gear from neutral or park conditions to drive or reverse conditions.

Shift feel of a manual or automatic transmission includes its ergonomics (i.e., the location of the gear stick or lever with respect to the driver and the pattern or spacing between the gear position, shape, size, and other factors). Other evaluations include testing for the clutch pedal forces through the range of operation, shifting characteristics, smoothness of engagement and disengagement, the point of engagement (i.e., the friction point), the motion required to engage and disengage the clutch, and others.

Evaluations also can be made on the ease of locking or unlocking of the four-by-four (4×4) or all-wheel-drive system, smoothness and noise and vibration of the driveline systems, effectiveness on various types of surfaces (especially those with low coefficients of friction), and evaluation in off-road driving conditions. Additional evaluations include trailer tow performance.

9.2 Steering and Braking

We have noticed (see Chapter 6) that load transfers from the inside wheels to the outside wheels when the vehicle undergoes cornering or steering. On the other hand, we also have noticed (see Chapter 7) that load transfers from the rear to the front axles when braking. In many circumstances, braking happens while the vehicle is under steering conditions (not a straight-ahead condition). This will add significant complexity to the mechanics of the vehicle and load transfer characteristics. The issue of braking and steering was discussed as early as 1974 by Limpert *et al.* (1974) and in a publication by the U.S. Department of the Army in 1976. We will consider first the braking of a vehicle on a horizontal road (i.e., no grade) and ignore the other forces (e.g., aerodynamic, powertrain resistance).

9.2.1 Simple Braking and Steering on a Horizontal Road

The equations derived in Chapter 6 (Eq. 6.29) that describe the load transfer from the inside to the outside wheels are repeated here

$$F_{zli} = \frac{F_{zl}}{2} - \Delta F_{zl}$$

$$F_{zlo} = \frac{F_{zl}}{2} + \Delta F_{zl} \tag{9.1}$$

$$\Delta F_{zl} = \frac{F_{yl}h_r + K_\phi \phi}{t}$$

where, as was found in Chapter 6 (Eq. 6.89).

$$F_{yl} = F_{zl}\frac{V^2}{2Rg} \tag{9.2}$$

In these equations, F_{z1} is the total vertical force on the front axle, F_{y1} is the total lateral force on the front axle, R is the radius of the turn, K_ϕ is the roll stiffness, ϕ is the roll angle, t is the wheel tread or track width, h_r is the roll center height, and V is the vehicle speed.

EXAMPLE E9.1

Consider a midsize sedan that weighs 14,234.3 N (3200 lb) under a certain load. The wheelbase of the vehicle is 2.74 m (9 ft, or 108 in.), and the center of gravity is 1.067 m (3.5 ft) from the front axle and 0.508 m (20 in.) above the ground. The vehicle speed is 96.56 km/hr (60 mph), the turning radius is 182.88 m (600 ft), the wheel tread is 1.219 m (4 ft), K_ϕ = 10157 Nm (90,000 in.-lb), the roll center height is 0.457 m (18 in.), and the roll angle is 5°. Determine the load on each wheel (F_{z1i}, F_{z1o}, F_{z2i}, F_{z2o}) if the vehicle is under no acceleration and is decelerating at 3.048 m/s² (10 ft/sec²).

Given in the problem are the following:

$$W = 3200 \text{ lb (thus, } M = \frac{3200}{32.2} = 99.38 \text{ slugs)}$$

$$c_1 + c_2 = 108 \text{ in.}$$

$$h \quad = \quad 20 \text{ in.}$$

$$c_1 \quad = \quad 42 \text{ in. (thus, } c_2 = 108 - 42 = 66 \text{ in.)}$$

$$V \quad = \quad 60 \text{ mph (88 ft/sec)}$$

$$R \quad = \quad 600 \text{ ft}$$

$$t \quad = \quad 4 \text{ ft (48 in.)}$$

$$K_\phi \quad = \quad 90,000 \text{ in.-lb}$$

$$h_r \quad = \quad 18 \text{ in.}$$

$$\phi \quad = \quad 5° \text{ (0.08727 radians)}$$

Solution:

Apply Eq. 7.25 to obtain the loading under no acceleration (i.e., a = 0) as

$$F_{z1} = mg \left(\frac{c_2}{c_1 + c_2} \right) - \frac{h}{c_1 + c_2} Ma$$

$$= 99.38 \times 32.2 \left(\frac{66}{108} \right) - \frac{20}{108} \times 99.38 \times 0$$

$$= 1955.56 \text{ lb}$$

$$F_{y1} = F_{z1} \frac{V^2}{2Rg}$$

$$= 1955.56 \times \frac{(88)^2}{2 \times 600 \times 32.2}$$

$$= 391.92 \text{ lb}$$

$$\Delta F_{z1} = \frac{F_{y1} h_r + K_\phi \phi}{t}$$

$$= \frac{391.92 \times 18 + 90,000 \times 0.08727}{48}$$

$$= 310.60 \text{ lb}$$

$$F_{z1i} = \frac{F_{z1}}{2} - \Delta F_{z1}$$

$$= \frac{1955.56}{2} - 310.60$$

$$= 667.18 \text{ lb}$$

$$F_{zlo} = \frac{F_{zl}}{2} + \Delta F_{zl}$$

$$= \frac{1955.56}{2} + 310.60$$

$$= 1288.38 \text{ lb}$$

Similarly for the rear axle,

$$F_{z2} = mg\left(\frac{c_1}{c_1 + c_2}\right) + \frac{h}{c_1 + c_2}Ma$$

$$= 99.38 \times 32.2\left(\frac{42}{108}\right) + \frac{20}{108} \times 99.38 \times 0$$

$$= 1244.44 \text{ lb}$$

$$F_{y2} = F_{z2}\frac{V^2}{2Rg}$$

$$= 1244.44 \times \frac{(88)^2}{2 \times 600 \times 32.2}$$

$$= 249.40 \text{ lb}$$

$$\Delta F_{z2} = \frac{F_{y2}h_r + K_\phi\phi}{t}$$

$$= \frac{249.40 \times 18 + 90,000 \times 0.08727}{48}$$

$$= 257.16 \text{ lb}$$

$$F_{z2i} = \frac{F_{z2}}{2} - \Delta F_{z2}$$

$$= \frac{1244.44}{2} - 257.16$$

$$= 365.06 \text{ lb}$$

$$F_{z2o} = \frac{F_{z2}}{2} + \Delta F_{z2}$$

$$= \frac{1244.44}{2} + 257.16$$

$$= 879.38 \text{ lb}$$

For a = −10 ft/sec^2,

$$F_{zl} = 99.38 \times 32.2 \left(\frac{66}{108} \right) - \frac{20}{108} \times 99.38 \times (-10)$$

$$= 2139.59 \text{ lb}$$

$$F_{yl} = F_{zl} \frac{V^2}{2 \, Rg} = 2139.59 \times \frac{(88)^2}{2 \times 600 \times 32.2}$$

$$= 428.80 \text{ lb}$$

$$\Delta F_{zl} = \frac{F_{yl} h_r + K_\phi \phi}{t}$$

$$= \frac{428.80 \times 18 + 90,000 \times 0.08727}{48}$$

$$= 324.43 \text{ lb}$$

$$F_{zli} = \frac{F_{zl}}{2} - \Delta F_{zl}$$

$$= \frac{2139.59}{2} - 324.43$$

$$= 745.36 \text{ lb}$$

$$F_{zlo} = \frac{F_{zl}}{2} + \Delta F_{zl}$$

$$= \frac{2139.59}{2} + 324.43$$

$$= 1394.22 \text{ lb}$$

$$F_{z2} = 99.38 \times 32.2 \left(\frac{42}{108} \right) + \frac{20}{108} \times 99.38 \times (-10)$$

$$= 1060.41 \text{ lb}$$

$$F_{y2} = F_{z2} \frac{V^2}{2Rg}$$

$$= 1060.41 \times \frac{(88)^2}{2 \times 600 \times 32.2}$$

$$= 212.52 \text{ lb}$$

$$\Delta F_{z2} = \frac{F_{y2}h_r + K_\phi\phi}{t}$$

$$= \frac{212.52 \times 18 + 90,000 \times 0.08727}{48}$$

$$= 243.32 \ lb$$

$$F_{z2i} = \frac{F_{z2}}{2} - \Delta F_{z2}$$

$$= \frac{1060.41}{2} - 243.32$$

$$= 286.88 \ lb$$

$$F_{z2o} = \frac{F_{z2}}{2} + \Delta F_{z2}$$

$$= \frac{1060.41}{2} + 243.32$$

$$= 773.52 \ lb$$

Note that the total load on the axles remains at 3200 lb, which is the weight of the vehicle. However, the load transferred from the rear axle to the front axle as the vehicle was decelerating and from the inner wheels to the outer wheels. Figure 9.1 shows the load distribution as a function of deceleration. Figure 9.2 shows the load distribution on each wheel as a function of velocity. Note that at high speeds, the load reduces significantly at the inner wheels. Figure 9.3 shows the wheel load as a function of radius of the turn. Note here that at a low radius of 100 ft, the front and rear inner wheels have negative forces, which is not possible. At these drifting conditions, the forces at the wheels essentially become zero, and the possibility of rollover becomes real.

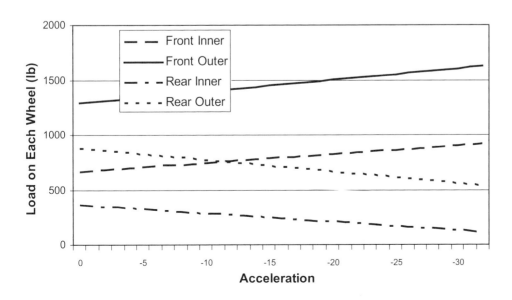

Figure 9.1 Load distribution under braking and cornering as a function of deceleration.

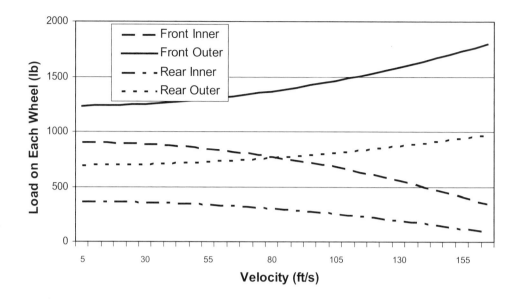

Figure 9.2 Load distribution under braking and cornering as a function of velocity.

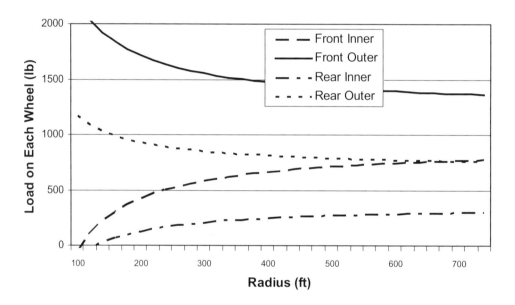

Figure 9.3 Load distribution under braking and cornering as a function of radius of the turn.

9.2.2 Optimal Braking Performance Under Steering

Braking performance under single axle braking conditions is not discussed here. We will consider braking with both axles, whereas single axle braking conditions are merely special cases. As mentioned, the brake engineer designs for the maximum braking force that can be obtained from the braking system at hand. Consideration is made for front lock-up to occur first under various loading and environmental conditions. Let us assume that, based on certain emergency conditions, the brake designer chooses to distribute the braking load such that a certain ratio (K_f) of the total braking force goes to the front brakes. The remainder that goes to the rear brakes is $(1 - K_f)$ of the total braking force. Note here that in conventional braking systems, the force distributed to the front axle does not take into consideration that the wheels of the front axle will have different loads when steering or cornering. On the other hand, anti-lock braking systems (ABS) distribute the braking force according to the wheel load.

The braking force of the front and rear axles can be written in terms of each other as

$$F_{x1} = \frac{K_f}{1 - K_f} F_{x2}$$

$$F_{x2} = \frac{1 - K_f}{K_f} F_{x1}$$

(9.3)

Note that front axle braking (single axle) is obtained when $K_f = 1$, and rear axle braking is obtained when $K_f = 0$. The total weight under no steering is distributed as

$$F_{z1} = \frac{1}{c_1 + c_2}\left(Mgc_2\cos\theta - h_a R_a - hMa - hMg\sin\theta\right)$$

$$F_{z2} = \frac{1}{c_1 + c_2}\left(Mgc_1\cos\theta + h_a R_a + hMa + hMg\sin\theta\right)$$

(9.4)

Manipulating Eqs. 9.1 and 9.2,

$$F_{z1i} = \frac{F_{z1}}{2} - \Delta F_{z1}$$

$$= \frac{F_{z1}}{2} - \frac{\left(F_{z1}\dfrac{V^2}{2Rg}\right)h_r + K_\phi\phi}{t}$$

$$= F_{z1}\left(\frac{1}{2} - \frac{V^2 h_r}{2Rgt}\right) - \frac{K_\phi\phi}{t}$$

(9.5)

$$= \left\{\frac{1}{c_1 + c_2}\left(Mgc_2\cos\theta - h_a R_a - hMa - hMg\sin\theta\right)\right\}\left(\frac{1}{2} - \frac{V^2 h_r}{2Rgt}\right) - \frac{K_\phi\phi}{t}$$

$$F_{z1o} = \left\{\frac{1}{c_1 + c_2}\left(Mgc_2\cos\theta - h_a R_a - hMa - hMg\sin\theta\right)\right\}\left(\frac{1}{2} + \frac{V^2 h_r}{2Rgt}\right) + \frac{K_\phi\phi}{t}$$

Similar derivation can be made for the rear axle as

$$F_{z2i} = \left\{\frac{1}{c_1 + c_2}\left(Mgc_1\cos\theta + h_a R_a + hMa + hMg\sin\theta\right)\right\}\left(\frac{1}{2} - \frac{V^2 h_r}{2Rgt}\right) - \frac{K_\phi\phi}{t}$$

$$F_{z2o} = \left\{\frac{1}{c_1 + c_2}\left(Mgc_1\cos\theta + h_a R_a + hMa + hMg\sin\theta\right)\right\}\left(\frac{1}{2} + \frac{V^2 h_r}{2Rgt}\right) + \frac{K_\phi\phi}{t}$$

Note also from Chapter 7 (Eq. 7.31) that by summing the forces in the direction of motion, we obtain

$$F_{x1} + F_{x2} - R_a - R_r - Mg\sin(\theta) = Ma$$

or from Eq. 9.3,

$$\left(1 + \frac{K_f}{1 - K_f}\right)F_{x2} - R_a - R_r - Mg\sin(\theta) = Ma$$

$$\left(1 + \frac{1 - K_f}{K_f}\right)F_{x1} - R_a - R_r - Mg\sin(\theta) = Ma$$

(9.6)

In Eq. 9.6, a is the longitudinal acceleration, to be distinguished from the lateral acceleration that results from cornering. Observe the simplifications

$$1 + \frac{K_f}{1 - K_f} = \frac{1 - K_f}{1 - K_f} + \frac{K_f}{1 - K_f} = \frac{1}{1 - K_f}$$

$$1 + \frac{1 - K_f}{K_f} = \frac{K_f}{K_f} + \frac{1 - K_f}{K_f} = \frac{1}{K_f}$$

(9.7)

9.2.2.1 Front Lock-Up

Under front lock-up, the maximum force at the front axle will be equal to twice that of the wheel where lock-up is introduced for conventional brakes. The inner wheel will always have less force; thus, the total axial force at the front axle will be $F_{x1} = -2\mu F_{z1i}$. Substituting this into Eq. 9.6 yields

$$-\frac{2\mu}{K_f}\left(\left\{\frac{1}{c_1 + c_2}\left(Mgc_2\cos\theta - h_aR_a - hMa_F - hMg\sin\theta\right)\right\}\left(\frac{1}{2} - \frac{V^2h_r}{2Rgt}\right) - \frac{K_\phi\phi}{t}\right)$$

$$- R_a - R_r - Mg\sin(\theta)$$

$$= Ma_F$$

(9.8)

Rearranging for a_F,

$$-\frac{2\mu}{K_f}\left(\left\{\frac{1}{c_1 + c_2}\left(Mgc_2\cos\theta - h_aR_a - hMg\sin\theta\right)\right\}\left(\frac{1}{2} - \frac{V^2h_r}{2Rgt}\right) - \frac{K_\phi\phi}{t}\right)$$

$$- R_a - R_r - Mg\sin(\theta)$$

(9.9)

$$= Ma_F\left[1 - \frac{\mu h}{(c_1 + c_2)K_f}\left(1 - \frac{V^2h_r}{Rgt}\right)\right]$$

or simply

$$a_F = \cfrac{-\cfrac{2\mu}{K_f}\left(\left\{\cfrac{1}{c_1+c_2}\left(Mgc_2\cos\theta - h_aR_a - hMg\sin\theta\right)\right\}\left(\cfrac{1}{2} - \cfrac{V^2h_r}{2Rgt}\right) - \cfrac{K_\phi\phi}{t}\right) - R_a - R_r - Mg\sin(\theta)}{M\left[1 - \cfrac{\mu h}{K_f(c_1+c_2)}\left(1 - \cfrac{V^2h_r}{Rgt}\right)\right]} \quad (9.10)$$

Note that the acceleration a has a subscript F to indicate that the acceleration obtained by using Eq. 9.10 is valid only when the front tires are locked up. Also note the degree of complexity of this equation. Ignoring aerodynamic drag, grade, and rolling resistance, the deceleration is

$$a_F = \cfrac{-\cfrac{2\mu}{K_f}\left[\left(\cfrac{Mgc_2}{c_1+c_2}\right)\left(\cfrac{1}{2} - \cfrac{V^2h_r}{2Rgt}\right) - \cfrac{K_\phi\phi}{t}\right]}{M\left[1 - \cfrac{\mu h}{K_f(c_1+c_2)}\left(1 - \cfrac{V^2h_r}{Rgt}\right)\right]} \quad (9.11)$$

Note that this expression may yield a small denominator of zero. This will yield a high number for a_F, which will mean that rear lock-up must happen first. It is worth comparing this equation with Eq. 7.43.

9.2.2.2 Rear Lock-Up

Under rear lock-up, the maximum force at the rear axle will equal twice that of the wheel where lock-up is introduced for conventional brakes. The inner wheel will always have less force; thus, the total axial force at the rear axle will be $F_{x2} = -2\mu F_{z2i}$. Substituting this into the Eq. 9.6 yields

$$-2\mu\left(1 + \cfrac{K_f}{1-K_f}\right)\left(\cfrac{1}{c_1+c_2}\left(Mgc_1\cos\theta + h_aR_a + hMa_R + hMg\sin\theta\right)\right)\left(\cfrac{1}{2} - \cfrac{V^2h_r}{2Rgt}\right) - \cfrac{K_\phi\phi}{t}$$
$$- R_a - R_r - Mg\sin(\theta) \quad (9.12)$$
$$= Ma_R$$

Rearranging for a_R,

$$-\cfrac{2\mu}{1-K_f}\left(\left\{\cfrac{1}{c_1+c_2}\left(Mgc_2\cos\theta + h_aR_a + hMg\sin\theta\right)\right\}\left(\cfrac{1}{2} - \cfrac{V^2h_r}{2Rgt}\right) - \cfrac{K_\phi\phi}{t}\right)$$
$$- R_a - R_r - Mg\sin(\theta) \quad (9.13)$$
$$= Ma_R\left[1 + \cfrac{\mu h}{(c_1+c_2)(1-K_f)}\left(1 - \cfrac{V^2h_r}{Rgt}\right)\right]$$

Simplify

$$a_R = \cfrac{-\cfrac{2\mu}{1-K_f}\left(\left\{\cfrac{1}{c_1+c_2}(Mgc_2\cos\theta + h_aR_a + hMg\sin\theta)\right\}\left(\cfrac{1}{2}-\cfrac{V^2h_r}{2Rgt}\right) - \cfrac{K_\phi\phi}{t}\right) - R_a - R_r - Mg\sin(\theta)}{M\left[1 + \cfrac{\mu h}{(1-K_f)(c_1+c_2)}\left(1 - \cfrac{V^2h_r}{Rgt}\right)\right]} \quad (9.14)$$

Ignoring aerodynamic drag, grade, and rolling resistance, the deceleration is

$$a_R = \cfrac{-\cfrac{2\mu}{1-K_f}\left(\left\{\cfrac{Mgc_2}{c_1+c_2}\right\}\left(\cfrac{1}{2}-\cfrac{V^2h_r}{2Rgt}\right) - \cfrac{K_\phi\phi}{t}\right)}{M\left[1 + \cfrac{\mu h}{(1-K_f)(c_1+c_2)}\left(1 - \cfrac{V^2h_r}{Rgt}\right)\right]} \quad (9.15)$$

It is worthwhile to compare Eq. 9.15 to Eq. 7.48 in Chapter 7.

The front tires will lock up first if $|a_F| < |a_R|$; otherwise, the rear tires will lock up first.

EXAMPLE E9.2

A 21,351.46-N (4800-lb) passenger car has a wheelbase of 2844.80 mm (112 in.) and a center of gravity 1270.00 mm (50 in.) behind the front axle and 508.00 mm (20 in.) above the ground. The braking effort distribution gives the front axle 60% of the total braking force. The car is moving on a horizontal plane. The car is going around a turning radius of 182.88 m (600 ft) at a speed of 96.56 km/hr (60 mph). The wheel tread is 1.219 m (4 ft), $K_\phi = 10,157$ Nm (90,000 in.-lb), the roll center height is 0.457 m (18 in.), and the roll angle is 5°. Determine which tires will lock up first if the car is moving on a surface with first a 0.75 coefficient of friction and second a 0.25 coefficient of friction. Ignore drag and rolling resistance.

Solution:

$$M = 4800 \text{ lb} = 149.07 \text{ slugs}$$

$$c_1 + c_2 = 112 \text{ in.}$$

$$c_1 = 50 \text{ in.}$$

$$c_2 = 62 \text{ in.}$$

$$K_f = 0.6$$

$$h = 20$$

$$\theta = 0$$

$$R_a = 0$$

$$R_r = 0$$

$$V = 60 \text{ mph } (88 \text{ ft/s}) = 1056 \text{ in./s}$$

$$R = 600 \text{ ft} = 7200 \text{ in.}$$

$$t = 4 \text{ ft } (48 \text{ in.})$$

$$K_\phi = 90{,}000 \text{ in.-lb}$$

$$h_r = 18 \text{ in.}$$

$$\phi = 5° \ (0.08727 \text{ radians})$$

$$g = 32.2 \text{ ft/s}^2 = 386.4 \text{ in./s}^2$$

First surface: $\mu = 0.75$

Acceleration under front-lock-up is

$$a_F = \dfrac{-\dfrac{2\mu}{K_f}\left[\left(\dfrac{Mgc_2}{c_1 + c_2}\right)\left(\dfrac{1}{2} - \dfrac{V^2 h_r}{2Rgt}\right) - \dfrac{K_\phi \phi}{t}\right]}{M\left[1 - \dfrac{\mu h}{K_f(c_1 + c_2)}\left(1 - \dfrac{V^2 h_r}{Rgt}\right)\right]}$$

$$a_F = \dfrac{-\dfrac{2 \times 0.75}{0.6}\left[\left(\dfrac{149.07 \times 386.4 \times 62}{112}\right)\left(\dfrac{1}{2} - \dfrac{(1056)^2 \times 18}{2 \times 7200 \times 386.4 \times 48}\right) - \dfrac{90{,}000 \times 0.08727}{48}\right]}{149.07\left[1 - \dfrac{0.75 \times 20}{0.6 \times 112}\left(1 - \dfrac{(1056)^2 \times 18}{7200 \times 386.4 \times 48}\right)\right]}$$

$$= -0.7168g$$

Acceleration under rear lock-up is

$$a_R = \dfrac{-\dfrac{2\mu}{1 - K_f}\left(\left\{\dfrac{Mgc_2}{c_1 + c_2}\right\}\left(\dfrac{1}{2} - \dfrac{V^2 h_r}{2Rgt}\right) - \dfrac{K_\phi \phi}{t}\right)}{M\left[1 + \dfrac{\mu h}{(1 - K_f)(c_1 + c_2)}\left(1 - \dfrac{V^2 h_r}{Rgt}\right)\right]}$$

$$a_R = \dfrac{-\dfrac{2 \times 0.75}{1 - 0.6}\left[\left(\dfrac{149.07 \times 386.4 \times 62}{112}\right)\left(\dfrac{1}{2} - \dfrac{(1056)^2 \times 18}{2 \times 7200 \times 386.4 \times 48}\right) - \dfrac{90{,}000 \times 0.08727}{48}\right]}{149.07\left[1 + \dfrac{0.75 \times 20}{(1 - 0.6) \times 112}\left(1 - \dfrac{(1056)^2 \times 18}{7200 \times 386.4 \times 48}\right)\right]}$$

$$= -0.6783g$$

Because $|a_F| > |a_R|$, the rear tires will lock up first.

Second surface: $\mu = 0.25$

Acceleration under front-lock-up is

$$a_F = \frac{-\dfrac{2 \times 0.25}{0.6}\left[\left(\dfrac{149.07 \times 386.4 \times 62}{112}\right)\left(\dfrac{1}{2} - \dfrac{(1056)^2 \times 18}{2 \times 7200 \times 386.4 \times 48}\right) - \dfrac{90,000 \times 0.08727}{48}\right]}{149.07\left[1 - \dfrac{0.25 \times 20}{0.6 \times 112}\left(1 - \dfrac{(1056)^2 \times 18}{7200 \times 386.4 \times 48}\right)\right]}$$

$$= -0.2067g$$

Acceleration under rear lock-up is

$$a_R = \frac{-\dfrac{2 \times 0.25}{1 - 0.6}\left[\left(\dfrac{149.07 \times 386.4 \times 62}{112}\right)\left(\dfrac{1}{2} - \dfrac{(1056)^2 \times 18}{2 \times 7200 \times 386.4 \times 48}\right) - \dfrac{90,000 \times 0.08727}{48}\right]}{149.07\left[1 + \dfrac{0.25 \times 20}{(1 - 0.6) \times 112}\left(1 - \dfrac{(1056)^2 \times 18}{7200 \times 386.4 \times 48}\right)\right]}$$

$$= -0.2653g$$

Because $|a_F| < |a_R|$, the front tires will lock up first.

Figure 9.4 shows how the acceleration under front and rear lock-up changes with the coefficient of friction for the parameters described in example E9.2. It clearly shows that rear lock-up is quite possible at surfaces with high coefficients of friction. Figure 9.5 shows that if the braking effort distribution is changed to 0.7, front brake lock-up will occur for all types of surfaces. However, this compromises the maximum braking force.

The K_f factor cannot be altered straightforwardly to account for the different coefficients of friction that various surfaces may offer in conventional brakes. An advanced anti-lock braking system provides the maximum brake force per wheel without lock-up.

9.3 Steering and Acceleration

Load transfers from the inside wheels to the outside wheels when the vehicle undergoes cornering or steering, and it transfers from the front to the rear axles when the vehicle accelerates. Acceleration can happen while the vehicle is under steering conditions. First we will consider acceleration of a vehicle on a horizontal road (i.e., no grade) and ignore the other forces (e.g., aerodynamic, powertrain resistance).

9.3.1 Simple Acceleration and Steering on a Horizontal Road

If we use Eq. 8.3 from Chapter 8 with Eq. 9.3, we can obtain the distribution of the weight of the vehicle under steering and acceleration. Consider the parameters of

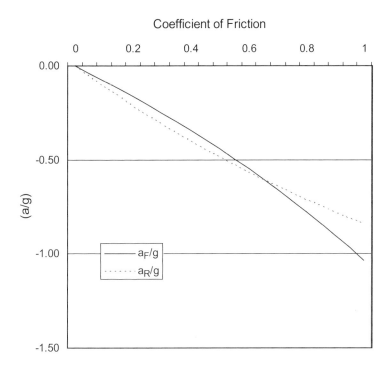

Figure 9.4 The values of a_F and a_R for different friction for the parameters of Example E9.2, where $K_f = 0.6$.

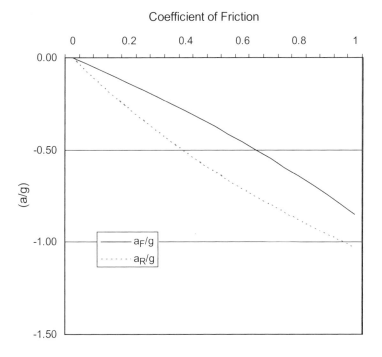

Figure 9.5 The values of a_F and a_R for different friction for the parameters of Example E9.2, where $K_f = 0.7$.

Example E9.1, and change the deceleration to an acceleration of 2 ft/s² (0.6096 m/s²). Figure 9.6 shows the load distribution as a function of acceleration. Figure 9.7 shows the load distribution on each wheel as a function of velocity. Note that at high speeds, the load reduces significantly at the inner wheels. Figure 9.8 shows the wheel load as a function of radius of the turn. Note here that at a low radius of 100 ft, the front and rear inner wheels have negative forces, which is not possible. At these drifting

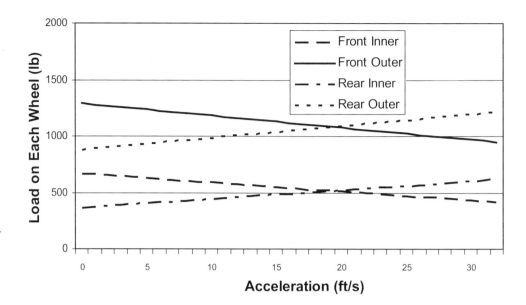

Figure 9.6 Load distribution under acceleration and cornering as a function of acceleration.

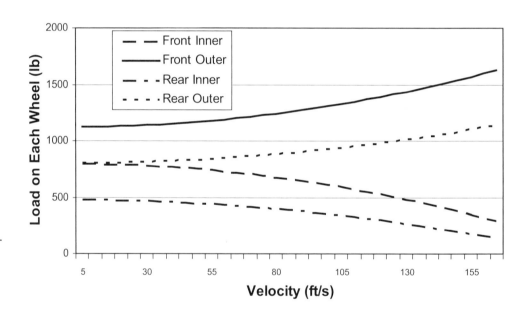

Figure 9.7 Load distribution under acceleration and cornering as a function of velocity.

conditions, the forces at the wheels essentially become zero, and the possibility of rollover becomes real.

9.3.2 Optimal Acceleration Performance Under Steering

Acceleration performance under a single axle drive condition (front-wheel drive or rear-wheel drive) is not discussed here. We will consider the cases of four-by-four or all-wheel-drive systems because the other front-wheel-drive and rear-wheel-drive

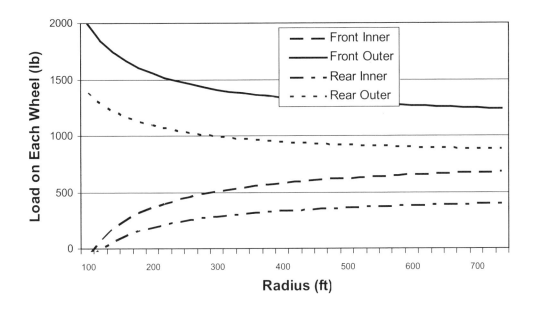

Figure 9.8 Load distribution under acceleration and cornering as a function of radius of the turn without skid.

conditions are special cases. The drive system is to be designed such that it delivers the maximum tractive forces. Let us distribute the tractive torque such that a certain ratio K_f of the total torque goes to the front axle, and the remainder goes to the rear axle. Note here that in conventional drive systems, the torque distributed to the front axle does not take into consideration the different loads on the right and left wheels when steering or cornering.

Note that front-wheel drive (single axle) is obtained when $K_f = 1$, and rear-wheel drive is obtained when $K_f = 0$. The total weight is distributed to the front and rear axle according to Eq. 9.4. The wheel loads are distributed according to Eq. 9.5. Also note that by summing the forces in the direction of motion, we obtain Eq. 9.6.

9.3.2.1 Front Skid

The maximum force at the front axle will be equal to twice that of the wheel where it is first felt for conventional traction systems. The inner wheel will always have less force; thus, the total axial force at the front axle will be $F_{x1} = 2\mu F_{z1i}$. Substituting this into Eq. 9.6 and using Eqs. 9.5 and 9.7 yields

$$\frac{2\mu}{K_f}\left(\left\{\frac{1}{c_1 + c_2}\left(Mgc_2\cos\theta - h_aR_a - hMa_F - hMg\sin\theta\right)\right\}\left(\frac{1}{2} - \frac{V^2h_r}{2Rgt}\right) - \frac{K_\phi\phi}{t}\right)$$

$$- R_a - R_r - Mg\sin(\theta)$$

$$= Ma_F$$

(9.16)

Rearranging for a_F,

$$\frac{2\mu}{K_f}\left(\left\{\frac{1}{c_1+c_2}\left(Mgc_2\cos\theta - h_aR_a - hMg\sin\theta\right)\right\}\left(\frac{1}{2} - \frac{V^2h_r}{2Rgt}\right) - \frac{K_\phi\phi}{t}\right)$$

$$- R_a - R_r - Mg\sin(\theta) \qquad (9.17)$$

$$= Ma_F\left[1 + \frac{\mu h}{(c_1+c_2)K_f}\left(1 - \frac{V^2h_r}{Rgt}\right)\right]$$

or simply

$$a_F = \frac{\frac{2\mu}{K_f}\left(\left\{\frac{1}{c_1+c_2}\left(Mgc_2\cos\theta - h_aR_a - hMg\sin\theta\right)\right\}\left(\frac{1}{2} - \frac{V^2h_r}{2Rgt}\right) - \frac{K_\phi\phi}{t}\right) - R_a - R_r - Mg\sin(\theta)}{M\left[1 + \frac{\mu h}{K_f(c_1+c_2)}\left(1 - \frac{V^2h_r}{Rgt}\right)\right]} \qquad (9.18)$$

The subscript F indicates that the acceleration obtained is valid when the front tires skid. Ignoring aerodynamic drag, grade, and rolling resistance, the deceleration is

$$a_F = \frac{\frac{2\mu}{K_f}\left[\left(\frac{Mgc_2}{c_1+c_2}\right)\left(\frac{1}{2} - \frac{V^2h_r}{2Rgt}\right) - \frac{K_\phi\phi}{t}\right]}{M\left[1 + \frac{\mu h}{K_f(c_1+c_2)}\left(1 - \frac{V^2h_r}{Rgt}\right)\right]} \qquad (9.19)$$

9.3.2.2 Rear Skid

The maximum force at the rear axle will be equal to twice that of the wheel where skid occurs first for conventional drive systems. Because the inner wheel has less force, the total axial force at the rear axle will be $F_{x2} = 2\mu F_{z2i}$. Substituting this into the Eq. 9.6 yields

$$2\mu\left(1 + \frac{K_f}{1-K_f}\right)\left(\frac{1}{c_1+c_2}\left(Mgc_1\cos\theta + h_aR_a + hMa_R + hMg\sin\theta\right)\right)\left(\frac{1}{2} - \frac{V^2h_r}{2Rgt}\right) - \frac{K_\phi\phi}{t}$$

$$- R_a - R_r - Mg\sin(\theta) \qquad (9.20)$$

$$= Ma_R$$

Rearranging for a_R,

$$\frac{2\mu}{1-K_f}\left(\left\{\frac{1}{c_1+c_2}\left(Mgc_2\cos\theta + h_aR_a + hMg\sin\theta\right)\right\}\left(\frac{1}{2} - \frac{V^2h_r}{2Rgt}\right) - \frac{K_\phi\phi}{t}\right)$$

$$- R_a - R_r - Mg\sin(\theta) \tag{9.21}$$

$$= Ma_R\left[1 - \frac{\mu h}{(c_1+c_2)(1-K_f)}\left(1 - \frac{V^2h_r}{Rgt}\right)\right]$$

Simplify as

$$a_R = \frac{\frac{2\mu}{1-K_f}\left(\left\{\frac{1}{c_1+c_2}\left(Mgc_2\cos\theta + h_aR_a + hMg\sin\theta\right)\right\}\left(\frac{1}{2} - \frac{V^2h_r}{2Rgt}\right) - \frac{K_\phi\phi}{t}\right) - R_a - R_r - Mg\sin(\theta)}{M\left[1 - \frac{\mu h}{(1-K_f)(c_1+c_2)}\left(1 - \frac{V^2h_r}{Rgt}\right)\right]} \tag{9.22}$$

Ignoring aerodynamic drag, grade, and rolling resistance, the acceleration is

$$a_R = \frac{\frac{2\mu}{1-K_f}\left(\left\{\frac{Mgc_2}{c_1+c_2}\right\}\left(\frac{1}{2} - \frac{V^2h_r}{2Rgt}\right) - \frac{K_\phi\phi}{t}\right)}{M\left[1 - \frac{\mu h}{(1-K_f)(c_1+c_2)}\left(1 - \frac{V^2h_r}{Rgt}\right)\right]} \tag{9.23}$$

The front tires will skid first if $|a_F| < |a_R|$; otherwise, the rear tires will skid first.

Example E9.3

Consider the vehicle in Example E9.2. The torque distribution gives the front axle 60% of the total torque delivered to the wheels. Determine which tires will skid first if the car is moving on a surface with first a 0.75 coefficient of friction and second a 0.25 coefficient of friction.

Solution:

$$M \quad = 4800\text{ lb} = 149.07\text{ slugs}$$

$$c_1 + c_2 = 112\text{ in.}$$

$$c_1 \quad = 50\text{ in.}$$

$$c_2 \quad = 62\text{ in.}$$

$$K_f \quad = 0.6$$

$$h \quad = 20\text{ in.}$$

$$\theta = 0$$

$$R_a = 0$$

$$R_r = 0$$

$$V = 60 \text{ mph } (88 \text{ ft/s}) = 1056 \text{ in./s}$$

$$R = 600 \text{ ft} = 7200 \text{ in.}$$

$$t = 4 \text{ ft } (48 \text{ in.})$$

$$K_\phi = 90,000 \text{ in.-lb}$$

$$h_r = 18 \text{ in.}$$

$$\phi = 5° (0.08727 \text{ radians})$$

$$g = 32.2 \text{ ft/s}^2 = 386.4 \text{ in./s}^2$$

First surface: $\mu = 0.75$

Acceleration under front skid is

$$a_F = \frac{\dfrac{2 \times 0.75}{0.6}\left[\left(\dfrac{149.07 \times 386.4 \times 62}{112}\right)\left(\dfrac{1}{2} - \dfrac{(1056)^2 \times 18}{2 \times 7200 \times 386.4 \times 48}\right) - \dfrac{90,000 \times 0.08727}{48}\right]}{149.07\left[1 + \dfrac{0.75 \times 20}{0.6 \times 112}\left(1 - \dfrac{(1056)^2 \times 18}{7200 \times 386.4 \times 48}\right)\right]}$$

$$= 0.7815g$$

Acceleration under rear skid is

$$a_R = \frac{\dfrac{2 \times 0.75}{1 - 0.6}\left[\left(\dfrac{149.07 \times 386.4 \times 62}{112}\right)\left(\dfrac{1}{2} - \dfrac{(1056)^2 \times 18}{2 \times 7200 \times 386.4 \times 48}\right) - \dfrac{90,000 \times 0.08727}{48}\right]}{149.07\left[1 - \dfrac{0.75 \times 20}{(1 - 0.6) \times 112}\left(1 - \dfrac{(1056)^2 \times 18}{7200 \times 386.4 \times 48}\right)\right]}$$

$$= 0.6290g$$

Because $|a_F| > |a_R|$, the rear tires will skid first.

Second surface: $\mu = 0.25$

Acceleration under front-lock-up is

$$a_F = \frac{\dfrac{2 \times 0.25}{0.6}\left[\left(\dfrac{149.07 \times 386.4 \times 62}{112}\right)\left(\dfrac{1}{2} - \dfrac{(1056)^2 \times 18}{2 \times 7200 \times 386.4 \times 48}\right) - \dfrac{90,000 \times 0.08727}{48}\right]}{149.07\left[1 + \dfrac{0.25 \times 20}{0.6 \times 112}\left(1 - \dfrac{(1056)^2 \times 18}{7200 \times 386.4 \times 48}\right)\right]}$$

$$= 0.2117g$$

Acceleration under rear lock-up is

$$a_R = \frac{\dfrac{2 \times 0.25}{1 - 0.6}\left[\left(\dfrac{149.07 \times 386.4 \times 62}{112}\right)\left(\dfrac{1}{2} - \dfrac{(1056)^2 \times 18}{2 \times 7200 \times 386.4 \times 48}\right) - \dfrac{90,000 \times 0.08727}{48}\right]}{149.07\left[1 - \dfrac{0.25 \times 20}{(1 - 0.6) \times 112}\left(1 - \dfrac{(1056)^2 \times 18}{7200 \times 386.4 \times 48}\right)\right]}$$

$$= 0.2574g$$

Because $|a_F| < |a_R|$, the front tires will skid first.

Figure 9.9 shows how the acceleration under front and rear lock-up changes with the coefficient of friction for the parameters described in Example E9.3. Rear skid is possible at surfaces with high coefficients of friction. Figure 9.10 shows that by changing the torque distribution to 0.7, front skid will occur for all types of surfaces. The K_f factor cannot be altered straightforwardly in conventional drive systems. An advanced traction control provides the maximum force per wheel without skid.

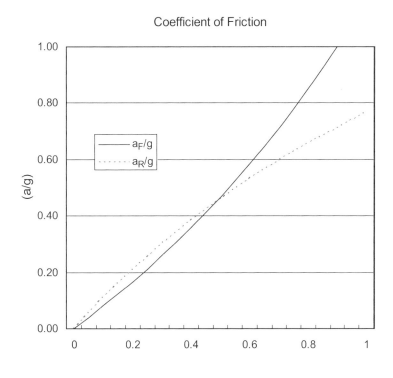

Figure 9.9 The values of a_F and a_R for different friction for the parameters of Example E9.3, where $K_f = 0.6$.

Coefficient of Friction

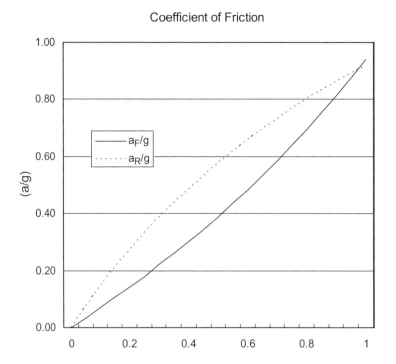

Figure 9.10 The values of a_F and a_R for different friction for the parameters of Example E9.2, where $K_f = 0.7$.

9.4 Vehicle Critical Speed

Vehicle critical speed (V_c) is the speed at which the tires of a cornering vehicle reach the limit of side friction and side slip begins. Slidge and Marshek (1997) reviewed formulas for estimating vehicle critical speed from yaw marks.

Figure 9.11 shows the vehicle forces in a banked plane. Noting that the lateral acceleration $a_y = \dfrac{V^2}{R}$ and applying Newton's laws, forces normal to the road surface are summed to give the normal force

$$F_z = Mg\cos(\theta) + \frac{MV^2\sin(\theta)}{R} \tag{9.24}$$

Note that the angle θ is the embankment angle and not the grade. Summing the forces in the lateral direction yields the maximum friction force

$$F_f = \frac{MV^2\cos(\theta)}{R} - Mg\sin(\theta) \tag{9.25}$$

Because $F_f = \mu F_z$ at the critical speed, further manipulation of Eq. 9.25 yields the equation of the critical speed (Slidge and Marshek, 1997)

$$V_c = \frac{\sqrt{gR(\mu + \tan\theta)}}{\sqrt{1 - \mu\tan\theta}} \tag{9.26}$$

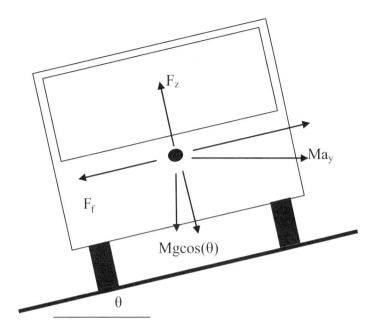

Figure 9.11 Vehicle forces in a banked plane.

EXAMPLE E9.4

Determine the critical speed of a vehicle traveling on a 5% embankment at a radius of 50 m. The coefficient of friction between the tires and the road is 0.8. What would be the critical speed if the coefficient of friction is 0.2?

Solution:

Given

$$G \quad = 9.81 \text{ m/s}^2$$

$$\tan \theta \ = 0.05$$

$$R \quad = 50 \text{ m}$$

First, consider $\mu = 0.8$.

The critical speed is

$$V_c = \frac{\sqrt{gR(\mu + \tan\theta)}}{\sqrt{1 - \mu \tan\theta}}$$

$$= \frac{\sqrt{9.81 \times 50(0.8 + 0.05)}}{\sqrt{1 - 0.8 \times 0.05}}$$

$$= 20.84 \text{ m/s}$$

Second, consider $\mu = 0.2$.

The critical speed is

$$V_c = \frac{\sqrt{gR\left(\mu + \tan\theta\right)}}{\sqrt{1 - \mu\tan\theta}}$$

$$= \frac{\sqrt{9.81 \times 50\left(0.2 + 0.05\right)}}{\sqrt{1 - 0.2 \times 0.05}}$$

$$= 11.13 \text{ m/s}$$

Figure 9.12 shows the critical speed dependence on the coefficient of friction. Note that at lower friction values, the critical speed is relatively small and can be encountered in normal driving conditions.

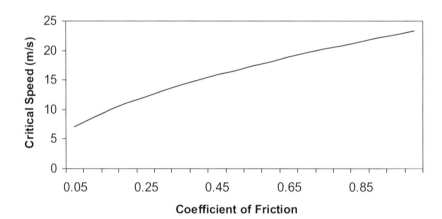

Figure 9.12 Vehicle critical speed as a function of friction.

Calculation of the critical speed of a vehicle is of great importance in determining the steering characteristics of the vehicle. As soon as the critical speed is reached, slippage starts laterally at the wheels. This action will cause the vehicle to oversteer. In a vehicle without electronic stability systems, this behavior will cause the driver to countersteer, but the vehicle will not respond because of skidding. This leads to hazardous situations and may cause accidents.

Equations of critical speed also are presented for various other conditions. In particular, critical speed can be obtained for vehicles turning while braking or accelerating. Interested readers are encouraged to pursue these conditions as discussed in the paper by Slidge and Marshek (1997).

9.5 Vehicle Stability

We already learned that rear lock-up is not desirable because it introduces instability to the vehicle. Rear lock-up will cause the vehicle to spin around its vertical axis. Typical drivers apply a large brake force when faced with unexpected emergency conditions, which will force wheel lock-up to occur. Figure 9.13 illustrates the initiation of vehicle instability. Assume that rear wheel lock-up occurs first, as shown in Figure 9.13(a), while the front wheels are still rolling (i.e., no lock-up). Any lateral disturbance due to wind, road irregularity, grade, embankment, or others will cause the front wheels to develop side or lateral forces (F_y) to resist this disturbance. The rear wheels will not produce such a force because they are still sliding. Thus, the side forces developed at the front wheels will cause a yaw moment at the vehicle (the magnitude of which is $F_y c_1$). This moment will cause the vehicle to rotate around its vertical axis. In turn, rotation causes the yawing moment to increase, thus causing instability.

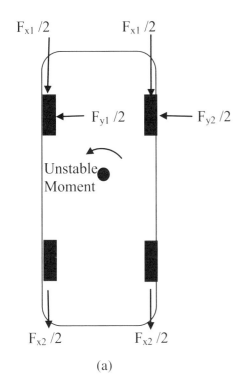

(a)

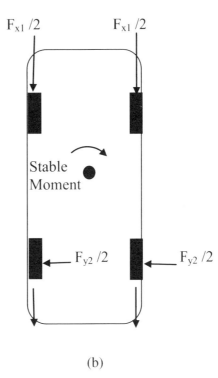

(b)

Figure 9.13 Vehicle stability under rear and front lock-up: (a) rear wheel lock-up, and (b) front wheel lock-up.

On the other hand, if the front wheels are to lock up first, lateral forces will develop on the rear wheels, causing the yaw moment in an opposite direction to that developed when the rear lock-up happens first. This yaw moment is balanced with the vehicle friction forces, causing it to diminish and thus regaining vehicle stability.

Note here that applying front brakes that are different at each of the front wheels can be engineered to balance any of the yawing forces. This will enable the driver to regain control of the vehicle.

Advancement of anti-lock braking and traction control systems led to integrating these two systems into one electronic unit. Furthermore, it is found that the same unit can be used together with the engine control unit to enhance the vehicle stability, particularly when steering and cornering.

With electronic stability, the brakes can be used to enhance the steering of the vehicle to the point of stability. In situations where the vehicle starts to understeer or oversteer due to yaw instability, brakes on certain wheels (left or right) are applied to steer the vehicle in a direction that corrects the oversteering or understeering. Braking can be directed to the left rear wheel to counter understeer, or to the right front wheel to counter oversteer. Advanced electronic stability systems also with the engine electronic control unit can cause the vehicle to accelerate at a certain moment to regain stability.

Situations that cause vehicle instability can be encountered in real driving. Rapid steering and countersteering can occur to avoid an obstacle that suddenly becomes apparent to the driver. Such behavior will generate substantial lateral forces on the front wheels. If the wheels were turned suddenly to the right, this would cause a clockwise rotation of the vehicle around its axis. Development of lateral forces at the rear wheel would be delayed, and typical drivers tend to countersteer; however, the vehicle will not respond. In this type of situation, the electronic stability will take charge and will apply braking on the left front wheel to counter the yawing moment produced by the lateral forces. Other actual maneuvers include fast lane changes, acceleration or deceleration during cornering, and rapid steering and countersteering required by certain roads (e.g., S-shaped roads).

The reader may refer to the book by Bosch (1999).

9.6 Summary

This chapter overviewed aspects of vehicle dynamics from a customer's perspective. It covered the different tests and evaluations that can be made to assess the dynamics of a particular vehicle. Emphasis was made on evaluations that involve more than one maneuver by the vehicle driver occurring simultaneously.

More detailed analysis was offered on the behavior of the vehicle under steering and braking conditions and under steering and acceleration conditions. Both treatments were made from a general perspective, where special cases could be derived from the general equations developed. For example, maximum tractive force for front-wheel-drive or rear-wheel-drive single-drive systems can be obtained from the general equations derived for all-wheel-drive systems.

Other subjects related to vehicle critical speed and vehicle stability were introduced.

9.7 References

1. Bosch Corporation, *Driving Safety Systems*, 2nd Ed., Society of Automotive Engineers, Warrendale, PA, 1999.

2. Genta, G., *Motor Vehicle Dynamics,* World Scientific Publishing, River Edge, NJ, 1997.

3. Gillespie, T.D., *Fundamentals of Vehicle Dynamics*, Society of Automotive Engineers, Warrendale, PA, 1992.

4. Kluger, M.A. and Long, D.M., "An Overview of Current Automatic, Manual and Continuously Variable Transmission Efficiencies and Their Projected Future Improvements," *SAE Transactions*, SAE Paper No. 1999-01-1259, Society of Automotive Engineers, Warrendale, PA, 1999.

5. Limpert, R., Gamero, F., and Boyer, R., "An Investigation of Brake Force Distribution for Straight and Curved Braking," *SAE Transactions*, SAE Paper No. 741086, Society of Automotive Engineers, Warrendale, PA, 1974.

6. Milliken, W.F. and Milliken, D.L., *Race Car Vehicle Dynamics,* Society of Automotive Engineers, Warrendale, PA, 1995.

7. SAE J670e, "Vehicle Dynamics Terminology," Society of Automotive Engineers, Warrendale, PA, 1978.

8. Slidge, N. and Marshek, K., "Formulas for Estimating Vehicle Critical Speed from Yaw Marks," *SAE Transactions*, SAE Paper No. 971147, Society of Automotive Engineers, Warrendale, PA, 1997.

9. Taborek, J.J., "Mechanics of Vehicles," *Machine Design,* May 30–December 26, 1957.

10. U.S. Department of the Army, *Engineering Design Handbook: Analysis and Design of Automotive Brake Systems*, DARCOM Pamphlet No. 706-358, 1976.

Chapter 10

Accident Reconstruction

10.1 Introduction and Objectives

An accident is an event in which one or more vehicles undergo unexpected action(s), generally involving physical contact with another vehicle or other object, producing injury, death, and/or property damage. The detailed process of observation, acquisition, and documentation of physical evidence and other information involving an accident or crash is known as accident investigation. The accident scene is the place where a vehicle accident occurs, both during and immediately following the accident, and before the vehicles and participants or persons involved have departed. The accident site refers to the place where an accident occurred, after the vehicles and participants or persons involved have departed the accident scene.

Accident reconstruction is a process conducted with the specific objective of estimating in both qualitative and quantitative manners how an accident occurred, using engineering, scientific, and mathematical principles based on evidence obtained through accident investigation.

The general objective of accident reconstruction is to describe the events of the accident and to determine what happened at a particular point in time. It includes the investigation of the circumstances of an accident. These circumstances include the effect of human, vehicle, and environmental factors on the pre-impact, impact, and post-impact events. Specific objectives of accident reconstruction include the following: collision analysis, injury analysis, accident avoidance analysis, injury avoidance analysis, and accident causation analysis. The ultimate objective of accident reconstruction is product or process improvement for future risk reduction.

The accident investigator observes or studies by close examination, and he or she preserves and documents the evidence discovered. The accident reconstructionist takes the evidence and all other data from various sources and rebuilds the accident one microsecond at a time. The data involve the objective data considered as facts, the engineering analysis, and witness statements. The objective data include facts such as vehicles, drivers, and the roadway involved in the accident, scene photographs, and measurements. The engineering analysis includes all engineering equations, analyses, testing, mathematical models, and computer simulation techniques that clearly and accurately describe the motions and associated vehicle(s) parameters and the occupants involved in the accident.

An uninvolved witness generally provides more accurate accident data for the accident reconstruction.

Webster's *New Universal Unabridged Dictionary* defines a reconstructionist as "A person who builds up from remaining parts and other evidence, an image of what something was in its original and complete form." The definition does not say recreate. In general, it is impossible to recreate even the simplest of accidents.

Accident reconstruction requires most or all of the following information:

- Color photographs taken at the accident scene

- Photographs of the vehicles taken at the accident scene or at a later date

- Police reports

- Witness statements

- Depositions

- Medical data (emergency medical service) and emergency room personnel

- Fire department reports

- TV station or independent videotapes taken at the accident scene or at a later date

- Newspaper clippings about the accident, people, scene, or vehicles

- Weather reports from the local National Oceanic and Atmospheric Administration (NOAA) office or other agencies and sources

Additional information or data are required, depending on the type of accident case. If the case involves a product defect, more specific information or data may be required.

10.2 Basic Equations of Motion

Kinematics is the study of the geometry of motion. Kinematics is used to relate displacement, velocity, acceleration, and time, without reference to the cause of the motion. Kinetics is the study of the relation existing between the forces acting on a body, the mass of the body, and the motion of the body. Kinetics is used to predict the motion caused by given forces or to determine the forces required to produce a given motion. Dynamics includes kinematics and kinetics.

Newton's laws of motion. Newton's three laws of motion are stated as follows:

1. Every body continues to be in its state of rest or of uniform motion unless disturbed by an extended force.

2. The rate of change of momentum of a body is equal to the external force and acts in the direction of force. In other words, the time rate of change of the product of the mass and velocity of a body is proportional to the force acting on the body, $F = ma$, where m is the mass of the body and V is the velocity of the body.

3. For every action, there is an equal and opposite reaction.

Table 10.1 provides a summary of the basic equations of motion useful for solving many problems found in accident reconstruction.

TABLE 10.1
BASIC EQUATIONS OF MOTION

S. No.	To Find	Given	Equation to Use
1.	Acceleration, a (ft/s² or m/s²)	t, v_i, v_e t, v_i, d v_i, v_e, d	$a = \dfrac{v_e - v_i}{t}$ $a = \dfrac{2d - 2v_i t}{t^2}$ $a = \dfrac{v_e^2 - v_i^2}{2d}$
2.	Distance, d (ft or m)	t, a, v_i a, v_i, v_e t, v_i, v_e	$d = v_i t + a\dfrac{t^2}{2}$ $d = \dfrac{v_e^2 - v_i^2}{2a}$ $d = \dfrac{t(v_i + v_e)}{2}$
3.	End velocity, v_e (ft/s or m/s)	t, a, v_i a, v_i, d	$v_e = v_i + at$ $v_e = \sqrt{v_i^2 + 2ad}$
4.	Initial velocity, v_i (ft/s or m/s)	t, a, v_e t, a, d a, v_e, d	$v_i = v_e - at$ $v_i = \dfrac{d}{t} - \dfrac{at}{2}$ $v_i = \sqrt{v_e^2 - 2ad}$
5.	Time, t (second)	a, v_i, v_e	$t = \dfrac{v_e - v_i}{a}$

Note:

a	=	acceleration, m/s (ft/s)
d	=	distance, m (ft)
t	=	time (sec)
v_i	=	initial velocity, m/s (ft/s)
v_e	=	end velocity, m/s (ft/s)
f	=	drag factor = a/g
g	=	acceleration due to gravity, 9.81 m/s² (32.2 ft/s²)

EXAMPLE E10.1

A road vehicle skids to a stop over two surfaces. The vehicle has skidded 27.43 m (90 ft) over the first surface and 15.24 m (50 ft) over the second surface. Assuming the first surface has a drag factor of 0.80 and the second surface has a drag factor of 0.70, determine the following:

a. The speed of the vehicle when the skidding first occurred

b. The time taken for the vehicle to skid over both surfaces

c. The speed of the vehicle after skidding 33.53 m (110 ft)

Solution:

a. From Table 10.1, the initial velocity v_i is obtained from

$$v_i = \sqrt{v_e^2 - 2ad}$$

$$= \sqrt{0^2 - 2(-6.87)(15.24)}$$

$$= 14.47 \text{ m/s } (47.48 \text{ ft/s})$$

where $a = fg = -0.70(9.81) = 6.87 \text{ m/s}^2 \left(-22.54 \text{ ft/s}^2\right)$

Here, 14.47 m/s (47.48 ft/s) is the velocity where the vehicle starts to skid on the second surface. It is the end velocity after skidding on the first surface. Therefore, on the first surface,

$$v_e = 14.47 \text{ m/s } (47.48 \text{ ft/s})$$

$$d = 27.43 \text{ m } (90 \text{ ft})$$

$$a = fg = -0.80(9.81) = 7.85 \text{ m/s}^2 \left(-25.76 \text{ ft/s}^2\right)$$

Now

$$v_i = \sqrt{v_e^2 - 2ad}$$

$$= \sqrt{14.47^2 - 2(-7.85)(27.43)}$$

$$= 25.30 \text{ m/s } \left(83.01 \text{ ft/s}\right)$$

$$= 90.86 \text{ km/h } \left(56.47 \text{ mph}\right)$$

b. Surface 1:

$$v_i = 25.30 \text{ m/s } (83.01 \text{ ft/s})$$

$$v_e = 14.47 \text{ m/s } (47.48 \text{ ft/s})$$

$$a = -7.85 \text{ m/}^2 \, (-25.76 \text{ ft/s}^2)$$

From Table 10.1, we have

$$t = \frac{v_e - v_i}{a}$$

$$= \frac{14.47 - 25.30}{-7.85}$$

$$= 1.38 \text{ s}$$

Surface 2:

$$v_i = 14.47 \text{ m/s (47.48 ft/s)}$$

$$v_e = 0$$

$$a = -6.87 \text{ m/s}^2 \; (-22.54 \text{ ft/s}^2)$$

and

$$t = \frac{v_e - v_i}{a}$$

$$= \frac{0 - 14.47}{-6.87}$$

$$= 2.11 \text{ s}$$

Total skidding time is $1.38 + 2.11 = 3.49$ s.

c. The end velocity can be obtained by knowing the initial velocity at the beginning of the second surface. An additional distance of 6.10 m (20 ft) is required to reach a total skidding distance of 33.53 m (110 ft).

Here,

$$v_i = 14.47 \text{ m/s (47.48 ft/s)}$$

$$a = -6.87 \text{ m/s}^2 \; (-22.54 \text{ ft/s}^2)$$

$$d = 6.10 \text{ m (20 ft)}$$

and

$$v_e = \sqrt{v_i^2 - 2ad}$$

$$= \sqrt{14.47^2 + 2(-6.87)(6.10)}$$

$$= 17.12 \text{ m/s } (56.18 \text{ ft/s})$$

$$= 64.50 \text{ km/h } (38.22 \text{ mph})$$

10.3 Drag Factor and Coefficient of Friction

The acceleration or deceleration of a vehicle is related to its drag factor f and the coefficient of friction μ. The drag factor is related to the acceleration by

$$a = fg \qquad (10.1)$$

where

a = acceleration, m/s^2 (ft/s^2)

f = drag factor (no dimension)

g = acceleration due to gravity, 9.81 m/s^2 (32.2 ft/s^2)

Drag factors and coefficients of friction generally are similar for all types of four-tired vehicles. The drag factors for motorcycles and heavy vehicles often can differ from those of four-tired vehicles.

The acceleration factor is the ratio of an acceleration rate to the acceleration of gravity. The deceleration rate, as related to friction between a skidding or sliding tire and the roadway surface due to the force of gravity, usually is less than the acceleration rate of gravity; therefore, it usually is expressed as a decimal fraction of gravity. This fraction is known as the drag factor f or, on a level surface, the coefficient of friction value.

Friction can be thought of as the resisting force to motion between two surfaces at their interface (contact). Static friction between two bodies is the tangential force that opposes the sliding of one body relative to the other. The limiting friction f′ is the maximum value of static friction that occurs when motion is impending. Kinetic friction is the tangential force between two bodies after motion begins. It is less than the static friction. The angle of friction is the angle between the action of the line of the total reaction of one body on another and the normal to the common tangent between the bodies when the motion is impending. The coefficient of static friction is the ratio of the limiting friction f′ to the normal force N

$$\mu = \frac{f'}{N} \qquad (10.2)$$

The coefficient of kinetic friction is the ratio of the kinetic friction to the normal force. The rolling friction (rolling resistance) refers to the resisting forces that come into play when a vehicle is rolling with no braking. Generally, these values are very low and are assumed to be insignificant in most accident reconstruction problems. However, rolling friction is important in tire design and other vehicle design considerations.

Laws of friction. The following laws of friction came into play:

1. The coefficient of friction is independent of the normal force; however, the limiting friction and kinetic friction are proportional to the normal force.

2. The coefficient of friction is independent of the area of contact.

3. The coefficient of kinetic friction is less than that of static friction.

4. At low speeds, friction is independent of speed. At high speeds, a decrease in friction has been noticed.

The drag factor f is defined as the force required for acceleration (or deceleration) in the direction of the acceleration (or deceleration) divided by the weight of the object (vehicle). Thus, we have

$$f = \frac{F}{W}$$ (10.3)

where

f = drag factor (no dimension)

F = force, N (lb)

W = weight of the object (vehicle), kg (lb)

For the coefficient of friction μ, the object must be sliding across the surface. This is not the case for the drag factor f. The drag factor and the coefficient of friction will be equal only in cases where all wheels are locked and are sliding on a level surface. Drag factors refer to conditions where a vehicle or object is slowing. The word "drag" implies slowing. Several methods are available for determining the drag factor and coefficient of friction for an actual accident, as follows:

1. Test-skit the accident vehicle or an exemplar vehicle

2. Slide a test tire (not the whole vehicle) to obtain the friction coefficient

3. Use existing highway department skid numbers for the road in question

4. Look up coefficients of friction in a table, and apply the appropriate adjustments to them for the case at hand

Drag factor on a grade. The drag factor on a grade is given by the formula

$$f_G = \frac{\mu + G}{\sqrt{1 + G^2}}$$ (10.4)

where

f_G = drag factor on a grade

μ = coefficient of friction between the tires and pavement

G = the percent grade expressed as a decimal (positive (+) for an uphill grade and negative (−) for a downhill grade)

Equation 10.4 applies for a case where a vehicle is skidding with all wheels locked on a non-level surface (that is, up or down grade). The drag factor is not equal to the coefficient of friction.

Table 10.2 gives a range of coefficients of friction for several road surface descriptions for passenger cars and pickup trucks equipped with typical tires. It can be seen from Table 10.2 that high friction coefficients are suggested for the low speed range. Generally, it is rare to see a coefficient of friction much larger than 0.90.

TABLE 10.2
COEFFICIENTS OF FRICTION FOR VARIOUS ROADWAY SURFACES

Description of Road Surface	Dry				Wet			
	Less than 30 mph		More than 30 mph		Less than 30 mph		More than 30 mph	
	From	To	From	To	From	To	From	To
Portland Cement								
New, Sharp	0.80	1.20	0.70	1.00	0.50	0.80	0.40	0.75
Traveled	0.60	0.80	0.60	0.75	0.45	0.70	0.45	0.65
Traffic Polished	0.65	0.75	0.50	0.65	0.45	0.65	0.45	0.60
Asphalt or Tar								
New, Sharp	0.80	1.20	0.65	1.00	0.50	0.80	0.45	0.75
Traveled	0.60	0.80	0.55	0.70	0.45	0.70	0.40	0.65
Traffic Polished	0.55	0.75	0.45	0.65	0.45	0.65	0.40	0.60
Excess Tar	0.50	0.60	0.35	0.60	0.30	0.60	0.25	0.55
Gravel								
Packed, Oiled	0.50	0.85	0.50	0.80	0.40	0.80	0.40	0.60
Loose	0.40	0.70	0.40	0.70	0.45	0.75	0.45	0.75
Cinders								
Packed	0.50	0.70	0.50	0.70	0.65	0.75	0.65	0.75
Rock								
Crushed	0.50	0.75	0.55	0.75	0.55	0.75	0.55	0.75
Ice								
Smooth	0.10	0.25	0.07	0.20	0.05	0.10	0.05	0.10
Snow								
Packed	0.30	0.55	0.35	0.55	0.30	0.60	0.30	0.60
Loose	0.10	0.25	0.10	0.20	0.30	0.60	0.30	0.60

Unequal drag factors on axles. For passenger cars during hard braking, the load is shifted to the front axle. Because of this load shift, no correction needs to be made when all the wheels are locked. However, if one axle has a drag factor that differs from that of another axle, then the drag factor for the vehicle is given by

$$f_R = \frac{f_f - x_f(f_f - f_r)}{1 - z(f_f - f_r)} \qquad (10.5)$$

where

f_R = drag factor on the vehicle

f_f = drag factor on the front axle

f_r = drag factor on the rear axle

x_f = horizontal distance of the center of mass from the front axle as a decimal fraction of the wheelbase

z = height of the center of mass as a decimal fraction of the wheelbase

Example E10.2

A vehicle is sliding down a 5% grade, and the coefficient of friction is 0.75. Determine the effective drag factor on the grade.

Solution:

The drag factor on a grade is obtained from Eq. 10.4 as

$$f_G = \frac{\mu + G}{\sqrt{1 + G^2}}$$

$$= \frac{0.75 + (-0.05)}{\sqrt{1 + (-0.05)^2}}$$

$$= \frac{0.70}{1.00125}$$

$$= 0.699$$

$$\approx 0.70$$

10.4 Work, Energy, and the Law of Conservation of Energy

Work. Work (W) for a force F, constant in magnitude and with its direction along a straight line, is given by

$$W = Fd \qquad\qquad (10.6)$$

where

W = work done, Nm (ft-lb)

F = constant force, N (lb)

d = displacement during the motor along the straight line, m (ft)

If the force and displacement (distance) are not in the same direction (collinear), then

$$W = Fd \cos\theta \qquad\qquad (10.7)$$

where θ is the angle between the force vector and the distance of travel.

Work is a scalar quantity.

Two forms of mechanical energy are important in understanding momentum and speed analysis: potential energy, and kinetic energy. Potential energy is the capacity of a mass or body to do work due to the position of the body.

Kinetic energy. The kinetic energy (KE) of a body with mass m and moving with speed v is given by

$$KE = \frac{1}{2}mv^2$$

$$= \frac{1}{2}\left(\frac{w}{g}\right)v^2$$

(10.8)

where

$$KE = \text{kinetic energy, Nm (ft-lb)}$$

$$w = \text{weight, N (lb)}$$

$$v = \text{velocity, m/s (ft/s)}$$

$$g = \text{acceleration due to gravity, } 9.81 \text{ m/s}^2 (32.2 \text{ ft/s}^2)$$

The amount of kinetic energy that must be converted to another form to bring a vehicle to a stop is

$$W = \text{work, Nm (ft-lb)} = W \, f \, d$$

(10.9)

where

$$w = \text{weight, N (lb)}$$

$$f = \text{drag factor}$$

$$d = \text{distance, m (ft)}$$

The work done to compress a spring is

$$W = \text{work} = \frac{1}{2}kx^2$$

(10.10)

where

$$W = \text{work, Nm (ft-lb)}$$

$$k = \text{spring constant, N/m (lb/ft)}$$

$$x = \text{distance compressed/stretched, m (ft)}$$

Potential energy. The potential energy (V) of a weight w close to the surface of the earth and at a height y above a given datum is

$$V = wy$$

(10.11)

where

w = weight, N (lb)

y = height, m (ft)

Law of conservation of energy. When a body moves under the action of conservative forces, the sum of the kinetic energy and of the potential energy of the body remains constant. The sum KE + V is called the mechanical energy of the body and is denoted by E,

$$\text{Mechanical energy } (E) = KE + V = \text{ constant} \qquad (10.12)$$

Linear momentum. Replacing the acceleration a by the derivative $\dfrac{dV}{dt}$, we can write Newton's second law of motion as

$$\Sigma F = m\frac{dV}{dt} \qquad (10.13)$$

Because the mass m of a body is constant, we have

$$\Sigma F = \frac{d}{dt}(mV) \qquad (10.14)$$

The vector mV is called the linear momentum or simply the momentum of the body. Equation 10.14 expresses that the resultant of the forces acting on the body is equal to the rate of change of the linear momentum of the body. Denoting by L the linear momentum of the body

$$L = mV \qquad (10.15)$$

and by $\dot{L}$ its derivative with respect to t, we can write Eq. 10.14 as

$$\Sigma F = \dot{L} \qquad (10.16)$$

It follows from Eq. 10.14 that the rate of change of the linear momentum mV is zero when $\Sigma F = 0$. Hence, if the resultant force acting on a body is zero, the linear momentum of the body remains constant, in both magnitude and direction. This is the principle of conservation of linear momentum for a body, which can be recognized as an alternative statement of Newton's first law of motion.

10.5 Driver Perception and Response

In traffic accident reconstruction, perception and response values are used primarily to

a. Estimate the possible stopping distance of vehicles, generally to determine whether a hazard could have been avoided under specified circumferences

b. Compare actual points of perception with possible points of perception to estimate perception delay as an indication of attentiveness

c. Determine whether a roadway user's response to a situation hazard was the best possible under existing circumferences

Perception is a human factor, which means the general process of detecting some object or situation and comprehending its significance. Reaction or response is a person's voluntary or involuntary response to a hazard or other situation that has been perceived, the response to a sensory stimulus. Reaction time is the time required from perception to the start of vehicle control for tactical or strategic operations.

In traffic accident reconstruction, perception and reaction time mainly have two significant uses:

a. To compare the least possible reaction under the circumferences with evasive tactics actually undertaken in order to evaluate the knowledge and skill of a driver or pedestrian

b. To establish the possible slowing or stopping distance of a vehicle at a stipulated speed

McGee *et al.* (1983) attempted to deduce perception-response time by listing its individual components, surveying the literature to find the appropriate values for each, and summing these values to obtain totals. Table 10.3 shows their results.

TABLE 10.3
RESULTS OF AN EFFORT TO DEDUCE DRIVER PERCEPTION-RESPONSE TIME, BASED ON SUMMATION OF ASSUMED COMPONENTS TAKEN FROM RESEARCH LITERATURE

Element	Percentile of Drivers					
	50	75	85	90	95	99
1. Perception						
a. Latency	0.24	0.27	0.31	0.33	0.35	0.45
b. Eye moment	0.09	0.09	0.09	0.09	0.09	0.09
c. Fixation	0.20	0.20	0.20	0.20	0.20	0.20
d. Recognition	0.40	0.45	0.55	0.55	0.60	0.65
2. Decision	0.50	0.75	0.90	0.90	0.95	1.00
3. Brake reaction	0.85	1.11	1.42	1.42	1.63	2.16
Total A (1a–d + 2 + 3)	2.3	2.9	3.5	3.5	3.8	4.6
Total B (1c–d + 2 + 3)	2.0	2.5	2.8	3.1	3.4	4.1
Total C (1a–d + 3)	1.8	2.1	2.3	2.6	2.9	3.6

From the preceding discussion, it is extremely difficult to determine the time required for an individual to actually perceive a hazard. Tests conducted to assess the combined time lapse for perception and reaction time in altered subjects indicated a range of 0.4 to 1.7 seconds. Using the result from tests and adding 0.3 second to the test results, the perception reaction range becomes 0.7 to 2.0 seconds. Given a correction factor, the mean for perception and reaction time would be in the area of 1.5 seconds.

The safe stopping distance for vehicles, knowing the actual time required for the driver to perceive and react to a hazard, is given by

$$SSD = vt_p + \frac{v^2}{2a} \qquad (10.17)$$

where

SSD = safe stopping distance for the vehicle (ft or m)

t_p = perception/reaction time (seconds)

v = velocity, m/s (ft/s)

a = acceleration factor (the effective drag factor times the gravitational constant)

Most design standards for rural and urban roadways use a 2.5-second perception and reaction time to include nearly all drivers and conditions for computing a safe stopping sight distance.

10.6 Engineering Models and Animations

Fay, Robinette, and Larson (1988) emphasized the importance of engineering models and animations in vehicular accident studies. Vehicle accident analysis relies heavily on mathematics and the principles of conservation of energy and momentum, as well as Newton's laws of motion. To apply these principles, it first is necessary to know the approximate vehicle motions. The analytical procedure is interactive, using a combination of model analysis to determine linear and angular velocities and accelerations. Scale accident scene models combined with aerial photography to enhance realism have been used extensively in evaluation, analysis, and presentation of vehicular accident reconstructions to non-technical audiences. Slide and video accident animations have been produced directly from aerial photograph enhanced models and have been used successfully in courtroom presentations since the 1970s.

10.6.1 Function of Accident Scene Models

Motions that occur in vehicle accidents sometimes are extremely complex because they involve collisions between vehicles, collisions between vehicles and fixed structures, and collisions between vehicles and pedestrians. Sometimes, they also include vaulting, flipping, and rolling. Determining what happened in a vehicle accident requires an engineering analysis of vehicle dynamics and kinematics and frequently an additional analysis of the dynamics and kinematics of the vehicle occupants and pedestrians. Models are useful at all phases of the engineering analysis and have an important role in presenting the engineering results in ways that are easy for people to understand. Actual vehicle impacts have been conducted with scale models to study vehicle motions and interactive forces.

A reconstructionist usually will have to expend significant effort in studying the physical evidence such as accident scene photos, the accident site, the vehicles, and driver and witness statements before the mathematical engineering analysis can be started. One of the most important steps is the proper evaluation and interpretation of the available information. Models have an important role when evaluating this information.

The base model for vehicle accident reconstruction is a model of the accident scene. The simplest possible model is a line drawing of a section of roadway or an intersection. The use of an aerial photograph is preferred over a line drawing because of the introduction of realism, detail, relative scale, and credibility. The aerial photograph can be enlarged to an appropriate scale for use with commercially available scaled vehicle models. If the accident scene is flat, the aerial photograph can be mounted on a flat surface. Surrounding landmarks such as trees, bushes, buildings, and traffic control devices all are displayed directly in the photo. Superimposing three-dimensional models at the appropriate locations can highlight any of these. If the accident scene is hilly or if specific elevation changes are important to the analysis or need to be emphasized, a scale three-dimensional terrain model can be constructed from an accident scene engineering survey. The aerial photograph can be superimposed over all or parts of the three-dimensional terrain base. Frequently, it is sufficient to mount only the road surface from the aerial photograph onto a plaster of Paris model and to construct the surrounding landscape with commercially available model supplies or plant material.

If possible, the aerial photograph is flown shortly after the accident to show vehicle skid marks, gouges, furrow marks, fluid strains, and so forth. There usually is a time delay between the accident and the aerial photograph, and the reconstructionist will have to superimpose this information of the accident onto the model. Most experienced reconstructionists have developed techniques and special equipment to measure or reconstruct such information, which then can be plotted onto the model. In addition, special tools and computer programs that are commercially available can assist with this process.

10.6.2 Model Application

The first step in analyzing a vehicular accident is to catalog the available information about the accident and then to superimpose or plot this information on the scene model and the vehicle models. In some cases, features of the terrain influence vehicle motion. Full-size models sometimes are used in accident studies.

Most accidents can be split into three phases: pre-impact, impact, and post-impact. Each phase is analyzed separately and collectively. This activity is accomplished with the assistance of accident reconstruction software.

10.6.3 Reconstruction Animations

Accurate physical and mathematical construction of an accident is of critical importance to the reconstructionist. Of equal importance is the process of communicating the reconstructionist's findings to a non-technical client or jury in such a manner that they can comprehend it. This step frequently is a central part of the litigation team effort in which the attorney must rely on the reconstructionist to present and defend the reconstruction. The model is an important demonstrative tool in effecting this goal. A logical extension of the model is a visual display of the vehicle motions presented over small increments of time and edited together into a smooth animation. Aerial photograph enhanced models were first used in the mid-1970s. Video animations of the reconstruction in real time have been used since the early 1980s.

To produce an animation, the motions of the vehicles and occupants are determined by computer analysis and are plotted on the model. The vehicles are moved to the plotted positions and photographed with video equipment. The frames are edited together to produce a smooth-running animation that shows the accident in real time. A sequence of slides can be made with fewer positions to produce a step view of the accident.

Introduction of terrain elevation changes and vehicle rollover motions can increase significantly the complexity of the mathematical motion analysis. In the following example, a single vehicle left a divided highway and rolled down a hill. The vehicle exhibited multiple types of motion: linear displacement, yawing, vehicle rollover, and vaulting down the embankment. The mathematical motion analysis became three-dimensional with multiple degrees of freedom constrained by the physical evidence associated with the accident (i.e., skid marks, points embankment, rollover path, location and position of the impacts between the vehicle and the ground, and the final point and orientation at rest). Evaluation and identification of these constraints constituted a major effort in itself and employed research data.

In summary, engineering models are useful in understanding the motions of vehicles during accident sequences and are an important part of the interactive analysis, which involves accident data and the principles of engineering and physics. The aerial photographs and models are useful in producing slide sequences and video animations, which portray realistic and physically accurate vehicle motions.

10.7 Lane Change Maneuver Model (Araszewski *et al.*, 2002)

Araszewski *et al.* (2002) developed a lane change model for determining the expected vehicle dynamics, including lateral, longitudinal, and angular vehicle displacements during the lane change maneuver. This model is based on the vehicle speed, peak and average accelerations, and driver steering input. The model uses vehicle handling parameters during steady-state handling conditions by incorporating tire, suspension, and handling characteristics.

A vehicle may perform a range of varying vehicle speeds, accelerations, and corresponding times over a range of lateral and longitudinal distances. Ultimately, the parameters of the lane change maneuver will depend on the driver input and specifically how the steering wheel is turned during a maneuver. Therefore, a significant human factor is involved in a lane change analysis.

The formulae for the longitudinal lane change distance and the lane change time are given by

$$d = 2V\sqrt{\frac{S}{a_{AVG}}} \qquad (10.18)$$

$$t = 2\sqrt{\frac{S}{a_{AVG}}} \qquad (10.19)$$

During a real-world lane change, the acceleration profile varies and is not constant. Because the largest unknown factor of the equations is the acceleration profile and thus the effective average lateral acceleration during a lane change maneuver, the model was adapted to consider different acceleration profiles. Two of the considered acceleration profiles included those obtained from a sinusoidal and triangular steer input by the vehicle operator. These adjustments allowed the peak acceleration to be an input parameter such that a theoretical upper limit could be obtained. For a sinusoidal steer input, the lane change longitudinal distance is given as

$$d_{Sine} = V \sqrt{\frac{2 \pi s}{a_{PEAK}}} \tag{10.20}$$

where a_{PEAK} is the peak lateral acceleration in meters per seconds squared (m/s^2).

For a triangular steer input, the lane change longitudinal distance is given by

$$d_{Tri} = 2V \sqrt{\frac{2s}{a_{PEAK}}} \tag{10.21}$$

The vehicle angle during the lane change maneuver also is derived here. The course angle may be expressed as the sum of the heading angle due to the travel path and the sideslip angle

$$\Theta = \Psi + \beta \tag{10.22}$$

where

Θ = the course angle

Ψ = the heading angle

β = the sideslip angle

The heading angle component derivation first requires a theoretical average radius during the lane change maneuver to be calculated. The theoretical average radius has no real-world or practical reference but is a useful analysis, too. The theoretical average radius is given by

$$R_{AVG} = \frac{V^2}{a_{AVG}} \tag{10.23}$$

Using a geometrical relation for the ratio between the traveled distance over half a lane change and the average turn radius, the heading angle at the middle point of the lane change may be determined as

$$\Psi_{Middle} = \left(\frac{180}{\pi} \right) \frac{Vt}{R_{AVG}} \tag{10.24}$$

For a symmetrical lane change maneuver, the maximum course angle occurs halfway through the lane change. At this point, the lateral acceleration is nominally zero;

therefore, the sideslip effects are absent. For a sinusoidal steer input during the lane change maneuver, the maximum course angle is given as

$$\Theta_{Sine(MAX)} = \left(\frac{180}{\pi^2}\right)\frac{a_{PEAK}t}{V} \tag{10.25}$$

For a triangular steer input, the maximum course angle is given by

$$\Theta_{Tri(MAX)} = \left(\frac{45}{\pi}\right)\frac{a_{PEAK}t}{V} \tag{10.26}$$

The maximum sideslip angle will occur when the peak lateral acceleration is experienced. For a symmetrical lane change maneuver, this will occur at the one-quarter and three-quarter points of the lane change. At these points, the heading angle will have an absolute value of half the maximum course angle, and the maximum sideslip angle is defined as

$$\beta_{MAX} = a_{PEAK}\left[k_\beta \frac{b}{V^2}\right] \tag{10.27}$$

For most accident reconstruction investigations, further analyses of the input parameters may not be required. However, in some cases, the steering angle for a lane change maneuver may be useful. This allows the required driver input to be calculated and assessed. The evaluation of the steering wheel input requires knowledge of the ratio of the steering wheel to the tire angle for the specific vehicle, and the understeer gradient of the vehicle must be known. Although the ratio of the steering wheel to the vehicle angle is often part of the vehicle specification data list, the understeer gradient is not readily available.

Additional relationships were derived for the peak steering angle required by the vehicle operator to conduct a lane change maneuver. The peak steering angle occurs when the lateral acceleration also is a peak value. Therefore, the peak steering angle is given by

$$\delta_{SW(MAX)} = G\left[k_U a_{PEAK} + \arctan\left(\frac{a_{PEAK}L}{V^2}\right)\right] \tag{10.28}$$

Other formulae that may be useful for accident reconstruction purposes are the lateral acceleration profile, the longitudinal and lateral displacement profiles, and the steering angle profile. These theoretical relationships are presented as follows:

Time:

t_i =   instantaneous time

t_{Tot} =   total lane change maneuver time

Instantaneous lateral acceleration:

$$a_{i(sine)} = a_{PEAK}\left[\sin\left(\frac{2\pi t_i}{t_{TOT}}\right)\right] \quad \text{sinusoidal steer input } (a_{AVG} = 0.64a_{PEAK}) \tag{10.29}$$

$$a_{i(Tri)} = \begin{cases} a_{PEAK}\left[\dfrac{4t_i}{t_{TOT}}\right] & 0 \le t_i < \dfrac{1}{4}t_{TOT} \\[3mm] a_{PEAK}\left[2 - \dfrac{4t_i}{t_{TOT}}\right] & \text{for} & \dfrac{1}{4}t_{TOT} \le t_i < \dfrac{3}{4}t_{TOT} \\[3mm] a_{PEAK}\left[\dfrac{4t_i}{t_{TOT}} - 4\right] & \dfrac{3}{4}t_{TOT} \le t_i \le t_{TOT} \end{cases} \tag{10.30}$$

triangular steer input $\left(a_{AVG} = 0.5a_{PEAK}\right)$

where

$\qquad a_{PEAK}$ = peak lateral acceleration

$\qquad a_i$ = instantaneous lateral acceleration

Instantaneous turn radius:

$$r_i = \frac{V_2}{a_i} \tag{10.31}$$

Instantaneous steer angle:

$$\delta_{Ki} = \arctan\left(\frac{L}{r_i}\right)$$

where L is the wheelbase, and

$$\delta_{Ui} = K_U a_i$$
$$\delta_i = \delta_{Ki} + \delta_{Ui} \tag{10.32}$$

where K_U is the understeer gradient.

Instantaneous steering wheel angle:

$$\delta_{SWi} = G\delta_i \tag{10.33}$$

where G is the steering-to-tire-angle ratio.

Longitudinal displacement: Stepwise incremental calculation of

$$d_{long} = d_{Vi} = d_{V(i-1)} + V\left[t_i - t_{(i-1)}\right]\cos(\psi_i) \tag{10.34}$$

Lateral displacement: Stepwise incremental calculation of

$$d_{lat} = d_{Xi} = d_{X(i-1)} + V\left[t_i - t_{(i-1)}\right]\sin(\Psi_i) \tag{10.35}$$

Instantaneous heading angle: Stepwise incremental calculation of

$$\Psi_i = \Psi_{(i-1)} + \left(\frac{180}{\pi}\right)\frac{V \mid t_i - t_{(i-1)} \mid}{r_i} \tag{10.36}$$

The staged lane change tests resulted in limited data over a range of vehicle speeds, lateral lane change distances, longitudinal lane change distances, and lane change times. Correspondingly, the peak and average accelerations of the staged lane change maneuvers also varied. The acceleration curve during the lane change maneuver was a direct relation of the steer input angle of the vehicle operator. Thus, as expected, the resulting acceleration profile varied from run to run and varied significantly from operator to operator. This was observed to be a factor of the driving style of the operator. Assessment of the steering wheel angle during the lane change maneuver suggests that the typical lane change maneuver likely falls within the range of results defined by the sinusoidal and triangular steer input. However, this assessment was made on a limited number of staged tests with a limited number of vehicle operators.

The test results were compared to the Limpert formulae that define the lane change distance and time in terms of average lateral acceleration. The Limpert formulae are summarized as follows:

Limpert (imperial):

$$d_{[ft]} = 0.458V_{[ft/sec]}\sqrt{\frac{s_{[ft]}}{a_{[g]}}}$$

$$d_{[ft]} = \frac{d_{[m]}}{0.3048}$$

$$s_{[ft]} = \frac{s_{[m]}}{0.3048 \text{ lim}}$$

$$V_{[ft/sec]} = \frac{V_{[m/s]}}{0.3048} \tag{10.37}$$

$$a_{[g]} = \frac{a_{[m/s^2]}}{9.81}$$

$$\left(\frac{d_{[m]}}{0.3048}\right) = 0.458\left(\frac{V_{[m/s]}}{0.3048}\right)\sqrt{\frac{\left(\frac{s_{[m]}}{0.3048}\right)}{\left(\frac{a_{[m/s^2]}}{9.81}\right)}}$$

Limpert (metric):

$$\Rightarrow d_{[m]} = 2.60 V_{[m/s]} \sqrt{\frac{s_{[m]}}{a_{\left[m/s^2\right]}}} \qquad (10.38)$$

where

d = lane change distance

V = speed

s = lateral lane change distance

a = average lateral acceleration

A comparison of the staged lane change maneuvers with the formulae presented in the study by Limpert (1989) suggests that the Limpert formulae consistently overestimate the lane change time and lane change distance. This resulted in an average overestimate of about 29% when the Limpert formula was applied using the average acceleration as recorded during the lane change maneuver.

The formulae provided by Dailey (1988) also were used in a comparison with the staged lane change maneuvers. The Dailey formulae assume a constant turn radius and therefore a constant lateral acceleration for each half of the lane change. The Dailey formulae are given as follows:

Dailey (imperial):

$$d_{[ft]} = 0.732\, V_{[mph]} \sqrt{\frac{m_{[ft]}}{a_{[g]}}}$$

$$m_{[ft]} = \left[\frac{1}{2}\right] s_{[ft]} = \left[\frac{1}{2}\right] \frac{s_{[m]}}{0.3048}$$

$$d_{[ft]} = \frac{d_{[m]}}{0.3048}$$

$$V_{[mph]} = \frac{V_{[km/h]}}{1.609} = \left(\frac{3.6}{1.609}\right) V_{[m/s]} \qquad (10.39)$$

$$a_{[g]} = \frac{a_{\left[m/s^2\right]}}{9.81}$$

$$\left(\frac{d_{[m]}}{0.3048}\right) = 0.732 \left(\frac{3.6}{1.609}\right) V_{[m/s]} \sqrt{\frac{\left[\frac{1}{2}\right]\left(\frac{s_{[m]}}{0.3048}\right)}{\left(\frac{a_{\left[m/s^2\right]}}{9.81}\right)}}$$

Dailey (metric):

$$\Rightarrow d_{[m]} = 2.00\, V_{[m/s]} \sqrt{\frac{s_{[m]}}{a_{[m/s^2]}}} \tag{10.40}$$

where

d = longitudinal lane change distance

V = speed

m = half the lateral lane change distance

a = average lateral acceleration

Note that although the Dailey formula accurately predicts the lane change distance and time if the average acceleration is known, it does not accurately model the travel path, the peak acceleration, or the acceleration profile. Due to the constant turn radius model, the vehicle path may deviate as much as about 30% at the quarter point of the full lane change maneuver. Also note that the Dailey formula likely is not accurate if peak or threshold acceleration values are used.

As expected, the average acceleration of the tests could be used to accurately evaluate the lane change distance, time, and angle of the test vehicle. The test results were predicted with nominally zero percentage average difference and a standard deviation of about 5%.

The application of the peak acceleration values resulted in calculated values that were near the actual value. The sinusoidal input condition yielded results that were on average about 5% below the actual value, with a standard deviation of about 6%. The triangular input condition yielded results that were on average about 7% above the actual value, with a standard deviation of about 7%. Therefore, the sinusoidal and triangular models appear to provide an envelope of the expected results. A significant portion of the resulting deviation in the calculated values likely originated from the operator input effect and the fact that the input was not perfectly symmetrical or smooth. However, this effect also likely would exist in real-world driving conditions.

Evaluation of the test data results yielded empirical coefficients to be determined. Considering the peak acceleration values generated during the lane change maneuvers, the empirical coefficient was determined to be 2.7, with a standard deviation of 0.2. Considering the average acceleration values generated during the lane change maneuvers, the empirical coefficient was determined to be 2.0, with a standard deviation of 0.1. Therefore, the peak and average accelerations can be used in the following empirical relations for a lane change maneuver as

$$d = 2.7V\sqrt{\frac{s}{a_{PEAK}}} \tag{10.41}$$

$$d = 2.0V\sqrt{\frac{s}{a_{AVG}}} \tag{10.42}$$

where

$$a_{AVG} = \text{average acceleration, m/s}^2$$

$$a_{PEAK} = \text{peak acceleration, m/s}^2$$

The ratio of the average acceleration to the peak acceleration during the test was about 0.58, with a standard deviation of about 0.04. Triangular and sinusoidal steering inputs result in average to peak acceleration ratios of 1.5 and about 0.64, respectively. This implies that the typical lane change maneuver was conducted with a steer input between a sinusoidal and triangular steer input.

The formulae presented here allow manual calculations to be conducted for lane change modeling. Alternatively, similar results may be possible if a validated simulation program is available and if data are applied correctly.

The models presented here appear to model the lane change maneuver with adequate accuracy and more completely define the lane change maneuver as compared to other manual techniques.

10.8 Speed Estimates for Fall, Flip, or Vault

When vehicles travel through the air in accidents, the motion can be described as a fall, flip, or vault. A fall occurs when a vehicle is traveling forward and is no longer in contact with the surface over which it is moving. Flip is the movement of a vehicle from a place where the forward velocity of a part of the vehicle suddenly is stopped by an object below its center of gravity such as a curb, rail, or furrow, with the result that the ensuing rotational motion lifts the vehicle from the ground. A flip occurs when a vehicle is moving sideways and the resistance at the tires is sufficient to cause the vehicle to raise and move through the air. Vault is a roll or pitch motion of a vehicle made following the loss of ground contact. A vault is similar to a flip, except a vault is an end-for-end flip.

10.8.1 Fall

The velocity of a vehicle at the point of takeoff is given by

$$v = d \sqrt{\frac{g}{2\cos\theta(d\sin\theta - h\cos\theta)}} \tag{10.43}$$

where

v = velocity of the vehicle at takeoff, m/s (ft/s)

d = horizontal distance traveled by the center of mass of the vehicle from takeoff to landing, m (ft)

g = acceleration due to gravity, 9.81 m/s^2 (32.2 ft/s^2)

h = distance traveled vertically by the center of mass of the vehicle from takeoff to landing (m or ft). h is positive if the landing point is above the takeoff point, or negative if the landing point is below the takeoff point.

G = percent grade

θ = angle of takeoff as measured relative to a horizontal plane (deg)

For a small angle, Eq. 10.43 becomes

$$v = d\sqrt{\frac{g}{2(dG - h)}} \qquad (10.44)$$

where $G = \sin\theta$ and $\cos\theta \approx 1$.

Figures 10.1 to 10.4 show how the horizontal distance d and the vertical distance h are measured for different situations. Figure 10.1 shows a vehicle falling through the air and landing at a higher ground than takeoff. Figure 10.2 shows the measurements required for the speed estimate. Figure 10.3 shows the landing position after a fall, which may not be the point where the vehicle finally comes to rest. Figure 10.4 shows the vehicle landing on a slope. A connection should be made on the basis of marks made where the vehicle first struck the ground after a fall. Figure 10.5 shows the takeoff slopes where an adjustment must be made to the vertical distance of the fall. This adjustment is equal to the slope times the horizontal distance. This adjustment is added to the vertical fall for a slope upward and subtracted for a slope downward.

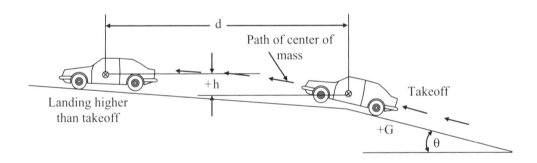

Figure 10.1 Vehicle landing at a higher ground than takeoff.

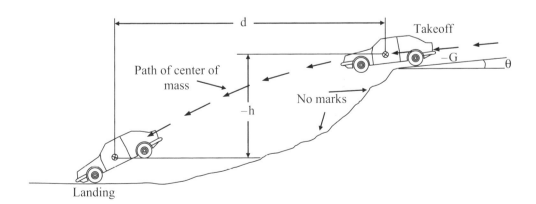

Figure 10.2 Measurements required for the speed estimate.

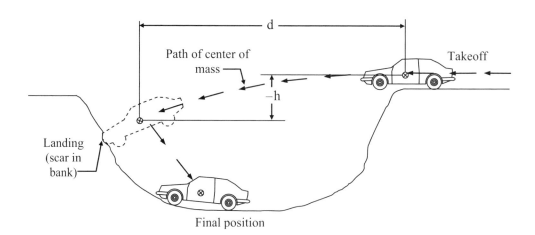

Figure 10.3 Landing position after a fall.

Figure 10.6 Slope of the ▮ in a fall.

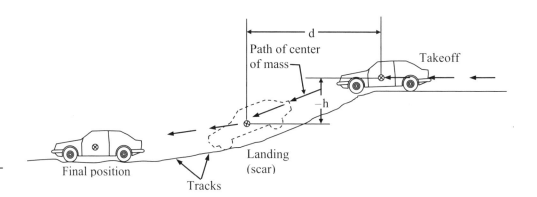

Figure 10.4 Vehicle landing on a slope.

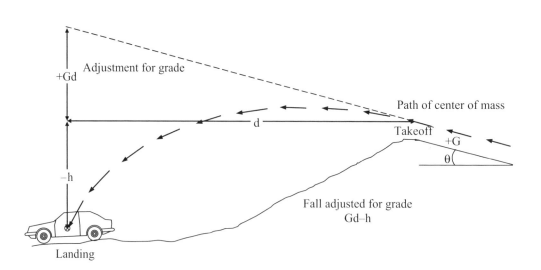

Figure 10.5 Takeoff slopes.

10.8.2 Flip

Equation 10.43 is the general equation for a flip. The distance is obtained by taking the shortest distance between any sign of where the vehicle took off (a curb or dig-in at the end of a furrow) and any sign of where it first landed, and adding the width of the vehicle (Figure 10.6). Figure 10.6 shows the slope of the takeoff in a fall. It is the slope of the exact path that the vehicle followed. A slope upward is positive, and a slope downward is negative.

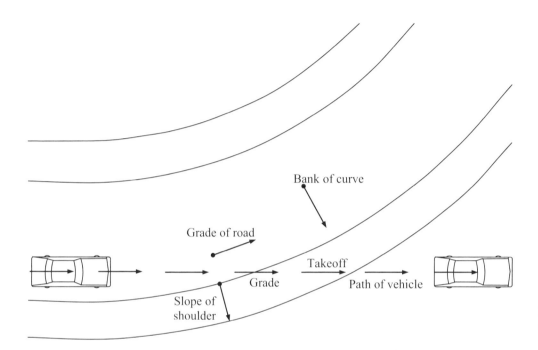

Figure 10.6 Slope of the takeoff in a fall.

The minimum speed is obtained when θ (the takeoff angle) is approximately 45°. The angle θ for the minimum speed estimate is given by

$$\theta = \frac{1}{2}\cos^{-1}\left[\frac{-h}{\sqrt{d^2 - h^2}}\right] \qquad (10.45)$$

where θ = 45°. Equation 10.43 gives the minimum speed at the takeoff point

$$v = d\sqrt{\frac{g}{d - h}} \qquad (10.46)$$

10.8.3 Vault

A vault is an end-for-end flip. Figure 10.7 shows the vault of a vehicle. A vault occurs when the front of a vehicle strikes an object in which further movement at the ground

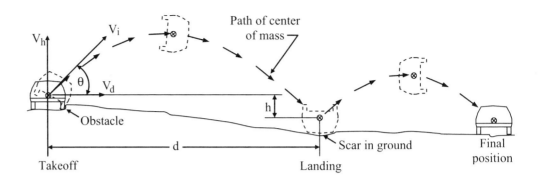

Figure 10.7 Vault of a vehicle.

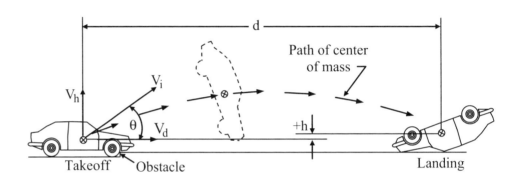

Figure 10.8 Path of the center of mass of a vehicle.

level stops. As with falls and flips, a vehicle that vaults likely will continue moving after it lands. It may roll or slide to its final position. The three equations that are used for a vault are given by Eqs. 10.43, 10.45, and 10.46. The variables v, d g, θ, and h are the same variables used for a flip. Figure 10.8 shows the path of the center of mass of the vehicle.

10.9　Speed Estimates from Yaw Marks

Yaw is the action of a vehicle revolving around its center of mass. This action is a sideslip movement and is commonly observed in a vehicle traveling around a curve whose rear end breaks away and appears to want to move in a direction other than that in which the vehicle is headed. Yaw commences at the time the rear tires start to sideslip more than the front tires and develops as the rear tires depart from their normal tracking path and cross over the front tire paths.

A yaw mark is a tire mark caused by a sideslipping tire often showing a striped pattern called striations. Yaw marks are the result of an avoidance maneuver, where a driver perceives an object or potential hazard in front of the vehicle and swerves suddenly to the left or right or from a sudden yaw velocity. During extreme steering maneuvers, the available tire friction can be saturated by the side forces on the tire, and the vehicle then will yaw and leave a peculiar type of tire mark. The speed at the beginning of the maneuver often can be calculated from yaw marks and often is referred to as the critical speed.

To calculate the speed to sideslip of a vehicle, one must know the radius of the curved path that is followed by the vehicle. The radius of curvature of the yaw mark is calculated using the formula

$$R = \frac{C^2}{8M} + \frac{M}{2}$$
(10.47)

where

R = radius, m (ft)

C = chord, m (ft)

M = middle ordinate, m (ft)

A chord measurement of 30.48 m (100 ft) is recommended. The speed of the vehicle is higher at the beginning of the yaw mark because the vehicle will decelerate as it sideslips. The critical speed is given by

$$V_{cr} = \sqrt{gRf}$$
(10.48)

where

V_{cr} = critical speed, m/s (ft/sec)

g = acceleration due to gravity, 9.81 m/s² (32.2 ft/s²)

R = radius, m (ft)

f = coefficient of friction

Equation 10.48 also can be written as

$$S = \sqrt{15Rf}$$
(10.49)

where

S = speed, mph

R = radius, ft

f = coefficient of friction

Also in S.I. units,

$$S = \sqrt{127Rf}$$

where

S = speed, km/h

R = radius, m

f = coefficient of friction

For a roadway with super elevation, the critical speed is given by

$$V_{cr} = \sqrt{\frac{gR\,(f + \tan\gamma)}{(1 - f\tan\gamma)}} \qquad (10.50)$$

where

g = acceleration due to gravity, 9.81 m/s^2 (32.2 ft/s^2)

f = coefficient of friction

R = radius, m (ft)

γ = bank angle (super elevation)

EXAMPLE E10.3

The measured chord and middle ordinates from a test of a passenger road vehicle of a sudden steer maneuver are 15.24 m (50 ft) and 0.37 m (1.2 ft). The test site has a zero grade and zero super elevation. Determine the following:

a. The speed of the vehicle over the measured arc of yaw marks. Assume an average frictional drag coefficient of 0.8.

b. The frictional drag coefficient that gives the measured speed of 80.45 km/h (50 mph).

Solution:

a. From Eq. 10.47, the radius of the circular arc is

$$R = \frac{C^2}{8M} + \frac{M}{2}$$

$$= \frac{15.24^2}{8(0.37)} + \frac{0.37}{2}$$

$$= 78.65 \text{ m } (258.04 \text{ ft})$$

The speed of the vehicle is given by Eq. 10.49 as

$$V_{cr} = \sqrt{127Rf}$$

$$= \sqrt{127(78.65)(0.8)}$$

$$= 89.39 \text{ km/h } (55.56 \text{ mph})$$

b. From Eq. 10.49,

$$f = \frac{V_{cr}^2}{127R}$$

$$= \frac{89.39^2}{127(78.65)}$$

$$= 0.80$$

10.10 Impact Analysis

A collision between two bodies that occurs in a very short interval of time and during which the two bodies exert relatively large forces on each other is called an impact. The common normal to the surfaces in contact during the impact is called the line of impact. The impact is the transmission of force from one body to another within a very short interval of time. The colliding bodies experience an elastic and/or plastic deformation.

When the amount of force between two objects in collision is sufficient to cause permanent deformation, the damage caused produces a loss of energy. An elastic collision between two bodies is one in which no permanent deformation takes place in either body, and both momentum and kinetic energy are conserved. In a perfectly elastic collision, one says that the sum of the translational kinetic energies of the object is not changed. A non-elastic collision, also referred to as an inelastic collision, is a collision between two bodies in which there is permanent deformation (i.e., the objects lose their original shapes). Some kinetic energy is transformed into internal energy in this type of collision. When two bodies collide, and they stick and travel together with a common final velocity after the collision, the collision is said to be completely inelastic.

If the centers of mass of the two colliding bodies are located on the line of impact, the impact is called a central impact. Otherwise, the impact is said to be eccentric. If the velocities of the two bodies are directed along the line of impact, the impact is said to be a direct impact. If either or both bodies move along a line other than the line of impact, the impact is said to be an oblique impact. Direct central impact occurs if the mass centers of the two bodies are also along the line of impact. Direct eccentric impact occurs if the initial velocities are parallel to the normal to the striking surfaces but are not collinear. An oblique impact occurs if the initial velocities are not along the line of impact.

10.10.1 Straight Central Impact

For the straight central impact of two bodies of mass m_1 and m_2, the law of conservation of momentum gives

$$m_1v_1 + m_2v_2 = m_1V_1 + m_2V_2 \qquad (10.51)$$

where v_1 and v_2 are the velocities of the two bodies before impact, and V_1 and V_2 are the velocities of the bodies after impact, respectively (Figure 10.9).

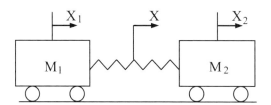

Figure 10.9 Central collisions.

In direct central impact of the two bodies, the coefficient of restitution is the ratio of the relative velocity of separation of the two bodies to their relative velocity of approach. That is,

$$e = -\frac{\left(V_2 - V_1\right)}{\left(v_2 - v_1\right)}$$ (10.52)

When the impact is oblique, the normal components of the velocities are used in Eq. 10.52.

In the case of an elastic impact $e = 1.0$, for a plastic impact $e = 0$, and both bodies remain in contact and have a common velocity V after impact. The coefficient of restitution for a particular case is determined from crash tests. The difference in velocities of the bodies before and after the collision indicates the change in kinetic energy into thermal energy, strain, or deformation energy, and the creation of sound. The change in kinetic energy without rotation is

$$E_A = \left(\frac{m_1}{2}\right)\left(v_1 - V_1\right)^2 + \left(\frac{m_2}{2}\right)\left(v_2 - V_2\right)^2$$ (10.53)

where E_A is the total energy dissipated during the approach period.

Equation 10.53 does not indicate how the crash energy is distributed to both bodies and vehicles. The distribution is a function of the deformation stiffness and the mass of each vehicle or body.

Using Newton's second law of motion and conservation of linear momentum, the equations for changes in velocities of vehicles 1 and 2 are given by

$$\Delta V_1' = \sqrt{\frac{2E_A m_2}{m_1\left(m_1 + m_2\right)}}$$ (10.54)

$$\Delta V_2' = \sqrt{\frac{2E_A m_1}{m_2\left(m_1 + m_2\right)}}$$ (10.55)

where

$\Delta V_1'$ = change in velocity of vehicle 1 during the approach period

$\Delta V_2'$ = change in velocity of vehicle 2 during the approach period

E_A = total energy dissipated during the approach period

m_1 = mass of vehicle 1

m_2 = mass of vehicle 2

The approach period referred to in the preceding discussion is the time between initial contact and the time when maximum crush occurs. The velocities of the two vehicles are assumed to be equal at the moment of maximum crush. Similarly, the separation period is the period between the maximum crush and complete separation of the vehicles. The total change in velocity is the change during the approach period added to the change during the separation period. The current model calculates only the approach velocity change. The total changes in velocity are given by

$$\Delta V_1 = \sqrt{\frac{2\left(1 + e^2\right)E_A m_2}{m_1\left(m_1 + m_2\right)}} \qquad (10.56)$$

$$\Delta V_2 = \sqrt{\frac{2\left(1 + e^2\right)E_A m_1}{m_2\left(m_1 + m_2\right)}} \qquad (10.57)$$

where

ΔV_1 = total change in velocity of vehicle 1

ΔV_2 = total change in velocity of vehicle 2

e = coefficient of restitution

10.10.2 Noncentral Collisions

In the more general case of noncentral collisions, a common velocity is assumed to be achieved at some point P in the region of contact (Figure 10.10). The changes in velocity at the center of gravity during the approach period are given by

$$\Delta V_1' = \sqrt{\frac{2E_A}{r_1 m_1\left[1 + \dfrac{r_1 m_1}{r_2 m_2}\right]}} \qquad (10.58)$$

$$\Delta V_2' = \sqrt{\frac{2E_A}{r_1 m_2\left[1 + \dfrac{r_2 m_2}{r_2 m_2}\right]}} \qquad (10.59)$$

$$r_1 = \frac{k_1^2}{k_1^2 + h_1^2}$$

(10.60)

$$r_2 = \frac{k_2^2}{k_2^2 + h_2^2}$$

where k_1 and k_2 are the radius of gyration of vehicles 1 and 2, respectively, and h_1 and h_2 are the moment arm of impact force of vehicles 1 and 2, respectively. h_1 and h_2 can be obtained from geometrical considerations and depend on the location of the centroid of the damaged area relative to the vehicle center of gravity, as well as the principal direction of force (PDOF).

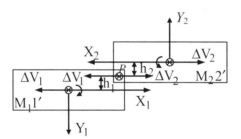

Figure 10.10 Noncentral collisions.

10.10.3 Crush Energy and ΔV

A number of crash tests were conducted to examine the crush behavior of vehicles. The crush energy E_A per unit width w of a crushed chassis can be expressed as a linear function of crush C as

$$\sqrt{\frac{2E_A}{w}} = d_0 + d_1C$$

(10.61)

where the residual crush C is measured normal (perpendicular) to and from the nominal undeformed surface. d_0 and d_1 are constants known as the crush stiffness coefficients, and they are obtained experimentally for a particular vehicle. They differ from vehicle to vehicle and among the front, side, and rear of each vehicle.

Prasad (1991) shows that if the crush profile is approximated by linear segments of crush between the six equally spaced points as shown in Figure 10.11, then the crush energy is obtained from

$$E_A = K_1 d_0^2 + K_2 d_0 d_1 + K_3 d_1^2$$

(10.62)

where

$$K_1 = \frac{w}{2}$$

$$K_2 = \frac{W\left[C_1 + 2\left(C_2 + C_3 + C_4 + C_5\right) + C_6\right]}{10}$$

and

$$K_3 = \frac{W\left[C_1^2 + 2\left(C_2^2 + C_3^2 + C_4^2 + C_5^2\right) + C_6^2 + C_1C_2 + C_2C_3 + C_4C_5 + C_5C_6\right]}{30} \quad (10.63)$$

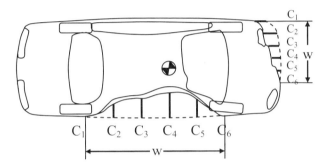

Figure 10.11 Linear segments of crush.

In these equations, the values of K_1, K_2, and K_3 are determined from the measured crush geometry, and d_0 and d_1 are vehicle crush parameters determined from a table of vehicle properties. This table is determined by crash test results or by estimating crash test performance using regression equations.

EXAMPLE E10.4

The following information is available for an accident involving two vehicles:

Weight of vehicle V-1 = w_1 = 1088.64 kg (2400 lb)

Weight of vehicle V-2 = w_2 = 2268 kg (5000 lb)

Travel direction at impact (vehicle V-1) is west

Travel direction at impact (vehicle V-2) is north

Travel direction after impact (vehicle V-1) is N 21° W

Travel direction after impact (vehicle V-2) is N 26° W

Post-impact travel distance of vehicle V-1 is d_1 = 15.85 m (52 ft)

Post-impact travel distance of vehicle V-2 is d_2 = 36.58 m (120 ft)

Post-impact drag factor of vehicle V-1 is f_1 = 0.28

Post-impact drag factor of vehicle V-2 is f_2 = 0.24

Travel angle of vehicle V-1 at impact is $\alpha_1 = 90°$

Travel angle of vehicle V-2 at impact is $\alpha_1 = 0°$

Travel angle of vehicle V-1 after impact is $\beta_1 = 21°$

Travel angle of vehicle V-2 after impact is $\beta_2 = 26°$

Determine:

a. The impact speed of vehicle V-1

b. The impact speed of vehicle V-2

c. The ΔV from impact of vehicle V-1

d. The ΔV from impact of vehicle V-2

Solution:

We will begin by considering that north is our $0°$ line. All angles measured counter-clockwise will be treated as positive.

	Direction	θ	$\sin \theta$	$\cos \theta$
α_1	West	$90°$	1	0
α_2	North	$0°$	0	1
β_1	N 21° W	$21°$	0.358	0.933
β_2	N 26° W	$26°$	0.438	0.898

To determine the weight ratios, let

$$w_1 = 1$$

Then

$$w_2 = \frac{2268}{1088.64} = 2.083$$

The post-impact speed of V-1 is

$$s_1 = \sqrt{254 d_1 f_1}$$

$$= \sqrt{254(15.85)(0.28)}$$

$$= 33.57 \text{ km/h} \ (20.87 \text{ mph})$$

The post-impact speed of V-2 is

$$s_2 = \sqrt{254 d_2 f_2}$$

$$= \sqrt{254(36.58)(0.24)}$$

$$= 47.22 \text{ km/h } (29.35 \text{ mph})$$

a. The law of conservation of momentum:

Let

$$S_1 = \text{speed of V-1 at impact}$$

$$S_2 = \text{speed of V-2 at impact}$$

Then

$$S_1 w_1 \sin \alpha_1 + S_2 w_2 \sin \alpha_2 = s_1 w_1 \sin \beta_1 + s_2 w_2 \sin \beta_2$$

$$S_1 = (1)(1) + S_2(2.083)(0)$$

$$= 33.57(1)(0.358) + 47.22(2.083)(0.438)$$

or

$$S_1 = 55.10 \text{ km/h } (34.24 \text{ mph})$$

b. Dissipation of energy equation:

$$S_1 w_1 \cos \alpha_1 + S_2 w_2 \cos \alpha_2 = s_1 w_1 \cos \beta_1 + s_2 w_2 \cos \beta_2$$

$$55.10(1)(0) + S_2(2.083)(1) = 33.57(1)(0.933) + 47.22(2.083)(0.898)$$

or

$$S_2(2.083) = 31.32 + 88.33$$

$$S_2 = 57.44 \text{ km/h } (35.70 \text{ mph})$$

c. ΔV determination:

V-1 speed/direction at impact = west @ 55.10 km/h

V-2 speed/direction after impact = N 21° W @ 33.57 km/h

Breaking the post-impact moment into components:

North component: $33.57 \sin 21° = 33.57(0.933) = 31.32$ km/h

West component: $33.57 \sin 21° = 33.57(0.358) = 12.02$ km/h

ΔV north: $0 + 31.32 = 31.32$ km/h (speed gain)

ΔV west: $12.02 - 55.10 = -43.08$ km/h (speed loss)

ΔV of vehicle V-1 $= \sqrt{31.32^2 + (-43.08)^2} = 53.26$ km/h $(33.10$ mph$)$

d.

Vehicle V-2 direction/speed at impact = north @ 57.44 km/h

Vehicle V-2 direction/speed after impact = N 26° W @ 47.22 km/h

Breaking the post-impact moment into components:

North component: $47.22 \sin 26° = 47.22(0.898) = 42.40$ km/h

West component: $47.22 \sin 26° = 47.22(0.438) = 20.68$ km/h

ΔV north: $42.40 - 57.44 = -15.04$ km/h (speed loss)

ΔV west: $20.68 - 0 = 20.68$ km/h (speed gain)

ΔV of vehicle V-2 $= \sqrt{(-15.04)^2 + (20.68)^2} = 25.57$ km/h $(15.89$ mph$)$

Check to be sure the ΔVs are in inverse proportion to the weights. Hence,

$$\frac{\text{Vehicle V-1} \, \Delta V}{\text{Vehicle V-2} \, \Delta V} = \frac{53.26}{25.57} = 2.083$$

$$\frac{w_2}{w_1} = \frac{2268}{1088.64} = 2.083$$

10.11 Vehicle–Pedestrian Collisions

Several mathematical and empirical models are available for the analysis of vehicle–pedestrian collisions. The main objective of the technical analysis is to determine the vehicle impact speed. In the real world, post-impact data from vehicle–pedestrian collisions often is insufficient to enable the use of pure mathematical models. However, mathematical models can be utilized successfully with the aid of some empirical data or reasonable assumptions. Conversely, the empirical models overcome the difficulty of insufficient data but must be applied within the model-defined limits.

An accident investigation can be divided into three areas: technical data, witness statements, and interpretation and analysis of the technical data.

Technical data. The first professionals to arrive at any serious traffic collisions are the emergency personal. Their main objective is to attend to the accident victims. However, in that process, they also contaminate technical evidence. Fortunately, a considerable amount of the physical evidence survives to facilitate scientific analysis.

In a real-world vehicle/pedestrian collision, the pedestrian trajectory data are limited. In fact, only the beginning and the end of the trajectory data often are available (i.e., only the point of impact and the final rest locations are known). Furthermore, in many cases, even the point of impact is difficult to establish due to lack of physical evidence. The pedestrian airborne distance (i.e., launch to initial landing distance) is another parameter that often cannot be evaluated from the physical evidence.

In some cases, secondary evidence may be available in the form of vehicle tire or skid marks. The skid marks can be used to determine the speed of the vehicle and to provide an independent method for checking the throw distance calculations. These real-world cases are special because they provide rare opportunity to either enhance or develop empirical solutions.

Witness statements: Witness statements can be a mixture of opinions and observations. Therefore, witness statements must be interpreted with a reasonable degree of caution. However, reliable witness observations of pedestrian movement prior to the collision can establish the pre-impact path of the pedestrian with some degree of accuracy.

Interpretation and analysis of the technical data: In cases where the physical evidence does not yield the point of impact, the point of impact can be assumed as the intersection of the pre-impact vehicle and pedestrian paths. The damage to the front end of the impacting vehicle can locate the point of impact within the width of the path of the vehicle.

Obviously, the total pedestrian throw distance is the distance between the point of impact and the pedestrian's center of gravity at rest.

10.11.1 Pedestrian Trajectories

During the motor vehicle accident, the pedestrian often is struck on the lower limbs by the leading edge of the vehicle. The initial contact frequently is followed by a more pronounced pedestrian–vehicle interaction. The nature of the subsequent contacts is dependent on the vehicle geometry and the pedestrian stature.

If the center of gravity (CG) of the pedestrian is lower than the upper edge of the leading face of the vehicle, then a forward projection trajectory ensues. In the forward projection trajectory, the upper body of the pedestrian may partially wrap over the hood of the vehicle; however, secondary contact between the pedestrian's head and the vehicle does not occur. If the pedestrian's center of gravity is propelled forward with the respect to the vehicle, then it is classified as a forward trajectory. Figure 10.12 illustrates a forward trajectory.

Figure 10.12 Example of a forward projection trajectory.

Conversely, if the pedestrian's center of gravity is above the upper leading edge of the vehicle, then the wrap trajectory occurs. For a wrap trajectory, the upper body will always wrap over the upper leading edge of the vehicle. After the initial impact, the pedestrian's center of gravity moves rearward with respect to the vehicle. If the pedestrian does not fall off to the side of the vehicle, then a secondary contact with the windshield or the upper surface of the hood takes place. After the secondary contact, the pedestrian's center of gravity will begin to move forward with respect to the vehicle. Figure 10.13 illustrates a wrap trajectory.

Figure 10.13 Example of a wrap trajectory.

During most pedestrian impacts, the impacting vehicle is decelerating prior to or immediately after impact. Upon impact, the pedestrian is accelerated rapidly to a speed that is nearly equal to or less than the impact speed of the vehicle. At the same time, the vehicle decelerates and continues decelerating, primarily due to braking. The vehicle and the pedestrian separate from each other when the pedestrian's speed exceeds the speed of the vehicle. After the separation, the pedestrian's trajectory will include a free flight through the air at some launch angle, an impact with the ground, and then sliding or tumbling to rest. If the vehicle is not braking, then the pedestrian is more likely to be carried on the vehicle for some distance or to vault over the roof of the vehicle.

If the pedestrian strikes near the corner of the vehicle and the pedestrian's center of gravity is above the upper leading edge of the vehicle, then after the initial contact, the pedestrian can simply slide off the fender. This trajectory is termed a fender vault. A fender vault also can occur if the pedestrian is walking quickly and the impact occurs just prior to the pedestrian clearing the path of the vehicle.

A vault trajectory occurs if the vehicle is traveling at sufficient velocity and is not decelerating. In this case, the pedestrian vaults over the vehicle and comes to rest behind the vehicle, primarily due to the fact that the vehicle does not stop. Analysis of vault trajectories must be conducted with extreme caution.

A list of variables is as follows:

$$d_{hood} = \text{height of the upper leading edge of the vehicle (m)}$$

$$h = \text{height of the pedestrian center of gravity (m)}$$

$$h^+ = \text{change in the height of the pedestrian's center of gravity (m)}$$

$$PE = \text{projection efficiency}$$

$$S = \text{pedestrian throw distance (m)}$$

$$V_v = \text{impact speed of the vehicle (m/s)}$$

$$v_p = \text{pedestrian throw speed (m/s)}$$

$$v_v = \text{impact speed of the vehicle (m/s)}$$

$$x_L = \text{pedestrian's displacement between initial and secondary contact (m)}$$

$$\Phi = \text{lateral displacement angle of the pedestrian (deg)}$$

μ = coefficient of friction between the pedestrian and the landing surface

μ_v = coefficient of friction between the vehicle tires and the road surface

θ = pedestrian's launch angle at impact (deg)

10.11.2 Mathematical and Hybrid Models

There are many published mathematical models (Toor, 2003). The traditional mathematical approach has been to treat the pedestrian as a projectile. Thus, either the projectile theory or some variation of the projectile theory is applied to develop mathematical models. There are simply too many published models to consider all models in this book. The mathematical models considered in this study were chosen for their user-friendliness. The user-friendliness of the models was assessed subjectively.

Collins (1979) took a simple approach by neglecting the pedestrian launch angle. Collins assumed that at impact, the pedestrian trajectory is horizontal in the direction of the impacting vehicle. The pedestrian's vertical acceleration is due only to gravity. Upon landing, the vertical velocity is exhausted, and the entire horizontal velocity is dissipated in the form of work done due to friction. This approach yielded the following quadratic equation incorporating the pedestrian's center of gravity height as

$$S = v_p \sqrt{\frac{2h}{g}} + \frac{v_p^2}{2\mu g} \tag{10.64}$$

Searle (1993) pointed out that the Collins method ignores any loss of horizontal speed due to impact with the ground on landing. The net result of this incomplete consideration is that Eq. 10.64 will slightly underpredict the average pedestrian launch speed. In addition, Collins indicates that a projection efficiency of 100% is applicable below impact speeds of about 50 km/h and references a pedestrian to road friction value of 0.8. Based on the data set available for this current study, both of these values probably are overestimates. Because the vehicle speed is inversely proportional to the projection efficiency but proportional to the frictional value, these two overestimates tend to cancel each other to some degree. Also, note that because the Collins model considers a horizontal throw angle and 100% projection efficiency, this suggests that the Collins model may be more appropriate for forward projection trajectories. If an appropriate friction value and a lower projection efficiency value are incorporated, then Eq. 10.64 also may be appropriate for wrap trajectories.

Searle considered an idealized particle and evaluated the following relationship for the particle launch speed as

$$v_p = \frac{\sqrt{2\mu g \left(S + \mu h^+\right)}}{\cos\theta + \mu\sin\theta} \tag{10.65}$$

where h^+ is the gain in elevation. For a pedestrian on level surface, this value will be negative.

Searle eliminated the throw angle term by further developing Eq. 10.65 to yield a minimum pedestrian impact speed. The minimum value is given by differentiating with respect to the throw angle and equating this term to zero (i.e., $\frac{\partial v}{\partial \theta} = 0$). Thus, the minimum pedestrian throw speed is given by

$$v_{p(min)} = \sqrt{\frac{2\mu g\left(S + \mu h^+\right)}{1 + \mu^2}} \qquad (10.66)$$

Searle reported that the pedestrian throw speed would be slightly lower than the ideal particle speed. To yield more accurate throw speed values, Searle suggests the incorporation of a correction factor that is dependent on the average coefficient of friction. Searle indicates that for u = 0.5 to 0.7, the correction factor is approximately –5%.

Because Eq. 10.65 yields the minimum throw speed, factors also must be incorporated to yield the average vehicle impact speed. The ratio of the pedestrian launch speed and the vehicle impact speed is defined as the projection efficiency. Searle indicates that the pedestrian throw speed is about 20% and 10% below the vehicle impact speed for adult and child pedestrians, respectively. The corresponding projection efficiency factors suggest that a wrap trajectory was referenced by Searle. Searle also indicates that the average vehicle impact speed is about 6% higher than the minimum.

Eubanks et al. (1994) published a three-phase model: pedestrian wrap on the hood, fall to the ground, and slide to rest. The Eubanks approach is similar to that of Collins as

$$S = \frac{v_p^2}{2\mu g S} + \left(\sqrt{\frac{2 d_{hood}}{g}} + \frac{d_{hood}}{v_p \sin \Phi}\right) \qquad (10.67)$$

If the hood height is substituted for the pedestrian's center of gravity height, and the pedestrian speed is considered to be negligible, then the Eubanks and Collins equations are equivalent.

Another theoretical model was presented by Han and Brach (2001). The Han-Brach model also considered distance traveled over the three phases of the pedestrian collision: the initial vehicle–pedestrian contact interaction, the pedestrian airborne trajectory, and the pedestrian slide/tumble/roll to rest. The Han-Brach model incorporates the launch angle, sliding friction, and the elevation gain or loss to determine the pedestrian launch speed. To assess the impact speed of the vehicle, the masses and a velocity ratio factor (to account for projection efficiency) are required. The airborne trajectory and the pedestrian slide/tumble/roll-to-rest components of the pedestrian impact are theoretically considered thoroughly; however, the initial impact phase and the projection efficiency factor must be explored further. For a road grade angle of zero, the Han-Brach model equations are

$$v_V = A_p \left[\frac{m_v + m_c}{\alpha m_c}\right] \sqrt{S - \left(x_L + \mu h\right)}$$

where

$$A_P = \sqrt{\frac{2\mu g}{\mu^2 \sin^2\theta + \mu \sin 2\theta + \cos^2\theta}}$$

Wood (1988), Wood and Simms (2000), and Wood and Walsh (2002) presented various models, both theoretical and empirical in origin. The Wood studies generally are thorough and comprehensive; however, the discussion of the Wood models will be limited for this study. The Wood formulae simplify considerably for single contact impacts that represent vault or forward projection trajectories. For wrap trajectories with secondary head contact, the formulae remain complex and require careful consideration of parameters.

Stcherbatcheff *et al.* (1975) also published a quadratic model similar to that of Collins; however, Stcherbatcheff *et al.* use as parameters the vehicle deceleration rate rather than the pedestrian/road friction and the vehicle impact speed rather than the pedestrian launch speed. Stcherbatcheff *et al.* also introduced an empirical coefficient based on test data. Their final equation took the form of

$$S = \frac{v_v^2}{2\left|\mu_{v(mean)}\right|} + 0.03\mu_{v(mean)}v_v \tag{10.68}$$

where $\mu_{v\,(mean)}$ is the mean vehicle deceleration.

The inclusion of the empirical coefficient may overcome the deficiencies of the Collins model. This model is useful only if the vehicle deceleration can be established with a reasonable degree of accuracy. In other words, unless some form of tire marks are present at the impact scene, the rate of vehicle deceleration cannot be established independently.

For a vehicle implementing full wheel lock-up on dry roads, an average deceleration rate of 0.7g may be appropriate. For a deceleration rate of 0.7g, the Stcherbatcheff *et al.* model may be simplified as

$$S = \frac{v_v^2}{1.4g} + 0.02\,gv_v \tag{10.69}$$

or

$$v_v = 6.81\sqrt{0.0424 + 0.291S} - 1.41 \tag{10.70}$$

Empirical models. In a real-world pedestrian collision, a human body is not an ideal projectile. The pedestrian's trajectory during and after the collision is complex. Many researchers have made excellent efforts to mathematically describe this behavior; however, the mathematical equations often are difficult to apply to real-world collisions. In other cases, additional assumptions must be made to solve these equations. The assumptions typically are based on empirical data.

Empirical models consider a black box approach, fitting a simple mathematical expression to the real data. These models incorporate many parameters necessary for solving mathematical equations into one or more constants. The outcome usually is a simple and practical mathematical expression. However, the disadvantage of the empirical models is that they typically model the average value of the considered data. If a limited sample size is considered, or if the data are skewed due to testing or recording techniques, then the empirical results may not be representative of real-world collisions.

Appel *et al.* (1975) derived the vehicle speed/throw distance relationship from the analysis of 137 real-world accidents. This study categorized separate equations for vehicle front-end geometry as low and high fronted, and the pedestrian size as adults and children. By definition, the leading edge of the hood of a low- and high-fronted vehicle is below the center of gravity of a 50th percentile adult; the leading edge of the hood of a blunt vehicle is above the center of gravity of a 50th percentile adult. Note that a child or an adult impacted by a blunt vehicle would likely produce a forward projection-type trajectory. The Appel paper published the following relationships:

$$\text{High-fronted vehicle: } S = 0.084v_v^2$$

$$\text{Low-fronted vehicle: } S = 0.065v_v^2$$

$$\text{Adults: } \qquad S = 0.070v_v^2$$

$$\text{Children: } \qquad S = 0.088v_v^2$$

(10.71)

Sturtz *et al.* (1976) reanalyzed the Appel data and defined children as persons aged 15 years or younger. They regrouped the data and found that the following cubic relationships were the best fit to the Appel data:

High-fronted vehicles:

$$\text{Children: } S = 0.6 + 0.76v_v + 0.0021v_v^2 \qquad (10.72)$$

$$\text{Adults: } \quad S = 1.0 + 0.61v_v + 0.0018v_v^2$$

Low-fronted vehicles:

$$\text{Children: } S = 0.6 + 0.0665_v^2 \qquad (10.73)$$

$$\text{Adults: } \quad S = 1.0 + 0.654v_v + 0.00145v_v^2$$

These equations can be solved easily for vehicle impact and speed in terms of throw distance.

Wood (1988), Wood and Simms (2000), and Wood and Walsh (2000) presented a hybrid model to yield minimum, maximum, and average vehicle impact speeds from pedestrian throw distance. Wood's equations are

$$v_{v(min)} = 2.5\sqrt{S}$$

$$v_{v(mean)} = 3.6\sqrt{S} \qquad (10.74)$$

$$v_{v(max)} = 4.5\sqrt{S}$$

Wood also has published numerous other excellent papers that extensively considered and contributed to the theory of vehicle–pedestrian collisions. The preceding empirical model was chosen for its ease of use.

Fugger and Randies (2000) present an empirical model that considers a power law relation between throw distance and vehicle impact speed. The Fugger model is applicable for wrap trajectories and is defined as

$$V_V = 8.3604\ S^{0.6046} \qquad (10.75)$$

where S is the throw distance measured in meters (m) and V_V is the vehicle impact speed in kilometers per hour (km/h).

Toor (2003) and Toor *et al.* (2000 and 2002) revised and validated an earlier empirical model. The Toor empirical model was based on an extensive data set of 359 collisions from a variety of sources. This model was separated according to wrap or forward projection trajectories. In addition, a combined trajectory model was presented for those cases where the trajectory is not known.

Forward projection trajectory model:

$$V_V = 11.3\sqrt{S} - 0.3 \qquad (10.76)$$

15th and 85th percentile prediction interval: ±10.5 km/h

5th and 95th percentile prediction interval: ±14.0 km/h

Wrap trajectory model:

$$V_V = 13.3\sqrt{S} - 4.6 \qquad (10.77)$$

15th and 85th percentile prediction interval: ±9 km/h

5th and 95th percentile prediction interval: ±12 km/h

Combined throw distance model:

$$V_V = 12.8\sqrt{S} - 3.6 \qquad (10.78)$$

15th and 85th percentile prediction interval: ±9.5 km/h

5th and 95th percentile prediction interval: ±12.9 km/h

The new model. The vehicle–pedestrian collision data used to formulate Eqs. 10.76, 10.77, and 10.79 were reevaluated to test the sensitivity of empirical models.

In the literature, a linear relationship between the square root of the throw distance and the impact speed was assumed. In this new model, a simple power law relationship has been considered. The following relationships were found to provide the best correlation.

Forward projection trajectory model:

$$V_V = 8.25 \, S^{0.61} \tag{10.79}$$

15th and 85th percentile prediction interval: ± 7.7 km/h

5th and 95th percentile prediction interval: ± 12.2 km/h

Wrap trajectory model:

$$V_V = 9.84 \, S^{0.57} \tag{10.80}$$

15th and 85th percentile prediction interval: ± 5.8 km/h

5th and 95th percentile prediction interval: ± 9.2 km/h

Combined (wrap and forward) throw distance model:

$$V_V = 9.19 \, S^{0.59} \tag{10.81}$$

15th and 85th percentile prediction interval: ± 6.5 km/h

5th and 95th percentile prediction interval: ± 10.2 km/h

To compare the theoretical models with the empirical models, the coefficient of friction for the pedestrian's post-impact trajectory is required. Published literature indicates that the coefficient of friction varies according to the pedestrian sliding on the ground or tumbling.

However, because these regions cannot be identified individually from real-world collision data, the most useful factor is the average coefficient of friction. Based on the published literature, the average friction value varies between 0.37 and 0.8.

Equations 10.76 through 10.78 were derived by performing linear regression analysis to the vehicle–pedestrian data. The aim was to achieve the best fit to calculate the vehicle impact speed. Using the same data but forcing the intercept through the origin yields

$$\begin{aligned}
\text{Forward trajectory:} \quad & V_V = 11.23\sqrt{S} \\
\text{Wrap trajectory:} \quad & V_V = 12.17\sqrt{S} \\
\text{Combined trajectory:} \quad & V_V = 11.90\sqrt{S}
\end{aligned} \tag{10.82}$$

where S is the throw distance measured in meters (m), and V_v is the vehicle impact speed in kilometers per hour (km/h).

Once the projection efficiency is incorporated to relate the pedestrian throw speed with the vehicle impact speed ($V_p = [PE]V_v$), these equations become directly comparable with the slide-to-stop equation of

$$v_p^2 = 2\mu gs \tag{10.83}$$

Correcting for units, the formula is expressed as

$$v_p^2 = [PE_{AVG}]^2 \begin{bmatrix} 3.12 \\ 3.38 \\ 3.31 \end{bmatrix}^2 S \tag{10.84}$$

$$\mu = [PE_{AVG}]^2 \begin{bmatrix} 0.50 \\ 0.58 \\ 0.56 \end{bmatrix} \tag{10.85}$$

Note that the average projection efficiency is different for wrap and forward projection trajectories. The average projection efficiency for forward projection, wrap, and combined trajectories can be defined as

$$PE_{AVG} = \begin{bmatrix} 0.95 \\ 0.80 \\ 0.85 \end{bmatrix} \tag{10.86}$$

Hence, the average effective coefficient of friction calculated from the collision data for forward projection, wrap, and combined trajectories is

$$\mu = \begin{bmatrix} 0.45 \\ 0.37 \\ 0.40 \end{bmatrix} \tag{10.87}$$

Thus, the average effective pedestrian coefficient of friction is calculated to be about 0.41; however, this is likely a lower bound based on the considered data due to its inclusion of the airborne portion of throw distance and the ground contact slide/tumble/roll distance.

The mathematical and empirical models predict vehicle impact speed with a reasonable degree of accuracy when each model is applied appropriately.

The advantage of empirical models is that they are simple to apply to real-world collision, particularly when the collision circumstances can be modeled by average or typical parameter values.

The mathematical models may provide useful insight when compared to empirical models and can provide trends if parameters vary significantly from the average or typical collision scenario values.

10.12 Accident Reconstruction Software

The success of the accident reconstruction methods is due largely to the availability of high-speed electronic digital computers. Several computer programs have been developed that are able to solve a wide variety of problems. There also are many accident reconstruction software programs, which were developed for the analysis of a variety of specific types of problems.

We present some examples of this software of general purpose and of special purpose in nature on accident reconstruction methods. Some of the special or key features of these programs, acronyms, contact details, and a few selected case studies are included.

10.12.1 Software Acronyms: REC-TEC with DRIVE³ and MSMAC^{RT}

REC-TEC key features. Angular Momentum * Finite Difference (Error) Analysis * AutoStats Lite (2005–4N6XPRT Systems) * Time/Distance * Acceleration-Deceleration Factor * Acceleration Deceleration (Swerve Comparison Option) * Multiple Surfaces (10) * Multiple Events (30) * Multiple Vehicles (2) * Motion Analysis (24 Phases) * V-TRAX (4-Vehicle) Animation * Collision Avoidance * Analysis * Braking * Passing * Turning * Angular Momentum * 360 Linear Momentum Fragmented (4) Vehicles * Spinout Trajectories * Vector Sum Analysis * Energy-Momentum Analysis * 360LM-PDOF Lite *Vector Momentum * Crush and Crash3 * Stiffness Coefficients * Energy * Break Fracture * Conservation * Reduced Mass * Kinetic * Fall/Vault * Optimum Angle * 3 Points on Arc * Pedestrian Vault * Vault-Slide Integration [minimum speed for throw distance] * Yaw/Critical Speed * Motorcycle Lean Angle * Photogrammetry * AASHTO (RR) Sight Triangle * Rotation Factor * Speed from RPM * Weight Shift [3 Axis] * Triangle Solver * Quadratic Solver * Commercial Vehicles * S-CAM Air/Air Disc Brake [48 Brakes] * Rollover (Static and Dynamic) * Maximum Off Tracking * Weight and Balance * LoadCheck [weights based on load for T/trailers or straight trucks and buses] * Animation * Simulation * Graphics * Sensitivity Tables * Saves and Loads Files * Imperial and/or Metric * Right- or Left-Hand Coordinate Systems * Selectable Decimal Precision * Selectable word processors, CAD/drawing programs, and Internet browsers * VideoTutor—More than 40 specialized training and help videos covering all aspects of the program are available from the website * Internet links to government and commercial sites * AutoStats Lite, Canadian Vehicle Specs plus Sisters and Clones * Free phone and e-mail support and (website) upgrades. Optimized for Windows™ 98, 98-2, 2000, ME, NT, and XP with seamless integration of Word/WordPad for creating reports.

DRIVE³ key features. The various sections in this program can be utilized for estimating a driver's response time. They present an accurate alternative to the rule-of-thumb estimation method that is without an empirical basis and, in fact, has been proven to be ineffective as an estimator of driver response times (DRT). The driver response time equations and adjustment-to-baseline methods of estimating response times are the only methods that have been based on empirical research and have estimated the response times of real-world drivers with statistical accuracy. However, the user must understand that this program is a tool, and as with any tool, it is only as good if it is utilized for the jobs for which it was created. Application of this or any driver response estimation method to completely dissimilar events will then show the predicted response times were derived will not likely produce an accurate estimate of response time. This program was not meant to predict the response time of any individual but to report how drivers have performed under various circumstances.

MSMAC^{RT} key features. This is the original SMAC program with many enhanced features, including the ability to overlay CAD drawings and bitmapped aerial photographs. The REC-TEC interface permits drag/drop positioning of vehicles and user-friendly data entry.

Contact information:

> George M. Bonnett, JD
> REC-TEC LLC
> P.O. Box 561031
> Rockledge, FL 32956-1031
> Phone: 1-321-639-7783
> Email: rec-tec@rec-tec.com
> Website: http://www.rec-tec.com

10.12.2 VCRware

VCRware—Vehicle Crash Reconstruction Software by Brach Engineering, LLC is a suite of computer programs based on Microsoft® Excel that provides the professional accident reconstructionist with the capability to reconstruct and analyze the vast majority of vehicular accidents. This software suite is unique in the industry, in that the theoretical basis for each program is covered in detail in the book *Vehicle Accident Analysis and Reconstruction Methods,* by Raymond M. Brach and R. Matthew Brach, published by SAE International in 2005.

The programs also benefit greatly from the availability of optimization utilities resident in Microsoft® Excel. These utilities include Goal Seek and Solver. Goal Seek allows the user to determine a value of an input parameter to achieve a specified or desired value of some spreadsheet output. For instance, the user can use Goal Seek with the Stopping Distance of a Vehicle spreadsheet to easily determine in one step the drag factor needed to stop a vehicle in 200 ft when traveling at an initial speed of 40 mph. Solver is similar in many ways to Goal Seek but can handle more complex problems such as maximization and minimization with multiple variables.

Where appropriate, the VCRware spreadsheets conveniently provide diagrams of the systems being analyzed. Calculations can be performed in either English or S.I. units. Printing of the results is handled using the Print command in the spreadsheet. Users also can copy and paste portions of some or all of the spreadsheets into other documents as a graphic for inclusion in technical reports and presentations.

In all, ten spreadsheets are provided. Interested users are encouraged to download the fully functional thirty-day trial version of the program for evaluation at www.brachengineering.com. Users wishing to purchase a copy of the software can do so conveniently and securely at www.brachengineering.com. The current pricing is available at the website. Orders will be shipped within ten working days of receipt of the order.

The following programs are included in the suite.

Critical speed formula. This program determines the (critical) speed of a single vehicle making yaw marks over a circular path following a sudden steer maneuver for a given tire–roadway frictional drag coefficient and measured curve coordinates.

Collision analysis for two point masses. This program uses point mass collision mechanics to calculate the initial speeds of two vehicles traveling in known directions and with known, straight-line post-impact paths over a flat surface with known frictional drag coefficients.

Planar impact mechanics analysis. For two vehicles with known physical parameters, orientations, collision centers, and initial velocity components, this spreadsheet uses planar impact mechanics to calculate the final velocity components, ΔV values, energy loss, principal direction of force, and more.

Low-speed vehicle impact analysis. This program enables analysis of a low-speed front-to-front or front-to-rear collision of two vehicles using point-mass collision mechanics, including the collision coefficient of restitution, contact duration, and vehicle roadway drag, when appropriate.

Crush stiffness coefficients from vehicle to barrier test. This program determines CRASH3 crush stiffness coefficients A, B, and G (and d_0 and d_1) using data from a vehicle-to-barrier test.

Energy loss and speed change from CRASH3. This program calculates the energy loss and ΔV values using the CRASH3 algorithm for a two-vehicle collision.

Vehicle–pedestrian collision analysis. This program calculates the times, velocities, pedestrian throw distance, and vehicle braking distance corresponding to the various events and sub-events of a vehicle–pedestrian (or vehicle–bicycle rider) collision.

Planar photogrammetric analysis. For a flat surface, this program calculates unknown locations of specific points on a photograph containing those points using the known locations of at least four calibration points from the site and contained on the same photograph.

Vehicle and semi-trailer dynamics program. This program calculates as a function of time the planar motion of a vehicle alone or a vehicle pulling a semi-trailer. Driver control modes include locked-wheel braking, lane change maneuver, or arbitrary steering. Individually locked wheels, drivetrain drag, and uniform acceleration can be simulated.

Stopping distance of a vehicle. This program calculates the stopping distance (reaction distance plus braking distance) from the speed or the speed from the distance.

Technical support for VCRware. While the software is quite intuitive to use, particularly to users familiar with Microsoft® Excel, technical support is available at the Brach Engineering website through a list of frequently asked questions. Support on other specific problems also is available through electronic mail at support@brachengineering.com.

Computer requirements. VCRware will run on a PC running Microsoft® Windows™ 98, 2000, or XP. The computer must be equipped with Microsoft® Excel. The program requires approximately 2 Mb of hard disk space for installation.

Contact information:

Brach Engineering, LLC
50515 Mercury Drive
Granger, IN 46530-8501
Phone: 574-273-8805
Fax: 574-272-8903
E-mail: matt_brach@brachengineering.com
Website: www.brachengineering.com

10.12.3 CRASHEX

Table 10.4 summarizes the key features of CRASHEX software.

TABLE 10.4
CRASHEX SOFTWARE KEY FEATURES

Acronym	Software Key Features	Contact Details
CRASHEX— Computerized Reconstruction of Approach Speeds on the Highway, Extended	• A classical planar conservation-of-momentum solution, refined by iterative inclusion of momentum loss due to (simulated) tire forces during impact. • Speed changes found by momentum are checked against those due to damage energy. • Animated plan view of vehicle motions and tire marks during approach, impact, and travel to rest. • The Method of Finite Differences is automated for disturbance of the inputs to a base case, one at a time, according to juried studies of likely errors of measurement, followed by statistical summation of the likely combined effects of all the likely errors of measurement. Upper and lower bounds of approach speed thus are found in a pervasive "what-if" study.	Albert G Fonda, P.E. Organization: Fonda Engineering Associates Address: 649 S. Henderson Rd. #C307 King of Prussia, PA 19406 Phone: 610-337-3311 E-mail: afonda7@crashex.com/ agfonda@gmail.com Website: www.crashex.com

Software case study: A.G. Fonda., CRASHEX: Computerized Reconstruction of Approach Speeds on the Highway Extended, 2006.

10.12.4 *ARSoftware*

Table 10.5 summarizes the ARSoftware product information.

TABLE 10.5
KEY FEATURES OF ARSOFTWARE

No.	Software Acronym	Software Key Features
1.	AITools WinSMAC	This is a prediction-simulation program based on the original SMAC model. Users can test assumptions and validate solutions obtained by traditional reconstruction techniques.
2.	AITools WinCRASH	WinCRASH is a Windows™ version of the CRASH III model. With scene data and damage patterns for the vehicles involved in a collision, a linear momentum and damage momentum solution can be calculated for the impact speeds of the vehicles. If only damage data are available, collision velocity changes can be evaluated.
3.	AITools Linear Momentum	This program was developed to provide the classical solution to the two-vehicle collision problem using conservation of linear momentum. The solution includes the velocity changes from the collision and a vector diagram.
4.	AITools Equations	This program provides solutions to formulae most commonly used in collision reconstruction. These include kinematic formulae for constant acceleration that relate speed, time, and distance. Problems involving airborne vehicles, critical cornering speeds, linear momentum, and total stopping distance also are addressed. Tables of results can be generated quickly and easily.
5.	AITools Brake Efficiency	This program analyzes the overall braking efficiency of vehicles with S-Cam airbrake systems. The model relies on the pioneer work of Ron Heusser but adapts and improves it for the computer.

Contact information:

Timothy A. Moebes, B.A.Sc, P.E.
ARSoftware
21108 77th PI W #104
Edmonds, WA 98026
Phone: 425-861-4666
Fax: 877-334-0905
E-mail: info@arsoftware.com
Website: arsoftware.com

10.12.5 Engineering Dynamics Corporation

Table 10.6 summarizes the various software acronyms and descriptions of the Engineering Dynamics Corporation (EDC).

TABLE 10.6
ENGINEERING DYNAMICS CORPORATION SOFTWARE DESCRIPTION

Topic	Acronym	Description
Vehicle Simulation	SIMON	SIMON (Simulation Model Non-Linear)—HVE-compatible 3-D dynamic simulation of the response of one or more unit vehicles to driver inputs and factors related to the environment. Takes advantage of HVE features, including HVE Brake Designer, Driver Model, and Tire Blow-Out Model.
	SIMON—articulation option	Articulated Vehicle option in SIMON to allow the study of articulated multiple-vehicle trains of virtually any vehicle/trailer configuration, including B-connections between trailers.
	SIMON—collision option	DyMESH (Dynamic MEchanical SHell) 3-D collision model for use in SIMON to allow the study of inter-vehicle collision(s). (U.S. Patent 6,195,625)
	EDVSM	EDVSM (Engineering Dynamics Vehicle Simulation Model)—HVE-compatible 3-D simulation analysis of the response of a unit vehicle to a driver's inputs. Typically used to study vehicle handling and vehicle rollover.
	EDVDS	EDVDS (Engineering Dynamics Vehicle Dynamics Simulator)—HVE-compatible 3-D simulation analysis of the dynamic response of a vehicle towing up to three trailers. Typically used to study heavy trucks or combination vehicles. (The first connection must be a fifth-wheel/kingpin connection, so the program cannot be used to model a car towing a utility trailer.)
	EDSMAC4	EDSMAC4 (Engineering Dynamics Simulation Model of Automobile Collisions)—HVE-compatible 2.5-D simulation analysis for studying complex crashes including simultaneous multiple-vehicle collisions, articulated vehicle crashes, and vehicle-to-barrier crashes.
	EDSVS	EDSVS (Engineering Dynamics Single Vehicle Simulation)—HVE-compatible 2.5-D simulation analysis of a four-wheeled automobile or truck having tandem axles and dual tires. It typically is used to study single vehicle loss of control.
	EDVTS	EDVTS (Engineering Dynamics Vehicle-Trailer System)—HVE-compatible 2.5-D simulation analysis of an articulated vehicle (vehicle towing single trailer). It typically is used to study simple articulated vehicle loss of control.

TABLE 10.6 *(Cont.)*

Vehicle Simulation *(cont.)*	EDGEN	EDGEN (Engineering Dynamics Corporation GENeral Analysis Tool)—HVE-compatible 3-D kinematics spreadsheet. The program typically is used to perform time-versus-distance studies, to move a human or vehicle between two or more known positions, or to move a camera car.
Accident Reconstruction	EDCRASH	EDCRASH (Engineering Dynamics Corporation Reconstruction of Accident Speeds on the Highway)—HVE-compatible analysis used to reconstruct single- and two-vehicle crashes. Determines conditions of impact using information obtained from vehicle (crush) and accident site.
Human Simulation	EDHIS	EDHIS (Engineering Dynamics Human Impact Simulator)—HVE-compatible 3-D analysis of the response of a human occupant during a crash. Typically used to study restraint system effectiveness. The human is represented by three inertial segments (head, torso, and lower extremities).
	GATB*	GATB (Graphical Articulated Total Body)—HVE-compatible 3-D human simulation. This is useful for studying complex occupant/pedestrian motions resulting from vehicle crashes or rollovers, including human-to-human contacts. Human is represented by 15 inertial segments. Multiple humans may be involved in a GATB simulation.

* Developed, distributed, and supported by Collision Engineering Associates, Inc.

HVE software case studies. Several white papers on the web page are excellent references and can be found at http://www.edccorp.com/library/whitepaper.html.

A few of them are listed here as follows:

WP-2003-2   "Downhill Commercial Vehicle Simulations—Part B (Intercity Bus Equipped with an Engine Data Recorder)"

WP-2003-3   "Reconstruction of Real World Side Impact Vehicle Collisions Using HVE—A Case Series of Pediatric Pelvic Fracture"

WP-2003-4   "Investigating the Use of Simulation Model Non-Linear (SIMON) for the Virtual Testing of Road Humps"

WP-2005-1   "HVE Data Inputs Based on Testing for a Wet Pavement Accident Involving an Intercity Bus and an SUV"

WP-2005-3   "SIMON and EDVDS Validation Study: Steady State and Transient Handling"

Contact information:

Engineering Dynamics Corporation
8625 S.W. Cascade Avenue, Suite 200
Beaverton, OR 97008
Phone: 1-503-644-4500
Fax: 1-503-526-0905
E-mail: info@edccorp.com
Website: www.edccorp.com

10.12.6 *Macinnis Engineering Associates (MEA) and MEA Forensic Engineers & Scientists*

PC Crash™ is a Windows™ collision and trajectory simulation tool that enables the accurate analysis of a variety of motor vehicle collisions and other incidents. Results are viewed as 3-D animations and as detailed reports, tables, and graphs.

PC Crash™ has an innovative collision model that efficiently balances simplicity and accuracy in reconstructing vehicle collisions. The collision optimizers find the best solution in a short time.

PC Rect™ converts oblique scene photographs pixel by pixel into scaled plan views. From the plan view, users then can measure in-plane distances and angles of accident scene evidence.

Features of PC Crash include the following:

- The ability to import 2-D or 3-D scenes or to create them within

- Collision Optimizer algorithms that reconstruct collisions using scene data and/or crush energy.

- Analysis of rollovers and vaults

- Steering with driver input sequences defined paths, or independently on wheels bent by damage

- Simultaneous simulation of up to 32 vehicles and objects

- 2-D or 3-D collision model based on conservation of linear and angular momentum, restitution, and contact friction (for sliding impacts)

- Collision and trajectory analysis of occupants, two-wheeled vehicles, and other objects

- Variable suspension and tire parameters

- Anti-lock braking system (ABS) and normal braking (with front–rear distribution valve for 3-D vehicles), and individual wheel braking

- Built-in 3-D animation with a fixed or moving camera

Validation studies include the following:

- Automatic Optimization of Pre-Impact Parameters Using Post-Impact Trajectories and Rest Positions

- The Trailer Simulation Model of PC Crash

- The Pedestrian Model in PC Crash™—The Introduction of Multi-Body System and Its Validation

- A New Approach to Occupant Simulation Through the Coupling of PC Crash™ and MADYMO®

- Validation of the Coupled PC Crash™—MADYMO® Occupant Simulation Model

- Validation of PC Crash™ Pedestrian Model:

  1. Data from Five Staged Car-to-Car Collisions and Comparison with Simulations.

  2. Reconstruction of Twenty Staged Collisions with PC Crash's™ Optimizer Application of Monte Carlo Methods for Stability Analysis within the Accident Reconstruction Software PC Crash™.

  3. An Evaluation of Rectified Bitmap 2D Photogrammetry with PC Rect™.

Sample animations, screen shots, software updates, software demos, and validation studies are available on the website.

Contact information:

> Macinnis Engineering Associates (MEA)
> Forensic Engineers & Scientists
> 23281 Vista Grande Dr., Suite A
> Laguna Hills, CA 92653
> Phone: 949-855-4632 or 1-877-855-5322
> Fax: 949-855-3340
> Website: http://www.maceng.com

10.12.7 Maine Computer Group

Accident Reconstruction Professional 7 (AR Pro). This program has been well endowed with formulae, options, and capabilities. It has some other formulae for truck braking efficiency, a multiple departing section 360° momentum formula (with up to eight departing sections), photogrammetry, and low-speed formulas.

AR Pro has a built-in Vehicle Database that includes virtually every car and light truck on the road since 1971. The database has information such as curb weights, overall length, overall height, and overall width.

Recall 7. Recall 7 is the entire National Highway Traffic Safety Administration (NHTSA) Recall Database. The database contains recalls from 1949 through today and covers items such as cars, trucks, motorcycles, semi-tractors, semi-trailers, motor homes, pop-up campers, boat trailers, utility trailers, helmets, child safety seats, tires, passenger buses, school buses, travel trailers, and fifth wheels.

TSB 7. The database contains TSBs from 1952 through today and covers items such as cars, trucks, motorcycles, semi-tractors, motor homes, passenger buses, and school buses. The program creates professional reports that can be printed and can include a single TSB, all TSBs for a particular make and model for a specific year, all TSBs for a particular make and model for any year, and all TSBs for a particular make and model from one specific year to another.

Contact information:

Maine Computer Group
117 Central Street
Hallowell, ME 04347
Phone: 207-623-1995
Fax: 207-623-0913
Website: http://www.mainecomputergroup.com
E-mail: mcg@mainecomputergroup.com

10.12.8 McHenry Software, Inc.

m-smac. Simulation Model of Automobile Collisions (SMAC) is a time-domain mathematical model in which the vehicles are represented by differential equations derived from Newtonian mechanics combined with empirical relationships for some components (e.g., crush properties, tires) that are solved for successive time increments by digital integration. Each vehicle is limited to the three degrees of freedom associated with the plane motion (i.e., two translations, one rotation). The tire forces are modeled by a nondimensional side force function, and the friction circle concept is included for the interaction between the side and circumferential tire forces. The collision force simulation is achieved by means of the modeling of each vehicle as a rigid mass surrounded by an isotropic, homogeneous periphery that exhibits elastic plastic behavior.

The SMAC computer model is an open form of reconstruction procedure wherein the user specifies the dimensional, inertial, crush, and tire properties of the vehicles, the initial speeds, the angles, and the driver-control inputs. The program, through stepwise integration of the equations of motion, produces detailed time histories of the vehicle trajectories, including the collision responses. The user compares the SMAC-predicted trajectories and collision deformations with the physical evidence to determine the degree of correlation. Iterative runs then can be performed, varying initial speeds, heading angles, and control inputs until an acceptable match of the physical evidence is achieved.

McHenry Software, Inc. (MSI) has ported the m-smac program to run on a PC in native 32-bit code. The company also has integrated the management of m-smac projects, including the creation, editing, submission of runs, and evaluation of output (including animation) into the m-edit environment, which runs in the Microsoft® Windows™ environment.

m-hvosm. This program, m-hvosm, stands for McHenry Highway Vehicle Object Simulation Model. The HVOSM mathematical model consists of up to 15 degrees of freedom: six for the sprung masses, and up to nine for the unsprung masses. The model is based on the fundamental laws of physics (i.e., Newtonian dynamics of rigid bodies), combined with empirical relationships derived from experimental test data (i.e., tire and suspension characteristics, load deflection properties of the vehicle structure). The balance of forces occurring within and applied to components of the system is defined in the form of a set of differential equations that constitute the mathematical model of the system. The HVOSM includes the general three-dimensional motions resulting from vehicle control inputs, traversals of terrain irregularities, and collisions with certain types of roadside obstacles.

In addition to the substantial effort in validation of the mathematical model, the HVOSM also was tested uniquely by designing an automobile stunt that was used both in a traveling auto-stunt thrill show and in the 1974 James Bond movie, *Man with the Golden Gun*, produced by United Artists Corporation.

M-crash. M-crash is an acronym for McHenry CRASH Reconstruction of Accident Speeds on the Highway. The program constitutes a major extension and refinement of the CRASH3 program (e.g., Brach, 2003 and 2004; Campbell, 1974; Casteel and Moss, 1999; Cheng *et al.*, 1987; Collins, 1979; Cooperrider *et al.*, 1990; Dailey, 1988; Day and Garvey, 2000; Dickerson, 1995; and Dixon, 1996), with the attained objective of significantly improving the reliability and accuracy of results.

The general approach to trajectory analysis in the CRASH programs (i.e., CRASH2, CRASH3, EDCRASH) initially has been to approximate the individual vehicle speeds at separation by means of work-energy relationships applied to the distances and characteristics of the travel of each vehicle from impact to rest. The calculated separation speeds then are included in an application of the principle of conservation of linear momentum to the collision to approximate the vehicle impact speeds.

The approach to damage analysis for the CRASH programs has been to approximate the total amount of energy absorption (to the point of common velocity) on the basis of empirical crush resistance properties. A relationship for the virtual crush properties of the vehicle(s) are derived from full-scale crash tests, from which the energy dissipated (excluding restitution) is determined as a function of the residual crush. (The term "virtual" is used to emphasize the fact that the crush energy is dissipated during the dynamic crushing of the vehicles and that equating the residual [restituted] crush to the energy dissipated is a virtual relationship, that is, they do not occur simultaneously.) The virtual crush properties then are applied to measured dimensions of structural crush for each vehicle in an individual accident to approximate the resulting speed change (excluding restitution effects). The inherent limitations of linearized virtual crush properties combined with limited crash test data for individual makes, models, years, and speeds limits the usefulness of the damage only to solutions to individual accident reconstructions. (Please see a brief review of the CRASH damage analysis algorithm for additional information.)

The published error levels associated with various versions of CRASH programs must be viewed in terms of the intended applications. In large-scale statistical studies (e.g., NCSS, NASS, FARS), the average error is the important consideration. However, in an application to an individual accident reconstruction, the maximum probable error in the given application must be defined as a part of any related conclusions. The widespread distribution of the EDCRASH program unfortunately has led to many inappropriate applications that frequently are accompanied by overstated claims of reliability and accuracy. Note that the EDCRASH validation calculation of percent error for the RICSAC test runs on the basis of the combined impact speed that does not appear to serve any useful purpose, other than yielding smaller percentages.

m-edit. The McHenry-Edit Environment (m-edit) user interface is an integrated analysis environment (IAE) that runs with the Microsoft® Windows™ operating system.

It integrates the following components into a cohesive collection that provides for complete accident reconstruction analysis, including:

- Field-sensitive input editor for the creation of m-smac and m-hvosm inputs, which includes several important additional auxiliary calculation routines.

- Input and output units may be specified in either English (smac original), metric, or EDSMAC compatible units.

- Execution of McHenry Software, Inc. simulation programs directly from the m-edit environment, where m-smac runs in native 32-bit PharLap DOS-Extended code for optimal speed and performance for Windows™ 3.1. For Windows™ 95 and NT 4.0, the simulation programs run in native 32-bit.

- Provisions for point-and-click commands for screen display and HP LaserJet printing of graphical outputs, both static (m-smacgr) and animated (m-smacan, m-hvgr).

- Provisions for point-and-click commands for review and printing of all tabular output data, including damage tables and the time history of vehicle positions, velocities, and accelerations for m-smac and all output datasets for vehicle positions, velocities, angles, suspension movements, and so forth for m-hvosm.

- Project files that organize with point-and-click control all input files related to a project by user-specified project names.

- Simplified organization of all output files for m-smac and m-hvosm runs related to an input file.

Other integrated analysis environment functions include the following:

- Informative status bars

- Pop-up forms for editing m-smac and m-hvosm inputs, which provide additional help on inputs.

- Optional in-place editing of m-smac and m-hvosm inputs for added convenience.

- Text files editor/viewer for viewing and editing any text file.

- Simple menu choices for utilizing other accident reconstruction tools by McHenry Software, Inc., including the following:

  - Customized Windows-based NHTSA vehicle specifications, NHTSA crash test, and Canadian SPECS database browsers

  - Launch, an occupant trajectory speed calculation routine

  - Marquard, a vehicle spinout speed calculation routine

- Simple procedures for the addition of user-defined tools to the menu to permit a customized environment to suit individual needs and preferences.

McHenry software demos and animations are available on the company website at http://www.mchenrysoftware2.com.

Contact information:

> McHenry Software, Inc.
> P.O. Box 5694
> Cary, NC 27512
> Phone: 919-468-9266
> Website: www.mchenrysoftware.com
> e-mail: McHenry@mchenrysoftware.com

The use of accident reconstruction software also can provide a learning experience because in the normal process of investigating and analyzing a crash, the investigator typically gains a great amount of knowledge about accident reconstruction. As a direct result, the investigator's ability to communicate his findings to others is vastly improved. The end result is a better understanding of what causes accidents, which is the primary reason for accident reconstruction.

10.12.9 Software Acronym: VDANL

VDANL is the acronym for Vehicle Dynamics and Analysis Program. Dynamics Simulation Model VDANL (Vehicle Dynamics Analysis, Nonlinear) is a comprehensive vehicle dynamics simulation program that runs under Windows™ 95, 98, NT, or 2000 and is intended for the analysis of passenger cars, light trucks, articulated vehicles, and multipurpose vehicles. The simulation model is designed to permit analysis of virtually all driver-induced maneuvering up through limit performance conditions defined by tire saturation characteristics, and it includes driver feedback control features.

The vehicle model. Model equations cover the full range of lateral/directional and longitudinal motions up through large angles experienced in spinout and rollover. The vehicle model includes components for sprung and unsprung masses, suspension, steering, braking, powertrain, drivetrain, and tires. The model includes a comprehensive tire model and properly accounts for the effects of maneuver-induced load transfer. The vehicle and tire models are based on past research and have been validated extensively.

Tires. Tire characteristics play a dramatic role in vehicle dynamics because they respond to vehicle maneuvering. The tire model generates lateral and longitudinal tire forces and aligning moments as functions of normal load, slip, and camber angle and includes appropriate interactions among these input variables, including force saturation.

Suspension. Composite suspension characteristics are designed to represent wheel steer and camber motions relative to the sprung mass and squat/lift forces resulting from tire ground plane forces acting on the suspension geometry. Wheel steer also arises from compliance in response to tire side force and aligning torque.

Steering. The steering model includes Ackerman steer effects and compliance, and a composite second-order characteristic to simulate steering dynamics in response to steering and aligning torque inputs.

Powertrain and drivetrain. The powertrain and drivetrain model includes the engine, transmission, differentials, and torque splitting between the front and rear axles. Front-, rear-, and four-wheel drive can be accommodated.

Brakes. The brake model includes simulation of vacuum boost runout and a nonlinear proportioning valve between the front and rear axles. It also includes a generic anti-lock braking system. Truck brakes include trailer pneumatic lag and fade due to overheating.

Capabilities and features. VDANL can be used in various ways to analyze vehicle maneuvering motions, handling, and stability. Input control commands (e.g., steering, braking, throttle, aerodynamics) can be applied directly to the open-loop vehicle. Under driver model closed-loop control, the simulation can be excited with road curvature, lane position, and speed commands. A large number of vehicle input and response variables can be saved to a file for subsequent analysis and/or plotted using time history and XY formats. An external Windows™-based plotting utility (WinEP by Mechanical Simulation Corporation, Ann Arbor, MI) is provided for data display and analysis.

Program control is achieved through a Windows™ point-and-click interface. The program also can be controlled with command files containing sequences of program commands. Built-in help features provide online display of the user's guide and technical reference.

Applications include the following:

- Dynamic response and stability

- Low-g and limit performance handling

- Traction control and braking

- Suspension design and load transfer

- Closed-loop driver/vehicle response problems

- Accident reconstruction

- Trailer towing

Additional VDANL features include the following:

- Common vehicle dynamics with the STISIM Drive™ driving simulator. Vehicle models can be developed using VDANL and then run in the real-time, driver-in-the-loop simulator.

- A 3-D terrain model allows a multi-lane roadway to be defined using simple commands. Horizontal and vertical curves can be defined with roadway cross-slopes. Surface attributes can be assigned to portions of the roadway, which are recognized by the vehicle tire model, allowing changing tire/roadway characteristics based on the position on the roadway. This 3-D terrain is common with STISIM Drive™.

- An Open Module option allows users to write code to be integrated into the VDANL simulation, enabling custom simulations to be created. Hardware-in-the-loop simulation is possible using the Open Module and STISIM Drive™.

- MATLAB$^®$ binary file output allows post-processing and plotting within MATLAB.

Hardware requirements. VDANL is designed to run on a PC with an Intel$^®$ Pentium$^®$ class processor or is compatible with Windows™ 95, 98, or NT/2000.

Systems Technology, Inc. (STI) has been involved in research and consulting in vehicle dynamics, manual and automatic controls, and human factors of aerospace and ground vehicles for more than 35 years. Specific experience with the analysis and testing of ground vehicles has resulted in the development of the vehicle model that is the basis for VDANL. Over the last two decades, the company's ground vehicle research has focused on the development and validation of comprehensive vehicle dynamics simulations and the development of enhancements to maximize the efficiency of VDANL applications.

Further information on VDANL applications has been documented in the literature, and a demonstration version of the program is available. Background information can be obtained by contacting STI at http://www.systemstech.com/content/view/32/39.

10.12.10 Expert AutoStats®—Vehicle Dimension-Weight-Performance Data

Expert AutoStats® (http://www.4n6xprt.com/4n6as.htm) is a program that includes more than 35,000 cars, pickup trucks, vans, and utility vehicles that range in years from the 1940s to the present. Expert AutoStats® has specifications that can assist in reconstructing accidents when the data for the vehicle are unavailable or if the vehicle is too severely damaged to obtain correct measurements.

For many vehicles from the mid-1960s to the present, data such as bumper height, front and rear overhang, hood height, and so forth also are included.

As of April 1995, the 4N6XPRT Systems® programs Expert AutoStats®, Expert Qwic Calcs®, Expert TireStuf®, and Expert VIN DeCoder® are accessible from within RECTEC.

System requirements. Expert AutoStats® has been tested on a variety of IBM laptop and desktop clones ranging from 8088 through Pentium® chips. A math coprocessor chip is not required. Expert AutoStats® also has been tested under the various versions of MS-DOS 3.0 through 7.0, DrDOS 6.0, and PC DOS 7.0. It also works as a DOS program under Windows™ 3.x, 95, 98, NT, Me, 2000, XP, OS/2 2.x, and OS/2 Warp, and various versions of LINUX.

A variety of dot matrix printers emulating the EPSON series have been used with no difficulty. The output also is compatible with Hewlett-Packard II, IIP, and III printers and Hewlett-Packard Desk Jet inkjet printers. Expert AutoStats® works with monochrome and color monitors.

4N6XPRT StifCalcs. This program (http://www.4n6xprt.com/4n6sc.htm) puts the NHTSA Crash Test database on your computer and calculates stiffness values from the tests when sufficient data are available to perform the calculations. The company also has several other programs offered for sale. See the company website (www.4n6xprt.com) for descriptions.

10.12.11 Other Accident Reconstruction Software Sites

The following are other useful sites:

- http://www.collisionrecon.com/products/products.html offers some Excel spreadsheet and TI calculator calculation programs.

- http://www.applied-kinematics.com/ak/products.asp offers a program called AI Momentum, which allows the user to set ranges of values for particular momentum calculations.

- http://www.visualstatement.com/products/ offers a visual statement software that is categorized as diagramming software but also is calculation software based on the diagram that the user builds.

- http://www.rudydegger.com/wecare-products/accident-reconstruction-calculator.asp offers a basic calculation software package.

Contact information:

4N6XPRT Systems
8387 University Avenue
La Mesa, CA 91941-3842
Phone: 619-464-3478 or 1-800-266-9778
Fax: 619-464-2206
E-mail: dv3@4n6xprt.com
Website: http://www.4n6xprt.com/

10.13 Low-Speed Sideswipe Collisions (Toor *et al.*, 2000)

Sideswipe collisions are characterized by prolonged sliding contact and with very little structural deformation. A sideswipe collision is considered as a collision in which the angle of impact is shallow and the vehicle interaction in the longitudinal direction is frictional. Sideswipes are characterized by prolonged sliding between the contact surfaces.

The dynamics of sideswipe collisions was described by Bailey *et al.* (1995), who conducted eleven shallow-angle (<30°) vehicle-to-vehicle sideswipe tests at approach speeds varying from 4.5 to 27 km/h. In some tests, vehicle sliding contact involved snagging the two interacting surfaces at points in which the contact surface geometry was irregular. Snagging was recognized as a different contact mechanism from sliding. Damage from snagging involved forward or rearward deformation of a structure. Levy (2000) presented a method for determining impact velocity in sideswipes in which a cycloidal truck tire mark has been imprinted on the side of the struck vehicle. This method is applicable to a small percentage of sideswipe collisions. Woolley (1987) modeled sideswipes in the IMPAC accident reconstruction computer program as frictional contact along a defined slip plane. SMAC (McHenry and McHenry, 1997) and CRASHEX (Fonda, 2000) also incorporate frictional inter-vehicular sliding forces. All these models compute the inter-vehicular forces from stiffness coefficients based on vehicle crush profiles measured after crash tests. These stiffness coefficients may not be accurate when modeling sideswipe collisions with little or no crash. No crash test data have been presented to evaluate the accuracy of these models in analyzing collisions that are dominated by inter-vehicular sliding rather than crush.

Toor *et al.* (2000) calculated lateral inter-vehicular force based on damage using vehicle side stiffness coefficients. Longitudinal force then was calculated assuming an inter-vehicular sliding coefficient of friction of 0.6. Average accelerations were calculated based on these forces with additional analysis to determine regions where tire slippage occurred. Snagging was not considered, the vehicle was assumed to be a rigid body, all tires were assumed to be steered straight and free rolling, and inward crush and yaw rotations were assumed to be small. They reported on 14 vehicle-to-barrier tests and four vehicle-to-vehicle tests used to validate the analytical method. The vehicle-to-barrier tests were conducted by steering the surface vehicle so that its side would contact a trunk lid mounted to a barrier. Similarly, in the vehicle-to-vehicle tests, the surface vehicle was steered so that its side would contact the bumper of the contact vehicle, which also was fixed to a barrier. Approach speeds varied from 4 to 10 km/h. Their method was reported to have estimated accurately the speed changes in their vehicle tests.

The method provides a useful starting point for evaluating sideswipe collision severity. The force calculation relies entirely on stiffness coefficients that are derived primarily from high-speed crash tests with substantial crush. There is little or no data to validate the accuracy of the force calculation at the intended low levels of crush (0 to 4 cm). The method does not take into account the approach angle between the vehicles. The approach angle strongly influences the length of sliding contact damage, which in turn strongly influences the calculated velocity change and acceleration of the vehicles.

10.13.1 Funk-Cormier-Bain Model (Funk et al., 2004)

This model is similar to SMAC (McHenry and McHenry, 1997), in that it calculates vehicle kinematics utilizing a forward timewise integration of the equations of rigid body motion. The model includes a novel non-crush-based method for calculating inter-vehicular force. An impact configuration is assumed in which the front corner of a bullet vehicle strikes the side of a stationary target vehicle at a given closing velocity and approach angle (Figure 10.14). The bullet and target vehicles were modeled as rectangular rigid bodies having the dimensions, masses, and inertial properties of the subject vehicles.

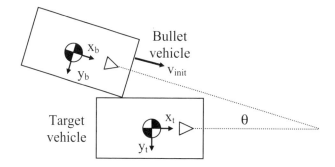

Figure 10.14 Schematic of a vehicle test setup.

The model was based on the following assumptions and limitations:

a. The combined lateral compliance of both vehicles due to bumper and door panel deformation, suspension, tire sidewall stiffness, and so forth was lumped into a single force-deflection relationship.

b. Longitudinal interaction between the sliding surfaces was assumed to be frictional. The model may not be valid when snagging or substantial inward crush occurs because this may cause the longitudinal force to exceed the force expected from pure friction.

c. All tires on both vehicles were assumed to be steered straight and free rolling with lateral slippage governed by Coulomb friction.

An inertial coordinate frame was established to track the positions of both vehicles. The inertial frame was aligned with the initial position of the target vehicle frame such that its origin was coincident with the center of gravity (CG) of the target vehicle. The bullet vehicle was assigned an initial position based on its relative approach speed and angle such that its trajectory would cause the bumper of the bullet vehicle to strike the side of the target vehicle at a predetermined point of initial contact. Equations are presented assuming the right front corner of the forward-moving bullet vehicle strikes the left side of the initially stationary target vehicle. Vehicle positions were tracked with respect to an inertial coordinate frame. Vehicle forces, velocities, and accelerations were tracked in the local vehicle coordinate frame originating at the center of gravity of each vehicle (Figure 10.14). Vehicle velocities were coordinate transformed to the inertial frame before being integrated to obtain positions.

10.13.2 Modeling Procedure

Modeling of the vehicle dynamics was performed by solving the equations of motion at a given point in time and successively incrementing in time to obtain results over a desired time duration. Because inter-vehicular and tire slip conditions could change the governing equations of motion several times during a simulation, the model is considered as an initial value problem. Initial values for all parameters were established at the time in which vehicle contact initiated. At each subsequent time increment, the following calculations were performed:

1. Vehicle positions were calculated by integrating the vehicle velocity using the trapezoidal rule. Based on the vehicle dimensions, a lateral overlap distance (d_{lat}) was calculated as the perpendicular distance between the side of the target vehicle and the front corner of the bullet vehicle. The longitudinal point of contact also was calculated as the perpendicular distance between the rear axle of the target vehicle and the front corner of the bullet vehicle.

2. The lateral contact force applied to the target vehicle was calculated as a function of the lateral overlap distance (d_{lat}). The force-deflection relationship was modeled as a linear elastic spring as

$$F_{yt} = k \cdot d_{lat} \tag{10.88}$$

where F_{yt} was the y axis force applied to the target vehicle, and k was the spring stiffness.

3. The longitudinal contact force applied to the target vehicle was calculated assuming frictional contact as

$$F_{xt} = \mu_{slide} \cdot F_{yt} \tag{10.89}$$

where F_{xt} is the x axis force applied to the target vehicle, and μ_{slide} is the intervehicular sliding coefficient of friction. The maximum value for μ_{slide} was assumed to be 0.5. After a common contact velocity was reached, μ_{slide} was calculated to equilibrate the inter-vehicular shear force such that no relative longitudinal acceleration occurred between the bumper of the bullet vehicle and the side of the target vehicle at the contact point. Later in the event, if this calculated shear force exceeded $\pm 0.5 \, F_{yt}$, then inter-vehicular sliding resumed.

4. The forces applied to the front corner of the bullet vehicle were calculated from Newton's third law by coordinate transforming the target vehicle forces to the bullet vehicle frame as

$$\begin{bmatrix} F_x \\ F_y \end{bmatrix}_b = \begin{bmatrix} \cos\theta & \sin\theta \\ -\sin\theta & \cos\theta \end{bmatrix} \begin{bmatrix} F_x \\ F_y \end{bmatrix}_t \tag{10.90}$$

where F_{xb} and F_{yb} were the contact forces applied to the bullet vehicle along its x axis and y axis, respectively, and θ was the angle between the vehicles (Figure 10.15).

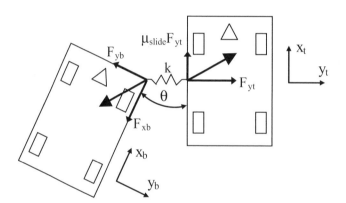

Figure 10.15 Inter-vehicular force interaction.

5. Once the contact forces were calculated, the front and rear lateral tire forces (F_f and F_r), lateral acceleration (a_y), and angular acceleration (α) were calculated for each vehicle. Only two dynamic equilibrium equations are available to solve for these four unknowns (Figure 10.16). These two equations may be obtained by summing the forces in the y direction and summing the moments about the vehicle center of gravity or the center of the front or rear axle (points f and r in Figure 10.16). Two additional equations are available if the lateral slip conditions of the tires are known. If no tires are slipping laterally, the lateral acceleration and angular acceleration of the vehicle are equal to zero. If the front and/or rear tires are slipping laterally, then the lateral tire force is equal to the maximum available tire force,

$$F_{f\,max} = \mu_{tire} \cdot W \cdot \frac{d_{cg-r}}{d_{wb}} \tag{10.91}$$

$$F_{r\,max} = \mu_{tire} \cdot W \cdot \frac{d_{cg-f}}{d_{wb}} \tag{10.92}$$

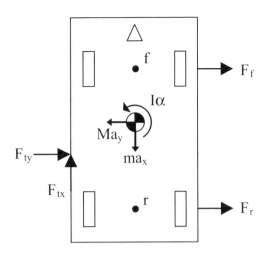

Figure 10.16 Free body diagram of the target vehicle.

where μ_{tire} is the coefficient of friction between the tires and the ground, W is the weight of the vehicle, d_{cg-r} is the distance from the center of gravity to the rear axle, d_{cg-f} is the distance from the center of gravity to the front axle, and d_{wb} is the wheelbase.

If the front tires are slipping laterally but the rear tires are gripping, then the lateral acceleration at the center of gravity and the angular acceleration of the vehicle are related kinematically as if the vehicle were rotating about the center of the rear axle,

$$a_y = \alpha \cdot d_{cg-r} \qquad (10.93)$$

Similarly, when the rear tires are slipping laterally and the front tires are gripping, then the lateral acceleration at the center of gravity and the angular acceleration of the vehicle are related kinematically as if the vehicle were rotating about the center of the front axle,

$$a_y = \alpha \cdot d_{cg-f} \qquad (10.94)$$

6. After calculating the lateral tire forces and the lateral and angular accelerations assuming a particular lateral slip condition for the tires, it was necessary to determine whether the assumed tire slip conditions were still valid. The calculation to determine whether the tire slip conditions had changed depended on the previous tire slip conditions. If a set of tires was not slipping, then slippage would initiate when the calculated lateral tire force exceeded the maximum allowable value (Eqs. 10.91 and 10.92). If a set of tires was already slipping, then slippage would stop when the lateral velocity reached zero. Lateral tire velocities were obtained by first calculating the lateral accelerations of the vehicle at the center of the front (a_f) and rear (a_r) axles as

$$a_f = a_y + \alpha \cdot d_{cg-f} \qquad (10.95)$$

$$a_r = a_y + \alpha \cdot d_{cg-r} \qquad (10.96)$$

The lateral velocities of the front (v_f) and rear (v_r) tires then were calculated by numerically integrating the lateral accelerations. If the tire slip conditions were determined to have changed, it was necessary to repeat step 5 using a new set of governing equations determined by the new tire slip conditions (Table 10.7).

TABLE 10.7
GOVERNING EQUATIONS FOR VEHICLE DYNAMICS PARAMETERS THAT DEPEND ON TIRE SLIP CONDITIONS

Front tires	Grip	Slip	Grip	Slip
Rear tires	Grip	Grip	Slip	Slip
a_y	0	$\alpha \cdot d_{cg-r}$	$\alpha \cdot d_{cg-r}$	ΣF_y
α	0	ΣM_r	ΣM_f	ΣM_{cg}
F_t	ΣM_r	$F_{t\,max}$	ΣF_y	$F_{t\,max}$
F_r	ΣM_t	ΣF_y	$F_{r\,max}$	$F_{r\,max}$

7. Next, all remaining parameters of interest describing the dynamics of the center of gravity of each vehicle were calculated. The longitudinal acceleration (a_x) of the center of gravity was calculated from Newton's second law. The linear velocity components (v_x and v_y) of the vehicle center of gravity in the local vehicle coordinate frame were calculated by numerical integration of the acceleration components using the trapezoidal rule. Angular velocity and angular displacement likewise were calculated by integrating and double-integrating angular acceleration, respectively. The relative angle between the vehicles (θ) then was given by

$$\theta = \theta_b - \theta_t \tag{10.97}$$

To calculate linear displacements, the velocity vector for each vehicle first was coordinate transformed to the inertial frame as

$$\begin{bmatrix} v_x \\ v_y \end{bmatrix}_{inertial} = \begin{bmatrix} \cos\theta_{veh} & -\sin\theta_{veh} \\ \sin\theta_{veh} & \cos\theta_{veh} \end{bmatrix} \begin{bmatrix} v_x \\ v_y \end{bmatrix}_{veh} \tag{10.98}$$

The position of the center of gravity of the vehicle then was calculated by integrating the velocity vector for each direction in the inertial frame.

8. Once the calculations were complete, then the data were output, the time was incremented, and steps 1 through 7 were repeated for the desired time duration.

To validate the model, seven low-speed (3 to 10 km/h), shallow-angle (15°) sideswipe collisions were staged with instrumented vehicles. These sideswipe collisions were characterized by long contact durations (~1 s) and low accelerations (<0.4g). The experimental collisions also were simulated with EDSMAC. EDSMAC overpredicted the peak longitudinal vehicle acceleration by an average of 83% and underpredicted the length of contact damage by an average of 50%. In contrast, the linear spring model accurately predicted the peak longitudinal vehicle acceleration (5% error) when the stiffness parameter was tuned to match the length of contact damage. These results suggest that a non-crush-based linear spring model for calculating inter-vehicular force could significantly improve the accuracy of reconstructions of low-speed sideswipe collisions compared to existing methods such as SMAC. The analytical model developed here also demonstrated that it is not possible to fully reconstruct a sideswipe collision based only on observations of vehicle damage.

10.14 Summary

Every vehicle accident reconstruction is unique, and no two accidents are the same. Accident reconstruction of these vehicle accidents often requires the use of a variety of methodologies and software based on the variation of physical evidence and accident investigative information. There is ample room for ingenuity and insight for the application of methods and software. Each accident case must be evaluated on its own merits, although it is not uncommon to have similarities among some cases. Appendix G presents a summary of basic formulae used in accident reconstruction.

10.15 References

1. Appel, H., Sturtz, G., and Gotzen, L., "Influence of Impact Speed and Vehicle Parameter on Injuries of Children and Adults in Pedestrian Accidents," Proceedings of the Second International Research Council on Biomechanics Injury (IRCOBI) Conference, 1975, pp. 83–100.

2. Araszewski, M., Toor, A., Overgaard, R., and Johal, R., "Lane Change Maneuver Modeling for Accident Reconstruction Applications," SAE Paper No. 2002-01-0817, Society of Automotive Engineers, Warrendale, PA, 2002.

3. Atkinson, D.R., Hennessy, C.J., and Watts, A.J., "Low Speed Automobile Accidents and Occupant Kinematics, Dynamics, and Biomechanics," Lawyers and Judges Publishing Company, Tuscon, AZ, 1996.

4. Backaitis, Stanley H., "Accident Reconstruction Technologies—Pedestrians and Motorcycles in Automotive Collisions," PT-35, Society of Automotive Engineers, Warrendale, PA, 1990.

5. Bailey, M.N., Wong, B.C., and Lawrence, J.M., "Data and Methods for Estimating the Severity of Minor Impacts," SAE Paper No. 950352, Society of Automotive Engineers, Warrendale, PA, 1995.

6. Baker, J.S. and Fricke, L.B., *The Traffic Accident Investigation Manual*, North-Western University Traffic Institute, Evanston, IL, 1986.

7. Bartlett, W.D. and Fonda, A.G., "Evaluating Uncertainty in Accident Reconstruction with Finite Differences," SAE Paper No. 2003-01-0469, Society of Automotive Engineers, Warrendale, PA, 2003.

8. Bartlett, W.D., Wright, W., Masory, O., Brach, R., Baxter, A., Schmidt, B., Navin, F., and Stanard, T., "Quantifying the Uncertainty in Various Measurement Tasks Common to Accident Reconstruction," SAE Paper No. 2002-01-0546, Society of Automotive Engineers, Warrendale, PA, 2003.

9. Bernard, J., Shannon, J., and Vanderploeg, M., "Vehicle Rollover on Smooth Surfaces," SAE Paper No. 891991, Society of Automotive Engineers, Warrendale, PA, 1989.

10. Bohan, T.L. and Damask, A.C., *Forensic Accident Investigation: Motor Vehicles*, Michie Butterworth Law Publishers, Charlottesville, VA, 1995.

11. Brach, Raymond M. and Brach, Matthew R., "Crush Energy and Planar Impact Mechanics for Accident Reconstruction," SAE Paper No. 980025, Society of Automotive Engineers, Warrendale, PA, 1998.

12. Brach, R.M., "Energy Loss in Vehicle Collisions," SAE Paper No. 871993, Society of Automotive Engineers, Warrendale, PA, 1987.

13. Brach, R.M., "Impact Analysis of Two-Vehicle Collisions," SAE Paper No. 830468, Society of Automotive Engineers, Warrendale, PA, 1987.

14. Brach, R.M., "Modeling of Low-Speed, Front-to-Rear Vehicle Impacts," SAE Paper No. 2003-01-0491, Society of Automotive Engineers, Warrendale, PA, 2003.

15. Brach, R.M. and Dunn, P.F., *Uncertainty Analysis for Forensic Science*, Lawyers and Judges Publishing Co., Tucson, AZ, 2004.

16. Campbell, K., "Energy as a Basis for Accident Severity," SAE Paper No. 74056, Society of Automotive Engineers, Warrendale, PA, 1974.

17. Casteel, D.A. and Moss., S.D., *Basic Collision Analysis and Scene Documentation*, 2nd Ed., Lawyers and Judges Publishing Company, Tucson, AZ, 1999.

18. Cheng, P.H., Sens, M.J., Weichel, J.F., and Guenther, D.A., "An Overview of the Evolution of Computer Assisted Motor Vehicle Accident Reconstruction," SAE Paper No. 871991, in *Reconstruction of Motor Vehicle Accidents: A Technical Compendium*, PT-34, Society of Automotive Engineers, Warrendale, PA, 1987.

19. Collins, J.C., *Accident Reconstruction,* Charles C. Thomas Publisher, Springfield, MA, November 1979.

20. Cooperrider, N.K., Thomas, T.M., and Hammond, S.A., "Testing and Analysis of Vehicle Rollover Behavior," SAE Paper No. 900366, Society of Automotive Engineers, Warrendale, PA, 1990.

21. Dailey, J., *Fundamentals of Traffic Accident Reconstruction*, Institute of Police Technology and Management, University of North Florida, Jacksonville, FL, 1988.

22. Day, T.D. and Garvey, J.T., "Applications and Limitations of 3-Dimensional Vehicle Rollover Simulation," SAE Paper No. 2000-01-0852, Society of Automotive Engineers, Warrendale, PA, 2000.

23. Dickerson, C.P., Arndt, M.W., Arndt, S.M., and Mowry, G.A., "Evaluation of Vehicle Velocity Predictions Using the Critical Speed Formula," SAE Paper No. 950137, Society of Automotive Engineers, Warrendale, PA, 1995.

24. Dixon, J., *Tires, Suspension and Handling*, 2nd Ed., Society of Automotive Engineers, Warrendale, PA, 1996.

25. Erdogan, L., Guenther, D., and Heydinger, G., "Suspension Parameter Measurement Using Side-Pull Test to Enhance Modeling of Vehicle Roll," SAE Paper No. 1999-01-1323, Society of Automotive Engineers, Warrendale, PA, 1999.

26. Eubanks, J.J., *Pedestrian Accident Reconstruction*, Lawyers and Judges Publishing Company, Tucson, AZ, 1994.

27. Fay, R.J., Robinette, R.D., and Larson, V.D., "Engineering Models and Animations in Vehicular Accident Studies," SAE Paper No. 880719, Society of Automotive Engineers, Warrendale, PA, 1988.

28. Fonda, A.G., "CRASH Extended for Desk and Handheld Computers," SAE Paper No. 870044, Society of Automotive Engineers, Warrendale, PA, 1987.

29. Fonda, A.G., "The Effects of Measurement Uncertainty on the Reconstruction of Various Vehicular Collisions," SAE Paper No. 2004-01-418, Society of Automotive Engineers, Warrendale, PA, 2004.

30. Fonda, A.G., "Nonconservation of Momentum During Impact," SAE Paper No. 950355, Society of Automotive Engineers, Warrendale, PA, 1995.

31. Fonda, A.G., "Partially Braked Impact and Trajectory Benchmarks, and Their Application to CRASH3 and CRASHEX," SAE Paper No. 2000-01-1315, Society of Automotive Engineers, Warrendale, PA, 2000.

32. Fonda, A.G., "Principles of Crush Energy Determination," SAE Paper No. 1999-01-0106, Society of Automotive Engineers, Warrendale, PA, 1999.

33. Fricke, L.B., *Traffic Accident Reconstruction, Traffic Accident Investigation Manual, Vol. 2,* Northwestern University Traffic Institute, Evanston, IL, 1990.

34. Fugger, T.F. and Randies, B.C., "Comparison of Pedestrian Accident Reconstruction Models to Experimental Test Data for Wrap Trajectories," C567/031/2000, Institute of Mechanical Engineers (IMechE), London, England, 2000.

35. Funk, F.R., Cormier, J.M., and Bain, C.E., "Analytical Model for Investigating Low-Speed Sideswipe Collisions," SAE Paper No. 2004-01-1185, Society of Automotive Engineers, Warrendale, PA, 2004.

36. Han, I. and Brach, R.M., "Throw Model for Frontal Pedestrian Collisions," SAE Paper No. 2001-01-0898, Society of Automotive Engineers, Warrendale, PA, 2001.

37. Happer, A., Araszewski, M., Toor, A., Overgaard, R., and Johal, R., "Comprehensive Analysis Method for Vehicle/Pedestrian Collision," SAE Paper No. 2000-10-0846, Society of Automotive Engineers, Warrendale, PA, 2000.

38. Haung, M., *Vehicle Crash Mechanics*, CRC Press, Boca Raton, FL, 2002.

39. Hill, G.S., "Calculations of Vehicle Speed from Pedestrian Throw," *Impact,* Institute of Traffic Accident Investigators, Shrewsbury, U.K., 1994, pp. 18–20.

40. Ishikawa, H., "Computer Simulation of Automobile Collision—Reconstruction of Accidents," SAE Paper No. 851729, Society of Automotive Engineers, Warrendale, PA, 1985.

41. Ishikawa, H., "Impact Center and Restitution Coefficients for Accident Reconstruction," SAE Paper No. 940564, Society of Automotive Engineers, Warrendale, PA, 1994.

42. Ishikawa, H., "Impact Model for Accident Reconstruction—Normal and Tangential Restitution Coefficients," SAE Paper No. 930654, Society of Automotive Engineers, Warrendale, PA, 1993.

43. Jones, I.S. and Baum, A.S., "Research Input for Computer Simulation of Automobile Collisions," Vol. IV, Staged Collision Reconstructions, U.S. DOT HS 805-040, December 1978.

44. Lambourn, R.F., "The Calculation of Motor Car Speeds from Curved Tire Marks," *Journal of the Forensic Science Society (U.K.)*, Vol. 29, 1989, pp. 371–386.

45. Levy, R.A., "Speed Determination in Car-Truck Sideswipe Collisions," SAE Paper No. 2000-01-0463, Society of Automotive Engineers, Warrendale, PA, 2000.

46. Limpert, R., *Motor Vehicle Accident Reconstruction and Cause Analysis*, 3rd Ed., Michie Company, Charlottesville, VA, 1989.

47. Lund, Y.I. and Bernard, J.E., "Analysis of Simple Rollover Metrics," SAE Paper No. 950306, Society of Automotive Engineers, Warrendale, PA, 1995.

48. Martinez, J.E. and Schlueter, R.J., "A Primer on the Reconstruction and Presentation of Rollover Accidents," SAE Paper No. 960647, Society of Automotive Engineers, Warrendale, PA, 1996.

49. Martinez, L., "Estimating Speed from Yaw Marks—An Empirical Study," *Accident Reconstruction Journal,* May/June 1993.

50. McGee, H.W., Hooper, K.G., Hughes, W.E., and Benson, W., "Highway Design and Operations Standards Affected by Driver Characteristics," Vol. II, Final Report, Bellomo-McGee, Inc., Report No. FHWA-RD-83-015, Vienna, VA, 1983.

51. McHenry, B.G. and McHenry, R.R., "SMAC-97: Refinement of the Collision Algorithm," SAE Paper No. 970947, Society of Automotive Engineers, Warrendale, PA, 1997.

52. McHenry, R.R., "A Comparison of Results Obtained with Different Analytical Techniques for Reconstruction of Highway Accidents," SAE Paper No. 750983, Society of Automotive Engineers, Warrendale, PA, 1975.

53. Meyer, S.E., Davis, M., Forrest, S., Cheng, D., and Herbst, B., "Accident Reconstruction of Rollovers—A Methodology," SAE Paper No. 2000-01-0853, Society of Automotive Engineers, Warrendale, PA, 2000.

54. Noon, R.K., *Engineering Analysis of Vehicle Accidents*, CRC Press, Boca Raton, FL, 1994.

55. NCS, "Manual on Classification of Motor Vehicle Traffic Accidents," 6th Ed., ANSI D16.1, National Safety Council, Chicago, IL, 1996.

56. Olson, P.L., *Forensic Aspects of Driver Perception and Response*, Lawyers and Judges Publishing, Tucson, AZ, 1996.

57. Orlowski, K.R., Moffatt, E.A., Bundorf, R.T., and Holcomb, M.P., "Reconstruction of Rollover Collisions," SAE Paper No. 890857, Society of Automotive Engineers, Warrendale, PA, 1989.

58. Peters, G.A. and Peters, B.J., *Automotive Vehicle Safety*, Society of Automotive Engineers, Warrendale, PA, 2002.

59. Prasad, A.K., "Energy Absorbed by Vehicle Structures in Side Impacts," SAE Paper No. 910599, Society of Automotive Engineers, Warrendale, PA, 1991.

60. Rivers, R.W., *Speed Analysis for Traffic Accident Investigation Manual*, Institute of Police Technology and Management (IPTM), University of North Florida, Jacksonville, FL, 1997.

61. Rivers, R.W., *Tire Failures and Evidence Manual*, Charles C. Thomas Publisher, Springfield, MA, November 2001.

62. Rivers, R.W., *Traffic Accident Investigator's Lamp Analysis Manual*, Charles C. Thomas Publisher, Springfield, MA, November 2001.

63. SAE International, "Collision Deformation Classification," SAE Standard J224, Society of Automotive Engineers, Warrendale, PA, March 1980.

64. SAE International, "Motor Vehicle Dimensions," SAE Standard J1100, Society of Automotive Engineers, Warrendale, PA, July 2002.

65. Searle, J.A., "The Physics of Throw Distance in Accident Reconstruction," SAE Paper No. 930659, Society of Automotive Engineers, Warrendale, PA, 1993.

66. Shelton, Sgt. Thomas, "Validation of the Estimation of Speed from Critical Speed Scuffmarks," *Accident Reconstruction Journal*, January/February 1995.

67. Sledge, N.H. and Marshek, K.M., "Formulas for Estimating Vehicle Critical Speed from Yaw Marks—A Review," SAE Paper No. 971147, Society of Automotive Engineers, Warrendale, PA, 1997.

68. Sledge, N.H. and Marshek, K.M., "Vehicle Critical Speed Formula—Values for the Coefficient of Friction—A Review," SAE Paper No. 971148, Society of Automotive Engineers, Warrendale, PA, 1997.

69. Stcherbatcheff, G., Tarrieere, C., Duclos, P., and Fayon, A., "Simulation of Collisions Between Pedestrians and Vehicles Using Adult and Child Dummies," SAE Paper No. 751167, Society of Automotive Engineers, Warrendale, PA, 1975.

70. Stevens, D.C., "Passenger Vehicle Rollover Reconstruction: Investigation, Analysis and Presentation of Results," Passenger Vehicle Rollover TOPTEC: Causes, Prevention, and Injury Prevalence, April 22–23, Scottsdale, AZ, Society of Automotive Engineers, Warrendale, PA, 2002.

71. Sturtz, G., Suren, E.G., Gotzen, L., Behrens, S., and Richter, K., "Biomechanics of Real Child Pedestrian Accidents," SAE Paper No. 760814, Society of Automotive Engineers, Warrendale, PA, 1976.

72. Toor, A., "Theoretical Versus Empirical Solutions for Vehicle/Pedestrian Collisions," SAE Paper No. 2003-01-0883, Society of Automotive Engineers, Warrendale, PA, 2003.

73. Toor, A., Araszewski, M., Johal, R., Overgaard, R., and Happer, A., "Revision and Validation of Vehicle/Pedestrian Collision Analysis Method," SAE Paper No. 2002-01-0550, Society of Automotive Engineers, Warrendale, PA, 2002.

74. Toor, A., Roenitz, E., Johal, R., Overgaard, R., Happer, A., and Araszewski, M., "Practical Analysis Technique for Quantifying Sideswipe Collisions," SAE Paper No. 1999-01-0094, Society of Automotive Engineers, Warrendale, PA, 2000.

75. Townes, Harry, "Photogrammetry in Accident Reconstruction," SAE Professional Development Seminar, Society of Automotive Engineers, Warrendale, PA, July 1998.

76. Wood, D. and Simms, C.K., "Coefficient of Friction in Pedestrian Throw," *Impact*, Vol. 9, No. 1, Institute of Traffic Accident Investigators, Shrewsbury, U.K., 2000, pp. 12–15.

77. Wood, D. and Simms, C.K., "A Hybrid Model for Pedestrian Impact and Projection," *International Journal of Crashworthiness*, Vol. 5, No. 4, Woodhead Publishing, Ltd., Cambridge, U.K., 2000, pp. 393–403.

78. Wood, D.P., "Impact and Movement of Pedestrians in Frontal Collisions with Vehicles," Proceedings of the Institution of Mechanical Engineers, Vol. 202, No. D2, London, U.K., 1988, pp. 101–110.

79. Wood, D.P. and Walsh, D.G., "Pedestrian Forward Projection Impact," *International Journal of Crashworthiness*, Vol. 7, No. 3, Woodhead Publishing, Ltd., Cambridge, U.K., 2002.

80. Woolley, R.L., "The IMPAC Program for Collision Analysis," SAE Paper No. 8700476, Society of Automotive Engineers, Warrendale, PA, 1987.

Vector Algebra

A.1 Real and Complex Vectors

An n-dimensional complex vector A is an ordered n-tuple of complex numbers $(a_1, a_2, ..., a_n)$ that form an n-dimensional vector space. If the ordered n-tuple numbers $(a_1, a_2, ..., a_n)$ admit only real numbers, then we define an n-dimensional real vector A.

The vector A can be represented as

$$A = \begin{Bmatrix} a_1 \\ a_2 \\ . \\ . \\ . \\ a_n \end{Bmatrix}$$

Now this is a vector (in the matrix sense) in which the components are distinguished by positioning in the column.

The vectors $A = (a_1, a_2, ..., a_n)$ and $B = (b_1, b_2, ..., b_n)$ are equal if and only if

$$a_1 = b_1, \ a_2 = b_2, \ ..., \ a_n = b_n$$

Vector addition and subtraction. The sum of the vectors $A = (a_1, a_2, ..., a_n)$ and $B = (b_1, b_2, ..., b_n)$ is the vector

$$C = A + B = (a_1 + b_1, \ a_2 + b_2, \ ..., \ a_n + b_n)$$

Vector multiplication by a scalar. The product of a vector $A = (a_1, a_2, ..., a_n)$ and a scalar number c is the vector

$$B = cA = (ca_1, ca_2, ..., ca_n)$$

A.2 Laws of Vector Operation

The following rules are satisfied by vector operations:

1. Commutative law: $A + B = B + A$

2. Associative law: $(A + B) + C = A + (B + C)$

3. Distributive law: $c(A + B) = cA + cB$ and $(c + d)A = cA + dA$

where c and d are real numbers

$$c(dA) = (cd)A$$

$0A = 0$, where 0 is called the zero vector or null vector

$c0 = 0$

The equality $cA = 0$ holds if and only if $c = 0$ or $A = 0$.

$$-(cA) = (-c)A = c(-A)$$

A.3 Linear Dependence

If we consider a vector U in which $u_1, u_2, ..., u_k$ are the components and $c_1, c_2, ..., c_k$ are k scalars, then the vector U given by

$$U = c_1 u_1 + c_2 u_2 + ... + c_k u_k$$

is called a linear combination of $u_1, u_2, ..., u_k$ with coefficients $c_1, c_2, ..., c_k$. The totality of linear combination of $u_1, u_2, ..., u_k$ obtained by letting $c_1, c_2, ..., c_k$ vary over R is a vector space. If the relation

$$c_1 u_1 + c_2 u_2 + ... + c_k u_k = 0$$

can be satisfied only for the trivial case, namely, when all the coefficients $c_1, c_2, ..., c_k$ are identically zero, then the vectors $u_1, u_2, ..., u_k$ are said to be linearly independent. If at least one of the coefficients $c_1, c_2, ..., c_k$ is different from zero, the vectors $u_1, u_2, ..., u_k$ are said to be linearly dependent, implying that one vector is a linear combination of the remaining vectors.

Linear combination. A vector A is said to be a linear combination of vectors $A_1, A_2, ..., A_k$ if the numbers $c_1, c_2, ..., c_k$ exist such that

$$A = c_1 A_1 + c_2 A_2 + \dots + c_k A_k$$

Thus, vectors $A_1, A_2, \dots, A_k$ are linearly dependent if and only if at least one of them can be expressed as a linear combination of the others.

Length of vector. The length of a vector $A = (a_1, a_2, \dots, a_n)$ is the non-negative number $\left[a_1^2 + a_2^2 + \dots + a_n^2 \right]^{1/2}$ and is denoted by $|A|$ or a. A vector whose length equals unity is called a unit vector. The terms modulus, magnitude, norm, or absolute value of a vector also are used for the length of a vector.

A.4 Three-Dimensional Vectors

In a rectangular Cartesian coordinate system, we employ x, y, and z to represent the three coordinates. The linearly independent unit vectors $i = (1, 0, 0)$, $j = (0, 1, 0)$, and $k = (0, 0, 1)$ in the directions x, y, and z, respectively, as shown in Figure A.1 are called coordinate vectors.

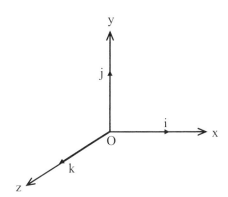

Figure A.1 Coordinate vectors.

In three-dimensional space, any four vectors are linearly dependent. Thus, every vector $A = (a_1, a_2, a_3)$ can be expressed as a linear combination of the coordinate vectors i, j, and k as

$$A = a_1 i + a_2 j + a_3 k$$

Two linearly dependent vectors are called collinear (parallel). Two vectors A and B are linearly dependent (parallel) if and only if one of them is a multiple of the other, that is, there is a number c such that $A = cB$. Three linearly dependent vectors are called coplanar.

Angle between vectors. The angle between two nonzero vectors A and B is the angle θ $(0 \leq \theta \leq \pi)$ between the directed segments representing both vectors.

Scalar product of two vectors. The scalar product, inner product, or dot product of two vectors $A = (a_1, a_2, a_3)$ and $B = (b_1, b_2, b_3)$ written as $A \cdot B$ is defined by

$$A \cdot B = a_1 b_1 + a_2 b_2 + a_3 b_3$$

It also can be shown that

$$A \cdot B = |A||B|\cos\theta$$

where $|A|$ and $|B|$ are the magnitudes, respectively, and θ is the angle between the vectors.

Perpendicular vectors. Two nonzero vectors A and B are perpendicular if and only if $A \cdot B = 0$.

A.5 Properties of the Scalar Product of Vectors

The scalar product of vectors satisfies the relation

1. $A \cdot B = B \cdot A$ (i.e., it is commutative)

2. $(A + B) \cdot C = A \cdot C + B \cdot C$ (i.e., it is distributive)

3. $A \cdot A = |A|^2$

A.6 Direction Angles

The angles that a nonzero vector makes with the coordinate vectors and thus with the coordinate axes are called the direction angles, and their cosines are the direction cosines of the given vector.

Let α, β, and γ be the direction angles of a nonzero vector $A = (a_1, a_2, a_3)$. The direction cosines obey the following rules:

1. $\cos\alpha = \dfrac{a_1}{|A|}$, $\cos\beta = \dfrac{a_2}{|A|}$, $\cos\gamma = \dfrac{a_3}{|A|}$

2. $\cos^2\alpha + \cos^2\beta + \cos^2\gamma = 1$

A.7 Vector Product

The vector product or cross product or outer product of vectors $A = (a_1, a_2, a_3)$ and $B = (b_1, b_2, b_3)$, denoted by $A \times B$, is a vector C defined by

$$A \times B = C$$

where C is a vector perpendicular to both A and B, is in the sense of a right-hand screw, and has a magnitude $|A||B|\sin\theta$, as shown in Figure A.2. Here, θ is the angle between A and B. The vector product $A \times B$ may be written as

$$A \times B = (a_2 b_3 - a_3 b_2)i + (a_3 b_1 - a_1 b_3)j + (a_1 b_2 - a_2 b_1)k$$

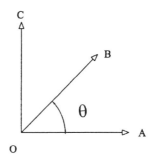

Figure A.2 Vector product of two vectors.

This result also is obtained by the expansion of the determinant

$$A \times B = \begin{vmatrix} i & j & k \\ a_1 & a_2 & a_3 \\ b_1 & b_2 & b_3 \end{vmatrix}$$

Properties of the vector product. The vector product satisfies the following relations:

1. It is not commutative (i.e., $A \times B \neq B \times A$, but $A \times B = -B \times A$)

2. It is distributive (i.e., $A \times (B + C) = A \times B + A \times C$)

3. It is not associative (i.e., $A \times (B \times C) \neq (A \times B) \times C$)

4. $k\, A \times B = k(A \times B)$, where k is a scalar number

Mixed product. The mixed product or triple scalar product of three vectors A, B, and C is a scalar denoted by $A \cdot (B \times C)$. The mixed product of three vectors sometimes also is called a trivector.

Properties of a mixed product. The mixed product satisfies the following relation:

$$A \cdot (B \times C) = B \cdot (C \times A) = C \cdot (A \times B)$$

$$-A \cdot (C \times B) = -C \cdot (B \times A) = -B \cdot (A \times C)$$

$$= \begin{vmatrix} a_1 & a_2 & a_3 \\ b_1 & b_2 & b_3 \\ c_1 & c_2 & c_3 \end{vmatrix}$$

A.8 Derivative of a Vector

The components of vectors are functions of a scalar variable t. Thus, for every value of t in the domain under consideration, we obtain, in general, a different vector. We deal with a vector field and denote the vector by $A(t)$. The derivative of $A(t)$ with respect to t is defined by

$$\frac{d}{dt} \lim_{\Delta t \to 0} A(t) = \frac{A(t + \Delta t) - A(t)}{\Delta t} = \dot{A}(t)$$

Similarly, we can define higher derivatives. For example,

$$\frac{d^2}{dt^2} A(t) = \lim_{\Delta t \to 0} \frac{A(t + \Delta t) - A(t)}{\Delta t} = \ddot{A}(t)$$

In general, the components of a vector can be functions of several variables. The corresponding partial derivatives can be defined in a similar manner.

Rules of vector differentiation. It can be shown that

1. $\dfrac{d}{dt}(c\,A) = c\,\dot{A} + \dot{c}\,A$, where c is a scalar function of t

2. $\dfrac{d}{dt}(A + B) = \dot{A} + \dot{B}$

3. $\dfrac{d}{dt}(A \cdot B) = \dot{A} \cdot B + A \cdot \dot{B}$

4. $\dfrac{d}{dt}(A \times B) = \dot{A} \times B + A \times \dot{B}$

A.9 References

1. Dukkipati, R.V., *Advanced Engineering Analysis*, Narosa Publishing House, New Delhi, India, 2006.

2. Greenberg, M.D., *Foundations of Applied Mathematics*, Prentice-Hall, Englewood Cliffs, NJ, 1978.

3. Hildebrand, F.B., *Advanced Calculus for Applications*, Prentice-Hall, Englewood Cliffs, NJ, 1978.

4. Jain, R.K. and Iyengar, S.R.K., *Advanced Engineering Mathematics*, Narosa Publishing Company, New Delhi, India, 2002.

5. Jeffrey, A., *Advanced Engineering Mathematics,* Harcourt/Academic Press, New York, 2002.

6. Kreyszig, E., *Advanced Engineering Mathematics*, 8th Ed., Wiley, New York, 1998.

7. Lipschutz, S., *Linear Algebra*, Schaum's Solved Problems Series, McGraw-Hill, New York, 1989.

8. Lipson, M., *Linear Algebra*, Schaum's Outlines, McGraw-Hill, New York, 2001.

9. Lopez, R.J., *Advanced Engineering Mathematics*, Addison-Wesley, New York, 2001.

10. O'Neil, P.V., *Advanced Engineering Mathematics*, 4th Ed., Brooks/Cole Publishing Company, New York, 1995.

11. Siegel, M., *Advanced Mathematics for Engineers and Scientists*, Schaum's Outlines, McGraw-Hill, New York, 1971.

12. Siegel, M., *Vector Analysis and Introduction to Tensor Analysis*, Schaum's Outlines, McGraw-Hill, New York, 1959.

13. Strang, G., *Linear Algebra and Its Applications*, Academic Press, New York, 1980.

14. Wylie, C.R. and Barret, L.C., *Advanced Engineering Mathematics*, 5th Ed., McGraw-Hill, New York, 1982.

Appendix B

Matrix Analysis

B.1 Introduction

The dynamic equations of vehicle systems generally lead to a set of second-order simultaneous differential equations. It often is desirable to express these equations in compact matrix form. In this appendix, we summarize some important results from matrix algebra that are useful in this book. The discussion of linear algebra presented here is relatively modest in nature, and its main objective is to familiarize the reader with some fundamental concepts of particular interest in dynamic systems.

B.2 Definitions of Matrices

A matrix is a collection of elements arranged in a specific order in rows and columns, such as

$$
A = \begin{bmatrix}
a_{11} & a_{12} & a_{13} & \cdots & a_{1n} \\
a_{21} & a_{22} & a_{23} & \cdots & a_{2n} \\
a_{31} & a_{32} & a_{33} & \cdots & a_{3n} \\
\vdots & \vdots & \vdots & \cdots & \vdots \\
a_{m1} & a_{m2} & a_{m3} & \cdots & a_{mn}
\end{bmatrix}
\tag{B.1}
$$

where A is called an $m \times n$ matrix because it contains m rows and n columns. The dimensions of A are $m \times n$. Each element a_{ij} (i = 1, 2, ..., m ; j = 1, 2, ..., n) of the matrix A represents a scalar. The position of the element a_{ij} in the matrix A is in the ith row and jth column, so that i is referred to as the row index and j as the column index. A square matrix has the same number of rows and columns (that is, m = n).

Column matrix. A matrix having a single column is said to be a column matrix and is written as

$$B = \begin{Bmatrix} b_{11} \\ b_{12} \\ b_{13} \\ \vdots \\ b_{m1} \end{Bmatrix} \qquad (B.2)$$

Row matrix. A matrix having a single row is called a row matrix and is represented as

$$C = \begin{bmatrix} c_{11} & c_{12} & \cdots & c_{1n} \end{bmatrix} \qquad (B.3)$$

Square matrix. A square matrix is one in which the number of rows is equal to the number of columns. It is referred to as $n \times n$ matrix or a matrix of order n.

Diagonal matrix. A diagonal matrix is a square matrix having all its elements as zero except those on the leading diagonal. For example, if D is a diagonal matrix of order n, then

$$D = \begin{bmatrix} a_{11} & 0 & \cdots & 0 \\ 0 & a_{22} & \cdots & 0 \\ 0 & 0 & a_{33} & 0 \\ \vdots & \vdots & \cdots & \vdots \\ 0 & 0 & \cdots & a_{nn} \end{bmatrix} \qquad (B.4)$$

Unit matrix. A unit matrix is a diagonal matrix whose diagonal elements are each equal to unity and is denoted as I, where

$$I = \begin{bmatrix} 1 & 0 & \cdots & 0 \\ 0 & 1 & \cdots & 0 \\ \vdots & \vdots & \cdots & \vdots \\ 0 & 0 & \cdots & 1 \end{bmatrix}_{n \times n} \qquad (B.5)$$

Null matrix (zero matrix). A null matrix or zero matrix has all its elements equal to zero and is written simply as zero. If $[0]$ is of order 2×4, it is given by

$$[0] = \begin{bmatrix} 0 & 0 & 0 & 0 \\ 0 & 0 & 0 & 0 \end{bmatrix}$$

Real matrix. A real matrix consists of mn numbers arranged in m rows and in n columns and whose elements are real quantities. For example, a 3×4 matrix would be

$$A = \begin{bmatrix} a_{11} & a_{12} & a_{13} & a_{14} \\ a_{21} & a_{22} & a_{23} & a_{24} \\ a_{31} & a_{32} & a_{33} & a_{34} \end{bmatrix} \quad (B.6)$$

Complex matrix. All elements of a matrix may be complex. A complex matrix has the form

$$\begin{bmatrix} a_{11} + ib_{11} & a_{12} + ib_{12} & a_{13} + ib_{13} \\ a_{21} + ib_{21} & a_{22} + ib_{22} & a_{23} + ib_{23} \\ a_{31} + ib_{31} & a_{32} + ib_{32} & a_{33} + ib_{33} \end{bmatrix} \quad (B.7)$$

Transpose of a matrix. The transpose of a matrix denoted as A^T is a matrix with the rows and columns interchanged from the original matrix A of Eq. B.1,

$$A^T = \begin{bmatrix} a_{11} & a_{21} & \cdots & a_{ml} \\ a_{12} & a_{22} & \cdots & a_{m2} \\ a_{13} & a_{23} & \cdots & a_{m3} \\ \vdots & \vdots & \cdots & \vdots \\ a_{1n} & a_{2n} & \cdots & a_{mn} \end{bmatrix} \quad (B.8)$$

For example, if

$$A = \begin{bmatrix} 3 & 6 & 5 \\ 2 & 1 & 7 \end{bmatrix}$$

$$A^T = \begin{bmatrix} 3 & 2 \\ 6 & 1 \\ 5 & 7 \end{bmatrix}$$

Symmetrical matrix. A symmetrical matrix is a square matrix whose elements are symmetrical about its leading diagonal. A symmetrical matrix is equal to its transpose. For example,

$$a_{ij} = a_{ji}$$

$$A = \begin{bmatrix} 4 & -2 & -5 \\ -2 & 0 & 8 \\ -5 & 8 & 5 \end{bmatrix} \quad (B.9)$$

Asymmetrical matrix. An asymmetrical matrix is a square matrix whose elements are symmetrical but with an opposite sign about its leading diagonal

$$a_{ij} = -a_{ji} \quad (B.10)$$

Triangular matrix. A triangular matrix is a square matrix that has zero elements either below or above the leading diagonal.

Trace. The sum of the main diagonal elements of square matrix $A = \left[a_{ij} \right]$ is called the trace of A and is given by

$$\text{Trace } A = a_{11} + a_{12} + \dots + a_{nn} \tag{B.11}$$

Minor. If the ith row and jth column of determinant A are deleted, the remaining (m–1) rows and (n–1) columns form a determinant M_{ij}. This determinant is called the minor of the element a_{ij}.

Principal minor. A minor of $|A|$ whose diagonal elements also are diagonal elements of A, is called a principal minor of $|A|$.

Cofactor. The cofactor c_{ij} of element a_{ij} of the matrix A is defined as

$$c_{ij} = \left(-1 \right)^{i+j} M_{ij} \tag{B.12}$$

where M_{ij} is the minor of the element a_{ij}.

For example, the cofactor of the element a_{32} of

$$\det |A| = \begin{vmatrix} a_{11} & a_{12} & a_{13} \\ a_{21} & a_{22} & a_{23} \\ a_{31} & a_{32} & a_{33} \end{vmatrix}$$

is given by

$$c_{32} = \left(-1 \right)^5 M_{32} = -\begin{vmatrix} a_{11} & a_{13} \\ a_{21} & a_{23} \end{vmatrix} \tag{B.13}$$

Adjoint matrix. The adjoint of a square matrix is found by replacing each element a_{ij} of matrix A by its cofactor c_{ij} and then transposing.

EXAMPLE EB.1

Consider the matrix A as follows:

$$A = \begin{bmatrix} -1 & 1 & 2 \\ 3 & -1 & 1 \\ -1 & 3 & 4 \end{bmatrix}$$

Solution:

First, calculate the determinant using the first row as

$$|A| = -\begin{vmatrix} -1 & 1 \\ 3 & 4 \end{vmatrix} - \begin{vmatrix} 3 & 1 \\ -1 & 4 \end{vmatrix} + 2\begin{vmatrix} 3 & -1 \\ -1 & 3 \end{vmatrix} = 10$$

Next, identify the minor corresponding to each entry to form the adjoint matrix. Thus, we have

$$\text{adj}(A) = \begin{bmatrix} -7 & 2 & 3 \\ -13 & -2 & 7 \\ 8 & 2 & -2 \end{bmatrix}$$

Singular and non-singular matrices. A square matrix is called singular if its associated determinant is zero. It is said to be non-singular if its associated determinant is non-zero.

Banded matrix. Matrices relating to engineering problems are of a banded nature. Bandedness means non-zero elements have a definite band around the principal diagonal. Because A is a symmetrical matrix, we can state this condition as $a_{ij} = 0$ for $j > i + m_A$ where $2m_A + 1$ is the bandwidth of A. The following matrix is a symmetrical banded matrix of order 5. The half-bandwidth m_A is 2.

$$A = \begin{bmatrix} 3 & 2 & 1 & 0 & 0 \\ 2 & 3 & 4 & 1 & 0 \\ 1 & 4 & 5 & 6 & 1 \\ 0 & 1 & 6 & 7 & 4 \\ 0 & 0 & 1 & 4 & 3 \end{bmatrix}$$

If the half bandwidth of a matrix is zero, the banded matrix reduces to a diagonal matrix.

Tridiagonal matrix. A banded matrix, which has only one non-zero element on either side of the principal diagonal, is said to be tridiagonal, that is, a half bandwidth in this case is 1.

Rank of a matrix. A matrix A is said to have rank or if there exists an r × r submatrix of A that is non-singular, and all other Q × Q submatrices (where $Q \geq r + 1$) are singular.

EXAMPLE EB.2

Find the determinant of

$$A = \begin{bmatrix} 1 & 5 & -3 \\ 4 & -1 & 1 \\ 2 & 0 & 1 \end{bmatrix}$$

Solution:

The determinant may be computed using the third row because it contains a zero; hence, fewer calculations will be involved. Therefore,

$$|A| = 2(-1)^{3+1} M_{31} + 0.M_{32} + 1(-1)^{3+3} M_{33}$$

$$= 2\begin{vmatrix} 5 & -3 \\ -1 & 1 \end{vmatrix} + \begin{vmatrix} 1 & 5 \\ 4 & -1 \end{vmatrix}$$

$$= 2(5-3) + (-1-20)$$

$$= -17$$

indicating that A is non-singular. The same result also can be obtained using the second column, which likewise contains a zero.

Conjugate matrix. The conjugate of a matrix A, denoted as A^*, is the matrix in which each element is the complex conjugate of the corresponding element of A.

B.3 Matrix Operations

Equality of matrices. Two matrices of the same order are equal if their corresponding elements are equal. Thus, two matrices A and B are equal if, for every j and k,

$$a_{jk} = b_{jk} \tag{B.14}$$

Hence,

$$A - B = 0$$

Matrix addition and subtraction. Addition or subtraction of matrices of the same order is performed by adding or subtracting corresponding elements. Thus, $A + B = C$, if for every j and k,

$$a_{jk} + b_{jk} = c_{jk} \tag{B.15}$$

EXAMPLE EB.3

Given

$$A = \begin{bmatrix} 7 & 3 & 2 \\ 5 & 4 & 6 \end{bmatrix}$$

and

$$B = \begin{bmatrix} 4 & 1 & 3 \\ 2 & 6 & 5 \end{bmatrix}$$

find $A + B$ and $A - B$.

Solution:

$$A + B = \begin{bmatrix} 7+4 & 3+1 & 2+3 \\ 5+2 & 4+6 & 6+5 \end{bmatrix} = \begin{bmatrix} 11 & 4 & 5 \\ 7 & 10 & 11 \end{bmatrix}$$

$$A - B = \begin{bmatrix} 7-4 & 3-1 & 2-3 \\ 5-2 & 4-6 & 6-5 \end{bmatrix} = \begin{bmatrix} 3 & 2 & -1 \\ 3 & -2 & 1 \end{bmatrix}$$

Multiplication of a matrix by a scalar. Multiplication of a matrix A by a scalar c multiples each element of the matrix by c, that is,

$$cA = c\left[a_{jk}\right] = \left[c\,a_{jk}\right] \tag{B.16}$$

In particular, the negative of a matrix has the sign of every element changed.

Matrix multiplication. If A is a matrix of order (m,n) and B is a matrix of order (n,p), then their matrix product AB = C is defined to be a matrix C of order (m, p) where, for every j and k,

$$c_{jk} = \sum_{r=1}^{n} a_{jr}\,b_{rk} \tag{B.17}$$

Consider the two matrices of

$$[A] = \begin{bmatrix} a_{11} & a_{12} & a_{13} \\ a_{21} & a_{22} & a_{23} \end{bmatrix}$$

$$[B] = \begin{bmatrix} b_{11} & b_{12} & b_{13} & b_{14} \\ b_{21} & b_{22} & b_{23} & b_{24} \\ b_{31} & b_{32} & b_{33} & b_{34} \end{bmatrix}$$

Then

$$[A][B] = \begin{bmatrix} a_{11} & a_{12} & a_{13} \\ a_{21} & a_{22} & a_{23} \end{bmatrix} \begin{bmatrix} b_{11} & b_{12} & b_{13} & b_{14} \\ b_{21} & b_{22} & b_{23} & b_{24} \\ b_{31} & b_{32} & b_{33} & b_{34} \end{bmatrix}$$

$$= \begin{bmatrix} (a_{11}b_{11} + a_{12}b_{21} + a_{13}b_{31}) & (a_{11}b_{12} + a_{12}b_{22} + a_{13}b_{32}) \\ (a_{21}b_{11} + a_{22}b_{21} + a_{23}b_{31}) & (a_{21}b_{12} + a_{22}b_{22} + a_{23}b_{32}) \end{bmatrix} \tag{B.18}$$

$$\begin{matrix} (a_{11}b_{13} + a_{12}b_{23} + a_{13}b_{33}) & (a_{11}b_{14} + a_{12}b_{24} + a_{13}b_{34}) \\ (a_{21}b_{13} + a_{22}b_{23} + a_{23}b_{33}) & (a_{21}b_{14} + a_{22}b_{24} + a_{23}b_{34}) \end{matrix} = [C]$$

From the foregoing, the general rule for multiplication of matrices may be stated as

$$c_{ij} = \sum_{k=1}^{n} a_{ik} b_{kj} \tag{B.19}$$

where k is a dummy index.

The product of two matrices can be obtained only if they are conformable, that is, if the number of columns in A is equal to number of rows in B. The symbolic equation

$$(m,n) \times (n,p) = (m,p) \tag{B.20}$$

indicates the orders of the matrices involved in a matrix product.

The following rules hold true for matrix multiplication:

1. Pre-multiplication of A by B does not equal post-multiplication of A by B

$$A \times B \neq B \times A \tag{B.21}$$

2. The distributive law applies as

$$A(B + C) = AB + AC \tag{B.22}$$

3. The associative law applies as

$$A(BC) = AB(C) \tag{B.23}$$

4. The product of two transposed matrices is equal to the transpose of the product of the original matrices in reverse order

$$A^T B^T = (BA)^T \tag{B.24}$$

5. Any matrix A multiplied by a unit matrix I gives a product identical with A

$$AI = A \tag{B.25}$$

EXAMPLE EB.4

Given

$$A = \begin{bmatrix} 1 & 8 \\ 2 & 1 \\ 4 & 3 \end{bmatrix}$$

and

$$B = \begin{bmatrix} 2 & 1 & 5 \\ 1 & 4 & 1 \end{bmatrix}$$

show that $AB \neq BA$.

Solution:

$$AB = \begin{bmatrix} 2 & 1 & 5 \\ 5 & 6 & 11 \\ 11 & 16 & 23 \end{bmatrix}$$

$$BA = \begin{bmatrix} 2 & 1 & 5 \\ 1 & 4 & 1 \end{bmatrix} \begin{bmatrix} 1 & 8 \\ 2 & 1 \\ 4 & 3 \end{bmatrix} = \begin{bmatrix} 24 & 16 \\ 13 & 7 \end{bmatrix}$$

Clearly, $AB \neq BA$.

B.4 Matrix Inversion

The inverse of a square matrix A, denoted as A^{-1}, is defined by the relation

$$A^{-1} A = A A^{-1} = I \tag{B.26}$$

A^{-1} is determined by the relation

$$A^{-1} = \frac{\text{adj } \mathbf{A}}{\det \mathbf{A}} \tag{B.27}$$

EXAMPLE EB.5

Determine the inverse of the matrix

$$A = \begin{bmatrix} -1 & 1 & 2 \\ 3 & -1 & 1 \\ -1 & 3 & 4 \end{bmatrix}$$

Solution:

First, calculate the determinant using the first row as

$$|A| = -|A| = -\begin{vmatrix} -1 & 1 \\ 3 & 4 \end{vmatrix} - \begin{vmatrix} 3 & 1 \\ -1 & 4 \end{vmatrix} + 2\begin{vmatrix} 3 & -1 \\ -1 & 3 \end{vmatrix} = 10$$

Next, identify the minor corresponding to each entry to form the adjoint matrix. Thus, we have

$$\text{adj}(A) = \begin{bmatrix} -7 & 2 & 3 \\ -13 & -2 & 7 \\ 8 & 2 & -2 \end{bmatrix}$$

Hence,

$$A^{-1} = \frac{1}{10}\text{adj}(A) = \begin{bmatrix} -0.7 & 0.2 & 0.3 \\ -1.3 & -0.2 & 0.7 \\ 0.8 & 0.2 & -0.2 \end{bmatrix}$$

B.5 Determinants

A determinant is a square array of numbers enclosed between two vertical lines. The numbers are called the elements of the determinant. Consider

$$\begin{vmatrix} 5 & 3 \\ 8 & 2 \end{vmatrix}$$

The numbers 5, 3, 8, and 2 are the elements. Because this determinant has two rows and two columns, it is called a 2×2 determinant or a second-order determinant.

The value of a 2×2 determinant is calculated as

$$\begin{vmatrix} a_{11} & a_{12} \\ a_{21} & a_{22} \end{vmatrix} = a_{11}a_{22} - a_{21}a_{12} \tag{B.28}$$

The value of a third-order determinant (3×3 determinant) can be calculated as

$$\begin{vmatrix} a_{11} & a_{12} & a_{13} \\ a_{21} & a_{22} & a_{23} \\ a_{31} & a_{32} & a_{33} \end{vmatrix} = a_{11}(\text{minor of } a_{11}) - a_{21}(\text{minor of } a_{21}) + a_{31}(\text{minor of } a_{31})$$

$$\tag{B.29}$$

$$= a_{11}\begin{vmatrix} a_{22} & a_{23} \\ a_{32} & a_{33} \end{vmatrix} - a_{21}\begin{vmatrix} a_{12} & a_{13} \\ a_{32} & a_{33} \end{vmatrix} + a_{31}\begin{vmatrix} a_{12} & a_{13} \\ a_{22} & a_{23} \end{vmatrix}$$

Third-order determinants. To calculate the value of a 3×3 determinant, we use minors. The minor of an element in a 3×3 determinant is a 2×2 determinant that results when both the row and the column that contain that element are deleted. For example,

$$\text{Minor } a_{11} \text{ of } \begin{vmatrix} a_{11} & a_{12} & a_{13} \\ a_{21} & a_{22} & a_{23} \\ a_{31} & a_{32} & a_{33} \end{vmatrix} = \begin{vmatrix} a_{22} & a_{23} \\ a_{32} & a_{33} \end{vmatrix}$$

$$\text{Minor } a_{21} \text{ of } \begin{vmatrix} a_{11} & a_{12} & a_{13} \\ a_{21} & a_{22} & a_{23} \\ a_{31} & a_{32} & a_{33} \end{vmatrix} = \begin{vmatrix} a_{12} & a_{13} \\ a_{32} & a_{33} \end{vmatrix} \qquad \text{(B.30)}$$

$$\text{Minor } a_{31} \text{ of } \begin{vmatrix} a_{11} & a_{12} & a_{13} \\ a_{21} & a_{22} & a_{23} \\ a_{31} & a_{32} & a_{33} \end{vmatrix} = \begin{vmatrix} a_{12} & a_{13} \\ a_{22} & a_{23} \end{vmatrix}$$

Properties of determinants. Let A be a square matrix.

1. If a multiple of one row of A is added to another row to produce a matrix B, then

$$\det B = \det A$$

2. If two rows of A are interchanged to produce B, then

$$\det B = -\det A$$

3. If one row of A is multiplied by k to produce B, then

$$\det B = k \det A$$

4. The determinant remains the same if the matrix is transposed

$$A = \begin{vmatrix} a_{11} & a_{12} \\ a_{21} & a_{22} \end{vmatrix}$$

$$A^{T} = \begin{vmatrix} a_{11} & a_{21} \\ a_{12} & a_{22} \end{vmatrix}$$

$$|A| = \left(a_{11}a_{22} - a_{21}a_{12} \right) = \left| A^{T} \right|$$

5. The determinant changes the sign if the two rows (or two columns) of the matrix are interchanged

$$A = \begin{vmatrix} a_{11} & a_{12} \\ a_{21} & a_{22} \end{vmatrix}$$

$$B = \begin{vmatrix} a_{21} & a_{22} \\ a_{11} & a_{12} \end{vmatrix}$$

$$|B| = \left(a_{11}a_{22} - a_{21}a_{12}\right) = -|A|$$

6. The determinant is zero if the two rows (or two columns) of the matrix are multiples of (or identical with) one another

$$A = \begin{bmatrix} a_{11} & a_{12} \\ 2a_{11} & 2a_{12} \end{bmatrix}$$

$$|A| = 2a_{11}a_{12} - 2a_{11}a_{12} = 0$$

7. The determinant is zero if the matrix has a zero row (or a zero column)

$$A = \begin{bmatrix} a_{11} & a_{12} \\ 0 & 0 \end{bmatrix}$$

$$|A| = \left[a_{11}(0) - (0)a_{12}\right] = 0$$

8. The determinant of a diagonal, an upper triangular matrix, and a lower triangular matrix is the product of the entries along the main diagonal.

9. $|A||B| = |AB|$

10. If any two rows (or columns) of A are linearly dependent, then

$$|A| = 0$$

11. If the ith row of A is multiplied by a scalar k and then is added to the jth row, and the jth row is replaced by the result, then the determinant of the new matrix is equal to the determinant of A.

General rule for determinants. Only a square matrix has a determinant. For a rectangular matrix, the determinant does not exist.

The determinant of a square n × n matrix A is

$$A = \begin{bmatrix} a_{ij} \end{bmatrix} \begin{bmatrix} a_{11} & a_{12} & \cdots & a_{1n} \\ a_{21} & a_{22} & \cdots & a_{2n} \\ \vdots & \vdots & \cdots & \vdots \\ a_{n1} & a_{n2} & \cdots & a_{nn} \end{bmatrix}$$

(B.31)

can be obtained by expanding in any one row or one column.

The determinant is given by

$$|A| = \sum_{j=1}^{n} a_{ij}A_{ij} = a_{i1}A_{i1} + a_{i2}A_{i2} + \ldots + a_{in}A_{in}$$

(B.32)

where A_{ij} are the elements of the cofactor matrix, cofactor A, or the cofactor matrix of A

$$\text{cofactor } A = \left| A_{ij} \right|$$

and its elements are given as

$$A_{ij} = \left(-1\right)^{i+j} \left| M_{ij} \right|$$

(B.33)

where after deleting the ith row and the jth column of A, the remainder matrix is called the minor matrix, M_{ij}. If the first row is selected, then we have

$$|A| = \sum_{j=1}^{n} a_{1j}A_{1j} = a_{11}A_{11} + a_{12}A_{12} + \ldots + a_{1n}A_{1n}$$

(B.34)

Similarly, if a particular jth column is chosen, then the determinant is given by

$$|A| = \sum_{j=1}^{n} a_{ij}A_{ij} = a_{1j}A_{1j} + a_{2j}A_{2j} + \ldots + a_{nj}A_{nj}$$

(B.35)

For a square 2 × 2 matrix,

$$A = \begin{bmatrix} a_{11} & a_{12} \\ a_{21} & a_{22} \end{bmatrix}$$

the determinant is given by

$$|A| = \sum_{j=1}^{n} a_{1j}A_{1j} = a_{11}A_{11} + a_{12}A_{12}$$

$$|A| = (-1)^{1+1} a_{11}a_{22} + (-1)^{1+2} a_{12}a_{21} \tag{B.36}$$

$$|A| = a_{11}a_{22} - a_{12}a_{21}$$

The plus and minus signs are determined by evaluating

$$(-1)^{i+j}$$

EXAMPLE EB.6

Given the two 2×2 matrices,

$$A = \begin{bmatrix} 6 & 0 \\ 2 & 1 \end{bmatrix}$$

and

$$B = \begin{bmatrix} 5 & 2 \\ 1 & 3 \end{bmatrix}$$

show that det $AB = (\det A)(\det B)$.

Solution:

$$\det A = 6 - 0 = 6$$

$$\det B = 15 - 2 = 13$$

$$AB = \begin{bmatrix} 6 & 0 \\ 2 & 1 \end{bmatrix}\begin{bmatrix} 5 & 2 \\ 1 & 3 \end{bmatrix} = \begin{bmatrix} 30 & 10 \\ 9 & 7 \end{bmatrix}$$

$$\det AB = 30(7) - 11(12) = 78$$

$$(\det A)(\det B) = (6)(13) = 78 = \det AB$$

EXAMPLE EB.7

Evaluate

$$\begin{vmatrix} 3 & -1 & -2 \\ 5 & 7 & -4 \\ 2 & -3 & 6 \end{vmatrix}$$

Solution:

Expand by minors about the first column

$$\begin{vmatrix} 3 & -1 & -2 \\ 5 & 7 & -4 \\ 2 & -3 & 6 \end{vmatrix} = 3\begin{vmatrix} 7 & -4 \\ -3 & 6 \end{vmatrix} - 5\begin{vmatrix} -1 & 2 \\ -3 & 6 \end{vmatrix} + 2\begin{vmatrix} -1 & -2 \\ 7 & -4 \end{vmatrix}$$

$$= 3\left[7(6) - (-3)(-4)\right] - 5\left[(-1)6 - (-3)(-2)\right] + 2\left[(-1)(-4) - 7(-2)\right] = 186$$

B.6 More on Matrix Inversion

If A and B are m × n matrices such that

$$AB = BA = I \qquad\qquad (B.37)$$

then B is said to be the inverse of A and is denoted by

$$B = A^{-1} \qquad\qquad (B.38)$$

To find the inverse of A^{-1}, provided the matrix A is given, let us consider the product

$$A \cdot \text{adj}A = \begin{bmatrix} a_{11} & a_{12} & \cdots & a_{1n} \\ a_{21} & a_{22} & \cdots & a_{2n} \\ a_{n1} & a_{n2} & \cdots & a_{nn} \end{bmatrix}$$

$$\times \begin{bmatrix} |M_{11}| & -|M_{12}| & \cdots & (-1)^{1+n}|M_{n1}| \\ -|M_{12}| & |M_{22}| & \cdots & (-1)^{2+n}|M_{n2}| \\ (-1)^{1+n}|M_{n1}| & (-1)^{2+n}|M_{n2}| & \cdots & |M_{nn}| \end{bmatrix} \qquad (B.39)$$

$$= \left[\sum_{j=1}^{n}(-1)^{i+j}a_{kj}|M_{ij}|\right]$$

An element of the matrix on the right side of Eq. B.39 has the value

$$\sum_{j=1}^{n}(-1)^{i+j}a_{kj}|M_{ij}| = \begin{vmatrix} a_{11} & a_{12} & \cdots & a_{1n} \\ a_{21} & a_{22} & \cdots & a_{2n} \\ a_{n1} & a_{n2} & \cdots & a_{nn} \end{vmatrix} = |a| \text{ if } i = k \qquad (B.40)$$

If $i \neq k$, the determinant possesses two identical rows because the determinant corresponding to $i \neq k$ is obtained from the matrix $[a]$ by replacing the ith row by the kth row and keeping the kth row intact. Therefore, if $i \neq k$, the value of the element is zero.

Equation B.39 can be written as

$$A \text{ adj} A = |A| \text{ I} \qquad (B.41)$$

Pre-multiplying Eq. B.41 throughout by A^{-1} and dividing the result by $|a|$, we obtain

$$A^{-1} = \frac{\text{adj } A}{\det A} \qquad (B.42)$$

so that the inverse of a matrix A is obtained by dividing its adjoint matrix by its determinant $|A|$.

If det A is equal to zero, then the elements of A^{-1} approach infinity (or are indeterminant at best), in which case the inverse A^{-1} is said not to exist, and the matrix A is said to be singular. The inverse of a matrix exists only if the determinant is not zero, that is, the matrix must be non-singular.

Transpose, inverse, and determinant of a product of matrices. If A is an $m \times n$ matrix and B is an $n \times p$ matrix, then $[C] = [A][B]$ is an $m \times p$ matrix with its elements given by

$$C_{ij} = \sum_{k=1}^{n} A_{ik}B_{kj} \qquad (B.43)$$

Now consider the product $B^T A^T$. Because to any element A_{ik} in A there corresponds the element A_{ki} in A^T, and to any element B_{kj} in B there corresponds the element B_{jk} in B^T, we can write

$$\sum_{k=1}^{n} B_{jk}A_{ki} = C_{ji} \qquad (B.44)$$

from which we obtain

$$C^T = B^T A^T \qquad (B.45)$$

or the transpose of a product of matrices is equal to the product of the transposed matrices in reverse order matrices. Therefore,

$$C = A_1 A_2 \ldots A_{s-1} A_s \tag{B.46}$$

Then

$$C^T = A_s^T A_{s-1}^T \ldots A_2^T A_1^T \tag{B.47}$$

Now consider

$$C = AB \tag{B.48}$$

where A and B are square matrices of order n. Then, pre-multiplying Eq. B.48 by $B^{-1}A^{-1}$, and post-multiplying the result by C^{-1}, we obtain simply

$$C^{-1} = B^{-1}A^{-1} \tag{B.49}$$

or the inverse of a product of matrices is equal to the product of the inverse matrices in reversed order. Equation B.49 can be generalized by considering the product in which all matrices $[A]_i$ (i = 1, 2, ..., s) are square matrices of order n and can be written as

$$C^{-1} = A_s^{-1} A_{s-1}^{-1} \ldots A_2^{-1} A_1^{-1} \tag{B.50}$$

Hence, the determinant of a product of matrices is equal to the product of the determinants of the matrices.

There is no direct division of matrices. The operation of division is performed by inversion. If

$$AB = C$$

then

$$B = A^{-1} C$$

where A^{-1} is called the inverse of matrix A.

The requirements for obtaining a unique inverse of a matrix are

1. The matrix is a square matrix.

2. The determinant of the matrix is not zero (the matrix is non-singular).

3. The inverse of a matrix also is defined by the relationship

$$A^{-1} A = I$$

The following are the properties of an inverted matrix:

1. The inverse of a matrix is unique.

2. The inverse of the product of two matrices is equal to the product of the inverse of the two matrices in reverse order

$$(AB)^{-1} = B^{-1}A^{-1}$$

3. The inverse of a triangular matrix is a triangular matrix of the same type.

4. The inverse of a symmetrical matrix is a symmetrical matrix.

5. The negative powers of a non-singular matrix are obtained by raising the inverse of the matrix to positive powers.

6. The inverse of the transpose of A is equal to the transpose of the inverse of A

$$\left(A^{T}\right)^{-1} = \left(A^{-1}\right)^{T}$$

B.7 System of Algebraic Equations

Cramer's rule. Consider the system of n simultaneous equations that are to be solved for the n unknowns $x_1, x_2, \ldots, x_n$,

$$a_{11}x_1 + a_{12}x_2 + \ldots a_{1n}x_n = y_1$$

$$a_{21}x_1 + a_{22}x_2 + \ldots a_{2n}x_n = y_2$$

$$\vdots$$

$$\vdots$$

$$a_{n1}x_1 + a_{n2}x_2 + \ldots a_{nn}x_n = y_n$$

(B.51)

Using the definitions of matrix addition and matrix multiplication, the system of Eq. B.51 can be written in matrix form as

$$Ax = y$$

(B.52)

where

$$A = \begin{bmatrix} a_{11} & a_{12} & \cdots & a_{1n} \\ a_{21} & a_{22} & \cdots & a_{2n} \\ \vdots & \vdots & \ddots & \vdots \\ a_{n1} & a_{n2} & \cdots & a_{nn} \end{bmatrix}$$

$$x = \begin{bmatrix} x_1 \\ x_2 \\ \vdots \\ x_n \end{bmatrix}$$

$$y = \begin{bmatrix} y_1 \\ y_2 \\ \vdots \\ y_n \end{bmatrix}$$

Cramer's rule can be used to solve for the components of x

$$x_i = \frac{|B_i|}{|A|} \tag{B.53}$$

where B_i is the matrix obtained by replacing the ith column of A with y. The matrix A cannot be singular.

EXAMPLE EB.8

Solve the following set of simultaneous linear equations for x, y, and z using Cramer's rule.

$$2x + 3y - z = -10$$

$$-x + 4y + 2z = -4$$

$$2x - 2y + 7z = 35$$

Solution:

Using Cramer's rule to solve the simultaneous linear equations, the coefficient matrix is

$$D = \begin{bmatrix} 2 & 3 & -1 \\ -1 & 4 & 2 \\ 2 & -2 & 7 \end{bmatrix}$$

The determinant is

$$|D| = 2\begin{vmatrix} 4 & 2 \\ -2 & 7 \end{vmatrix} + 1\begin{vmatrix} 3 & -1 \\ -2 & 7 \end{vmatrix} + 2\begin{vmatrix} 3 & -1 \\ 4 & 2 \end{vmatrix} = 103$$

The determinants of the substitutional matrices are

$$|A_1| = \begin{vmatrix} -10 & 3 & -1 \\ -4 & 4 & 2 \\ 35 & -2 & 7 \end{vmatrix} = 106$$

$$|A_2| = \begin{vmatrix} 2 & -10 & -1 \\ -1 & -4 & 2 \\ 2 & 35 & 7 \end{vmatrix} = -279$$

$$|A_3| = \begin{vmatrix} 2 & 3 & -10 \\ -1 & 4 & -4 \\ 2 & -2 & 35 \end{vmatrix} = 405$$

$$x = \frac{106}{103} = 1.0291$$

$$y = \frac{-279}{103} = -2.7087$$

$$z = \frac{405}{103} = 3.9320$$

Inversion of a matrix. Consider a set of three simultaneous linear algebraic equations

$$a_{11}x_1 + a_{12}x_2 + a_{13}x_3 = y_1$$
$$a_{21}x_1 + a_{22}x_2 + a_{23}x_3 = y_2 \quad \text{(B.54)}$$
$$a_{31}x_1 + a_{32}x_2 + a_{33}x_3 = y_3$$

Equation B.54 can be expressed in the matrix form

$$Ax = y \quad \text{(B.55)}$$

Pre-multiplying by the inverse of A^{-1}, we obtain the solution of x as

$$x = A^{-1}y \quad \text{(B.56)}$$

The term A^{-1} can be identified by Cramer's rule as follows. The solution for x_1 is

$$x_1 = \frac{1}{|A|} \begin{vmatrix} y_1 & a_{12} & a_{13} \\ y_2 & a_{22} & a_{23} \\ y_3 & a_{32} & a_{33} \end{vmatrix}$$

$$= \frac{1}{|A|} \left\{ y_1 \begin{vmatrix} a_{22} & a_{23} \\ a_{32} & a_{33} \end{vmatrix} - y_2 \begin{vmatrix} a_{12} & a_{13} \\ a_{32} & a_{33} \end{vmatrix} + y_3 \begin{vmatrix} a_{12} & a_{13} \\ a_{22} & a_{23} \end{vmatrix} \right\}$$

$$= \frac{1}{|A|} \left\{ y_1 C_{11} + y_2 C_{21} + y_3 C_{31} \right\}$$

where A is the determinant of the coefficient matrix A, and C_{11}, C_{21}, and C_{31} are the cofactors of A corresponding to element 11, 21, and 31. We also can write similar expressions for x_2 and x_3 by replacing the second and third columns by the y column, respectively. Hence, the complete solution can be written in matrix form as

$$\begin{Bmatrix} x_1 \\ x_2 \\ x_3 \end{Bmatrix} = \frac{1}{|A|} \begin{bmatrix} C_{11} & C_{21} & C_{31} \\ C_{12} & C_{22} & C_{32} \\ C_{13} & C_{23} & C_{33} \end{bmatrix} \begin{Bmatrix} y_1 \\ y_2 \\ y_3 \end{Bmatrix} \tag{B.57}$$

or

$$\{x\} = \frac{1}{|A|} \left[C_{ji} \right] \{y\} = \frac{1}{|A|} \left[\text{adj } A \right] \{y\}$$

Hence,

$$A^{-1} = \frac{1}{|A|} \text{adj } A \tag{B.58}$$

EXAMPLE EB.9

Solve the following system of linear equations for x, y, and z using the matrix inversion method:

$$10x + 3y + 10z = 5$$

$$8x - 2y + 9z = 11$$

$$8x + 4y - 10z = 7$$

Solution:

There are several ways to solve this problem.

$$AX = B$$

$$\begin{bmatrix} 10 & 3 & 10 \\ 8 & -2 & 9 \\ 8 & 4 & -10 \end{bmatrix} \begin{bmatrix} x \\ y \\ z \end{bmatrix} = \begin{bmatrix} 5 \\ 11 \\ 7 \end{bmatrix}$$

$$A A^{-1} X = A^{-1} B$$

$$I X = A^{-1} B$$

or

$$X = A^{-1}B$$

$$A^{-1} = x = \begin{bmatrix} 1.3131 \\ -1.6624 \\ -0.3144 \end{bmatrix}$$

Hence, the solution is $x = 1.3131$, $y = -1.6624$, and $z = -0.3144$.

B.8 Eigenvalues and Eigenvectors

Consider a set of linear simultaneous equations in the form

$$AX = \lambda X \tag{B.59}$$

where A is a square matrix, X is a column matrix, and λ is a number. We can rewrite Eq. B.59 as

$$\begin{bmatrix} a_{11} - \lambda & a_{12} & \cdots & a_{1n} \\ a_{21} & a_{22} - \lambda & \cdots & a_{2n} \\ a_{31} & a_{32} & \cdots & a_{3n} \\ \vdots & \vdots & \cdots & \vdots \\ a_{n1} & a_{n2} & \cdots & a_{nn} - \lambda \end{bmatrix} \begin{Bmatrix} x_1 \\ x_2 \\ x_3 \\ \vdots \\ x_n \end{Bmatrix} = 0$$

or

$$(A - \lambda I)X = 0 \tag{B.60}$$

A nontrivial solution of Eq. B.60 can exist only when the determinant of $(A - \lambda I)$ vanishes, or

$$|A - \lambda I| = 0 \tag{B.61}$$

Equation B.61 is called the characteristic equation. The roots λ_1, λ_2, ..., λ_n of the characteristic equation are called the eigenvalues of matrix A. When each root is substituted back into Eq. B.59, we obtain a set of linear equations that are not all independent. By assuming a value of one x, say x_1, and discarding one equation, we can solve for the values of other x's. The column matrix obtained by this procedure is called a characteristic vector or eigenvector. Thus, there is one characteristic vector for each eigenvalue.

Determination of eigenvectors. The eigenvector X_i corresponding to the eigenvalue λ_i can be determined from the cofactors of any row of the characteristic equation.

Let $\left[A - \lambda_i I\right]X_i = 0$ be written for a third-order system as

$$\begin{bmatrix} (a_{11} - \lambda_i) & a_{12} & a_{13} \\ a_{21} & (a_{22} - \lambda_i) & a_{23} \\ a_{31} & a_{32} & (a_{33} - \lambda_i) \end{bmatrix} \begin{Bmatrix} x_1 \\ x_2 \\ x_3 \end{Bmatrix} = 0 \tag{B.62}$$

The characteristic equation $\left[A - \lambda_i I\right] = 0$ is

$$\begin{bmatrix} (a_{11} - \lambda_i) & a_{12} & a_{13} \\ a_{21} & (a_{22} - \lambda_i) & a_{23} \\ a_{31} & a_{32} & (a_{33} - \lambda_i) \end{bmatrix} = 0 \tag{B.63}$$

Expanding Eq. B.63 in terms of the cofactors of the first row gives

$$(a_{11} - \lambda_1)C_{11} + a_{12}C_{12} + a_{13}C_{13} = 0 \tag{B.64}$$

Now replace the first row of the determinant by the second row, leaving the other two rows unchanged. The value of the determinant is zero because there are two identical rows

$$\begin{vmatrix} a_{21} & (a_{22} - \lambda_i) & a_{23} \\ a_{21} & (a_{22} - \lambda_i) & a_{23} \\ a_{31} & a_{32} & (a_{33} - \lambda_i) \end{vmatrix} = 0 \tag{B.65}$$

Similarly, if we expand in terms of the cofactors of the first row, which are identical to the cofactors of the previous determinant, we obtain

$$a_{21}C_{11} + (a_{22} - \lambda_i)C_{12} + a_{23}C_{13} = 0 \tag{B.66}$$

Finally, replace the first row by the third row and expand in terms of the first row of the new determinant

$$\begin{vmatrix} a_{31} & a_{32} & (a_{33} - \lambda_i) \\ a_{21} & (a_{22} - \lambda_i) & a_{23} \\ a_{31} & a_{32} & (a_{33} - \lambda_i) \end{vmatrix} = 0 \tag{B.67}$$

$$a_{31}C_{11} + a_{32}C_{12} + (a_{33} - \lambda_i)C_{13} = 0 \tag{B.68}$$

Equations B.64, B.66, and B.68 now can be assembled in a single matrix equation as

$$\begin{bmatrix} (a_{11} - \lambda_i) & a_{12} & a_{13} \\ a_{21} & (a_{22} - \lambda_i) & a_{23} \\ a_{31} & a_{32} & (a_{33} - \lambda_i) \end{bmatrix} \begin{Bmatrix} C_{11} \\ C_{12} \\ C_{13} \end{Bmatrix} = 0 \tag{B.69}$$

Hence, the eigenvector X_i can be obtained from the cofactors of the characteristic equation with $\lambda = \lambda_i$. Because the eigenvectors are relative to a normalized coordinate, the column of cofactors can differ by a multiplying factor, β

$$\begin{Bmatrix} x_1 \\ x_2 \\ x_3 \end{Bmatrix} = \beta \begin{Bmatrix} C_{11} \\ C_{12} \\ C_{13} \end{Bmatrix} \tag{B.70}$$

EXAMPLE EB.10

Determine the eigenvalues and eigenvectors of the matrix

$$A = \begin{bmatrix} 2 & -1 & 0 \\ -1 & 3 & -2 \\ 0 & -2 & 3 \end{bmatrix}$$

Solution:

The eigenvalues of A are determined by finding the values of λ satisfying

$$\begin{vmatrix} 2 - \lambda & -1 & 0 \\ -1 & 3 - \lambda & -2 \\ 0 & -2 & 3 - \lambda \end{vmatrix} = 0$$

Expansion of the determinant gives

$$(2 - \lambda) \begin{vmatrix} 3 - \lambda & -2 \\ -2 & 3 - \lambda \end{vmatrix} - (-1) \begin{vmatrix} -1 & -2 \\ 0 & 3 - \lambda \end{vmatrix} = 0$$

When the 2×2 determinants are expanded, the following cubic equation is obtained:

$$\lambda^3 + 8\lambda^2 - 16\lambda + 7 = 0$$

The eigenvalues are the roots of the cubic equation, which are 0.609, 2.227, and 5.164. The eigenvector corresponding to the smallest eigenvalue is obtained by solving

$$
\begin{bmatrix} 1.391 & -1 & 0 \\ -1 & 2.391 & -2 \\ 0 & -2 & 2.391 \end{bmatrix} \begin{bmatrix} x_1 \\ x_2 \\ x_3 \end{bmatrix} = \begin{bmatrix} 0 \\ 0 \\ 0 \end{bmatrix}
$$

The first equation gives $x_1 = 0.719x_2$. The third equation gives $x_3 = 0.836x_2$. When these relationships are substituted into the second equation, it is identically satisfied. Thus, x_2 remains arbitrary, and the eigenvector of A corresponding to $\lambda = 0.609$ is

$$
C_1 \begin{bmatrix} 0.719 \\ 1 \\ 0.836 \end{bmatrix}
$$

where C_1 is an arbitrary constant. The same procedure is followed, yielding the eigenvectors corresponding to the second and third eigenvalues. These are, respectively,

$$
C_2 \begin{bmatrix} -4.41 \\ 1 \\ 2.59 \end{bmatrix}
$$

$$
C_3 \begin{bmatrix} -0.316 \\ 1 \\ -0.924 \end{bmatrix}
$$

If A is an $n \times n$ singular matrix, then one of its eigenvalues is zero. If A is non-singular, then the eigenvalues of A^{-1} are the reciprocals of the eigenvalues of A. The eigenvectors of A^{-1} are the same as the eigenvectors of A.

Properties of eigenvalues and eigenvectors. The following properties apply to eigenvalues and eigenvectors.

1. If a real matrix A has eigenvalues λ_i and characteristic vectors X_i, then A^T has the same eigenvalues λ_i but with characteristic vectors Y_i orthogonal to X_i as

$$
Y_j^T X_i = \begin{cases} 1 & \text{when} \quad i = j \\ 0 & \text{when} \quad i \neq j \end{cases} \tag{B.71}
$$

where the vectors X and Y are normalized.

2. If a matrix A is symmetric and all its elements are real numbers, the eigenvalues and characteristic vectors are real. Moreover, the characteristic vectors are orthogonal to each other as

$$X_i^T X_j = \begin{cases} 1 & \text{when} \quad i = j \\ 0 & \text{when} \quad i \neq j \end{cases} \tag{B.72}$$

3. The determinant of a matrix is equal to the product of all its eigenvalues as

$$|A| = \lambda_1 \lambda_2 \dots \lambda_n \tag{B.73}$$

4. The trace of a matrix, which is the sum of the elements on the leading diagonal of a matrix, is equal to the sum of its eigenvalues

$$a_{11} + a_{22} + \dots + a_{nn} = \lambda_1 + \lambda_2 + \dots + \lambda_n \tag{B.74}$$

5. If the eigenvalues and characteristic vectors of matrix A are λ_i and X_i, then a matrix $B = TAT^{-1}$ has the same eigenvalue λ_i but characteristic vectors equal to TX_i, T being any non-singular square matrix.

B.9 Quadratic Forms

A function of n variables $F\left(x_1, x_2, \dots, x_n\right)$ is called a quadratic form if

$$F\left(x_1, x_2, \dots, x_n\right) = \sum_{i=1}^{n} \sum_{j=1}^{n} q_{ij} \, x_i \, x_j = X^T Q X \tag{B.75}$$

where $Q = \left[q_{ij}\right]_{n \times n}$ and $X^T = \left(x_1, x_2, \dots, x_n\right)$ without any loss of generality, and Q can always be assumed symmetric. Otherwise, Q may be replaced by the symmetrical matrix $\dfrac{\left(Q + Q^T\right)}{2}$ without changing the value of the quadratic form.

B.10 Positive Definite Matrix

A matrix A is positive definite if and only if the quadratic form $X^t A X > 0$ for all $X \neq 0$. A matrix A is positive semi-definite if and only if the quadratic form $X^T A X \geq 0$ for all X, and there exists an $X \neq 0$ such that $X^T A X = 0$.

Tests for positive definite matrices. The following are true for all positive definite matrices:

1. All the eigenvalues must be positive.

2. All diagonal elements must be positive.

3. All the leading principal determinants must be positive, that is, if

$$A = \begin{bmatrix} a_{11} & a_{12} & \cdots & a_{1n} \\ a_{21} & a_{22} & \cdots & a_{2n} \\ \vdots & \vdots & \cdots & \vdots \\ a_{n1} & a_{n2} & \cdots & a_{nn} \end{bmatrix}_{n \times n} \tag{B.76}$$

then

$$A_1 = a_{11}$$

$$A_2 = \begin{vmatrix} a_{11} & a_{12} \\ a_{21} & a_{22} \end{vmatrix}$$

$$A_3 = \begin{vmatrix} a_{11} & a_{12} & a_{13} \\ a_{21} & a_{22} & a_{23} \\ a_{31} & a_{32} & a_{33} \end{vmatrix} \tag{B.77}$$

$$\vdots$$

$$A_n = \begin{vmatrix} a_{11} & a_{12} & \cdots & a_{1n} \\ a_{21} & a_{22} & \cdots & a_{2n} \\ \vdots & \vdots & \cdots & \vdots \\ a_{n1} & a_{n2} & \cdots & a_{nn} \end{vmatrix}$$

A is said to be positive definite if all $A_1, A_2, ..., A_n$ are positive.

EXAMPLE EB.11

Consider the matrix A given by

$$A = \begin{bmatrix} 5 & 3 & 4 \\ 3 & 2 & 1 \\ 2 & 3 & 10 \end{bmatrix}$$

The principal minors of A are

$$A_1 = 5 > 0$$

$$A_2 = \begin{vmatrix} 5 & 3 \\ 3 & 2 \end{vmatrix} = 1 > 0$$

$$A_3 = \begin{vmatrix} 5 & 3 & 4 \\ 3 & 2 & 1 \\ 2 & 3 & 10 \end{vmatrix} = 21 > 0$$

Because all the principal minors of A are positive, the matrix A is a positive definite matrix.

B.11 Negative Definite Matrix

A matrix A is negative definite if and only if A is positive definite. In other words, A is negative definite when $X^T A X < 0$ for all $X \neq 0$.

A matrix A is said to be negative semi-definite if $X^T A X \leq 0$ for all $X \neq 0$.

Tests for negative definite matrices. Consider Eqs. B.76 and B.77. The matrix A is said to be negative definite if

$$A_j = \left(-1^j \right) \qquad j = 1, 2, ..., n \qquad\qquad (B.78)$$

EXAMPLE EB.12

Consider the matrix A given by

$$A = \begin{vmatrix} -9 & 4 \\ 1 & -5 \end{vmatrix}$$

The principal minors of A are

$$A_1 = -9 < 0$$

$$A_2 = \begin{vmatrix} -9 & 4 \\ 1 & -5 \end{vmatrix} = 41 > 0$$

Because the principal minor of A_1 is negative and A_2 is positive, the matrix A is a negative definite matrix.

B.12 Indefinite Matrix

A matrix A is indefinite if X^TAX is positive for some X and negative for other X.

For example, consider Eqs. B.75 and B.76. The matrix A is said to be indefinite if

$$A_j = (-1)^{j+1} \qquad j = 1, 2, ..., n \qquad (B.79)$$

EXAMPLE EB.13

Consider the matrix A given by

$$A = \begin{bmatrix} 3 & 4 \\ 1 & -8 \end{bmatrix}$$

The principal minors of A are

$$A_1 = 3 > 0$$

$$A_2 = \begin{bmatrix} 3 & 4 \\ 1 & -8 \end{bmatrix} = -28 > 0$$

Because the principal minor of A_1 is positive and A_2 is negative, the matrix A is an indefinite matrix.

B.13 Norm of a Vector

A norm is any function that assigns to every vector X of real numbers, a real number K satisfying the following four properties:

$$\|X\| \geq 0 \qquad (B.80)$$

$$\|X + Y\| \leq \|X\| + \|Y\| \qquad (B.81)$$

$$\|\beta X\| = \beta \|X\| \text{ for every scalar } \beta \qquad (B.82)$$

$$\|X\| = 0 \text{ if and only if } X = 0 \qquad (B.83)$$

The most commonly used norm is the Euclidean norm and is given by

$$\|X\| = \left[\sum_{i=1}^{n} X_i^2 \right]^{1/2} \qquad (B.84)$$

The other norms are

$$\|X\| = \sum_{i=1}^{n} |X_i| \qquad X \text{ is an } n \times 1 \text{ vector} \qquad (B.85)$$

$$\|X\| = \max \left\{ |X_i| \right\} \qquad (B.86)$$

$$\|X\| = \left(X^T X \right)^{1/2} \qquad (B.87)$$

$$\|X\| = \left(X^T A X \right)^{1/2} \qquad (B.88)$$

where A is a positive definite matrix.

B.14 Partitioning of Matrices

When two matrices are multiplied together, we may subdivide the work into separate compartments by drawing horizontal and vertical lines called partitions through the matrices

$$AB = \begin{bmatrix} a_{11} & a_{12} & \vdots & a_{13} & a_{14} \\ a_{21} & a_{22} & \vdots & a_{23} & a_{24} \\ \cdots & \cdots & \cdots & \cdots & \cdots \\ a_{31} & a_{32} & \vdots & a_{33} & a_{34} \\ a_{41} & a_{42} & \vdots & a_{43} & a_{44} \end{bmatrix} \begin{bmatrix} b_{11} & b_{12} & b_{13} \\ b_{21} & b_{22} & b_{23} \\ \cdots & \cdots & \cdots \\ b_{31} & b_{32} & b_{33} \\ b_{41} & b_{42} & b_{43} \end{bmatrix} \qquad (B.89)$$

$$= \begin{bmatrix} A_1 & A_2 \\ A_3 & A_4 \end{bmatrix} \begin{bmatrix} B_1 \\ B_2 \end{bmatrix}$$

where

$$A_1 = \begin{bmatrix} a_{11} & a_{12} \\ a_{21} & a_{22} \end{bmatrix}$$

$$A_2 = \begin{bmatrix} a_{13} & a_{14} \\ a_{23} & a_{24} \end{bmatrix}$$

$$A_3 = \begin{bmatrix} a_{31} & a_{32} \\ a_{41} & a_{42} \end{bmatrix}$$

$$A_4 = \begin{bmatrix} a_{33} & a_{34} \\ a_{43} & a_{44} \end{bmatrix}$$

$$B_1 = \begin{bmatrix} b_{11} & b_{12} & b_{13} \\ b_{21} & b_{22} & b_{23} \end{bmatrix}$$

$$B_2 = \begin{bmatrix} b_{31} & b_{32} & b_{33} \\ b_{41} & b_{42} & b_{43} \end{bmatrix}$$

The matrices A_1, A_2, A_3, A_4, B_1, and B_2 are called submatrices, and a matrix whose elements are matrices is called a super matrix. By applying the rules of matrix multiplication to the super matrices of Eq. B.89, we can write

$$AB = \begin{bmatrix} A_1B_1 + A_2B_2 \\ A_3B_1 + A_4B_2 \end{bmatrix} \qquad (B.90)$$

By expanding Eq. B.89, it will be found that Eq. B.90 is correct. In general, we can partition matrices in any way desired as long as all submatrices, which are to be multiplied together, are conformable.

B.15 Augmented Matrix

A system of linear equations in matrix notation takes the form $AX = b$, where A is of the order $m \times n$, X is of the order $n \times 1$, and b is of the order $m \times 1$. The augmented matrix A_b can be obtained by adjoining column b to matrix A. In terms of partitioned matrices, we have

$$A_b = \begin{bmatrix} A \vdots b \end{bmatrix}$$

As an example, let us consider a set of linear equations as

$$X_1 + X_2 + X_3 = 6$$

$$X_1 + X_2 - X_3 = 0$$

$$X_1 - X_2 + X_3 = 2$$

Then we have

$$A = \begin{bmatrix} 1 & 1 & 1 \\ 1 & 1 & -1 \\ 1 & -1 & 1 \end{bmatrix}$$

$$b = \begin{Bmatrix} 6 \\ 0 \\ 2 \end{Bmatrix}$$

B.16 Matrix Calculus

Integration. If $X(t)$ is an $m \times n$ matrix, with $a_{11}(t), a_{12}(t), ..., a_{mn}(t)$ as the elements, then

$$\int X(t)\, dt = \begin{bmatrix} \int a_{11}(t)dt & \int a_{12}(t)dt & \cdots & \int a_{1n}(t)dt \\ \int a_{21}(t)dt & \int a_{22}(t)dt & \cdots & \int a_{2n}(t)dt \\ \vdots & \vdots & \cdots & \vdots \\ \int a_{m1}(t)dt & \int a_{m2}(t)dt & \cdots & \int a_{mn}(t)dt \end{bmatrix} \qquad (B.91)$$

Differentiation. If $X(t)$ is an $m \times n$ matrix, with $a_{11}(t), a_{12}(t), ..., a_{mn}(t)$ as the elements, then

$$\frac{d}{dt}(X) = \begin{bmatrix} \dfrac{da_{11}}{dt} & \dfrac{da_{12}}{dt} & \cdots & \dfrac{da_{1n}}{dt} \\ \dfrac{da_{21}}{dt} & \dfrac{da_{22}}{dt} & \cdots & \dfrac{da_{2n}}{dt} \\ \vdots & \vdots & \cdots & \vdots \\ \dfrac{da_{m1}}{dt} & \dfrac{da_{m2}}{dt} & \cdots & \dfrac{da_{mn}}{dt} \end{bmatrix} \qquad (B.92)$$

Derivatives of product and sum.

$$\begin{array}{ccc} A = A(t) & B = B(t) & C = AB \\ (m \times n \text{ matrix}) & (n \times p \text{ matrix}) & (m \times p \text{ matrix}) \end{array}$$

$$\frac{dC}{dt} = \frac{dA}{dt}B + A\frac{dB}{dt} \qquad (B.93)$$

$$\begin{array}{ccc} A = A(t) & B = B(t) & C = A + B \\ (m \times n \text{ matrix}) & (m \times n \text{ matrix}) & (m \times n \text{ matrix}) \end{array}$$

$$\frac{dC}{dt} = \frac{dA}{dt} + \frac{dB}{dt} \qquad (B.94)$$

Derivatives of linear and quadratic forms.

$$\frac{dX^{T}}{dX} = I \qquad X \text{ is an } n \times 1 \text{ vector} \qquad (B.95)$$

$$\frac{d(X^{T}b)}{dX} = b \qquad b \text{ is an } n \times 1 \text{ vector} \qquad (B.96)$$

$$\frac{d\left(X^T X\right)}{dX} = 2X \qquad\qquad (B.97)$$

B.17 Summary

A matrix is a rectangular array of elements in rows and columns. The elements of a matrix can be numbers, coefficients, terms, or variables. This appendix provided the relevant and useful elements of matrix analysis for the study of dynamics and control. Topics covered included matrix definitions, matrix operations, determinants, matrix inversion, trace, transpose, system of algebraic equations and solution, eigenvalues and eigenvector determination, definite matrix, partitioning of matrices, augmented matrix, and matrix calculus.

B.18 References

1. Barnett, C., *Matrix Methods for Engineers and Scientists*, McGraw-Hill, New York, 1982.

2. Churchill, R.V., *Operational Mathematics*, 3rd Ed., McGraw-Hill, New York, 1972.

3. Dukkipati, R.V., *Advanced Engineering Analysis*, Narosa Publishing House, New Delhi, India, 2006.

4. Gantmacher, F.R., *Theory of Matrices, Vols. I and II*, Chelsea Publishing Company, New York, 1959.

5. Halmos, P.R., *Finite Dimensional Vector Spaces*, Van Nostrand Reinhold, New York, 1958.

6. Jain, R.K. and Iyengar, S.R.K., *Advanced Engineering Mathematics*, Narosa Publishing Company, New Delhi, India, 2002.

7. Jeffrey, A., *Advanced Engineering Mathematics*, Harcourt/Academic Press, New York, 2002.

8. Kreyszig, E., *Advanced Engineering Mathematics*, 8th Ed., Wiley, New York, 1998.

9. Lay, D.C., *Linear Algebra and Its Applications*, 2nd Ed., Addison-Wesley, Reading, MA, 2000.

10. Lipschutz, S., *Linear Algebra*, Schaum's Solved Problems Series, McGraw-Hill, New York, 1989.

11. Lipson, M., *Linear Algebra*, Schaum's Outlines, McGraw-Hill, New York, 2001.

12. Lopez, R.J., *Advanced Engineering Mathematics*, Addison-Wesley, New York, 2001.

13. Noble, B. and Daniel, J., *Applied Linear Algebra*, 2nd Ed., Prentice-Hall, Upper Saddle River, NJ, 1977.

14. O'Neil, P.V., *Advanced Engineering Mathematics*, 4th Ed., Brooks/Cole Publishing Company, New York, 1995.

15. Siegel, M., *Advanced Mathematics for Engineers and Scientists*, Schaum's Outlines, McGraw-Hill, New York, 1971.

16. Siegel, M., *Vector Analysis and Introduction to Tensor Analysis*, Schaum's Outlines, McGraw-Hill, New York, 1959.

B.19 Glossary of Terms

Adjugate (or classical adjoint): The matrix adj A formed from a square matrix A by replacing the (i, j)-entry of A by the (i, j)-cofactor, or all i and j, and then transposing the resulting matrix.

Associative law of multiplication: $A(BC) = (AB)C$ for all A, B, and C.

Augmented matrix: A matrix made up of a coefficient matrix for a linear system and one or more columns to the right. Each extra column contains the constants from the right side of a system with the given coefficient matrix.

Band matrix: A matrix whose nonzero entries lie within a band along the main diagonal.

Basic variable: A variable in a linear system that corresponds to a pivot column in the coefficient matrix.

Characteristic equation (of A): $\det(A - \lambda I) = 0$.

Characteristic polynomial (of A): $\det(A - \lambda I)$ or $\det(\lambda I - A)$.

Coefficient matrix: A matrix whose entries are the coefficients of a system of linear equations.

Cofactor: A number $C_{ij} = (-1)^{i+j} \det A_{ij}$, called the (i, j)-cofactor of A, where A_{ij} is the submatrix formed by deleting the ith row and the jth column of A.

Cofactor expansion: A formula for det A using cofactors associated with one row or one column, such as for row 1: $\det A = a_{11}C_{11} + \ldots + a_{1n}C_{1n}$.

Column-row expansion: The expression of a product AB as a sum of outer products: $\text{col}_1(A) \text{ row } 1(B) + \ldots + \text{col}_n(A) \text{ row } n(B)$, where n is the number of columns of A.

Column sum: The sum of the entries in a column of a matrix.

Column vector: A matrix with only one column, or a single column of a matrix that has several columns.

Commuting matrices: Two matrices A and B, such that AB = BA.

Complex eigenvalue: A nonreal root of the characteristic equation of an $n \times n$ matrix.

Complex eigenvector: A nonzero vector x in C^n, such that $Ax = \lambda x$, where A is an $n \times n$ matrix and λ is a complex eigenvalue.

Cramer's rule: A formula for each entry in the solution x of the equation $Ax = B$ when A is an invertible matrix.

Determinant (of a square matrix A): The number det A defined inductively by a cofactor expansion along the first row of A. Also, $(-1)^r$ times the product of the diagonal entries in any echelon form U obtained from A by row replacements and r row interchanges (but no scaling operations).

Diagonal matrix: A square matrix whose entries not on the main diagonal are all zero.

Distributive laws: (left) $A(B + C) = AB + AC$, and (right) $(B + C)A = BA + CA$, for all A, B, and C.

Dynamical system: *See* Discrete linear dynamical system.

Echelon form (or row echelon form, of a matrix): An echelon matrix that is row equivalent to the given matrix.

Echelon matrix (or row echelon matrix): A rectangular matrix that has three properties: (1) All nonzero rows are above any row of all zeros; (2) Each leading entry of a row is in a column to the right of the leading entry of the row above it; and (3) All entries in a column below a leading entry are zero.

Eigenspace (of A corresponding to λ): The set of all solutions of $Ax = \lambda x$, where λ is an eigenvalue of A. Consists of the zero vector and all eigenvectors corresponding to λ.

Eigenvalue (of A): A scalar λ such that the equation $Ax = \lambda x$ has a solution for some nonzero vector x.

Eigenvector (of A): A nonzero vector x such that $Ax = \lambda x$ for some scalar λ.

Elementary matrix: An invertible matrix that results by performing one elementary row operation on an identity matrix.

Elementary row operations: (1) (Replacement) Replace one row by the sum of itself and a multiple of another row. (2) Interchange two rows. (3) (Scaling) Multiply all entries in a row by a nonzero constant.

Free variable: Any variable in a linear system that is not a basic variable.

Full rank (matrix): An $m \times n$ matrix whose rank is the smaller of m and n.

Homogeneous equation: An equation of the form $Ax = 0$, possibly written as a vector equation or as a system of linear equations.

Ill-conditioned matrix: A square matrix with a large (or possibly infinite) condition number; a matrix that is singular or can become singular if some of its entries are changed ever so slightly.

Inverse (of an $n \times n$ matrix A): An $n \times n$ matrix A^{-1} such that $AA^{-1} = A^{-1}A = I_n$.

Invertible matrix: A square matrix that possesses an inverse.

Linear equation (in the variables $x_1, ..., x_n$): An equation that can be written in the form $a_1x_1 + a_2x_2 + ... + a_nx_n = b$, where b and the coefficients $a_1, ..., a_n$ are real or complex numbers.

Linear system: A collection of one or more linear equations involving the same variables, say, $x_1, \ldots, x_n$.

Lower triangular matrix: A matrix with zeros above the main diagonal.

Lower triangular part (of A): A lower triangular matrix whose entries on the main diagonal and below agree with those in A.

LU factorization: The representation of a matrix A in the form $A = LU$, where L is a square lower triangular matrix with ones on the diagonal (a unit lower triangular matrix) and U is an echelon form of A.

Main diagonal (of a matrix): The entries with equal row and column indices.

Matrix: A rectangular array of numbers.

Matrix equation: An equation that involves at least one matrix, such as $Ax = B$.

Nonsingular (matrix): An invertible matrix.

Nonzero (matrix or vector): A matrix (with possibly only one row or column) that contains at least one nonzero entry.

Positive definite matrix: A symmetrical matrix A such that $x^T A x > 0$ for all $x \neq 0$.

Positive semi-definite matrix: A symmetrical matrix A such that $x^T A x \geq 0$ for all x.

Rank (of a matrix A): The dimension of the column space of A, denoted by rank A.

Row-column rule: The rule for computing a product AB in which the (i, j)-entry of AB is the sum of the products of corresponding entries from row i of A and column j of B.

Row equivalent (matrices): Two matrices for which there exists a (finite) sequence of row operations that transforms one matrix into the other.

Row sum: The sum of the entries in a row of a matrix.

Row vector: A matrix with only one row, or a single row of a matrix that has several rows.

Scalar: A (real) number used to multiply a vector or matrix.

Submatrix (of A): Any matrix obtained by deleting some row and/or columns of A; also, A itself.

Symmetrical matrix: A matrix A such that $A^T = A$.

System of linear equations (or a linear system): A collection of one or more linear equations involving the same set of variables, say, $x_1, \ldots, x_n$.

Trace (of a square matrix A): The sum of the diagonal entries in A, denoted by tr A.

Transpose (of A): An $n \times m$ matrix A^T whose columns are the corresponding rows of the $m \times n$ matrix A.

Upper triangular matrix: A matrix U with zeros below the diagonal entries $u_{11}, u_{22}, \ldots$

Appendix C

Laplace Transforms

The Laplace transformation is an operational method for solving linear, time-invariant, differential equations and corresponding initial and boundary value problems. The process of solution consists mainly of three steps:

1. The given problem is transformed into a simple equation, often called a subsidiary equation.

2. The subsidiary equation is solved by purely algebraic manipulations.

3. The solution of the subsidiary equation is transformed back to obtain the solution of the given problem.

The Laplace transformation reduces the problem of solving a differential equation to an algebraic problem. In comparison with the classical method of solving linear differential equations, the Laplace transform method has the advantage that the initial conditions are automatically taken care of, and the homogeneous equation and the particular integral are solved in one operation.

C.1 Laplace Transformation

Let us consider a function $f(t)$ defined for all values of time larger than zero, $t > 0$, and define the (one-sided) Laplace transformation of $f(t)$ by the definite integral

$$L\left[f(t)\right] = F(s)$$

$$= \int_0^\infty e^{-st} f(t) dt$$

(C.1)

where e^{-st} is known as the kernel of the transformation, and s is referred to as a subsidiary variable. L is an operational symbol indicating that the quantity that it prefixes is to be transformed by the Laplace integral $\int_0^\infty e^{-st} dt$.

$F(s)$ is the Laplace transform of $f(t)$ and $f(t) = 0$ for $t < 0$. The variable s is, in general, a complex quantity (complex variable), and the associated complex plane $s = x + iy$ is called the s plane and at times the Laplace plane. Because Eq. C.1 is defined in terms of an integral, it is said to be an integral transformation, also commonly referred to as an integral transform.

The function $f(t)$ must be such that integral Eq. C.1 exists, which places on $f(t)$ the restriction

$$\left[e^{-st} f(t) \right] < C e^{-(s-a)t} \qquad \text{Re } s > a \qquad (C.2)$$

where C is a constant. The condition in Eq. C.2 implies that $f(t)$ must not increase with time more rapidly than the exponential function Ce^{-st}. Another restriction on $f(t)$ is that it must be piecewise continuous. Most functions describing physical phenomena satisfy these conditions.

C.2 Existence of Laplace Transform

The necessary conditions for the existence of the integral in Eq. C.1 are as follows:

a. $f(t)$ is piecewise continuous on every finite (time) interval of $(0, \infty)$

b. The magnitude of $f(t)$ is bounded by an exponential function

$$\left| f(t) \right| \le M\, e^{at} \qquad (C.3)$$

for some real constants a and M and all $t \in (0, \infty)$.

EXAMPLE EC.1

Find the Laplace transform of $f(t)$ defined by

$$f(t) = 0 \qquad t < 0$$

$$= te^{-5t} \qquad t \ge 0$$

Solution:

Because

$$L(t) = G(s) = \frac{1}{s^2}$$

and

$$L\left[e^{-\alpha t}f(t)\right]\int_0^\infty e^{-\alpha t}f(t)e^{-5t}dt = F(s+\alpha)$$

$$F(s) = L\left[te^{-5t}\right]$$

$$= G(s+5)$$

$$= \frac{1}{(s+5)^2}$$

C.3 Inverse Laplace Transform

The operation of obtaining $f(t)$ from the Laplace transform $F(s)$ is termed the inverse Laplace transformation. The inverse Laplace transform of $F(s)$ is denoted by

$$f(t) = L^{-1}\left[F(s)\right] \tag{C.4}$$

and is given by the inverse Laplace transform integral

$$f(t) = \frac{1}{2\pi j}\int_{c-j\infty}^{c+j\infty}F(s)e^{st}ds \tag{C.5}$$

where c is a real constant that is greater than the real parts of all the singularities of $F(s)$. Equation C.5 represents a line integral that is to be evaluated in the s plane. However, for most engineering purposes, the inverse Laplace transform operation can be accomplished simply by referring to the Laplace transform table.

C.4 Properties of the Laplace Transform

The applications of the Laplace transform in many instances are simplified by the utilization of the properties of the transform. These properties are presented in the following theorems.

C.4.1 Multiplication by a Constant

The Laplace transform of the product of a constant k and a time function $f(t)$ is the constant k multiplied by the Laplace transform of $f(t)$, that is,

$$L\left[kf(t)\right] = kF(s) \tag{C.6}$$

where $F(s)$ is the Laplace transform of $f(t)$.

C.4.2 Sum and Difference

The Laplace transform of the sum (or difference) of two time functions is the sum (or difference) of the Laplace transforms of the time functions, that is,

$$L\left[f_1(t) \pm f_2(t)\right] = F_1(s) \pm F_2(s) \tag{C.7}$$

where $F_1(s)$ and $F_2(s)$ are the Laplace transforms of $f_1(t)$ and $f_2(t)$, respectively.

C.5 Special Functions

C.5.1 Exponential Function

Consider the exponential function

$$
\begin{aligned}
f(t) &= 0 && \text{for } t < 0 \\
&= A\,e^{-at} && \text{for } t \ge 0
\end{aligned}
\tag{C.8}
$$

where A and a are constants.

The Laplace transform of Eq. C.8 is obtained as

$$
\begin{aligned}
L\left[A\,e^{-at}\right] &= \int_0^\infty A e^{-at} e^{-st} dt \\
&= A \int_0^\infty e^{-(a+s)t} dt \\
&= \frac{A}{s+a}
\end{aligned}
\tag{C.9}
$$

In obtaining Eq. C.9, we assumed that the real part of s is greater than $-a$, so that the integration converges. The Laplace transform of $F(s)$ of any Laplace transformable function $f(t)$ obtained this way is valid throughout the entire s plane except at the poles of $F(s)$.

C.5.2 Step Function

The step function can be visualized as a constant force of magnitude A applied to a system at time $t \ge 0^+$, as shown in Figure C.1.

Consider the step function

$$
\begin{aligned}
f(t) &= 0 && \text{for } t < 0 \\
&= A && \text{for } t \ge 0
\end{aligned}
\tag{C.10}
$$

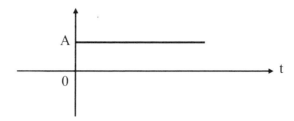

Figure C.1 Step function.

where A is a constant. It is a special case of the exponential function Ae^{-at}, where $a = 0$. The step function is undefined (finite) at $t = 0$. The Laplace transform of Eq. C.10 is given by

$$L[A] = \int_0^\infty A e^{-st} dt = \frac{A}{s} \qquad \text{(C.11)}$$

The step function when $A = 1$ is called the unit step function. The step function whose height is A can be written $A1(t)$.

The Laplace transform of the unit step function that is described by

$$\begin{aligned} 1(t) &= 0 && \text{for } t < 0 \\ &= 1 && \text{for } t \geq 0 \end{aligned} \qquad \text{(C.12)}$$

is

$$\frac{1}{s} \quad \text{or} \quad L[1(t)] = \frac{1}{s} \qquad \text{(C.13)}$$

Equation C.13 shows that a step function occurring at $t = t_0$ corresponds to a constant signal suddenly applied to the system at time $t = t_0$.

C.5.3 Ramp Function

Consider the ramp function

$$\begin{aligned} f(t) &= 0 && \text{for } t < 0 \\ &= At && \text{for } t \geq 0 \end{aligned} \qquad \text{(C.14)}$$

where A is a constant. This ramp function is shown in Figure C.2. The Laplace transform of this ramp function Eq. C.14 is given by

$$L[At] = A\int_0^\infty t e^{-st} dt \qquad \text{(C.15)}$$

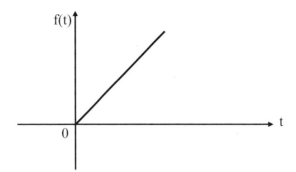

Figure C.2 Ramp function.

Applying the formula for integration by parts to Eq. C.15, we get

$$\int_a^b u \, dv = uv \Big|_a^b - \int_a^b v \, du \qquad \text{(C.16)}$$

Here,

$$u = t$$

$$dv = e^{-st} dt$$

$$v = \frac{e^{-st}}{(-s)}$$

Therefore,

$$L[At] = A \int_0^\infty t \, e^{-st} \, dt$$

$$= A \left(t \frac{e^{-st}}{-s} \Big|_0^\infty - \int_0^\infty \frac{e^{-st}}{-s} dt \right) \qquad \text{(C.17)}$$

$$= \frac{A}{s} \int_0^\infty e^{-st} dt = \frac{A}{s^2}$$

when $A = 1$, the ramp function is known as the unit ramp function.

C.5.4 Pulse Function

Figure C.3 shows a pulse where

$$f(t) = \frac{A}{t_0} \qquad \text{for } 0 \le t \le t_0 \qquad \qquad \text{(C.18)}$$

$$= 0 \qquad \text{for } t < 0 \text{ and } t_0 < t$$

in which A and t_0 are constants.

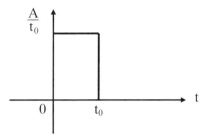

Figure C.3 Pulse function.

The pulse function is a step function of height $\dfrac{A}{t_0}$ that starts at $t = 0$ and that is superimposed by a negative function of height $\dfrac{A}{t_0}$ beginning at $t = t_0$. That is,

$$f(t) = \frac{A}{t_0} 1(t) - \frac{A}{t_0} 1(t - t_0) \tag{C.19}$$

The Laplace transform of $f(t)$ in Eq. C.19 is given by

$$
\begin{aligned}
L\left[f(t)\right] &= L\left[\frac{A}{t_0}1(t)\right] - L\left[\frac{A}{t_0}1(t - t_0)\right] \\[2mm]
&= \frac{A}{t_0 s} - \frac{A}{t_0 s}e^{st_0} \tag{C.20} \\[2mm]
&= \frac{A}{t_0 s}\left(1 - e^{-st_0}\right)
\end{aligned}
$$

C.5.5 Impulse Function

The impulse function is a special limiting case of the pulse function. The impulse function is defined by

$$
\begin{aligned}
f(t) &= \lim_{t_0 \to 0} \frac{A}{t_0} && \text{for } 0 < t < t_0 \\[2mm]
&= 0 && \text{for } t \le 0 \text{ and } t_0 \le t
\end{aligned}
\tag{C.21}
$$

The impulse function is shown in Figure C.4. The area under the impulse is equal to A, the height of the impulse function is $\dfrac{A}{t_0}$, and the duration is t_0. As $t_0 \to 0$, $\dfrac{A}{t_0} \to \infty$; however, the area under the impulse remains as A.

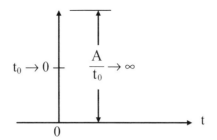

Figure C.4 Impulse function.

Now, the Laplace transform of Eq. C.21 is given by

$$L\left[f\left(t\right)\right] = \lim_{t_0 \to 0}\left[\frac{A}{t_0 s}\left(1 - e^{-st_0}\right)\right]$$

$$= \lim_{t_0 \to 0}\frac{\dfrac{d}{dt_0}\left[A\left(1 - e^{-st_0}\right)\right]}{\dfrac{d}{dt_0}\left(t_0 s\right)} \qquad \text{(C.22)}$$

$$= \frac{As}{s}$$

$$= A$$

Equation C.22 shows that the Laplace transform of the impulse function is equal to the area under the impulse, A.

C.5.6 Dirac Delta Function

The impulse function whose area is equal to unity is called the unit impulse function or the Dirac delta function. A unit impulse function, denoted by $\delta\left(t\right)$, can be visualized as a force that is very large in magnitude and is applied for a very short period of time.

The Laplace transform of a unit impulse function is given by

$$L\left[\delta\left(t\right)\right] = \Delta\left(s\right)$$

$$= \lim_{t_1 \to 0}\left[\frac{1}{st_1}\left(1 - e^{-st_1}\right)\right] \qquad \text{(C.23)}$$

$$= \lim_{t_1 \to 0}\left[\frac{se^{-st_1}}{s}\right] = 1$$

The area under the unit impulse function is unity. An impulse function usually is expressed as $A\delta\left(t\right)$, where A represents the area. Also, $\delta\left(t - \tau\right)$ denotes a unit impulse applied at time $t = \tau$. It has the property

$$\int_{-\infty}^{\infty} f(\tau)\delta(t - \tau)\,d\tau = f(t) \qquad\qquad \text{(C.24)}$$

provided that $f(t)$ is continuous at $t = \tau$.

At $t = 0$, Eq. C.24 gives

$$\int_{-\infty}^{\infty} f(\tau)\delta(t - \tau)\,d\tau = f(0) \qquad\qquad \text{(C.25)}$$

From the definition of Laplace transform, we have

$$L\left[\delta(t - \tau)\right] = e^{-\tau s} \qquad\qquad \text{(C.26)}$$

C.5.7 Sinusoidal Function

The Laplace transform of the sinusoidal function

$$
\begin{aligned}
f(t) &= 0 && \text{for } t < 0 \\
&= A\,\sin\omega t && \text{for } t \geq 0
\end{aligned}
\qquad\qquad \text{(C.27)}
$$

where A and ω are constants is given by

$$
\begin{aligned}
\frac{A}{2j} L\left[A\sin\omega t\right] &= \frac{A}{2j}\int_0^{\infty}\left(e^{j\omega t} - e^{-j\omega t}\right)e^{-st}\,dt \\
&= \frac{A}{2j}\frac{1}{s - j\omega} - \frac{A}{2j}\frac{1}{s + j\omega} \qquad\qquad \text{(C.28)} \\
&= \frac{A\omega}{s^2 + \omega^2}
\end{aligned}
$$

where we have used the identities

$$
\begin{aligned}
e^{j\omega t} &= \cos\omega t + j\sin\omega t \\
e^{-j\omega t} &= \cos\omega t - j\sin\omega t
\end{aligned}
\qquad\qquad \text{(C.29)}
$$

and

$$\sin\omega t = \frac{1}{2j}\left(e^{j\omega t} - e^{-j\omega t}\right)$$

The Laplace transforms of $A \cos \omega t$ are given by

$$L[A \cos \omega t] = \frac{As}{s^2 + \omega^2} \qquad (C.30)$$

If $A = 1$, then $f(t)$ given by Eq. C.27 is known as a unit sinusoidal function.

For a unit sinusoidal function,

$$L[\sin \omega t] = \frac{\omega}{s^2 + \omega^2}$$

Similarly,

$$L[\cos \omega t] = \frac{s}{s^2 + \omega^2} \qquad (C.31)$$

C.6 Multiplication of $f(t)$ by e^{-at}

If $f(t)$ is Laplace transformable and its Laplace transform is $F(s)$, then the Laplace transform of $e^{-at}f(t)$ is given by

$$L\left[e^{-at}f(t)\right] = \int_0^\infty e^{-at}f(t)e^{-st}dt$$
$$= F(s + a) \qquad (C.32)$$

Here, a may be real or complex. Equation C.32 is useful in finding the Laplace transforms of functions such as $e^{-at} \sin \omega t$ and $e^{-at} \cos \omega t$. For instance,

$$L[\sin \omega t] = \frac{\omega}{s^2 + \omega^2}$$
$$= F(s)$$
$$L\left[e^{-at} \sin \omega t\right] = F(s + a) \qquad (C.33)$$
$$= \frac{\omega}{(s + a)^2 + \omega^2}$$

Similarly,

$$L[\cos \omega t] = \frac{s}{s^2 + \omega^2}$$

and

$$L\left[e^{-at}\cos\omega t\right] = G(s + a)$$

$$= \frac{s + a}{(s + a)^2 + \omega^2}$$

(C.34)

C.7 Differentiation

The Laplace transform of the derivative of a function $f(t)$ is given by

$$L\left[\frac{d}{dt}f(t)\right] = sF(s) - f(0)$$

(C.35)

where $f(0)$ is the initial value of $f(t)$ evaluated at $t = 0$.

The values of $f(0+)$ and $f(0-)$ may not be the same, as shown in Figure C.5. The distance between $f(0+)$ and $f(0-)$ is important when $f(t)$ has a discontinuity at $t = 0$ because, in such a case, $\dfrac{df(t)}{dt}$ will involve an impulse function at $t = 0$. If $f(0+) \neq f(0-)$, Eq. C.35 can be written as

$$L + \left[\frac{d}{dt}f(t)\right] = sF(s) - f(0+)$$

$$L - \left[\frac{d}{dt}f(t)\right] = sF(s) - f(0-)$$

(C.36)

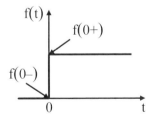

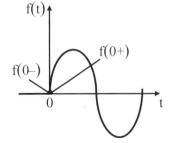

Figure C.5 Step function and sine function indicating initial values at $t = 0-$ and $t = 0+$.

To prove this equation, we integrate the Laplace integral by parts, which gives

$$\int_0^\infty f(t)\, e^{-st} dt = f(t)\frac{e^{-st}}{-s}\bigg|_0^\infty - \int_0^\infty \left[\frac{d}{dt}f(t)\right]\frac{e^{-st}}{-s} dt$$

Therefore,

$$F(s) = \frac{f(0)}{s} + \frac{1}{s}L\left[\frac{d}{dt}f(t)\right] \tag{C.37}$$

Hence,

$$L\left[\frac{d}{dt}f(t)\right] = sF(s) - f(0) \tag{C.38}$$

Similarly, the following relationship can be obtained for the second derivative of $f(t)$ as

$$L\left[\frac{d^2}{dt^2}f(t)\right] = s^2F(s) - sf(0) - \dot{f}(0) \tag{C.39}$$

where $\dot{f}(0)$ is the value of $\dfrac{df(t)}{dt}$ evaluated at $t = 0$.

Let

$$\frac{d}{dt}f(t) = g(t)$$

Then

$$L\left[\frac{d^2}{dt^2}f(t)\right] = L\left[\frac{d}{dt}g(t)\right]$$

$$= sL[g(t)] - g(0) \tag{C.40}$$

$$= sL\left[\frac{d}{dt}f(t)\right] - \dot{f}(0) = s^2F(s) - sF(0) - \dot{f}(0)$$

Extending for the nth derivative of $f(t)$, we get

$$L\left[\frac{d^n}{dt^n}f(t)\right] = s^nF(s) - s^{n-1}f(0) - s^{n-2}\dot{f}(0) - \ldots - f^{(n-1)}(0) \tag{C.41}$$

where $f(0)$, $\dot{f}(0)$, ..., $f^{(n-1)}(0)$ represent the values of $f(t)$, $\dfrac{df(t)}{dt}$, ..., $\dfrac{d^{n-1}f(t)}{dt^{n-1}}$, respectively, evaluated at $t = 0$. If all the initial values of $f(t)$ and its derivatives are equal to zero, then the Laplace transform of the nth derivative of $f(t)$ is given by $s^nF(s)$.

C.8 Integration

If $f(t)$ is of exponential order, then the Laplace transform of $\int f(t)dt$ exists and is given by

$$L\left[\int f(t)dt\right] = \frac{F(s)}{s} + \frac{f^{-1}(0)}{s} \qquad (C.42)$$

where $F(s) = L\left[f(t)\right]$ and $f^{-1}(0) = \int f(t)dt$, evaluated at $t - 0$.

Table C.1 shows the Laplace transforms of time functions that frequently appear in linear dynamic systems and control systems.

TABLE C.1
TABLE OF LAPLACE TRANSFORMS

	Function Transform $f(t)$	Laplace Transform $F(s)$
6	$e^{at}f(t)$	$F(s-a)$
7	$f(t-b)$ when $t > b$ $0 \qquad$ when $t < b$	$e^{-bs}F(s)$
12	e^{-at}	$\dfrac{1}{s+a}$
13	e^{at}	$\dfrac{1}{s-a}$
14	t^n	$\dfrac{n!}{s^{n+1}}\quad n = 1, 2, \ldots$
15	$\dfrac{t^{n-1}}{(n-1)!}$	$\dfrac{1}{s^n}\quad n = 1, 2, \ldots$
16	$\dfrac{1}{(n-1)!}t^{n-1}e^{-at}$	$\dfrac{1}{(s+a)^n}$
17	t	$\dfrac{1}{s^2}$
18	$1 - e^{-at}$	$\dfrac{a}{s(s+a)}$
19	$\sin at$	$\dfrac{a}{s^2 + a^2}$
20	$\sinh at$	$\dfrac{a}{s^2 - a^2}$
21	$t\,e^{-at}$	$\dfrac{1}{(s+a)^2}$
22	$t\,e^{at}$	$\dfrac{1}{(s-a)^2}$

TABLE C.1 *(cont.)*

	Function Transform $f(t)$	Laplace Transform $F(s)$
23	$\dfrac{1}{a}\sinh at$	$\dfrac{1}{s^2 - a^2}$
24	$\cosh at$	$\dfrac{s}{s^2 - a^2}$
25	$\dfrac{1}{a^2}\left(1 - \cosh at\right)$	$\dfrac{1}{s\left(s^2 + a^2\right)}$
26	$\dfrac{e^{at} - e^{bt}}{a - b}$	$\dfrac{1}{(s - a)(s - b)}$
27	$e^{-at}\sin \omega t$	$\dfrac{\omega}{\left(s + a\right)^2 + \omega^2}$
28	$e^{-at}\cos \omega t$	$\dfrac{\left(s + a\right)}{\left(s + a\right)^2 + \omega^2}$
29	$\dfrac{1}{(b - a)}\left(e^{-at} - e^{-bt}\right)$	$\dfrac{1}{(s + a)(s + b)}$
30	$\dfrac{1}{(b - a)}\left[(\alpha - a)e^{-at} - (\alpha - b)e^{-bt}\right]$	$\dfrac{s + \alpha}{(s + a)(s + b)}$
31	$\dfrac{1}{\omega}\sin \omega t$	$\dfrac{1}{s^2 + \omega^2}$
32	$\dfrac{e^{-d\omega t}\sin\left(\omega t\sqrt{1 - d^2}\right)}{\omega\sqrt{1 - d^2}}\quad d^2 < 1$	$\dfrac{1}{s^2 + 2d\omega s + \omega^2}$
33	$\dfrac{e^{-d\omega t}\sin\left(\omega t\sqrt{d^2 - 1}\right)}{\omega\sqrt{d^2 - 1}}\quad d^2 > 1$	$\dfrac{1}{s^2 + 2d\omega s + \omega^2}$
34	$t\,e^{-\omega t}$	$\dfrac{1}{s^2 + 2\omega s + \omega^2}$
35	$\cos hat$	$\dfrac{s}{s^2 - a^2}$
36	$\cos at$	$\dfrac{s}{s^2 + a^2}$
37	$\dfrac{1}{a}\sin at$	$\dfrac{1}{s^2 + a^2}$

TABLE C.1 *(cont.)*

	Function Transform $f(t)$	Laplace Transform $F(s)$
38	$\dfrac{t}{2a}\sin at$	$\dfrac{s}{\left(s^2 + a^2\right)^2}$
39	$t\cos at$	$\dfrac{s^2 - a^2}{\left(s^2 + a^2\right)^2}$
40	$e^{at}\cos kt$	$\dfrac{s + a}{\left(s + a\right)^2 + k^2}$

C.9 Final Value Theorem

The final value theorem relates the steady-state behavior of $f(t)$ to the behavior of $sF(s)$ in the neighborhood of $s = 0$. This theorem is applicable only if $\lim_{t \to \infty} f(t)$ exists. If all poles of $sF(s)$ lie in the left half of the s plane, $\lim_{t \to \infty} f(t)$ exists. But if $sF(s)$ has poles on the imaginary axis or in the right half of the s plane, $f(t)$ will contain oscillating or exponentially increasing time functions, respectively, and $\lim_{t \to \infty} f(t)$ will not exist.

The final value theorem may be stated as follows. If $f(t)$ and $\dfrac{df(t)}{dt}$ are Laplace transformable, if $F(s)$ is the Laplace transform of $f(t)$, and if $\lim_{t \to \infty} f(t)$ exists, then

$$\lim_{t \to \infty} f(t) = \lim_{s \to \infty} sF(s) \qquad\qquad (C.43)$$

The final value theorem is a useful relation in the analysis and design of feedback control systems because it gives the final value of a time function by determining the behavior of its Laplace transform as s tends to zero. However, the final value theorem is not valid if $sF(s)$ contains any poles whose real part is zero or positive, which are equivalent to the analytic requirement of $sF(s)$ stated in the theorem.

C.10 Initial Value Theorem

The initial value theorem does not give the value of $f(t)$ at exactly $t = 0$ but at a time slightly greater than zero. If $f(t)$ and $\dfrac{df(t)}{dt}$ are both Laplace transformable and if $\lim_{s \to \infty} sF(s)$ exists, then

$$f(0+) \lim_{s \to \infty} sF(s) \qquad\qquad (C.44)$$

The initial value theorem is valid for the sinusoidal function.

C.11 Shift in Time

The Laplace transform of $f(t)$ delayed by time T is equal to the Laplace transform of $f(t)$ multiplied by e^{-Ts}, that is,

$$L\left[f(t-\tau)u_s(t-\tau)\right] = e^{-Ts}F(s) \qquad (C.45)$$

where $u_s(t-\tau)$ denotes the unit step function, which is shifted in time to the right by T.

C.12 Complex Shifting

The Laplace transform of $f(t)$ multiplied by $e^{\mp at}$, where a is a constant, is equal to the Laplace transform $F(s)$ with s replaced by $s \pm a$, that is,

$$L\left[e^{\mp at}f(t)\right] = F(s \pm a) \qquad (C.46)$$

C.13 Real Convolution (Complex Multiplication)

If the function $f_1(t)$ and $f_2(t)$ have the Laplace transforms $F_1(s)$ and $F_2(s)$, respectively, and $f_1(t)=0$ and $f_2(t) = 0$ for $t < 0$, then Eq. C.46 shows that multiplication of two transformed functions in the complex s domain is equivalent to the convolution of the two corresponding real functions of t in the t domain

$$\begin{aligned}
F_1(s)F_2(s) &= L\left[\int_0^t f_1(\tau)f_2(t-\tau)d\tau\right] \\
&= L\left[\int_0^t f_2(\tau)f_1(t-\tau)d\tau\right] \qquad (C.47) \\
&= L\left[f_1(t)*f_2(t)\right]
\end{aligned}$$

In Eq. C.47, the symbol * denotes convolution in the t domain. An important fact to remember is that the inverse Laplace transform of the product of two functions in the s domain is not equal to the product of the two corresponding real functions in the t domain, that is, in general,

$$L^{-1}\left[F_1(s)F_2(s)\right] \neq f_1(t)f_2(t) \qquad (C.48)$$

There also is a dual relation to the real convolution theorem, called the complex convolution or real multiplication. Essentially, the theorem states that multiplication in the real t domain is equivalent to convolution in the complex s domain, that is,

$$L\left[f_1(t)f_2(t)\right] = F_1(s) * F_2(s) \qquad \text{(C.49)}$$

where * denotes complex convolution.

Table C.2 summarizes the properties of Laplace transforms.

TABLE C.2
PROPERTIES OF LAPLACE TRANSFORMS

1	$$L\left[A\,f(t)\right] = A\,F(s)$$
2	$$L\left[f_1(t) \pm f_2(t)\right] = F_1(s) \pm F_2(s)$$
3	$$L_\pm\left[\frac{d}{dt}f(t)\right] = sF(s) - f(0\pm)$$
4	$$L_\pm\left[\frac{d^2}{dt^2}f(t)\right] = s^2F(s) - s\,f(0\pm) - \dot{f}(0\pm)$$
5	$$L_\pm\left[\frac{d^n}{dt^n}f(t)\right] = s^nF(s) - \sum_{k=1}^{n} s^{n-k}f^{(k-1)}(0\pm)$$ where $f^{(k-1)}(t) = \dfrac{d^{k-1}}{dt^{k-1}}f(t)$
6	$$L\left[\int f(t)dt\right] = \frac{F(s)}{s} + \frac{1\left[\int f(t)dt\right]_{t=0\pm}}{s}$$
7	$$L_\pm\left[\int\dots\int f(t)(dt)^n\right] = \frac{F(s)}{s^n} + \sum_{k=1}^{n}\frac{1}{s^{n-k+1}}\left[\int\dots\int f(t)(dt)^k\right]_{t=0\pm}$$
8	$$L\left[\int_0^t f(t)dt\right] = \frac{F(s)}{s}$$
9	$$\int_0^\infty f(t)dt = \lim_{s\to 0} F(s) \quad \text{if } \int_0^\infty f(t)\,dt \text{ exists}$$
10	$$L\left[e^{-at}f(t)\right] = F(s+a)$$
11	$$L\left[f(t-a)1(1-a)\right] = e^{-as}F(s) \quad a \geq 0$$
12	$$L\left[t\,f(t)\right] = -\frac{dF(s)}{ds}$$
13	$$L\left[t^2 f(t)\right] = \frac{d^2}{ds^2}F(s)$$
14	$$L\left[t^n f(t)\right] = (-1)^n\frac{d^n}{ds^n}F(s) \quad n = 1, 2, 3, \dots$$
15	$$L\left[\frac{1}{t}f(t)\right] = \int_0^\infty F(s)ds \quad \text{if } \lim_{t\to 0}\frac{1}{t}f(t) \text{ exists}$$

TABLE C.2 *(cont.)*

16	$$L\left[f\left(\frac{1}{a}\right)\right] = a\,f(as)$$
17	$$l\left[\int_0^t f_1(t-\tau)f_2(\tau)d\tau\right] = F_1(s)F_2(s)$$

C.14 Inverse Laplace Transformation

Generally, inverse Laplace transforms are obtained through the partial fraction expansion procedure.

C.14.1 Partial Fraction Expansions

In obtaining the solution of linear differential equations with constant coefficients or analysis of linear time-invariant systems by the Laplace transform method, we encounter rational algebraic fractions that are the ratio of two polynomials in s, such as

$$F(s) = \frac{P(s)}{Q(s)}$$

$$= \frac{b_0 s^m + b_1 s^{m-1} + \ldots + b_m}{a_0 s^n + a_1 s^{n-1} + \ldots + a_n} \tag{C.50}$$

In practical systems, the order of the polynomial in the numerator is equal to or less than that of the denominator. In terms of the orders m and n, rational algebraic fractions are subdivided as follows:

   i. Improper fraction if $m \geq n$

   ii. Proper fraction if $m < n$

An improper fraction can be separated into a sum of a polynomial in s and a proper fraction, that is,

$$F(s) = \underset{\text{Improper}}{\frac{P(s)}{Q(s)}}$$

$$= d(s) + \underset{\text{Proper}}{\frac{P(s)}{Q(s)}} \tag{C.51}$$

This can be achieved by performing a long division. To obtain the partial fraction expansion of a proper fraction, first we factorize the polynomial $Q(s)$ into n first-order factors. The roots may be real, complex, distinct, or repeated. Various cases are discussed here.

C.14.2 Case I—Partial Fraction Expansion When $Q(s)$ Has Distinct Roots

In this case, Eq. C.66 may be written as

$$F(s) = \frac{P(s)}{Q(s)} = \frac{P(s)}{(s + p_1)(s + p_2) \cdots (s + p_k) \cdots (s + p_n)} \tag{C.52}$$

which, when expanded, gives

$$F(s) = \frac{A_1}{s + p_1} + \frac{A_2}{s + p_2} + \cdots + \frac{A_k}{s + p_k} + \cdots + \frac{A_n}{s + p_n} \tag{C.53}$$

where A_k ($k = 1, 2, \ldots, n$) are constants.

The coefficient A_k is called the residue at the pole $s = -p_k$. To evaluate A_k, multiply $F(s)$ in Eq. C.52 by $(s + p_k)$ and let $s = -p_k$. This gives

$$A_k = (s + p_k) \frac{P(s)}{Q(s)} \bigg|_{S=-p_k}$$

$$= \frac{P(s)}{\dfrac{d}{ds} Q(s)} \bigg|_{S=-p_k} \tag{C.54}$$

$$= \frac{P(-p_k)}{(p_1 - p_k)(p_2 - p_k) \cdots (p_{k-1} - p_k) \cdots (p_n - p_k)} \tag{C.55}$$

From Eq. C.53, we get

$$f(t) = L^{-1}\big[F(s)\big]$$

$$= A_1 e^{-p_1 t} + A_2 e^{-p_2 t} + \ldots + A_n e^{-p_n t} \tag{C.56}$$

EXAMPLE EC.2

Find the initial value of $\dfrac{df(t)}{dt}$ when the Laplace transform of $f(t)$ is given by

$$F(s) = L\big[f(t)\big]$$

$$= \frac{2s + 1}{s^2 + s + 1}$$

Solution:

Using the initial value theorem,

$$\lim_{t \to 0+} f(t) = \lim_{s \to 0} sF(s) = \lim_{s \to 0} \frac{s(2s+1)}{s^2 + s + 1} = 2$$

Because the L_+ transform of $\dfrac{df(t)}{dt} = g(t)$ is

$$L_+\left[g(t)\right] = sF(s) - f(0+)$$

$$= \frac{s(2s+1)}{s^2 + s + 1} - 2$$

$$= \frac{-s - 2}{s^2 + s + 1}$$

the initial value of $\dfrac{df(t)}{dt}$ is

$$\lim_{t \to 0} \frac{df(t)}{dt} = g(0+) = \lim_{s \to 0} s\left[sF(s) - f(0+)\right]$$

$$= \lim_{s \to 0} \frac{-s^2 - 2s}{s^2 + s + 1} = -1$$

C.14.3 Case II—Partial Fraction Expansion When Q(s) Has Complex Conjugate Roots

Suppose a pair of complex conjugate roots in $Q(s)$ is given by

$$s = -a - j\omega$$

and

$$s = -a + j\omega$$

Then $F(s)$ may be written as

$$F(s) = \frac{P(s)}{Q(s)}$$

$$= \frac{P(s)}{(s + a + j\omega)(s + a - j\omega)(s + p_3)(s + p_4)\ldots(s + p_n)}$$

(C.57)

which when expanded gives

$$F(s) = \frac{A_1}{(s + a + j\omega)} + \frac{A_2}{(s + a - j\omega)}$$

$$+ \frac{A_3}{s + p_3} + \frac{A_4}{s + p_4} + \cdots + \frac{A_n}{s + p_n}$$

(C.58)

where A_1 and A_2 are residues at the poles $s = -(a + j\omega)$ and $s = -(a - j\omega)$, respectively. These residues form a complex conjugate pair. From Eq. C.58, the inverse Laplace transform of $F(s)$ may be obtained as

$$f(t) = L^{-1}\left[F(s)\right] = L^{-1}\left[\frac{A_1}{(s + a + j\omega)} + \frac{A_2}{(s + a - j\omega)}\right]$$

$$+ A_3 e^{-p_3 t} + \cdots + A_n e^{-p_n t}$$

(C.59)

$$= 2\operatorname{Re}\left[A_1 e^{-(a+j\omega)t}\right] + A_3 e^{-p_3 t} + \cdots + A_n e^{-p_n t}$$

(C.60)

From Eq. C.54, the residue A_1 is given by

$$A_1 = \frac{P(s)}{Q(s)}(s + a + j\omega)\Bigg|_{s=-(a+j\omega)}$$

(C.61)

C.14.4 Case III—Partial Fraction Expansion When Q(s) Has Repeated Roots

Assume that root p_1 of $Q(s)$ is of multiplicity r and that other roots are distinct. The function $F(s)$ may be written as

$$F(s) = \frac{P(s)}{Q(s)} = \frac{P(s)}{(s + p_1)^r (s + p_{r+1})(s + p_{r+2}) \cdots (s + p_n)}$$

(C.62)

which when expanded gives

$$F(s) = \frac{A_{1(r)}}{(s + p_1)^r} + \frac{A_{1(r-1)}}{(s + p_1)^{r-1}} + \frac{A_{1(r-2)}}{(s + p_1)^{r-2}} + \cdots + \frac{A_{12}}{(s + p_1)^2}$$

$$+ \frac{A_{11}}{(s + p_1)} + \frac{A_{r+1}}{(s + p_{r+1})} + \frac{A_{r+2}}{(s + p_{r+2})} + \cdots + \frac{A_n}{(s + p_n)}$$

(C.63)

The coefficients of repeated roots may be obtained using the relation

$$A_{1(r-i)} = \frac{1}{i!}\left[\frac{d^i}{ds^i}\left\{(s+p_1)^r\frac{P(s)}{Q(s)}\right\}\right]_{s=-p_1} \qquad i = 0,1,2,\cdots,r-1 \qquad (C.64)$$

From Eq. C.63, we obtain

$$f(t) = L^{-1}\left[F(s)\right] = \left[\frac{A_{1(r)}}{(r-1)!}t^{r-1} + \frac{A_{1(r-1)}}{(r-2)!}t^{r-2} + \cdots + A_{12}t + A_{11}\right]e^{-p_1 t}$$

$$+ A_{r+1}e^{-p_{r+1}t} + \cdots + A_n e^{-p_n t} \qquad (C.65)$$

EXAMPLE EC.3

Find the inverse Laplace transform of $F(s)$, where

$$F(s) = \frac{1}{s\left(s^2 + 4s + 5\right)}$$

Solution:

Because

$$s^2 + 4s + 5 = (s + 2 + jl)(s + 2 - jl)$$

We note that $F(s)$ involves a pair of complex-conjugate poles; hence, we expand $F(s)$ into the form

$$F(s) = \frac{1}{s\left(s^2 + 4s + 5\right)}$$

$$= \frac{a_1}{s} + \frac{a_2 s + a_3}{s^2 + 4s + 5}$$

where a_1, a_2, and a_3 are obtained

$$1 = a_1\left(s^2 + 4s + 5\right) + \left(s_2 s + a_3\right)s$$

Comparing the coefficients of the s^2, s, and s^0 terms on both sides of Eq. 3, respectively, we obtain

$$a_1 + a_2 = 0$$

$$4a_1 + a_3 = 0$$

$$5a_1 = 1$$

from which

$$a_1 = \frac{1}{5}$$

$$a_2 = -\frac{1}{5}$$

$$a_3 = -\frac{4}{5}$$

Hence,

$$F(s) = \frac{1}{5}\frac{1}{s} + \frac{-\frac{1}{5}s - \frac{4}{5}}{\left(s^2 + 4s + 5\right)} = \frac{1}{5}\frac{1}{s} - \frac{1}{5}\frac{s+4}{s^2 + 4s + 5}$$

$$= \frac{1}{5}\frac{1}{s} - \frac{1}{5}\frac{2}{(s+2)^2 + 1^2} - \frac{1}{5}\frac{s+2}{(s+2)^2 + 1^2}$$

The inverse Laplace transform of $F(s)$ is

$$f(t) = \frac{1}{5} - \frac{2}{5}e^{-2t}\sin 2t - \frac{1}{5}e^{-2t}\cos t$$

C.15 Solution of Differential Equations

The solution of a linear differential equation with constant coefficients (pertaining to a linear system) can be obtained conveniently through the Laplace transform technique by applying the following steps:

1. Using the real differentiation and the linearity properties and the transform pairs as given in Table C.1, transform the differential equation into an algebraic equation in the complex variable s.

2. Through algebraic manipulation, solve for the unknown in terms of the complex variable s.

3. The time domain solution of the unknown is obtained by inverse Laplace transforming the s domain solution.

This procedure is best illustrated by means of examples.

EXAMPLE EC.4

Find the solution of the differential equation

$$\frac{dx}{dt} + bx = A\sin\omega t \qquad x(0) = a$$

Solution:

$$\frac{dx}{dt} + bx = A\sin\omega t \qquad x(0) = a \qquad (EC4.1)$$

Taking the Laplace transform on both sides of Eq. EC5.1, we obtain

$$\left[sX(s) - x(0)\right] + bX(s) = A\frac{\omega}{s^2 + \omega^2} \qquad (EC4.2)$$

or

$$(s + n)X(s) = \frac{A\omega}{s^2 + \omega^2} + a \qquad (EC4.3)$$

Solving Eq. EC4.3 for $X(s)$, we obtain

$$X(s) = \frac{A\omega}{(s + b)(s^2 + \omega^2)} + \frac{a}{s + b}$$

$$= \frac{A\omega}{b^2 + \omega^2}\left[\frac{1}{s + b} - \frac{s - b}{s^2 + \omega^2}\right] + \frac{a}{s + b}$$

$$= \left(a + \frac{A\omega}{b^2 + \omega^2}\right)\frac{1}{s + b} + \frac{Ab}{b^2 + \omega^2}\frac{\omega^2}{s^2 + \omega^2} - \frac{A\omega}{b^2 + \omega^2}\frac{s}{s^2 + \omega^2} + \frac{1}{s + b}$$

Hence, the inverse Laplace transform of $X(s)$ is

$$x(t) = L^{-1}\left[X(s)\right] = \left(a + \frac{A\omega}{b^2 + \omega^2}\right)e^{-bt} + \frac{Ab}{b^2 + \omega^2}\sin\omega t \qquad t \geq 0$$

C.16 Summary

The Laplace transform method is an important mathematical tool for linear dynamic system analysis and design. In this appendix, we defined the Laplace transformation and gave Laplace transforms of several common time functions and important Laplace transform theorems. This appendix also dealt with the inverse Laplace transformations. In dynamic system analysis and design, we often must solve time-invariant, linear differential equations of second or higher orders. One convenient method for solving the differential equations is to use the Laplace transformation method, in which the differential equation is transformed into an algebraic equation in terms of the transfer function of the unknown quantity to be determined. The transform function is a function of a complex variable, denoted by s. A systematic approach for the solution of the linear, time-invariant differential equations was presented.

C.17 References

1. Dukkipati, R.V., *Advanced Engineering Analysis*, Narosa Publishing House, New Delhi, India, 2006.

2. Greenberg, M.D., *Foundations of Applied Mathematics*, Prentice-Hall, Englewood Cliffs, NJ, 1978.

3. Hildebrand, F.B., *Advanced Calculus for Applications*, Prentice-Hall, Englewood Cliffs, NJ, 1978.

4. Jain, R.K. and Iyengar, S.R.K., *Advanced Engineering Mathematics*, Narosa Publishing House, New Delhi, India, 2002.

5. Jeffrey, A., *Advanced Engineering Mathematics*, Harcourt/Academic Press, New York, 2002.

6. Kreyszig, E., *Advanced Engineering Mathematics*, 8th Ed., Wiley, New York, 1998.

7. Lipschutz, S., *Linear Algebra*, Schaum's Solved Problems Series, McGraw-Hill, New York, 1989.

8. Lipson, M., *Linear Algebra*, Schaum's Outlines, McGraw-Hill, New York, 2001.

9. Lopez, R.J., *Advanced Engineering Mathematics*, Addison-Wesley, New York, 2001.

10. O'Neil, P.V., *Advanced Engineering Mathematics*, 4th Ed., Brooks/Cole Publishing Company, New York, 1995.

11. Siegel, M., *Advanced Mathematics for Engineers and Scientists*, Schaum's Outlines, McGraw-Hill, New York, 1971.

12. Siegel, M., *Vector Analysis and Introduction to Tensor Analysis*, Schaum's Outlines, McGraw-Hill, New York, 1959.

13. Spiegel, M.R., *Theory and Problems of Laplace Transforms*, Schaum's Outline Series, McGraw-Hill, New York, 1965.

14. Wylie, C.R. and Barret, L.C., *Advanced Engineering Mathematics*, 5th, Ed., McGraw-Hill, New York, 1982.

Appendix D

Glossary of Terms

Terminology used frequently in the field of road vehicle dynamics is compiled here from various sources, including the document SAE J670e, Vehicle Dynamics Terminology, SAE International, Warrendale, PA, July 1976.

A, B, C, and D pillars and posts: The vertical pillars and posts of a light vehicle forming the major vertical structural members of a body. Pillars typically are at window height; posts are below window height. From front to rear, the A post/pillar is the most forward number, the B post/pillar is the second most forward vertical member, and so forth (Figure D.1).

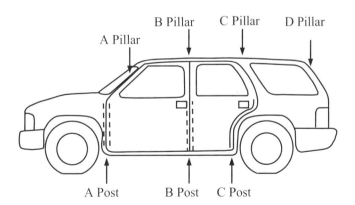

Figure D.1 A, B, C, and D pillars and posts.

AASHTO: American Association of State Highway and Transport Officials.

ABS: *See* Anti-lock braking system.

Acceleration: Acceleration of a point is the time rate of change of the velocity of the point.

Acceleration scuff: A scuffmark made when sufficient power is applied to the driving wheels to make at least one of them spin on the road surface.

Accelerometer: An instrument used to measure acceleration, usually linear acceleration.

Accident: An unintended event that results in damage or injury. *See also* Crash; Collision.

Accident investigation: The process of observation, acquisition, and documentation of physical evidence and other information regarding an accident or crash.

Accident reconstruction: A procedure carried out with the specific purpose of estimating in both a qualitative and quantitative manner how an accident occurred, using engineering, scientific, and mathematical principles and based on evidence obtained through accident investigation.

Accident scene: A place where a traffic accident occurs, both during and immediately following the accident, and before vehicles and participants have departed. *See* Accident site.

Accident site: A place where a traffic accident occurred, after vehicles and participants have departed the scene. *See* Accident scene.

Ackerman steer angle: The angle whose tangent is the wheelbase divided by the radius of turn.

Ackerman steer angle gradient: The rate of change of Ackerman steer angle with respect to the change in steady-state lateral acceleration on a level road at a given trim and test conditions.

Acute angle: An angle of less than 90° and more than 0°.

Aerodynamic angle of attack: The angle between the vehicle x axis and the trace of the resultant air velocity vector on a vertical plane containing the vehicle x axis.

Aerodynamic sideslip angle: The angle between the traces on the vehicle x-y plane of the vehicle x axis and the resultant air velocity vector at some specified point in the vehicle.

Aggressivity: The properties and characteristics of a vehicle, such as mass and structural stiffness, that cause a vehicle to provide more protection for participants than another vehicle in a crash.

Agricultural commodity trailer: A trailer designed to transport bulk commodities from harvest sites to process or storage sites.

Air resistance: The force-opposing moment of a body through the air surrounding it. Due to the relatively low speeds and the short airborne time periods dealt with in vehicle collisions, the effects of wind and air resistance are negligible.

Airbag: A device in the interior of a vehicle that inflates and acts between an occupant and an interior vehicle surface to prevent injury in a crash. *See* Supplemental restraint system.

Aligning moment: *See* Aligning torque (Aligning moment).

Aligning stiffness (Aligning torque stiffness): The rate of change of aligning torque with respect to change in slip angle, usually evaluated at zero slip angle.

Aligning torque (Aligning moment): The component of the tire moment vector tending to rotate the tire about the Z' axis, positive clockwise when looking in the positive direction of the Z' axis.

Aligning torque stiffness: *See* Aligning stiffness (Aligning torque stiffness).

Amplification factor: *See* Magnification factor (Amplitude ratio, Amplification factor).

Amplitude: The amplitude of displacement at a point in a vibrating system is the largest value of displacement that the point attains with reference to its equilibrium position.

Amplitude ratio: The amplitude ratio or magnification factor is the ratio of the maximum force developed in the spring of a mass-spring-dashpot system to the maximum value of the exciting force. *See also* Magnification factor (Amplitude ratio, Amplification factor).

Angle collision: A collision between two traffic units approaching on separate roadways or other paths that intersect.

Angular acceleration: The time rate of change of angular velocity.

Angular frequency (Circular frequency): The angular frequency of a periodic quantity, in radians per unit time, is the frequency multiplied by 2π.

Angular impact: When two vehicles collide in such a manner that their respective directions of force are not parallel.

Angular impulse: The angular impulse of a constant torque T acting for a time t is the product Tt.

Angular mechanical impedance (Rotational mechanical impedance): The impedance involving the ratio of torque to angular velocity. *See* Impedance.

Angular momentum: The angular momentum of a body about its axis of rotation is the moment of its linear momentum about the axis.

Angular motion: *See* Rotational particle motion (Angular motion, Circular motion).

Angular velocity: The time rate of change of rotational displacement.

Animation: The process by which movement of objects is illustrated.

ANSI: American National Standards Institute.

Anti-lock braking system (ABS): A portion of the service brake system of a vehicle that automatically controls the degree of rotational slip of one or more wheels, allowing for braking near the peak coefficient and preventing wheel lock-up.

Anti-resonance: For a system in forced oscillation, anti-resonance exists at a point when any change, however small, in the frequency of excitation causes an increase in the response at this point.

AOI: *See* Area of impact.

Apogee: The highest vertical point a body in flight reaches, occurring at half the horizontal distance.

Approach speed: The speed of a vehicle just prior to the first significant event, such as contact in an accident. *See* Closing speed.

Aquaplaning: *See* Hydroplaning.

Arc: Part of a curve between two points on the curve.

Area of impact (AOI): Area encompassed by the interface between colliding objects projected onto the road. *See* Point of impact (POI); Impact center.

Articulated vehicle: A vehicle jointed or joined together permanently or semi-permanently by means of a pivot connection for operating separate segments as a unit.

Asphalt: *See* Bituminous pavement.

Asphalted concrete: Concrete cemented with a bituminous material such as tar or asphalt.

Asymptotic stability: Asymptotic stability exists at a prescribed trim if, for any small temporary change in disturbance or control input, the vehicle will approach the motion defined by the trim.

At-scene investigation: Examining and recording the results of a traffic accident and obtaining additional information that supplements data obtained for the accident report.

Attitude angle: *See* Sideslip angle (Attitude angle).

Attribute: Any inherent characteristic of a road, a vehicle, or a person that affects the probability of a traffic accident.

Auxiliary mass damper (Damped vibration absorber): A system consisting of a mass, spring, and damper, which tend to reduce vibration by the dissipation of energy in the damper as a result of relative motion between the mass and the structure to which the damper is attached.

Average acceleration: Calculated by dividing the change in velocity of an object by the time that is required to gain that velocity.

Axis: A straight line around which an object rotates. Also a reference line from which or along which distances or angles are measured in a system of coordinates.

Axle fore-and-aft shake: The oscillatory motion of an axle, which consists purely of longitudinal displacement.

Axle side shake: The oscillatory motion of an axle, which consists of transverse displacement.

Axle windup: The oscillatory motion of an axle about the horizontal transverse axis through its center of gravity.

Axle yaw: The oscillatory motion of an axle around the vertical axis through its center of gravity.

BAC: Blood alcohol concentration.

Backlite: The rear or back window that spans from the driver's side to the passenger's side of the vehicle.

Backlite header: The structural body member, which connects the upper portions of the rearmost driver and passenger pillars and forms the top edge of the backlite (back window).

Barrier equivalent velocity (BEV): The forward speed and corresponding kinetic energy with which a vehicle contacts a flat fixed rigid barrier at 90° with no rebound. *See also* Energy equivalent speed (EES).

Bead: The portion of the tire that fits onto the rim of the wheel.

Bead base: The approximately cylindrical portion of the bead of a tire that forms its inside diameter.

Bead toe: That portion of the bead of a tire that joins the bead base and the inside surface of the tire.

Beaming: A mode of vibration involving predominantly bending deformations of the sprung mass about the vehicle y axis.

Beats: Periodic variations that result from the superposition of two simple harmonic quantities of different frequencies f_1 and f_2. They involve the periodic increase and decrease of amplitude at the beat frequency.

BEV: *See* Barrier equivalent velocity (BEV).

Bicycle model: A two-wheeled vehicle used conceptually in vehicle dynamics studies to represent a four-wheeled vehicle where the side-to-side extent of the vehicle is neglected for simplicity.

Biomechanics: The study of mechanical motion in a biological system.

Bituminous pavement: A pavement comprising an upper layer or layers of aggregate with a bituminous binder (e.g., asphalt, coal tars, natural tars) and surface treatment such as chip seals, slurry seals, sand seals, and cape seals.

Black box: *See* Event data recorder.

Blacktop: *See* Bituminous pavement.

Bobtail: A term used to refer to a truck tractor being driven without a semi-trailer.

Boom: A high-intensity vibration (25 to 100 Hz) perceived audibly and characterized as a sensation of pressure by the ear.

Bounce: *See* Vertical (Bounce).

Brake fade: Repeated application of the brakes, causing a loss of friction ability due to heat and resulting in the impairment of braking efficiency.

Brake hop: An oscillatory hopping motion of a single wheel or of a pair of wheels, which occurs when brakes are applied in forward or reverse motion of the vehicle.

Brake slip: *See* Wheel slip.

Braking coefficient: *See* Braking force coefficient (Braking coefficient).

Braking distance: The distance to bring a vehicle to rest during brake application in straightforward motion. *See also* Stopping distance.

Braking (driving) squeal: The squeal resulting from longitudinal slip.

Braking (driving) stiffness: The rate of change of longitudinal force with respect to change in longitudinal slip, usually evaluated at zero longitudinal slip.

Braking (driving) stiffness coefficient: The ratio of braking (driving) stiffness of a straight free-rolling tire to the vertical load.

Braking force: (1) The force over the contact surface between a tire and a road in the direction of heading of the braked wheel that develops as a result of brake application. (2) The negative longitudinal force resulting from braking torque application.

Braking force coefficient (Braking coefficient): The ratio of the braking force to the vertical load.

Braking force, peak: The largest force that can be developed during brake application as wheel slip is varied over the range of free-rolling slip to locked wheel slip.

Braking time: The time required to traverse the braking distance.

Braking torque: The negative wheel torque.

Braking traction coefficient: The maximum of the braking force coefficient that can be reached without locking a wheel on a given tire and road surface for a given environment and operating condition.

BTS: Bureau of Transportation Statistics.

Bump stop: An elastic member that increases the wheel rate toward the end of the compression travel.

Bus: A vehicle designed to transport more than 15 passengers, including the driver.

Camber angle: The inclination of the wheel plane to the vertical. It is considered positive when the wheel leans outward at the top and negative when the wheel leans inward.

Camber coefficient: *See* Camber stiffness coefficient (Camber coefficient).

Camber force (Camber thrust): The lateral force when the slip angle is zero and the ply steer and conicity forces have been subtracted.

Camber stiffness: The rate of change of the lateral force with respect to the change in the inclination angle, usually evaluated at zero inclination angles.

Camber stiffness coefficient (Camber coefficient): The ratio of the camber stiffness of a straight free-rolling tire to the vertical load.

Camber thrust: *See* Camber force (Camber thrust).

Caster angle: The angle in side elevation between the steering axis and the vertical. It is considered positive when the steering axis is inclined rearward (in the upward direction) and negative when the steering axis is inclined forward.

Caster offset: The distance in side elevation between the point where the steering axis intersects the ground and the center of tire contact. The offset is considered positive when the intersection point is forward of the tire contact center and negative when it is rearward.

CDC: *See* Collision deformation classification.

Center of gravity (CG): (1) That point of a body through which the resultant force of gravity (weight) acts irrespective of the orientation of the body. (2) That point of a body from which it could be suspended or supported and be in balance.

Center of impact: *See* Impact center.

Center of mass: *See* Center of gravity (CG).

Center of parallel wheel motion: The center of curvature of the path along which each of a pair of wheel centers moves in a longitudinal vertical plane relative to the sprung mass when both wheels are equally displaced.

Center of thrust: A thrust directed toward the center of mass of the vehicle or other object in collision.

Center of tire contact: The intersection of the wheel plane and the vertical projection of the spin axis of the wheel onto the road plane.

Centered force: A force generated in a collision that is directed through the center of gravity of the vehicle. *See also* Eccentric force.

Centerline: *See* Longitudinal axis (Roll axis, Centerline).

Central force: The component of the tire force vector in the direction perpendicular to the direction of travel of the center of tire contact. The central force is equal to the lateral force times the cosine of the slip angle minus the longitudinal force times the sine of the slip angle.

Central impact: An impact in which the contact impulses pass through the center of gravity. *See* Oblique impact.

Centrifugal caster: The unbalance moment about the steering axis produced by a lateral acceleration equal to gravity acting at the combined center of gravity of all the steerable parts. It is considered positive if the combined center of gravity is forward of the steering axis and negative if rearward of the steering axis.

Centrifugal force: (1) The force of a body in motion that tends to keep it continuing in the same direction rather than following a curve path. (2) If a body rotates at the end of an arm, the force is provided by the tension in the arm. The reaction to this force acts at the center of rotation and is called the centrifugal force. It represents the inertia force of the body, resisting the change in the direction of its motion.

Centrifugal skid: *See* Critical speed scuff.

Centripetal acceleration: (1) Acceleration toward the center of a curve or circle. (2) The component of the vector acceleration of a point in the vehicle perpendicular to the tangent to the path of that point and parallel to the road plane. (3) The acceleration that is directed toward the center of rotation.

Centripetal force: A center-seeking force that causes an object to move toward the center.

CFR: Code of Federal Regulations.

CG: *See* Center of gravity (CG).

Change in momentum: (1) The difference of the momentum (product mass and velocity) of a mass from one time to another. (2) The difference of momentum of a mass between the beginning and end of contact with another mass. (3) The difference in the momentum of a system of bodies. *See also* Conservation of momentum.

Change of velocity: The difference between velocity vectors at two points in time. *See also* Delta v (ΔV).

Characteristic speed: That forward speed for an understeer vehicle at which the steering sensitivity at zero lateral acceleration trim is one-half the steering sensitivity of a neutral steer vehicle.

Chop: (1) A broad shallow gouge in a road surface, beginning with an even, regular, deeper side and terminating in scratches and striations on the opposite shallower side. (2) A depression in pavement made by a strong, sharp metal edge moving under heavy pressure, more commonly occurring at an impact event as opposed to post-impact trajectory.

Chop mark: A broad shallow gauge, even and regular on the deeper side and terminating in scratches and striations on the opposite side.

Chord: A straight line connecting two points of an arc or two points of a curve.

Circular frequency: *See* Angular frequency (Circular frequency).

Circular motion: *See* Rotational particle motion (Angular motion, Circular motion).

Clearance lamp: A light used on the front or rear of a motor vehicle to indicate overall width or height.

Closing speed: (1) The magnitude of the relative velocity between two vehicles at a given point in time as they approach each other. (2) The magnitude of relative velocity between two vehicles as they approach each other at the beginning of an accident. (3) A normal component of the closing velocity. *See* Approach speed.

Closing velocity: (1) The magnitude of the relative velocity between two vehicles at a given point in time as they approach each other. (2) The magnitude of the relative velocity between two vehicles at the beginning of a crash. (3) The vector difference between the velocity of the vehicle and the velocity/object struck immediately before the impact.

Coefficient of friction (μ): (1) A number ratio representing the resistance to sliding of two surfaces contacting each other. The coefficient of friction is the same as the drag factor if all wheels of a vehicle are sliding (locked) and the surface is level, $\mu = \dfrac{F}{w}$.

(2) A number representing the resistance to sliding of two flat surfaces in contact, defined as the ratio of the resistance force to the normal force between the surfaces. *See* Friction drag coefficient.

Coefficient of restitution: (1) A number between 0 (most plastic/permanent damage) and 1 (most elastic) that represents the tendency for a damaged object to try to resume its original shape. (2) The ratio of the relative normal velocity at the time of separation to the relative normal velocity at the time of initial contact between the impact centers of two colliding bodies.

Coefficient of rolling resistance: The ratio of the force of resistance to rolling with zero slip to the vertical load of a wheel or vehicle. *See* Rolling resistance; Rolling resistance force coefficient (Coefficient of rolling resistance).

Collinear collision: A collision between two vehicles in which their respective directions of travel are parallel to one another, either as a rear-end or head-on collision.

Collision: A dynamic event consisting of the contact between two or more bodies, resulting in a change of momentum of a least one participating body. *See also* Accident; Crash; Impact.

Collision deformation classification (CDC): A classification of the extent of deformation to an automobile, utility vehicle, pickup, or van from a crash.

Collision scrub: A short, usually broad skid mark or yaw mark made during the engagement of collision vehicles.

Common contact point: *See* Impact center.

Common velocity conditions: Two independent conditions applicable to a collision where at the time of separation, the relative normal velocity component is zero (no restitution) and the relative tangential component of velocity is zero (i.e., sliding has ended).

Complex damping: Damping in which the force opposing the vibratory motion is variable, but not proportional, to the velocity.

Complex function: A function having real and imaginary parts.

Complex vibration: A vibration whose components are sinusoids not harmonically related to one another. *See* Harmonic.

Compliance: The reciprocal of stiffness. *See* Stiffness.

Compliance camber: The camber motion of a wheel, resulting from compliance in suspension linkages and produced by forces and/or moments applied at the tire-road contact.

Compliance camber coefficient: The rate of change in wheel inclination angle with respect to the change in forces or moments applied at the tire-road contact.

Compliance oversteer: Compliance steer, which decreases vehicle understeer or increases vehicle oversteer.

Compliance steer: The change in the steer angle of the front or rear wheels, resulting from compliance in suspension and steering linkages and produced by forces and/or moments applied at the tire-road contact.

Compliance steer coefficient: The rate of change in compliance steer with respect to change in forces or moments applied at the tire-road contact.

Compliance understeer: Compliance steer, which increases vehicle understeer or decreases vehicle oversteer.

Compression: The relative displacement of sprung and unsprung masses in the suspension system in which the distance between the masses decreases from that at static condition. *See also* Metal-to-metal position (Compression).

Concrete pavement: A solidified pavement with an upper layer of aggregate (such as sand and stone) mixed with Portland cement paste binder.

Conicity force: The component of lateral force offset, which changes the sign (with respect to the tire axis system) with a change in the direction of rotation (positive away from the serial number or toward the whitewall). The force is positive when it is directed away from the serial number on the right-side tire and negative when it is directed toward the serial number on the left-side tire.

Conservation of angular momentum: The total angular momentum of a system of masses about any one axis remains constant unless acted upon by an external torque about that axis.

Conservation of energy: (1) Energy can be neither created nor destroyed. (2) The principle of physics that states that the amount of energy in a closed system remains constant (conserved), regardless of the changes in the form of that energy.

Conservation of linear momentum: The total momentum of a system of masses in any one direction remains constant unless acted upon by an external force in that direction.

Conservation of momentum: The principle of physics that states that the sum of the momentum existing prior to a collision is equal to the sum of the momentum existing after the collision.

Conservative system: A system in which there is no mechanism for dissipating or adding energy.

Contact damage: Deformation sustained in a vehicle from physical engagement with another vehicle or object. *See* Direct damage; Dynamic crush; Induced damage; Residual crush.

Contact path: The area or region of mutual contact between a tire and the surface over which it moves or rests.

Contact point: The point of intersection of the resultant contact impulse with the intervehicular contact surface of each of two colliding vehicles. *See* Impact center.

Contact surface: *See* Intervehicular contact surface.

Control gain: *See* Steering sensitivity (Control gain.)

Coordinate system, vehicle: *See* Three-axis vehicle coordinate system.

Cornering effect: *See* Sideslip coefficient.

Cornering limit: The maximum lateral acceleration of a vehicle.

Cornering squeal: The squeal produced by a free-rolling tire, resulting from the slip angle.

Cornering stiffness: The negative of the rate of change of lateral force with respect to the change in the slip angle, usually evaluated at zero slip angles.

Cornering stiffness coefficient (Cornering coefficient): The ratio of the cornering stiffness of a straight free-rolling tire to the vertical load.

Coulomb damping: Damping that occurs due to dry friction when two surfaces slide against one another.

Coulomb friction: *See* Static friction (Coulomb friction, Fluid friction).

Coupled modes: Modes of vibration that are not independent but influence one another because of energy transfer from one mode to the other. *See* Mode of vibration.

Coupling: The term used in mechanical vibration to indicate a connection between equations of motion.

Crash: An event in which one or more vehicles make unintended contact with another vehicle or other object, producing injury, death, and/or property damage. *See* Collision; Impact.

Crash reconstruction: *See* Accident reconstruction.

CRASH3: An acronym for Calspan Reconstruction of Speeds on the Highway, Version 3, a method of reconstruction that uses the calculation of the crush energy of a collision and an approximate post-impact trajectory spinout simulation.

Crashworthiness: The characteristics of a motor vehicle that represent the occupant protection of the vehicle in a specific collision.

Critical damping: The minimum amount of viscous damping required in a linear system to prevent the displacement of the system from passing the equilibrium position upon returning from an initial displacement.

Critical speed: (1) That forward speed for an oversteer vehicle at which the steering sensitivity at zero lateral acceleration trim is infinite. (2) The maximum speed at which a vehicle can negotiate a curve.

Critical speed formula: A formula, $v_{cr} = \sqrt{fgR}$ that calculates the speed of a vehicle from its radius of curvature R, friction drag coefficient f, and acceleration of gravity g.

Critical speed scuff: A tire mark left by a rotating wheel that is sliding or slipping sideways due to the centrifugal force exceeding the peak lateral friction. The skid mark normally will have striation marks. Also referred to as centrifugal skid or inertial mark. *See also* Sideslip; Yaw.

Critical velocity: The minimum velocity at the highest point of the loop to complete a cycle.

Critically damped system: A system is said to be critically damped if the amount of damping is such that the resulting motion is on the border between the two cases of underdamped and overdamped systems.

Cross slope: *See* Super elevation.

Crown: The lateral slope of a roadway where the center is higher than the outside edge, usually for drainage purposes.

Crumple zone: That portion of the front or rear of a vehicle designed to absorb energy of a collision for the protection of the occupants.

Crush area: An area defined by the original vehicle exterior and a crush profile.

Crush distance: The permanent damage or deformation that a vehicle sustains as a result of impact. "C" measurements are taken from the original bodyline at the frame height and from left to right and from rear to front. To determine the five space intervals for C1–C6 measurements, double the damage width (in inches) and divide by 10.

Crush equivalent speed: *See* Energy equivalent speed (EES).

Crush stiffness: *See* Crush stiffness coefficient.

Crush stiffness coefficient: An empirical quantity used in the calculation of the energy dissipated in a collision and associated with the velocity change V of each vehicle. *See* CRASH3.

Curb weight: The weight of a motor vehicle with standard equipment and maximum fuel capacity.

Cycle: The complete sequence of variations in displacement that occur during a period.

D'Alembert's principle: The virtual work performed by the effective forces through infinitesimal virtual displacements compatible with the system constraints is zero.

Damped natural frequency: The frequency of free vibration of a damped linear system. The free vibration of a damped system may be considered periodic in the limited sense that the time interval between zero crossings in the same direction is constant, even though successive amplitudes decrease progressively. The frequency of the vibration is the reciprocal of this time interval.

Damped systems: Those systems in which energy is dissipated by forces opposing the vibratory motion.

Damped vibration absorber: *See* Auxiliary mass damper (Damped vibration absorber).

Damping: The process of energy dissipation generally is referred to in the study of vibrations as damping.

Damping devices: As distinct from specific types of damping, damping devices refer to the actual mechanisms used to obtain damping of suspension systems.

Damping ratio: The ratio of the amount of viscous damping present in a system to that required for critical damping.

Dashpot: *See* Viscous damper (Dashpot).

Debris: Accumulation of broken or detached matter.

Deflection (static): The radial difference between the undeflected tire radius and the static loaded radius, under specified loads and inflation.

Deformation: The alteration of form or shape as a result of a collision.

Degree of freedom: The number of degrees of freedom of a vibrating system is the sum total of all ways in which the masses of the system can be displaced independently from their respective equilibrium.

Delta t (Δt): (1) A time interval associated with an event such as vehicle-to-vehicle contact. (2) The time duration during impulse.

Delta v (Δv): (1) The difference or change of velocity vector over a time interval. (2) The difference in the velocity vector of the center of gravity of a vehicle between separation and first contact in a crash.

Departure velocity: *See* Velocity; Separation.

Dependent variables: The variables that describe the physical behavior of a system.

Deposition: (1) The pretrial questioning of a witness, under oath, subject to cross-examination for the purpose of discovering evidence or perpetuating testimony. (2) The document in which this is recorded that describes vehicle damage resulting from an impact.

Deterministic excitation: If the excitation force is known at all instants of time, the excitation is said to be deterministic.

DGPS: Differential global positioning system.

Direct damage: Damage indicating which portions of the vehicles directly contacted each other.

Direction of principal force: *See* Principal direction of force (PDOF).

Discrete system: A system with a finite number of degrees of freedom.

Displacement (Linear displacement): The net change in the position of a particle as determined from the position function.

Distributed system: A system in which mass and elasticity are considered to be distributed parameters.

Disturbance response: The vehicle motion resulting from unwanted force or displacement inputs applied to the vehicle. Examples of disturbances are wind forces or vertical road displacements.

Divergent instability: Divergent instability exists at a prescribed trim if any small temporary disturbance or control input causes an ever-increasing vehicle response without oscillation.

Divot: A piece of turf or sod torn up by dynamic contact.

DoT (DOT): United States Department of Transportation. *See* NHTSA.

Double amplitude: *See* Peak-to-peak amplitude (Double amplitude).

Drag factor (f): A number representing the ratio of the acceleration or deceleration of a vehicle or other body to the acceleration of gravity. In a skid, f is the average of the coefficient of friction during the entire skid. $F = \dfrac{a}{g}$. *See* Friction drag coefficient.

Drag force: The negative tractive force.

Drag sled: A weighted device (whose bottom surface is covered with a portion of tire thread) that is pulled along a roadway surface and provides a sliding friction coefficient of that device and roadway surface by computing the ratio of pull force to its weight.

Driver: Any person who drives or who is in physical control of a vehicle.

Driving force: The longitudinal force resulting from the driving torque application.

Driving force coefficient: The ratio of the driving force to the vertical load.

Driving point impedance: The impedance involving the ratio of force to velocity when both the force and velocity are measured at the same point and in the same direction. *See* Impedance.

Driving torque: The positive wheel torque.

Driving traction coefficient: The maximum value of the driving force coefficient, which can be reached on a given tire and road surface for a given environment and operating condition.

Dry friction damping: *See* Coulomb damping.

Dwt: Deadweight tons.

Dynamic crush: The deformation formed by the external surface of the vehicle at any time during an impact, usually measured relative to the corresponding as-manufactured unreformed surface. *See* Crush area; Crush profile; Residual crush.

Dynamic index: (k^2/ab ratio) is the square of the radius of gyration (k) of the sprung mass about a transverse axis through the center of gravity, divided by the product of the two longitudinal distances (a and b) from the center of gravity to the front and rear wheel centers.

Dynamic rate: The dynamic rate of an elastic member is the rate measured during rapid deflection where the member is not allowed to reach static equilibrium.

Dynamic vibration absorber (Tuned damper): An auxiliary mass-spring system that tends to neutralize the vibration of a structure to which it is attached. The basic principle of operation is vibration out-of-phase with the vibration of such structure, thereby applying a counteracting force.

Dynamically coupled: If an equation contains cross products of velocity, or if the kinetic energy contains cross products of velocity, that equation of motion is dynamically coupled.

Dynamics: The branch of mechanics that studies the motion of bodies (kinematics) and the action of forces that produce or change the motion.

Earth-fixed axis system (X, Y, Z): A right-hand orthogonal axis system fixed on the earth. The trajectory of the vehicle is described with respect to this earth-fixed axis system. The X axis and Y axis are in a horizontal plane, and the Z axis is directed downward.

EBS: *See* Equivalent barrier speed (EBS).

Eccentric force: A force generated in a collision that is not directed through the center of gravity of a vehicle. *See also* Centered force.

Eccentric impact: *See* Oblique impact.

EES: *See* Energy equivalent speed (EES).

Effective rolling radius (Re): The ratio of the linear velocity of the wheel center in the X′ direction to the spin velocity.

Effective static deflection: The effective static deflection of a loaded suspension system equals the static load divided by the spring rate of the system at that load.

Elastic collision: A collision between two bodies in which no permanent (plastic) deformation takes place, and both momentum and kinetic energy are conserved. *See also* Coefficient of restitution.

Elastic deformation: Deformation that is fully recovered after an applied force is removed.

Elastic impact: An idealized impact where the kinetic energy at separation equals the kinetic energy at the initiation of contact. A fully elastic impact is an impact where the coefficient of restitution is equal to one.

Elasticity: A material property that causes that material to return to its natural state after being compressed.

Encroachment: Occupancy of a highway right-of-way by non-highway structures or objects.

Energy: The capacity to do work. Mechanical energy is equal to the work done on a body in altering either its position or its velocity.

Energy equivalent speed (EES): The speed and corresponding kinetic energy with which the vehicle must contact a fixed rigid object with no rebound for equivalence to conditions of another collision. For example, the energy may be equal to a specified level of residual crush. Energy equivalent speed is the preferred term, broader than barrier equivalent velocity (BEV), equivalent barrier speed (EBS), and equivalent test speed (ETS). *See also* Barrier equivalent velocity (BEV); Equivalent barrier speed (EBS); Equivalent test speed (ETS).

Energy equivalent velocity: *See* Energy equivalent speed (EES); *see also* Equivalent barrier speed (EBS).

Equilibrium: The state of a body where the body has no net force acting on it.

Equivalent barrier speed (EBS): A vehicle velocity at which the kinetic energy of the vehicle would equal the energy that was absorbed in plastic (permanent) deformation.

Equivalent system: A system that may be substituted for another system for the purpose of analysis. Many types of equivalence are common in vibration and shock technology:

(1) equivalent stiffness, (2) equivalent damping, (3) torsional system equivalent to a translational system, (4) electrical or acoustical system equivalent to a mechanical system, and so forth.

Equivalent test deformation: *See* Equivalent energy speed (EES).

Equivalent viscous damping: A value of viscous damping assumed for the purpose of analysis of a vibratory motion, such that the dissipation of energy per cycle at resonance is the same for either the assumed or actual damping force.

ETS: *See* Equivalent test speed (ETS).

Excitation (Stimulus): An external force (or other input) applied to a system that causes the system to respond in some way.

Exciting frequency: The frequency of variation of the exciting force.

Expert: A person with special knowledge, skill, experience, training, or education.

External forces: The actions of other bodies on a rigid body.

Factor: Any circumstance contributing to a result without which the result could not have occurred.

Fall: The downward movement of an object in air under the force of gravity, where the landing point is lower than the takeoff point. *See also* Vault.

Fatigue: In reference to the strength of metals, fatigue is the deterioration of the metallic structure caused by repeated loading.

FCP: *See* First contact position (FCP).

FHWA: Federal Highway Administration.

Final rest position: *See* Point of rest.

First contact position (FCP): The position or location at an accident scene (measured relative to a coordinate system fixed to the earth) of a vehicle, pedestrian, or other object at the time it first has contact with another body in a collision.

First contact velocity: The velocity of the center of gravity of a vehicle, pedestrian, or other object at its first contact position.

First order lateral force variation: The peak-to-peak amplitude of the fundamental frequency component of the Fourier series-representing lateral force variation. Its frequency is equal to the rotational frequency of the tire.

First order radial force variation: The peak-to-peak amplitude of the fundamental frequency component of the Fourier series representing radial force variation. Its frequency is equal to the rotational frequency of the tire.

Fixed control: That mode of vehicle control wherein the position of some point in the steering system (front wheels, Pitman arm, steering wheel) is held fixed. This is a special case of position control.

Fixed object: A stationary object such as a guardrail, bridge railing, abutment, construction barricade, impact attenuator, tree, embedded rock, utility pole, ditch side, steep earth or rock slope, culvert, fence, or building.

Flat-tire mark: A scuff mark made by an over-deflected tire; a mark made by a tire that is seriously under-inflated or overloaded.

Flat tire radius: The distance from the spin axis to the road surface of a loaded tire on a specified rim at zero inflation.

Flip: (1) A sudden upward and downward movement off the ground when the horizontal movement of an object is obstructed below its center of mass by an obstacle on the surface supporting the object. The rotation during the flip is rapid, and the object normally lands upside down. Sometimes also called vault. (2) Movement of a vehicle from a place where the forward velocity of a part of the vehicle suddenly is stopped by an object below its center of gravity such as a curb, rail, or furrow, with the result that the ensuing rotation lifts the vehicle from the ground. (3) Sideways movement of a vehicle in the air where its forward velocity suddenly is stopped below the center of mass, with the result that the ensuing rotation lifts the vehicle off the ground. *See also* Vault.

Fluid friction: *See* Static friction (Coulomb friction, Fluid friction).

FMCSA: Federal Motor Carrier Safety Administration.

FMCSR: Federal Motor Carrier Safety Regulations.

FMVSS: Federal Motor Vehicle Safety Standard.

Force: A push or a pull that one body exerts on another and includes gravitational, electrostatic, magnetic, and contact influences.

Force control: That mode of vehicle control wherein inputs or restraints are placed on the steering system in the form of forces, independent of the displacement required.

Forced vibration: Forced vibration of a system is vibration during which variable forces outside the system determine the period of the vibration.

Forward projection pedestrian collision: A frontal collision of a vehicle and a pedestrian or cyclist in which the initial contact area is at or above the height of the center of gravity of the pedestrian or cyclist and in which a single impact with the frontal geometry of the vehicle causes the pedestrian or cyclist to be projected forward relative to the vehicle.

Forward velocity: Forward velocity of a point in the vehicle is the component of the vector velocity perpendicular to the y axis and parallel to the road plane.

Foundation (Support): A structure that supports the gravity load of a mechanical system. It may be fixed in space, or it may undergo a motion that provides excitation for the supported system.

Four point transformation: A photogrammetric technique whereby points positioned on a surface reasonably approximated by a plane with unknown locations can be located through the use of four additional points whose locations are known. *See* Photogrammetry.

Fraction of critical damping: The fraction of critical damping (damping ratio) for a system with viscous damping is the ratio of the actual damping coefficient c to the critical damping coefficient cc.

Free body diagram method: One method of deriving the differentiated equations of motion that involves applying conservation laws to free body diagrams of the system drawn at an arbitrary instant.

Free control: That mode of vehicle control wherein no restraints are placed on the steering system. This is a special case of force control.

Free-rolling tire: A loaded rolling tire operated without application of driving or braking torque.

Free vibration: The vibration of a system during which no variable force is applied externally to the system.

Frequency, angular: *See* Angular frequency (Circular frequency).

Frequency ratio: The ratio of the exciting frequency to the natural frequency.

Frequency of vibration: The number of periods occurring in unit time.

Friction: The resistance to motion between two bodies in contact with each other.

Friction coefficient: *See* Coefficient of friction (μ).

Friction drag coefficient: An average, uniform (constant) value of a friction coefficient applied to a specific sliding event such as when an object slides from an initial speed to stop over a distance d or during a speed change of delta V (ΔV).

Friction drag factor: *See* Friction drag coefficient.

Friction mark: A tire mark made when a slipping or sliding tire rubs the surface of the road or other surface.

Frontal impact: An impact or collision involving the front of a vehicle.

Full impact: An impact during which motion momentarily ceases between some areas of the colliding bodies while they are in contact with each other. When the colliding bodies do not separate after collision, the impact is complete.

Full trailer: A towed vehicle with a fixed rear axle and a front axle that pivots and is made to be pulled by a powered tow vehicle.

Fundamental frequency: Natural frequencies can be arranged in order of increasing magnitude, and the lowest frequency is referred to as the fundamental frequency.

Fundamental mode of vibration: In a system, this is the mode having the lowest natural frequency.

Furrow: A ditch dug by a tire, wheel, or body part sliding in dirt or a loose material surface.

Gap skid: A skid mark in which there exists a gap of ten feet or more between segments from the release and reapplication of brakes. Each segment is measured separately. *See also* Skip skid.

GAW: *See* Gross axle weight (GAW).

GAWR: *See* Gross axle weight rating (GAWR).

GCW: *See* Gross combined weight (GCW).

GCWR: *See* Gross combined weight rating (GCWR).

Generalized coordinates: A set of independent coordinates that properly and completely defines the configuration of a system and whose number is equal to the number of degrees of freedom.

Generalized forces: These usually are not actual or observable forces acting on the system but some component of a combination of such forces.

Gore point: The triangular area of land formed by the shoulders of two diverging roadways.

Gouge (Gouge mark): A pavement or ground scar deep enough to be felt easily with the fingers, produced when a protruding metal part of a vehicle digs into the pavement.

Gouge mark: *See* Gouge (Gouge mark).

GPS: Global positioning system.

Grade: The change in elevation in unit distance in a specified direction along the centerline of a roadway or path of travel of a vehicle. It is the difference between the levels of two points divided by the level distance between the points.

Grade factor: The decimal equivalent of the grade of a slope by dividing the vertical distance (rise) by the horizontal distance (run).

Gravity, acceleration due to: The ratio between the weight of a body and the mass of the body. The earth's gravitational constant is 32.2 ft/sec/sec (32.22) or 9.81 m/sec/sec (9.812).

Groove: A long, narrow pavement gouge or a channel in a pavement.

Gross axle vehicle rating: The value specified by the manufacturer as the maximum loaded weight over a single axle.

Gross axle weight (GAW): The total weight carried by an individual axle including the vehicle weight and cargo.

Gross axle weight rating (GAWR): The maximum allowable weight that can be carried by a single axle (front or rear).

Gross combined weight (GCW): The weight of a loaded vehicle plus the weight of a fully loaded semi-trailer.

Gross combined weight rating (GCWR): The maximum allowable weight of a vehicle and loaded semi-trailer.

Gross contact area: The total area enclosing the pattern of the tire tread in contact with a flat surface, including the area of grooves or voids.

Gross vehicle weight (GVW): The combined weight of a vehicle and its cargo.

Gross vehicle weight rating (GVWR): (1) The upper limit of combined weight and cargo for a vehicle, established by design, regulation, or both. (2) The value specified by the manufacturer as the maximum loaded weight of a vehicle.

GVW: *See* Gross vehicle weight (GVW).

GVWR: *See* Gross vehicle weight rating (GVWR).

Harmonic: A sinusoidal quantity having a frequency that is an integral multiple of the frequency of a periodic quantity to which it is related.

Harmonic excitation: If the excitation force is periodic, the excitation is said to be harmonic.

Harmonic response: Harmonic response is the periodic response of a vibrating system exhibiting the characteristics of resonance at a frequency that is a multiple of the excitation frequency.

Harshness: Vibrations (15 to 100 Hz) perceived tactually and/or audibly, produced by the interaction of the tire with road irregularities.

Heading angle: The angle between a reference axis fixed in the vehicle and a reference axis fixed in the roadway, giving a measure of a vehicle yaw rotation or directional orientation relative to the roadway.

Head-on impact: Frontal impact where the principle direction of force (PDOF) is at or near zero degrees.

Holonomic coordinates: If each coordinate is independent of the others, the coordinates are known as holonomic coordinates.

Holonomic system: A system having equations of constraint containing only coordinates or coordinates and time.

Homogenous differential equation: A differential equation in which all terms contain the unknown function or its derivative.

Hop: The vertical oscillatory motion of a wheel between the road surface and the sprung mass.

Hydroplaning: The effect that occurs when a tire rolling on a wet surface of a road reaches a speed where the tire becomes detached from the road and skis along the top of a film of water.

Hysteric damping: The existence of a hysterics loop leads to energy dissipation from the system during each cycle, which causes this type of natural damping.

Impact: A single collision of one mass in motion with a second mass that may be either in motion or at rest.

Impact center: The point of intersection of the contact impulse and the intervehicular contact surface for an impact. *See* Contact point.

Impact force (lever arm) movement arm: *See* Impulse moment arm.

Impact velocity: The velocity of the center of gravity of an object relative to a coordinate system fixed in the earth during an impact.

Impedance: Mechanical impedance is the ratio of a force-like quantity to a velocity-like quantity when the arguments of the real (or imaginary) parts of the quantities increase linearly with time. Examples of force-like quantities are force, sound pressure, voltage, and temperature. Examples of velocity-like quantities are velocity, volume velocity, current, and heat flow. Impedance is the reciprocal of mobility.

Imprint: An area of contact damage that clearly shows the shape of the object creating the damage.

Impulse: A change in momentum that takes place along the principle direction of force (thrust) during a collision. Force multiplied by time.

Impulse moment arm: The perpendicular distance from the center of gravity of an object to the line of action of an impulse.

Impulse ratio: The ratio of the tangential and normal impulse components in planar impact mechanics.

Impulsive force: A force that has a large magnitude and acts during a short time duration such that the time integral of the force is finite.

Impulsive torque: A torque that acts for a short time.

Independent variables: The variables with which the dependent variable changes.

Induced damage: Noncontact damage to a vehicle caused by some other part of the same vehicle or by the shock of a collision, indicated by crumpling, distortion, buckling, and breaking.

Inertia: The tendency of a body to resist a change in acceleration or motion. *See* Newton's first law of motion.

Inertial mark: *See* Critical speed scuff.

Influence coefficient: An influence coefficient, denoted by α_{12}, is defined as the status deflection of the system at position 1 due to a unit force applied at position 2 when the unit force is the only force acting.

Initial contact: (1) The point in time and space when two objects begin to touch or interact with no significant force. (2) The beginning of an impact.

Internal forces: The forces that hold together the parts of a rigid body.

Intrusion: Reduction of the pre-crash space within the passenger space compartment.

ISO: International Organization for Standardization, in Geneva, Switzerland.

Isolation: A reduction in the capacity of a system to respond to an excitation, attained by the use of a resilient support. In steady-state forced vibration, isolation is expressed quantitatively as the complement of transmissibility.

Jackknife: The behavior of a tractor-trailer system, wherein the trailer centerline assumes a large angle to the tractor centerline.

Jerk: A concise term used to denote the time rate of change of acceleration of a point.

Kinematics: The study of geometry of motion, or how things move, with little or no reference to the forces that cause the motion.

Kinetic energy: The kinetic energy of a body is the energy it possesses due to its velocity. If a body of mass m attains a velocity v from rest under the influence of a force P and moves a distance s, then work done by P is Ps, and the kinetic energy of the body is $\left[\dfrac{1}{2}mv^2\right]$.

Kinetics: The study of motion and the forces that cause motion.

Kingpin inclination: The angle in front elevation between the steering axis and the vertical.

Kingpin offset: Kingpin offset at the ground is the horizontal distance in front elevation between the point where the steering axis intersects the ground and the center of tire contact.

kph: Kilometers per hour; also km/h.

Lagrange's method: The technique known as Lagrange's method utilizes both the principle of virtual displacements and the D'Alembert principle to derive the equations of motion of a vibrating system.

Lagrangian or Lagrangian function: The difference between the kinetic energy and the potential energy of a system.

Lateral acceleration: The component of the vector acceleration of a point in the vehicle perpendicular to the vehicle x axis and parallel to the road plane.

Lateral axis (Pitch axis): The axis that extends from the left to the right and through the center of mass of a vehicle, perpendicular to the longitudinal axis.

Lateral force: That portion of the friction force on the tires that is directed perpendicular to the vehicle centerline.

Lateral force coefficient: The ratio of the lateral force to the vertical load.

Lateral force offset: The average lateral force of a straight free-rolling tire.

Lateral force variation: The periodic variation of lateral force of a straight free-rolling tire that repeats each revolution, at a fixed loaded radius, a given mean normal force, a constant speed, a given inflation pressure, and a test surface curvature.

Lateral traction coefficient: The maximum value of the lateral force coefficient that can be reached on a free-rolling tire for a given road surface, environment, and operating condition.

Lateral velocity: The lateral velocity of a point in the vehicle is the component of the vector velocity perpendicular to the x axis and parallel to the road plane.

Law of acceleration: *See* Newton's second law of motion (Law of acceleration).

Law of action and reaction: *See* Newton's third law of motion (Law of action and reaction).

Law of inertia: *See* Newton's first law of motion (Law of inertia).

Leading edge: The foremost part of a vehicle with respect to the motion and attitude of that vehicle.

Light vehicle: An automobile, passenger van, pickup truck, or sport utility vehicle.

Line spectrum: A spectrum whose components occur at a number of discrete frequencies.

Linear acceleration: The acceleration, whether uniform or nonuniform, of a vehicle or object that moves in a straight line.

Linear damping: With linear damping, the damping force is proportional to the velocity.

Linear differential equation: An equation that contains no products of the solution function and/or its derivatives.

Linear displacement: *See* Displacement (Linear displacement).

Linear mechanical impedance: The impedance involving the ratio of force to linear velocity. *See* Impedance.

Linear system (Rectilinear system): A system in which particles move only in a straight line.

Loaded radius: The distance from the center of tire contact to the wheel center measured in the wheel plane.

Locked-wheel skid mark: A skid mark left by a braked, nonrotational wheel sliding in contact with a road surface.

Logarithmic decrement: The rate of decay of amplitude, expressed as the natural logarithm of the amplitude ratio.

Longitudinal acceleration: The component of the vector acceleration of a point in the vehicle in the x direction.

Longitudinal axis (Roll axis, Centerline): The axis extending from the rear to the front and through the center of mass of a vehicle, parallel to its sides.

Longitudinal force: The component of the force vector in the x direction.

Longitudinal slip (Percent slip): The ratio of the longitudinal slip velocity to the spin velocity of the straight free-rolling tire, expressed as a percentage.

Longitudinal slip velocity: The difference between the spin velocity of the driven or braked tire and the spin velocity of the straight free-rolling tire. Both spin velocities are measured at the same linear velocity at the wheel center in the X′ direction. A positive value results from driving torque.

Longitudinal velocity: The longitudinal velocity of a point in a vehicle is the component of the vector velocity in the x direction.

Lumped mass systems: Systems that can be modeled as a combination of distinct mass and elastic elements, which possess many degrees of freedom.

Magnification factor (Amplitude ratio, Amplification factor): The ratio of the steady-state vibration amplitude to the pseudo-static deflection.

Mass: The mass of a body is determined by comparison with a standard mass, using a beam-type balance.

Maximum crush depth: The deepest part of a crush profile. *See* Dynamic crush. *See also* Residual crush.

Maximum engagement: (1) The point in time when the maximum dynamic crush occurs. (2) The point of a collision at which the greatest force and contact area between the vehicles has occurred.

Mean square value: The mean square value of a time function is found from the average of the squared values integrated over some time interval.

Mechanical impedance: *See* Impedance.

Mechanical shock: A nonperiodic excitation (e.g., a motion of the foundation or an applied force) of a mechanical system that is characterized by suddenness and severity and usually causes significant relative displacements in the system.

Mechanical system: An aggregate of matter comprising a defined configuration of mass, stiffness, and damping.

Median: That portion of a divided highway separating the roadways for traffic traveling in opposite directions.

Metal-to-metal position (Compression): The point of maximum compression travel limited by interference of substantially rigid members. *See also* Compression.

Metal-to-metal position (Rebound): The point of maximum rebound travel limited by interference of substantially rigid members. *See also* Rebound.

Middle ordinate: A line perpendicular to a chord that connects the midpoint of the chord to a point on the arc of a curve or circle.

Modal analysis: The procedure of solving a system of simultaneous differential equations of motion by transforming them into a set of independent equations by means of the modal matrix.

Mode of vibration: In a system undergoing vibration, a mode of vibration is a characteristic pattern assumed by the system in which the motion of every particle is simple harmonic with the same frequency. Two or more modes may exist concurrently in a multiple degrees-of-freedom system.

Moment of inertia: A physical property of a body that represents its resistance to rotational acceleration.

Momentum: The momentum of a body is the product of its mass and velocity.

mpg: Miles per gallon.

mph: Miles per hour.

Multiple degrees-of-freedom system: A system for which two or more coordinates are required to define completely the position of the system at any instant.

Natural frequency: The frequency of free vibration of a system. For a multiple degrees-of-freedom system, the natural frequencies are the frequencies of the normal modes of vibration.

Natural modes: The eigenvectors also are referred to as modal vectors and represent physically so-called natural modes.

Natural motions: The free vibration problem admits special independent solutions in which the system vibrates in any one of the natural modes. These solutions are referred to as natural motions.

Natural vibration: If the oscillating motion about an equilibrium point is the result of a disturbing force that is applied once and then removed, the motion is known as natural (or free) vibration.

Net contact area: The area enclosing the pattern of the tire tread in contact with a flat surface, excluding the area of grooves or other depressions.

Neutral stability: Neutral stability exists at a prescribed trim if, for any small temporary change in disturbance or control input, the resulting motion of the vehicle remains close to, but does not return to, the motion defined by the trim.

Neutral steer: A vehicle is neutral steer at a given trim if the ratio of the steering wheel angle gradient to the overall steering ratio equals the Ackerman steer angle gradient.

Neutral steer line: The set of points in the x-z plane at which external lateral forces applied to the sprung mass produce no steady-state yaw velocity.

Newton's first law of motion (Law of inertia): An object at rest or in uniform motion will remain at rest or in uniform motion until acted upon by an unbalanced or external force.

Newton's second law of motion (Law of acceleration): The acceleration a of a body is directly proportional to the force F acting upon the body and is inversely proportional to the mass m of the body where F = ma; also $a = \dfrac{F}{m}$.

Newton's third law of motion (Law of action and reaction): For every action, there is an equal and opposite reaction. Or, whenever one body exerts force on another, the second body always exerts a force on the first body that is equal in magnitude, opposite in direction, and has the same line of action.

NHTSA: National Highway Traffic Safety Administration.

Nonholonomic systems: Systems having equations of constraint-containing velocities.

Nonlinear damping: Damping due to a damping force that is not proportional to velocity.

Nonlinear system: A system in which its motion is governed by nonlinear differential equations.

Nonlinear vibrating system: A system in which any of the variable forces are not directly proportional to the displacement, or to its derivatives, with respect to time.

Normal acceleration: The component of the vector acceleration of a point in the vehicle in the z direction.

Normal mode of vibration: A mode of vibration that is uncoupled from (i.e., can exist independently of) other modes of vibration of a system. When vibration of the system is defined as an eigenvalue problem, the normal modes are the eigenvectors, and the normal mode frequencies are the eigenvalues. The term "classical normal mode" sometimes is applied to the normal modes of a vibrating system, characterized by vibration of each element of the system at the same frequency and phase. In general, classical normal modes exist only in systems having no damping or having particular types of damping.

Normal modes: The process of adjusting the elements of the natural modes to render their amplitude is called normalization, and the resulting vectors are referred to as normal modes.

Normalization: The process in which it often is convenient to choose the magnitude of the modal vectors so as to reduce the matrix [m] to the identity matrix, which automatically reduces the matrix [k] to the diagonal matrix of the natural frequencies squared.

NTSB: National Transportation Safety Board.

Number of degrees of freedom: The equivalent to the number of coordinates required to completely specify the state of an object.

OAW: *See* Vehicle width (Overall width [OAW]).

Oblique collision: A collision between two vehicles in which the direction of the vehicles prior to the collision was neither parallel nor perpendicular. *See also* Angular impact.

Oblique impact: An impact in which the contact impulse does not pass through the center of gravity.

Occupant compartment: That portion of the interior of a vehicle designed for the accommodation of passengers.

Occupant kinematics: An analysis of the motion of occupants during a collision for correlating occupant injuries to interior vehicle contact damage and the determination of relative seating position.

Offset: The distance between the longitudinal heading axes of two vehicles in frontal contact. *See also* Overlap.

Off-tracking: The tendency of a towed trailer to follow a path that is different from the path of the towing vehicle as the combination negotiates a turn.

Orthonormal: Modes that are normalized.

Oscillation: The variation, usually with time, of the magnitude of a quantity with respect to a specified reference when the magnitude is alternately greater and smaller than the reference.

Oscillatory instability: This type of instability exists if a small temporary disturbance or control input causes an oscillatory vehicle response of ever-increasing amplitude about the initial trim.

Outside diameter: The maximum diameter of the new unloaded tire inflated to the normal recommended pressure and mounted on a specified rim.

Over-deflection: An overloaded or under-inflated tire.

Overall length: *See* Vehicle length (Overall length).

Overall steering ratio: The rate of change of the steering wheel angle at a given steering wheel trim position, with respect to change in the average steer angle of a pair of steered wheels, assuming an infinitely stiff steering system with no roll of the vehicle.

Overall width (OAW): *See* Vehicle width (Overall width [OAW]).

Overdamped system: If the damping is heavy, the motion is non-oscillatory, and the system is said to be overdamped.

Overhang, front/rear: The longitudinal dimensions of a vehicle from the center of the front/rear wheels to the foremost/rearmost point on the vehicle, including the bumper, bump guards, tow hooks, and/or rub strips if standard equipment.

Overlap: The measurement of engagement of two objects in contact with one another.

Override: A condition in a collision where the main structural members, such as a bumper, of the striking vehicle are above the main structural members, such as the frame rails, of the struck vehicle. *See also* Underride.

Oversteer: (1) A characteristic of a vehicle that results in a tendency to steer toward the inside of a curve. A vehicle is oversteer at a given trim if the ratio of the steering wheel angle gradient to the overall steering ratio is less than the Ackerman steer angle gradient. *See also* Understeer.

Overturning couple: The overturning moment on the vehicle with respect to a central, longitudinal axis in the road plane due to lateral acceleration and roll acceleration.

Overturning couple distribution: The distribution of the total overturning couple between the front and rear suspensions expressed as a percentage of the total.

Overturning moment: The component of the tire moment vector tending to rotate the tire about the X' axis, positive clockwise when looking in the positive direction of the X' axis.

Parallel hop: The form of wheel hop in which a pair of wheels hops in phase.

Parallel springing: A description of the suspension of a vehicle in which the effective static deflections of the two ends are equal; that is, the spring center passes through the center of gravity of the sprung mass.

Partial node: The point, line, or surface in a standing-wave system where some characteristic of the wave field has minimum amplitude differing from zero. The appropriate modifier should be used with the words "partial node" to signify the type that is intended (e.g., displacement partial node, velocity partial node, pressure partial node).

PDOF: *See* Principal direction of force (PDOF).

PDR time: *See* Perception-decision-reaction time (PDR time).

Peak-to-peak amplitude (Double amplitude): The peak-to-peak amplitude of displacement at a point in a vibrating system is the sum of the extreme values of displacement in both directions from the equilibrium position.

Peak-to-peak lateral tire run-out: The difference between the maximum and minimum indicator readings, measured parallel to the spin axis at the point of maximum tire section, on a true running wheel (measured separately for each sidewall).

Peak-to-peak lateral wheel run-out: The difference between the maximum and minimum indicator readings, measured parallel to the spin axis on the inside vertical portion of a rim flange (measured separately for each flange).

Peak-to-peak loaded radial tire run-out: The difference between the maximum and minimum values of the loaded radius on a true running wheel.

Peak-to-peak radial wheel run-out: The difference between the maximum and minimum values of the wheel bead seat radius, measured in a plane perpendicular to the spin axis (measured separately for each bead seat).

Peak-to-peak (total) lateral force variation: The difference between the maximum and minimum values of the lateral force during one revolution of the tire.

Peak-to-peak (total) radial force variation: The difference between the maximum and minimum values of the normal force during one revolution of the tire.

Peak-to-peak unloaded radial tire run-out: The difference between the maximum and minimum undetected values of the tire radius, measured in a plane perpendicular to the spin axis on a true running wheel.

Peak-to-peak value: The peak-to-peak value of a vibrating quantity is the algebraic difference between the extremes of the quantity.

Peak value: Generally the maximum stress that the vibrating part is undergoing.

Pedestrian center of gravity (pedestrian CG): To locate the center of gravity for an adult pedestrian, multiply the person's height (in inches, without shoes) by 0.57 and than add the shoe heel height.

Percent deflection: The static deflection expressed as a percentage of the unloaded section height above the top of the rim flange.

Percent slip: *See* Longitudinal slip (Percent slip).

Perception: The process of detecting some object or situation and comprehending its significance, typically 0.75 of a second for the average driver. *See also* Reaction.

Perception-decision-reaction time (PDR time): The time required by a person to complete a response to an event or stimulus.

Period: The period of an oscillation is the smallest increment of time in which one complete sequence of variation in displacement occurs.

Periodic quantity: An oscillating quantity whose values recur for certain increments of the independent variable.

Periodic vibration: This type of vibration exists in a system when the recurring cycles take place in equal time intervals.

Phase of a periodic quantity: For a particular value of the independent variable, the fractional part of a period through which the independent variable has advanced, measured from an arbitrary reference.

Photogrammetry: The science of obtaining reliable measurement from photographs.

Pickup: *See* Transducer (Pickup).

Pitch: The angular component of ride vibrations of the sprung mass about the vehicle y axis. *See* Pitch, roll, and yaw.

Pitch axis: *See* Lateral axis (Pitch axis).

Pitch, roll, and yaw: Terms that distinguish the rotations of a vehicle about three perpendicular axes with origin at the center of gravity of the vehicle. Pitch is rotation about the horizontal side-to-side axis, roll is rotation about a horizontal front-to-rear axis, and yaw is rotation about the vertical axis (Figure D.2).

Pitch velocity: The angular velocity about the y axis.

Pitching moment: The component of the moment vector tending to rotate the vehicle about the y axis, positive clockwise when looking in the positive direction of the y axis.

Planar impact: An impact in which all forces, moments, and motion take place in a plane.

Plastic damage: Permanent damage to a vehicle that occurs during a collision.

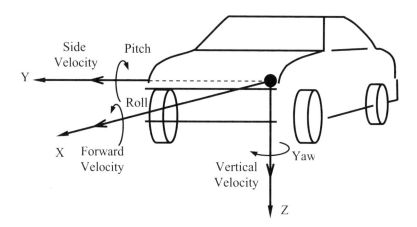

Figure D.2 Pitch, roll, and yaw.

Plastic impact: An impact with little or no rebound at the end of impact. A perfectly plastic impact occurs when the coefficient of restitution equals zero.

Ply steer force: The component of lateral force offset that does not change sign (with respect to the tire axis system) with a change in the direction of rotation (positive along the Y′ axis). The force remains positive when it is directed away from the serial number on the right side tire and toward the serial number on the left side tire.

POI: *See* Point of impact (POI).

Point of actual perception: The point at which an object is detected and recognized as a potential hazard.

Point of contact: The point of intersection of the contact impulse and the intervehicular contact surface during an impact. *See also* First contact position (FCP); Impact center; Principal direction of force (PDOF).

Point of impact (POI): (1) The collision sequence where involved objects first come into one another or a surface; can be referred to as the area of impact (AOI) and the initial contact. (2) The location on the road or other surface where objects such as vehicles collided. *See also* Area of impact (AOI).

Point mass: An idealized concept from mechanics in which an object is considered to have mass but no extent, no finite dimensions, and which, as a consequence, its rotation is irrelevant. *See* Rigid body.

Point of no escape: The perception and reaction distance plus the slide-to-stop distance in which a harmful event cannot be avoided.

Point of rest: Controlled or uncontrolled post-collision position of vehicles or persons.

Point of separation: *See* Separation.

Portland cement concrete: A concrete of an aggregate mixed with Portland cement.

Position control: That mode of vehicle control wherein inputs or restraints are placed on the steering system in the form of displacements at some control point in the steering system (front wheels, Pitman arm, steering wheel), independent of the force required.

Post-collision trajectory: *See* Post-impact trajectory.

Post-crash: The time in a collision sequence from the point of separation to the last event that significantly influences the vehicles or occupants of those vehicles.

Post-crash damage: Damage existing to a vehicle after it comes to rest, including damage that may result during rescue, towing, and salvage operations.

Post-impact speed: The magnitude of the velocity of an object in a collision at the time of separation or the end of contact. *See* Post-impact velocity; Separation speed.

Post-impact trajectory: The path of the center of gravity of a vehicle from impact to rest.

Post-impact velocity: The velocity of an object in a collision at the time of separation or the end of contact.

Potential energy: The potential energy of a body is the energy it possesses due to its position and is equal to the work done in raising it from some datum level. Thus, the potential energy of a body of mass m at a height h above the datum level is equal to mass times gravitational constant times height (mgh).

Power hop: An oscillatory hopping motion of a single wheel or of a pair of wheels, which occurs when tractive force is applied in forward or reverse motion of the vehicle.

Pre-crash: The period of time in a collision sequence from the first significant event associated with a collision to initial contact.

Pre-impact velocity: The velocity of a vehicle in a collision at the instant of its initial contact.

Principal direction of force (PDOF): (1) Determined from vehicle damage, this indicates the magnitude and direction of the opposing force acting on a vehicle during collision. (2) The direction of the line of action of the contact impulse in a planar collision expressed in degrees, measured positive clockwise from the longitudinal axis of a vehicle. Also known as thrust and the direction of principal force.

Radial force variation: The periodic variation of the normal force of a loaded straight free-rolling tire, which repeats each revolution at a fixed loaded radius, a given mean normal force, a constant speed, a given inflation pressure, and a test surface curvature.

Radius: A line segment that joins the center of a circle with any point on its circumference.

Radius of gyration: The square root of the quotient of the momentum of inertia and the mass of a rigid body.

Random excitation: This type of excitation is said to be random if the excitation force is unknown but the average and standard deviations are known.

Random vibration: Random vibration exists in a system when the oscillation is sustained but irregular both as to period and amplitude.

Rate of camber change: The change of camber angle per unit vertical displacement of the wheel center relative to the sprung mass.

Rate of caster change: The change in caster angle per unit vertical displacement of the wheel center relative to the sprung mass.

Rate of track change: The change in wheel track per unit vertical displacement of both wheel centers in the same direction relative to the sprung mass.

Ratio of critical damping: *See* Fraction of critical damping.

Reaction: The time when a driver first perceives a reason to stop his or her vehicle, and the instant that driver takes action to stop to vehicle, typically 0.75 of a second for the average driver. *See also* Perception.

Reaction distance: The distance that a vehicle travels during driver reaction time.

Reaction time: *See* Perception-decision-reaction time (PDR time).

Rebound: The relative displacement of the sprung and unsprung masses in a suspension system in which the distance between the masses increases from that at static condition. *See also* Metal-to-metal position (Rebound).

Rebound clearance: The maximum displacement in rebound of the sprung mass relative to the wheel center permitted by the suspension system, from the normal load position.

Rebound stop: An elastic member that increases the wheel rate toward the end of the rebound travel.

Reconstruction: A systematic process of evaluating evidence associated with a collision sequence and applying accepted physical principles to determine how the collision occurred.

Rectilinear system: *See* Linear system (Rectilinear system).

Reference line: A line, often the edge of a roadway, from which measurements are made.

Reference point: A point from which measurements are made to locate spots in an area, described in terms of the relation to permanent landmarks.

Residual crush: The permanent deformation formed by the nominal external surface of a vehicle caused by an impact, usually measured relative to the corresponding as-manufactured undeformed surface. *See* Crush area; Crush profile.

Resonance: A forced vibration phenomenon that exists if any small change in frequency of the applied force causes a decrease in the amplitude of the vibrating system.

Resonant frequency (Resonance frequency): The frequency at which resonance exists.

Response: For a device or system, the motion (or other output) that results from an excitation (stimulus) under specified conditions.

Response spectrum: *See* Shock spectrum (Response spectrum).

Rest position: The location of the center of gravity of a vehicle following an accident, measured relative to a coordinate system fixed in the earth.

Restitution: *See* Coefficient of restitution.

Resultant air velocity vector: The vector difference of the ambient wind velocity vector and the projection of the velocity vector of the vehicle on the X-Y plane.

Resultant force: The combined force acting on colliding vehicles to give a single resultant force.

Reverse projection photogrammetry: The photogrammetric procedure of inserting a transparency, which contains outlines of transient and fixed objects of a scene, into a camera for the purpose of determining the position and orientation of the camera at the time the original photograph was taken and to facilitate the relocation of the transient information.

Ride: The low-frequency (up to 5 Hz) vibrations of the sprung mass as a rigid body.

Ride clearance: The maximum displacement in compression of the sprung mass relative to the wheel center permitted by the suspension system, from the normal load position.

Ride rate: The change of wheel load, at the center of tire contact, per unit vertical displacement of the sprung mass relative to the ground at a specified load.

Ride time: The period of a time that begins when the bumper of a vehicle strikes a pedestrian or bicyclist and ends when the pedestrian or bicyclist strikes the ground. *See also* Throw distance.

Right angle: An angle of 90° formed by two lines perpendicular to each other.

Rim diameter: The diameter at the intersection of the bead seat and the flange.

Rim width: The distance between the inside surfaces or the rim flanges.

Roll: Rotation about the longitudinal axis of the vehicle. *See* Pitch, roll, and yaw; Yaw angle.

Roll axis: The line joining the front and rear roll centers. *See also* Longitudinal axis (Roll axis, Centerline).

Roll camber: The camber displacements of a wheel resulting from suspension roll.

Roll camber coefficient: The rate of change in the wheel inclination angle with respect to change in the suspension roll angle.

Roll center: The point in the transverse vertical plane through any pair of wheel centers at which lateral forces may be applied to the sprung mass without producing suspension roll.

Roll out: Part or all of a post-impact trajectory in which little or no sideslip of the wheels of a vehicle occurs. *See* Spinout.

Roll oversteer: Roll steer that decreases vehicle understeer or increases vehicle oversteer.

Roll steer: The change in the steer angle of the front or rear wheels due to suspension roll.

Roll steer coefficient: The rate of change in roll steer with respect to the change in the suspension roll angle at a given trim.

Roll stiffness distribution: The distribution of the vehicle roll stiffness between the front and rear suspension, expressed as a percentage of the vehicle roll stiffness.

Roll understeer: Roll steer that increases vehicle understeer or decreases vehicle oversteer.

Roll velocity: The angular velocity about the x axis.

Rolling friction: Resistance to motion of the rolling of one body over another.

Rolling moment: The component of the moment vector tending to rotate the vehicle about the x axis, positive clockwise when looking in the positive direction of the x axis.

Rolling resistance: The retarding force of a freely rolling wheel due to interaction with a contact surface, parallel to the heading axis of a wheel of a moving vehicle.

Rolling resistance force: The negative longitudinal force resulting from energy losses due to deformations of a rolling tire.

Rolling resistance force coefficient (Coefficient of rolling resistance): The ratio of the rolling resistance to the vertical load. *See also* Coefficient of rolling resistance.

Rolling resistance moment: The component of the tire moment vector tending to rotate the tire about the Y′ axis, positive clockwise when looking in the positive direction of the Y′ axis.

Rollover: Vehicle motion in which the wheels of the vehicle leave the road surface and at least one side or top of the vehicle contacts the ground. *See* Flip; Vault.

Rollover threshold: The lateral acceleration rate at which the vehicle will begin rotation about its longitudinal axis.

Root mean square value (rms): The square root of the mean square value.

ROR: Run off the road.

Rotational mechanical impedance: *See* Angular mechanical impedance (Rotational mechanical impedance).

Rotational particle motion (Angular motion, Circular motion): The motion of a particle around a circular path.

Roughness: Vibration (15 to 100 Hz) perceived tactually and/or audibly, generated by a rolling tire on a smooth road surface and producing the sensation of driving on a coarse or irregular surface.

Run-off: Fluid that flows from a ruptured vehicle reservoir down a sloping surface.

SAE: SAE International, formerly known as the Society of Automotive Engineers.

SAE coordinated system: Three-axis vehicle coordinated system.

Scalar: Any quantity that requires only magnitude, irrespective of direction.

Scene: *See* Accident scene.

Scrape: An area covered with scratches or striations made by a sliding object passing over the surface.

Scratch: A light and usually irregular scar mode on a hard surface, such as paving, by a sliding metal part without great pressure. Scratches are visible but normally are not distinguishable to the touch.

Scrub (Side skid mark): A skid mark left by tires pushed sideways or kept from rotating by forces of a collision, indicating the movement of the tire during impact between the vehicle and another vehicle or object.

Scuff mark: A relatively short mark made by a moving tire on a road or other surface in an erratic fashion with no specific, consistent features, such as acceleration scuff, impact scuff, or a flat-tire mark. *See also* Acceleration scuff; Flat-tire mark; Skid mark; Yaw mark.

Second impact: An impact between an occupant and an interior surface of a vehicle caused by and following an impact between the vehicle and another object.

Secondary impact: A second or subsequent impact between the same two vehicles during a crash.

Self-induced (self-excited) vibration: The vibration of a mechanical system is self-induced if it results from conversion, within the system, of non-oscillatory excitation to oscillatory excitation.

Semi: *See* Farm tractor; Semi-trailer; Truck tractor.

Semi-trailer: A towed vehicle equipped with one or more axles to the rear of its laden center of gravity and whose front end forms part of a pivot joint attached to a truck tractor or other powered tow vehicle. Examples are cargo, recreational, boat, and livestock trailers.

Separation: The point in the collision sequence where the collision forces have been dissipated, and the objects in contact have the opportunity to physically separate from each other, whether they actually separate or not.

Separation speed: The speed at the time of loss of contact of two vehicles in a collision. This also can refer to the speed of the centers of gravity or at the contact point.

Separation velocity: The vector velocity at the time of loss of contact of two vehicles in a collision. It can refer to the speed of the centers of gravity or at the contact point.

Service brake system: The primary brake system used for slowing and stopping a vehicle.

Shake: The intermediate frequency (5 to 25 Hz) vibrations of the sprung mass as a flexible body.

Shallow angle collision: A collision between two objects in which the angle between the directions of travel is less than $15°$.

Shimmy: A self-excited oscillation of a pair of steerable wheels about their steering axes, accompanied by appreciable tramp.

Shock: A transient phenomenon that results in a sharp, nearly sudden change in velocity.

Shock absorber: A generic term commonly applied to hydraulic mechanisms for producing damping of suspension systems.

Shock impulse: *See* Shock pulse (Shock impulse).

Shock isolator (Shock mount): A resilient support that tends to isolate a system from a shock motion.

Shock motion: An excitation involving the motion of a foundation. *See* Foundation (Support); Mechanical shock.

Shock mount: *See* Shock isolator (Shock mount).

Shock-pulse duration: *See* Duration of shock pulse.

Shock pulse (Shock impulse): A disturbing force characterized by a rise and subsequent delay of acceleration in a short period of time.

Shock spectrum (Response spectrum): A shock spectrum is a plot of the maximum response experienced by a single degree-of-freedom system, as a function of its own natural frequency, in response to an applied shock. The response may be expressed in terms of acceleration, velocity, or displacement.

Shoulder: The portion of a highway adjoining the outside traffic lanes of a roadway.

S.I. System of Units: Metric system (Système International d'Unités, or International System of Units).

Side acceleration: The component of the vector acceleration of a point in the vehicle in the y direction.

Side force: The component of the force vector in the y direction.

Side rail: The outermost edge on the side of the roof of a vehicle, connecting the upper ends of the A, B, C, and D pillars.

Side skid mark: *See* Scrub (Side skid mark).

Side velocity: The side velocity of a point in a vehicle is the component of the vector velocity in the y direction.

Sideslip: Lateral/transverse translation of a vehicle perpendicular to its heading (Figure D.3). *See also* Wheel slip.

Sideslip angle (Attitude angle): The angle between the traces on the X-Y plane of the vehicle x axis and the vehicle velocity vector at some specified point in the vehicle. In Figure D.3, the sideslip angle is shown as a negative angle.

Sideslip angle gradient: The rate of change of the sideslip angle with respect to the change in steady-state lateral acceleration on a level road at a given trim and test conditions.

Sideslip angle, tire: *See* Wheel sideslip angle.

Sideslip angle, vehicle: The angle between the heading of the vehicle and its velocity vector. *See* Sideslip.

Sideslip coefficient: The slope of the initial linear portion of the side force-sideslip curve of a tire.

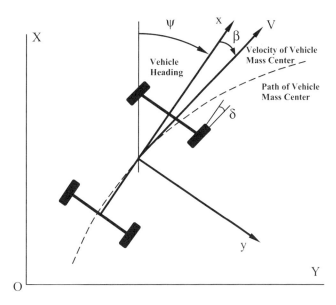

Figure D.3 Sideslip.

Sideslip mark: *See* Critical speed scuff.

Sideslip stiffness: *See* Sideslip coefficient.

Sideswipe collision: A collision of a vehicle where sliding (relative tangential motion) over the intervehicular contact surface does not end at or before separation. *See* Common velocity conditions.

Sidewall: The portion of either side of the tire that connects the bead with the tread.

Sidewall rib: A raised circumferential rib located on the sidewall.

Simple harmonic vibration: Vibration at a point in a system is simple harmonic when the displacement with respect to time is described by a simple sine function.

Simulation: The use of mathematics and mechanics, usually done using a computer to represent, reproduce, or model a physical process.

Single-degree-of-freedom system: A system for which only one coordinate is required to define completely the configuration of the system at any instant.

Sinusoidal motion: *See* Simple harmonic motion.

Sizzle: A tread noise (up to 4000 Hz) characterized by a soft frying sound, particularly noticeable on a smooth road surface.

Skid mark: A friction mark left on a road or any surface by a braked or locked tire.

Skid number: A number representing tire-pavement frictional drag determined by measurements made according to standard equipment, conditions, and procedures and usually stated as 100 times a friction coefficient.

Skip skid: A skid mark with short gaps, usually three feet or less from the tire bouncing off the roadway due to a hole, bump, or object, measured as one continuous skid. *See also* Gap skid.

Slap: An airborne smacking noise produced by a tire traversing road seams such as tar strips and expansion joints.

Sliding braking traction coefficient: The value of the braking force coefficient of a tire obtained on a locked wheel on a given tire and road surface for a given environment and operating condition.

Sliding friction coefficient: *See* Coefficient of friction (μ).

Slip: *See* Sideslip; Wheel slip.

Slip angle: The difference between the steered direction of a tire and the direction followed.

Slip angle force: The lateral force when the inclination angle is zero and the ply steer and conicity forces have been subtracted.

Slip stiffness: *See* Wheel slip coefficient.

Slope: The ratio of a vertical distance change to the horizontal distance change. *See* Grade.

Snubber: A generic term commonly applied to mechanisms that employ dry friction to produce damping of suspension systems.

Spatter: This occurs when a fluid container is ruptured by collision, and the liquid is spattered on the road or vehicle parts.

Spectrum: The magnitude of the frequency components that constitute a quantity.

Speed: A change in distance with respect to time, regardless of direction.

Spin axis: The axis of rotation of a wheel.

Spin-down: *See* Impending skid mark.

Spin velocity: The angular velocity of a wheel on which the tire is mounted, about its spin axis.

Spinout: A descriptive term for post-impact vehicle motion, including significant yaw rotation. *See* Post-impact trajectory.

Spring: A flexible mechanical line between two particles in a mechanical system.

Spring center: The vertical line along which a vertical load applied to the sprung mass will produce only uniform vertical displacement.

Spring rate: The change of load of a spring per unit deflection, taken as a mean between loading and unloading at a specified load.

Spring stiffness: A linear spring obeys a force-displacement law of $F = {*}x$, where $*$ is called the spring stiffness or spring constant and has the dimensions of force for length, and x is the displacement of the spring.

Sprung mass: Considered to be a rigid body having equal mass, the same center of gravity, and the same moments of inertia about identical axes as the total sprung weight.

Sprung weight: All weight supported by the suspension, including portions of the weight of the suspension members.

Squeal: Narrow-band airborne tire noise (150 to 800 Hz), resulting from either longitudinal slip or slip angle, or both.

Stability: The property of a system that causes it to return to a condition of equilibrium or steady motion when disturbed.

Standard loads and inflations: Those combinations of loads and inflations up to the maximum load and inflation recommended by the Tire & Rim Association and published in the yearly editions of the *Tire & Rim Association Yearbook*.

Static amplitude: The static amplitude in forced vibration at a point in a system is that displacement of the point from its specified equilibrium position that would be produced by a static force equal to the maximum value of exciting force.

Static crush: *See* Residual crush.

Static deflection: The deflection of a mechanical system due to gravitational force alone.

Static friction (Coulomb friction, Fluid friction): The frictional force exerted on a stationery body.

Static loaded radius: The loaded radius of a stationary tire inflated to normal recommended pressure.

Static margin: The horizontal distance from the center of gravity to the neutral steer line divided by the wheelbase. It is positive if the center of gravity is forward of the neutral steer line.

Static rate: For an elastic member, the rate measured between successive stationary positions at which the member has settled to substantially equilibrium condition.

Static stability factor: A dimensionless number meant to indicate the resistance of the vehicle to rollover. The number is calculated using the formula $SSF = \dfrac{T}{2h}$, where T is the vehicle track width and h is the height of the center of gravity of the vehicle.

Static toe: Static toe-in or toe-out of a pair of wheels, at a specified wheel load or relative position of the wheel center with respect to the sprung mass, is the difference in the transverse distances between the wheel planes, taken at the extreme rear and front points of the tire treads. When the distance at the rear is greater, the wheels are toed-in by this amount; when the distance is smaller, the wheels are toed-out.

Static toe angle: The static toe angle (in degrees) of a wheel, at a specified wheel load or relative position of the wheel center with respect to the sprung mass, is the angle between a longitudinal axis of the vehicle and the line of intersection of the wheel plane and the road surface. The wheel is toed-in if the forward portion of the wheel is turned toward a central longitudinal axis of the vehicle, and toed-out if turned away.

Statically coupled: A description of an equation of motion that contains cross products of coordinates.

Station line: Reference line from where measurements are obtained from predetermined reference points and from which measurements are made at right angles.

Steady state: A condition that exists when periodic (or constant) vehicle responses to periodic (or constant) control and/or disturbance inputs do not change over an arbitrarily long time. The motion responses in steady state are referred to as steady-state responses. This definition does not require the vehicle to be operating in a straight line or on a level road surface. It also can be in a turn of constant radius or on a cambered road surface.

Steady-state response: The system response after the transient motion has decayed sufficiently.

Steady-state response gain: The ratio of change in the steady-state response of any motion variable with respect to change in input at a given trim.

Steady-state vibration: Steady-state vibration exists in a system if the displacement at each point recurs for equal increments of time.

Steer angle: The angle between the projection of a longitudinal axis of the vehicle and the line of intersection of the wheel plane and the road surface.

Steering response: The vehicle motion resulting from an input to the steering (control) element.

Steering sensitivity (Control gain): The change in steady-state lateral acceleration on a level road with respect to the change in the steering wheel angle at a given trim and test conditions.

Steering wheel angle: Angular displacement of the steering wheel measured from the straight-ahead position (position corresponding to zero average steer angle of a pair of steered wheels).

Steering wheel angle gradient: The rate of change in the steering wheel angle with respect to the change in steady-state lateral acceleration on a level road at a given trim and test conditions.

Steering wheel torque: The torque applied to the steering wheel about its axis of rotation.

Steering wheel torque gradient: The rate of change in the steering wheel torque with respect to the change in steady-state lateral acceleration on a level road at a given trim and test conditions.

Stiffness: The ratio of the change of force (or torque) to the corresponding change in translational (or rotational) deflection of an elastic element.

Stiffness coefficient: *See* Crush stiffness coefficient.

Stimulus: *See* Excitation (Stimulus).

Stopping distance: The distance taken by a driver to bring a vehicle to rest in straight-forward motion by braking, including the distance traveled during perception decision reaction time prior to brake application. *See also* Braking distance.

Straight free-rolling tire: A free-rolling tire moving in a straight line at zero inclination angle and zero slip angle.

Strain energy: The strain energy of a body is the energy stored when the body is deformed. If an elastic body of stiffness S is extended a distance x by a force P, then the work done is equal to the strain energy that is equal to $\left[\dfrac{1}{2}Sx^2\right]$.

Striations: Narrow, parallel stripes or marks usually made by friction or abrasion on the roadway or on vehicle parts.

Structural damping: Damping that results from within the structure due to energy loss in the material or at joints.

Subharmonic: A sinusoidal quantity having a frequency that is integral submultiples of the fundamental frequency of a periodic quantity to which it is related.

Subharmonic response: A term sometimes used to denote a particular type of harmonic response that dominates the total response of the system. It frequently occurs when the excitation frequency is submultiples of the frequency of the fundamental resource.

Super elevation: A side-to-side slope of a road measured in degrees or percent.

Supplemental restraint system: An interior vehicle device that inflates when actuated by accelerometers and/or crash sensors and acts between an occupant and an interior vehicle surface to prevent injury due to sudden contact. *See* Airbag.

Suspension rate (Wheel rate): The change of wheel load, at the center of tire contact, per unit vertical displacement of the sprung mass relative to the wheel at a specified load.

Suspension roll: The rotation of the vehicle sprung mass about the x axis with respect to a transverse axis joining a pair of wheel centers.

Suspension roll angle: The angular displacement produced by suspension roll.

Suspension roll gradient: The rate of change in the suspension roll angle with respect to change in steady-state lateral acceleration on a level road at a given trim and test conditions.

Suspension roll stiffness: The rate of change in the restoring couple exerted by the suspension of a pair of wheels on the sprung mass of the vehicle with respect to the change in the suspension roll angle.

SUV: Sport utility vehicle.

Swing-arm radius: The horizontal distance from the swing center to the center of tire contact.

Swing center: That instantaneous center in the transverse vertical plane through any pair of wheel centers about which the wheel moves relative to the sprung mass.

Synchronous: Two harmonic oscillations are said to be synchronous if they have the same frequency (or angular velocity).

Takeoff angle (Fall, Vault): The angle between the trajectory or path of the car and the horizontal in a fall or vault. *See* Fall; Vault.

Takeoff speed: The speed of a vehicle that leaves a surface and launches off a precipice, embankment, or terminal point in the load and travels through the air.

Tangent: A line that touches the curve at only one point and is perpendicular to the radius at that point; also a straight section of a roadway.

TDC: *See* Truck deformation classification (TDC).

Three-axis vehicle coordinate system: A three-dimensional vehicle coordinate system.

Throw distance: For a pedestrian or bicyclist impact, the distance from the point of impact to the final uncontrolled point of rest, including impact, airborne, and sliding distance. *See also* Ride time.

Thump: A periodic vibration and/or audible sound generated by the tire and producing a pounding sensation that is synchronous with wheel rotation.

Tire axis system: The origin of the tire axis system is the center of tire contact.

Tire lateral load transfer: The vertical load transfer from one of the front tires (or rear tires) to the other that is due to acceleration, rotational, or inertial effects in the lateral direction.

Tire lateral load transfer distribution: The distribution of the total tire lateral load transfer between the front and rear tires, expressed as the percentage of the total.

Tire longitudinal load transfer: The vertical load transferred from a front tire to the corresponding rear tire that is due to acceleration, rotational, or inertial effects in the longitudinal direction.

Tire marks: The general term for marks on a surface generated by tires. Tire marks can be scuffs, skids, yaw marks, prints, and so forth.

Tire overall width: The width of unloaded new tires, mounted on a specified rim, inflated to the normal recommended pressure, and including the protective rib, bars, and decorations.

Tire rate (static): The static rate measured by the change of wheel load per unit vertical displacement of the wheel relative to the ground at a specified load and inflation pressure.

Tire section height: Half the difference between the tire outside diameter and the nominal rim diameter.

Tire section width: The width of an unloaded new tire mounted on a specified rim, inflated to the normal recommended pressure, and including the normal sidewalls, but excluding the protective rib, bars, and decorations.

Tire sideslip angle: *See* Wheel sideslip angle.

Torque: A rotational effort produced by a force times a moment arm.

Torque-arm center in braking: The instantaneous center in a vertical longitudinal plane through the wheel center about which the wheel moves relative to the sprung mass when the brake is locked.

Torque-arm center in drive: The instantaneous center in a vertical longitudinal plane through the wheel center about which the wheel moves relative to the sprung mass when the drive mechanism is locked at the power source.

Torque-arm radius: The horizontal distance from the torque-arm center to the wheel center.

Torsional shake: A mode of vibration involving twisting deformations of the sprung mass about the x axis of the vehicle.

Torsional spring: A link in a mechanical system where application of torque leads to angular displacement between the ends of the torsional spring.

Torsional vibration: The vibration of a rigid body about a specific reference axis. The displacement is measured in terms of an angular coordinate.

Total static deflection: For a loaded suspension system, the overall deflection under the static load from the position at which all elastic elements are free of load.

Track change: The change in wheel track resulting from vertical suspension displacements of both wheels in the same direction.

Track width: The distance between the center of the lateral width of the wheel on one side of the vehicle to the center of the wheel on the opposite side.

Tractive force: (1) The component of the tire force vector in the direction of travel of the center of tire contact. Tractive force is equal to lateral force times the sine of the slip angle plus the longitudinal force times the cosine of the slip angle. (2) The part of the friction force on the tires that is directed along the vehicle centerline. *See also* Lateral force.

Tractor: *See* Farm tractor; Truck tractor.

Tractor, semi-trailer: A truck tractor (cab) with two or more axles pulling a semi-trailer.

Tractor trailer: A truck tractor (cab) with two or more axles pulling a trailer.

Trailer: A towed vehicle that is equipped with two axles. The front axle is attached to the tow vehicle and is pivoted for turning, whereas the rear axle is fixed.

Trailing edge: That portion of a vehicle component (e.g., door, window, fender, quarter) that is closest to the rear of the vehicle; the rearmost part of a vehicle with respect to the motion and attitude of the vehicle.

Trajectory: (1) The curve that a body follows as it moves through space. (2) The path of the center of gravity of a body as it moves through space, usually associated with coordinates of the center of gravity as a function of time.

Tramp: The form of wheel hop in which a pair of wheels hops in an opposite phase.

Transducer (Pickup): A device that converts shock or vibratory motions into an optical, mechanical, or, most commonly, electrical signal that is proportional to a parameter of the experienced motion.

Transfer impedance: Transfer impedance between two points is the impedance involving the ratio of force to velocity when the force is measured at one point and the velocity is measured at the other point; also denotes the ratio of force to velocity measured at the same point but in different directions. *See* Impedance.

Transient state: The state that exists when the motion responses, the external forces relative to the vehicle, or the control positions are changing with time.

Transient vibration: Temporarily sustained vibration of a mechanical system, which may consist of forced or free vibration, or both.

Transmissibility: In forced vibration, the ratio of the transmitted force to the applied force.

Tread: The peripheral portion of the tire, the exterior of which is designed to contact the road surface.

Tread arc width: The distance measured along the tread contour of an unloaded tire between one edge of the tread and the other. For tires with rounded tread edges, the point of measurement is that point in space that is at the intersection of the tread radius extended until it meets the prolongation of the upper sidewall contour.

Tread chord width: The distance measured parallel to the spin axis of an unloaded tire between one edge of the tread and the other. For tires with rounded tread edges, the point of measurement is that point in space that is at the intersection of the tread radius extended until it meets the prolongation of the upper sidewall contour.

Tread contact length: The perpendicular distance between the tangent to the edges of the leading and following points of road contact and parallel to the wheel plane.

Tread contact width: The distance between the extreme edges of road contact at a specified load and pressure measured parallel to the Y' axis at zero slip angles and zero inclination angles.

Tread contour: The cross-sectional shape of the tread surface of an inflated unloaded tire, neglecting the tread pattern depressions.

Tread depth: The distance between the bases of a tire tread groove and a line tangent to the surface of the two adjacent tread ribs or rows.

Tread noise: Airborne sound (up to 5000 Hz) except squeal and slap produced by the interaction between the tire and the road surface.

Tread pattern: The molded configuration on the face of the tread, generally composed of ribs, rows, grooves, bars, lugs, and the like.

Tread radius: The radius or combination of radii describing the tread contour.

Trim: The steady-state (that is, equilibrium) condition of the vehicle with constant input, which is used as the reference point for analysis of dynamic vehicle stability and control characteristics.

Trip point: That location along a ground surface at which the motion of a vehicle component suddenly is halted followed by a flip, rollover, or vault.

Truck deformation classification (TDC): A classification system used to approximately describe a collision-damaged truck, consisting of seven alphanumeric characters arranged in specific order to form a descriptive composite of the vehicle damage.

Truck tractor: A motor designed for pulling semi-trailers. Basic types are cab-over-engine and conventional.

Tuned damper: *See* Dynamic vibration absorber (Tuned damper).

Uncoupled mode: An uncoupled mode of vibration is a mode that can exist in a system concurrently with and independently of other modes.

Undamped natural frequency: For a mechanical system, the frequency of free vibration resulting from only elastic and inertial forces of the system.

Undamped system: A system in which no forces are opposing the vibratory motion to dissipate energy.

Underdamped system: For the motion of a system where the displacement is a harmonic function having amplitude that decays exponentially with time, the system is said to be underdamped, and the damping is below critical.

Underride: A motor in a collision in which the main structural components of one vehicle are below the main structural components of the other vehicle.

Understeer: A vehicle characteristic that results in a tendency to steer toward the outside of a curve. *See also* Oversteer.

Understeer/oversteer gradient: The quantity obtained by subtracting the Ackerman steer angle gradient from the ratio of the steering wheel angle gradient to the overall steering ratio.

Uniform acceleration: Acceleration that changes at a constant rate.

Uniform mass damping: A system is said to possess uniform mass damping if the damping that acts on each mass is proportional to the magnitude of the mass.

Uniform motion: A term meaning uniform velocity.

Unladen weight: The maximum weight of a vehicle with no payload.

Unrestrained system: A system that has a rigid body mode corresponding to a natural frequency of zero.

Unsprung masses: The equivalent masses that reproduce the inertia forces that are produced by the motions of the corresponding unsprung parts.

Unsprung weight: All weight that is not carried by the suspension system but is supported directly by the tire or wheel and is considered to move with it.

Unstretched length: The length of a spring when it is not subjected to external forces.

Variable acceleration: Acceleration that varies at a nonconstant rate.

Variance: The mean of the squares of the deviations from the mean value of a vibrating quantity.

Vault: A vehicle or object launched into the air where the takeoff is uphill. *See* Flip; *see also* Fall.

Vector: Any quantity that has both magnitude and direction such as force, velocity, acceleration, and momentum.

Vehicle acceleration: The vector quantity expressing the acceleration of a point in the vehicle relative to the earth-fixed axis system.

Vehicle area: The projected frontal area, including the tires and underbody parts.

Vehicle axis system: A right-hand orthogonal axis system fixed in a vehicle such that with the vehicle moving steadily in a straight line on a level road, the x axis is substantially horizontal, points forward, and is in the longitudinal plane of symmetry. The y axis points to the driver's right, and the z axis points downward.

Vehicle coordinate system: *See* Three-axis vehicle coordinate system.

Vehicle length (Overall length): The maximum dimension measured longitudinally between the foremost point and the rearmost point in the vehicle, including the bumper, bumper guards, tow hooks, and/or rub strips, if standard equipment.

Vehicle pitch angle: The angle between the vehicle x axis and the ground plane.

Vehicle response: The vehicle motion resulting from some internal or external input to the vehicle. Response tests can be used to determine the stability and control characteristics of a vehicle.

Vehicle roll angle: The angle between the vehicle y axis and the ground plane.

Vehicle roll gradient: The rate of change in vehicle roll angle with respect to the change in steady-state lateral acceleration on a level road at a given trim and test conditions.

Vehicle roll stiffness: The sum of the separate suspension roll stiffness.

Vehicle velocity: The vector quantity expressing the velocity of a point in the vehicle relative to the earth-fixed axis system.

Vehicle wheelbase: The characteristic length upon which aerodynamic moment coefficients are based.

Vehicle width (Overall width [OAW]): The maximum dimension measured between the widest point on the vehicle, excluding exterior mirrors, flexible mud flaps, and marker limos, but including bumpers, moldings, sheet metal protrusions, or dual wheels if standard equipment.

Velocity: The rate of change of displacement with both a magnitude and direction. The magnitude of velocity is referred to as speed.

Velocity-time curve: A graphical depiction of velocity of the center of gravity of a vehicle as it changes over time.

Vertical axis (Yaw axis): The axis perpendicular to the plane of the horizon and passing through the center of mass of the object.

Vertical (Bounce): The translational component of ride vibrations of the sprung mass in the direction of the vehicle z axis.

Vertical load: The normal reaction of the tire on the road, which is equal to the negative of the normal force.

Vibration: An oscillation where the quantity is a parameter that defines the motion of a mechanical system. *See* Oscillation.

Vibration control: The use of vibration analysis to develop methods to eliminate or reduce unwanted vibrations or to use vibrations to protect against unwanted force or motion transmission.

Vibration damper: An auxiliary system composed of an inertia element and a viscous damper that is connected to a primary system as a means of vibration control.

Vibration, general (Oscillation): The variation with time of the displacement of a body with respect to a specified reference dimension when the displacement is alternately greater and smaller than the reference.

Vibration isolator: A resilient support that tends to isolate a system from steady-state excitation.

Vibratory motion: *See* Vibration.

Viscous damping: The dissipation of energy that occurs when a particle in a vibrating system is resisted by a force that has a magnitude proportional to the magnitude of the velocity of the particle and direction opposite to the direction of the particle.

Weight: The weight of a body is the force of attraction that the earth exerts upon it and is determined by a suitably calibrated spring-type balance.

Weight shift: When the weight of a vehicle shifts in reaction to a change in direction speed.

Wheel center: The point at which the spin axis of the wheel intersects the wheel plane.

Wheel fights: A rotary disturbance of the steering wheel produced by forces acting on the steerable wheels.

Wheel flutter: Forced oscillation of steerable wheels about their steering axes.

Wheel plane: The central plane of the tire, normal to the spin axis.

Wheel rate: *See* Suspension rate (Wheel rate).

Wheel sideslip angle: The angle between a heading axis of a wheel and the direction of the velocity vector of the center of the wheel.

Wheel skid: The occurrence of sliding between the tire and road interface that takes place within the entire contact area. Skid can result from braking, driving, and/or cornering.

Wheel slip: The ratio of the forward velocity of a tire at the road contact patch to the forward velocity at the center of the wheel (for braking), or the ratio of the forward velocity of a tire at the center of the wheel to the forward velocity at the road contact patch (for traction or acceleration).

Wheel slip coefficient: The slope of the initial linear portion of the forward force wheel slip curve of a tire.

Wheel torque: The external torque applied to the tire from the vehicle about the spin axis.

Wheel track (Wheel tread): The lateral distance between the centers of tire contact of a pair of wheels. For vehicles with dual wheels, it is the distance between the points centrally located between the centers of tire contact of the inner and outer wheels.

Wheel tread: *See* Wheel track (Wheel tread).

Wheel wobble: A self-excited oscillation of steerable wheels about their steering axes, occurring without appreciable tramp.

Wheelbase: (1) The distance from the center of the front wheel of a vehicle to the center of the rear wheel of that vehicle. (2) The perpendicular distance between axes through the front and rear wheel centerlines of a vehicle. In the case of dual axles, the distance is to the midpoint of the centerlines of the dual axles.

Wind resistance: *See* Air resistance.

Windshield header: The structural body member that connects the upper portions of the left and right A pillars and is above the top edge of the windshield.

Work: The product of the average force and the distance moved in the direction of the force by its point of application.

Wrap pedestrian collision: A frontal collision of a vehicle and pedestrian or cyclist, in which the initial contact occurs at a point below the center of gravity of the pedestrian or cyclist and where the frontal geometry of the vehicle allows the pedestrian to move rearward relative to the vehicle and to strike another portion of the vehicle such as a windshield. The latter impact causes the pedestrian or cyclist to develop an airborne trajectory followed by an impact with the ground.

Yaw: (1) The rotation of a body about its vertical axis. (2) A steering-based sideways movement of a turning vehicle, resulting from exceeding critical speed. *See* Pitch, roll, and yaw; Yaw angle.

Yaw angle: The angle between the heading of a vehicle and a fixed reference.

Yaw axis: *See* Vertical axis (Yaw axis).

Yaw mark: A tire mark caused by a sideslipping tire, often showing a striped pattern called striations. *See* Critical speed scuff.

Yaw moment of inertia: The moment of inertia about a vertical axis of a vehicle. *See* Moment of inertia; Radius of gyration.

Yaw rate: The angular velocity about the vertical axis.

Yaw velocity: The angular velocity about the z axis.

Yawing moment: The component of the moment vector tending to rotate the vehicle about the z axis, positive clockwise when looking in the positive direction of the z axis.

Appendix E

Direct Numerical Integration Methods

E.1 Introduction

The equations of dynamic equilibrium for a multiple-degrees-of-freedom vehicle system in motion can be written as

$$[M]\{\ddot{X}\} + [C]\{\dot{X}\} + [K]\{X\} = \{F(t)\} \tag{E.1}$$

where

$$[M] \quad = \text{ mass matrix}$$

$$[C] \quad = \text{ damping matrix}$$

$$[K] \quad = \text{ stiffness matrix}$$

$$\{F(t)\} = \text{ external load vector}$$

$$[X] \quad = \text{ displacement vector}$$

$$[\dot{X}] \quad = \text{ velocity vector}$$

$$[\ddot{X}] \quad = \text{ acceleration vector}$$

In dynamic analysis, the effect of acceleration-dependent inertia forces, velocity-dependent damping forces, and displacement-dependent elastic forces are considered.

Here, we will consider the direct numerical integration methods for the solution of Eq. E.1. In the direct integration method, Eq. E.1 is integrated using a numerical step-by-step procedure, with the term "direct" meaning that prior to the numerical integration, no transformation of the equation into a different form is carried out.

E.2 Single-Degree-of-Freedom System

The general equation of a viscously damped single-degree-of-freedom dynamical system can be written in the general form as

$$M\ddot{X} + C\dot{X} + KX = F(t) \qquad (E.2)$$

where for the system

$$
\begin{aligned}
M &= \text{mass} \\
C &= \text{damping} \\
K &= \text{stiffness} \\
F(t) &= \text{applied force} \\
X &= \text{displacement} \\
\dot{X} &= \text{velocity} \\
\ddot{X} &= \text{acceleration}
\end{aligned}
$$

E.2.1 Finite Difference Method

If the equilibrium relation (Eq. E.2) is regarded as an ordinary differential equation with constant coefficients, it follows that any convenient finite difference expressions to approximate the velocities and accelerations in terms of displacements can be used. The central idea in the finite difference method is to use approximations to derivatives. Hence, the general differential equation such as Eq. E.2 and the associated boundary condition are replaced by the corresponding finite difference equations. The continuous variable t is replaced by the discrete variable t_i, and the differential equation is solved progressively in time increments $h = \Delta t$, starting from known initial conditions. The solution obtained is approximate, but the accuracy of the solution can be improved by suitably selecting the time increment.

In this method, we replace the solution domain with a finite number of points, known as mesh or grid points, and obtain the solution at these points. The mesh or grid points generally are equally spaced along the independent coordinate, as shown in Figure E.1. The central difference method is based on the Taylor's series expansion of X_{i+1} and X_{i-1} about the grid point i

$$X_{i+1} = X_i + h\dot{X}_i + \frac{h^2}{2}\ddot{X}_i + \frac{h^3}{6}\dddot{X}_i + \dots \qquad (E.3)$$

$$X_{i-1} = X_i - h\dot{X}_i + \frac{h^2}{2}\ddot{X}_i - \frac{h^3}{6}\dddot{X}_i + \dots \qquad (E.4)$$

where $X_i = X(t = t_i)$ and the time interval $h = t_{i+1} - t = \Delta t$. By taking only the first two terms and subtracting Eq. E.4 from Eq. E.3, we obtain the central difference approximation to the first derivative of x at $t = t_i$ as

$$\dot{X}_i = \frac{dx}{dt}\bigg|_{t_i} = \frac{1}{2h}\left(X_{i+1} - X_{i-1}\right) \tag{E.5}$$

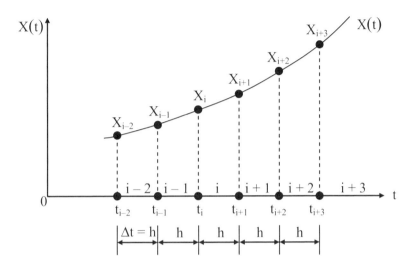

Figure E.1 Mesh or grid points in the finite difference method.

Adding Eqs. E.4 and E.3, we obtain the central difference formula for the second derivative as

$$\ddot{X}_i = \frac{d^2x}{dt^2}\bigg|_{t_i} = \frac{1}{h^2}\left(X_{i+1} - 2X_i + X_{i-1}\right) \tag{E.6}$$

Although a number of finite difference schemes exist, here we consider only two methods selected for their simplicity. They are the central difference method and the Runge-Kutta method.

E.2.2 Central Difference Method

Let the duration over which the numerical solution of Eq. E.2 is required be divided into n equal parts of interval $h = \Delta t$ each. The initial conditions are assumed to be given by $X(t = 0) = X_0$ and $\dot{X}(t = 0) = \dot{X}_0$.

The accuracy of the solution always depends on the size of the time step. The critical time step is given by $\Delta t_{cri} = \dfrac{\tau_n}{\pi}$, where τ_n is the natural period of the system. If Δt is selected to be larger than Δt_{cri}, the method becomes unstable, that is, the truncation of higher-order terms or rounding-off in the computer causes errors to grow and makes the dynamic response calculations meaningless. Numerical methods that require the use of time step Δt smaller than the critical time step Δt_{cri} are said to be conditionally stable. A safe rule to use is to choose $h \leq \dfrac{\tau_n}{10}$. Substituting Eqs. E.6 and E.5 for $\dot{X}_i$ and $\ddot{X}_i$, respectively, in Eq. E.2 at mesh or grid point i gives

$$M\left\{\frac{X_{i+1} - 2X_i + X_{i-1}}{(\Delta t)^2}\right\} + C\left\{\frac{X_{i+1} - X_{i-1}}{2\Delta t}\right\} + KX_i = F_i \qquad (E.7)$$

where $X_i = X(t_i)$ and $F_i = F(t_i)$. Solving Eq. E.7 for X_{i+1} gives

$$X_{i+1} = \left\{\frac{1}{\dfrac{M}{(\Delta t)^2} + \dfrac{C}{2\Delta t}}\right\}\left[\left\{\frac{2M}{(\Delta t)^2} - K\right\}X_i + \left\{\frac{C}{2\Delta t} - \frac{M}{(\Delta t)^2}\right\}X_{i-1} + F_i\right] \qquad (E.8)$$

This is known as the recurrence formula.

Thus, the displacement of the mass (X_{i+1}) can be calculated using Eq. E.8 if we know the previous displacements X_i, X_{i-1}, and the present external force F_i. Repeated application of Eq. E.8 gives us the response time history of the system. Because the solution of X_{i+1} in Eq. E.8 is based on the previous displacement X_i in Eq. E.7, the integration procedure is known as an explicit integration method. Note that to compute X_i, both X_0 and X_{-1} are required, but the initial conditions provide only the values of X_0 and $\dot{X}_0$. Therefore, the central difference method is not self-starting. However, the value of X_{-1} can be obtained from Eqs. E.5 and E.6 as follows. First, calculate the value of $\ddot{X}_0$ by substituting the given values of X_0 and $\dot{X}_0$ in Eq. E.2 as

$$\ddot{X}_0 = \frac{1}{M}\left[F(t = 0) - C\dot{X}_0 - KX_0\right] \qquad (E.9)$$

The value of X_{-1} then is obtained by the application of Eqs. E.5 and E.6 at i = 0

$$\ddot{X}_{-1} = X_0 - \Delta t\dot{X}_0 + \frac{(\Delta t)^2}{2}\ddot{X}_0 \qquad (E.10)$$

EXAMPLE E.1

Find the response of a viscously damped single-degree-of-freedom system subjected to a force

$$F(t) = F_0\left(1 - \sin\frac{\pi}{t_0}\right)$$

with the following data:

$$F_0 = 2 \text{ N}$$

$$t_0 = \pi,$$

$$M = 2 \text{ kg}$$

$$C = 0.3$$

$$K = 1 \text{ N/m}$$

Assume the value of the displacement and velocity of the mass at $t = 0$ to be zero.

Solution:

The differential equation of motion is

$$M\ddot{X} + C\dot{X} + KX = F(t) = F_0\left(1 - \sin\frac{\pi}{t_0}\right) \qquad (E.11)$$

Because the initial conditions are $X_0 = \dot{X}_0 = 0$, we have from Eq. E.9

$$\ddot{X}_0 = \frac{1}{M}\left[F(t = 0) - C\dot{X}_0 - KX_0\right] = \frac{1}{2}[2 - 0 - 0] = 1$$

Hence, Eq. E.10 gives

$$\ddot{X}_{-1} = X_0 - \Delta t\dot{X}_0 + \frac{(\Delta t)^2}{2}\ddot{X}_0 = \frac{(\Delta t)^2}{2}$$

Therefore, Eq. E.11 can be written as a recurrence relation of

$$X_{i+1} = \frac{1}{\left[\dfrac{M}{(\Delta t)^2} + \dfrac{C}{2\Delta t}\right]}\left[\left\{\dfrac{2M}{(\Delta t)^2} - K\right\}X_i + \left\{\dfrac{C}{2\Delta t} - \dfrac{M}{(\Delta t)^2}\right\}X_{i-1} + F_i\right] \qquad i = 0, 1, 2, \dots \; (E.12)$$

with

$$X_0 = 0$$

$$X_{-1} = \frac{(\Delta t)^2}{2}$$

$$X_i = X(t_i) = X(i\,\Delta t)$$

$$F_i = F(t_i) = F_0\left(1 - \sin\frac{i\pi\Delta t}{t_0}\right)$$

The undamped natural frequency and the natural period of the system are given by

$$\omega_n = \left(\frac{K}{M}\right)^{\frac{1}{2}} = 1$$

$$\tau_n = \frac{2\pi}{\omega_n} = 2\pi$$

Generally, the time step Δt must be less than $\frac{\tau_n}{\pi} = 2.0$. Let us find the solution of Eq. E.11 by using the time steps $\Delta t = \frac{\tau_n}{20}$. The values of the response X_i obtained according to Eq. E.12 are shown in Table E.1 for ten time steps.

TABLE E.1
RESPONSE X OF THE SYSTEM IN EXAMPLE E.1
OBTAINED USING THE CENTRAL DIFFERENCE METHOD
AND THE RUNGE-KUTTA METHOD

Time (t_i)	Central Difference Method	Runge-Kutta Method	
	X	$X_1 = X$	$X_2 = \dot{X}$
0	0	0	0
$\pi/10$	0.0493	0.0433	0.2564
$2\pi/10$	0.1606	0.1489	0.3981
$3\pi/10$	0.2988	0.2826	0.4372
$4\pi/10$	0.4347	0.4154	0.3977
$5\pi/10$	0.5481	0.5276	0.3116
$6\pi/10$	0.6298	0.6101	0.2146
$7\pi/10$	0.6821	0.6649	0.1413
$8\pi/10$	0.7175	0.7044	0.1207
$9\pi/10$	0.7564	0.7484	0.1723
π	0.8234	0.8212	0.3039

E.2.3 Runge-Kutta Method

In the Runge-Kutta method, an approximation to $X_{t+\Delta t}$ is obtained from X_t in such a way that the power series expansion of the approximation coincides, up to terms of a certain order $(\Delta t)^N$ in the time interval Δt, with the actual Taylor series expansion of $(t + \Delta t)$ in powers of Δt. The method is based on the assumption that the higher derivatives and partial derivatives exist at points required.

The Runge-Kutta method is self-starting and has the advantage that no initial values are needed beyond the prescribed values. A brief discussion of its basis is presented here. In the Runge-Kutta method, the second-order differential equation is first reduced to two first-order equations. Consider the differential equation for the single-degree-of-freedom system of Eq. E.2. Equation E.2 can be rewritten as

$$\ddot{X} = \frac{1}{M}\left[F(t) - C\dot{X} - KX\right] = f\left(X, \dot{X}, t\right) \tag{E.13}$$

By letting $X_1 = X$ and $X_2 = \dot{X}$, Eq. E.13 can be reduced to the following two first-order equations of

$$\begin{aligned} \dot{X}_1 &= X_2 \\ \dot{X}_2 &= f\left(X_1, X_2, t\right) \end{aligned} \tag{E.14}$$

By defining

$$X(t) = \begin{Bmatrix} X_1(t) \\ X_2(t) \end{Bmatrix}$$

$$F(t) = \begin{Bmatrix} X_2 \\ f\left(X_1, X_2, t\right) \end{Bmatrix}$$

the following recurrence formula is obtained to find the values of $X(t)$ at mesh or grid points t_i according to the fourth-order Runge-Kutta method. We omit the details of the derivation of the method.

$$X_{i+1} = X_i + \frac{1}{6}\left[K_1 + 2K_2 + 2K_3 + K_4\right]$$

where

$$K_1 = h\,F\left(X_i, t_i\right)$$

$$K_2 = h\,F\left(X_i + \frac{1}{2}K_1, t_i + \frac{1}{2}h\right)$$

$$K_3 = h\,F\left(X_i + \frac{1}{2}K_2, t_i + \frac{1}{2}h\right) \tag{E.15}$$

$$K_4 = h\,F\left(X_i + \frac{1}{2}K_3, t_{i+1}\right)$$

Although the Runge-Kutta method does not require the computation of derivatives beyond the first, its higher accuracy is obtained by four evaluations of the first derivatives to obtain agreement with the Taylor series solution through terms of order h^4. Because the fourth-order Runge-Kutta method is an explicit method, the maximum time step usually is governed by stability considerations. The change in time step can be implemented easily between iterations; hence, the method can be considered as an inherently stable method. The main drawback of the method is that each forward step requires several computations of the functions, thus increasing the computational cost. The Runge-Kutta method is applicable and extendable to a system of differential equations.

EXAMPLE E.2

Calculate the displacement response of the system considered in Example E.1 using the Runge-Kutta method.

Solution:

Denoting $X_1 = X$ and $X_2 = \dot{X}$, Eq. E.12 can be written as

$$\ddot{X} = \frac{1}{M}\Big[F(t) - C\dot{X} - KX\Big] = f\big(X, \dot{X}, t\big)$$

or

$$\dot{X}_1 = X_2$$

$$\dot{X}_2 = f\big(X_1, X_2, t\big)$$

Also by defining

$$\{X(t)\} = \begin{Bmatrix} X_1(t) \\ X_2(t) \end{Bmatrix}$$

and

$$\{F(t)\} = \begin{Bmatrix} X_2 \\ f\big(X, \dot{X}, t\big) \end{Bmatrix}$$

or

$$\{F(t)\} = \begin{Bmatrix} \dot{X}(t) \\ \dfrac{1}{M}\Big[F_0\Big(1 - \sin\dfrac{\pi t}{t_0}\Big) - C\dot{X}(t) - KX(t)\Big] \end{Bmatrix}$$

with the initial conditions $\{X(0)\} = \begin{Bmatrix} 0 \\ 0 \end{Bmatrix}$. The values of $\{X_{i+1}\}$, where $i = 0, 1, 2, ..., 10$ obtained are shown in Table E.1.

E.3 Multiple-Degrees-of-Freedom System

Several numerical direct integration schemes are available to determine the approximate solution of a system of equations of motion given by Eq. E.1. For a linear dynamic system, matrices $[M]$, $[C]$, and $[K]$ are independent of time and therefore remain unchanged during the integration procedure. These matrices vary with time for a nonlinear dynamic system and must be modified during the integration of equations of motion. For the solution of equations of motion for a linear dynamic system, either the normal mode superposition method of dynamic analysis or the direct numerical integration method can be used. However, for the solution of nonlinear equations of motion, direct numerical integration methods generally are recommended.

Basically, two principal approaches are used in the direct integration method: explicit and implicit schemes. In an explicit scheme, the response quantities are expressed in terms of previously determined values of displacement, velocity, and acceleration. In an implicit scheme, the difference equations are combined with the equations of motion, and the displacements are calculated directly by solving the equations.

Here, only selected numerical integration schemes widely used for linear and nonlinear dynamics analyses are considered. Three explicit and four implicit direct integration schemes are examined. A brief description of these schemes is presented, and their application is illustrated. The explicit schemes presented are the central difference method and the fourth-order Runge-Kutta method. The implicit schemes include the Houbolt, Wilson-θ, and Newmark-β methods.

E.4 Explicit Schemes

As mentioned, the response quantities in an explicit formulation are expressed explicitly in terms of previously determined values of displacement, velocity, and acceleration.

E.4.1 Central Difference Method

The procedure indicated for the case of a single-degree-of-freedom system can be extended directly to this case. Consider a displacement time history curve, as shown in Figure E.2.

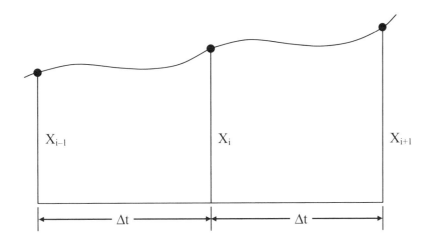

Figure E.2 Displacement versus time (central difference method).

At the middle of the time interval Δt, the velocity is given by

$$\dot{X}_{i+\frac{1}{2}} = \frac{X_{i+1} - X_i}{\Delta t} \qquad (E.16)$$

$$\dot{X}_{i-\frac{1}{2}} = \frac{X_i - X_{i-1}}{\Delta t} \qquad (E.17)$$

The acceleration is

$$\ddot{X}_i = \frac{\dot{X}_{i+\frac{1}{2}} - \dot{X}_{i-\frac{1}{2}}}{\Delta t} \tag{E.18}$$

Substituting $\dot{X}_{i+\frac{1}{2}}$ and $\dot{X}_{i-\frac{1}{2}}$ from Eqs. E.16 and E.17 into Eq. E.18, we obtain

$$\ddot{X}_i = \frac{1}{\Delta t^2}\left(X_{i+1} - 2X_i + X_{i-1}\right) \tag{E.19}$$

The difference formulas in the central difference method for velocity and acceleration are written in terms of displacement as

$$\left\{\dot{X}_t\right\} = \frac{1}{2\Delta t}\left[\left\{X_{t+\Delta t}\right\} - \left\{X_{t-\Delta t}\right\}\right] \tag{E.20}$$

$$\left\{\ddot{X}_t\right\} = \frac{1}{\Delta t^2}\left[\left\{X_{t+\Delta t}\right\} - 2\left\{X_t\right\} + \left\{X_{t-\Delta t}\right\}\right] \tag{E.21}$$

Substituting $\left\{\dot{X}_t\right\}$ and $\left\{\ddot{X}_t\right\}$ from Eqs. E.20 and E.21, respectively, into Eq. E.15, we get

$$\left[\bar{M}\right]\left\{X_{t+\Delta t}\right\} = \left\{\bar{F}_t\right\} \tag{E.22}$$

where $\left[\bar{M}\right]$, which is the effective mass matrix, and $\left\{\bar{F}_t\right\}$, which is the effective force vector, are given by

$$\left[\bar{M}\right] = \frac{1}{\Delta t^2}[M] + \frac{1}{2\Delta t}[C] \tag{E.23}$$

$$\left\{\bar{F}_t\right\} = \left\{F_t\right\} - \left([K] - \frac{2}{\Delta t^2}[M]\right)\left\{X_t\right\} - \left(\frac{1}{\Delta t^2}[M] - \frac{1}{2\Delta t}[C]\right)\left\{X_{t-\Delta t}\right\} \tag{E.24}$$

At time $t + \Delta t$, the displacements $\left\{X_{t+\Delta t}\right\}$ can be computed by solving Eq. E.22, and the velocities and accelerations at time t are determined by substituting these values of $\left\{X_{t+\Delta t}\right\}$ into Eqs. E.20 and E.21. Note that the calculation of $\left\{X_{t+\Delta t}\right\}$ involves $\left\{X_t\right\}$ and $\left\{X_{t-\Delta t}\right\}$. Hence, to obtain the solution at time Δt, a special starting procedure is needed.

The local truncation error of this method is of the order of Δt^2. An important consideration in the central difference method is that the integration requires a time step Δt that must be smaller than a critical value Δt_{cr}. This critical value is limited by the highest frequency of the discrete system ω_{max}, where

$$\Delta t \le \Delta t_{cr} \le \frac{2}{\omega_{max}} \tag{E.25}$$

If the time step is larger than Δt_{cr}, the integration is unstable, meaning that any errors resulting from the numerical integration or round-off in the computer grow and make the dynamic response calculations questionable.

E.4.2 Fourth-Order Runge-Kutta Method

In the fourth-order Runge-Kutta method, the system of second-order differential equations (Eq. E.1) is converted into state variable form. That is, both the displacements and velocities are treated as unknowns $\{y\}$ defined by

$$\{y\} = \left\{ \begin{matrix} \{\dot{X}\} \\ \{X\} \end{matrix} \right\} \tag{E.26}$$

Using Eq. E.24, Eq. E.1 can be rewritten as

$$\{\ddot{X}\} = -[M]^{-1}[K]\{\Delta X_t\}\{X\} - [M]^{-1}[C]\{\Delta \dot{X}_t\} + [M]^{-1}\{F(t)\} \tag{E.27}$$

Using the identity

$$\{\dot{y}\} = \{\dot{y}\} \tag{E.28}$$

we obtain from Eqs. E.26 and E.27

$$\{\dot{y}\} = \left\{ \begin{matrix} \{\dot{X}\} \\ \{\ddot{X}\} \end{matrix} \right\} \left[\begin{array}{c|c} [0] & [I] \\ \hline -[M]^{-1}[K] & -[M]^{-1}[C] \end{array} \right] \left\{ \begin{matrix} \{X\} \\ \{\dot{X}\} \end{matrix} \right\} + \left\{ \begin{matrix} \{0\} \\ -[M]^{-1}\{F(t)\} \end{matrix} \right\} \tag{E.29}$$

or

$$\{\dot{y}\} = [E]\{y\} + \{F*(t)\} \tag{E.30}$$

That is,

$$\{\dot{y}\} = \{f(t,y)\} \tag{E.31}$$

In the Runge-Kutta method, an approximation to $\{y_{t+\Delta t}\}$ is obtained from $\{y_t\}$ in such a way that the power series expansion of the approximation coincides, up to the terms of a certain order $(\Delta t)^N$ in the time interval Δt, with an actual Taylor series expansion of $(t + \Delta t)$ in powers of Δt. This method has the advantage that no initial values are required beyond the prescribed ones. The general fourth-order algorithms are based on formulae of the form

$$\{y_{t+\Delta t}\} = \{y_t\} + \Delta t\left(a\,K_1 + b\,K_2 + c\,K_3 + d\,K_4\right)$$

where a, b, c, and d are constants, and K_1, K_2, K_3, and K_4 are the appropriate derivative values computed in the interval $t_K < t < t_{K+\Delta t}$. Several fourth-order algorithms have been proposed. The following is due to Runge-Kutta, and we omit the details of its derivation:

$$\{y_{t+\Delta t}\} = \{y_t\} + \frac{\Delta t}{6}\left(K_1 + 2K_2 + 2K_3 + K_4\right)$$

in which

$$K_1 = f\left(t,\, y_t\right)$$

$$K_2 = f\left(t + \frac{\Delta t}{2}, y_t + K_1\frac{\Delta t}{2}\right)$$

$$K_3 = f\left(t + \frac{\Delta t}{2},\, y_t + K_2\frac{\Delta t}{2}\right)$$ (E.32)

$$K_4 = f\left(t + \Delta t,\, y_t + K_3\Delta t\right)$$

The Runge-Kutta algorithm does not require the calculation of higher derivatives. The method is completely self-starting, and the step-size can be changed easily between iterations. Hence, the method can be considered inherently stable. The truncation error e_t for the fourth-order Runge-Kutta scheme is of the form

$$e_t = c\left(\Delta t\right)^5$$ (E.33)

where c is a constant, which depends on $f\left(t,\, y\right)$ and its higher-order partial derivatives. The Runge-Kutta method generates an artificial damping that unduly suppresses the amplitude of the response of a dynamic system.

E.5 Implicit Schemes

In an implicit scheme, the difference equations are combined with the equations of motion, and the response to be evaluated is related implicitly to previously determined values. Hence, the response quantities are calculated by directly solving the difference equations.

E.5.1 Houbolt Method

The Houbolt method is based on third-order interpolation of displacements X_t and the multistep implicit formulae for $\dot{X}_t$, and $\ddot{X}_t$ is obtained in terms of X_t by using backward differences. The difference formulae are summarized in the following with reference to Figure E.3.

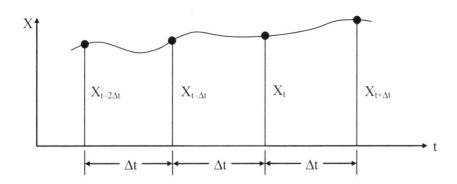

Figure E.3 Displacement versus time history curve.

By considering a cubic curve that passes through the four successive ordinates, we obtain

$$X_t = X_{t+\Delta t} - \Delta t \dot{X}_{t+\Delta t} + \frac{\Delta t^2}{2}\ddot{X}_{t+\Delta t} - \frac{\Delta t^3}{6}\dddot{X}_{t+\Delta t} \qquad (E.34)$$

$$X_{t-\Delta t} = X_{t+\Delta t} - (2\Delta t)\dot{X}_{t+\Delta t} + \left(\frac{2\Delta t}{2}\right)^2\ddot{X}_{t+\Delta t} - \left(\frac{2\Delta t}{6}\right)^3\dddot{X}_{t+\Delta t} \qquad (E.35)$$

$$X_{t-2\Delta t} = X_{t+\Delta t} - (3\Delta t)\dot{X}_{t+\Delta t} + \left(\frac{3\Delta t}{2}\right)^2\ddot{X}_{t+\Delta t} - \left(\frac{3\Delta t}{6}\right)^3\dddot{X}_{t+\Delta t} \qquad (E.36)$$

Solving Eqs. E.34 to E.36 for $\ddot{X}_{t+\Delta t}$ and $\dot{X}_{t+\Delta t}$, we obtain

$$\ddot{X}_{t+\Delta t} = \frac{1}{\Delta t^2}\left(2X_{t+\Delta t} - 5X_t + 4X_{t-\Delta t} - X_{t-2\Delta t}\right) \qquad (E.37)$$

$$\dot{X}_{t+\Delta t} = \frac{1}{6\Delta t}\left(11X_{t+\Delta t} - 18X_t + 9X_{t-\Delta t} - 2X_{t-2\Delta t}\right) \qquad (E.38)$$

The difference formulae in the Houbolt algorithm therefore are given by

$$\left\{\ddot{X}_{t+\Delta t}\right\} = \frac{1}{\Delta t^2}\left[2\left\{X_{t+\Delta t}\right\} - 5\left\{X_t\right\} + 4\left\{X_{t-\Delta t}\right\} - \left\{X_{t-2\Delta t}\right\}\right] \qquad (E.39)$$

$$\left\{\dot{X}_{t+\Delta t}\right\} = \frac{1}{6\Delta t}\left(11\left\{X_{t+\Delta t}\right\} - 18\left\{X_t\right\} + 9\left\{X_{t-\Delta t}\right\} - 2\left\{X_{t-2\Delta t}\right\}\right) \qquad (E.40)$$

By substituting the expressions for $\ddot{X}_{t+\Delta t}$ and $\dot{X}_{t+\Delta t}$ from Eqs. E.39 and E.40, respectively, into Eq. E.1, we obtain

$$\left[\bar{M}\right]\left\{X_{t+\Delta t}\right\} = \left\{\bar{F}_{t+\Delta t}\right\} \qquad (E.41)$$

where $\left[\overline{M}\right]$ is the effective mass matrix and $\left\{\overline{F}_{t+\Delta t}\right\}$ is the effective force vector

$$\left[\overline{M}\right] = \frac{2}{\Delta t^2}[M] + \frac{11}{6\Delta t}[C] + [K] \tag{E.42}$$

$$\left\{\overline{F}_{t+\Delta t}\right\} = \left\{F_{t+\Delta t}\right\} + \left(\frac{5}{\Delta t^2}[M] + \frac{3}{\Delta t}[C]\right)\left\{X_t\right\}$$

$$-\left(\frac{4}{\Delta t^2}[M] + \frac{3}{2\Delta t}[C]\right)\left\{X_{t-\Delta t}\right\} \tag{E.43}$$

$$+\left(\frac{1}{\Delta t^2}[M] + \frac{1}{3\Delta t}[C]\right) + \left\{X_{t-2\Delta t}\right\}$$

Note that the equilibrium equation at time $t + \Delta t$, which is given by Eq. E.41, is used in finding the solution for $\left\{X_{t+\Delta t}\right\}$. For this reason, this method is called an implicit integration method. It can be seen that the velocities and accelerations at time $t + \Delta t$ are obtained by substituting for $\left\{X_{t+\Delta t}\right\}$ in Eqs. E.40 and E.39, respectively. Also, a knowledge of X_t, $X_{t-\Delta t}$, and $X_{t-2\Delta t}$ is needed to find the solution for $\left\{X_{t+\Delta t}\right\}$. Because no direct method is available to find $\left\{X_{t-\Delta t}\right\}$ and $\left\{X_{t-2\Delta t}\right\}$, we use the central difference method initially to find the solution at time Δt and $2\Delta t$. This makes the method non-self-starting. The method also requires a large amount of computer storage to store displacements for the previous time steps. A basic difference between the Houbolt method and the central difference method is the presence of the stiffness matrix K as a factor to the required displacements $X_{t+\Delta t}$ in the Houbolt method. The term $KX_{t+\Delta t}$ appears because the equilibrium is considered at time $t + \Delta t$ and not at time t as in the central difference method. There is no critical time-step limit, and Δt in general can be larger than that given for the central difference method.

E.5.2 Wilson-θ Method

The Wilson-θ method assumes that the acceleration of the system varies linearly between two instants of time. Referring to Figure E.4, the acceleration is assumed to be linear from time $t_1 = i\Delta t$ to time $t_{i+\theta} = t_i + \theta\Delta t$, where $\theta \geq 1.0$. For this reason, the method is known as the Wilson-θ method.

If τ is the time varying between t and $t + \theta\Delta t$, then for the time interval t to $t + \theta\Delta t$, we can assume that

$$\ddot{X}_{t+\tau} = \ddot{X}_t + \frac{\tau}{\theta\Delta t}\left(\ddot{X}_{t+\theta\Delta t} - \ddot{X}_t\right) \tag{E.44}$$

Successive integration for Eq. E.44 gives the following expressions for $\dot{X}_{t+\tau}$ and $\ddot{X}_{t+\tau}$

$$\ddot{X}_{t+\tau} = X_t + \dot{X}_t\tau + \frac{\tau^2}{2\theta\Delta t}\left(\ddot{X}_{t+\theta\Delta t} - \ddot{X}_t\right) \tag{E.45}$$

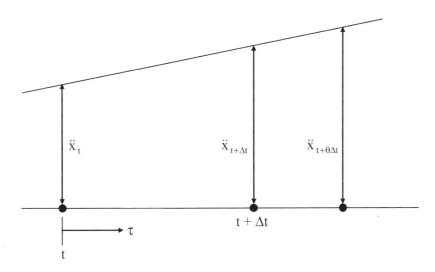

Figure E.4 Linear acceleration assumption in the Wilson-θ method.

$$\dot{X}_{t+\tau} = X_t + \dot{X}_t\tau + \frac{1}{2}\ddot{X}_t\tau + \frac{\tau^3}{6\theta\Delta t}\left(\ddot{X}_{t+\theta\Delta t} - \ddot{X}_t\right) \tag{E.46}$$

Substituting $\tau = \theta\Delta t$ into Eqs. E.45 and E.46, we obtain the following expressions for $\dot{X}$ and X at time $t + \theta\Delta t$:

$$\dot{X}_{t+\theta\Delta t} = \dot{X}_t + \frac{\theta\Delta t}{2}\left(\ddot{X}_t + \ddot{X}_{t+\theta\Delta t}\right) \tag{E.47}$$

$$X_{t+\theta\Delta t} = X_t + \theta\Delta t\dot{X}_t + \frac{1}{2}\ddot{X}_t\tau + \frac{\theta^2\Delta t^2}{6}\left(\ddot{X}_{t+\theta\Delta t} - 2\ddot{X}_t\right) \tag{E.48}$$

Solving Eqs. E.47 and E.48 for $\ddot{X}_{t+\theta\Delta t}$ and $\dot{X}_{t+\theta\Delta t}$ in terms of $X_{t+\theta\Delta t}$, we obtain

$$\ddot{X}_{t+\theta\Delta t} = \frac{6}{\theta^2\Delta t^2}\left(X_{t+\theta\Delta t} - X_t\right) - \frac{6}{\theta\Delta t}\left(\dot{X}_t - 2\ddot{X}_t\right)$$

$$\dot{X}_{t+\theta\Delta t} = \frac{3}{\theta\Delta t}\left(X_{t+\theta\Delta t} - X_t\right) - 2X_t - 2\dot{X}_t - \frac{\theta\Delta t}{2}\ddot{X}_t \tag{E.49}$$

The difference formulae in the Wilson-θ algorithm then are given by

$$\left\{\ddot{X}_{t+\theta\Delta t}\right\} = \frac{6}{\theta^2\Delta t^2}\left(\left\{X_{t+\theta\Delta t}\right\} - \left\{X_t\right\}\right) - \frac{6}{\theta\Delta t}\left(\left\{\dot{X}_t\right\} - 2\left\{\ddot{X}_t\right\}\right) \tag{E.50}$$

$$\left\{\dot{X}_{t+\theta\Delta t}\right\} = \frac{3}{\theta\Delta t}\left(\left\{X_{t+\theta\Delta t}\right\} - \left\{X_t\right\}\right) - 2\left\{\dot{X}_t\right\} - \frac{\theta\Delta t}{2}\left\{\ddot{X}_t\right\} \tag{E.51}$$

We employ Eq. E.1 at time t + Δt to obtain a solution for displacement, velocity, and acceleration at time t + Δt. Because accelerations vary linearly, a linearly projected force vector is used, such that

$$[M]\{\ddot{X}_{t+\theta\Delta t}\} + [C]\{\dot{X}_{t+\theta\Delta t}\} + [K]\{X_{t+\theta\Delta t}\} = \{F_{t+\theta\Delta t}\} \tag{E.52}$$

where

$$\{F_{t+\theta\Delta t}\} = \{F_t\} + \theta(\{F_{t+\theta\Delta t}\} - \{F_t\})$$

By substituting the expressions for $\{\ddot{X}_{t+\theta\Delta t}\}$ and $\{\dot{X}_{t+\theta\Delta t}\}$ from Eqs. E.50 and E.51, respectively, into Eq. E.52, we obtain

$$[\bar{M}]\{X_{t+\theta\Delta t}\} = \{\bar{F}_{t+\theta\Delta t}\} \tag{E.53}$$

where the effective mass matrix $[\bar{M}]$ and the effective force vector $\{\bar{F}_{t+\theta\Delta t}\}$ are given by

$$[\bar{M}] = \frac{6}{\theta^2\Delta t^2}[M] + \frac{3}{\theta\Delta t}[C] + [K] \tag{E.54}$$

$$\{\bar{F}_{t+\theta\Delta t}\} = \{F_{t+\theta\Delta t}\} + \left(\frac{6}{\theta^2\Delta t^2}[M] + \frac{3}{\theta\Delta t}[C]\right) + \{X_t\}$$

$$+ \left(\frac{6}{\theta\Delta t}[M] + 2[C]\right)\{\dot{X}_t\} + \left(2[M] + \frac{\theta\Delta t}{2}[C]\right)\{\ddot{X}_t\} \tag{E.55}$$

The solution of Eq. E.53 gives $\{X_{t+\theta\Delta t}\}$, which then is substituted into the following relationships to obtain the displacements, velocities, and accelerations at time t + Δt

$$\{\ddot{X}_{t+\Delta t}\} = \frac{6}{\theta^2\Delta t^2}(\{X_{t+\theta\Delta t}\} - \{X_t\}) - \frac{6}{\theta^2\Delta t}\{\dot{X}_t\} + \left(1 - \frac{3}{\theta}\right)\{\ddot{X}_t\} \tag{E.56}$$

$$\{\dot{X}_{t+\Delta t}\} = \{\dot{X}_t\} + \frac{\Delta t}{2}\{\ddot{X}_{t+\Delta t}\} - \{\ddot{X}_t\} \tag{E.57}$$

$$\{X_{t+\Delta t}\} = \{X_t\} + \Delta t\{\dot{X}_t\} + \frac{\Delta t^2}{6}(\{\ddot{X}_{t+\Delta t}\} - 2\{\ddot{X}_t\}) \tag{E.58}$$

When θ = 1.0, the method reduces to the linear acceleration scheme. The method is unconditionally stable for linear dynamic systems when θ ≥ 1.37, and a value of θ = 1.4 often is used for nonlinear dynamic systems. Also note that no special starting procedures are needed, because X, Ẋ, and Ẍ are expressed at time t + Δt in terms of the same quantities at time t only.

E.5.3 Newmark-β Method

The Newmark-β integration method also is based on the assumption that the acceleration varies linearly between two instants of time. In this method, the two parameters α and β can be changed to suit the requirements of a particular problem. The expressions for velocity and displacements are given by Figure E.5 as

$$\left\{\dot{X}_{t+\Delta t}\right\} = \left\{\dot{X}_t\right\} + \left[(1-\alpha)\left\{\ddot{X}_t\right\} + \alpha\left\{\ddot{X}_{t+\Delta t}\right\}\right]\Delta t \qquad (E.59)$$

$$\left\{X_{t+\Delta t}\right\} = \left\{X_t\right\} + \left\{\dot{X}_t\Delta t\right\} + \left[\left(\frac{1}{2}-\beta\right)\left\{\ddot{X}_t\right\} + \beta\left\{\ddot{X}_{t+\Delta t}\right\}\right]\Delta t^2 \qquad (E.60)$$

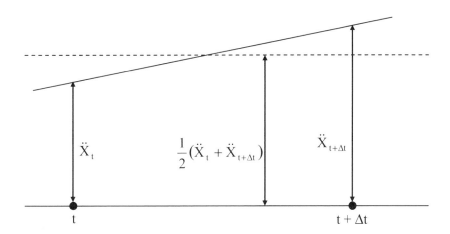

Figure E.5 Newmark's constant average acceleration scheme.

The parameters α and β signify the influence of acceleration at the end of the interval on the velocity and displacement, respectively, at the end of the interval Δt. Therefore, α and β are chosen to obtain the desired integration accuracy and stability. When $\alpha = \frac{1}{6}$ and $\beta = \frac{1}{2}$, Eqs. E.59 and E.60 correspond to the linear acceleration method (which also can be obtained using $\theta = 1$ in the Wilson-θ method). When $\alpha = \frac{1}{2}$ and $\beta = 0$, the acceleration is constant and equal to $\ddot{X}_t$ during each time interval. If $\alpha = \frac{1}{2}$ and $\beta = \frac{1}{8}$, the acceleration is constant from the beginning as $\ddot{X}_t$ and then changes to $\ddot{X}_{t+\Delta t}$ in the middle of the time interval Δt. When $\alpha = \frac{1}{2}$ and $\beta = \frac{1}{4}$, this corresponds to the assumption that the acceleration remains constant at an average value of $\frac{\left(\ddot{X}_t + \ddot{X}_{t+\Delta t}\right)}{2}$. The finite difference formulae for the Newmark-β scheme are

$$\left\{\ddot{X}_{t+\Delta t}\right\} = \frac{1}{\beta\Delta t^2}\left(\left\{X_{t+\Delta t}\right\} - \left\{X_t\right\}\right) - \frac{1}{\beta\Delta t}\left\{\dot{X}_t\right\} - \left(\frac{1}{2\beta}-1\right)\left\{\ddot{X}_t\right\} \qquad (E.61)$$

$$\left\{\dot{X}_{t+\Delta t}\right\} = \frac{\alpha}{\beta \Delta t}\left(\left\{X_{t+\Delta t}\right\} - \left\{X_t\right\}\right) - \left(\frac{\alpha}{\beta} - 1\right)\left\{\dot{X}_t\right\} - \Delta t\left(\frac{\alpha}{2\beta} - 1\right)\left\{\ddot{X}_t\right\} \qquad (E.62)$$

Equation E.1 can be employed to obtain a solution for displacements, velocities, and accelerations at time $t + \Delta t$. Therefore, by substituting the expressions for $\left\{\ddot{X}_{t+\Delta t}\right\}$ and $\left\{\ddot{X}_t\right\}$ from Eqs. E.61 and E.62, respectively, into Eq. E.1, we obtain

$$\left[\bar{M}\right]\left\{X_{t+\Delta t}\right\} = \left\{\bar{F}_{t+\Delta t}\right\} \qquad (E.63)$$

where the effective mass matrix $\left[\bar{M}\right]$ and the effective force vector $\left\{\bar{F}_{t+\Delta t}\right\}$ are given by

$$\left[\bar{M}\right] = \frac{1}{\beta \Delta t^2}[M] + \frac{\alpha}{\beta \Delta t}[C] + [K] \qquad (E.64)$$

$$\left\{\bar{F}_{t+\Delta t}\right\} = \left\{F_{t+\Delta t}\right\} + \left[\left(\frac{1}{2\beta} - 1\right)[M] + \Delta t\left(\frac{\alpha}{2\beta} - 1\right)[C]\right]\left\{\ddot{X}_t\right\}$$
$$+ \left[\frac{1}{\beta \Delta t}[M] + \left(\frac{\alpha}{\beta} - 1\right)[C]\right]\left\{\dot{X}_t\right\} + \left[\frac{1}{\beta \Delta t^2}[M] + \frac{\alpha}{\beta \Delta t}[C]\right]\left\{X_t\right\} \qquad (E.65)$$

Solution of Eq. E.63 gives $\left\{X_{t+\Delta t}\right\}$, which then is substituted into Eq. E.61 and E.62 to obtain the accelerations and velocities at time $t + \Delta t$. One of the features of the Newmark-β method is that for linear systems, the amplitude is conserved, and the response is unconditionally stable, provided that $\alpha \geq \frac{1}{2}$ and $\beta \geq \frac{1}{4}\left(\alpha + \frac{1}{2}\right)^2$. For values of $\alpha = \frac{1}{2}$ and $\beta = \frac{1}{4}$, the largest truncation errors occur in the frequency of the response as opposed to other β values. Also, note that unless $\beta = \frac{1}{2}$, a spurious damping is introduced, proportional to $\left(\beta - \frac{1}{2}\right)$. If $\beta = 0$, a negative damping results, which involves a self-excited vibration arising solely from the numerical procedure. Likewise, if $B > \frac{1}{2}$, a positive damping is introduced, which reduces the magnitude of the response even without the presence of real damping in the system. For a multiple-degrees-of-freedom system in which a number of modes constitute the total response, the peak amplitude may not be correct.

E.6 Case Studies

E.6.1 Linear Dynamic System

To illustrate the performance of various integration methods and to compare the computational time, efficiency, stability, and accuracy, a linear dynamic system with two degrees of freedom was used, as shown in Figure E.6. The equations of motion for this system are

$$\begin{bmatrix} M_1 & 0 \\ 0 & M_2 \end{bmatrix} \begin{bmatrix} \ddot{X}_1 \\ \ddot{X}_2 \end{bmatrix} + \begin{bmatrix} C_1 + C_2 & -C_2 \\ -C_2 & C_2 \end{bmatrix} \begin{bmatrix} \dot{X}_1 \\ \dot{X}_2 \end{bmatrix} + \begin{bmatrix} K_1 + K_2 & -K_2 \\ -K_2 & K_2 \end{bmatrix} \begin{bmatrix} X_1 \\ X_2 \end{bmatrix} = \begin{bmatrix} F_1(t) \\ F_2(t) \end{bmatrix} \quad \text{(E.66)}$$

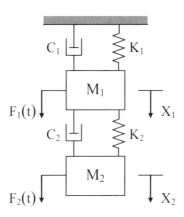

Fig. E. 6 Two degree of freedom linear dynamic system.

The numerical values of the mass, damping, and stiffness matrices are given as

$$[M] = \begin{bmatrix} 1 & 0 \\ 0 & 10 \end{bmatrix}$$

$$[K] = \begin{bmatrix} 21 & -1 \\ -1 & 1 \end{bmatrix}$$

$$[C] = \begin{bmatrix} 0.2 & -0.1 \\ -0.1 & 0.1 \end{bmatrix}$$

All the initial conditions are assumed to be zero, and the forcing vector is assumed as

$$\begin{Bmatrix} F_1(t) \\ F_2(t) \end{Bmatrix} = \begin{Bmatrix} 0 \\ 4 \end{Bmatrix} \qquad \text{for } t > 0$$

and

$$\begin{Bmatrix} F_1(t) \\ F_2(t) \end{Bmatrix} = \begin{Bmatrix} 0 \\ 0 \end{Bmatrix} \qquad \text{for } t < 0$$

The results obtained by various integration methods are given in Dukkipati (2001). For all of the methods, the time step of 0.025 sec was chosen. A value of $\theta = 1.5$ was used in the Wilson-θ method, and the parameters $\alpha = \dfrac{1}{2}$ and $\beta = \dfrac{1}{6}$ were used for the

Newmark-β method. All the integration methods gave stable and accurate results. The implicit methods were started by using the fourth-order Runge-Kutta method for the initial time steps. The accuracy of these various integration methods is evident from the comparison with the analytical solutions that also are given in Dukkipati (2001). The comparison of computational time of various integration methods used in this case study is summarized in Dukkipati (2001).

For a linear dynamic system, the matrices $[M]$, $[C]$, and $[K]$ in Eq. E.66 are independent of time and remain unchanged during the integration procedure. However, for a non-linear dynamic system, these matrices vary with time and therefore should be modified (i.e., updated) during the integration of the equations of motion.

E.6.2 Nonlinear Dynamic System

To illustrate the performance of various explicit and implicit integration methods, the methods are applied to a nonlinear dynamic system of a railway wheel-axle set, as shown in Figure E.7 and described by two nonlinear equations of motion.

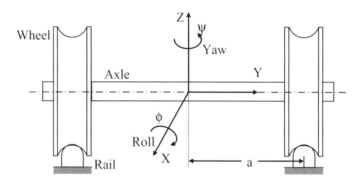

Figure E.7 Railway wheel-axle set.

The two nonlinear equations of motion for this wheel-axle set, for the lateral and yaw degrees of freedom, are in a matrix form of

$$
\begin{bmatrix} M & 0 \\ 0 & I\omega \end{bmatrix} \begin{Bmatrix} \ddot{y} \\ \ddot{\psi} \end{Bmatrix} + \begin{bmatrix} \dfrac{2f_{33}}{V} + C_y & \dfrac{2f_{12}}{V} \\ -\dfrac{2f_{12}}{V} & C_\psi + \dfrac{2a^2}{V}f_{11} + \dfrac{2f_{12}}{V} \end{bmatrix} \begin{bmatrix} \dot{y} \\ \dot{\psi} \end{bmatrix}
$$

$$
+ \begin{bmatrix} K_y & -2f_{33} \\ 0 & 2f_{12} - aW_A\delta_0 + K_\psi \end{bmatrix} \begin{bmatrix} y \\ \psi \end{bmatrix} \qquad (E.67)
$$

$$
= \begin{Bmatrix} F_y(t) + \dfrac{2f_{12}}{r_0}\Delta_2(y) - W_A\Delta_L(y) \\ F_\psi(t) + \dfrac{2f_{12}}{r_0}\Delta_1(y) - \dfrac{2af_{11}}{r_0}\left(\dfrac{r_L - r_R}{2}\right) \end{Bmatrix}
$$

where

M = mass of the wheel set: 5253 Ns^2/m (30 lb $s^2/in.$)

I_ω = yaw moment of inertia of the wheel: 1864 kg m s^2 (16,500 lb in. s^2)

a = half the distance between contact points on two rails: 0.762 m (30 in.)

r_0 = radius of the wheel: 0.508 m (20 in.)

f_{33} = longitudinal creep coefficient: 1.633×10^3 kg (3.6×10^6 lb)

f_{12} = lateral/spin creep coefficient: 0.209×10^6 kg (0.46×10^6 lb)

f_{22} = spin creep coefficient: 19.3 kg m^2 (66,000 $in.^2$ lb)

f_{11} = lateral creep coefficient: 1.769×10^6 kg (3.9×10^6 lb)

W_A = axle load: 29,937 kg (66,000 lb)

K_y = lateral stiffness: 87.5×10^4 N/m (5000 lb/in.)

C_y = lateral damping: 45.3 kg s/rad (100 lb sec/rad)

K_ψ = yaw stiffness: 21.15×10^6 kg m/rad (187,200,000 in. lb/rad)

C_ψ = yaw damping: 3525 kg m s/rad (31,200 lb in. s/rad)

V = axle speed: 382.4 m/s (15,056 in./s)

δ_0 = wheel conicity: 0.05

$F_y(t)$ = lateral input force (Figure E.8)

$F_\psi(t)$ = yaw input moment: 0

Figure E.8 shows the relationship of lateral impact force versus time.

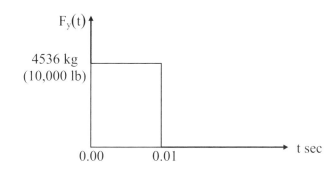

Figure E.8 Lateral impact force versus time relationship.

The initial conditions are at time $t = 0$, and $y = \dot{y} = \psi = \dot{\psi} = 0$. The variations of the difference in rolling radii $\Delta_1(y)$, $\Delta_2(y)$, and $\Delta_L(y)$ with respect to the lateral displacement y are given in Figures E.9 and E.10.

Equation E.1 was solved numerically by using the following direct integration methods: (1) central difference method, (2) fourth-order Runge-Kutta method, (3) Houbolt method,

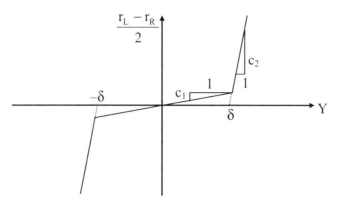

Figure E.9 Variation of the difference in the rolling radii with the lateral displacement of the wheel-axle set ($c_1 = 0.005$, $c_2 = 10$, and $\delta_1 = 0.889$ cm (0.35 in.)].

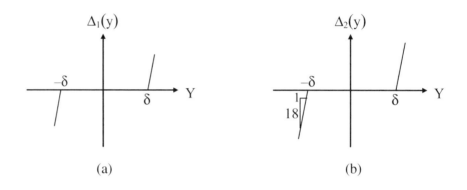

Figure E.10 (a) Variation of $\Delta_1(y)$ with the lateral displacement y, (b) variation of $\Delta_2(y)$ with the lateral displacement y, and (c) variation of $\Delta_L(y)$ with the lateral displacement y.

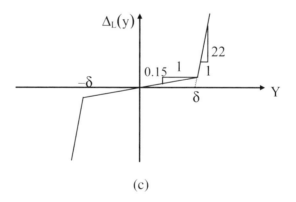

(4) Wilson-θ method, and (5) Newmark-β method. All of these integration methods were run using time steps of 0.001 sec for a total duration of 1 sec, by which time the lateral (y) and yaw (ψ) displacements of the wheel-axle set had reached steady-state values. Time histories for the two parameters y and ψ also were plotted, as shown in Dukkipati (2001). The results were compared with respect to computing time and stability of solution. Of the two explicit and three implicit integration methods studied, both of the explicit methods (central difference method and fourth-order Runge-Kutta method) and one of the implicit methods (Newmark-β method with $\alpha = \dfrac{1}{2}$ and $\beta = \dfrac{1}{6}$) methods were found to be stable. The central difference and Runge-Kutta methods introduced a

certain amount of damping. The Runge-Kutta method required the maximum amount of computational time, whereas all of the other methods required approximately the same amount of computational time.

Note that it is not possible to establish the stability of a single numerical direct integration method for all possible applications. The selection of a particular integration method depends greatly on the nature of the problem to be solved and often is dictated by the desired solution accuracy and system nonlinearities.

E.7 Summary

In this appendix, we briefly reviewed the direct numerical integration methods for the solution of one or a system of differential equations. The integration schemes considered were two explicit and three implicit methods. The explicit schemes are the central difference method and the fourth-order Runge-Kutta method. The implicit schemes are the Houbolt method, the Wilson-θ method, and the Newmark-β method. Application of these direct numerical integration methods is discussed with a case study of linear and nonlinear dynamic systems.

When these five integration methods were used to compute the response of a two-degrees-of-freedom nonlinear dynamic system, it was found that the Houbolt and Wilson-θ methods were unstable. The displacement responses from the central difference method and the fourth-order Runge-Kutta method were highly damped. It is difficult to determine the stability, accuracy, and efficiency of a single numerical integration method for all possible applications. The use of a particular integration method is dependent mainly on the nature of the problem and often is dictated by the desired accuracy of the solution.

E.8 References

1. Ali, R., "Finite Difference Methods in Vibration Analysis," *Shock and Vibration Digest*, Vol. 15, March 1983, pp. 3–7.

2. Bathe, K.J. and Wilson, E.L., *Numerical Methods in Finite Element Analysis*, Prentice-Hall, Englewood Cliffs, NJ, 1976.

3. Belytschko, T., "Explicit Time Integration of Structure-Mechanical Systems," in Donea, J. (ed.), *Advanced Structural Dynamics*, Applied Science Publishers, London, 1980, pp. 97–122.

4. Belytschko, T. and Mullen, R., "Stability of Explicit-Implicit Mesh Partitions in Time Integration," *International Journal for Numerical Methods in Engineering*, Vol. 12, 1978, pp. 1575–1586.

5. Belytschko, T. and Schoeberle, D.F., "On the Unconditional Stability of an Implicit Algorithm for Nonlinear Structural Dynamics," *Journal of Applied Mechanics*, Vol. 42, 1975, pp. 865–869.

6. Belytschko, T., Yen, H.J., and Mullen, R., "Mixed Methods for Time Integration," *Journal of Computer Methods in Applied Mechanics and Engineering*, Vol. 17/18, 1979, pp. 259–275.

7. Biggs, J.M., *Introduction to Structural Dynamics*, McGraw-Hill, New York, 1964.

8. Clough, R.W. and Penzien, J., *Dynamics of Structures*, McGraw-Hill, New York, 1974.

9. Conte, S.D. and DeBoor, C.W., *Elementary Numerical Analysis: An Algorithmic Approach*, 2nd Ed., McGraw-Hill, New York, 1972.

10. Denman, E.D., "A Numerical Method for Coupled Differential Equations," *International Journal for Numerical Methods in Engineering*, Vol. 4, 1972, pp. 587–596.

11. Destefano, G.P., "Cause of Instabilities in Numerical Integration Techniques," *International Journal of Computational Mathematics*, Vol. 2, 1968, pp. 123–142.

12. Dukkipati, R.V., *Vehicle Dynamics*, Narosa Publishing House, New Delhi, India, 2001.

13. Dukkipati, R.V., Ananda Rao, M., and. Bhat, R.B., *Computer Aided Analysis and Design of Machine Elements*, New Age International Limited, New Delhi, India, 1999.

14. Garg, V.K. and Dukkipati, R.V., *Dynamics of Railway Vehicle Systems*, Academic Press, New York, 1984.

15. Gerald, C.F. and Wheatley, P.O., *Applied Numerical Analysis*, 3rd Ed., Addison-Wesley, Reading, MA, 1984.

16. Houbolt, J.C., "A Recurrence Matrix Solution for the Dynamic Response of Elastic Aircraft," *Journal of Aeronautical Science*, Vol. 17, 1950, pp. 540–550.

17. Hughes, T.J.R., "A Note on the Stability of Newmark's Algorithm in Nonlinear Structural Dynamics," *International Journal for Numerical Methods in Engineering*, Vol. 11, 1976, pp. 383–386.

18. Hurty, W.C. and Rubinstein, M.F., *Dynamics of Structures*, Prentice-Hall, Englewood Cliffs, NJ, 1970.

19. Jennings, A. and Orr, D.R.L., "Application of the Simultaneous Iteration Method to Undamped Vibration Problems," *International Journal for Numerical Methods in Engineering*, Vol. 31, 1971, pp. 13–24.

20. Donea, J. (ed.), *Advanced Structural Dynamics*, Applied Science Publishers, London, 1980.

21. Krieg, R.D., "Unconditional Stability in Numerical Time Integration Methods," *Journal of Applied Mechanics*, Vol. 40, 1973, pp. 417–421.

22. Lau, P.C.M., "Finite Difference Approximation for Ordinary Derivatives," *International Journal for Numerical Methods in Engineering*, Vol. 17, 1981, pp. 663–678.

23. Leech, J.W., Hsu, T., and Mack, E.W., "Stability of a Finite-Difference Method for Solving Matrix Equations," *AIAA Journal*, Vol. 3, 1965, pp. 2172–2173.

24. Nakamura, S., *Computational Methods in Engineering and Science*, John Wiley, New York, 1977.

25. Newmark, N.M., "A Method of Computation for Structural Dynamics," *American Society of Civil Engineers Journal of Engineering Mechanics Division*, Vol. 85, 1959, pp. 67–94.

26. Nickell, R.E., "Direct Integration in Structural Dynamics," *American Society of Civil Engineers Journal of Engineering Mechanics Division*, Vol. 99, 1973, pp. 303–317.

27. Park, K.C., "An Improved Stiffly Stable Method for Direct Integration of Non-linear Structural Dynamic Equations," *American Society of Civil Engineers Journal of Applied Mechanics*, Vol. 42, 1975, pp. 464–470.

28. Romanelli, M.I., "Runge-Kutta Method for the Solution of Ordinary Differential Equations, " in *Mathematical Methods for Digital Computers*, Ralston, A. and Wilf, H.S. (eds.), John Wiley, New York, 1965.

29. Wang, P.C., *Numerical and Matrix Methods in Structural Mechanics*, John Wiley, New York, 1966.

Appendix F

Units and Conversion

F.1 The S.I. System of Units

The International System of Units, whose name in French is Système International d'Unités (S.I.), has been recognized universally and adopted by most countries of the world. The Unites States is still in the process of making the transition from the U.S./English system to the S.I. system. Hence, both systems of units are followed throughout this book and in worked examples.

Units. A physical quantity can be measured only by a comparison with a like quantity. A distinct amount of a physical quantity is called a unit. Any physical quantity of the same kind can be compared with it, and its value can be stated in terms of a ratio number and the unit used.

Basic units and derived units. The general unit of a physical quantity is defined as its dimension. A unit system can be developed by selecting the basic dimension of the system, a specific unit (for instance, meter for length, kilogram for mass, and second for time). Such a unit is called a basic unit, and the corresponding physical quantity is called a basic quantity. Units that are not basic are called derived units.

Systematic units. Systematic units are units derived systematically within a unit system. They can be obtained by replacing the general units (dimensions) by the basic units of the system.

If we define the dimensions of length, mass, and time as $[L]$, $[M]$, and $[T]$, respectively, then the physical quantities may be expressed as $[L]^x[M]^y[T]^z$. For instance, the dimension of acceleration is $[L][T]^{-2}$, and that of force is $[L][M][T]^{-2}$.

Absolute systems of units and gravitational systems of units. Systems of units in which the mass is taken as a basic unit are called absolute systems of units, whereas those in which the force rather than the mass is taken as a basic unit are called gravitational systems of units.

British Engineering System of Units (B.E.S.). The British Engineering System of Units (abbreviated B.E.S.) is a gravitational system of units and is based on the foot, pound force, and second as the basic units. This is the system used in the United States. The derived unit of mass is lb_f-s^2/ft and is called a slug (1 slug = 1 lb_f -s^2/ft).

International System of Units (S.I.). The International System of Units (abbreviated S.I.) is the internationally agreed system of units for expressing the values of physical quantities (Table F.1). In this system, four basic units are added to the customary three basic units (meter, kilogram, second) of the metric (mks) absolute system of units. The four added basic units are the ampere as the unit of electric current, the kelvin as the unit of thermodynamic temperature, the candela as the unit of luminous intensity, and the mole as the unit of amount of substance. Thus, in S.I. units, the meter, kilogram, second, ampere, kelvin, candela, and mole constitute the seven basic units. There are two auxiliary units in the S.I. units: the radian, which is the unit of a plane angle, and the steradian, which is the unit of a solid angle. Table F.1 lists the seven basic units, two auxiliary units, and some of the derived units of the International System of Units (S.I.). (Multiples and submultiples of the units are indicated by a series of 16 prefixes for various powers of 10 [Table F.2].)

TABLE F.1
INTERNATIONAL SYSTEM OF UNITS (S.I.)

	Quantity	Unit	Symbol	Dimension
Basic units	Length	meter	m	
	Mass	kilogram	kg	
	Time	second	s	
	Electric current	ampere	A	
	Temperature	kelvin	K	
	Luminous intensity	candela	cd	
	Amount of substance	mole	mol	
Auxiliary units	Plane angle	radian	rad	
	Solid angle	steradian	sr	
Derived units	Acceleration	meter per second squared	m/s^2	
	Activity (of radioactive source)	1 per second	s^{-1}	
	Angular acceleration	radian per second squared	rad/s^2	
	Angular velocity	radian per second	rad/s	
	Area	square meter	m^2	
	Density	kilogram per cubic meter	kg/m^3	

TABLE F.1 *(cont.)*

	Quantity	Unit	Symbol	Dimension
Derived units *(cont.)*	Dynamic viscosity	newton second per square meter	N-s/m^2	m^{-1} kg s^{-1}
	Electric capacitance	farad	F	m^{-2} kg^{-1} s^4 A^2
	Electric charge	coulomb	C	As
	Electric field strength	volt per meter	V/m	m kg s^{-3} A^{-1}
	Electric resistance	ohm	Ω	m^2 kg s^{-3} A^{-2}
	Entropy	joule per kelvin	J/K	m^2 kg s^{-2} K^{-1}
	Force	newton	N	m kg s^{-2}
	Frequency	hertz	Hz	s^{-1}
	Illumination	lux	lx	m^{-2} cd sr
	Inductance	henry	H	m^2 kg s^{-2} A^{-2}
	Kinematic viscosity	square meter per second	m^2/s	
	Luminance	candela per square meter	cd/m^2	
	Luminous flux	lumen	lm	cd sr
	Magnetic field strength	ampere per meter	A/m	
	Magnetic flux	weber	Wb	m^2 kg s^{-2} A^{-1}
	Magnetic flux density	tesla	T	kg s^{-2} A^{-1}
	Magnetomotive force	ampere turn	A	A
	Power	watt	W	m^2 kg s^{-3}
	Pressure	pascal (newton per square meter)	Pa (N/m^2)	m^{-1} kg s^{-3}
	Radiant intensity	watt per steradian	W/sr	m^2 kg s^{-3} sr^{-1}
	Specific heat	joule per kilogram kelvin	J/kg-K	m^2 s^{-2}K^{-1}
	Thermal conductivity	watt per meter kelvin	W/m-K	m kg s^{-3} K^{-1}
	Velocity	meter per second	m/s	
	Voltage	volt	V	m^2 kg s^{-3} A^{-1}
	Volume	cubic meter	m^3	
	Wave number	1 per meter	m^{-1}	
	Work, energy, quantity of heat	joule	J	m^2 kg s^{-2}

TABLE F.2
PREFIXES AND ABBREVIATED PREFIXES FOR MULTIPLES
AND SUBMULTIPLES IN POWERS OF 10

Multiplication Factor		Prefix	Abbreviated Prefix
1,000,000,000,000,000,000	$= 10^{18}$	exa	E
1,000,000,000,000,000	$= 10^{15}$	peta	P
1,000,000,000,000	$= 10^{12}$	terra	T
1,000,000,000	$= 10^{9}$	giga	G
1,000,000	$= 10^{6}$	mega	M
1,000	$= 10^{3}$	kilo	k
100	$= 10^{2}$	hecto	h
10	$= 10^{1}$	deka	da
0.1	$= 10^{-1}$	deci	d
0.01	$= 10^{-2}$	centi	c
0.001	$= 10^{-3}$	milli	m
0.000001	$= 10^{-6}$	micro	μ
0.000000001	$= 10^{-9}$	nano	n
0.000000000001	$= 10^{-12}$	pico	p
0.000000000000001	$= 10^{-15}$	femto	f
0.000000000000000001	$= 10^{-18}$	atto	a

The seven basic S.I. units are defined as follows:

- **Meter.** The meter is the length equal to 1,650,763.73 wavelengths of radiation in vacuum corresponding to the unperturbed transition between levels $2P_{10}$ and $5d_5$ of the atom of krypton 86, the orange-red line.

- **Kilogram.** The kilogram is the mass of a particular cylinder (of diameter 39 mm and height 39 mm) of platinum-iridium alloy, called the International Prototype Kilogram, which is preserved in a vault at Sèvres, France, by the International Bureau of Weights and Measures.

- **Second.** The second is the duration of 9,192,631,770 periods of the radiation corresponding to the transition between the two hyperfine levels of the fundamental state of the atom of cesium 133.

- **Ampere.** The ampere is a constant current that—if maintained in two straight, parallel conductors of infinite length, of negligible circular cross sections, and placed 1 meter apart in a vacuum—will produce between these conductors a force equal to 2×10^{-7} newtons per meter of length.

- **Kelvin.** The kelvin is the fraction $\dfrac{1}{273.16}$ of the thermodynamic temperature of the triple point of water. (Note that the triple point of water is 0.0l°C.)

- **Candela.** The candela is the luminous intensity, in the direction of the normal, of a black-body surface $\dfrac{1}{600,000}$ square meter in area, at the temperature of solidification of platinum under a pressure of 101,325 newtons per square meter.

- **Mole.** The mole is the amount of substance of a system that contains as many elementary entities as there are atoms in 0.012 kilogram of carbon 12.

The two auxiliary S.I. units are defined as follows:

- **Radian.** The radian is a unit of plane angular measurement equal to the angle at the center of a circle subtended by an arc equal in length to the radius. (The dimension of the radian is zero because it is a ratio of the quantities of the same dimension.)

- **Steradian.** The steradian is a unit of measure of solid angles that is expressed as the solid angle subtended at the center of the sphere by a portion of the surface whose area is equal to the square of the radius of the sphere. (The dimension of the steradian also is zero because it is a ratio of the quantities of the same dimension.)

The major differences between the two systems of units are attributed to the basic units and derived units established by Newton's second law of motion. These are summarized in Table F.1

F.2 S.I. Unit Prefixes

Table F.2 gives the common prefixes for multiples and submultiples of S.I. units.

F.3 S.I. Conversion

To convert the units of a given quantity from one system to another, the equivalence of units and the conversion factors are given in Table F.3. A simple procedure is to write the desired S.I. unit equal to the U.S./English units and then put in canceling unit factors. The following example illustrates the procedure.

EXAMPLE EF.1

Express stress in S.I. units in terms of stress in U.S./English units.

TABLE F.3
CONVERSION OF UNITS

Quantity	S.I. Equivalence	English Equivalence
Acceleration	$1 \text{ ft/s}^2 = 0.3048 \text{ m/s}^2$	$1 \text{ m/s}^2 = 0.3281 \text{ ft/s}^2$
Angles	$1 \text{ rad} = 57.2958$ degrees $1 \text{ rpm} = 0.1667 \text{ rev/sec}$ $\quad = 0.1047 \text{ rad/sec}$	$1 \text{ degree} = 0.0175 \text{ rad}$ $1 \text{ rad/sec} = 9.5493 \text{ rpm}$
Angular velocity	$1 \text{ rpm} = 0.1047 \text{ rad/sec}$	$1 \text{ rad/sec} = 9.5511 \text{ rpm}$
Area	$1 \text{ in.}^2 = 645.16 \times 10^{-6} \text{ m}^2$ $1 \text{ ft}^2 = 0.0929 \text{ m}^2$	$1 \text{ m}^2 = 1550 \text{ in.}^2$ $\quad = 10.7639 \text{ ft}^2$
Area moment of inertia or second moment of area	$1 \text{ in.}^4 = 41.6231 \times 10^{-8} \text{ m}^4$ $1 \text{ ft}^4 = 86.3097 \times 10^{-4} \text{ m}^4$	$1 \text{ m}^4 = 240.251 \times 10^4 \text{ in.}^4$ $\quad = 115.862 \text{ ft}^4$
Damping constant:		
Translatory	$1 \text{ lb}_f \cdot \text{sec/in.} = 175.1268 \text{ N s/m}$	$1 \text{ N} \cdot \text{s/m} = 5.7102 \times 10^{-3} \text{ lb}_f \text{ sec/in.}$
Torsional	$1 \text{ in.} \cdot \text{lb}_f - \text{sec/rad} = 0.1130 \text{ m N s/rad}$	$1 \text{m} \cdot \text{N} \cdot \text{s/rad} = 8.8507 \text{ lb}_f.\text{in.sec/rad}$
Force or weight	$1 \text{ lb}_f = 4.4482 \text{ N}$	$1 \text{ N} = 0.2248 \text{ lb}_f$
Length	$1 \text{ in.} = 0.0254 \text{ m}$ $1 \text{ ft} = 0.3048 \text{ m}$ $1 \text{ mile} = 5280 \text{ ft} = 1.6093 \text{ km}$	$1 \text{ m} = 39.370 \text{ in} = 3.2808 \text{ ft}$ $1 \text{ km} = 3280.84 \text{ ft} = 0.6214 \text{ mile}$
Mass	$1 \text{lb}_f \cdot \text{sec}^2/\text{ft (slug)} = 14.5939 \text{ kg}$ $\quad = 32.174 \text{ lb}_m$ $1 \text{ lb}_m = 0.4536 \text{ kg}$ $1 \text{ short ton} = 907.185 \text{ kg}$	$1 \text{ kg} = 2.2046 \text{ lb}_m$ $\quad = 0.0685 \text{ slugs (lb}_f \cdot \text{sec}^2/\text{ft})$
Mass density	$1 \text{ lb}_m/\text{in.}^3 = 27.6799 \times 10^3 \text{ kg/m}^3$ $1 \text{ lb}_m/\text{ft}^3 = 16.0185 \text{ kg/m}^3$	$1 \text{ kg/m}^3 = 36.127 \times 10^{-6} \text{ lb}_m/\text{in.}^3$ $\quad = 62.428 \times 10^{-3} \text{ lb}_m/\text{ft}^3$
Mass moment of inertia	$1 \text{ in.} \cdot \text{lb}_f - \text{sec}^2 = 0.1130 \text{ m}^2 \text{ kg}$	$1 \text{ m}^2 \text{ kg} = 8.8507 \text{ in.} \cdot \text{lb}_f \cdot \text{sec}^2$
Moment	$1 \text{ ft.} \cdot \text{lb} = 1.3558 \text{ N.m}$	$1 \text{ N-m} = 0.7376 \text{ ft-lb}$
Power	$1 \text{ in.} \cdot \text{lb}_f = 0.1129 \text{ W}$ $1 \text{ ft.} \cdot \text{lb}_f = 1.3558 \text{ W}$ $\quad = 1818.2 \times 10^{-6} \text{ hp}$ $1 \text{ hp} = 747.7 \text{ W}$	$1 \text{ W} = 8.8507 \text{ in.} \cdot \text{lb}_f/\text{sec}$ $\quad = 0.7376 \text{ ft.} \cdot \text{lb}_f/\text{sec}$ $\quad = 1.3410 \times 10^{-3} \text{ hp}$
Spring constant:		
Translatory	$1 \text{ lb}_f/\text{in.} = 175.1268 \text{ N/m}$ $1 \text{ lb}_f/\text{ft} = 14.5939 \text{ N/m}$	$1 \text{ N/m} = 5.7102 \times 10^{-3} \text{ lb}_f/\text{in.}$ $\quad = 68.522 \times 10^{-3} \text{ lb}_f/\text{ft}$
Torsional	$1 \text{ in.} \cdot \text{lb}_f/\text{rad} = 0.1130 \text{ m N/rad}$	$1 \text{ m N/rad} = 8.8507 \text{ in.} \cdot \text{lb}_f/\text{rad}$ $\quad = 0.7376 \text{ lb}_f \cdot \text{ft/rad}$

TABLE F.3 *(cont.)*

Quantity	S.I. Equivalence	English Equivalence
Stress, pressure, or elastic modulus	1 $lb_f/in.^2$ (psi) = 6894.757 Pa 1 lb_f/ft^2 = 47.88 Pa	1 Pa = 1.4504 × 10^{-4} $lb_f/in.^2$ (psi) = 208.8543 × 10^{-4} lb_f/ft^2
Torque or moment	1 $lb_f \cdot$in. = 0.1130 N m 1 $lb_f \cdot$ft = 1.3558 N m	1 N m = 8.8507 lb_f in. = 0.7376 lb_f – ft
Velocity	1 ft/sec = 0.682 mph = 0.3048 m/sec 1 mph = 1.467 ft/sec = 1.609 km/h	1 m/s = 3.2808 ft/s = 2.237 mph 1 km/h = 0.622 mph
Volume	1 $in.^3$ = 16.3871 × 10^{-6} m^3 1 ft^3 = 28.3168 × 10^{-3} m^3 1 U.S. gallon = 3.7853 liters = 3.7853 × 10^{-3} m^3	1 m^3 = 61.0237 × 10^3 $in.^3$ = 35.3147 ft^3 = 10^3 liters = 0.2642 U.S. gallon
Work or energy	1 $lb_f \cdot$in. = 0.1130 J 1 $lb_f \cdot$ft = 1.3558 J 1 Btu = 1055.056 J 1 kWh = 3.6 × 10^6 J	1 J = 8.8507 in.$\cdot lb_f$ = 0.7376 ft. lb_f = 0.9478 × 10^{-3} Btu = 0.2778 kWh

Solution:

Stress:

(Stress in S.I. units) = (Stress in English units) × (Multiplication factor)

$$(Pa) = \left(\frac{N}{m^2}\right) = \left(\frac{N}{lb_f} lb_f\right)\left(\frac{\frac{1}{m}}{in.} in.\right)^2 = \left(\frac{N}{lb_f}\right)\left(\frac{\frac{1}{m}}{in.}\right)^2 \left(\frac{lb_f}{in.^2}\right)$$

$$= \left(\frac{N \text{ per } 1\, lb_f}{m \text{ per } 1\, in.^2}\right)\left(\frac{lb_f}{in.^2}\right)$$

$$= \left(\frac{4.4482}{0.0254^2}\right)\left(\frac{lb_f}{in.^2}\right)$$

$$= 6894.757\left(\frac{lb_f}{in.^2}\right)$$

Temperature conversion. To convert from Fahrenheit t_f to Celsius t_c,

$$t_c = \frac{5}{9}\left(t_f - 32\right)$$

To convert from Celsius t_c to Fahrenheit t_f,

$$t_f = \frac{9}{5}\left(t_c + 32\right)$$

F.4 References

1. IBM, *SI Metric Reference Manual*, 3rd Ed., New York, March 1974.

2. IEEE/ASTM-SI-10, "Standard for Use of the International System of Units (SI): The Modern Metric System," ASTM International, West Conshohocken, PA.

3. Mechtly, E.A., "The International System of Units," 2nd revision, NASA SP-7012, 1973.

4. National Bureau of Standards, "The International System of Units (SI)," July 1974.

5. National Institute for Science and Technology, "Guide for the Use of the International System of Units (SI), Special Publication 811," abridged, Gaithersburg, MD, 1995.

6. Wandmacher, C., *Metric Units in Engineering—Going SI*, Industrial Press, New York, 1978.

Appendix G

Accident Reconstruction Formulae

This appendix contains a summary of formulae commonly used in road vehicle accident reconstruction.

G.1 Center of Mass

The following notation is used (Figures G.1 to G.4):

a = horizontal distance between the center of mass and the center of the tire, cm (in.)

b = horizontal distance between the center of the front axle and the center of mass, cm (in.)

CM = center of mass

H = height of the center of mass above the roadway surface, measured, cm (in.)

H_1 = height to the center of the rear axle with the rear raised, cm (in.)

L = length of the wheelbase, cm (in.)

LF = left front wheel

LR = left rear wheel

RF = right front wheel

RR = right rear wheel

S_1 = weight supported by the right wheels, kg (lb)

T_1 = track width of the front tires (center to center), cm (in.)

T_2 = track width of the rear tires (center to center), cm (in.)

T_r = front tire radius, cm (in.)

W = total vehicle weight, kg (lb)

W_1 = weight supported by the front wheels, kg (lb)

W_2 = weight supported by the rear wheels, kg (lb)

W_3 = weight on the front wheels with the rear wheels raised, kg (lb)

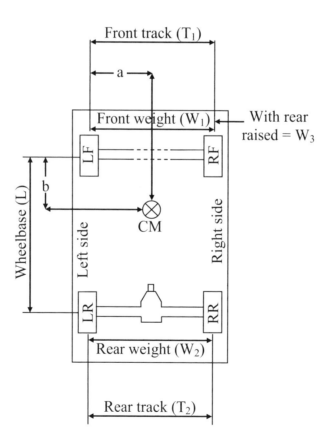

Figure G.1 Location of the center of mass.

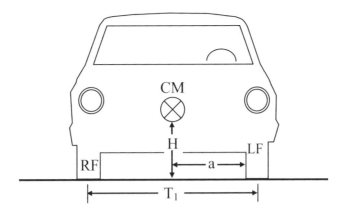

Figure G.2 The vertical location H.

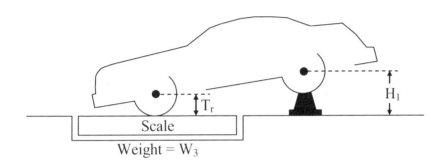

Figure G.3 Weighting position to locate the height of the center of mass.

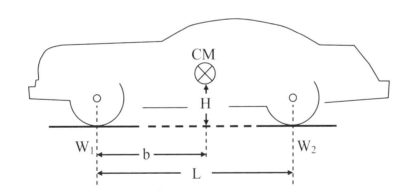

Figure G.4 Location of distance b.

Distance a:

$$a = \frac{S_1 T_1}{W}$$

Distance b:

$$b = \frac{W_2 L}{W}$$

The vertical location, H:

$$H = \frac{T_r + (W_3 - W_1)L\sqrt{L^2 - (H_1 - T_r)^2}}{W(H_1 - T_r)}$$

G.2 Slide-to-a-Stop Speed

The following notation is used:

d $\quad=\quad$ length of a slide or skid, m (ft)

$d_1 \quad=\quad$ skid distance on the first surface, m (ft)

$d_2 \quad=\quad$ skid distance on the second surface, m (ft)

e $\quad=\quad$ super elevation

f $\quad=\quad$ drag factor (level surface)

$f_1 \quad=\quad$ drag factor of the first surface

$f_2 \quad=\quad$ drag factor of the second surface

m $\quad=\quad$ grade or slope

n $\quad=\quad$ percent braking efficiency or capability

$n_1 \quad=\quad$ percent braking on the first surface

$n_2 \quad=\quad$ percent braking on the second surface

S $\quad=\quad$ minimum speed in km/h (mph) at the beginning of a slide of skid marks

$\theta \quad=\quad$ slope angle in degrees; use (+) for uphill grades and (–) for downhill grades

$\mu \quad=\quad$ coefficient of friction

Speed based on the length of skid marks or the length of a slide by an object over a surface:

English	S.I.
$S = \sqrt{30\,df}$	$S = \sqrt{254\,df}$

S is the minimum speed in kilometers per hour, km/h (mph), at the beginning of a slide or skid marks.

Slide-to-stop speed involving braking efficiency or capability:

English	S.I.
$S = \sqrt{30\,dfn}$	$S = \sqrt{254\,dfn}$

S is the minimum.

Slide-to-stop speed involving braking efficiency and grade or super elevation:

English | S.I.

$$S = \sqrt{30\,d\left(fn \pm e\right)}$$

$$S = \sqrt{254\,d\left(fn \pm e\right)}$$

or

$$S = \sqrt{30\,d\left(\mu n \pm m\right)}$$

$$S = \sqrt{254\,d\left(\mu n \pm m\right)}$$

Slide-to-stop speed involving an uphill grade of 10% or less:

English | S.I.

$$S = \sqrt{30\,d\left(f + m\right)}$$

$$S = \sqrt{254\,d\left(f + m\right)}$$

Slide-to-stop speed involving an uphill grade greater than 10%:

English | S.I.

$$S = \sqrt{30\,d\left(f\cos\theta + \sin\theta\right)}$$

$$S = \sqrt{254\,d\left(f\cos\theta + \sin\theta\right)}$$

Slide-to-stop speed involving a downhill grade of 10% or less:

English | S.I.

$$S = \sqrt{30\,d\left(f - m\right)}$$

$$S = \sqrt{254\,d\left(f - m\right)}$$

Slide-to-stop speed involving a downhill grade greater than 10% ($\pm$):

English | S.I.

$$S = \sqrt{30\,d\left(f\cos\theta \pm \sin\theta\right)}$$

$$S = \sqrt{254\,d\left(f\cos\theta \pm \sin\theta\right)}$$

Speed for continuous skid across various types of surfaces:

English | S.I.

$$S = \sqrt{30\left(f_1 d_1 + f_2 d_2\right)}$$

$$S = \sqrt{254\left(f_1 d_1 + f_2 d_2\right)}$$

Skidding to a stop when each side of the vehicle is on different types of surfaces:

English | S.I.

$$S = \sqrt{15\left(f_1 + f_2\right)d}$$

$$S = \sqrt{127\left(f_1 + f_2\right)d}$$

where f_1 and f_2 are drag factors on sides 1 and 2, respectively.

Skid-to-a-stop where the right and left wheels are on different types of roadway surfaces where there are different braking efficiencies:

English

$$S = \sqrt{30\left(f_1 n_1 + f_2 n_2\right)d}$$

S.I.

$$S = \sqrt{254\left(f_1 n_1 + f_2 n_2\right)d}$$

Skid-to-a-stop where the right and left wheels are on different types of roadway surfaces with a grade and where there are different braking efficiencies:

English

$$S = \sqrt{30\left[\left(f_1 n_1 + f_2 n_2\right) \pm m\right]d}$$

S.I.

$$S = \sqrt{254\left[\left(f_1 n_1 + f_2 n_2\right) \pm m\right]d}$$

Speed attained after accelerating from a stop or decelerating to a stop:

English

$$S = \sqrt{30\,df}$$

S.I.

$$S = \sqrt{254\,df}$$

where

d = distance, m (ft)

f = acceleration/deceleration factor

S = speed, km/h (mph)

G.3   Yaw, Sideslip, and Critical Curve Speed

For a level roadway surface:

English

$$S = 3.86\sqrt{R\,f}$$

S.I.

$$S = 11.27\sqrt{R\,f}$$

When e is ±10% or less:

English

$$S = 3.86\sqrt{R\left(f \pm e\right)}$$

S.I.

$$S = 11.27\sqrt{R\left(f \pm e\right)}$$

When e exceeds +10%:

English

$$S = \frac{3.86\sqrt{R\left(f + e\right)}}{\sqrt{1 - fe}}$$

S.I.

$$S = \frac{11.27\sqrt{R\left(f + e\right)}}{\sqrt{1 - fe}}$$

When e exceeds −10%:

English

$$S = \frac{3.86\sqrt{R(f - e)}}{\sqrt{1 + fe}}$$

S.I.

$$S = \frac{11.27\sqrt{R(f - e)}}{\sqrt{1 + fe}}$$

where

e = super elevation

f = drag factor

R = radius, m (ft)

S = speed, km/h (mph)

G.4 Combined Speeds

Combined speed where the initial speed is based on a combination of speeds:

English:

$$S_c = \sqrt{S_1^2 + S_2^2 + \cdots + S_n^2}$$

$$S_c = \sqrt{30(d_1f_1 + d_2f_2 + \cdots + d_nf_n)}$$

$$S_c = \sqrt{30(d_1f_1 + d_2f_2 + \cdots + d_nf_n) + (\text{speed at impact})^2}$$

S.I.:

$$S_c = \sqrt{S_1^2 + S_2^2 + \cdots + S_n^2}$$

$$S_c = \sqrt{254(d_1f_1 + d_2f_2 + \cdots + d_nf_n)}$$

$$S_c = \sqrt{254(d_1f_1 + d_2f_2 + \cdots + d_nf_n) + (\text{speed at impact})^2}$$

Combined speed involving a vehicle skid on various surfaces and one additional speed:

English

$$S_c = \sqrt{30(d_1f_1 + d_2f_2) + S_3^2}$$

S.I.

$$S_c = \sqrt{254(d_1f_1 + d_2f_2) + S_3^2}$$

where

d_1 = skid distance on the first surface

d_2 = skid distance on the second surface

d_n = skid distance(s) on subsequent surface(s)

f_1 = drag factor on the first surface

f_2 = drag factor on the second surface

f_n = drag factor(s) on subsequent surface(s)

S_1 = speed of the vehicle on the first surface

S_2 = speed of the vehicle on the second surface on the second event

S_c = speeds combined (initial speed; speed at the start of the first event)

S_n = subsequent speed or calculation of speed

Combined speed involving vehicle skid on various surfaces and one additional speed:

<table>
<tr><td align="center">English</td><td align="center">S.I.</td></tr>
<tr><td align="center">$$S_c = \sqrt{30(d_1f_1 + d_2f_2) + S_3^2}$$</td><td align="center">$$S_c = \sqrt{254(d_1f_1 + d_2f_2) + S_3^2}$$</td></tr>
</table>

where

d = distance, m (ft)

f = drag factor

S_3 = speed, km/h (mph)

S_c = combined speed, km/h (mph)

Combination of two separately calculated speeds:

<table>
<tr><td align="center">English</td><td align="center">S.I.</td></tr>
<tr><td align="center">$$S_c = \sqrt{S_1^2 + S_2^2}$$</td><td align="center">$$S_c = \sqrt{S_1^2 + S_2^2}$$</td></tr>
</table>

where

S_1 = speed on the first surface, km/h (mph)

S_2 = speed on the second surface or speed from the second event, km/h (mph)

S_c = speeds combined, km/h (mph)

G.5 360-Degree Momentum Speed Analysis

The following notation is used (Figures G.5 to G.7):

S_1 = impact speed of vehicle 1, km/h (mph)

S_2 = impact speed of vehicle 2, km/h (mph)

S_3 = post-impact speed of vehicle 1, km/h (mph) (departure speed)

S_4 = post-impact speed of vehicle 2, km/h (mph) (departure speed)

W_1 = weight of vehicle 1, kg (lb)

$W_1 S_1$ = approach, vehicle 1

$W_1 S_3$ = departure, vehicle 1

W_2 = weight of vehicle 2, kg (lb)

$W_2 S_2$ = approach, vehicle 2

$W_2 S_4$ = departure, vehicle 2

α = approach angle for vehicle 1

θ = departure angle of vehicle 1, degrees

ψ = approach angle of vehicle 2, degrees

ϕ = departure angle of vehicle 2, degrees

Impact speed of vehicle 2:

$$S_2 = \frac{W_1 S_3 \sin\theta}{W_2 \sin\psi} + \frac{S_4 \sin\phi}{\sin\psi}$$

Impact speed of vehicle 1:

$$S_1 = S_3 \cos\theta + \frac{W_2 S_4 \cos\phi}{W_1} - \frac{W_2 S_2 \cos\psi}{W_1}$$

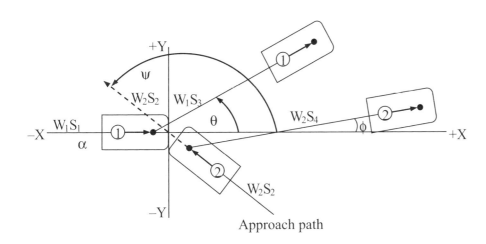

Figure G.5 Angular collision of two vehicles.

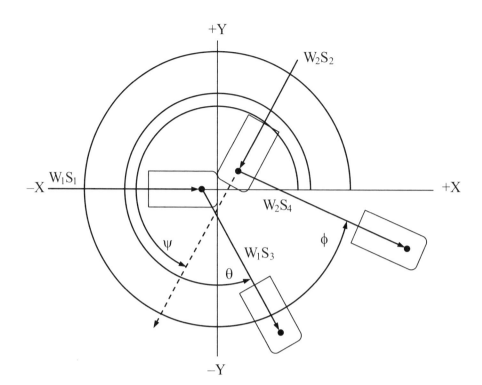

Figure G.6 Two-vehicle collision with negative sine and cosine values.

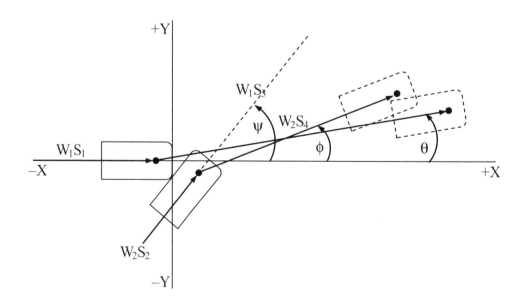

Figure G.7 Two-vehicle collision with angles less than 90°.

G.6 Tip and Rollover Speed

Level roadway:

The critical velocity or speed at which a vehicle could negotiate a curve involving a level roadway without overturning is given by

Velocity:
$$V = \sqrt{\frac{Rg}{2}\left(\frac{t_w}{h}\right)}$$

Speed:

English

$$S = 0.682\sqrt{\frac{Rg}{2}\left(\frac{t_w}{h}\right)}$$

S.I.

$$S = 3.6\sqrt{\frac{Rg}{2}\left(\frac{t_w}{h}\right)}$$

where

g = acceleration due to gravity, 9.81 m/s^2 (32.2 ft/sec^2)

h = height of the center of mass, cm (in.)

R = radius, m (ft)

S = speed, km/h (mph)

t_w = track width, cm (in.)

V = velocity, m/s (ft/sec)

Curve with bank or super elevation:

The critical velocity or speed at which a vehicle could negotiate a curve involving super elevation without overturning is given by

Velocity:
$$V = \sqrt{\frac{Rg(0.5t_w + he)}{(h - 0.5t_w)e}}$$

Speed:

English

$$S = 0.682\sqrt{\frac{Rg(0.5t_w + he)}{(h - 0.5t_w)e}}$$

S.I.

$$S = 3.6\sqrt{\frac{Rg(0.5t_w + he)}{(h - 0.5t_w)e}}$$

where e is the super elevation or bank.

Lateral acceleration factor:

The lateral acceleration factor where the vehicle will sideslip but not tip over is given by

$$f_L = \frac{t_w}{2h}$$

where

f_L = lateral acceleration factor

t_w = track width, cm (in.)

h = height of center of mass, m (in.)

The angular velocity at which the vehicle will overturn rather than spin-out on a level surface is given by

$$\omega = \sqrt{\frac{gt}{Rh}}$$

where

g = acceleration due to gravity, 9.81 m/s^2 (32.2 ft/sec^2)

h = height of center of mass, m (ft)

R = radius of turn, m (ft)

t = half track width of vehicle, m (ft)

ω = angular velocity, rad/sec

G.7 Weight Shift and Speed

Weight shift:

The weight shift from the rear axle to the front axle during braking or deceleration is given by

$$W_S = \frac{\dfrac{h}{L}\left[W_F f_F + W_R f_R\right]}{\left[1 - \left(\dfrac{h}{L} f_F\right)\right] + \left(\dfrac{h}{L} f_R\right)}$$

Weight shift speed:

The speed involving weight shift during braking on deceleration is given by

Level surface:

English:

$$S = \sqrt{30\left[\frac{f_F d_F\left(W_F + W_S\right) + f_R d_R\left(W_R - W_S\right)}{W_T}\right]}$$

S.I.:

$$S = \sqrt{254\left[\frac{f_F d_F\left(W_F + W_S\right) + f_R d_R\left(W_R - W_S\right)}{W_T}\right]}$$

Grade or slope:

English:

$$S = \sqrt{30\left[\frac{f_F d_F\left(W_F + W_S\right) + f_R d_R\left(W_R - W_S\right)}{W_T}\right] \pm de}$$

S.I.:

$$S = \sqrt{254\left[\frac{f_F d_F\left(W_F + W_S\right) + f_R d_R\left(W_R - W_S\right)}{W_T}\right] \pm de}$$

where

d = distance, m (ft)

d_F = skid distance for the front tires, m (ft)

d_R = skid distance for the rear tires, m (ft)

e = grade or slope

f_F = drag factor for the front tires

f_R = drag factor for the rear tires

h = height of the center of mass, cm (in.)

L = length of the wheelbase, cm (in.)

S = speed, km/h (mph)

W_F = weight of the front wheels, kg (lb)

W_R = weight of the rear wheels, kg (lb)

W_S = weight shift from the rear axle to the front axle, kg (lb)

W_T = total weight of the vehicle, kg (lb)

G.8 Kinetic Energy and Speed

The following notation is used:

g = acceleration due to gravity, 9.81 m/s^2 (32.2 ft/s^2)

KE = kinetic energy, joules (ft-lb)

m = mass of vehicle, kg (slug)

S = speed, km/h (mph)

V = velocity, m/s (ft/s)

W = weight of the vehicle, kg (lb)

English	S.I.

$$KE = \frac{1}{2}mV^2 \qquad\qquad KE = \frac{1}{2}mV^2$$

$$KE = \frac{WS^2}{30} \qquad\qquad KE = \frac{WS^2}{26}$$

$$V = \sqrt{\frac{2g\,KE}{W}} \qquad\qquad V = \sqrt{\frac{2\,KE}{W}}$$

$$S = \sqrt{\frac{30\,KE}{W}} \qquad\qquad S = \sqrt{\frac{KE}{0.0385\,m}}$$

G.9 Fall, Slip, and Vault Speeds

Takeoff and fall speed when the takeoff area is level:

English	S.I.

$$S = \frac{2.73d}{\sqrt{h}} \qquad\qquad S = \sqrt{\frac{7.97d}{\sqrt{h}}}$$

where

d = horizontal distance

h = vertical distance

S = speed at takeoff

Takeoff and fall speed when the takeoff area has a ± grade (m) not exceeding 10%:

English	S.I.

$$S = \frac{2.73d}{\sqrt{h \pm dm}} \qquad\qquad S = \sqrt{\frac{7.97d}{\sqrt{h \pm dm}}}$$

where

d = horizontal distance

m = % grade

S = speed at takeoff

Takeoff and fall speed involving any downward takeoff angle, including those having a grade exceeding –10%:

English

S.I.

$$S = \cos\theta\, \frac{2.73d}{\sqrt{h - d\tan\theta}} \qquad\qquad S = \cos\theta\, \frac{7.97d}{\sqrt{h - d\tan\theta}}$$

where

   d = horizontal distance

   h = vertical distance

   S = speed at the takeoff point

   θ = takeoff angle, degrees

General vault formula:

English

S.I.

$$S = \frac{2.73d}{\sqrt{d\sin\theta\cos\theta \pm h\cos^2\theta}} \qquad\qquad S = \frac{7.97d}{\sqrt{d\sin\theta\cos\theta \pm h\cos^2\theta}}$$

where

   S = speed

   d = horizontal distance

   h = vertical distance

   θ = takeoff angle, deg

Flip and vault takeoff speed when the takeoff and landing are at the same level:

English

S.I.

$$S = \sqrt{15d} \qquad\qquad S = \sqrt{127d}$$

where

   d = horizontal distance between the centers of mass

   S = speed at the point of takeoff

Flip and vault takeoff speed when the landing is either above or below the takeoff point:

English

S.I.

$$S = \frac{3.86d}{\sqrt{d \pm h}} \qquad\qquad S = \frac{11.27d}{\sqrt{d \pm h}}$$

where

d = horizontal distance between the center of mass positions

h = vertical distance (rise or fall) between the center of mass positions

S = speed at the point of takeoff

Speed of the vehicle or other airborne object at the time it becomes airborne:

English

S.I.

$$S = \frac{\sqrt{30\,fd}}{\cos\theta + f\sin\theta}$$

$$S = \frac{\sqrt{254\,fd}}{\cos\theta + f\sin\theta}$$

where

d = distance, m (ft)

f = deceleration factor

S = speed, km/h (mph)

Bibliography

Several outstanding text and reference books on road vehicle dynamics merit consultation for those readers who wish to pursue these topics further. Also, several publications are devoted to presenting research results and in-depth case studies in road vehicle dynamics. The following list is only a representative sample of the many excellent references that include journals and periodicals on road vehicle dynamics.

Books

1.   Artamonov, M.D., Ilarionov, V.A., and Morin, M.M., *Motor Vehicles, Fundamentals and Design*, MIR, Moscow, 1976.

2.   Bastow, D., Howard, G.P., and Whitehead, J.P., *Car Suspension and Handling,* 4th Ed., Society of Automotive Engineers, Warrendale, PA, 2004.

3.   Bencini, M., *Diminica Del Veicolo*, Tamburini, Milano, 1956.

4.   Bhat, R. and Dukkipati, R.V., *Advanced Dynamics*, Narosa Publishing House, New Delhi, India, 2001.

5.   Blundell, M. and Harty, D., *The Multibody Systems Approach to Vehicle Dynamics*, Society of Automotive Engineers, Warrendale, PA, 2004.

6.   Brach, Raymond M. and Brach, R. Matthew, *Vehicle Accident Analysis and Reconstruction Methods,* Society of Automotive Engineers, Warrendale, PA, 2005.

7.   Breuer, B. and Dausend, U., *Advanced Brake Technology*, Society of Automotive Engineers, Warrendale, PA, 2003.

8.   Bussien, N., *Automobiltechnisches Handbuch*, Technischer Verlag Berlin, 1953.

9. Campbell, C., *New Directions in Suspension Design*, Robert Bentley, Inc., Cambridge, MA, 1981.

10. Costin, M. and Phillips, D., *Racing and Sports Car Chassis Design*, 2nd Ed., Robert Bentley, Inc., Cambridge, MA, 1965.

11. Dixon, J.C., *The Shock Absorber Handbook*, Society of Automotive Engineers, Warrendale, PA, 1999.

12. Dixon, J.C., *Tires, Suspension and Handling*, Society of Automotive Engineers, Warrendale, PA, 1996.

13. Dixon, J.C., *Tyres, Suspension and Handling*, Cambridge University Press, Cambridge, 1991.

14. Dukkipati, R.V., *Advanced Engineering Analysis*, Narosa Publishing House, New Delhi, India, 2006.

15. Dukkipati, R.V., *Advanced Mechanical Vibrations*, Narosa Publishing House, New Delhi, India, 2006.

16. Dukkipati, R.V., *Analysis and Design of Control Systems Using MATLAB*, Narosa Publishing House, New Delhi, India, 2005.

17. Dukkipati, R.V., *Control Systems*, Narosa Publishing House, New Delhi, India, 2005.

18. Dukkipati, R.V., *Engineering System Dynamics*, Narosa Publishing House, New Delhi, India, 2004.

19. Dukkipati, R.V., *MATLAB for Mechanical Engineers*, New Age International Publishers, New Delhi, India, 2007.

20. Dukkipati, R.V., *Solving Engineering Mechanics Problems Using MATLAB*, New Age International Publishers, New Delhi, India, 2007.

21. Dukkipati, R.V., *Solving Engineering System Dynamics Problems with MATLAB*, New Age International Publishers, New Delhi, India, 2006.

22. Dukkipati, R.V., *Solving Vibration Analysis Problems Using MATLAB*, New Age International Publishers, New Delhi, India, 2006.

23. Dukkipati, R.V., *Spatial Mechanisms: Analysis and Synthesis*, Narosa Publishing House, New Delhi, India, 2001.

24. Dukkipati, R.V., *Vehicle Dynamics*, Narosa Publishing House, New Delhi, India, 2000.

25. Dukkipati, R.V., *Vibration Analysis*, Narosa Publishing House, New Delhi, India, 2004.

26. Dukkipati, R.V. and Amyot, J.R., *Computer Aided Simulation in Railway Dynamics*, Marcel Dekker, New York, 1988.

27. Dukkipati, R.V., Ananda Rao, M., and Bhat, R., *Computer Aided Analysis and Design of Machine Elements*, New Age International Publishers, New Delhi, India, 2000.

28. Dukkipati, R.V. and Ray, P.K., *Product and Process Design for Quality, Economy and Reliability*, New Age International Publishers, New Delhi, India, 2007.

29. Dukkipati, R.V. and Ray, P.K., *Quality Control, Improvement and Management*, New Age International Publishers, New Delhi, India, 2007.

30. Dukkipati, R.V. and Srinivas, J.S., *Mechanical Vibrations*, Prentice-Hall of India, New Delhi, India, 2005.

31. Dukkipati, R.V. and Srinivas, J.S., *Vibrations: Problem Solving Companion*, New Age International Publishers, New Delhi, India, 2007.

32. Dukkipati, R.V., Srinivas, J.S., and RaviKanth, A., *Numerical Methods*, New Age International Publishers, New Delhi, India, 2008.

33. Ellis, J.R., *Road Vehicle Dynamics*, John R. Ellis, Inc., Akron, OH, 1988.

34. Ellis, J.R., *Vehicle Dynamics*, Business Books Ltd., London, 1969.

35. Elsley, G.H. and Devereux, A.J., *Hovercraft Design and Construction*, Cornell Maritime Press, Centerville, MD, 1968.

36. Firch, J.W., *Motor Truck Engineering Handbook*, 3rd Ed., James W. Fitch, Anacortes, WA, 1984.

37. Garg, V.K. and Dukkipati, R.V., *Dynamics of Railway Vehicle Systems*, Academic Press, New York, 1984.

38. Geary, P.J., *Magnetic and Elastic Suspensions—A Survey of Their Design, Construction and Use*, Taylor and Francis, London, England, 1964.

39. Genta, G., *Meccanica Dell'Autoveicolo*, Levrotto & Bella, Torino, 1993.

40. Giles, J.G., *Steering, Suspension and Tyres*, Iliffe Books Ltd., London, 1968.

41. Gillespie, T.D., *Fundamentals of Vehicle Dynamics*, Society of Automotive Engineers, Warrendale, PA, 1992.

42. Gillespie, T.D., *Fundamentals of Vehicle Dynamics Software*, Society of Automotive Engineers, Warrendale, PA, 2004.

43. Goodsell, D., *Dictionary of Automotive Engineering*, Butterworths, London, 1989.

44. Hay, W.W., *An Introduction to Transportation Engineering*, Wiley, New York, 1961.

45. Hennes, R.G. and Ekse, M., *Fundamentals of Transportation Engineering*, McGraw-Hill, New York, 1969.

46. Huang, M., *Vehicle Crash Mechanics*, CRC Press, Boca Raton, FL, 2002.

47. Hucho, W.H., "The Aerodynamic Drag of Cars. Current Understanding, Unresolved Problems, and Future Prospects," in *Aerodynamic Drag Mechanics of Bluff Bodies and Road Vehicles*, Plenum Press, New York, 1978.

48. ISO, "Road Vehicles—Lateral Transient Response Test Methods, ISO 7401-1988," International Organization for Standardization, 1988.

49. ISO, "Road Vehicles—Steady State Circular Test Procedure, ISO 4138-1982 (E)," International Organization for Standardization, 1982.

50. Johnson, E., ed., *Tires and Handling,* Society of Automotive Engineers, Warrendale, PA, 1996.

51. Jones, B., *Elements of Practical Aerodynamics*, Wiley, New York, 1942.

52. Koenig, E. and Fachsenfield, R., *Aerodynamik des Kraftfahrzeuge*, Verlag der Motor, Rundshou-Umsha Verlag, Frankfurt A.M., 1951.

53. Limpert, R., *Brake Design and Safety*, 2nd Ed., Society of Automotive Engineers, Warrendale, PA, 1999.

54. Lucas, G.G., *Road Vehicle Performance*, Gordon & Breach, London, 1986.

55. Magnus, K., ed., *Dynamics of Multibody Systems,* Springer-Verlag, Berlin, 1978.

56. Milliken, W.F. and Milliken, D.L., *Chassis Design: Principles and Analysis*, Society of Automotive Engineers, Warrendale, PA, 2002.

57. Milliken, W.F. and Milliken, D.L., *Race Car Vehicle Dynamics*, Society of Automotive Engneers, Warrendale, PA, 1995.

58. Mitschke, M., *Dinamik der Kraftfahzeuge*, Springer, Berlin, 1972.

59. Morelli, A., "Costruzioni Automobilistiche," in *Enciclopedia dell'Ingegneria*, ISEDI, Milano, 1972.

60. Newton, K., Steeds, W., and Garrett, T.K., *The Motor Vehicle*, 10th Ed., Butterworths, London, 1983.

61. Pacejka, H.B., ed., *The Dynamics of Vehicles on Roads and Railway Tracks*, Swets & Zeitlinger, Lisse, The Netherlands, 1976.

62. Pernau, F., *Die Entscheidenden Reifeneigen Shaften*, Vortragstext, Eszter, 1967.

63. Pollone, G., *Il Veicolo*, Lecrotto & Bella, Torino, 1970.

64. Puhn, F., *How to Make Your Car Handle*, H.P. Books, Los Angeles, CA, 1981.

65. Rao, J.S. and Dukkipati, R.V., *Mechanism and Machine Theory,* Wiley Eastern Ltd. [now New Age International], New Delhi, India, 1992.

66. Reimpell, J. and Betzler, J.W., *The Automotive Chassis Engineering Principles*, Society of Automotive Engineers, Warrendale, PA, 2001.

67. Richardson, H.H., *et al.*, "Special Issue on Ground Transportation," *ASME Journal of Dynamic Systems, Measurement and Control*, Vol. 96, No. 2, 1974.

68. Rill, Georg, "Vehicle Dynamics—Lecture Notes," Fachhochschule Regensburg, University of Applied Sciences, Hochschule fur Technik Wirtschaft Soziales, October 2003.

69. Roberts, P., *Collector's History of the Automobile*, Bonanza Books, New York, 1978.

70. SAE International, "2004 SAE Vehicle Dynamics Technology Collection on CD-ROM," Society of Automotive Engineers, Warrendale, PA, 2004.

71. SAE International, "ABS/TCS, Brake Technology and Foundation Brake NVH, and Tire and Wheel Technology," Society of Automotive Engineers, Warrendale, PA, 2004.

72. SAE International, *Automobile Handbook*, 2nd Ed., Robert Bosch GmbH, Stuttgart, 1994.

73. SAE International, "Fundamentals of Vehicle Dynamics—A Technical Video Series," Society of Automotive Engineers, Warrendale, PA, 2004.

74. SAE International, "Laboratory Testing Machines for Measuring the Steady-State Force and Moment Properties of Passenger Car Tires, SAE J1107," Society of Automotive Engineers, Warrendale, PA, September 1975.

75. SAE International, "Proposed Passenger Car and Light Truck Directional Control Response Test Procedures, SAEXJ 266," Society of Automotive Engineers, Warrendale, PA, 1985.

76. SAE International, "Vehicle Aerodynamics 2004," Society of Automotive Engineers, Warrendale, PA, 2004.

77. SAE International, "Vehicle Aerodynamics Terminology, SAE J1594," Society of Automotive Engineers, Warrendale, PA, June 1987.

78. SAE International, "Vehicle Dynamics, Braking, Steering and Suspension," Society of Automotive Engineers, Warrendale, PA, 2004.

79. SAE International, "Vehicle Dynamics and Simulation 2004," Society of Automotive Engineers, Warrendale, PA, 2004.

80. SAE International, "Vehicle Dynamics and Simulation, Steering and Suspensions on CD," Society of Automotive Engineers, Warrendale, PA, 2004.

81. SAE International, "Vehicle Dynamics Terminology, SAE J670e," Society of Automotive Engineers, Warrendale, PA, July 1976.

82. Scibor Ryilski, A.J., *Road Vehicle Aerodynamics*, Pentech Press, London, 1975.

83. Segel, L., "Research in the Fundamentals of Automobile Control and Stability," *SAE Transactions*, Society of Automotive Engineers, Warrendale, PA, 1985.

84. Shi, H. and Sheng G., *Mechanical Vibration System: Analysis, Measurement, Modeling and Control*, HUST Press, Wuhan, China, 1991 (National Award Winner Book in 1996).

85. Slibar, A. and Springer, H., *The Dynamics of Vehicles on Roads and Tracks*, Swets & Zeitlinger, Lisse, The Netherlands, 1978.

86. Steeds, W., *Mechanics of Road Vehicles*, Iliffe and Sons, Ltd., London, 1960.

87. Taborek, J.J., *Mechanics of Vehicles*, Towmotor Corporation, Cleveland, OH, 1957.

88. Terry, L. and Baker, A., *Racing Car Design and Development*, Robert Bentley, Inc., Cambridge, MA, 1973.

89. Tidbury, G.H., *Advances in Automobile Engineering*, Pergamon Press, London, 1965.

90. Van Eldik Thieme, H.C.A. and Pacejka, H.B., "The Tire as a Vehicle Component," Delft University of Technology, Delft, The Netherlands, 1971.

91. Whitehead, J.P. and Bastow, D., *Car Suspension and Handling*, 4th Ed., Society of Automotive Engineers, Warrendale, PA, 2004.

92. Willumeit, H.P., ed., *The Dynamics of Vehicles on Roads and Tracks*, Swets & Zeitlinger, Lisse, The Netherlands, 1980.

93. Wong, J.Y., *Theory of Ground Vehicles*, 3rd Ed., Wiley, New York, 2001.

Journals and Periodicals

General Journals

Most of the information on road vehicle dynamics is found in papers published in journals or presented at conferences. Relevant papers can be found in journals covering mechanical systems, dynamics, or mechanical engineering in general (e.g., *Dynamics & Control*, Kluwer Academic Publication, Dordrecht, The Netherlands; *Journal of Guidance, Control and Dynamics*, American Institute of Aeronautics and Astronautics (AIAA), New York, USA; and several journals of the American Society of Mechanical Engineers (ASME) and other similar societies).

Specialized Journals on Road Vehicle Dynamics

SAE International publishes specialized journals and conference proceedings, organizes conferences, and promotes road vehicle dynamics in numerous ways. The following are specialized journals on road vehicle dynamics:

1. *Accident Reconstruction Journal* (ARJ), Waldorf, MD.

2. *Associazione Tecnica dell'Automobile* (ATA), Torino, Italy.

3. *Automobiltechnische Zeitschrift* (ATZ), Berlin, Germany.

4. *Automotive Engineering International*, SAE International, Warrendale, PA.

5. *International Journal of Vehicle Design*, Interscience, Geneva, Switzerland.

6. *Vehicle System Dynamics*, Swets & Zeitlinger, Lisse, The Netherlands.

List of Symbols

Chapter 1

A   arbitrary vector

c   damping coefficient

c_c critical damping

c_i damping constant of ith damper

E   modulus of elasticity

eq   equivalent value

F   applied force

f_f general friction force

f_k kinetic friction force

f_s static friction force

F_d damping force

g   acceleration due to gravity

G   modulus of rigidity

H_G angular momentum of a rigid body about its mass center G

H_o angular momentum

H_p angular momentum of a rigid body about point p

$(H_x)_p$ $(H_y)_p$ $(H_z)_p$	three Cartesian components of the angular moment vector H_p
i	ith value
I_p	mass moment of inertia of the rigid body about point p
J	polar moment of inertia
k	spring constant
k_i	spring constant of ith spring
L	linear momentum
m	mass
m_i	ith mass
n	total number of rigid bodies
N	normal force
q_i	generalized coordinates
Q_i	generalized force in the direction of the ith generalized coordinate
r	position vector
t	time, sec
T	kinetic energy
U	potential energy
V	velocity
x_1	displacement of mass m_1
x_2	displacement of mass m_2
α	angular acceleration
δW	virtual work
μ_k	kinetic friction coefficient
μ_s	static friction coefficient
ω	angular velocity
ω_n	natural frequency
$(\cdot)$	$\frac{d}{dt}(\ )$

$(\ddot{\ })$	$\dfrac{d^2}{dt^2}(\ )$
$[\ ]$	matrix
$[\ ]^{-1}$	inverse of $[\ ]$
$[\ ]^{T}$	transpose of $[\ ]$

Chapter 2

a	acceleration
A	wavelength
a_o, a_n, b_n	complex Fourier coefficients
c	damping coefficient
c_c	critical damping
c_i	damping constant of ith damper
$[c]$	damping matrix
$[D]$	dynamic matrix
eq	equivalent value
$E(x)$	expected value of x
f	linear frequency
f_0	critical frequency
$F(t)$	exciting force
F_0	amplitude of force
$F_d, F_d(t)$	damping force
$F_s(t)$	spring force
g	acceleration due to gravity
$H(\omega)$	complex frequency response function
i	ith value
$[I]$	unit matrix or identity matrix
k	spring constant
k_i	spring constant of ith spring
$[k]$	stiffness matrix

m	mass
m_i	ith mass
$p(x)$	probability density function
$p(x,y)$	joint probability density function of variables x and y
PSD	power spectral density
r	frequency ratio
$R_{xy}(\tau)$	cross-correlation function of the time history records of $x(t)$ and $y(t)$
t	time
T_p	period
X	amplitude
$x(t)$	instantaneous value at time t
$\bar{x}(t)$	mean value
$\bar{x}^2(t)$	mean square value
δ	logarithmic decrement
$\delta(t-\tau)$	impulse response function
ϕ	phase angle
$[\bar{\phi}]$	weighted modal matrix
λ	eigenvalues
ξ	damping ratio
σ_x	standard deviation of x
σ_x^2	variance of x
τ	period of oscillation
ψ	modal matrix
ω	frequency of excitation
ω_d	damped frequency
ω_n	natural frequency
$(\dot{\ })$	$\dfrac{d}{dt}(\ )$
$(\ddot{\ })$	$\dfrac{d^2}{dt^2}(\ )$
$[\ ]$	matrix

$[\ ]^{-1}$ inverse of $[\ ]$

$[\ ]^{T}$ transpose of $[\ ]$

$L(\ )$ Laplace transform of $(\ )$

$L^{-1}(\ )$ inverse Laplace transform of $(\ )$

Chapter 3

a	distance from the center point to the sum of normal force
A	parameter can be obtained through two points
C	damping coefficient of the tire
C_r	cornering stiffness of the tire
D_x	longitudinal trail
D_y	lateral trail
$\overline{D}_x$	nondimensional longitudinal trail
$\overline{D}_y$	nondimensional lateral trail
f_R	rolling resistance coefficient
f_δ	toe-in resistance coefficient
$F_c(x)$	function of the radial force
F_x	longitudinal force between the tire and road surface
F_y	lateral force between the tire and road surface
F_{yf}	cornering force of the front tire to balance the centrifugal force of the vehicle
F_z	normal force of the tire on the contact patch
$\overline{F}$	relative resultant tangential force
$\overline{F}_x$	nondimensional longitudinal force
$\overline{F}_y$	nondimensional lateral force
$F_{\alpha x}$	driving force applied to the wheel
$F_{\delta v}$	side force due to the tire lateral deformation caused by the angle
F_φ	maximum force of a tire on hard surfaces
$h(X)$	function of the surface irregularity

$\dot{h}(X)$	function of the slope of the surface irregularity just under the center of the tire
$h(X - x)$	function of the surface irregularity of the road with a distance apart from the contact point
K	nominal stiffness of the tire
k_i	distributed stiffness coefficient of the tire corresponding to the site x
k_x	tire longitudinal stiffness per unit area in the contact patch
k_y	lateral distributed stiffness per length of the tire tread
k_{yr}	lateral stiffness of the tire tread
k_z	radial stiffness of the tire
K_a	tire lateral stiffness
K_{by}	lateral carcass stiffness
K_c	camber stiffness of the tire carcass
K_d	dynamic stiffness of the tire
K_r	camber stiffness
K_s	static stiffness of the tire
K_x	total longitudinal stiffness
K_{x0}	longitudinal carcass deformation stiffness
K_y	slip stiffness of the tire
K_{y0}	lateral carcass deformation stiffness
ℓ, ℓ_a	length of the tire contact patch
ℓ_x	longitudinal relaxation length
ℓ_y	lateral relaxation length
M_b	braking torque of the wheel
M_f	moment produced by the normal force about the axis of rotation of the tire
M_t	driving torque of the wheel
M_y	rolling resistance moment
p	tire/road contact pressure
$q(x)$	resultant tangential stress in the slide region
q_z	vertical pressure distribution

r, R_e	effective rolling radius of the tire, m
R	free radius of the tire
r_d	vertical distance from the center of the wheel to the tire-ground contact point
s	slip ratio
s_a, s_b	longitudinal skid
S_b	braking slip ratio
S_c	critical value to divide the adhesion region and the sliding region
S_d	driving slip ratio
$S_{s\alpha}$	associated parameter of slip ratio
S_x	effective longitudinal slip ratio
S_y	effective lateral slip ratio
T_z	total aligning torque
$\overline{T}_z$	nondimensional aligning moment
v	horizontal velocity
v_c	circumference speed of the wheel
v_x	longitudinal velocity of the vehicle
v_y	lateral velocity of the vehicle
V_r	component speed of the tire center relative to the ground
V_{sx}	effective longitudinal slip velocity
V_{sxn}	tire nominal longitudinal slip velocity of the contact patch center
V_{sy}	effective lateral slip velocity
V_{syn}	tire nominal slip velocity of the contact patch center
V_x	velocity of the tire center in the longitudinal direction
y_0	lateral deformation
y_b	lateral bending deformation
y_r	tread deformation
α	slip angle, deg
α_f	slip angle of the front tire
α_r	slip angle of the rear tire
γ	camber angle of the tires

γ_c	roll angle
δ	radial deformation of the tire
δ_0	steering angle
δ_{v0}	toe-in angle of the front wheel on one side
Δ	distance between the normal force and wheel centerline
$\zeta(X)$	function of the displacement of the center of the axis arm
$\dot{\zeta}(X)$	function of the velocity of the center of the axis arm
$\eta(\mu)$	load distribution function
η_γ	influence coefficient of the camber angle
ξ	tire longitudinal deformation
σ_ξ	longitudinal stress of the tire tread
φ	adhesion coefficient that varies with the state of the tire rolling or slipping
φ_0	road/tire friction coefficient when no sliding takes place
φ_1	road/tire friction coefficient when the slip ratio is S1
φ_p	average peak values of the coefficient of road adhesion
φ_s	average sliding values of the coefficient of road adhesion
φ_x	road/tire friction coefficient
ϕ	relative resultant slip ratio
ϕ_x	longitudinal slip ratio
ϕ_y	lateral slip ratio
ω	angular speed of a rolling tire; vertical velocity

Chapter 4

a	acceleration
a_{active}	active acceleration
$a_{passive}$	passive acceleration
$a_w(t)$	weighted acceleration
C	damping coefficient
$[C]$	damping matrix
E	modal energy matrix

$E_i(f)$	frequency weighting function
f	temporal frequency or vehicle internal excitation force
f_{firing}	firing frequency
f_{roll}	roll frequency
Δf	temporal frequency range
F	force
F_x, F_y, F_z	summation of all forces acting on the body in the x, y, and z directions, respectively
H	transmissibility
I_{xx}, I_{yy}, I_{zz}	moments of inertia
I_{xy}, I_{xz}, I_{yz}	products of inertia
K	stiffness
k_f, k_r	stiffness of the front ride and rear ride, respectively
K_e	stiffness of the engine bracket
K_i	stiffness of the engine mount
K_{LP}, K_{LI}, K_{LD}	control gains
K_s	suspension stiffness
K_t	tire stiffness
K_v	stiffness of the body bracket
$[K]$	stiffness matrix
ℓ_1	distance between the front axle and the center of gravity
ℓ_2	distance between the rear axle and the center of gravity
L_i	wavelength
m, M	mass
M_x	moments acting around the x axis
M_y	moments acting around the y axis
M_z	moments acting around the z axis
$[M]$	mass matrix
N	frequency index
P	pressure
r	distance from the imbalance mass center to the shaft rotation center

R	gear ratio
r_y^2	radius of gyration
t	time
T	transmissibility
TE	transmission error
V	vehicle speed
VDV	vibration dose value
$W(f)$	power spectral density
$W_z(\Omega)$	power spectral density of the road profile
x_c, y_c, z_c	coordinates of the powerplant center of gravity
z	displacement; sprung mass displacement input
$\dot{z}$	velocity
$\ddot{z}$	acceleration
z_r	road displacement input
z_u	unsprung mass displacement input
$\overline{z_{\Delta\Omega}^2}$	mean-square value within $\Delta\Omega$
α_c	rotational angular acceleration around the x axis
β_c	rotational angular acceleration around the y axis
γ_c	rotational angular acceleration around the z axis
δ	relative displacement
η	acceleration reduction ratio
θ	angular
λ	frequency ratio
ξ	damping ratio
ρ	fluid density
$\Phi, \tilde{\Phi}$	eigenvector matrix; actual modes; target modes
Θ	angular amplitutide
ψ	surrogate actual modes
ω	excitation frequency
ω_n	natural frequency

ω_s	sprung system frequency
ω_u	unsprung system frequency
Ω	spatial frequency
Ω_0	reference spatial frequency
$\Delta\Omega$	frequency bandwidth

Chapter 5

a	acceleration
b	distance between two suspension springs
d	vehicle track width
E_k	instantaneous rotational potential energy
E_v	instantaneous rotational kinetic energy
F_{1z}, F_{2z}	vertical forces acting on tires
F_{k1}, F_{k2}	spring forces
h	vehicle center of gravity height
h_1	distance from the rollover center to the ground
h_2	distance from the vehicle center of gravity to the rollover center
I	moment of inertia
K	suspension spring stiffness at one side
K_{tire}	tire stiffness
K_{tire}^*	dimensionless stiffness for the tire
K_{θ_2}	roll stiffness of the suspension
$K_{\theta_2}^*$	dimensionless stiffness for the suspension
m	mass
m_s	sprung mass
m_u	unsprung mass
r	tire height
Δr	tire deflection
T_0	vehicle lateral kinetic energy
T_1	vehicle lateral rotational energy

V	velocity
θ	road slope angle
θ_1	deflection angle between the two tires
θ_2	deflection angle between two suspension springs
φ	vehicle rotation angle
ϕ	table angle

Chapter 6

a, b	distances of the front and rear axles from the center of gravity
a_y	lateral acceleration (g)
$bC_r - aC_f$	understeer gradient
C_α	cornering stiffness
F_w	wind gust excitation force
F_x	longitudinal force
F_{xf}	longitudinal force at the front axle
F_{xr}	longitudinal force at the rear axle
F_y	lateral force
$F_{yf} = C_f\alpha_f$	front cornering force
$F_{yr} = C_r\alpha_r$	rear cornering force
F_z	vertical force
I	moment of inertia of the vehicle about its yaw axis
K	understeer gradient
L	$L = a + b$
M	vehicle mass
M_w	wind gust excitation moment
M_x	tilting torque
M_y	rolling resistance torque
M_z	self-aligning and bore torque
$r = \dot{\psi}$	yaw rate or yaw velocity
R	radius of the turn
v	forward velocity

V_x	longitudinal component of the center of gravity velocity
V_y	lateral component of the center of gravity velocity
W_f	load on the front axle (N)
W_r	load on the rear axle (N)
α	slip angle
β	sideslip angle
γ	vehicle heading angle
δ	steer angle
ψ	direction angle of velocity

Chapter 7

a	vehicle acceleration
A	friction surface area of contact between the brake shoes and the driven disk
a_1	radius between the center of the brake to the brake pivot
a_F	acceleration (deceleration) when only the front brakes are applied (front lock-up)
a_R	acceleration (deceleration) when only the rear brakes are applied
A_F	projected area of the front of the vehicle
b	width of the brake shoe
C	slope of the linear portion of the relation between the braking effort and the skid
c_1	horizontal (parallel to the road surface) distance between the contact point of the front wheel and the center of gravity
c_2	horizontal distance between the contact point of the rear wheel and the center of gravity
C_D	drag coefficient
C_L	coefficient of aerodynamic lift
C_M	coefficient of aerodynamic pitching moment obtained from wind tunnel testing
CG	center of gravity
e	vertical distance between the pivot and the contact with the brake cylinder

e_1	vertical distance between the pivot and the center of application of the friction force
e_2	horizontal distance between the pivot and the center of application of the friction force
F	brake cylinder "actuation" force in drum brakes, applied force on disk brakes
F_A, F_B	friction forces at the friction pads in drum brakes
F_b	total braking force on both axles
F_x	components of the actuation force F, in the brake x direction
F_{x1}	friction force on the front wheels
F_{x2}	friction force on the rear wheels
F_{xc}	critical value at which the braking effort ends being linear with the skid
F_y	components of the actuation force F, in the brake y direction
F_{z1}	vertical force on the front wheels
F_{z2}	vertical force on the rear wheels
g	acceleration of gravity
h	vertical (perpendicular to the road surface) distance between the contact points of the contact of the wheel with the ground and the center of gravity
h_a	height at which the aerodynamic force R_a is acting
i	skid as a percentage of maximum skid
i_{SAE}	SAE recommended equation for skid
K_f	brake distribution factor, the ratio of the total braking force that goes to the front brakes
K_{fmax}	brake distribution factor that achieves the maximum braking force
KE	kinetic energy
L_C	characteristic length of the vehicle
m	mass of the braking system with the drum
M	mass of the vehicle
M_c	aerodynamic pitching moment on the vehicle
M_F	moment resulting from the frictional forces over the arc length of the shoe contact
M_N	moment produced by the normal forces (pressure) about the pivot

M_{Pivot}	moment around the pivot of the brake
N_A, N_B	normal forces at the friction pads in drum brakes
p	pressure at any angle θ on the lining, pressure on disk brakes
p_{max}	maximum pressure over the range of contact at the lining
r	radius between the center of the brake to the lining
R	reaction force resultant at the pivot
r_{ave}	average radius from the center of the moving element to the contacting surface
r_t	radius of the tire
R_a	aerodynamic drag force
R_L	aerodynamic lift force that acts upward on the vehicle
R_r	rolling resistance (force) between the tires and the drive axle
R_x	reaction force components at the pivot in the x direction
R_y	reaction force components at the pivot in the y direction
S	specific heat
t	time
T	torque applied to the drum as a result of the frictional forces ($T = T_L + T_R$)
T_L	torque on the left-hand shoe
T_R	torque on the right-hand shoe
$\Delta temp$	brake temperature rise
V	vehicle speed
V_f	vehicle final velocity
V_i	vehicle initial velocity
V_r	relative velocity of the vehicle with respect to the wind
W	weight of the vehicle
X_b	distance traveled during braking
θ	grade or slope of the road; angle between the line connecting the center of the brake with the pivot and the line connecting the center of the brake with the point of the pressure p
θ_1, θ_2	angles between the line connecting the center of the brake with the pivot and those connecting the center of the brake with the beginning and end of the lining, respectively

μ	coefficient of friction
μ_r	rolling coefficient
μ_p	peak value of the coefficient of friction
ρ	density of air
ω	angular speed of the tire

Chapter 8

a	vehicle acceleration
a_F	acceleration when tractive effort is applied to the front axle
a_R	acceleration when tractive effort is applied to the rear axle
A_F	projected area of the front of the vehicle
B_d	dry pressure
B_{do}	absolute air dry pressure
B_o	reference absolute pressure
c_1	horizontal (parallel to the road surface) distance between the contact point of the front wheel and the center of gravity
c_2	horizontal distance between the contact point of the rear wheel and the center of gravity
C_D	drag coefficient
F	force needed to accelerate the vehicle
F_{net}	net tractive force
F_x	total tractive force
F_{x1}	friction force on the front wheels
F_{x2}	friction force on the rear wheels
F_{z1}	vertical force on the front wheels
F_{z2}	vertical force on the rear wheels
g	acceleration of gravity
h	vertical (perpendicular to the road surface) distance between the contact points of the contact of the wheel with the ground and the center of gravity
h_a	height at which the aerodynamic force R_a is acting
i	skid/slip as a percentage of maximum skid

i_{SAE}	SAE recommended equation for skid
I_i	moments of inertia of the rotating components attached to the drive train
I_w	mass moment of inertia for a wheel
K_f	tractive torque distribution factor, the ratio of the total tractive torque that goes to the front axle
K_{fmax}	torque distribution factor that achieves the maximum tractive force
K_g	factor for finding gear ratios in a transmission
K_{tc}	capacity factor
M	mass of the vehicle
m_w	mass of the wheel system
n_e	engine speed corresponding to the maximum speed of the vehicle
n_{tc}	speed of the torque converter
P	power
P_o	reference power
P_{tcin}	power input of the torque converter
P_{tcout}	power output of the torque converter
r	rolling radius of the wheels
r_g	radius of gyration
R_a	aerodynamic drag force
R_r	rolling resistance (force) between the tires and the drive axle
S	distance travelled
t	time
T	torque applied to the wheels; temperature
T_e	engine torque
T_{emax}	maximum engine torque
T_{max}	maximum torque available at the wheels
T_o	inlet temperature
T_{tcin}	input torque of the torque converter
T_{tcout}	output torque of the torque converter
$\Delta temp$	brake temperature rise

V	vehicle speed
V_f	vehicle final velocity
V_i	vehicle initial velocity
V_{max}	maximum speed of the vehicle
V_r	relative velocity of the vehicle with respect to the wind
W	weight of the vehicle
X_b	distance traveled during acceleration
γ_m	mass factor
η_t	overall transmission efficiency
η_{tc}	efficiency of the torque converter
θ	grade or slope of the road
θ_{max}	maximum slope
μ	coefficient of friction
μ_p	peak value of the coefficient of friction
μ_r	rolling coefficient
ξ_1	gear ratio of the lowest gear
ξ_n	gear ratio of the highest gear
ξ_o	overall reduction ratio
ξ_x	gear ratio in the drive axle
ρ	density of air
ω	angular speed
ω_e	engine speed
ω_{tcin}	input speed of the torque converter
ω_{tcout}	output speed of the torque converter

Chapter 9

a	vehicle acceleration
a_F	acceleration when the tractive effort/braking force is applied to the front axle
a_R	acceleration when the tractive effort/braking force is applied to the rear axle
a_y	lateral acceleration

c_1	horizontal (parallel to the road surface) distance between the contact point of the front wheel and the center of gravity
c_2	horizontal (parallel to the road surface) distance between the contact point of the rear wheel and the center of gravity
F	force needed to accelerate the vehicle
F_f	maximum friction force in the lateral direction
F_{x1}	friction force on the front wheels
F_{x2}	friction force on the rear wheels
F_{y1}	total lateral force on the front axle
F_{y2}	total lateral force on the rear axle
F_{z1}	vertical force on the front wheels
F_{z1i}	vertical force on the front inner wheel
F_{z1o}	vertical force on the front outer wheel
F_{z2}	vertical force on the rear wheels
F_{z2i}	vertical force on the rear inner wheel
F_{z2o}	vertical force on the rear outer wheel
g	acceleration of gravity
h	vertical (perpendicular to the road surface) distance between the contact points of the contact of the wheel with the ground and the center of gravity
h_a	height at which the aerodynamic force R_a is acting
h_r	roll center height
K_f	tractive torque or brake force distribution factor, ratio of the total tractive torque or brake force that goes to the front axle
K_ϕ	roll stiffness
M	mass of the vehicle
R	radius of the turn
R_a	aerodynamic force
R_r	rolling resistance (force) between the tires and the drive axle
t	wheel tread or track width
V	vehicle speed
V_c	vehicle critical speed
θ	grade or slope of the road

| μ | coefficient of friction |
| ϕ | roll angle |

Chapter 10

a	acceleration of a body
a_{AVG}	average lateral acceleration
a_{PEAK}	peak lateral acceleration
a_y	lateral acceleration
c	regional crush
C	chord
d	distance
d_0, d_1	constants
d_{cg-f}	distance from the center of gravity to the front axle
d_{cg-r}	distance from the center of gravity to the rear axle
d_{hood}	height of the upper leading edge of the vehicle
d_{lat}	lateral overlap distance
d_{wb}	wheelbase
e	coefficient of restitution
E_A	total energy dissipated during the approach period
f	coefficient of friction; drag factor
F	force acting on a body
f'	limiting friction
f_f	drag factor on the front axle
f_G	drag factor on a grade
f_r	drag factor on the rear axle
f_R	drag factor on the vehicle
F_{xb}	contact force applied to the bullet vehicle along the x axis
F_{xt}	x axis force applied to the target vehicle
F_{yb}	contact force applied to the bullet vehicle along the y axis
F_{yt}	y axis force applied to the target vehicle
g	acceleration due to gravity

G	percent grade expressed as a decimal
h	distance traveled vertically by the center of mass of the vehicle from takeoff to landing
h^+	change in height of the pedestrian center of gravity
ⓗ	course angle
k	spring stiffness
k_1	radius of gyration of vehicle 1
k_2	radius of gyration of vehicle 2
k_U	understeer gradient
KE	kinetic energy of a body
L	wheelbase
ⓜ	mass of a body
M	middle ordinate
m_i	mass of the vehicle i
mV	linear momentum
N	normal force
PE	projection efficiency
R	radius
R_{AVG}	theoretical average radius
s	lateral lane change distance
S	speed
SSD	safe stopping distance for a vehicle
t	time
t_p	perception/reaction time
t_{TOT}	total lane change maneuver time
v	velocity of a body
v_e	end velocity
v_i	initial velocity
v_p	pedestrian throw speed
v_v	vehicle impact speed
V_{cr}	critical speed

ΔV_1	total change in velocity of vehicle 1
ΔV_2	total change in velocity of vehicle 2
w	weight of a body
W	work done
x	distance compressed or stretched
x_f	horizontal distance of the center of mass from the front axle as a decimal fraction of the wheelbase
x_L	pedestrian displacement between the initial and secondary contacts
y	height
z	height of the center of mass as a decimal fraction of the wheelbase
α	angular acceleration, rad/sec^2
β	sideslip angle, deg
β_{MAX}	maximum sideslip angle
γ	bank angle (super elevation)
$\delta_{SW(MAX)}$	peak steering angle
θ	pedestrian launch angle at impact; angle of takeoff as measured relative to a horizontal plane
μ	coefficient of static friction
μ_{slide}	intervehicular sliding coefficient of friction
μ_{tire}	coefficient of friction between the tires and the ground
μ_v	coefficient of friction between the vehicle tires and the road surface
Φ	lateral displacement angle of the pedestrian
ψ	heading angle
ψ_{Middle}	heading angle at the middle point of a lane change

Index

About the Authors

Rao Dukkipati, Ph.D., P.E., F.A.S.M.E., F.C.S.M.E.
Professor, Chair and Graduate Program Director
Department of Mechanical Engineering
Fairfield University, Fairfield, CT

Rao Dukkipati is a professor and chair of mechanical engineering, and the director of the graduate program at Fairfield University, as well as a consultant in the field of accident reconstruction. He is member of the Connecticut Academy of Sciences and Engineering (CASE) and a Fellow of both the American Society of Mechanical Engineers (ASME) and the Canadian Society for Mechanical Engineers (CSME). Dr. Dukkipati's other professional memberships include the American Society for Engineering Education (ASEE) and SAE International. He has a Ph.D. in Mechanical Engineering from Oklahoma State University (1973), an M.S. in Mechanical Engineering from both the Andhra University, India (1969) and the University of New Brunswick, Canada (1971), and a B.S. in Mechanical Engineering from Sri Venkateswara University, India (1966). Dr. Dukkipati's specialized areas of teaching and research include mechanical design, mechanics, vibrations, control systems, engineering system dynamics, economy and reliability, quality control, advanced engineering mathematics, vehicle dynamics, and accident reconstruction. Previously, Dr. Dukkipati worked at Pratt & Whitney Aircraft of Canada of the United Technologies Corporation in Montreal, Canada; the National Research Council of Canada in Ottawa, Canada; the University of Windsor in Windsor,

Canada; and the University of Toledo in Ohio. He is a licensed Professional Engineer in the province of Ontario, Canada. Dr. Dukkipati has published more than 300 papers in national and international journals and conferences and numerous technical reports. He has authored or co-authored more than 30 engineering books, spanning a range of topics that includes vibrations, control systems, engineering system dynamics, economy and reliability, quality control, engineering mathematics, robotics, numerical methods, kinematics and dynamics, railway vehicle dynamics, and road vehicle dynamics. Dr. Dukkipati was awarded the 2008 American Society for Engineering Education (ASEE) New England Section Outstanding Teacher Award.

Jian Pang, Ph.D.
Researcher and PD Engineer
Powertrain Programs
Ford Motor Company, Dearborn, MI

Jian Pang is a senior researcher and PD engineer with Ford Motor Company. He formerly was a technical specialist and senior engineer with Stewart & Steven Service, Inc. in Texas. He received his Ph.D. in Mechanical Engineering from the University of Oklahoma in 1996. He received his B.S. (1985) and M.S. (1991) degrees in Mechanical Engineering from Wuhan University of Technology and Shanghai Jiao Tong University, respectively. Dr. Pang has 20 years of diversified experience in vehicle and ship engineering. He has published more than 40 papers in international and national journals and conferences and more than 50 industrial technical reports in the area of automotive system dynamics, noise and vibration, durability, ship structure, and calculation method analysis. Dr. Pang is the leading co-author of a book on automotive noise, vibration, and harshness. He is a member of the American Society of Mechanical Engineers (ASME) and SAE International. Dr. Pang also has published a novel and more than 100 articles in magazines, newspapers, Internet sites, and so forth.

Mohamad Qatu, Ph.D., P.E.
Adjunct Professor
Oakland University, Rochester, MI
and
Leader, Advanced Research
Ford Motor Company, Dearborn, MI

Mohamad Qatu received his undergraduate engineering degree from Yarmouk University in Jordan (1985) and his M.S. and Ph.D. from Ohio State University in 1986 and 1989, respectively. His areas of specialization include solid mechanics, computer-aided engineering, and noise and vibration. Dr. Qatu's academic experience includes working as a professor and director of the mechanical engineering technology program at Franklin University, an associate professor of mechanical engineering at Lake Superior State University, a visiting professor at An-Najah National University, and an adjunct professor at Oakland University. Dr. Qatu's industrial experience includes working for Dresser Industries, Honda of America, Dana Corporation, and Ford Motor Company, where he currently leads a group on noise, vibration, and harshness. He is the author of one book and more than 70 research papers, review articles, and book reviews, and he is a registered Professional Engineer in both Ohio and Michigan. Dr. Qatu has received two patents on noise suppression and several technical awards, and he wrote a few manuals to his credit. Recently, he established a new journal, *International Journal of Vehicle Noise and Vibration*, and serves as its executive editor. Dr. Qatu also is on the editorial board of the *Journal of Composite Structures*. He is a Fellow of the American Society of Mechanical Engineers (ASME) and a member of SAE International.

Gang Sheng, Ph.D.
Research Scientist
Gates Corporation, Rochester Hills, Michigan

Gang Sheng received his undergraduate engineering degree and his M.S. from Shanghai Jiao Tong University in 1984 and 1987, respectively, and his Ph.D. from Singapore Nanyang Technological University in 1997. Presently, he is an assistant professor at the University of Alaska, Fairbanks. Previously, Dr. Sheng worked as a research scientist at Gates Corporation, an advisory scientist at IBM, a principal consultant for CMA in the United States, a principal engineer at Singapore National Data Storage Institute, a principal engineer/group leader at Sony in Singapore, and an assistant professor/lecturer in Huazhong University of Science and Technology in China. Dr. Sheng holds five patents and has received several academic and industrial awards. He has co-authored five books and has published more than 100 research papers covering topics such as dynamics, noise and vibration, and tribology.

Shuguang Zuo, Ph.D.
Professor of Mechanical Engineering
Automobile Engineering School
Tongji University, Shanghai, P.R. China

Shuguang Zuo is a professor of automotive engineering at Tongji University in the Peoples Republic of China. He received his B.S. in Mechanical Design in 1990 from Hunan Agricultural University, and his M.S. in 1993 and his Ph.D. in 1996 in Automotive Engineering from Jilin Institute of Technology (now called Jilin University). From 1996 to 1998, Dr. Zuo was the Post-Doctoral Researcher in Aerospace Technology at the Post-Doctoral Exchange Center of Nanjing University of Aeronautics and Astronautics and the Chunlan Enterprise Post-Doctoral Exchange Center. Since 1998, Dr. Zuo has worked as an Associate Professor, Director of Automotive Teaching–Research Office, and Deputy Director of the Vehicle Research Laboratory in Shanghai Tiedao University; and a Professor and Ph.D. Supervisor in the School of Automobiles at Tongji University. His achievements include being an 11th Shanghai Shuguang Scholar, an Accreditation Expert of High New Technology Achievement Transfer Projects in the Science and Technology Commission of Shanghai Municipality, an Accreditation Expert of the National Natural Science Foundation, and an Accreditation Expert of the National 863 Program Project. Dr. Zuo has many years of extensive research and teaching experience in automotive engineering. He has led and participated in more than 20 research projects and has applied for four patents, two of which have already been granted. Dr. Zuo's current major research projects include two projects for the National Natural Science Foundation, "Body Vibration and Noise Characteristic Research on Fuel Cell Electric Vehicle" and "Noise and Vibration Characteristics Experimental and Theoretical Analysis on Fuel Cell Electric Vehicle" sub-items of the National 863 Program Project, "Electric Vehicle–Fuel Cell Vehicle Complete Car Project." He has published more than 40 papers in Chinese and foreign core academic journals and is a member of SAE International.